Stochastic Processes: General Theory

Mathematics and Its Applications

Managing Editor:

M. HAZEWINKEL

Centre for Mathematics and Computer Science, Amsterdam, The Netherlands

Volume 342

Stochastic Processes: General Theory

by

M. M. Rao
University of California

KLUWER ACADEMIC PUBLISHERS
DORDRECHT / BOSTON / LONDON

Library of Congress Cataloging-in-Publication Data

Rao, M. M. (Malempati Madhusudana), 1929-
Stochastic processes : general theory / M.M. Rao.
p. cm. -- (Mathematics and its applications ; v. 342)
Includes bibliographical references and indexes.

1. Stochastic processes. I. Title. II. Series: Mathematics and its applications (Kluwer Academic Publishers) ; v. 342.
QA274.R37 1995
519.2--dc20 95-20902

ISBN 978-1-4419-4749-9

Published by Kluwer Academic Publishers,
P.O. Box 17, 3300 AA Dordrecht, The Netherlands.

Kluwer Academic Publishers incorporates
the publishing programmes of
D. Reidel, Martinus Nijhoff, Dr W. Junk and MTP Press.

Sold and distributed in the U.S.A. and Canada
by Kluwer Academic Publishers,
101 Philip Drive, Norwell, MA 02061, U.S.A.

In all other countries, sold and distributed
by Kluwer Academic Publishers Group,
P.O. Box 322, 3300 AH Dordrecht, The Netherlands.

Printed on acid-free paper

All Rights Reserved
© 1995 Kluwer Academic Publishers
Softcover reprint of the hardcover 1st edition 1995
No part of the material protected by this copyright notice may be reproduced or
utilized in any form or by any means, electronic or mechanical,
including photocopying, recording or by any information storage and
retrieval system, without written permission from the copyright owner.

To the memory

of my parents

CONTENTS

Preface

The following work represents a completely revised and enlarged version of my book *Stochastic Processes and Integration* (1979). The new material is so extensive that it was deemed appropriate to modify the title to the current one, which reflects the content and generality. Although the book follows the original format, the changes from the previous edition are everywhere. I shall briefly explain the differences and the additions here.

The present version contains a more detailed and complete treatment of Kolmogorov's existence theorem in terms of projective limits and various applications. The first five chapters are devoted to the general theory of processes, and the final two are largely new. To accommodate all the work and to keep the book in reasonable bounds, several parts of the original presentation have been shortened and some have been omitted.

A few words on the specific changes should contrast the original work with this revised edition. A major difference is the inclusion of a generalized version of Bochner's boundedness principle which enables a novel unification of all the currently used stochastic integrals. This plays a key role in Chapter VI, where both linear and nonlinear higher order stochastic differential equations are presented as applications of this idea. A special feature of these equations, compared to the first order case, is also stressed; and many new problems awaiting solutions are pointed out. This is not discussed in other books, as far as I know. Chapter VII continues the general theme, but for processes taking values in smooth manifolds or for multiparameters. The old Chapter IV is split into two in the present version, and much of the lifting theory of the original Chapter III is shortened to make room for the new work. A somewhat similar condensation occurs in Chapters I and II of the original version, but the Kolmogorov existence theory of processes is expanded. I tried, however, to present the new version more tightly and it is better focussed. Some suggestions of the reviewers of the original monograph are also taken into account. Since every chapter begins with an outline of its contents, I shall omit further detailed discussion, except noting that complements and exercises parts (often with hints) supplement the text in a number of ways.

Each chapter has a bibliographical notes section assigning proper credits to various contributors. Hopefully I have been successful in this attempt. There is an expanded bibliography as well as notation, author and subject indexes. The numbering system is standard. Thus an item such as VI.3.5 denotes the fifth in Section 3 of Chapter VI. In a given chapter only the section and item number are used, but in a section both the chapter and section numbers are also omitted, retaining only the item number. Further equations are numbered afresh in each section. However, chapter and section titles appear on simultaneous pages to allow easy location of the items.

The revision has taken much more time than originally envisaged. I would like to thank Prof. V.V. Sazonov for some helpful comments and especially for his collaboration on the projective limit theory. A large part of the preparation of the manuscript, using TEX, was accomplished with the great help of Ms. Jan Patterson. I am also grateful to Dr. Y. Kakihara for showing me by example that this TEX craft can be learned by someone like me who is not well-versed in typing. Finally I shall be happy if the subject covered and the problems raised here stimulate enough interest in researchers in furthering the subject.

Riverside, CA. M.M. Rao
May, 1995

Chapter I

Introduction and foundations

After introducing some terminology, and motivational remarks for the study of stochastic processes, this chapter is devoted to the basic Kolmogorov existence theorem, some of its extensions as projective limits, and a few applications. These results will form a foundation for the rest of the work in this book.

1.1 Concept of a stochastic process and types

(a) *Introduction.* From an empirical point of view, a stochastic process is the description of a physical phenomenon governed by certain laws of probability and evolving in time. A mathematical abstraction of it is that a *stochastic* (or *random*) *process* is any indexed collection of random variables $\{X_t, t \in T\}$ defined on a fixed probability space. To make this explicit, the axiomatic foundations of probability, due to A. N. Kolmogorov, will be employed. Other models (mainly due to B. de Finetti, L. J. Savage, and A. Rényi) involve conditional probability spaces and the analytical work here applies to them also. Thus a probability space is a triple (Ω, Σ, P) where Ω is a point set representing all possible *outcomes* of an experiment, Σ is a σ-algebra of subsets of Ω, the elements of which are called *events* comprising all relevant questions in which an experimenter may be interested, and P is a σ-additive non-negative function on Σ with $P(\Omega) = 1$. Then P is called a probability function assigning a unique number in $[0, 1]$ for each event signifying a measure of uncertainty. Let Ω' be another point set, Σ' be a σ-algebra of subsets of Ω' and $X : \Omega \to \Omega'$ be a mapping. Then X is called an *abstract* or *generalized* (or Ω'-valued) *random variable* if $X^{-1}(\Sigma') \subset \Sigma$, i.e., $A = \{\omega : X(\omega) \in A'\} = X^{-1}(A') \in \Sigma$, all $A' \in \Sigma'$. If $\Omega' = \mathbb{R}$, the real line, $\Sigma' = \mathcal{B}$, the Borel σ-algebra of $\mathbb{R}$, then X is simply called a random variable, omitting any reference to the pair $(\mathbb{R}, \mathcal{B})$. [If Ω' is a completely regular space and Σ' is the Baire σ-algebra, then X is sometimes called a weak random variable and it is a strong random variable if Ω' is also separable.] Thus a (real) stochastic process is a family $\{X_t, t \in T\}$ such that for each $t \in T, X_t$ is a random variable. It is thus clear that the measure theoretic concepts are basic in this study and the standard results from real analysis (cf., e.g., Sion [1], Halmos [1], Rao [11], or Royden [1]) will be freely used.

In many applications, the index set T is a real interval, and then it is identified as time so that we can think of X_t as the description of a phenomenon at the instant t. If ω is the outcome, $X_t(\omega)$ will be the value (or observation) in Ω' (or $\mathbb{R}$) of the experiment in progress. However, T may be a subset of the plane or of the Euclidean space $\mathbb{R}^n$. For instance $X_t(\omega)$ may be the force of turbulence when $t = (t_1, t_2)$, the components being velocity and time. To distinguish these cases $\{X_t, t \in T\}$ is sometimes called a *(random) process* if $T \subseteq \mathbb{R}$, and a *(random) field* if $T \subseteq \mathbb{R}^n, n \geq 2$. (We write $\mathbb{R}$ for $\mathbb{R}^1$, following custom.) The definition of a process can also be viewed somewhat differently. Thus $\{X_t, t \in T\}$ may be regarded as a function X on the product space $T \times \Omega$ into Ω' (or $\mathbb{R}$), and $X_t = X(t, \cdot)$, in the above notation. Then the mapping $X(t, \cdot)$ is a *random function* if each t-section $X(t, \cdot)$ is a random variable from (Ω, Σ) to (Ω', Σ'), and the ω-section $X(\cdot, \omega)$ from T into Ω' is called a *sample function, realization*, or *trajectory* (= *path*) of X (or of the process $\{X_t, t \in T\}$). If T is countable, then $X(\cdot, \omega)$ is a *sample sequence*. In general, the term stochastic process (or simply, process) is used in situations in which T is infinite.

(b) *Image probabilities and distribution functions.* In all the above definitions the probability measure P played no part. If Y is a random variable on (Ω, Σ, P) to (Ω', Σ') and $A' \in \Sigma'$, let $A = Y^{-1}(A')$. Then from the fact that the inverse relation preserves all set operations, we deduce that the function P uniquely induces a probability P' (also denoted P_Y) on Σ' by the equation: $P_Y(A') = P(Y^{-1}(A'))$, $A' \in \Sigma'$. The set function P', or P_Y, is called the *image probability* under Y. If $\Omega' = \mathbb{R}, \Sigma' = \mathcal{B}, A'_x = (-\infty, x) \in \mathcal{B}, x \in \mathbb{R}$, let $P_Y(A'_x) = F_Y(x)$. Then $F_Y(\cdot)$ is called the *distribution function of* Y. Since such intervals generate $\mathcal{B}, P_Y$ and hence F_Y are uniquely defined on $\mathcal{B}$ and $\mathbb{R}$, and they determine each other through the Stieltjes integral $P_Y(A) = \int_A dF_Y(x)$, $A \in \mathcal{B}$. If

$Y = \{Y_i, 1 \le i \le n\}$, the components Y_i being random variables, then Y maps Ω into $\Omega' = \mathbb{R}^n$ and we replace A'_x by an n-dimensional open interval $A'_x = (-\infty, x_1) \times \cdots \times (-\infty, x_n)$ in $\mathbb{R}^n$ and set $A = Y^{-1}(A'_x)$. The image probability $P_Y(A'_x) = P(A) = F_{Y_1,\ldots,Y_n}(x_1, \ldots, x_n)$ of such A, is similarly defined and $F_{Y_1,\ldots,Y_n}(\cdot, \ldots, \cdot)$ is called the *joint* (or *n-dimensional*, or *multivariate*) *distribution function* of the vector Y. The general definition of a stochastic process given at the beginning will now be made more concrete by relating it to distribution functions.

Consider the (real) stochastic process $X = \{X_t, t \in T\}$ on (Ω, Σ, P) and, for any n, let $\{t_1, \ldots, t_n\}$ be a set of n-points from T. Then the joint distribution of $\{X_{t_1}, \ldots, X_{t_n}\}$ is given, for $x_i \in \mathbb{R}$, by

$$\begin{aligned} F_{X_{t_1},\ldots,X_{t_n}}(x_1, \ldots, x_n) &= P\left\{\omega : (X_{t_1}, \ldots, X_{t_n})(\omega) \in \mathop{\times}_{i=1}^{n} (-\infty, x_i)\right\} \\ &= P\left[\bigcap_{i=1}^{n} \{\omega : X_{t_i}(\omega) \in (-\infty, x_i)\}\right]. \end{aligned} \tag{1}$$

This will be written for simplicity as:

$$F_{t_1,\ldots,t_n}(x_1, \ldots, x_n) = P\{\omega : X_{t_1}(\omega) < x_1, \ldots, X_{t_n}(\omega) < x_n\}. \tag{2}$$

Varying n and the t's in T, (2) generates a family $\{F_{t_1,\ldots,t_n}, n \ge 1\}$ of all *finite dimensional distributions* of the process X on (Ω, Σ, P). Since $\{\omega : X_{t_1}(\omega) < \infty\} = \Omega$, the defining equations (1) and (2) imply the following two properties of the family:

$$F_{t_1,\ldots,t_n,t_{n+1}}(x_1, \ldots, x_n, \infty) = F_{t_1,\ldots,t_n}(x_1, \ldots, x_n), \tag{3}$$

$$F_{t_{i_1},\ldots,t_{i_n}}(x_{i_1}, \ldots, x_{i_n}) = F_{t_1,\ldots,t_n}(x_1, \ldots, x_n), \tag{4}$$

where $(i_1, \ldots, i_n)$ is a permutation of $(1, \ldots, n)$. Equation (4) is a consequence of (1) and the result that the set intersection is a commutative operation. It also emphasizes the fact that the measure of the set in (2) does not depend on the order of the t-points in the index T.

A collection of finite dimensional distributions $\{F_{t_1,\ldots,t_n}, t_i \in T, n \ge 1\}$ satisfying (3) and (4) is said to be a *compatible family*, even if they are only given *a priori* and not necessarily by (2) for a stochastic process X. Then equations (3) and (4) are called the *compatibility conditions* of the family indexed by the set T. These concepts raise the following two natural questions: For any given stochastic process on a probability space, there is a uniquely determined family of finite dimensional distributions satisfying the compatibility conditions (3) and (4). Suppose now that, conversely, a compatible family of distribution functions is given. Does there exist a probability space and a stochastic process such that its (set of) finite dimensional distributions are the given family? The second question is: does there exist a compatible family of distribution functions for which the preceding question may be asked? If the answers to these questions were negative, the subject would have very little interest. We shall first settle the second question by exhibiting (nontrivial) compatible families, and then in the following sections treat the important first question in some detail. It will be found that the existence problem is related to the topological nature of the range space Ω' of the process or, in the present context, the spaces $\{\mathbb{R}^n, n \geq 1\}$. If the family $\{F_{t_1,\dots,t_n}, n \geq 1\}$ is replaced by a compatible set of (image) probability functions on a general space Ω', then a probability space (Ω, Σ, P) can fail to exist, and some restrictions on the probability functions will be needed. An understanding of this problem and of its solution will be of central importance for much of the work in this book and in the stochastic theory generally.

(c) *Some compatible families of distribution functions.* The relatively simple second question will be considered here. We first note that a distribution function $F(\cdot,\dots,\cdot)$ can be characterized by the following properties: $F(x_1,\dots,x_n)$ is nonnegative, nondecreasing in each variable x_i, continuous from the left, and such that (see (1))

(i) $\lim_{x_i \to -\infty} F(x_1,\dots,x_n) = 0$,

(ii) $\lim_{x_n \to \infty} F(x_1,\dots,x_n) = F(x_1,\dots,x_{n-1})$,

(iii) $\lim_{x_i \to \infty, i=1,\dots,n} F(x_1,\dots,x_n) = F(\infty,\dots,\infty) = 1$.

Moreover, the "increments" are nonnegative: i.e., if $x_i \leq y_i, i = 1,\dots,n$, then (iv) $\Delta F \geq 0$, where

$$\begin{aligned}\Delta F = F(y_1,\dots,y_n) &- \sum_{i=1}^{n} F(y_1,\dots,y_{i-1},x_i,y_{i+1},\dots,y_n)\\ &+ \sum_{i<j} F(y_1,\dots,y_{i-1},x_i,y_{i+1},\dots,y_{j-1},x_j,y_{j+1},\dots,y_n)\\ &+ \cdots + (-1)^n F(x_1,\dots,x_n).\end{aligned}$$

We now exhibit two compatible families in order of their frequent appearances in probability theory. So our first example will be the Gaussian (or normal) family of distributions, and the second the Poisson family. A function F_n on $\mathbb{R}^n$ to the unit interval is said to be a *Gaussian distribution function* if it has the form:

$$F_n(x_1, \ldots, x_n) = [(2\pi)^n \det(K)]^{-1/2} \times \int_{-\infty}^{x_1} \cdots \int_{-\infty}^{x_n} \exp\left[-\frac{1}{2}(K^{-1}(t-\alpha), (t-\alpha))\right] dt_n \cdots dt_1, \tag{5}$$

where $K = (k_{ij})$ is a symmetric positive definite matrix of order $n, \alpha = (\alpha_1, \ldots, \alpha_n)'$ is a vector of $\mathbb{R}^n$, and $(u, v) \left\{= \sum_{i=1}^{n} u_i v_i\right\}$ is the scalar product in $\mathbb{R}^n$ (with prime "′" for transpose), and $\det(K)$ is the determinant of K. To show that, for all $x_i \in \mathbb{R}$, (5) defines an n-dimensional distribution function, some computation is required. In fact all the properties are clear except for $F_n(\infty, \ldots, \infty) = 1$. To see the latter, diagonalize K as $U'D_\lambda U$ where $UU' = U'U =$ identity, and D_λ is the diagonal matrix with elements $\lambda_i > 0$ which are the eigenvalues of K. Thus $\det(K) = \prod_{i=1}^{n} \lambda_i$ and if $v = U(t-\alpha)$ then

$$(K^{-1}(t-\alpha), (t-\alpha)) = (D_\lambda^{-1} v, v) = \sum_{i=1}^{n} v_i^2/\lambda_i,$$

and

$$F_n(+\infty, \ldots, +\infty) = \prod_{i=1}^{n} \left[(2\pi\lambda_i)^{-1/2} \int_{-\infty}^{\infty} \exp\left(\frac{-v_i^2}{2\lambda_i}\right) dv_i\right] = 1.$$

A similar computation with two applications of Fubini's theorem shows the validity of the compatibility conditions (3) and (4). This provides an affirmative answer to the second question of the preceding paragraph.

An important reason for considering this very general form of the Gaussian family, rather than simpler examples, is that Gaussian distributions occur very frequently both in practical contexts and in the theory of probability. Poincaré reportedly said (attributing to Lippman) that there must be something mysterious about this distribution since mathematicians think it is a law of nature and physicists are convinced that it is a mathematical theorem.

Another important example, following the Gaussian case, is the *Poisson distribution.* It is defined on $\mathbb{R}^n$ with its points of increase on the integer lattice (= points with integer coordinates) of the positive orthant of $\mathbb{R}^n$ and is given by

$$F_n(x_1, \ldots, x_n) = \sum_{k_1=0}^{(x_1)} \sum_{k_2=0}^{(x_2)} \cdots \sum_{k_n=0}^{(x_n)} \exp(-\lambda_1 - \cdots - \lambda_n) \frac{\lambda_1^{k_1}}{k_1!} \cdots \frac{\lambda_n^{k_n}}{k_n!}, \tag{6}$$

where $\lambda_i > 0$ and (x_i) is the largest integer less than $x_i \in \mathbb{R}$, and F_n vanishes if any $(x_i) < 0$. It is clear that the set $\{F_n, n \geq 1\}$ of (6) is a compatible family. But (6) suggests the following simple class of compatible families. Let $f_k(\cdot)$ be any nonnegative integrable function on $\mathbb{R}$, relative to a measure μ on the Borel sets, whose integral is 1. If F_n is defined by

$$F_n(x_1, \ldots, x_n) = \int_{-\infty}^{x_1} \cdots \int_{-\infty}^{x_n} f_1(t_1) \cdots f_n(t_n) d\mu(t_n) \cdots d\mu(t_1), \tag{7}$$

then Fubini's theorem shows that $\{F_n, n \geq 1\}$ is a compatible family. Note that, in contrast to (6) and (7), the distributions in (5) cannot in general be expressed as $F_n(x_1, \ldots, x_n) = \prod_{i=1}^{n} F_n(+\infty, \ldots, +\infty, x_i, +\infty, \ldots, \infty), x_i \in \mathbb{R}, i = 1, \ldots, n$. *Thus the theory must be sufficiently general to include such families.*

Before considering the existence problem, we introduce some further concepts.

Since the distribution functions are not necessarily continuous (see (6)), sometimes it will be more convenient to work with their Fourier-Stieltjes transforms which are always (uniformly) continuous. Thus if F_n is a distribution function, define

$$\varphi_n(t_1, \ldots, t_n) = \int_{-\infty}^{\infty} \cdots \int_{-\infty}^{\infty} \exp[i(t_1 x_1 + \cdots + t_n x_n)] dF_n(x_1, \ldots, x_n), \tag{8}$$

for all $t_j \in \mathbb{R}, j = 1, \ldots, n$. Since F_n is bounded on $\mathbb{R}^n$, φ_n always exists and is called the *characteristic function* of F_n. The *uniqueness theorem* of the Fourier-Stieltjes transforms implies that φ_n and F_n uniquely determine each other and then many properties of F_n can be deduced from a study of the φ_n.

If F_n on $\mathbb{R}^n$ can be expressed, relative to a measure μ on the Borel sets $\mathcal{B}_n$ of $\mathbb{R}^n$, as

$$F_n(x_1, \ldots, x_n) = \int_{-\infty}^{x_1} \cdots \int_{-\infty}^{x_n} f_n(t_1, \ldots, t_n) d\mu(t_1, \ldots, t_n) \qquad (9)$$

where f_n is a nonnegative μ-integrable (necessarily μ-unique) function on $\mathbb{R}^n$, then f_n is called the *density function* of F_n relative to μ, and if μ is the Lebesgue measure, then f_n is simply termed the *density.* Thus F_n has a density in (5) and has a density relative to the counting measure μ on $\mathcal{B}_n$ in (6). By a classical result in Fourier analysis (the Riemann-Lebesgue lemma), if F_n has a density then $\varphi_n(t_1, \ldots, t_n) \to 0$ as $t_1^2 + \cdots + t_n^2 \to \infty$, but this is not true in the general case that μ does not vanish on Lebesgue null sets.

(d) *Types.* A stochastic process on a probability space will be called a *Gaussian, Poisson,* or *some other process* if its finite dimensional distributions constitute a Gaussian, Poisson or other family, respectively. There are numerous other classes, as will be seen later. Stochastic processes can be classified into various *types* not only by their finite dimensional distribution families, but also according to certain other "regularity" properties cutting across each of the above classes. Thus a process $\{X_t, t \in T\}, T \subseteq \mathbb{R}$ (or $T \subseteq \mathbb{R}^n$) is said to be *strictly stationary* if for any $t_i \in T, h \in \mathbb{R} (t \in T, h \in \mathbb{R}^n)$ with $t_i + h \in T, i = 1, \ldots, k (t_i + h \in T)$, and for any $x_i \in \mathbb{R}$, its finite dimensional distributions F_t satisfy:

$$F_{t_1+h, \ldots, t_k+h}(x_1, \ldots, x_k) = F_{t_1, \ldots, t_k}(x_1, \ldots, x_k), \quad n \geq 1. \qquad (10)$$

In words, this says that the finite dimensional distributions remain invariant under any change (or shift) of "time". Thus a Gaussian, Poisson, or other process may (or may not) be strictly stationary. In contrast to the above, another (in a sense weaker) stationarity concept can also be introduced. Let $\{X_t, t \in T\}, T \subseteq \mathbb{R}$ (or $T \subseteq \mathbb{R}^n$), be a stochastic process on (Ω, Σ, P) (with perhaps complex values so that the real and imaginary parts are random variables) and suppose that for each $t \in T, \int_\Omega |X_t|^2 dP < \infty$. Then the process is said to be *weakly stationary*, or *stationary in the wide sense*, if the functions $m(\cdot)$ and $K(\cdot,\cdot)$ defined by

$$m(t) = \int_{\Omega} X_t dP, \quad K(s,t) = \int_{\Omega} X_s X_t^* dP, \tag{11}$$

for s, t in T, (complex conjugates are denoted by asterisks) satisfy (i) $m(t)$ = constant, $t \in T$, (ii) $K(s,t) = \tilde{K}(s-t)$, so that $K(\cdot,\cdot)$ is a function of one variable only. (Some authors require only condition (ii), but we shall assume that both conditions hold.) In (11), $m(\cdot)$ is called the *mean function* and $K(\cdot,\cdot)$ the *second* (mixed) *moment function* of the process. If $C(\cdot,\cdot)$ and $V(\cdot)$ are defined by

$$C(s,t) = K(s,t) - m(s)m(t)^*, \quad V(t) = K(t,t) - |m(t)|^2, \tag{12}$$

then they are called the *covariance* and *variance functions* respectively. Since, by the Cauchy-Buniakowsky-Schwarz (CBS)-inequality

$$|K(s,t)|^2 \leq K(s,s)K(t,t), \quad |m(t)|^2 \leq K(t,t), \tag{13}$$

it follows that $V(t) \geq 0$, and $|C(s,t)|^2 \leq V(s)V(t)$, for all s, t in T. If $\tilde{C}(s,t) = C(s,t)[V(s)V(t)]^{-1/2}$, then $\tilde{C}(\cdot,\cdot)$ is known as the *correlation function*, and we evidently have $|\tilde{C}(s,t)| \leq 1$. The Lebesgue integral $\int_{\Omega} X_t dP$ is denoted by $E(X_t)$. It is called the *expectation* of X_t (relative to the measure P and is also denoted $E_P(X_t)$).

To connect the strict and wide (or weak) sense concepts, recall that if Y is a measurable mapping (= abstract random variable) on (Ω, Σ, P) into (Ω', Σ'), and P_Y is the image of P, i.e., $P_Y(A') = P(Y^{-1}(A'))$, $A' \in \Sigma'$, then for any real measurable function Z on Ω', we have the *fundamental relation*

$$\int_{\Omega} Z(Y(\omega))dP(\omega) = \int_{\Omega'} Z(\omega')dP_Y(\omega'), \tag{14}$$

in that if one side exists, so does the other, and equality holds. In fact, this is the defining equation of P_Y, if $Z = \chi_{A'}$, the *indicator* of A' (i.e., $\chi_{A'} = 1$ on A', $= 0$ off A', so that $Z(Y) = \chi_{Y^{-1}(A')}$), and thus (14) is true, by the linearity of the integral for any positive simple function. By the Monotone Convergence theorem, the equation follows for $Z \geq 0$, and then the general case is deduced. Taking $\Omega' = \mathbb{R}^n$, and P_Y as the finite dimensional distribution, (11) becomes (if $x = x_1 + ix_2, y = y_1 + iy_2, F_{t,t} = F_t$):

$$m(t) = \int_{-\infty}^{\infty}\int_{-\infty}^{\infty} x \, dF_t(x_1, x_2), \; K(s,t) = \int_{-\infty}^{\infty}\int_{-\infty}^{\infty} xy^* \, d^2 F_{s,t}(x_1, x_2, y_1, y_2). \tag{11'}$$

If the family $\{F_{t_1,\dots,t_n}, n \geq 1\}$ is invariant under time changes (i.e. shifts of time axis), it follows after a small computation that $m(t) =$ constant and $K(s,t) = \tilde{K}(s-t)$. Hence every strictly stationary process for which $K(t,t) < \infty (t \in T)$ is also stationary in the wide sense. The converse of this statement is clearly false, but a simple computation, based on the form of the density, shows that these two concepts coincide for Gaussian processes (another reason for the importance of the latter). Note that no moments are assumed to exist in the strict sense definition so that, in general, the wide sense concept need not be defined.

To consider other types, we have to introduce the notion of independence. If $X_1, \dots, X_n$ are n random variables on (Ω, Σ, P) to (Ω', Σ'), they are said to be *mutually independent* if for any sets $A_i' \in \Sigma', i = 1, \dots, n$, the following equations hold true:

$$P\left(\bigcap_{i=1}^{m} [X_i^{-1}(A_i')]\right) = \prod_{i=1}^{m} P[X_i^{-1}(A_i')], \quad 1 < m \leq n. \tag{15}$$

If $\Omega' = \mathbb{R}, \Sigma' = \mathcal{B}$, then the above equations and (14) provide an equivalent definition of independence, based on the use of distribution functions. If we take A_i' to be the generating family of open intervals of $\mathcal{B}$, as in subsection (b), then the X_i are mutually independent iff

$$P\left(\bigcap_{i=1}^{m} \{\omega : X_i(\omega) < x_i\}\right) = \prod_{i=1}^{m} F_{X_i}(x_i), \quad x_i \in \mathbb{R}, 1 < m \leq n. \tag{15'}$$

An infinite family of random variables $\{X_i, i \in I\}$ is said to be independent if every finite subfamily consists of mutually independent random variables. A stochastic process $\{X_t, t \in T\}, T \subset \mathbb{R}$, is said to have *independent increments* if for any $t_1 < t_2 < \cdots < t_n, t_i \in T$, the random variables $(X_{t_2} - X_{t_1}), (X_{t_3} - X_{t_2}), \dots, (X_{t_n} - X_{t_{n-1}})$ are mutually independent for all $n \geq 3$. The corresponding wide sense concept is that if $\int_\Omega |X_t|^2 dP < \infty$ for all $t \in T$, then the *increments are orthogonal*, i.e., for any $t_1 < t_2 \leq t_3 < t_4$, it is true that $(X_{t_2} - X_{t_1}) \perp (X_{t_4} - X_{t_3})$, meaning

$$E((X_{t_2} - X_{t_1})(X_{t_4} - X_{t_3})^*) = 0. \tag{16}$$

Again for the Gaussian processes with mean function zero, these two concepts coincide. Since the increments form a process in their own right, it is meaningful to talk about the processes with stationary (in either sense) increments which are moreover Gaussian, or Poisson, etc.

Two more types of processes that are fundamental for our purposes are Markov processes and Martingales. To define and understand these two families, it is necessary to introduce the concepts of conditional expectation and of conditional probability. This will be studied in the next chapter. We now consider the basic existence problem, and then the above two classes of processes in later chapters.

1.2 The Kolmogorov existence theorem

We begin with a classical result of Kolmogorov's (1933) which establishes the existence of a real stochastic process on a measure space having the given compatible family of distribution functions as its finite dimensional distributions. It was later shown by Bochner (1955) that Kolmogorov's idea applies to a much more general situation; what is essential for this is to find appropriate formulations of the relevant concepts in an abstract topological setting. Since the present day research and applications find abstract stochastic processes very useful, both these results and their generalizations will be considered in this and the following sections.

1. Theorem. (Kolmogorov) *Let T be a set of real numbers and let $t_1 < t_2 < \cdots < t_n$ be n points from it. Corresponding to each such set, let $F_{t_1,\cdots,t_n}$ be an n-dimensional distribution function on $\mathbb{R}^n$. Let the family $\{F_{t_1,\ldots,t_n}, n \geq 1\}$ of all such finite dimensional distribution functions satisfy the compatibility conditions (3) and (4) of the last section. Define Ω as the space of all extended real valued functions $\omega : T \to \bar{\mathbb{R}}$. Let $\mathcal{B}_T$ be the σ-algebra of Ω generated by sets of the form $\{\omega : \omega(t) < a\}, \{\omega : \omega(t) \leq \infty\}; t \in T, a \in \bar{\mathbb{R}}$. Then there exists a unique probability measure P on $\mathcal{B}_T$ such that, if $X_t(\omega) = \omega(t)$ is the tth coordinate function of $\omega \in \Omega, \{X_t, t \in T\}$ is the desired process* *on $(\Omega, \mathcal{B}_T, P)$ which has the given family of distribution functions as its finite dimensional distributions, so that for any $x_i \in \mathbb{R}, t_i \in T, i = 1, \ldots, n$ we have*

$$P\left(\bigcap_{i=1}^{n}\{\omega : X_{t_i}(\omega) < x_i\}\right) = F_{t_1,\ldots,t_n}(x_1, \ldots, x_n), \quad n \geq 1. \tag{1}$$

Proof. The idea of proof here is to define a suitable function, using the compatibility hypothesis, on sets of the form

$$I = \{\omega : a_i \leq \omega(t_i) < b_i, i = 1, \ldots, n\} \subset \Omega = \bar{\mathbb{R}}^T \tag{2}$$

(if $b_i = +\infty$ for an i, one can replace "$<$" by "$\leq$"), called *cylinders* with interval or rectangular *bases* in $\mathbb{R}^n$, which form a semi-ring $\mathcal{L}$, and then to show that the defined set function on $\mathcal{L}$ is σ-additive. The Carathéodory and Hahn extension theorems then give the unique P on $\mathcal{B}_T$, the σ-algebra generated by $\mathcal{L}$, and this proves the result. We establish these statements in steps for convenience.

I. The class $\mathcal{L}$ of cylinders in Ω is a semi-ring, i.e., $\emptyset \in \mathcal{L}, I_1, I_2 \in \mathcal{L} \Rightarrow I_1 \cap I_2 \in \mathcal{L}$, and $I_1 - I_2$ can be expressed as a finite disjoint union of cylinders. In fact, by taking $a_i = -\infty, b_i = +\infty$, or both, in the definition of I in (2), we may assume that $n_1 = n_2 = k$, in I_1 and I_2 from $\mathcal{L}$, and hence that the same t-points occur in both. Then

$$I_1 \cap I_2 = \{\omega \in \Omega : \max(a_i^1, a_i^2) \leq \omega(t_i) < \min(b_i^1, b_i^2), i = 1, \ldots, k\}$$

and $I_1 \cap I_2 \in \mathcal{L}$. Since $\Omega \in \mathcal{L}$, and $I_1 - I_2 = I_1 \cap (\Omega - I_2)$ it suffices to show that $(I)^c = \Omega - I$ is a finite disjoint union of cylinders. If I has the representation (2), then $(I)^c$ can be expressed as:

$$\begin{aligned}(I)^c = \{\omega \in \Omega : -\infty \leq \omega(t_1) < a_1, a_i \leq \omega(t_i) < b_i, i = 2, \ldots, k\} \\ \cup \{\omega \in \Omega : b_1 \leq \omega(t_1) \leq \infty, a_i \leq \omega(t_i) < b_i, i = 2, \ldots, k\} \cup \cdots \\ \cup \{\omega \in \Omega : a_i \leq \omega(t_i) < b_i, i = 1, \ldots, k-1, b_k \leq \omega(t_k) \leq \infty\}.\end{aligned}$$

Since the right side is a finite set of disjoint cylinders, we conclude that $\mathcal{L}$ is a semi-ring (cf. Halmos [1]).

II. Definition of P on $\mathcal{L}$: Let $A \subset \mathbb{R}^n$ be a rectangle, i.e., $A = \times_{i=1}^{n} [a_i, b_i)$. If $I = \{\omega \in \Omega : (\omega(t_1), \ldots, \omega(t_n)) \in A\} \in \mathcal{L}$, define P by the equation

$$P(I) = \int_A dF_{t_1,\ldots,t_n}(x_1, \ldots, x_n), \tag{3}$$

using the Lebesgue-Stieltjes integration. We show that P is finitely additive on $\mathcal{L}$. Let I, I_i be cylinders such that $I = \bigcup_{i=1}^{r} I_i$, where I_i are disjoint. Let A, A_i be the corresponding rectangular bases of these cylinders. As before, we may assume that the same t-points $(t_1, \ldots, t_n)$ are used for the rI_i's, and I, by allowing some $a_j = -\infty$ and (or) $b_j = +\infty$. The values of $P(I)$ and $P(I_i)$ given by (3) are unaltered because of the compatibility hypothesis of $\{F_{t_1,\ldots,t_n}, n \geq 1\}$, and the Fubini theorem. (Without compatibility, the proof breaks down at this point.) Thus $A = \bigcup_{j=1}^{r} A_j$ and A_j are disjoint rectangles in $\mathbb{R}^n$. This device, with compatibility, will be repeatedly used in the proof whenever we are dealing with a *finite* collection of cylinders. Hence

$$P(I) = \int_A dF_{t_1,\ldots,t_n} = \sum_{j=1}^{r} \int_{A_j} dF_{t_1,\ldots,t_n} = \sum_{j=1}^{r} P(I_j).$$

III. Let $\mathcal{R}$ be the ring generated by $\mathcal{L}$. It is then the class of all finite disjoint unions of sets of $\mathcal{L}$. Define $\bar{P}$ on $\mathcal{R}$ as follows. If $J \in \mathcal{R}, J = \bigcup_{i=1}^{m} I_i, I_i \in \mathcal{L}$, disjoint, let

$$\bar{P}(J) = \sum_{i=1}^{m} P(I_i). \tag{4}$$

If $J = \bigcup_{i=1}^{m'} I_i', I_i' \in \mathcal{L}$ (I_i' are disjoint) is another representation, then, as in Step II, we may assume that these $m + m'$ cylinders have the same t-points in their definitions and hence $J = \bigcup_{j=1}^{m} \bigcup_{i=1}^{m'} I_j \cap I_i', I_j \cap I_i' \in \mathcal{L}$. With the additivity of P on $\mathcal{L}$, we have

$$\begin{aligned}\bar{P}(J) &= \sum_{i=1}^{m} P(I_i) = \sum_{i=1}^{m} P\left(I_i \cap \bigcup_{j=1}^{m'} I_j'\right) = \sum_{i=1}^{m}\sum_{j=1}^{m'} P(I_i \cap I_j') \\ &= \sum_{j=1}^{m'} P\left(I_j' \cap \bigcup_{i=1}^{m} I_i\right) = \sum_{j=1}^{m'} P(I_j').\end{aligned}$$

Thus the definition of $\bar{P}$ by (4) is unambiguous.

We may use the same device and prove the additivity of $\bar{P}$ on $\mathcal{R}$. Briefly, if J_1, J_2 are disjoint sets of $\mathcal{R}$, and $J_i = \bigcup_{j=1}^{m_i} I_j^i, i = 1,2$ is a representation, and if we set $Q_j = I_j^1$ for $j = 1, \ldots, m_1$ and $= I_k^2$ for $j = m_1 + k, k = 1, \ldots, m_2$, then $J_1 \cup J_2 = \bigcup_{j=1}^{m_1+m_2} Q_j$ and

$$\bar{P}(J_1 \cup J_2) = \sum_{j=1}^{m_1+m_2} P(Q_j) = \sum_{j=1}^{m_1} P(I_j^1) + \sum_{j=1}^{m_2} P(I_j^2) = \bar{P}(J_1) + \bar{P}(J_2). \tag{5}$$

It is clear that $\bar{P} = P$ on $\mathcal{L}$ and we may use the same symbol on both $\mathcal{L}$ and $\mathcal{R}$ by the above proof.

IV. P is σ-additive on $\mathcal{L}$: Since additivity clearly implies σ-superadditivity of P, it suffices to show that it is σ-subadditive on $\mathcal{L}$. It is here that the topological properties of Ω are decisive.

Since $\Omega = \bar{\mathbb{R}}^T$, and $\bar{\mathbb{R}} = [-\infty, \infty]$ is compact, it follows by the Tychonov theorem that Ω is compact (in the product topology). Let $I_i \in \mathcal{L}, I \in \mathcal{L}$, and $I = \bigcup_{i=1}^{\infty} I_i$ where $I_i \cap I_j = \emptyset (i \neq j)$. We may take $I_i \neq \Omega, i \geq 1$. If $\varepsilon > 0$ is given, let H_i and K be cylinders such that the interior of $H_i, \mathrm{int}(H_i) \supset I_i$ and the closure of $K, \mathrm{cl}(K) \subset I$ and moreover,

$$(a) \qquad P(H_i) < P(I_i) + \frac{\varepsilon}{2^{i+1}}, \quad (b) \qquad P(I) < P(K) + \frac{\varepsilon}{2}. \tag{6}$$

This is possible because of the left continuity of the distribution functions F. For instance, replace a_i by $(a_i - \eta)$ and(or) b_i by $(b_i - \eta)$, for small enough $\eta > 0$, in the definitions of I_n and I. Thus $\mathrm{cl}(K) \subset I = \bigcup_{n=1}^{\infty} I_n \subset \bigcup_{n=1}^{\infty} \mathrm{int}(H_n)$. Since $\mathrm{int}(H_i) \subset \Omega$ and $\mathrm{cl}(K)$ is compact (Ω being compact), there is a finite subcover with, say N_0, H_i's. We may assume, by relabelling if necessary, that these H_i's to be the first N_0 of the collection. Hence, using (6), and the fact that $\bigcup_{i=1}^{N_0} H_i \in \mathcal{R}$ on which P is subadditive by Step III, we have:

$$\begin{aligned} P(I) - \frac{\varepsilon}{2} < P(K) \leq P\left(\bigcup_{i=1}^{N_0} H_i\right) &\leq \sum_{i=1}^{N_0} P(H_i) \\ &\leq \sum_{i=1}^{\infty} \left[P(I_i) + \frac{\varepsilon}{2^{i+1}}\right] < \sum_{n=1}^{\infty} P(I_n) + \frac{\varepsilon}{2} \,. \end{aligned} \tag{7}$$

Since $\varepsilon > 0$ is arbitrary, (7) proves the σ-subadditivity of P on $\mathcal{L}$ as asserted.

V. To complete the proof, let P^* on Ω be the outer measure generated by $(\mathcal{L}, P)$, i.e.,

$$P^*(E) = \inf\left\{\sum_{i=1}^{\infty} P(E_i) : E_i \in \mathcal{L}, E \subset \bigcup_{i=1}^{\infty} E_i\right\}, \quad E \subset \Omega. \tag{8}$$

Then the classical Carathéodory extension theorem says that the P^* measurable sets Σ_{P^*} form a σ-algebra containing $\mathcal{L}$ and that $P^* = P$ on $\mathcal{L}$. Let $\mathcal{B}_T = \sigma(\mathcal{L})$, the σ-algebra generated by $\mathcal{L}$. Then $\mathcal{B}_T \subset \Sigma_{P^*}$ and P^* on $\mathcal{B}_T$ (since it is so on Σ_{P^*}) is σ-additive. Now P is finite. So the σ-additive extension P_1 of P from $\mathcal{L}$ to $\mathcal{B}_T$ is also unique, by the classical Hahn extension theorem, and we again write P for P_1 on $\mathcal{B}_T$. Thus $(\Omega, \mathcal{B}_T, P)$ is the desired probability space and P on $\mathcal{L}$ is given by (3). (For these standard results, see, e.g., Rao [11], Sections 2.2 and 2.3.)

Define the coordinate function $X_t : \omega \mapsto \omega(t), t \in T$, i.e., $X_t(\omega) = \omega(t)$. Then it follows that

$$\begin{aligned} P\{\omega : a_i \le X_{t_i}(\omega) < b_i, i = 1, \dots, n\} \\ &= P\left\{\omega : (\omega(t_1), \dots, \omega(t_n)) \in \mathop{\times}_{i=1}^{n} [a_i, b_i)\right\} \\ &= \int_{a_1}^{b_1} \cdots \int_{a_n}^{b_n} dF_{t_1, \dots, t_n}. \end{aligned}$$

This establishes the theorem. $\square$

Remark. It should be noted that the stochastic process $\{X_t, t \in T\}$ constructed in the above theorem is extended-real-valued. This is only a technical device. The definition of a distribution function implies that in $\bar{\mathbb{R}}$ the two points "$+\infty$", and "$-\infty$" receive measure zero, since we have $\lim_{x_i \to -\infty} F_{t_i, \dots, t_n}(x_1, \dots, x_n) = 0$, and $\lim_{x_1, \dots, x_n \to +\infty} (1 - F_{t_1, \dots t_n}(x_1, \dots, x_n)) = 0$. If $\tilde{\Omega} = \mathbb{R}^T$, then $\tilde{\Omega} \subset \Omega$ and the latter is the space defined above. However, it can be shown in terms of Step V, that $P^*(\tilde{\Omega}) = 1$. Define $\tilde{\mathcal{B}}_t = (\mathcal{B}_T \cap \tilde{\Omega})$, the restriction or *trace* of $\mathcal{B}_T$ on $\tilde{\Omega}$. Then $\tilde{\mathcal{B}}_T$ is a σ-algebra, and if $\tilde{P}$ is the restriction of P to $\tilde{\mathcal{B}}_T$, then

$(\tilde{\Omega}, \tilde{\mathcal{B}}_T, \tilde{P})$ is a probability space (cf. also Theorem 3.4 later). If we let $\tilde{\omega}(t) = \omega(t)$ for $\omega \in \tilde{\Omega} \subset \Omega$, and $= 0$ for $\omega \in \Omega - \tilde{\Omega}$, and next define $\tilde{X}_t(\tilde{\omega}) = X_t(\tilde{\omega})$, then $\{\tilde{X}_t, t \in T\}$ is a real-valued stochastic process on $(\tilde{\Omega}, \tilde{\mathcal{B}}_T, \tilde{P})$ with the given compatible family of distributions as its finite dimensional distribution functions. Hence the above theorem can be reformulated as:

2. Theorem. *Let T be a set of real numbers and $t_1 < t_2 \cdots < t_n$ be n points from it. Corresponding to each such set let $F_{t_1,\dots,t_n}$ be an n-dimensional distribution function on $\mathbb{R}^n$. Let the family $\{F_{t_1,\dots,t_n}, n \geq 1\}$ be compatible in the sense of the last section. Let $\Omega = \mathbb{R}^T$, and let $\mathcal{B}_T$ be the σ-algebra generated by sets of the form $\{\{\omega : -\infty < \omega(t) < a\}, t \in T, a \in \mathbb{R}\}$. Then there exists a unique probability measure P on $\mathcal{B}_T$ such that, if $X_{(\cdot)}(\omega) = \omega(\cdot)$ is the coordinate function of ω in Ω, $\{X_t, t \in T\}$ is a stochastic process on $(\Omega, \mathcal{B}_T, P)$ which has the given family of distributuions as its finite dimensional distribution functions so that we have*

$$P\left(\bigcap_{i=1}^{n}\{\omega : X_{t_i}(\omega) < x_i\}\right) = F_{t_1,\dots,t_n}(x_1, \dots, x_n), \quad n \geq 1,$$

for any $x_i \in \mathbb{R}, t_i \in T, i = 1, \dots, n$.

3. Remarks. 1. The original result of Kolmogorov is formulated as in Theorem 2, and it can be proved directly without reference to Theorem 1. However, Step IV of the preceding proof does not carry over to this case since $\Omega = \mathbb{R}^T$ is not a compact space. To get around this difficulty Kolmogorov gave an indirect proof. We shall include later an extension of his argument for an abstract result (cf. Theorem 4 of Section 3) from which Theorem 2 is again a consequence. It is useful to have various methods of proof since they suggest different generalizations.

2. The left continuiuty of a distribution function F is precisely the regularity of the P it determines on $\mathbb{R}$; i.e., if P^* is the outer measure "generated by" the function F, then it has the following four properties, relating to (Baire and) Borel probabilities:

(a) $P^*(A) = \inf\{P^*(O) : O \supset A, O \text{ open}\}, A \subset \mathbb{R}$,

(b) $P^*(B) = \sup\{P^*(C) : C \subset B, C \text{ compact}\}, B \subset \mathbb{R}$ open,

(c) open sets are P^*-measurable,

(d) $P^*(C) < \infty$, for each compact $C \subset \mathbb{R}$.

Such properties used in the proof of Theorem 1 suggest that the existence problem admits an extension for probabilities on general topological spaces. Those possibilities will be examined below, since these results also form a good connecting link with the martingales introduced in Chapter II.

3. The above theorem says that a real stochastic process on a probability space (Ω, Σ, P) can be replaced by another process on some other probability space $(\mathbb{R}^T, \mathcal{B}_T, \bar{P})$ having the same finite dimensional distributions. In the first case, Ω is an abstract set; but in the second case it is a function space, $\mathbb{R}^T$. For this reason, the latter is often called the *function space* or *canonical representation* of a process. Since the space of sample functions and the space of "elementary events" are now the same, this identification is useful in many problems. It must be noted, however, that various (e.g., non-linear) operations on a stochastic process of function space type lead to new processes which are not of the same type. If for such a new process the same procedure is applied, a function space representation can again be obtained but now on a new probability space which may have little, if any, relation with the old one. Thus it is not possible to assume that all processes are of function space type in our study.

4. Another important consequence of Kolmogorov's theorem is that a stochastic process can be defined as a family of real measurable functions $\{X_t, t \in T\}$ on a probability space (Ω, Σ, P), or as a compatible family of finite dimensional distributions $\{F_{t_1,\dots,t_n}, n \geq 1\}$ for $t_i \in T$. This point of view emphasizes the fact that the distribution functions (or image probability laws) are the key objects of interest in a large part of probability theory. For this reason, it is advantageous to take either of the two forms as definitions of a stochastic process depending on the context. Notice that the image probability spaces usually have "finer" structure and are topological.

We now discuss an abstraction of the preceding result, following Bochner, and then treat some useful ramifications. Recall that a set $(D, <)$ is *directed* if it is partially ordered by a relation "$<$" and that for any d_1, d_2 in D, there is a d in D with $d_i < d, i = 1, 2$.

4. Definition. Let $\{(\Omega_\alpha, \Sigma_\alpha, P_\alpha), \alpha \in D\}$ be a family of measure spaces, where D is a directed set, and let $\{g_{\alpha\beta}, \alpha < \beta, \alpha, \beta \text{ in } D\}$ be a family of mappings such that: (i) $g_{\alpha\beta} : \Omega_\beta \to \Omega_\alpha$, and $g_{\alpha\beta}^{-1}(\Sigma_\alpha) \subset \Sigma_\beta$, i.e., each $g_{\alpha\beta}$ is measurable for the pair $(\Sigma_\beta, \Sigma_\alpha)$, (ii) for any $\alpha < \beta < \gamma, g_{\alpha\gamma} = g_{\alpha\beta} \circ g_{\beta\gamma}$ and $g_{\alpha\alpha}$ = identity, i.e., the mappings are *compatible*, and (iii) for every $\alpha < \beta, P_\alpha$ is the image measure of P_β under $g_{\alpha\beta}$, i.e., $P_\alpha = P_\beta \circ g_{\alpha\beta}^{-1}$ (sometimes this is also written as $P_\alpha = g_{\alpha\beta} \circ P_\beta$). Then the abstract collection $\{(\Omega_\alpha, \Sigma_\alpha, P_\alpha, g_{\alpha\beta})_{\alpha<\beta} : \alpha, \beta \text{ in } D\}$ is called a (*generalized*) *compatible family*, or *a projective system of measure* (*or probability*) *spaces* relative to the mappings $g_{\alpha\beta}$. If Ω_α is a topological space and Σ_α is the σ-algebra of Baire sets, then *the mappings $g_{\alpha\beta}$ are also required to be continuous.*

Sometimes the term "inverse system" is used for the concept of a projective system, introduced in the above definition. Moreover, if each P_α is regular in the sense that it satisfies conditions (a)-(d) of Remark 3 (2) above, then the family is called a *topological projective system.* When topology intervenes, it is always Hausdorff in our considerations, but this is not essential for many of the results below.

It will be instructive to note that, in Kolmogorov's theorem above, $\Omega_\alpha = \bar{\mathbb{R}}^\alpha$, or $= \mathbb{R}^\alpha$, where $\alpha \subset T$ is a finite set (D =all finite subsets of T ordered by inclusion), $g_{\alpha\beta}$'s are coordinate projections from $\mathbb{R}^\beta$ onto $\mathbb{R}^\alpha$, and Σ_α's are the Borel (= Baire) σ-algebras. The functions $\{P_\alpha, \alpha \in D\}$ are the Lebesgue-Stieltjes measures generated respectively by $\{F_\alpha, \alpha \in D\}$ which satisfy the conditions (i)-(iii) of the above definition, and then the collection $\{(\Omega_\alpha, \Sigma_\alpha, P_\alpha, g_{\alpha\beta})_{\alpha<\beta} : \alpha, \beta$ in $D\}$ becomes a topological projective system. This transcription also shows that the linear ordering of T, as a subset of $\mathbb{R}$, is not essential to the problem and the abstraction illuminates its structure. We shall consider this point in some detail in the following section as it is also needed for some work in later chapters. Sometimes it is convenient to identify $\mathbb{R}^\alpha$ and $\mathbb{R}^{\text{card}(\alpha)}, \alpha \in D$, which are linearly homeomorphic. This will be done without comment.

1.3 Some generalizations of the existence theorem: projective limits

Let $\{(\Omega_\alpha, \Sigma_\alpha, P_\alpha, g_{\alpha\beta})_{\alpha<\beta} : \alpha, \beta \text{ in } D\}$ be a projective system of probability spaces. If $\Omega_D = \times_{\alpha \in D} \Omega_\alpha$ is the cartesian product, let Ω be the subset consisting of those elements, called *threads*, $\omega = (\omega_\alpha, \alpha \in D) \in \Omega_D$, such that for each pair α, β satisfying $\alpha < \beta$

we have $g_{\alpha\beta}(\omega_\beta) = \omega_\alpha$. If $g_\alpha : \omega \mapsto \omega_\alpha, \omega \in \Omega$, then Ω is called the projective limit of the system $\{(\Omega_\alpha, g_{\alpha\beta}) : \alpha, \beta \text{ in } D\}$ and denoted $\Omega = \varprojlim(\Omega_\alpha, g_{\alpha\beta})$. In general Ω can be very small (or empty) even if each $g_{\alpha\beta}$ is an onto mapping. However, if each $\Omega_\alpha = \mathbb{R}^\alpha, g_{\alpha\beta} : \mathbb{R}^\beta \to \mathbb{R}^\alpha$ is the coordinate projection, as in the Kolmogorov system, then $\Omega \cong \mathbb{R}^T, g_\alpha : \Omega \to \Omega_\alpha$ is also onto ($\alpha \in D$, all finite subsets of T, cf., Exercise 15). In any case, if we let $\Sigma_0 = \bigcup_\alpha g_\alpha^{-1}(\Sigma_\alpha)$, then Σ_0 is an algebra of subsets of Ω. To proceed further, we can state the following simple result which holds without any real restrictions on a projective system of measures.

1. Proposition. *Let $\{(\Omega_\alpha, \Sigma_\alpha, P_\alpha, g_{\alpha\beta})_{\alpha<\beta} : \alpha, \beta \text{ in } D\}$ be a general projective system of measure spaces and $\Omega = \varprojlim(\Omega_\alpha, g_{\alpha\beta})$. If Σ_0 is the algebra defined above (it is generated by the "cylinder" sets of Ω with bases in $\Sigma_\alpha, \alpha \in D$), and if $g_\alpha : \Omega \to \Omega_\alpha, \alpha \in D$, is onto, then there exists a unique (finitely) additive function P on Σ_0 such that $P_\alpha = P \circ g_\alpha^{-1} : \Sigma_\alpha \to \mathbb{R}^+$ for each $\alpha \in D$.*

Proof. Let $A \in \Sigma_0 = \bigcup_\alpha g_\alpha^{-1}(\Sigma_\alpha)$. If $A \in g_\alpha^{-1}(\Sigma_\alpha) \cap g_\beta^{-1}(\Sigma_\beta)$, then there exist $B_1 \in \Sigma_\alpha, B_2 \in \Sigma_\beta$ such that $A = g_\alpha^{-1}(B_1) = g_\beta^{-1}(B_2)$. Since D is directed, there exists a $\gamma \in D$ such that $\gamma > \alpha, \gamma > \beta$, and by compatibility of the mappings, we have $g_\alpha = g_{\alpha\gamma} \circ g_\gamma, g_\beta = g_{\beta\gamma} \circ g_\gamma$. Hence

$$g_\gamma^{-1}(g_{\alpha\gamma}^{-1}(B_1)) = g_\alpha^{-1}(B_1) = A = g_\beta^{-1}(B_2) = g_\gamma^{-1}(g_{\beta\gamma}^{-1}(B_2)). \tag{1}$$

Since by hypothesis, $g_\gamma : \Omega \to \Omega_\gamma$ is onto, it follows that $g_\gamma^{-1} : \Sigma_\gamma \to \Sigma_0$ is one-to-one so that (1) implies $g_{\alpha\gamma}^{-1}(B_1) = g_{\beta\gamma}^{-1}(B_2)$. Thus

$$P_\alpha(B_1) = P_\gamma(g_{\alpha\gamma}^{-1}(B_1)) = P_\gamma(g_{\beta\gamma}^{-1}(B_2)) = P_\beta(B_2), \tag{2}$$

by the compatibility of the measures. If we set $P(A) = P_\alpha(B_1) = P_\beta(B_2), A \in \Sigma_0$, then by (2) P is unambiguously defined on Σ_0. Moreover, this prescription also implies $P_\alpha = P \circ g_\alpha^{-1}, \alpha \in D$. It is clear that, the thus defined P is nonnegative and finitely additive on Σ_0. □

If for a projective system of probability spaces, for which each $g_\alpha : \Omega \to \Omega_\alpha$ is onto, the additive set function P of the above proposition is in fact σ-additive, then it has a unique σ-additive extension, also denoted by P, to $\Sigma = \sigma(\Sigma_0)$, the σ-algebra generated by Σ_0. In this case the triple (Ω, Σ, P) is called the *projective limit* of the projective system $\{(\Omega_\alpha, \Sigma_\alpha, P_\alpha, g_{\alpha\beta})_{\alpha<\beta}\,; \alpha, \beta \text{ in } D\}$. It is not hard to show that the above definition is valid if each P_α is a σ-finite measure on Σ_α. We omit this verification here. One notes that the concepts "projective system" and "projective limit" are meaningful if *each P_α is a signed or vector valued additive set function.* This remark will be utilized later.

Also observe that, if $\Omega_\alpha = \mathbb{R}^\alpha, \Sigma_\alpha = \mathcal{B}_\alpha$, and $g_{\alpha\beta} = \pi_{\alpha\beta}$ is the coordinate projection, and if moreover each P_α is a Lebesgue-Stieltjes probability measure, then Kolmogorov's theorem asserts the existence of the projective limit for this system. In general, however, P on Σ_0 is *not* σ-additive. (For a counter-example, see Exercise 6 at the end of the chapter.) Therefore we need to analyze the situation further.

In order to overcome the initial hurdle of an empty projective limit set, we introduce the following useful (sufficient) condition, due to Bochner.

2. Definition. A projective system of spaces $\{(\Omega_\alpha, g_{\alpha\beta})_{\alpha<\beta} : \alpha, \beta$ in $D\}$ is said to satisfy the *sequential maximality* (s.m.) condition if for each sequence $\alpha_1 < \alpha_2 < \cdots$ in D, and any points ω_{α_i} in Ω_{α_i} such that $g_{\alpha_i\alpha_{i+1}}(\omega_{\alpha_{i+1}}) = \omega_{\alpha_i}, i \geq 1$, there is an $\omega \in \Omega$, for which $\omega_{\alpha_i} = g_{\alpha_i\infty}(\omega), i \geq 1$. (Consequently $g_\alpha(\Omega) = \Omega_\alpha, \alpha \in D$, where $g_\alpha = g_{\alpha\infty}$, as usual.)

It is obvious that the s.m. condition is automatic if the $g_{\alpha\beta}$ are coordinate projections, as in Kolmogorov's theorem. It is not very hard to verify that the s.m. condition holds if each Ω_α is a nonvoid compact Hausdorff space and $g_{\alpha\beta}$'s are onto mappings. In this case $\emptyset \neq \Omega \subset \Omega_D$ is compact. These observations imply that in most cases of interest the s.m. property will be present. However, Millington and Sion ([1], p. 649) showed the existence of projective systems admitting limits, but for which the s.m. condition is not satisfied. (See Complements and exercises on this point.) Nevertheless, we shall use this condition in the following because of its simplicity in abstract cases.

In the remainder of this section (and in later chapters) we present some characterizations of projective limits. The next definition and result will be useful for this purpose.

3. Definition. Let (Ω, Σ) be a measurable space and $\{\Sigma_\alpha, \alpha \in D\}$ be a directed family of σ-subalgebras of Σ such that, for $\alpha < \beta$, we have $\Sigma_\alpha \subset \Sigma_\beta$. Let $\mu_\alpha : \Sigma_\alpha \to \mathbb{R}$ be additive functions such that the restriction of μ_β to Σ_α is μ_α for $\alpha < \beta$, i.e., symbolically, $\mu_\beta|\Sigma_\alpha = \mu_\alpha$. Then $\{(\Omega, \Sigma_\alpha, \mu_\alpha), \alpha \in D\}$ is called an additive set martingale and $\{\Sigma_\alpha, \alpha \in D\}$ is its *base*. If each μ_α is σ-additive, then it is a (σ-additive, or simply) *set martingale*.

Evidently each set martingale defines a projective system (take $g_{\alpha\beta}$ = identity, $\Omega_\alpha \equiv \Omega$) of (signed) measures. Conversely, if $\{(\Omega_\alpha, \Sigma_\alpha, P_\alpha, g_{\alpha\beta})_{\alpha<\beta} : \alpha, \beta \text{ in } D\}$ is an arbitrary projective system satisfying the s.m. condition so that $\Omega = \underleftarrow{\lim}(\Omega, g_{\alpha\beta})$ is nontrivial (or simply assuming the condition that $g_\alpha(\Omega) = \Omega_\alpha$), then $\{(\Omega, \Sigma_\alpha^*, P_\alpha^*), \alpha \in D\}$ is a set martingale where $\Sigma_\alpha^* = g_\alpha^{-1}(\Sigma_\alpha)$ and $P_\alpha^* \circ g_\alpha^{-1} = P_\alpha$, for each $\alpha \in D$. If there is a measure μ on $\Sigma_0 = \bigcup_\alpha \Sigma_\alpha^*$ such that $\mu|\Sigma_\alpha^*$ is σ-finite for each α (or μ is finite) and each P_α^* is μ-continuous, then $\{f_\alpha, \Sigma_\alpha^*, \alpha \in D\}$ is called a *martingale* (*of point functions*) where f_α is the Radon-Nikodým derivative of P_α^* relative to $\mu|\Sigma_\alpha$. This simple looking connection between a martingale family and the projective system is of fundamental importance. We start a serious study of martingales in the next chapter, and several of its extensions, as well as applications, will occupy a large part of this book; indeed, they are a very important part in current research.

Let us consider a topological projective system, as this will be useful even in the study of abstract measure families. The first general result on projective limits is the following theorem, due essentially to Bochner (1955) and the last part to Choksi (1958).

4. Theorem. *Let $\{(\Omega_\alpha, \Sigma_\alpha, P_\alpha, g_{\alpha\beta})_{\alpha<\beta} : \alpha, \beta$ in $D\}$ be a projective system of Hausdorff topological probability spaces with the s.m. property. Let each Σ_α contain all the closed subsets of Ω_α and each P_α on Σ_α be inner regular (for the compact sets). Then the system admits a unique projective limit (Ω, Σ, P). Moreover P is inner regular on $\Sigma_0 = \bigcup_{\alpha\in D} g_\alpha^{-1}(\Sigma_\alpha)$ relative to the class $\mathcal{C} \subset \Sigma$ of all cylinders with compact bases, and hence P is inner regular on $\Sigma = \sigma(\Sigma_0)$ relative to the class $(\mathcal{C})_\delta = \left\{C \subset \Omega : C = \bigcap_{n=1}^{\infty} C_n, C_n \in \mathcal{C}\right\} \subset \Sigma$. (The $g_\alpha : \Omega \to \Omega_\alpha$ are the canonical mappings determined by $g_{\alpha\beta}$, as usual.) If each Ω_α is also compact, then Σ contains the Baire σ-algebra of the compact space Ω and thus P is a Baire measure. In fact, in this case the hypothesis may be weakened by replacing Σ_α (which are now Borel σ-algebras) to $\mathcal{B}_\alpha$, the Baire σ-algebras of Ω_α, so that each P_α is a Baire* *measure and (Ω, Σ, P) exists and is a Baire measure space. Hence P is (Baire) regular, i.e., P of every compact set can be approximated from above by the P of (Baire) open sets; similarly P of open sets can be approximated from below by P of compact (Baire) sets.* [The concepts of Baire (and Borel) regularity and related matters will be recalled in the proof. Cf., also Royden [1], Dinculeanu [1], or Rao [11].]

Proof. The first part is similar to Theorem 2.1, except that the $g_{\alpha\beta}$ need not be the natural projections now, but the s.m. condition implies that the projective limit Ω of $(\Omega_\alpha, g_{\alpha\beta})$ has sufficiently many points. Note that $\Omega \subsetneqq \Omega_D$ is possible, and Ω_α's are noncompact. So we give an indirect proof and it also provides an alternative argument for Theorems 2.1 and 2.2.

By Proposition 1 above, there is a unique (finitely) additive P on the algebra Σ_0 to $[0,1]$, and by the extension theory of measures, it suffices to show that P is σ-additive on Σ_0 so that it has the same property on Σ. For this, let $A_n \in \Sigma_0, A_n \supset A_{n+1}$ and $\bigcap_{n=1}^{\infty} A_n = \emptyset$. Then $P(A_n) \to 0$, as $n \to \infty$, is to be shown. Suppose this is false. Then there exists a cylinder sequence $\{A_n, n \geq 1\} \subset \Sigma_0$, as above, and an $\varepsilon > 0$ such that $P(A_n) \geq 2\varepsilon > 0$. But $A_n \in g_{\alpha_n}^{-1}(\Sigma_{\alpha_n}) = \Sigma_{\alpha_n}^*$ for some α_n, and since $\{\Sigma_{\alpha_n}^*\}_1^\infty$ is directed, we may and do choose the α_n satisfying $\alpha_n < \alpha_{n+1}$. Let $A_n = g_{\alpha_n}^{-1}(B_n)$. Then by the inner regularity of P_{α_n} on Σ_{α_n}, there is a compact $\tilde{C}_n \subset B_n \subset \Omega_{\alpha_n}$ such that

$$P_{\alpha_n}(\tilde{C}_n) > P_{\alpha_n}(B_n) - \frac{\varepsilon}{2^n}. \tag{3}$$

However, $\{g_{\alpha_n}^{-1}(\tilde{C}_n)\}_1^\infty$ need not be monotone. So let $\tilde{C}_n^* = g_{\alpha_n}^{-1}(\tilde{C}_n) \subset A_n$ and define $C_n^* = \bigcap_{i=1}^{n} \tilde{C}_i^*$. Note that $A_n \downarrow \emptyset$ implies $C_n^* \downarrow \emptyset$, and the latter are cylinders contained in A_n. Since g_{α_n} is continuous and $\tilde{C}_n$ is compact, C_n^* is closed. But C_n^* is a cylinder. So there exists a $C_n \in \Sigma_{\alpha_n}$ such that $C_n^* = g_{\alpha_n}^{-1}(C_n)$ and C_n is compact since C_n^* is the intersection of a finite number of cylinders each with a compact base. In fact, let $\tilde{C}_i \subset \Omega_{\alpha_i}$ be a compact base defining $\tilde{C}_i^* = g_{\alpha_i}^{-1}(\tilde{C}_i)$. If $\tilde{\tilde{C}}_{in} = g_{\alpha_i\alpha_n}^{-1}(\tilde{C}_i)$, then $\tilde{\tilde{C}}_{in} \subset \Omega_{\alpha_n}, i = 1, \ldots, n$, closed, since g_{α_i} and $g_{\alpha\beta}$ are continuous and consistent. Also $\tilde{\tilde{C}}_{nn} = \tilde{C}_n$. Then $C_n = \bigcap_{i=1}^{n} \tilde{\tilde{C}}_{in}$ is compact since $\tilde{C}_n$ is, and $C_n^* = g_{\alpha_n}^{-1}(C_n) = \bigcap_{i=1}^{n} g_{\alpha_n}^{-1}(\tilde{\tilde{C}}_{in}) = \bigcap_{i=1}^{n} \tilde{C}_i^*$.

Moreover, the compatibility of the mappings g_{α_n} and monotonicity of C_n^*'s imply, for each n, $C_n = g_{\alpha_n}(C_n^*) \supset g_{\alpha_n \alpha_{n+1}}(g_{\alpha_{n+1}}(C_{n-1}^*)) = g_{\alpha_n \alpha_{n+1}}(C_{n+1})$, and $P(C_n^*) = P_{\alpha_n}(C_n)$. With this reduction, we assert that (i) $C_n \neq \emptyset$ for all n, and then (ii) $\bigcap_{n=1}^{\infty} C_n^* \neq \emptyset$. Since $C_n^* \downarrow \emptyset$, (ii) provides the desired contradiction.

For (i) it suffices to show that $P_{\alpha_n}(C_n) = P(C_n^*) > 0$, for all n, using (3). But (3) can be expressed as

$$P(\tilde{C}_n^*) = P(g_{\alpha_n}^{-1}(\tilde{C}_n)) = P_{\alpha_n}(\tilde{C}_n) > P_{\alpha_n}(B_n) - \frac{\varepsilon}{2^n} = P(A_n) - \frac{\varepsilon}{2^n}, \quad (4)$$

and since $A_n \supset \tilde{C}_n^*$, this implies $P(A_n - \tilde{C}_n^*) < \frac{\varepsilon}{2^n}$. Also, $C_n^* = \bigcap_{i=1}^{n} \tilde{C}_i^* = \tilde{C}_n^* - \left[\bigcup_{i=1}^{n-1} (A_i - \tilde{C}_i^*)\right]$. Hence using the subtractive property of P, we have

$$P(C_n^*) = P(\tilde{C}_n^*) - P\left(\bigcup_{i=1}^{n-1} (A_i - \tilde{C}_i^*)\right) \geq P(\tilde{C}_n^*) - \sum_{i=1}^{n-1} \frac{\varepsilon}{2^i}. \quad (5)$$

Thus (4) and (5) yield

$$P_{\alpha_n}(C_n) = P(C_n^*) \geq P(A_n) - \sum_{i=1}^{n} \frac{\varepsilon}{2^i} \geq 2\varepsilon - \varepsilon = \varepsilon > 0. \quad (6)$$

To complete the result, note that C_n's are nonempty and compact in Ω_{α_n}'s. Let $\omega_n^n \in C_n$ and for $m > n, \omega_m^m \in C_m$. Since $g_{\alpha_i \alpha_m}(C_m) \subset C_n$, we let $\omega_n^m = g_{\alpha_n \alpha_m}(\omega_m^m) \in C_n$. Consider the set $\{\omega_n^r, r \geq n\} \subset C_n$. The compactness of C_n implies the existence of a convergent subsequence of this sequence (denoted by itself) with limit $\omega_n^0 \in C_n$, as $r \to \infty$. This is true for each n, and by the continuity of $g_{\alpha_n \alpha_m}$

$$\omega_n^0 = \lim_{r \to \infty} g_{\alpha_n \alpha_m}(\omega_m^r) = g_{\alpha_n \alpha_m}(\omega_m^0), \quad m > n .$$

Thus $\omega_n^0 \in \Omega_{\alpha_n}$ and $\omega_n^0 = g_{\alpha_n \alpha_{n+1}}(\omega_{n+1}^0)$ so that by the s.m. condition, there exists an $\omega \in \Omega$ such that $\omega_n^0 = g_{\alpha_n}(\omega)$, for all n. This implies that $\omega \in C_n^*$ for all n, and $\bigcap_{n=1}^{\infty} C_n^* \neq \emptyset$. This contradicts the choice of C_n^*, and hence P must be σ-additive on Σ_0. Moreover, by (6) we have (since $A_n \supset C_n^*$)

$$P(A_n) \geq P(C_n^*) \geq P(A_n) - \varepsilon. \quad (7)$$

Thus $P(A_n - C_n^*) \leq \varepsilon$, and the fact that $A_n \in \Sigma_0, C_n^* \in \mathcal{C}$ yield the inner regularity of P on Σ_0. This also follows directly after a simple computation using the inner regularity of each P_α. The uniqueness of the limit space is evident.

To see that P is inner regular on Σ, for $(\mathcal{C})_\delta$, let $A \in \Sigma$. Since $\Sigma \subset \Sigma_{P^*}$, the Carathéodory class of P^*-measurable sets, where P^* is the generated (outer) measure by (Σ_0, P), since $P(\Omega) < \infty$, we have for any $\varepsilon > 0$, a sequence $\{A_n\}_1^\infty, A_n \in \Sigma_0$ such that $A \subset \bigcup_{n=1}^{\infty} A_n$ and $P\left(\bigcup_n A_n - A\right) < \frac{\varepsilon}{2}$, or $P\left(A^c - \bigcap_{n=1}^{\infty} A_n^c\right) < \frac{\varepsilon}{2}$. But $A \in \Sigma$ implies $A^c \in \Sigma$ and $A_n^c \in \Sigma_0$, $\bigcap_{n=1}^{\infty} A_n^c \subset A^c$. Thus given $A \in \Sigma$ and $\varepsilon > 0$, there exist $B_n \in \Sigma_0$ such that $A \supset \bigcap_{n=1}^{\infty} B_n$ and $P\left(A - \bigcap_{n=1}^{\infty} B_n\right) < \varepsilon/2$. Now by the inner regularity of P on Σ_0, there exist $C_n^* \in \mathcal{C}, C_n^* \subset B_n$ such that

$$P(B_n - C_n^*) < \frac{\varepsilon}{2^{n+1}}. \tag{8}$$

Let $C_\infty^* = \bigcap_{n=1}^{\infty} C_n^* \in (\mathcal{C})_\delta$, so $C_\infty^* \subset \bigcap_{n=1}^{\infty} B_n = B \subset A$ and we have the asserted inner regularity:

$$P(A - C_\infty^*) \leq P(A - B) + \sum_{n=1}^{\infty} P(B_n - C_n) < \frac{\varepsilon}{2} + \sum_{n=1}^{\infty} \frac{\varepsilon}{2^{n+1}} = \varepsilon. \tag{9}$$

For the last part, some facts about Baire sets are needed. Recall that the Baire σ-algebra $\mathcal{B}_\alpha$ of the compact Ω_α is the σ-algebra generated by all the compact G_δ sets of Ω_α, or equivalently, the σ-algebra generated by all the real continuous functions on Ω_α. Since $\mathcal{B}_\alpha \subset \Sigma_\alpha$, and $P|\mathcal{B}_\alpha$ is a Baire measure in the first part (but P_α need not have this property when Ω_α are arbitrary noncompact spaces), the assumptions of the present part are weaker than the first case, and the same conclusion holds. In fact, since each Baire measure is (inner) regular for the class of compact G_δ-sets, the above proof holds verbatim and thus the system admits the projective limit (Ω, Σ, P) which is inner regular for the class of cylinder sets with compact bases in Ω_α's. Now it is necessary to show that every Baire set of Ω belongs to Σ. This is technical, and we include only an outline here to get a "feeling" referring the reader to Choksi [1]. (A more general situation is covered by Prokhorov's theorem below.)

We have noted after Definition 2 that Ω is a compact space and its topology is relativized from that of Ω_D which is compact in its product topology. But by the standard measure theory, it follows that the open Baire sets form a base for the topology of the spaces Ω, Ω_D, and Ω_α (cf. Halmos [1], Theorem D, p. 218). From this it is not hard to show that every Baire set of Ω_D is in $\mathcal{B}$, the σ-algebra generated by the cylinders with open Baire bases. However, every G_δ set $C \subset \Omega$ is of the form $C = \Omega \cap A$ for a G_δ set $A \subset \Omega_D$. It may now be deduced that every compact G_δ set $C \subset \Omega$ is of the form $C = \Omega \cap \bigcup_{n=1}^{\infty} U_n$, with $U_n \subset \Omega_D$, open Baire set (so $U_n \in \mathcal{B}$). Thus $C \in \mathcal{B}(\Omega)$, the trace algebra. Since the topology is also generated by $\{g_\alpha^{-1}(U_\alpha) : U_\alpha \subset \Omega_\alpha$ open Baire, $\alpha \in D\}$, it follows that for every open $V (\in \mathcal{B})$ there is an open $U_\alpha \in \Sigma_\alpha$ with $V \cap \Omega = g_\alpha^{-1}(U_\alpha)$ for some $\alpha \in D$. Hence each compact (and so each) Baire set of Ω is in $\mathcal{B}(\Omega)$, and $\mathcal{B}(\Omega) \subset \Sigma = \sigma\left(\bigcup_\alpha g_\alpha^{-1}(\Sigma_\alpha)\right)$. But the opposite inclusion is easy, because $\sigma\{g_\alpha^{-1}(U_\alpha), \alpha \in D\} = \Sigma$. □

5. Remarks. Even if each $(\Omega_\alpha, \Sigma_\alpha, P_\alpha)$ is a Borel space, we may not conclude that (Ω, Σ, P) is Borel. This is due to the relations between the topology of Ω_α's, and the continuous functions on them (and the continuous g_α mappings that connect the various spaces). There is no such relation for Borel classes in general. Hence even if $(\Omega_\alpha, \Sigma_\alpha, P_\alpha)$ are Borel spaces, we can only assert that (Ω, Σ, P) generally is a Baire space. In case D is countable, each Ω_α is separable and compact, then Borel and Baire are the same classes and the conclusion will then be that (Ω, Σ, P) is a Borel space. However, a Baire measure on a σ-compact space has a unique extension to a Borel measure on the same space and in this sense the conclusion of the last part of the above theorem is strong enough for all applications to follow. This comment will be used for a discussion of Prokhorov's result in Theorem 8 below. We note in passing that Theorem 4 is a generalization of the Kolmogorov existence theorem, and, moreover, the result and its proof hold if all measures are σ-finite.

In most studies, the probability spaces and thus the projective systems are not topological. So the preceding result is not directly applicable. Therefore we turn to abstract measure systems and prove the following fundamental result which is in the nature of a **representation theorem** of the system, in whose proof Theorem 4 will be used.

6. Theorem. *Let $\{(\Omega_\alpha, \Sigma_\alpha, P_\alpha, g_{\alpha\beta})_{\alpha<\beta} : \alpha, \beta \text{ in } D\}$ be an abstract projective system of probability spaces, each Σ_α separating points of Ω_α, i.e., ω_1, ω_2 in $\Omega_\alpha \Rightarrow$ there is $A \in \Sigma_\alpha$ with $\omega_1 \in A, \omega_2 \notin A$. Then there exists a Hausdorff topological system of (Borel) regular probability spaces $\{(S_\alpha, \mathcal{B}_\alpha, \mu_\alpha, \hat{g}_{\alpha\beta})_{\alpha<\beta} : \alpha, \beta \text{ in } D\}$ admitting a Borel regular projective limit $(S, \mathcal{B}, \mu)$ such that:*

(i) for each $\alpha \in D, S_\alpha$ is a totally disconnected compact Hausdorff space such that $(S_\alpha, \mathcal{B}_\alpha, \mu_\alpha)$ is a Borel regular probability space,

(ii) there exist mappings $u_\alpha : \Omega_\alpha \to S_\alpha, \alpha \in D$, such that $u_\alpha(\Omega_\alpha)$ is dense in S_α, and $\hat{g}_{\alpha\beta} : S_\beta \to S_\alpha$, compatible, continuous and onto,

(iii) the compatible family $\{\hat{g}_{\alpha\beta}; \alpha, \beta \text{ in } D\}$ is related to $g_{\alpha\beta}$ by $u_\alpha \circ g_{\alpha\beta} = \hat{g}_{\alpha\beta} \circ u_\beta, \alpha < \beta$, i.e., the following diagram is commutative,

$$\begin{array}{cccc} & \Omega_\beta & \stackrel{u_\beta}{\rightarrow} & S_\beta & \\ g_{\alpha\beta} & \downarrow & & \downarrow & \hat{g}_{\alpha\beta} \\ & \Omega_\alpha & \stackrel{u_\alpha}{\rightarrow} & S_\alpha & \end{array}$$

(iv) the measures P_α and μ_α are related by the equation for $\alpha \in D, \mu_\alpha = P_\alpha \circ u_\alpha^{-1}$, i.e., μ_α is the image measure of P_α under u_α, and

(v) if $\Omega = \varprojlim(\Omega_\alpha, g_{\alpha\beta}), S = \varprojlim(S_\alpha, \hat{g}_{\alpha\beta})$, and the s.m. condition holds for the given system, then there is a unique mapping $u : \Omega \to S$ such that, if $g_\alpha : \Omega \to \Omega_\alpha$ and $\hat{g}_\alpha : S \to S_\alpha$ are the canonical projections of the systems, one has $u_\alpha \circ g_\alpha = \hat{g}_\alpha \circ u, \alpha \in D$, i.e., the following diagram is commutative:

$$\begin{array}{cccc} & \Omega & \stackrel{u}{\rightarrow} & S & \\ g_\alpha & \downarrow & & \downarrow & \hat{g}_\alpha \\ & \Omega_\alpha & \stackrel{u_\alpha}{\rightarrow} & S_\alpha & \end{array}$$

The image system is unique to within a homeomorphism.

We shall present a useful result for which the above representation plays a key role following the proof. This gives an appreciation of the technical result contained in the theorem.

Proof. The representation spaces S_α are obtained using a form of the Stone space determined by Σ_α of Ω_α. Thus we take S_α to be the set of all finitely additive two valued (i.e. 0 and 1 valued) measures on Σ_α. The topology in S_α is given by the following neighborhood system: For each $A \in \Sigma_\alpha$ let $\mathcal{V}_\alpha(A)$ be defined by

$$\mathcal{V}_\alpha(A) = \{v \in S_\alpha : v(A) = 1\}. \tag{10}$$

Then for $A, B \in \Sigma_\alpha$ one has

$$\begin{aligned} \mathcal{V}_\alpha(A \cap B) &= \mathcal{V}_\alpha(A) \cap \mathcal{V}_\alpha(B) \\ \mathcal{V}_\alpha(A \cup B) &= \mathcal{V}_\alpha(A) \cup \mathcal{V}_\alpha(B), \end{aligned} \tag{11}$$

and, as is well-known, $\{\mathcal{V}_\alpha(A), A \in \Sigma_\alpha\}$ forms a base for the topology of S_α in which the latter space becomes compact Hausdorff, (since Σ_α separates points of Ω_α). Moreover, the class $\{\mathcal{V}_\alpha(A), A \in \Sigma_\alpha\}$ becomes the set of all clopen (= closed-open) sets in this topology and S_α becomes totally disconnected.

Next let us define the mappings $u_\alpha : \Omega_\alpha \to S_\alpha$ by setting $u_\alpha(\omega) = \delta_\omega$ for each $\omega \in \Omega_\alpha$ where $\delta_\omega(A) = 1$ if $\omega \in A, = 0$ if $\omega \notin A$. We claim that $u_\alpha(\Omega_\alpha)$ is dense in S_α. In fact let $\hat{\Omega}_\alpha = u_\alpha(\Omega_\alpha) \subset S_\alpha$ and $v \in S_\alpha - \hat{\Omega}_\alpha$. It suffices to show that each neighborhood of v in S_α intersects $\hat{\Omega}_\alpha$. Thus by definition of (10) for each neighborhood U of v in S_α, there exists an $A \in \Sigma_\alpha$ such that $v \in \mathcal{V}_\alpha(A) \subset U$. Hence for any $\omega \in A$, the set function $\delta_\omega \in \mathcal{V}_\alpha(A) \subset U$ so that U intersects $\hat{\Omega}_\alpha$, proving our claim.

To construct compatible mappings $\hat{g}_{\alpha\beta}$, let $\alpha < \beta$ be given and consider $g_{\alpha\beta} : \Omega_\beta \to \Omega_\alpha$ and $v \in S_\beta$. Then for each $A \in \Sigma_\alpha, g_{\alpha\beta}^{-1}(A) \in \Sigma_\beta$ and $v(g_{\alpha\beta}^{-1}(A))$ is well-defined, and $(v \circ g_{\alpha\beta}^{-1})(\cdot)$ is a finitely additive $\{0, 1\}$-valued function on Σ_α so that $v \circ g_{\alpha\beta}^{-1} \in S_\alpha$. Let this correspondence be denoted by $\bar{g}_{\alpha\beta}$ which is a well-defined mapping and $\bar{g}_{\alpha\beta} : \hat{\Omega}_\beta \to \hat{\Omega}_\alpha$. It is also onto by construction. These maps can be extended to S_α as follows. If $v \in S_\alpha$ then it defines a (finitely additive) $\tilde{v}$ on a subalgebra of $\Sigma_\beta(\alpha < \beta)$ given by $\Sigma_{0\beta} = \{g_{\alpha\beta}^{-1}(A), \ A \in \Sigma_\alpha\}$, where $\tilde{v}(g_{\alpha\beta}^{-1}(\cdot)) = v(\cdot)$. We can now extend $\tilde{v}$ from $\Sigma_{0\beta}$ to Σ_β as a finitely additive $\{0, 1\}$ valued function $\hat{v}$ (say). [This extension is similar to that of taking a filter to a maximal filter, cf., e.g., Davis [1], p. 10.] It is clear that $\hat{v} \circ g_{\alpha\beta}^{-1} = v$, and the mapping $\hat{g}_{\alpha\beta} : \hat{v} \mapsto v$ is an extension of $\bar{g}_{\alpha\beta}$ from S_β onto S_α. The construction also implies that for each $\omega \in \Omega_\beta$ we have

$$(u_\alpha \circ g_{\alpha\beta})(\omega) = (\hat{g}_{\alpha\beta} \circ u_\beta)(\omega) \quad , \quad \alpha < \beta, \tag{12}$$

so that these mappings are as in the commutative diagram of (iii).

To see that each $\hat{g}_{\alpha\beta} : S_\beta \to S_\alpha$ is continuous, let $v_0 \in S_\beta$ and consider a neighborhood U of $\hat{g}_{\alpha\beta} \circ v_0$ in S_α. Then U is of the form

$U = \{u \in S_\alpha : u(A) = 1\}$ with $A \in \Sigma_\alpha$ satisfying $(\hat{g}_{\alpha\beta} \circ v_0)(A) = 1$. If V is a neighborhood of v_0 in S_β so that $V = \{v \in S_\beta : v(g_{\alpha\beta}^{-1}(A)) = 1\}$ for A defining U, and if $v \in V$ then we have

$$(\hat{g}_{\alpha\beta} \circ v)(A) = v(g_{\alpha\beta}^{-1}(A)) = 1.$$

This implies that $\hat{g}_{\alpha\beta}^{-1}(V) \subset U$ so that each $\hat{g}_{\alpha\beta}$ is also continuous. Thus parts (ii) and (iii) are established, since the compatibility of $\{\hat{g}_{\alpha\beta}, \alpha < \beta$ in $D\}$ follows on $\hat{\Omega}_\alpha \subset S_\alpha$ from definition. In fact, for $\alpha < \beta < \gamma$, if $v \in \hat{\Omega}_\gamma$, then $\hat{g}_{\beta\gamma} \circ v \in S_\beta$ so that $\hat{g}_{\alpha\beta}(\hat{g}_{\beta\gamma} \circ v)$ is defined and we have

$$\begin{aligned} \hat{g}_{\alpha\beta}(\hat{g}_{\beta\gamma} \circ v) &= \hat{g}_{\alpha\beta}(v \circ g_{\beta\gamma}^{-1}) \\ &= (v \circ g_{\beta\gamma}^{-1}) \circ g_{\alpha\beta}^{-1} \\ &= v \circ (g_{\beta\gamma}^{-1} \circ g_{\alpha\beta}^{-1}) \\ &= v \circ g_{\alpha\gamma}^{-1} = \hat{g}_{\alpha\gamma} \circ v. \end{aligned} \tag{13}$$

Since $v \in \hat{\Omega}_\gamma$ is arbitrary, one has $\hat{g}_{\alpha\beta} \circ \hat{g}_{\beta\gamma} = \hat{g}_{\alpha\gamma}$ on $\hat{\Omega}_\gamma$. But these mappings are continuous and the $\hat{\Omega}_\alpha$ are dense in the S_α. Hence the same relations hold on the S_α. We next deduce (i) and (iv) from the above construction.

Let $\mathcal{B}_{0\alpha}$ be the class of all clopen subsets of S_α. Then it is an algebra and the σ-algebra generated by $\mathcal{B}_{0\alpha}$, denoted $\tilde{\mathcal{B}}_\alpha$, is the Baire σ-algebra of S_α. Define the (finitely) additive $\bar{\mu}_\alpha : \mathcal{B}_{0\alpha} \to \mathbb{R}^+$ by

$$\bar{\mu}_\alpha(\mathcal{V}_\alpha(A)) = P_\alpha(A), \quad A \in \Sigma_\alpha, \tag{14}$$

where $\mathcal{V}_\alpha(A)$ is given by (10). It is evident that $\bar{\mu}_\alpha$ is unambiguously defined and is additive on $\mathcal{B}_{0\alpha}$ by the properties of $\mathcal{V}_\alpha(\cdot)$ noted after (10). But since $\mathcal{B}_{0\alpha}$ is an algebra of clopen sets, the finite and σ-additivities are equivalent on it. Hence by the classical Hahn extension theorem (cf., e.g., Rao [11], p. 65), $\bar{\mu}_\alpha$ has a unique σ-additive extension to $\sigma(\mathcal{B}_{0\alpha})$, the Baire σ-algebra of S_α. Since S_α is compact, $\bar{\mu}_\alpha$ has a unique further extension, denoted $\hat{\mu}_\alpha$, to the Borel σ-algebra $\mathcal{B}_\alpha$ (cf., again Rao [11], p. 453, and the same conclusion also follows from the Henry extension theorem, *ibid* [11], p. 76 the uniqueness coming from the fact that the class $\{\mathcal{V}_\alpha(A), A \in \Sigma_\alpha\} \subset \mathcal{B}_{0\alpha}$ also forms a base for the topology of S_α as noted in Ex. 11 on p. 80 of the last reference). We next verify that $\{\hat{\mu}_\alpha, \hat{g}_{\alpha\beta}, \alpha < \beta$ in $D\}$ is a projective system.

If $A \in \Sigma_\alpha$, $\alpha < \beta$, then one has

$$\begin{aligned}\hat{\mu}_\alpha(\mathcal{V}_\alpha(A)) &= P_\alpha(A) = P_\beta(g_{\alpha\beta}^{-1}(A)), \quad \text{by hypothesis},\\ &= \hat{\mu}_\beta(\mathcal{V}_\beta(g_{\alpha\beta}^{-1}(A))), \quad \text{by definition},\\ &= \hat{\mu}_\beta(g_{\alpha\beta}^{-1}(\mathcal{V}_\alpha(A))), \quad \text{since} \quad \mathcal{V}_\beta \circ g_{\alpha\beta}^{-1} = \hat{g}_{\alpha\beta}^{-1} \circ \mathcal{V}_\alpha,\end{aligned} \tag{15}$$

and this implies that $\hat{\mu}_\alpha = \hat{\mu}_\beta \circ \hat{g}_{\alpha\beta}^{-1}$, and the system is projective. We now connect the u_α (point) mappings with the $\mathcal{V}_\alpha$ to complete the identification.

Consider $\tilde{\mathcal{V}}_\alpha$ of (10), restricted to $\hat{\Omega}_\alpha (\subset S_\alpha)$. Then we have

$$\begin{aligned}\tilde{\mathcal{V}}_\alpha(A) &= \{v \in \hat{\Omega}_\alpha : v(A) = 1\}, \quad A \in \Sigma_\alpha \quad ,\\ &= \{\delta_\omega : \omega \in A\} = \{u_\alpha(\omega), \omega \in A\} = u_\alpha(A).\end{aligned} \tag{16}$$

But for each $B \subset \hat{\Omega}_\alpha$, if $\tilde{B} = u_\alpha^{-1}(B)$, then $u_\alpha : \Omega_\alpha \to \hat{\Omega}_\alpha$ being onto, one has $u_\alpha(\tilde{B}) = B$ so that

$$\tilde{\mathcal{V}}_\alpha(\tilde{B}) = u_\alpha(\tilde{B}) = B. \tag{17}$$

Now for $B \in \mathcal{B}_{0\alpha}$, consider $\tilde{B} = u_\alpha^{-1}(B)$ and one has

$$\begin{aligned}\hat{\mu}_\alpha(B) &= \hat{\mu}_\alpha(u_\alpha(\tilde{B})) = \hat{\mu}_\alpha(\mathcal{V}_\alpha(\tilde{B})), \quad \text{by (17)},\\ &= P_\alpha(\tilde{B}), \quad \text{by definition of } \hat{\mu}_\alpha \text{ in (14)},\\ &= P_\alpha(u_\alpha^{-1}(B)).\end{aligned} \tag{18}$$

If Σ_α^* denotes the class of all P_α^*-measurable sets of Ω_α, then it is a σ-algebra on which (the outer measure P_α^* is generated by (Σ_α, P)) P_α^* is σ-additive, $P_\alpha^*|\Sigma_\alpha = P_\alpha \quad (\Sigma_\alpha \subset \Sigma_\alpha^*)$. Then (18) implies that u_α is $(\Sigma_\alpha^*, \mathcal{B}_\alpha)$-measurable. Since we may assume that P_α is Carathéodory regular, for convenience, this shows that Σ_α^* is P_α-completion of Σ_α and we can conclude that u_α is P_α-measurable. This gives (i) and (iv) of the theorem. Note that by Theorem 4, the above representation system $\{(S_\alpha, \mathcal{B}_\alpha, \mu_\alpha, \hat{g}_{\alpha\beta}) : \alpha < \beta \text{ in } D\}$ admits a Baire and then Borel regular projective system with Baire (and then Borel) regular projective limit $(S, \mathcal{B}, \mu)$ where $S = \varprojlim(S_\alpha, \hat{g}_{\alpha\beta})$ and $\mu = \varprojlim(\hat{\mu}_\alpha, \hat{\mathcal{B}}_\alpha)$. Let $\Omega = \varprojlim(\Omega_\alpha, g_{\alpha\beta})$.

We next construct $u : \Omega \to S$ as required in the theorem. Since now the given projective system is assumed to satisfy the s.m. condition,

$\Omega \neq \emptyset$ and the canonical mappings $g_\alpha = g_{\alpha\infty} : \Omega \to \Omega_\alpha$ and $\hat{g}_\alpha = \hat{g}_{\alpha\infty} : S \to S_\alpha$ are well-defined, and $g_\alpha = g_{\alpha\beta} \circ g_\beta$ and similarly for $\hat{g}_\alpha$. Set $h_\alpha = u_\alpha \circ g_\alpha : \Omega \to S_\alpha$. Then for $\alpha < \beta$.

$$h_\alpha = u_\alpha \circ (g_{\alpha\beta} \circ g_\beta) = (\hat{g}_{\alpha\beta} \circ u_\beta) \circ g_\beta = \hat{g}_{\alpha\beta} \circ h_\beta. \tag{19}$$

Consider a 'thread' $u(\omega) = \{\tilde{u}_\alpha(\omega) \in S_\alpha, \alpha \in D\} \in \times_{\alpha \in D} S_\alpha \supset S$. In order that $u(\omega) \in S$, we need to have $\tilde{u}_\alpha = g_{\alpha\beta} \circ \tilde{u}_\beta$ for $\alpha < \beta$. But (19) shows that this holds if $\tilde{u}_\alpha = h_\alpha$. Thus $u(\omega) = \{h_\alpha(\omega) : \alpha \in D\} \in S, \omega \in \Omega$ and $u : \Omega \to S$ is the desired function. The uniqueness is immediate since two such mappings are different only if at least one of the components is distinct. So (v) also holds. □

A family of mappings $\{u_\alpha, \alpha \in D\}$ between the systems of spaces $\{(\Omega_\alpha, g_{\alpha\beta})_{\alpha<\beta} : \alpha, \beta \text{ in } D\}$ and $\{(S_\alpha, \hat{g}_{\alpha\beta})_{\alpha<\beta} : \alpha, \beta \text{ in } D\}$ such that $u_\alpha \circ g_{\alpha\beta} = \hat{g}_{\alpha\beta} \circ u_\beta$ for $\alpha < \beta$ (everywhere) is said to be a *projective system* of the first set of spaces into the second. By Theorem 6(v), there is a unique mapping $u : \Omega \to S$ where $\Omega = \varprojlim(\Omega_\alpha, g_{\alpha\beta}), S = \varprojlim(S_\alpha, \hat{g}_{\alpha\beta})$. This u is called the *projective limit* of the family $\{u_\alpha, \alpha \in D\}$, and denoted $u = \varprojlim u_\alpha$. This implies

$$u(\Omega) \subset \varprojlim u_\alpha(\Omega_\alpha) \subset S. \tag{20}$$

The inclusion can be proper and $u(\Omega) \notin \mathcal{B} = \sigma\left(\bigcup_\alpha \hat{g}_\alpha^{-1}(\mathcal{B}_\alpha)\right)$ so that u need not be $(\Sigma, \mathcal{B})$-measurable. However, by Proposition 1, there is a unique finitely additive P on $\Sigma_0 = \bigcup g_\alpha^{-1}(\Sigma_\alpha), P(\Omega) = 1$, whenever s.m. holds and $g_\alpha(\Omega) = \Omega_\alpha$ for all $\alpha \in D$. The problem of σ-additivity of P on Σ_0 and the "size" of $u(\Omega)$ in S (as well as the essential equality in (10)) are interrelated. [Here "size" refers to the measure of the set.]

The following result gives a characterization of projective systems admitting limits, and also answers the above question about the "size" of $u(\Omega)$ in S. This plays an important part in many places – both in theory as well as applications.

7. Theorem. *Let $\{(\Omega_\alpha, \Sigma_\alpha, P_\alpha, g_{\alpha\beta})_{\alpha<\beta} : \alpha, \beta$ in $D\}$ be a projective system as in Theorem 6, and satisfying the s.m. condition. Let $\{(S_\alpha, \mathcal{B}_\alpha, \mu_\alpha, \hat{g}_{\alpha\beta})_{\alpha<\beta} : \alpha, \beta$ in $D\}$ be the (compact) representing projective system of the above family, guaranteed by Theorem 6, and let*

$u_\alpha : \Omega_\alpha \to S_\alpha, u = \varprojlim u_\alpha, \mu = \varprojlim u_\alpha$ which exist (and μ is Baire and then Borel regular) as in Theorem 6. Then $P = \varprojlim P_\alpha$ exists (i.e., P of Proposition 1 is σ-additive) iff $u(\Omega)$ is μ-full or thick in S (i.e., $\mu^*(u(\Omega)) = \mu(S) = 1$ where μ^* is the outer measure generated by $(\mathcal{B}, \mu)$ in S). Moreover, $\mu = P \circ u^{-1}$.

After giving a proof of this result we shall illustrate it at various places in the following work.

Proof. For the direct part, let $u(\Omega) \subset S$ be μ-full, and consider the trace σ-algebra $\tilde{\mathcal{B}} = \{B \cap u(\Omega) : B \in \mathcal{B}\}$, $(S, \mathcal{B}, \mu)$ being the (regular) projective limit of the representing system, given by Theorem 6. If $\tilde{\mu}(B \cap C) = \mu(B)$ where $C = u(\Omega)$, the full (or μ-thick) set, then $\tilde{\mu}$ is well-defined and, by a classical measure theory result, is a probability function on $\tilde{\mathcal{B}}$ (cf., e.g., Rao [11], p. 75). Define $\tilde{P} : \tilde{\Sigma} = u^{-1}(\mathcal{B}) \to [0,1]$, by the equation $\tilde{P}(u^{-1}(B)) = \mu(B)(= \tilde{\mu}(C \cap B))$. It is evident that $\tilde{P}$ is a probability measure on $\tilde{\Sigma}$. We now assert that $P = \tilde{P}|\Sigma_0$ where $\Sigma_0 = \bigcup_\alpha g_\alpha^{-1}(\Sigma_\alpha)$, so that P is σ-additive on the algebra Σ_0 and has a unique extension to $\Sigma = \sigma(\Sigma_0)$ which thus shows $P = \varprojlim(P_\alpha, g_\alpha)$ exists and the desired result follows.

Let $A \in \Sigma_0$ so that $A = g_\alpha^{-1}(A_\alpha)$ for some $\alpha \in D$, where $A_\alpha \in \Sigma_\alpha$ and $P(A) = P_\alpha(A_\alpha)$, by Proposition 1. Using the notation of the preceding theorem and its proof, let $B_\alpha = \tilde{\mathcal{V}}_\alpha(A_\alpha) \subset \Omega_\alpha$ where $\tilde{\mathcal{V}}_\alpha$ is the restriction to $\hat{\Omega}_\alpha$. Then, as in (17) and (18), $u_\alpha(u_\alpha^{-1}(B_\alpha)) = B_\alpha$ and $\tilde{\mathcal{V}}_\alpha(A_\alpha) = u_\alpha(A_\alpha)$ so that

$$\hat{\mu}_\alpha(B_\alpha) = \hat{\mu}_\alpha(u_\alpha(A_\alpha)) = P_\alpha(A_\alpha) = P_\alpha(u_\alpha^{-1}(B_\alpha)). \tag{21}$$

Using the fact that $u_\alpha \circ g_\alpha = \hat{g}_\alpha \circ u$ of Theorem 6(v), we have

$$\begin{aligned} P(A) &= P_\alpha(A_\alpha) = \hat{\mu}_\alpha(B_\alpha) \quad , \quad \text{by (21)}, \\ &= \hat{\mu}(\hat{g}_\alpha^{-1}(B_\alpha)) = \tilde{\mu}(\hat{g}_\alpha^{-1}(B_\alpha) \cap C) \\ &= \tilde{P}(u^{-1} \circ \hat{g}_\alpha^{-1}(B_\alpha)) = \tilde{P}(A), \quad A \in \Sigma_0. \end{aligned} \tag{22}$$

Thus P and $\tilde{P}$ agree on Σ_0. By the earlier reduction, this shows that P is σ-additive and the result follows in this direction.

For the converse, let $C = u(\Omega)$ and consider $\tilde{\mathcal{B}}_0 = \{C \cap B : B \in \mathcal{B}_0\}$ the trace of the algebra $\mathcal{B}_0$ on C, $(\mathcal{B}_0 = \bigcup_\alpha g_\alpha^{-1}(\mathcal{B}_\alpha))$. If $P =$

$\underleftarrow{\lim}(P_\alpha, g_{\alpha\beta})$, let $\tilde{\mu}(C \cap B) = P(u^{-1}(B)), B \in \tilde{\mathcal{B}}_0$. Then $\tilde{\mu}$ is well-defined since $C \cap B = \emptyset$ implies $u^{-1}(C \cap B) = \emptyset$ so that

$$\emptyset = u^{-1}(C \cap B) = u^{-1}(C) \cap u^{-1}(B) = \Omega \cap u^{-1}(B) = u^{-1}(B),$$

and $P(u^{-1}(B)) = 0$. The σ-additivity of P makes $\tilde{\mu}$ a measure on the algebra $\tilde{\mathcal{B}}_0$. If μ^* is the outer measure generated by $(\mathcal{B}_0, \hat{\mu})$, then we claim that $\tilde{\mu}$ and μ^* agree on $\tilde{\mathcal{B}}_0$ and hence on $\sigma(\tilde{\mathcal{B}}_0)$ and $\mu^*(C) = \tilde{\mu}(C) = 1$ which will complete the argument.

If $B \in \mathcal{B}_0$, then $B = \hat{g}_\alpha^{-1}(B_\alpha), B_\alpha \in \mathcal{B}_\alpha$ for some $\alpha \in D$. Hence we have

$$\begin{aligned}
\hat{\mu}(B) = \hat{\mu}_\alpha(B_\alpha) &= P_\alpha(u_\alpha^{-1}(B_\alpha)), \quad \text{as in (22)}, \\
&= P(g_\alpha^{-1}(u_\alpha^{-1}(B_\alpha))) \\
&= P(u^{-1}(\hat{g}_\alpha^{-1}(B_\alpha))), \quad \text{since } u_\alpha \circ g_\alpha = \hat{g}_\alpha \circ u, \\
&= P(u^{-1}(B)) = \tilde{\mu}(C \cap B) = \tilde{\mu}(C \cap g_\alpha^{-1}(B_\alpha)). \qquad (23)
\end{aligned}$$

If μ^* is the outer measure generated by $(\mathcal{B}_0, \tilde{\mu})$, then $\mu^*|\mathcal{B}_0 = \hat{\mu}$ and taking $B_\alpha = S_\alpha$ in (23), we get from $\mu^*(C \cap \tilde{g}_\alpha^{-1}(B_\alpha)) = \tilde{\mu}(C \cap \hat{g}_\alpha^{-1}(B_\alpha)$, that $\mu^*(C) = \hat{\mu}_\alpha(S_\alpha) = 1$ so that C is μ-full. Then the argument of the direct part of Theorem 6 shows that μ is essentially the image measure of P under u. □

The next result, due to Prokhorov [1], is an extension of the last part of Theorem 4 for compact spaces, used in Theorem 6. It gives a characterization of topological projective systems without imposing the s.m. condition explicitly. It is actually an extension (due to Bourbaki [1]) of the original statement of Prokhorov's, and the proof is somewhat involved. We omit it here refering to the detailed version given in (Rao [11], pp. 359-364).

8. Theorem. *Let $\{(\Omega_\alpha, \Sigma_\alpha, P_\alpha, g_{\alpha\beta})_{\alpha<\beta} : \alpha, \beta \text{ in } D\}$ be a projective system of probability spaces such that : (i) each Ω_α is Hausdorff, (ii) Σ_α contains the compact subsets of Ω_α and P_α is inner regular (for the compact sets) and (iii) $g_{\alpha\beta}$'s are continuous. Let $\Omega = \underleftarrow{\lim}(\Omega_\alpha, g_{\alpha\beta})$ be nontrivial and the canonical mappings $g_\alpha : \Omega \to \Omega_\alpha, \alpha \in D(g_\alpha = g_{\alpha\beta} \circ g_\beta$ for $\alpha < \beta)$ be continuous and separate points of Ω. Then $(\Omega, \Sigma, P) = \underleftarrow{\lim}(\Omega_\alpha, \Sigma_\alpha, P_\alpha, g_{\alpha\beta})$ exists and the probability P is inner regular (for the compact sets of Ω) iff the following condition holds:*

(*) For each $\varepsilon > 0$, there is a compact set $K_\varepsilon \subset \Omega$ (so $g_\alpha(K_\varepsilon) \in \Sigma_\alpha$) such that $P_\alpha(\Omega_\alpha - g_\alpha(K_\varepsilon)) < \varepsilon, \alpha \in D$.

If the limit measure is a Radon probability, then the condition (*) is clearly true, and so only the sufficiency is to be proved. By Theorem 4, $P = \underset{\leftarrow}{\lim} P_\alpha$ exists under the hypotheses (i)-(iii) even if the condition (*) is not valid (but then P has no such regularity property). If each Ω_α is compact, then again (*) follows by Theorem 4 and Remark 5. In the general case, one needs a different argument.

The following extension of the above theorem of Prokhorov, also due to the same author, is what we use in applications.

9. Theorem. *Let $\{(\Omega_\alpha, \Sigma_\alpha, P_\alpha, g_{\alpha\beta})_{\alpha<\beta} : \alpha, \beta \text{ in } D\}$ be a projective system of probability spaces satisfying the conditions (i)-(iii) of the above theorem. Let T be any Hausdorff topological space. Suppose there is a set $\{h_\alpha, \alpha \in D\}$ of mappings such that (a) $h_\alpha : T \to \Omega_\alpha$ is continuous, $\alpha \in D$, and (b) $h_\alpha = g_{\alpha\beta} \circ h_\beta$ for each $\alpha < \beta$ in D. Then there exists a Radon probability Q on the Borel algebra of T with $Q \circ h_\alpha^{-1} = P_\alpha, \alpha \in D$, iff the following condition holds:*

(′) For each $\varepsilon > 0$, there is a compact set $K_\varepsilon \subset T$ such that $P_\alpha(\Omega_\alpha - h_\alpha(K_\varepsilon)) < \varepsilon, \alpha \in D$.*

If the h_α's separate points of T, then Q is unique.

Proof. If there is such a Q, then (*′) clearly holds. Conversely, let $\Omega = \underset{\leftarrow}{\lim}(\Omega_\alpha, g_{\alpha\beta})$. Then, as in Theorem 6(v), one finds a unique $u : T \to \Omega$ such that $h_\alpha = g_\alpha \circ u$. In fact, the last equation symbolizes that $h_\alpha(t)$ equals the αth coordinate of $u(t) \in \underset{\alpha\in D}{\times} \Omega_\alpha$, and we need to show that $u(t) \in \Omega$ for all $t \in T$, or that $u(t) = (u_\alpha(t), \alpha \in D)$ is a thread. This means, for $\alpha \leq \beta, g_{\alpha\beta}(u_\beta(t)) = u_\alpha(t), t \in T$. But by (b) of the hypothesis, this holds if $u_\alpha = h_\alpha, \alpha \in D$. Thus $u : t \mapsto \{h_\alpha(t), \alpha \in D\}$ is the desired mapping since the uniqueness follows immediately. Note that since each component h_α is continuous by hypothesis, it is seen that u is continuous from the space T into Ω. Thus the following diagram is commutative with continuous mappings.

$$\begin{array}{ccccc} & T & \xrightarrow{u} & \Omega & \\ h_\beta & \downarrow & \searrow h_\alpha & \downarrow & g_\alpha \\ & \Omega_\beta & \xrightarrow{g_{\alpha\beta}} & \Omega_\alpha & \end{array}$$

If $\tilde{K}_\varepsilon = u(K_\varepsilon) \subset \Omega$, then $\tilde{K}_\varepsilon$ is compact and $(*')$ reduces to $(*)$ of Theorem 8. Hence $P = \lim\limits_{\leftarrow} P_\alpha$ is a Radon measure on Ω. Define Q on $u^{-1}(\Sigma)$ by the equation $Q \circ u^{-1} = P$. Then $Q(T) = 1$ and Q is the desired probability on T. Indeed, if $\ell(f) = \int_\Omega f \circ u^{-1}\, dP$, for $f \in C(K_\varepsilon)$, the space of real continuous functions on the compact set K_ε, then by the Riesz representation (cf., Royden [1], p. 310) we have $\ell(f) = \int_T f dQ$ with $Q(K_\varepsilon) = P(u(K_\varepsilon))$. Letting $\varepsilon \downarrow 0$ through a sequence, one deduces that Q is a regular probability on T, with a σ-compact support. The uniqueness, when the h_α separate points, is immediate since then u is one-to-one, as seen from its definition. □

Remark. This extension of Theorem 8 is particularly useful for applications of the next section where T is a Banach space and Ω_α's are finite dimensional subspaces of T. The projective limit space Ω is much bigger than the original space T. But the above theorem gives conditions for a Radon probability to live on T itself.

1.4 Applications of projective limits

In this section we illustrate the preceding general theory on two problems. The first application is an immediate consequence of Theorem 3.9 but its reformulation leads to a very useful theorem of Gross [2] on abstract Wiener spaces, and we use it later. The second application is to a construction of a Gaussian measure on $\mathbb{R}^\infty$, the space of real sequences endowed with the projective limit topology; equivalently, we obtain a Gaussian probability on $\mathbb{R}^\infty$ as a projective limit of uniform measures on n-spheres (of $\mathbb{R}^{n+1}, n \geq 1$).

Let $\mathcal{X}$ be a real topological vector space and $\mathcal{F}$ be the class of all closed subspaces of finite deficiency, i.e., $F \in \mathcal{F}$ iff the quotient space $\mathcal{X}/F$ (considered as a vector space) is finite dimensional. If α, β are in $\mathcal{F}$, then define an ordering "<" by the rule: $\alpha < \beta$ iff $\alpha \supset \beta$. Then $(\mathcal{F}, <)$ becomes a directed set. Let $\mathcal{X}_\alpha = \mathcal{X}/\alpha$ and $\varphi_\alpha : \mathcal{X} \to \mathcal{X}_\alpha$ be the (linear) canonical projection and similarly let $\varphi_{\alpha\beta} : \mathcal{X}_\beta \to \mathcal{X}_\alpha$ for $\alpha < \beta$ in $\mathcal{F}$. Then for $\alpha < \beta < \gamma$ we have the composition rules $\varphi_{\alpha\beta} \circ \varphi_{\beta\gamma} = \varphi_{\alpha\gamma}$ and $\varphi_{\alpha\alpha} =$ identity, which follow from the elementary properties of quotients in vector spaces. Thus the family $\{(\mathcal{X}_\alpha, \varphi_{\alpha\beta})_{\alpha<\beta} : \alpha, \beta \text{ in } \mathcal{F}\}$ forms a projective system of linear spaces with $\varphi_{\alpha\beta}$'s as continuous open mappings whenever $\mathcal{X}$ has a locally convex topology (e.g., a Banach space). Let P_α be a probability measure on the Borel σ-algebra Σ_α of $\mathcal{X}_\alpha$ such that for $\alpha < \beta$ we have $P_\alpha = P_\beta \circ \varphi_{\alpha\beta}^{-1}$. Then $\{(\mathcal{X}_\alpha, \Sigma_\alpha, P_\alpha, \varphi_{\alpha\beta})_{\alpha<\beta} : \alpha, \beta \text{ in } \mathcal{F}\}$ is a projective system of probability spaces.

Let $\Omega = \varprojlim(\mathcal{X}_\alpha, \varphi_{\alpha\beta})$. How are $\mathcal{X}$ and Ω related? If $\mathcal{X}$ is finite dimensional they may obviously be identified. However, in the infinite dimensional case they are essentially different. An easy computation shows that the projective limit topology of Ω is weaker than the given locally convex topology of $\mathcal{X}$ and, by the next result, $\mathcal{X} \subsetneqq \Omega$. If $\mathcal{X}$ is as above, let $\mathcal{X}^*$ be its topological adjoint and $\mathcal{X}'$ its algebraic adjoint vector spaces.

1. Proposition. *Let $\{(\mathcal{X}_\alpha, \varphi_{\alpha\beta})_{\alpha<\beta} : \alpha, \beta$ in $\mathcal{F}\}$ be the projective system of (finite dimensional quotient) vector spaces defined from a locally convex topological vector space $\mathcal{X}$, and let $\Omega = \varprojlim(\mathcal{X}_\alpha, \varphi_{\alpha\beta})$. Then Ω and $\mathcal{X}^{*'}$ can be identified in the sense that there is a bicontinuous linear bijection between these two spaces, where Ω is endowed with the projective limit topology and $\mathcal{X}^{*'}$ is the algebraic dual of $\mathcal{X}^*$ with $\sigma(\mathcal{X}^{*'}, \mathcal{X}^*)$-topology. Thus under the natural identifications we have the (generally proper) inclusions $\mathcal{X} \subset \mathcal{X}^{**} \subset \mathcal{X}^{*'} \cong \Omega$, where $\mathcal{X}^*$ and $\mathcal{X}^{**}$ are the first and second adjoint spaces of $\mathcal{X}$.*

We shall outline a proof of this proposition in the Complements section at the end of the chapter. A reason for the locally convex topology here is that, in the more general cases, there may not be enough continuous linear functionals on $\mathcal{X}$ in which case this result fails.

Under the hypothesis of the proposition, consider the projective system of probability spaces $\{(\mathcal{X}_\alpha, \Sigma_\alpha, P_\alpha, \varphi_{\alpha\beta})_{\alpha<\beta} : \alpha, \beta \text{ in } \mathcal{F}\}$. In this case (of vector spaces) the system is also called a "cylindrical measure," but as we know by Proposition 3.1, it only determines on (Ω, Σ) a finitely additive set function, so that the appellation "measure" is somewhat inappropriate. Mostly we adopt the former terminology. Let P be the additive set function on $\Sigma_0 = \bigcup_\alpha g_\alpha^{-1}(\Sigma_\alpha)$ where $g_\alpha : \Omega \to \mathcal{X}_\alpha$ (and the g_α is the same as φ_α when its domain is the natural image of $\mathcal{X}$ in $\mathcal{X}^{*'} \cong \Omega$) is the canonical mapping of Ω onto $\mathcal{X}_\alpha$ (the "onto" parts are automatic for all these quotient mappings). By Theorem 3.8 we can present the best condition for the σ-additivity of P

– namely, for each $\varepsilon > 0$, there exists a compact set $K_\varepsilon \subset \Omega$ such that $P_\alpha(\mathcal{X}_\alpha - \varphi_\alpha(K_\varepsilon)) < \varepsilon, \alpha \in \mathcal{F}$. However, if it is desired to have P supported by $\mathcal{X}$ itself, it is necessary (and obviously sufficient) that the $K_\varepsilon \subset \mathcal{X}$. For this we apply Theorem 3.9. Since the φ_α's separate points of $\mathcal{X}$, one has the following:

2. Proposition. *Let $\mathcal{X}$ be a (real) locally convex topological vector space, and consider a projective system $\{(\mathcal{X}_\alpha, \Sigma_\alpha, P_\alpha, \varphi_{\alpha\beta})_{\alpha<\beta} : \alpha, \beta$ in $\mathcal{F}\}$ where Σ_α is the Borel σ-algebra and P_α on Σ_α is an inner regular probability. Then the system admits a regular limit (so, $P = \underleftarrow{\lim} P_\alpha$) and the limit probability is supported by $\mathcal{X}$ iff for each $\varepsilon > 0$, there exists a compact set $K_\varepsilon \subset \mathcal{X}$ such that $(*) P_\alpha(\mathcal{X}_\alpha - g_\alpha(K_\varepsilon)) < \varepsilon$, for all $\alpha \in \mathcal{F}$ where g_α is the canonical mapping of Ω onto $\mathcal{X}_\alpha [g_\alpha | \mathcal{X} = \varphi_\alpha]$.*

In general $\mathcal{X}$ can be very small in Ω. It may even have P-measure zero. (Then, of course, the condition $(*)$ of Proposition 2 cannot hold.) On the other hand, one may consider the product space $\times_{\alpha\in\mathcal{F}} \hat{\mathcal{X}}_\alpha$, where $\hat{\mathcal{X}}_\alpha$ is the Stone-Čech compactification of $\mathcal{X}_\alpha$, and then construct an extension μ using the procedure of Kolmogorov's original theorem with coordinate projections π_α's, as the connecting maps. If P is the additive set function of the projective system of Proposition 2, which is not σ-additive, then $P \circ g_\alpha^{-1} = \mu \circ \pi_\alpha^{-1} = P_\alpha, \alpha \in \mathcal{F}$, and even Ω (and not only $\mathcal{X}$) can have μ-measure zero.

As it stands, the preceding result is a little too general. Actually one should know the class of locally convex vector spaces $\mathcal{X}$ for which the above criterion can be applied. A characterization of such spaces presents a formidable problem. For instance, if $\mathcal{X} = \mathcal{Y}^*$ (i.e., $\mathcal{X}$ is an adjoint space of a locally convex vector space $\mathcal{Y}$) then, by a result of R. A. Minlos, for the σ-additivity of P in $\mathcal{X}$ (i.e., in order that $P = \underleftarrow{\lim} P_\alpha$ exist), it is necessary that $\mathcal{Y}$ belong to a special class called "nuclear" spaces. (Cf., Gel'fand and Vilenkin [1], Ch. IV, for details.) This does *not* cover the cases where $\mathcal{X}$ is not an adjoint space. Even a determination of spaces $\mathcal{X}$ for which $\mathcal{X} \cong \Omega (= \underleftarrow{\lim}(\mathcal{X}_\alpha, \varphi_{\alpha\beta}))$ is nontrivial. For a detailed study of the spaces for which some of the preceding characterizations can be obtained, we have to refer the reader to the monograph by L. Schwartz [1].

A specialization of the above proposition when $\mathcal{X}$ is a Hilbert space leads to an interesting class, called the *abstract Wiener spaces.* We include here an account of this, and use it later in a study of certain stochastic differential equations.

In pursuing the stated specialization we first make the relations explicit between the linearity of the space $\mathcal{X}$ and the projective systems (or "cylindrical measures") on it.

3. Definition. Let (Ω, Σ, μ) be a probability space and $L^0(\mu)$ be the class of all (real) random variables on Ω. If $\mathcal{X}$ is a locally convex vector space and $\mathcal{X}^*$ its adjoint, consider a (not necessarily continuous) linear mapping $F : \mathcal{X}^* \to L^0(\mu)$. Let $\mathcal{F}$ be the set of equivalence classes of all such mappings, i.e., $\hat{F} \in \mathcal{F}$ iff $F_1, F_2 \in \hat{F} \Rightarrow \mu \circ T_1^{-1} = \mu \circ T_2^{-1}, T_j = (F_j(y_1), \dots, F_j(y_n))', n \geq 1, \{y_i\}_1^n \subset \mathcal{X}^*, j = 1, 2$. Then each member of $\mathcal{F}$ is called a *weak distribution* over $\mathcal{X}$ (or in $\mathcal{X}^*$), relative to μ.

Now $\{(\hat{F}(y), y \in \mathcal{X}^*\}$ is a stochastic process on (Ω, Σ, μ). Any member F of $\hat{F}$ is also called a *random linear mapping.* If $\mathcal{X} = C_c^\infty(\mathbb{R})$, the space of infinitely differentiable real functions vanishing off compact sets ($\{f_n, n \geq 1\} \subset C_c^\infty(\mathbb{R})$ is said to converge to f, if $(f_n, f)_{n \geq 1}$ vanish outside a fixed compact set, $D^p f = f^{(p)}$, and $f_n^{(p)}(x) \to f^{(p)}(x)$ uniformly in x, for each $p \geq 0$ so that the limit f is in $C_c^\infty(\mathbb{R})$ [this defines a locally convex topology; $C_c^\infty(\mathbb{R})$ is called a Schwartz space], and if $L^0(\mu)$ is topologized (nonlocally convex in general) by saying that $\{g_n, n \geq 1\} \subset L^0(\mu)$ converges when $g_n \to g$ in probability, then a random linear mapping $F : \mathcal{X}^* \to L^0(\mu)$ is a *generalized random process* or a *random Schwartz distribution* whenever F is also continuous. Clearly the concept makes sense if $\mathcal{X}$ is any locally convex vector space. The notion of a weak (respectively, random Schwartz) distribution is due to I. Segal (respectively I. Gel'fand, and independently to K. Itô).

We present a positive solution of the following natural existence problem: Given $\mathcal{X}$ (hence $\mathcal{X}^*$) as above, can one find some probability space (Ω, Σ, μ) so as to obtain a specified (type of) weak distribution on $\mathcal{X}^*$ into $L^0(\mu)$?

4. Theorem. *Let $\mathcal{C}$ be the class of all cylindrical probabilities (i.e., all projective systems of probability spaces) on $\mathcal{X}$ as defined after Proposition 1. Let $\mathcal{W}$ be the class of all weak distributions from $\mathcal{X}^*$ (each relative to some probability space). Then there exists a bijective correspondence between $\mathcal{C}$ and $\mathcal{W}$ and hence they can be identified.*

Proof. Let $f \in \mathcal{W}$. This implies the existence of a probability space

(Ω, Σ, μ) and $f : \mathcal{X}^* \to L^0(\mu)$ defines a random linear mapping where f is identified with its equivalence class (or a member of the weak distribution). If $T_n = (x_1^*, \ldots, x_n^*)', x_i^* \in \mathcal{X}^*$, then $T_n : \mathcal{X} \to \mathbb{R}^n$ is a continuous linear mapping and every linear operator between these spaces is of that form. If D is the family of all finite subsets of $\mathcal{X}^*$, directed $(<)$ by inclusion, then T_n (or $\mathrm{sp}(\{x_i^*\}_1^n)$) *identifies* as an $\alpha(\in D)$, and $f \circ T_n = f \circ \alpha = (f(x_1^*), \ldots, f(x_n^*))'$ is a random vector. If $\mathcal{X}_\alpha \subset \mathcal{X}$ is a finite dimensional subspace (i.e., α "lives" on $\mathcal{X}_\alpha$) with $\alpha \in D$, we show that there is a projective system $\{(\mathcal{X}_\alpha, \mathcal{B}_\alpha, P_\alpha, g_{\alpha\beta})_{\alpha<\beta} : \alpha, \beta \text{ in } D\}$ determined by f. Evidently the notation $\mathcal{X}_\alpha$ here, and in Propositions 1 and 2, is only different in appearance. Indeed, if $\mathcal{N}$ is the annihilator of $\mathrm{sp}(\{x_i^*\}_1^n) = \mathrm{sp}(\alpha)$, then $\mathcal{N}$ is of finite deficiency and $\mathcal{X}/\mathcal{N}$ is $\mathcal{X}_\alpha$. So the two definitions are equivalent. We can and will take the elements of α here as linearly independent.

Let $T_n \cong \alpha$ (identified) and $\beta \cong \tau_m = (y_1^*, \ldots, y_m^*)'$. So $\tau_m : \mathcal{X} \to \mathbb{R}^m$. Since $f(x^*) : \Omega \to \mathbb{R}$ is a random variable, if $\tilde{f} : \Omega \to \mathbb{R}^{\mathcal{X}^*}$ is defined by $\tilde{f}(\omega)|_{x^*} = f(x^*)(\omega)$, then $\tilde{f}$ denotes a canonical representation of the "process" f, in terms of Remark 2.3. We can write this as: $\tilde{f}(\omega)(x^*) = f(x^*)(\omega)$. Recalling that π_α is a coordinate projection on $\mathbb{R}^{\mathcal{X}^*} \to \mathbb{R}^\alpha, \alpha \in D$, we may express similarly $\{\pi_\alpha(\tilde{f}), \alpha \in D\}$ as a set of random vectors, since $\pi_\alpha(\tilde{f}) = f \circ \alpha$ is a measurable mapping of Ω into $\mathbb{R}^n$. If $\tilde{\mu}_\alpha = \mu \circ (f \circ \alpha)^{-1}$, then $\tilde{\mu}_\alpha$ is the image probability on the Borel algebra $\tilde{\mathcal{B}}_\alpha$ of $\mathbb{R}^\alpha, \alpha \in D$. To verify the compatibility of $\tilde{\mu}_\alpha$ and $\tilde{\mu}_\beta$, consider the linear mappings T_n and τ_m, representing α and β defined above, between $\mathcal{X}$ and $\mathbb{R}^n$, and $\mathcal{X}$ and $\mathbb{R}^m$. By an elementary analysis one can find an operator (coordinate projection) $u : \mathbb{R}^n \to \mathbb{R}^m$ such that $u \circ T_n = \tau_m$. In coordinate notation, if $u = (u_{ij})$, this gives $y_j^* = \sum_{i=1}^n u_{ij} x_i^*$, and hence for μ-almost all $\omega \in \Omega$, we have:

$$
\begin{aligned}
f(y_j^*)(\omega) = \tilde{f}(\omega)(y_j^*) &= \sum_{i=1}^n u_{ij} \tilde{f}(\omega)(y_j^*), \\
&= \sum_{i=1}^n u_{ij} f(x_i^*)(\omega), \quad 1 \le j \le m. \qquad (1)
\end{aligned}
$$

Thus $f o \beta = u o f o \alpha$ and $\tilde{\mu}_\beta = \mu o (f o \beta)^{-1} = \mu o (f o \alpha)^{-1} o u^{-1} = \tilde{\mu}_\alpha o u^{-1}$. Now let $g_{\alpha\beta} : \mathcal{X}_\beta \to \mathcal{X}_\alpha$ be the canonical projection for $\alpha < \beta$. We can find a bijection $u_\alpha : \mathcal{X}_\alpha \to \mathbb{R}^\alpha$ such that $g_{\alpha\beta} = u_\alpha^{-1} \circ u \circ u_\beta$. Let

$\mathcal{B}_\alpha = u_\alpha^{-1}(\tilde{\mathcal{B}}_\alpha)$. Then $\mathcal{P} = \{(\mathcal{X}_\alpha, \mathcal{B}_\alpha, \mu_\alpha, g_{\alpha\beta})_{\alpha<\beta} : \alpha, \beta \text{ in } D\}$, with $\mu_\alpha \circ u_\alpha^{-1} = \tilde{\mu}_\alpha$, is the desired system. If f and g are two equivalent members of the weak distributional definition, $\mu \circ (f \circ \alpha)$ and $\mu \circ (g \circ \alpha)$ define the same measures for $\alpha \in D$, and so the *same* projective system results. Thus $\mathcal{P} \in \mathcal{C}$.

Conversely, let an element $\mathcal{P} \in \mathcal{C}$ be given. To construct a weak distribution, we first need to show the existence of a probability space (Ω, Σ, μ). Let $\mathcal{P} = \{(\mathcal{X}_\alpha, \Sigma_\alpha, P_\alpha, g_{\alpha\beta})_{\alpha<\beta} : \alpha, \beta \text{ in } D\}$ where each $\mathcal{X}_\alpha$ is finite dimensional. Let $\mathbb{R}^\alpha$ be the Euclidean space of the same dimension as $\mathcal{X}_\alpha$ and $u_\alpha : \mathcal{X}_\alpha \to \mathbb{R}^\alpha$ be a bijection between these spaces (which exists by the elementary analysis as noted above), where $\alpha \in D$. These u_α's can be further chosen such that for $\alpha < \beta$ if $\pi_{\alpha\beta} : \mathbb{R}^\beta \to \mathbb{R}^\alpha$ is the coordinate projection, then the mappings $g_{\alpha\beta} : \mathcal{X}_\beta \to \mathcal{X}_\alpha$ and $\pi_{\alpha\beta}$ are again related by the commutative diagram:

$$\begin{array}{ccccc} & \mathcal{X}_\beta & \xrightarrow{u_\beta} & \mathbb{R}^\beta & \\ g_{\alpha\beta} & \downarrow & & \downarrow & \pi_{\alpha\beta}, \\ & \mathcal{X}_\alpha & \xrightarrow{u_\alpha} & \mathbb{R}^\alpha & \end{array} \qquad u_\alpha \circ g_{\alpha\beta} = \pi_{\alpha\beta} \circ u_\beta. \tag{2}$$

Let $\tilde{\mu}_\alpha$ be the image measure of P_α on the Borel σ-algebra $\tilde{\mathcal{B}}_\alpha$ of $\mathbb{R}^\alpha$ so $P_\alpha \circ u_\alpha^{-1} = \tilde{\mu}_\alpha$. Then $\{(\mathbb{R}^\alpha, \tilde{\mathcal{B}}_\alpha, \tilde{\mu}_\alpha, \pi_{\alpha\beta})_{\alpha<\beta} : \alpha, \beta \text{ in } D\}$ is a projective system of (Lebesgue-Stieltjes or) regular probability spaces. Hence by Theorem 2.2 (or 3.4) the latter system admits a unique limit $(\Omega, \mathcal{B}, \mu)$ where $\Omega = \mathbb{R}^D \cong \varprojlim(\mathbb{R}^\alpha, \pi_{\alpha\beta}), \mathcal{B} = \sigma(\bigcup_\alpha \pi_\alpha^{-1}(\tilde{\mathcal{B}}_\alpha))$, and $\mu = \varprojlim \tilde{\mu}_\alpha$. If $u = \varprojlim u_\alpha$ (as defined after the proof of Theorem 3.6), then $u(\varprojlim(\mathcal{X}_\alpha, g_{\alpha\beta})) \subset \Omega = \varprojlim(u_\alpha(\mathcal{X}_\alpha), \pi_{\alpha\beta})$ and the left is "full" in Ω (relative to the μ-measure) iff the given system $\mathcal{P}$ admits a limit. This follows by Theorem 3.7. Thus we have a probability space $(\Omega, \mathcal{B}, \mu)$ determined by $\mathcal{P} \in \mathcal{C}$. To define $f : \mathcal{X}^* \to L^0(\mu)$, which gives a weak distribution, let $\tilde{f} : \mathbb{R}^{\mathcal{X}^*} \to \mathbb{R}^{\mathcal{X}^*}$ be the identity mapping and define $f : \mathcal{X}^* \to L^0(\mu)$ by the equation $f(x^*) = \pi_{x^*}(\tilde{f}), x^* \in \mathcal{X}^*$, where π_{x^*} is the coordinate projection as in the first part of the proof. Since D consists of all finite subsets of $\mathcal{X}^*$, we may and do inject Ω into $\mathbb{R}^{\mathcal{X}^*}$. Then $f(x^*)(\omega) = \pi_{x^*}(\tilde{f}(\omega)) = \pi_{x^*}(\omega) = \omega(x^*) \in \mathbb{R}$ for each $\omega \in \Omega$. Clearly $f(x^*)$ is a random variable since $\{\omega : \pi_{x^*}(\tilde{f})(\omega) < a\} \subset \Omega$ is a cylinder set in $\mathcal{B}$, and similarly for each $\alpha \cong (x_1^*, \ldots, x_n^*)', f \circ \alpha$ is a random vector. If $\pi_\alpha : \Omega \to \mathbb{R}^\alpha$ is the canonical projection, it follows from the preceding analysis (cf., Theorem 2.1) that $P_\alpha \circ u_\alpha^{-1} = \mu_\alpha = \mu \circ \pi_\alpha^{-1} =$

$\mu \circ (f \circ \alpha)^{-1}, \alpha \in D$. Hence if $\mathcal{P}^f$ is the projective system determined by f, then $P_\alpha^f \circ u_\alpha^{-1}$ is given by $\mu \circ (f \circ \alpha)^{-1}$. Thus $P_\alpha = P_\alpha^f, \alpha \in D$, and $\mathcal{P}^f = \mathcal{P}$.

It remains to verify the linearity of f. Let $\alpha = (x_1^*, x_2^*, x_1^* + x_2^*)'$ so that $\alpha : \mathcal{X} \to \mathbb{R}^3$ and $P_\alpha^f \circ u_\alpha^{-1}$ is a probability on the Borel sets of $\mathbb{R}^3$. But the mapping $v : \mathbb{R}^3 \to \mathbb{R}$ defined by $v : (a, b, c)' \mapsto (a + b - c)$ is continuous and $v \circ \alpha = 0$. So $P_\alpha^f \circ u_\alpha^{-1}(v(A)) = 1$ or 0, according to whether $0 \in A$ or $0 \notin A$ where A is a Borel set of $\mathbb{R}$ (since u_α is a bijection). Hence $P_\alpha^f \circ u_\alpha^{-1}$ is supported by the plane $a + b = c$ in $\mathbb{R}^3$, and we deduce from the definition of f above that $\omega(x_1^*) + \omega(x_2^*) - \omega(x_1^* + x_2^*) = 0$ for μ-almost all ω, since $P_\alpha^f \circ u^{-1} = \mu \circ (f \circ \alpha)^{-1}$. Thus f is additive. On the other hand, for any $a \in \mathbb{R}$, $\alpha = (x^*, ax^*)' : \mathcal{X} \to \mathbb{R}^2, v : (p, q)' \mapsto ap - q$ gives by a similar argument that $f(ax^*) = af(x^*)$. Hence $f : \mathcal{X}^* \to L^0(\mu)$ is linear. Consequently its μ-equivalence class $\hat{f}$ is in $\mathcal{W}$. By the first part, we can conclude that $\hat{f}$ determines a projective system of probability spaces which coincides with $\mathcal{P}$. Hence $\hat{f}$ and $\mathcal{P}$ again correspond to each other uniquely. This completes the proof. □

In the above result the mapping f is generally not continuous, and hence does not define a generalized random process. However, if we restrict the spaces $\mathcal{X}$ considerably (but including Schwartz spaces), or restrict the classes of projective systems severely (but including Gaussian systems), then a certain correspondence can be established. Regarding the former, a basic result due to R. A. Minlos ([1], Theorem 4) states that a projective system of probability measures $\mathcal{P}$ in $\mathcal{X} = \mathcal{Y}^*$, where $\mathcal{Y}$ is a "nuclear" space, admits a limit iff the corresponding weak distribution f is continuous from $\mathcal{X}^*$ into $L^0(\mu)$ where μ is determined by $\mathcal{P}$. So f is a generalized random process. Here we omit the definition of "nuclearity" as well as the proof of the above statement. [See e.g., Gel'fand and Vilenkin [1], Chapter IV, Theorem 3 and the following discussion.] We now turn to the Gaussian case.

To use the notation and ideas of Proposition 2, let $\mathcal{X}$ be a Hilbert space and $\alpha \subset \mathcal{X}$, a closed subspace of finite deficiency. We may identify $\mathcal{X}_\alpha = \mathcal{X}/\alpha$ with the orthogonal complement $\alpha^\perp = \mathcal{X} \ominus \alpha$, and then the canonical mappings $\varphi_\alpha : \mathcal{X} \to \mathcal{X}_\alpha$ are orthogonal projections. Let $\mathcal{P} = \{(\mathcal{X}_\alpha, \Sigma_\alpha, P_\alpha, \varphi_{\alpha\beta})_{\alpha<\beta} : \alpha, \beta \text{ in } \mathcal{F}\}$ be a projective stystem of probability spaces (or a cylindrical probability) in $\mathcal{X}$ where $\mathcal{F}$ is the directed set (as before) of subspaces of finite deficiency in $\mathcal{X}$ and each P_α is inner regular on Σ_α. Then this $\mathcal{P}$ admits the limit $P = \lim\limits_{\leftarrow} P_\alpha$, and P is supported by $\mathcal{X}$ iff for each $\epsilon > 0$ there is a compact set $K_\epsilon \subset \mathcal{X}$ such that (by Theorem 3.9) (*) $P_\alpha(\varphi_\alpha(K_\epsilon)) \geq 1 - \epsilon$, for all $\alpha \in \mathcal{F}$. Let $\mathcal{X}$ be the closed linear span of K_ϵ. Suppose that $\mathcal{X}_\epsilon$ happened to be finite dimensional. [That this is very "nearly" the case for Gaussian processes, for many $\mathcal{X}$'s, is the content of Theorem 5 below.] If $\pi_0 : \mathcal{X} \to \mathcal{X}_\epsilon$ is an orthogonal projection and if Π_0 is the set of all orthogonal projections π of finite ranks satisfying $\pi\pi_0 = \pi_0\pi = 0$ (denoted $\pi \perp \pi_0$), then (*) implies that for each Borel set $A \subset \pi_1(\mathcal{X}) - \{0\}, \pi_1 \in \Pi_0, P_{\alpha_1}(A) < \epsilon$ since clearly $\{0\} = \pi_1(\mathcal{X}_\epsilon)$, and K_ϵ is projected into the zero element of $\pi_1(\mathcal{X}) = \alpha_1^\perp$. Let q be a function depending only on finitely many variables such that $q \circ \pi : \mathcal{X} \to \mathbb{R}^+$ is a homogeneous subadditive functional (a semi-norm) defined for each $\pi \in \Pi_0$. Such a q is a (positive) "tame" or a "cylindrical" function. Let q be continuous in the metric of $\mathcal{X}$ so that q is a weaker (semi-) norm than the given norm of $\mathcal{X}$. Thus if (*) holds, then ($P = \lim\limits_{\leftarrow} P_\alpha$ exists and) the set $\{x : q(\pi x) > a\}$ is P_α-measurable for all $a \in \mathbb{R}, \alpha = \pi(\mathcal{X})^\perp$, and

$$P_\alpha\{x : q(\pi x) > \epsilon\} = P\{x : q(\pi x) > \epsilon\} < \epsilon, \qquad \pi \in \Pi_0. \tag{3}$$

If such a q (in the general case) exists on $\mathcal{X}$, it is called a *measurable semi-norm* (m. s. n.) relative to P. On the other hand, if $\mathcal{P}$ is a Gaussian cylindrical probability (i.e., each $P_\alpha, \alpha \in \mathcal{F}$, is Gaussian), and if $(\Omega, \mathcal{B}, \mu)$ is the Gaussian probability space determined by $\mathcal{P}$ as shown in the proof of Theorem 4 (see equation (2), where $\Omega = \lim\limits_{\leftarrow} \mathbb{R}^\alpha$), let $f : \mathcal{X}^* \to L^0(\Omega, \mathcal{B}, \mu)$ be the associated weak distribution. Then using the isomorphism $u_\alpha : \mathcal{X}_\alpha \to \mathbb{R}^\alpha, \pi(\mathcal{X}) = \mathcal{X}_\alpha$ (here α in $\mathbb{R}^\alpha$ also denotes the dimension of $\mathcal{X}_\alpha$), and u_α is a "matrix" determined by a basis $(y_1^*, \ldots, y_n^*)$ of $\mathcal{X}_\alpha^*$ so that $q(x)$ can be expressed as $q(x) = h \circ u_\alpha(y_1^*(x), \ldots, y_n^*(x))$ where $h : \mathbb{R}^\alpha \to \mathbb{R}^+$ is a Baire function (it is continuous since $q(\cdot)$ is such a cylindrical function), $x \in \mathcal{X}_\alpha$. If $(\tilde{q} \circ \pi)(\omega) = h \circ u_\alpha(f(y_1^*)(\omega), \ldots, f(y_n^*)(\omega))$, then $\tilde{q} \circ \pi$ is a random variable on $(\Omega, \mathcal{B}, \mu)$. Moreover, by the proof of Theorem 4 (the correspondence between P_α on $\mathcal{X}_\alpha$ and μ_α on $\mathbb{R}^\alpha$) we have

$$\int_{\mathcal{X}_\alpha} q(x)dP_\alpha = \int_\Omega \tilde{q} \circ \pi(\omega) d\mu(\omega) = \int_{\mathbb{R}^\alpha} \tilde{q}(t) d\mu_\alpha(t). \tag{4}$$

Hence (3) can be expressed, for all $\pi \in \Pi_0$ with $\mathcal{X}_\alpha = \pi(\mathcal{X})$, as:

$$P_\alpha\{x \in \mathcal{X}_\alpha : q(x) > \epsilon\} = \mu_\alpha\{\omega : \tilde{q}(\omega) > \epsilon\} = \mu\{\omega : \tilde{q} \circ \pi(\omega) > \epsilon\} < \epsilon. \tag{5}$$

But since μ is always a measure, the right side of (5) can be used for a *definition* of the functional $q(\cdot)$ to be a measurable (semi-) norm on $\mathcal{X}$, if its representative $\tilde{q}$ by the weak distribution (through $\mathcal{P}$) on $(\Omega, \mathcal{B}, \mu)$ is such that $(+)\mu\{\omega : \tilde{q}(\pi\omega) > \epsilon\} < \epsilon$, for all $\pi \in \Pi_0$. Using the hypothesis that μ is Gaussian, at this point, one can deduce several properties of q *whenever* $(+)$ *holds*. We need the following facts, established by L. Gross [1] (recall that for $y \in \mathcal{X}^*$, $f(y)$ is Gaussian with mean zero and variance $\|y\|_*^2$ if f is a weak distribution on $\mathcal{X}^*$):

(i) q is continuous in the norm of $\mathcal{X}$, and is bounded on the unit ball of $\mathcal{X}$. The existence of such a q on $\mathcal{X}$ implies that $\mathcal{X}$ is separable.

(ii) $\{\tilde{q} \circ \pi(\cdot), \pi \in \mathbb{P}\}$ converges in probability to a random variable Y, as the net of finite rank orthogonal projections $\mathbb{P}$ tends to the identity. Moreover, if $q \circ \pi$ is replaced by q_n a monotone sequence, $\tilde{q}_n \to Y$ in μ-measure, and $\mu[Y \leq \epsilon] > 0$ for all $\epsilon > 0$, then the equation $q_0(x) = \lim_n q_n(x), x \in \mathcal{X}$, defines an m. s. n. on $\mathcal{X}$.

With these notions and properties, we can present the following comprehensive result, due to L. Gross [2].

5. Theorem. *Let $\mathcal{X}$ be a real Hilbert space and $\mathcal{P}$ be a projective system of Gaussian probability spaces. If there exists a measurable seminorm q on $\mathcal{X}$ so that there exists a $\pi_0 = \pi(\epsilon)$, an orthogonal projection of finite rank, such that for each finite rank orthogonal projection π satisfying $\pi \perp \pi_0$ (i.e., $\pi \in \Pi_0$), one has*

$$P\{x : q \circ \pi(x) > \epsilon\} < \epsilon, \qquad \pi \in \Pi_0, \tag{6}$$

then $P = \overleftarrow{\lim} P_\alpha$ exists, and the support B of P is given by $B = \bar{sp}(\mathcal{X})$, in the m. s. n. q, so that $(B, \mathcal{B}, P)$ is a Gaussian probability space where $\mathcal{B}$ is the Borel algebra of B (so P is regular).

Remark. Note that $\mathcal{X} \subset B$ and the embedding is continuous. If B^* is the adjoint of B, then $B^* \subset \mathcal{X}^* \cong \mathcal{X} \subset B$ and the triple $(B, \mathcal{B}, P)$ [or $(B^*, \mathcal{X}, B)$] is called an *abstract Wiener space*. It is also denoted by $(i, \mathcal{X}, B)$ where $i : \mathcal{X} \hookrightarrow B$ is the injection map.

Proof. How condition (6) arises has already been seen prior to the statement of the theorem. We therefore establish the important existence of the limit, assuming that the projective system is Gaussian. The idea of the proof is to show, with the properties (i) and (ii) above, that the apparently weaker condition (6) is now equivalent to Prokhorov's of Proposition 2, in the space B. We present the details in steps for clarity.

(I) If $\mathbb{P}$ is the set of all orthogonal projections of finite rank on $\mathcal{X}$, and $\{a_n > 0, n \geq 1\}$ is given, then there exists $\{\pi_n, n \geq 1\} \subset \mathbb{P}$ such that $\pi_n \perp \pi_m, m \neq n$, $\sum_{n=1}^{\infty} \pi_n =$ id., and $\sum_{n=1}^{\infty} a_n q(\pi_n x) = q_0(x) < \infty$, for each $x \in \mathcal{X}$, and q_0 is a measurable norm on $\mathcal{X}$. (Here again "id." denotes the identity on $\mathcal{X}$.)

Proof of (I). Let $(\Omega, \mathcal{B}, \mu)$ be the regular probability space associated with $\mathcal{P}$. Then $\tilde{q} \circ \pi$ is a random variable on this space for each $\pi \in \mathbb{P}$. Since q is an m. s. n. (so that (6) holds for each $\epsilon > 0$ with $\pi_\epsilon \in \mathbb{P}$), for each $a_n > 0$ there exists a $\tilde{\pi}_n \in \mathbb{P}$ such that for all $\pi \in \mathbb{P}, \pi \perp \tilde{\pi}_n$, one has

$$\mu\left\{\omega : (\tilde{q} \circ \pi)(\omega) > \frac{1}{a_n 2^n}\right\} < \frac{1}{2^n}, \quad n \geq 1. \tag{7}$$

We may assume that (by addition) $\tilde{\pi}_n \tilde{\pi}_{n+1} = \tilde{\pi}_n, \|\tilde{\pi}_n x - x\|_{\mathcal{X}} \to 0$. Let $\pi_1 = \tilde{\pi}_1$ and $\pi_n = \tilde{\pi}_{n+1} - \tilde{\pi}_n \in \mathbb{P}$. We claim that $\{\pi_n, n \geq 1\}$ satisfies the requirements.

Indeed, it is clear that $\sum_{n=1}^{\infty} \pi_n =$ id. (strong convergence), and (7) implies, on using the fact that $\pi_n \perp \tilde{\pi}_n$,

$$\mu\{\omega : a_n \tilde{q} \circ \pi_n(\omega) > 2^{-n}\} < 2^{-n}, \quad n \geq 1. \tag{8}$$

Hence, if for any $\epsilon > 0$ we choose $m \geq 1$ such that $2^{-m} < \epsilon$, then

$$\begin{aligned}\mu\left\{\omega : \sum_{k=m+1}^{n} a_k \tilde{q} \circ \pi_k(\omega) > \epsilon\right\} &\leq \mu\left[\bigcup_{k=m+1}^{n} \{\omega : a_k \tilde{q} \circ \pi_k(\omega) > 2^{-k}\}\right] \\ &\leq \sum_{k \geq m+1} \mu\{\omega : a_k \tilde{q} \circ \pi_k(\omega) > 2^{-k}\} \\ &\leq 2^{-m} < \epsilon.\end{aligned}$$

Consequently, $\left\{\sum_{k=1}^{m} a_k \tilde{q} \circ \pi_k, m \geq 1\right\}$ is Cauchy in $L^0(\mu)$ so that it converges to some Y in $L^0(\mu)$ (this means it converges in probability). We can now invoke property (ii) for the monotone sequence

$\left\{\sum_{k=1}^{n} a_k q \circ \pi_k, n \geq 1\right\}$ to conclude that $\sum_{k=1}^{\infty} a_k q \circ \pi_k(x) = q_0(x)$ defines an m. s. n. on $\mathcal{X}$, provided it is shown that $\mu[Y \leq \delta] > 0$ for any $\delta > 0$.

To see that the last condition holds for Y, since $\left\{\sum_{k=1}^{n} a_k \tilde{q} \circ \pi_k(\cdot)\right\}_{n=1}^{\infty}$ is Cauchy in probability, and q is a norm dominated by $\|\cdot\|_{\mathcal{X}}$, there exists an $n_0 = n_0(\delta) > 1$ such that if $f_{n_0} = \sum_{n > n_0} a_n \tilde{q} \circ \pi_n$, we have $\alpha = \mu[f_{n_0} \leq \delta/2] > 0$. Also if $f'_{n_0} = \sum_{k=1}^{n_0} a_k \tilde{q} \circ \pi_k$, then $\beta = \mu[f'_{n_0} \leq \delta/2] > 0$. In both these inequalities we used property (i). Next observe that for any $x \in \mathcal{X}, y_1 = \pi_n(x), y_2 = \pi_m(x)$ are mutually orthogonal. Since the weak distribution F, corresponding to the Gaussian cylinder probability $\mathcal{P}$, is linear and the $F(y_i)(\in L^0(\mu))$ are normally distributed with mean zero and variance $\|y_i\|_{\mathcal{X}}^2$ (by definition), one has

$$\begin{aligned}
\|y_1 + y_2\|_{\mathcal{X}}^2 &= \mathrm{Var}(F(y_1 + y_2)) = \mathrm{Var}(F(y_1) + F(y_2)) \\
&= \mathrm{Var}F(y_1) + \mathrm{Var}F(y_2) + \mathrm{Cov}(F(y_1), F(y_2))) \\
&= \|y_1\|_{\mathcal{X}}^2 + \|y_2\|_{\mathcal{X}}^2 + \mathrm{Cov}(F(y_1), F(y_2)).
\end{aligned} \tag{9}$$

Since $y_1 \perp y_2$, $\|y_1 + y_2\|_{\mathcal{X}}^2 = \|y_1\|_{\mathcal{X}}^2 + \|y_2\|_{\mathcal{X}}^2$, and hence $\mathrm{Cov}(F(y_1), F(y_2))$ is 0; so the $F(y_i)$ are uncorrelated. But in a Gaussian family this implies (Exercise 3) independence. Hence

$$\mu[Y \leq \delta] = \mu[f'_{n_0} + f_{n_0} \leq \delta] \geq \mu\left[\left(f'_{n_0} \leq \frac{\delta}{2}\right) \cap \left(f_{n_0} \leq \frac{\delta}{2}\right)\right] = \alpha\beta > 0, \tag{10}$$

because f_{n_0} and f'_{n_0} are independent being based on mutually independent random variables $F(\pi_k x), k \leq n_0$, and $F(\pi_r x), r \geq n_0 + 1$. Thus Y determines $q_0(x) = \sum_{n=1}^{\infty} a_n q \circ \pi_n(x)$, and q_0 is m. s. n. But $q_0(x) > 0$ if $x \neq 0$ since q is m. s. n. and $\pi_n(x) \neq 0$ for some $n (a_n > 0$ all $n)$. So q_0 is a measurable norm.

II. If B is the completion of $\mathcal{X}$ under $q(\cdot)$, then there exists a measurable norm $q_0(\cdot)$ on $\mathcal{X}$ such that the ball $U_r = \{x \in \mathcal{X} : q_0(x) \leq r\}$ is precompact in B for each $r > 0$.

Proof of (II). Using the notation of the above step, let $\{a_n, n \geq 1\}$ there be chosen subject to $\sum_{n=1}^{\infty} a_n^{-1} < \infty$, and let $q_0(\cdot)$ be the corresponding measurable norm where $q_0(x) = \sum_{n=1}^{\infty} a_n q \circ \pi_n(x)$. Let U_r be the ball in $\mathcal{X}$ as in the assertion, and consider a sequence $\{x_n, n \geq 1\} \subset U_r \subset B$. We need to show that there is a convergent (in B) subsequence $\{x_{n_j}, j \geq 1\}$. But by definition of q_0 and U_r, we have $q \circ \pi_k(x_n) \leq r a_k^{-1}$ for all $n \geq 1$, and $k \geq 1$. For each $k, \pi_k(\mathcal{X})$ is finite dimensional and $q(\cdot)$ is a norm on this subspace. So the bounded sequence $\{\pi_k(x_n), n \geq 1\}$ has a convergent subsequence $\{\pi_k(x_{n_j}), j \geq 1\}$ for each $k \geq 1$, by the Bolzano-Weierstrass theorem. Thus by the diagonalization procedure we can find a subsequence $\{x^m, m \geq 1\} \subset \{x_n, n \geq 1\}$ such that $\{\pi_k(x^m), m \geq 1\}$ converges in the $q(\cdot)$ norm for every $k \geq 1$. However, since $\sum_{k=1}^{\infty} \pi_k x = x$ (strongly) and

$$q(x^n - x^m) \leq \sum_{k=1}^{N} q(\pi_k(x^n - x^m)) + q\left(\sum_{k \geq N} \pi_k(x^n - x^m)\right), \tag{11}$$

we conclude the $\left\|\sum_{k \geq N} \pi_k(x^n - x^m)\right\| \to 0$ as $n \to \infty$ and then by property (i), $q\left(\sum_{k \geq N} \pi_k(x^n - x^m)\right) \to 0$ as $N \to \infty$. Hence (11) becomes $q(x^n - x^m) \leq \sum_{k=1}^{\infty} q(\pi_k(x^n - x^m)) \leq \sum_{k=1}^{\infty} 2r a_k^{-1} < \infty$. By the Dominated Convergence, we may let $n, m \to \infty$ for the middle series and (since each term tends to zero) conclude that the sequence $\{x^m, m \geq 1\}$ is Cauchy in B, as desired.

(III). The function $P = \lim\limits_{\leftarrow} P_\alpha$, of $\mathcal{P}$, exists and $\text{supp}(P) = B$.

Proof of (III). We reduce the result to that of Proposition 2. Since $q(\cdot)$ is an m.s.n. satisfying (6), let $q_0(\cdot)$ be the measurable norm defined to satisfy the conditions of Step II, the existence of which is shown in Step I. Then (6) holds for q_0 (by definition). Given $\epsilon > 0$, choose $\delta = \delta_\epsilon > 0$ such that $\mu\{\omega : \tilde{q}_0(\omega) > \delta\} < \epsilon$. Let $\tilde{K}_\epsilon = \{x \in \mathcal{X} : q_0(x) \leq \delta\}$. Then $\tilde{K}_\epsilon \in \Sigma$, and by Step II, $\tilde{K}_\epsilon \subset B$ is precompact (in $q(\cdot)$-norm) and convex. If K_ϵ is the closure of $\tilde{K}_\epsilon$ in B, then K_ϵ is compact in $B, g_\alpha(K_\epsilon)$ is measurable, and by (5)

$$P_\alpha(g_\alpha(K_\epsilon)) \geq P_\alpha(g_\alpha(\tilde{K}_\epsilon)) = \mu\{\omega : \tilde{q}_0(\omega) \leq \delta\} \geq 1 - \epsilon,$$

for all $\pi(\mathcal{X}) = \alpha^{\perp} \in \mathcal{F}$. Hence by Proposition 2, P is σ-additive and is supported in B. The argument leading to (10) shows that every open set of B has positive P-measure, so that supp$(P) = B$. But B is separable. So the Baire algebra $\mathcal{B}$ is also Borel. Since $\mathcal{X} \subset B$ and the inclusion being a continuous embedding (by property(i)) we deduce that $B^* \subset \mathcal{X}^* \cong \mathcal{X} \subset B$ when the Hilbert space $\mathcal{X}$ and its adjoint $\mathcal{X}^*$ are identified. Thus $(B, \mathcal{B}, P)$ is the desire projective limit. □

It is natural to ask for a classification of Hilbert spaces admitting a measurable norm (*relative to a Gaussian cylindrical probability*). This seems to be unknown, and we list some interesting spaces each admitting an m. s. n.

1. The first (and most important) example is the space $B_1 = C_0[0,1]$, the Banach space of continuous real functions on the unit interval, vanishing at the origin; and this is also historically the earliest space considered by N. Wiener (1923) and is often called *the Wiener space.* Here $\mathcal{X}_1 = \{f \in B_1 : f(u) = \int_0^u f'(t)dt, 0 \le u \le 1, \|f\|^2 = \int_0^1 |f'|^2 dt < \infty\}$ where f' is the derivative of f so that $f \in \mathcal{X}_1$ iff it is absolutely continuous on $(0,1)$ with a square integrable derivative. Then $\mathcal{X}_1$ is a Hilbert space in the norm $\|\cdot\|$. Let $q(x) = \sup\{|x(t)| : 0 \le t \le 1\}$. Then it can be shown (nontrivially) that $q(\cdot)$ is an m. s. n. on $\mathcal{X}$. Clearly C_0 is the completion of $\mathcal{X}_1$ for $q(\cdot)$, since $\mathcal{X}$ contains all the polynomials of C_0.

2. This is a generalization of the preceding example to functions of two variables. Thus we construct $B_2 \subset C([0,1] \times [0,1])$ as follows. Let

$$\mathcal{X}_2 = \left\{ f : f(t,s) = \int_0^t \int_0^s \frac{\partial^2 f}{\partial u \partial v}(u,v)du\, dv, 0 \le t \le 1, 0 \le s \le 1, \right.$$

$$\left. \text{and} \quad \|f\|^2 = \int_0^1 \int_0^1 \left| \frac{\partial^2 f}{\partial u \partial v} \right|^2 du\, dv < \infty \right\}.$$

Then $\mathcal{X}_2$ is a Hilbert space, and if $q(f) = \sup\{|f(u,v)| : 0 \le u \le 1, 0 \le v \le 1\}$, we get the space B_2 (= completion of $\mathcal{X}_2$ in q) to be a Wiener space with $q(\cdot)$ as an m. s. n. (relative to a Gaussian cylindrical measure, as always). If instead one considers the tensor products $\tilde{\mathcal{X}}_2 = \mathcal{X}_1 \otimes \mathcal{X}_1$ and $\tilde{B}_2 = B_1 \otimes B_1$ of the first example (with norms as the sums of norms in the individual spaces), then again the set

$(i, \tilde{\mathcal{X}}_2, \tilde{B}_2)$ is an abstract Wiener space, the sum of the uniform norms being the m. s. n. (These results admit generalization to spaces of n-variables. However, in each case the proof is involved, and numerous painful details have to be verified for the m. s. n. See Finlayson [1], [2] on this. Originally the space $(i, \mathcal{X}_2, B_2)$ was treated by J. Yeh [1].)

3. Let $\mathcal{X}_1 \subset C_0[0,1]$ be as in the first example and consider for $0 < \alpha < \frac{1}{2}$, $\mathrm{Lip}_\alpha(f) = \sup\{|f(t) - f(s)| \cdot |t-s|^{-\alpha} : 0 \le s,\ t \le 1, s \ne t\}$. If B_3^α is the completion of $\mathcal{X}_1$ for the norm $\mathrm{Lip}_\alpha(\cdot)$, then $(i, \mathcal{X}_1, B_3^\alpha)$ is again a Wiener space with $\mathrm{Lip}_\alpha(\cdot)$ as an m. s. n.

4. If $\mathcal{X}$ is a (separable) Hilbert space and $A : \mathcal{X} \to \mathcal{X}$ is a bounded linear operator, then it is said to be of a *trace class* (or *nuclear*) operator iff $\sum_{n=1}^{\infty} \|Ae_n\|_{\mathcal{X}} < \infty$ for an orthonormal basis $\{e_n, n \ge 1\}$. Let A be a positive definite nuclear operator on $\mathcal{X}$. This means we also require of $A : (x, Ax)_{\mathcal{X}} = (Ax, x)_{\mathcal{X}} > 0$ for all $x \in \mathcal{X}, x \ne 0$, in addition to nuclearity. If $q(x) = (Ax, x)_{\mathcal{X}}^{1/2}$ then $q(\cdot)$ is an m. s. n. on $\mathcal{X}$. (See Gross [1], [2] for details of these two examples.)

In view of the above discussion and examples it is fairly obvious that many Banach or even Hilbert spaces do not support a Gaussian measure. In fact, it was already noted by Gross ([2], p. 39) that essentially only the abstract Wiener spaces have this property. More explicitly we have the following result:

6. Proposition. *Let B be a separable Banach space with norm $q(\cdot)$, and let P be a Gaussian probability on the Borel algebra $\mathcal{B}$ of B. Then there exists a separable Hilbert space $\mathcal{X}_0$ (with norm $\|\cdot\|$) $\subset B$ such that $q(\cdot)$ is a measurable norm on $\mathcal{X}_0$ and the support of P is $B_0 \subset B$ where B_0 is the completion of $\mathcal{X}_0$ relative to $q(\cdot)$. Equivalently, if (Ω, Σ, μ) is a probability space and $X : \Omega \to B$ is a $(\Sigma, \mathcal{B})$ measurable mapping such that $P = \mu \circ X^{-1}$ (the image measure) is Gaussian in B, then $(i, \mathcal{X}_0, B_0)$ is an abstract Wiener space with $B_0 = \bar{\mathrm{sp}}(\mathcal{X}_0) \subset B, \mathcal{X}_0 = \bar{\mathrm{sp}}(X(\Omega))$ the last closure relative to an inner product.*

We omit a proof of this propostion. It will not be needed later. A reason for its presentation here is to understand the extent and importance of abstract Wiener spaces. It should be noted, however, that by Minlos' result stated above there is a fairly large class of infinite dimensional locally convex (nonnormable) conjugate vector spaces $\mathcal{Y}(= \mathcal{Z}^*)$, namely, "nuclear" spaces $\mathcal{Z}$, which support Gaussian measures. Thus both Minlos and Gross theories are important specializations of the general Theorem 3.9 of Prokhorov's. The point here (and of the above proposition) is that the geometrical *structure of a general topological vector space is intimately related to the (class of) probability measures it can support.* We do not treat vector valued processes *per se* in the present work.

Finally, we sketch the second example mentioned at the beginning of this section. This is not included in the abstract Wiener spaces, and involves construction of a Gaussian measure on an infinite dimensional sequence space. It will aid in understanding the earlier projective limit theory more "concretely." This problem was originally presented by Hida and Nomoto ([1], [2]). (The space turns out to be "nuclear.")

Consider the n-sphere of radius r in $\mathbb{R}^{n+1}$, i.e., the set $S_n(r) = \left\{(x_1, \dots, x_{n+1}) : \sum_{i=1}^{n+1} x_i^2 = r^2\right\}$. The uniform measure on $S_n(r)$ is the normalized Lebesgue measure and it can be expressed more conveniently in spherial polar coordinates as follows: $x_1 = r \prod_{i=1}^{n} \sin\theta_i$, and for $2 \le k \le n$, $x_k = r \cos\theta_{k-1} \prod_{i=k}^{n} \sin\theta_i$, and $x_{n+1} = r\cos\theta_n$, where $0 < r < \infty, 0 \le \theta_1 \le 2\pi$, and $0 \le \theta_i \le \pi, i \ge 2$. Then the uniform probability P_n on the Borel σ-algebra of $S_n(r)$ can be given by the known formula:

$$dP_n(\theta_1, \dots, \theta_n) = \frac{1}{2}\Gamma\left(\frac{n+1}{2}\right)\pi^{(-n+1)/2}\left[\prod_{i=2}^{n}\sin^{i-1}\theta_i\right] d\theta_1, \dots, d\theta_n. \tag{12}$$

(Because of the normalization, we see that in (12) r does not appear.) In this formulation we can set up a one-to-one correspondence between $S_n(r)$ and the open box $B_n = \{(\theta_1, \dots, \theta_n) : 0 < \theta_1 < 2\pi, 0 < \theta_i < \pi, 2 \le i \le n\}$ if we exclude the null set from $S_n(r)$ corresponding to the obvious boundary points of B_n. Thus (12) can be considered as a probability measure on the Borel sets of B_n, and since the component measures factor, the θ_i (being coordinate functions) can be considered as independent random variables on this space for the measure P_n. With this observation, we may go back to the original space $S_n(r)$ and let Ω_n be the (open) set of points on this sphere which is homeomorphic to B_n. The set of points deleted from $S_n(r)$ forms an $(n-2)$-dimensional subspace of $\mathbb{R}^{n+1}$, and has zero (Lebesgue or

P_n-) measure. Thus in $\mathbb{R}^2, \Omega_n$ is the circle with the point $(r,0)$ removed. [This allows for convenient computations and the degeneracy eliminated.] Let $g_{mn} : \Omega_n \to \Omega_m$ be the mapping induced by the coordinate projections on B_n to $B_m, m \le n$. It is clear that if $\mathcal{B}_n$ is the Borel σ-algebra of Ω_n, μ_n is the corresponding probability on $\mathcal{B}_n$ induced by the P_n of (12), with the above transformation between x's and θ's, then for $\ell < m < n$ we have $g_{\ell m} \circ g_{mn} = g_{\ell n}, g_{nn} =$ identity, and for any $A \in \mathcal{B}_m, \mu_m(A) = \mu_n(g_{mn}^{-1}(A))$, i.e., $\mu_m = \mu_n \circ g_{mn}^{-1}$. Thus $\{(\Omega_n, \mathcal{B}_n, \mu_n, g_{mn})_{m<n}, n \ge m \ge 1\}$ is a regular projective system of probability spaces, where Ω_n is a full subset of the n-sphere (i.e., $\mu_n(\Omega_n) = 1$), and g_{mn} are continuous. Since the family is denumerable (and linearly ordered), we can immediately use the proof of Theorem 2.1. More generally we can also apply Bochner's theorem (Theorem 3.4) to conclude that the system admits a projective limit $(\Omega, \mathcal{B}, \mu)$ provided we show that the sequential maximality holds. But this is immediate here since it is so for the sets B_n, and by identification (which involves only a finite number of operations at any time in this mapping) the same holds for Ω_n's. Hence $(\Omega, \mathcal{B}, \mu)$ is the limit since there exist $g_n : \Omega \to \Omega_n$ with $g_n = g_m \circ g_{mn}$ for $n \ge m$ and $\mu \circ g_n^{-1} = \mu_n, \mu$ being a Borel probability. (See also Exercise 10.)

We can identify μ explicitly, and show that $(\Omega, \mathcal{B}, \mu)$ is a Gaussian probability space and that it induces, through an appropriate mapping, a Gaussian measure on the space $\mathbb{R}^{\mathbb{N}}$, of infinite sequences. (Endowed with the projective limit topology, this is an infinite dimensional space and it becomes "nuclear.") Thus let $X^{(n)} \in \Omega_n \subset S_n$ so that $X^{(n)} = (X_i^{(n)}, i = 1, \ldots, n+1)$. Since $g_n = g_{n,\infty} : \Omega \to \Omega_n$ is onto, and if $\omega \in \Omega, g_n(\omega) \in \Omega_n$, we have $g_n(\cdot) = (X_1^{(n)}, \ldots, X_{(n+1)}^{(n)}) \in S_n$ and $X_j^{(n)} (= X_j^{(n)}(\cdot))$ can be regarded as the jth coordinate variable of $g_n(\cdot)$. The measurability of g_n implies that of $X_j^{(n)}(\cdot)$ for each j and n. We may alternatively regard $X_j^{(n)} : \Omega \to \mathbb{R}$, as measurable real functions on Ω (by an appropriate identification). We claim that $\{X_i^{(n)}/\sqrt{n}\}$ is a Cauchy sequence in $L^2(\mu)$ for each $i = 1, 2, \ldots$, and that the limit sequence $X_i(\cdot)$ (i.e., $n^{-1/2} X_i^{(n)} \to X_i$ in L^2) forms a Gaussian process, where $\mu \circ g_n^{-1} = \mu_n$ and where μ_n on $\mathcal{B}_n$ is identifiable with P_n of (12) on the Borel sets of S_n.

To prove this, notice that

$$\int_{\Omega} X_i^{(n)}(\omega)d\mu = \int_{\Omega_n} X_i^{(n)}\, d\mu_n = \int_{S_n(r)} x_i^{(n)}(\theta_1, \dots, \theta_n)dP_n = 0, \quad i \le n,$$

and for $m \le n$ (δ_{ij} is the Kronecker delta),

$$\int_{\Omega} X_i^{(m)}(\omega)X_j^{(n)}(\omega)d\mu = \int_{\Omega_n} X_i^{(m)}\, X_j^{(n)}\, d\mu_n = \delta_{ij}\left(\frac{n}{m}\right)^{1/2} \frac{\Gamma\left(\frac{m+1}{2}\right)\Gamma\left(\frac{n}{2}\right)}{\Gamma\left(\frac{m}{2}\right)\Gamma\left(\frac{n+1}{2}\right)}, \tag{13}$$

where we evaluated the n-fold elementary integral, omitting the intermediate computation, and where $\Gamma(\cdot)$ is, as usual, the gamma function. From (13) we get

$$\int_{\Omega} \left(\frac{X_i^{(n)}}{\sqrt{n}} - \frac{X_i^{(m)}}{\sqrt{m}}\right)^2 d\mu = 2\left[1 - \left(\frac{n}{m}\right)^{1/2}\right] \to 0, \text{ as } n \le m,\ n \to \infty. \tag{14}$$

Hence $X_i = \underset{n}{\text{l.i.m.}} \frac{X_i^{(n)}}{\sqrt{n}}$ is well-defined, $X_i \in L^2(\Omega, \mathcal{B}, \mu), i = 1, 2, \dots$. Note that $X_i(\omega) = X_i(\omega')$, for all i iff ω and ω' belong to a μ-null set, since the $X_i^{(n)}(\cdot) = (g_n(\cdot))_i$ separate points of Ω. It remains to show that μ induces a Gaussian measure, i.e., $\{X_i, i \ge 1\}$ is a Gaussian process. In fact, it suffices to prove that, for a finite subset $(t_1, \dots, t_k)$ of the integers, $(X_{t_1}, \dots, X_{t_k})$ is Gaussian. From the discussion of Gaussian densities in Section 1, it follows (cf., Exercises 3 and 5) that we only need to show, for any scalars $(a_1, \dots, a_k)$, that the functions $Y_k = \sum_{i=1}^{k} a_i X_{t_i} = \underset{n}{\text{l.i.m.}} \frac{1}{\sqrt{n}} \sum_{i=1}^{k} a_i X_{t_i}^{(n)}$ are Gaussian distributed. Because of the one-to-oneness of the characteristic functions and distribution functions and since clearly the above convergence implies the convergence of the corresponding distributions (by the above cited exercise), we consider the latter. Thus for $\lambda = (\lambda_1, \dots, \lambda_k)$,

$$\begin{aligned} \varphi_n(\lambda) &= \int_{\Omega} \exp\left[in^{-1/2}\left(\sum_{j=1}^{k} \lambda_j a_j X_{t_j}^{(n)}\right)\right] d\mu \\ &= \int_{S_n(r)} \exp\left[i \sum_{i=1}^{k} \lambda_j a_j n^{-1/2}\, x_{t_j}^{(n)}\right] dP_n \\ &\to \exp\left[-\frac{1}{2}\sum_{j=1}^{k} a_j^2 \lambda_j^2\right] = \varphi(\lambda). \end{aligned} \tag{15}$$

An explicit evaluation of the last integral involves Bessel functions, and the elementary but tedious computation is left to the reader. Since $\varphi(\cdot)$ is a product of Gaussian characteristic functions, it follows that the X_i's are independent and each has the same standard (= mean zero and variance 1) Gaussian distribution. Thus μ on Ω gives the image measure on $\mathbb{R}^{\mathbb{N}}$ to be a (product) Gaussian measure where the set $\mathbb{R}_0^{\mathbb{N}} = \{a = (a_1, a_2, \ldots) : a_i = X_i(\omega), \omega \in \Omega\} \subset \mathbb{R}^{\mathbb{N}}$, has measure one.

This result can be stated in a more abstract but equivalent form. It is a known fact that, if $\mathcal{O}(n)$ denotes the class of all $n \times n$ orthogonal matrices (or the orthogonal group) acting on (rotations of) the unit sphere S_{n-1}, then the subset of all of its elements which leave the north pole of S_{n-1} invariant is isomorphic to a subgroup $\mathcal{O}(n-1)$ of $\mathcal{O}(n)$, which is the group of Euclidean rotations of the $(n-2)$-dimensional subspace $x_n = 0$. Each element u of this subgroup can be expressed as the matrix $\begin{pmatrix} u_1 & 0 \\ 0 & 1 \end{pmatrix}$ where $u_1 \in \mathcal{O}(n-1)$ (as the group acting on S_{n-2}). It then follows that the quotient space $\mathcal{O}(n)/\mathcal{O}(n-1)$ can be identified with the space S_{n-1}. By induction, $\mathcal{O}(n)$ then is isomorphic with the cartesian product space $S_1 \times \cdots \times S_{n-1}$. Thus the image of the uniform measures on this product space (which accounts for the ultimate independence of X_i's above) corresponds to the invariant measure on $\mathcal{O}(n)$. This identification then gives (since the projective limit procedure is the same except that we now get $u^{(n)} = (u_{ij}^{(n)}) \in \mathcal{O}(n)$ and the range space is $\mathbb{R}^{\mathbb{Z}}$) the space of doubly infinite sequences. The measure μ is again Gaussian and we get this as the invariant measure on the infinite dimensional space $\Omega = \lim\limits_{\leftarrow} \mathcal{O}(n)$ which will be an orthogonal group. This interpretation is useful for some work in quantum field theory. (Cf., also Shale [1], and Yamasaki [1].) Shale uses direct limits and only gets a finitely additive "probability," whereas the second author uses the procedure indicated above.

1.5 Complements and exercises

1. Using the notation of Equation (1.1.5), let F_n be an n-dimensional Gaussian distribution function in $\mathbb{R}^n$, with parameters m and K, where m is an n-vector and K is an nth order positive definite symmetric

matrix. If φ_n is the characteristic function of F_n, show that

$$\varphi_n(t_1, \dots, t_n) = \exp[-\frac{1}{2}(Kt, t) + i(m, t)],$$

where $t = (t_1, \dots, t_n), t_i \in \mathbb{R}$, and (m, t) denotes the inner product. Find also the characteristic function of the Poisson distribution function (cf. Equation (1.1.6)).

2. Let $T \subset \mathbb{R}$, and, for each n, consider $t_1 < \cdots < t_n, t_i \in T$. Suppose $\{F_{t_1,\dots,t_n}\}_{n\geq 1}$ is the associated family of finite dimensional distribution functions satisfying the Kolmogorov compatibility conditions. If $\{\varphi_{t_1,\dots,t_n}\}_{n\geq 1}$ is the corresponding family of characteristic functions, show that the compatibility conditions on the F's are equivalent to the following on the φ's:

(i) $\lim_{u_i \to 0} \varphi_{t_1,\dots,t_n}(u_1, \dots, u_n) = \varphi_{t_1,\dots,t_{i-1},t_{i+1},\dots,t_n}(u_1, \dots, u_{i-1}, u_{i+1}, \dots, u_n)$,

(ii) $\varphi_{t_{i_1},\dots,t_{i_n}}(u_{i_1}, \dots, u_{i_n}) = \varphi_{t_1,\dots,t_n}(u_1, \dots, u_n)$ where $(i_1, \dots, i_n)$ is a permutation of $(1, \dots, n)$.

3. Let F_n be the distribution function of a random vector $X = (X_1, \dots, X_n)'$ on (Ω, Σ, P) to $\mathbb{R}^n$. Its mean vector, denoted $E(X)$, is $(E(X_1), \dots, E(X_n))'$ where ($'$ denotes transposition as before)

$$E(X_i) = \int_\Omega X_i dP = \int_{-\infty}^{\infty} \cdots \int_{-\infty}^{\infty} x_i dF_n(x_1, \dots, x_n),$$

and similarly the (second) moment matrix is $E(XX') = (E(X_iX_j), i, j = 1, \dots, n)$ where

$$E(X_iX_j) = \int_\Omega X_iX_j dP = \int_{-\infty}^{\infty} \cdots \int_{-\infty}^{\infty} x_ix_j dF_n(x_1, \dots, x_n).$$

The Lebesgue-Stieltjes integrals are assumed to exist here. Clearly $E[(X - E(X))(X - E(X))']$ is defined if the first two moments of X_i exist. This is the covariance matrix of X. If F_n is Gaussian, as in Exercise 1, show that its mean vector is m and covariance matrix is K. [Thus, by Exercise 1, a Gaussian distribution is uniquely determined by the pair of parameters: the mean vector and its covariance matrix.] If now the covariance matrix of a Gaussian random vector is diagonal, then deduce that the component random variables are mutually independent, and that orthogonality and independence are equivalent concepts for Gaussian random variables with zero means. Show also that a vector $X = (X_1, \ldots, X_n)'$ has an n-dimensional Gaussian distribution with mean m and covariance matrix K iff for each $a = (a_1, \ldots, a_n)' \in \mathbb{R}^n$, the random variable $a'X = \sum_{i=1}^{n} a_i X_i$ has a univariate Gaussian distribution with mean $a'm$ and variance (Ka, a).

4. Another useful class of finite dimensional distribution functions is the following. Let $T = (0, \infty)$, and for x, y in $\mathbb{R}$, define $f_t, t \in T$, by the inverse Fourier transform as follows:

$$f_t(x|y) = \frac{1}{2\pi} \int_{-\infty}^{\infty} \exp[iu(y-x) - t|u|^\alpha]du, \quad 0 < \alpha \le 2.$$

If $0 = t_0 < t_1 < t_2 < \cdots < t_n, t_i \in T$, let

$$F_{t_1, \ldots, t_n}(x_1, \ldots, x_n)$$
$$= \int_{-\infty}^{x_1} \cdots \int_{-\infty}^{x_n} f_{t_1}(0|u_1) f_{t_2 - t_1}(u_1|u_2) \cdots f_{t_n - t_{n-1}}(u_{n-1}|u_n) du_n \cdots du_1.$$

Show that $\{F_{t_1, \ldots, t_n}\}_{n \ge 1}$ is a consistent family of (continuous) distribution functions in $\mathbb{R}^n$. [If $\alpha = 2$ this is a subclass of a Gaussian family, called the *Brownian motion* or *Wiener* distribution family, and if $\alpha = 1$ it is called the *Cauchy* distribution family. If $0 < \alpha < 2$ this is called the P. Lévy distribution (or '*stable*') family with exponent α.]

5. If Y_n, Y are k-dimensional random vectors with characteristic functions φ_n, φ then we say that $Y_n \to Y$ in distribution iff $\varphi_n(t) \to \varphi(t)$ for all $t \in \mathbb{R}^k$. Show that $Y_n \to Y$ in distribution iff for each $\alpha \in \mathbb{R}^k$, the scalar sequence $a'Y_n \to a'Y$ in distribution where $a'Y = \sum_{i=1}^{k} a_i Y_i$. (This result is useful in reducing some computations for multidimensional limit theorems to one dimensional cases.)

6.(a) (Projective limits need not exist.) Let μ be a finite measure on a σ-algebra $\Sigma, \Sigma_n \subset \Sigma$ an increasing sequence of σ-algebras, of a set Ω, and let $f_n : \Omega \to \mathbb{R}$ be Σ_n-measurable such that $P_n(A) = \int_A f_n d\mu_n, A \in \Sigma_n, \mu_n = \mu|\Sigma_n$. Suppose that f_n satisfies $\int_A f_n d\mu_n = \int_A f_{n-1} d\mu_{n-1}, A \in \Sigma_{n-1}, f_n \to 0$ a. e., as $n \to \infty$, and $\int_\Omega f_n d\mu_n = 1$ for all n. Show that P on $\Sigma_0 = \bigcup_{n=1}^{\infty} \Sigma_n$ defined by $P(A) = \lim_{n\to\infty} P_n(A)$ is only finitely additive for an appropriate (unbounded) sequence $\{f_n\}_1^\infty$ of functions. For instance, let (Ω, Σ, μ) be the Lebesgue unit interval. For each n, let $\Sigma_n \subset \Sigma$ be generated by the intervals $[0, \frac{1}{2^{n+1}}]$, $(\frac{j}{2^{n+1}}, \frac{j+1}{2^{n+1}}], j = 1, \ldots, 2^n - 2, 2^n, \ldots, 2^{n+1} - 1$, and $I_n = (\frac{1}{2} - \frac{1}{2^{n+1}}, \frac{1}{2})$. Let $f_n = 2^{n+1} \chi_{I_n}, n \geq 1$, where χ_A is the indicator function of the set A. Then show that this sequence fulfills all the requirements.

(b) Let $\{\mu_n\}_1^\infty$ be a sequence of uniformly bounded σ-additive set functions on (Ω, Σ), where Σ is a σ-algebra of the set Ω, such that $\lim_{n\to\infty} \mu_n(A) = \mu(A), A \in \Sigma$. Then μ is σ-additive (and bounded). This is a corollary of the basic Vitali-Hahn-Saks theorem. Explain why this does not contradict the example of part (a).

7.(a) If $\mathcal{B}_n$ is the Borel algebra of $\mathbb{R}^n$, and $P : \mathcal{B}_n \to \mathbb{R}^+$ is any (probability) measure, then P is (inner) regular, i. e., $A \in \mathcal{B}_n$ implies $P(A) = \sup\{P(C) : C \subset A, \text{ compact}\}$. Thus every finite measure on $\mathcal{B}_n$ enjoys the regularity properties of the distribution functions used in Kolmogorov's theorem.

(b) Let $\mathcal{B}_T$ be the product σ-algebra $\bigotimes_{t\in T} \mathcal{B}_t, \mathcal{B}_t = \mathcal{B}$ in $\mathbb{R}_t = \mathbb{R}$. If $\pi_\alpha : \mathbb{R}^T \to \mathbb{R}^\alpha$ for $\alpha \subset T$, (α finite) is the coordinate projection $\left(\Omega = \mathbb{R}^T, \mathbb{R}^\alpha = \times_{t\in\alpha} R_t\right)$ and $\mathcal{B}_\alpha$ is the corresponding Borel algebra of $\mathbb{R}^\alpha$ let $\tilde{\mathcal{B}}_\alpha = \pi_\alpha^{-1}(\mathcal{B}_\alpha) \subset \mathcal{B}_T$. If D is the collection of all finite subsets of T, directed by inclusion, then the algebra $\mathcal{B}_0 = \bigcup_{\alpha\in D} \tilde{\mathcal{B}}_\alpha$ generates $\mathcal{B}_T$. Show that if $\{\tilde{\mathcal{B}}_\alpha, P_\alpha, \alpha \in D\}$ is an increasing family of probability measures, i. e., $P_\alpha(A) \leq P_\beta(A)$ for all $\alpha \leq \beta$, then the limit P on $\mathcal{B}_0$, given by $P(A) = \lim_\alpha P_\alpha(A), A \in \mathcal{B}_0$, is σ-additive and has a unique extension to $\mathcal{B}_T$. [*Hint:* Use (a) to conclude that P_α is (inner) regular, for each $\alpha \in D$. This corresponds to "set submartingales".]

8. In some applications of Kolmogorov-Bochner theorems, the space $\lim\limits_{\leftarrow}(\Omega_\alpha, g_{\alpha\beta}) = \Omega$ may be too large, or a subset Ω^0 will be relevant, but non-measurable for (Ω, Σ, P). Then we need a result about the σ-additivity of P on the trace σ-algeba $\Sigma(\Omega^0)$. The following procedure shows that, in many cases, a useful result can be obtained. Let $\{(\Omega_\alpha, \Sigma_\alpha, P_\alpha, g_{\alpha\beta})_{\alpha<\beta} : \alpha, \beta \text{ in } D\}$ be a projective system of probability spaces, and let $\Omega_\alpha^0 \subset \Omega_\alpha$ be P_α-*thick* for each $\alpha \in D$, i.e. the outer measure $\tilde{P}_\alpha$ of Ω_α^0 is unity, so that for each $E \in \Sigma_\alpha$ with $P_\alpha(E) > 0$ we have: $\tilde{P}_\alpha(\Omega_\alpha^0 \cap E) > 0$. Let $\Sigma_\alpha^0 = \Omega_\alpha^0 \cap \Sigma_\alpha = \{\Omega_\alpha^0 \cap E : E \in \Sigma_\alpha\}$,

and $P_\alpha^0(\Omega_\alpha^0 \cap E) = P_\alpha(f_\alpha^{-1}(\Omega_\alpha^0 \cap E)) = P_\alpha(E), E \in \Sigma_\alpha$ where $f_\alpha : E \to \Omega_\alpha^0 \cap E$, is an algebraic isomorphism on Σ_α to Σ_α^0. Show that (i) P_α^0 is well-defined, (ii) $g_{\alpha\beta}(\Omega_\beta^0) = \Omega_\alpha^0 \Rightarrow \{(\Omega_\alpha^0, \Sigma_\alpha^0, P_\alpha^0, g_{\alpha\beta})_{\alpha<\beta} : \alpha, \beta$ in $D\}$ is a projective system of probability spaces. Also if the original system admits the projective limit (Ω, Σ, P) then so does the new one under the following condition: If $\Omega^0 = \underleftarrow{\lim}(\Omega_\alpha^0, g_{\alpha\beta})$, then Ω^0 is P-thick. Let $\Sigma_0 = \bigcup_{\alpha\in D} g_\alpha^{-1}(\Sigma_\alpha), \Sigma^0 = \bigcup_{\alpha\in D} g_\alpha^{-1}(\Sigma_\alpha^0)$, where $g_\alpha = g_{\alpha D}$, and $\Sigma^0 = \Omega^0 \cap \Sigma_0$. Then $(\Omega^0, \tilde{\Sigma}^0, P^0)$ is the projective limit with $\tilde{\Sigma}^0$ as the σ-algebra generated by $\Sigma^0, P^0(\Omega^0 \cap E) = P(E)$ for $E \in \Sigma_0$. (Even if the old one is a topological (regular) projective limit, the new measure space need not be, in spite of the fact that it still satisfies the s. m. condition.) Conversely, if $g_{\alpha\beta}$'s are coordinate projections, the original system consists of Carathéodory outer regular probability spaces (i. e., no more extensions) admitting the limit (Ω, Σ, P) and $\Omega^0 \subset \Omega$ is P-thick, let $\Omega_\alpha^0 = g_\alpha(\Omega^0) \subset \Omega_\alpha$. Then Ω_α^0 is P_α-thick for each $\alpha \in D$, and $(\Omega^0, \tilde{\Sigma}^0, P^0)$, as in the above, is the projective limit of the new system where $\Sigma_\alpha^0 = \Omega_\alpha^0 \cap \Sigma_\alpha$, and $P_\alpha^0(\Omega_\alpha^0 \cap A) = P_\alpha(A)$. [This last result remains valid if $g_{\alpha\beta}$'s satisfy the compatibility conditions, g_α's separate points of Ω and P_α is such that each subset of Ω_α has a P_α-measurable cover, instead of outer regularity.]

This extension has key applications for many problems in stochastic processes. For instance, in the original Kolmogorov existence theorem, $\Omega = \mathbb{R}^T$, and if one considers the finite dimensional distributions of the Brownian motion (cf., Exercise 4), then for (Ω, Σ, P) the set Ω^0 of continuous functions in Ω is non-measurable and is P-thick. If $X_{(\cdot)}(\omega) = \omega(\cdot)$ is the sample function, it is continuous for almost all ω (shown later in Chapter III), so that the original procedure cannot give any information about the thick set Ω^0. The above extension shows how we may obtain a useful restatement of the problem and that P on $\Sigma(\Omega^0)$ is a probability. As another example, if $\Omega = \mathbb{R}^T$, and $\Omega^0 \subset \Omega$ is the set of left continuous functions without discontinuities of the second kind, and if the projective system is considered with the finite dimensional distributions of the stable processes of exponent $\alpha, 0 < \alpha < 2$, (cf., Exercise 4), then the projective limit (Ω, Σ, P) exists but the set $\Omega^\infty \subset \Omega$ of all continuous functions satisfies $\tilde{P}(\Omega^\infty) = 0$. But fortunately Ω^0 is P-thick. The result of this exercise can be used to obtain the projective limit $(\Omega^0, \tilde{\Sigma}^0, P^0)$ in which $\tilde{P}(\Omega^0) > 0$. [These two modifications for the above processes were discussed separately by J. L. Doob and P. Lévy.] (Regarding the first part, see also Kingman [1], Halmos [1], p. 75 on thick sets, and Mallory and Sion [1], for related results.)

9. This exercise explains the underlying reasons for the above cases. Let (Ω, Σ, P) be a probability space and $(E_t, \mathcal{E}_t)_{t\in T}$ be an abstract family of measurable spaces. Let D be the class of all finite subsets of T as usual. If $X_t : \Omega \to E_t$ is a $(\Sigma, \mathcal{E}_t)$-measurable mapping, then $\{X_t, t \in T\}$ is an "abstract" stochastic process (sometimes called a measurable "vector field" and then the vector field is a vector if $E_t = E_0$ for all t). (a) Let $\mathcal{L} \subset \Sigma$ be the smallest σ-algebra with respect to which all the X_t are measurable. If $J \subset T$ is countable, let $\mathcal{L}_J$ be the σ-algebra relative to which the $X_t, t \in J$, are measurable. Show that if $\mathcal{L}_0 = \cup\{\mathcal{L}_J : J \subset T, \text{ countable}\}$, then $\mathcal{L}$ is generated by $\mathcal{L}_0$. (b) For each $\alpha \in D$, let $(E_\alpha, \mathcal{E}_\alpha) = \underset{t\in\alpha}{\times} (E_t, \mathcal{E}_t)$ be the cartesian product and let P_α on $\mathcal{E}_\alpha$ be the image measure of P by the "process" or "vector-field" $\{X_t, t \in \alpha\}$. If $\pi_{\alpha\beta} : E_\beta \to E_\alpha (\alpha < \beta$ in $D)$ are the coordinate projections, then $\{(E_\alpha, \mathcal{E}_\alpha, P_\alpha, \pi_{\alpha\beta})_{\alpha<\beta} : \alpha, \beta \text{ in } D\}$ is a projective system of (image) probability spaces for which the projective limit exists and is denoted $(E_D, \mathcal{E}_D, \tilde{P})$, where $E_D = \underset{\leftarrow}{\lim} E_\alpha, \mathcal{E}_D = \sigma\left(\bigcup_\alpha \pi_\alpha^{-1}(\mathcal{E}_\alpha)\right)$. Moreover, $(\Omega, \mathcal{L}, P)$, $(E_D, \mathcal{E}_D, \tilde{P})$ are equivalent measure spaces in the sense that $P_\alpha = \tilde{P} \circ \pi_{\alpha D}^{-1} = P \circ X_\alpha^{-1}, \alpha \in D$. (c) In case the P_α of (b) are not image probabilities but the family is merely a projective system, then the projective limit need not exist. Give an example illustrating this last possibility. Find conditions on P_α's in order that the projective limit may exist. [Note that, by (b), when $E_t = E_0$, all $t \in T$, then the new process on E_D is the coordinate process: $Y_t(\omega) = \omega(t) \in E_0$. Also since by (a), $\mathcal{L}$ is generated by $\mathcal{L}_0$ in the equivalence, deduce that the only sets in $\mathcal{E}_D$ are those that depend on some *countably many* Y_t's, each such collection depending on the set of $\mathcal{E}_D$ under consideration.]

10. Simplify the proof of Theorem 3.4 if D has countable cofinal subset, say, $\{\alpha_n\}_1^\infty$. Thus we may extract inductively, by directedness of D, a subsequence $A = \{\alpha'_n\}_1^\infty$ such that $\alpha'_n < \alpha'_{n+1}$ for $n \geq 1$. Then $A \subset D$ is also cofinal and since the compatibility conditions imply, for $\alpha < \alpha'_n, g_\alpha = g_{\alpha\alpha'_n} \circ g_{\alpha'_n}$ we may use $\{g_{\alpha'_n}\}_1^\infty$ in place of $\{g_\alpha\}_{\alpha\in D}$ and it suffices to prove the result for $D = \mathbb{N}$, the natural numbers. (If

$P = \lim_{\overline{\mathbb{N}}} P_{\alpha'_n}$, then $P \circ g_\alpha^{-1} = P \circ (g_{\alpha'_n}^{-1} \circ g_{\alpha\alpha'_n}^{-1}) = P_{\alpha'_n} \circ g_{\alpha\alpha'_n}^{-1} = P_\alpha$. So $P = \lim_{\leftarrow} P_\alpha$.)

(a) The system with $D = \mathbb{N}$ satisfies the s. m. condition. (Let $\omega_1 \in \Omega_1$ and we choose inductively as before the elements $\omega_n \in \Omega_n \cap g_{(n-1)n}^{-1}(\omega_{n-1})$. Then $\omega_n = g_{nm}(\omega_m)$ for $m \geq n$, and the point $\{\omega_n\}_1^\infty \in \lim_{\leftarrow}(\Omega_n, g_n) = \Omega$. So the s. m. condition holds, and $g_n : \Omega \to \Omega_n$ is onto.)

(b) By the inner regularity, given $\epsilon > 0$, for each $k \geq 1$, we can find a compact set $C_k \subset \Omega_k$ such that $P_k(\Omega_k - C_k) \leq \epsilon$ and $C_{k+1} \subset g_{k(k+1)}^{-1}(C_k)$. $\Big($Proceed inductively by choosing $P_{k+1}(g_{k(k+1)}^{-1}(C_k) - C_{k+1}) \leq \frac{\epsilon}{2^{n+2}}$.$\Big)$

(c) For each n, $\{g_{nm}(C_m), m \geq n\}$ is a decreasing sequence of compact subsets of Ω_n, and $C = \bigcap_{n\geq 1} g_n^{-1}(C_n) \subset \Omega$ is compact, $g_n(C) = \bigcap_{m \geq n} g_{nm}(C_m)$. (Note that $C = \Omega \cap \times_{n\geq 1} C_n$, and $\Omega \subset \times_{\alpha\in D} \Omega_\alpha$ is closed, so C is compact.) Hence $P_n(\Omega_n - g_n(C)) = \lim_{m\geq n} P_n(\Omega_n - g_{nm}(C_m)) = \lim_{m\to\infty} [P_m \circ g_{nm}^{-1}(\Omega_n - g_{nm}(C_n))] \leq \lim_{m\to\infty} [P_m(\Omega_m - C_m)] \leq \epsilon$. [See also Bourbaki [1].]

11. Let $\{(\Omega_\alpha, \Sigma_\alpha, P_\alpha, g_{\alpha\beta})_{\alpha<\beta} : \alpha, \beta \text{ in } D\}$ be a projective system of probability spaces. The system is *almost sequentially maximal* (a. s. m.) iff for each $\epsilon > 0$, and any sequence $\alpha_1 < \alpha_2 < \cdots$ in D, there exists a sequence of sets $\{A_n\}_1^\infty, A_n \in \Sigma_{\alpha_n}$, with $g_{\alpha_n\alpha_{n+1}}^{-1}(A_n) \subset A_{n+1}$ (or, $A_n \subset g_{\alpha_n\alpha_{n+1}}(A_{n+1})$) such that (i) $P_{\alpha_n}(A_n) < \epsilon, n \geq 1$, and (ii) $\{(\Omega_{\alpha_n} - A_n, g_{\alpha_n\alpha_m}) : m \geq n \geq 1\}$ satisfies the s. m. condition in that for each $\omega_n \in \Omega_{\alpha_n} - A_n$ with $g_{\alpha_n\alpha_{n+1}}(\omega_{n+1}) = \omega_n$ we have $g_{\alpha_n}(\omega) = \omega_n, n \geq 1$, for some $\omega \in \Omega (= \lim_{\leftarrow}(\Omega_\alpha, g_\alpha))$. Thus with $A_n = \emptyset$, we have the s. m. property irrespective of any measure.

(a) Suppose (Ω, Σ, P) is the projective limit of the given system, and let $\Omega^0 \subset \Omega$ be a set for which the a. s. m. condition holds, i.e., for each $\omega_n \in \Omega_{\alpha_n} - A_n$ in the above definition, there is $\omega^0 \in \Omega^0$ such that $g_{\alpha_n}(\omega^0) = \omega_n, n \geq 1$. Show that the set function $\tilde{P}$ defined by $\tilde{P}(\Omega^0 \cap A) = P(A)$ for $A \in \Sigma_0 = \bigcup_{\alpha \in D} g_\alpha^{-1}(\Sigma_\alpha)$ is σ-additive, and has a unique extension to $\tilde{\Sigma} = \sigma(\Sigma_0(\Omega^0))$.

(b) If $\Omega^0 \subset \Omega$ is P-thick, show that it satisfies the a. s. m. condition (and hence (a) applies). Thus the result of Exercise 8 is extended. [*Hints:* Let $E = \{\omega \in \Omega : g_{\alpha_n}(\omega^0) = g_{\alpha_n}(\omega), n \geq 1 \text{ for no } \omega^0 \in \Omega^0\} = \bigcup_{\{\alpha_n\}} \bigcup_{\omega^0 \in \Omega^0} \bigcap_{n=1}^{\infty} \{\omega \in \Omega : g_{\alpha_n}(\omega) \neq g_{\alpha_n}(\omega^0)\}$. Then $E \subset \Omega - \Omega^0 = E_0$. Note that E has inner P-measure zero if a. s. m. holds and E_0 (and hence E) has the same property when Ω^0 is P-thick. In either case, show that if $\{B_n\}_1^\infty \subset \Sigma_0$ and $\Omega^0 \subset \bigcup_{n=1}^{\infty} B_n$, then $\Omega - \bigcup_{n=1}^{\infty} B_n \subset E$ and is P-null. Hence $\tilde{P}(\Omega^0 \cap A) = P(A)$ yields the desired probability function.]

(c) If the given system is topological (cf., Definition 1.2.4 ff) and if $\Omega = \varprojlim(\Omega_\alpha, g_\alpha)$ has the a. s. m. property, then the projective limit of the system exists. [This extends the first part of Theorem 3.4. The proof is an extension of the latter with a careful modification, using the a. s. m. in place of the stronger s. m. condition employed there. The result is actually true when the measures take values in an abelian topological group. For details and additional results, see Millington and Sion [1]. The reader will also find in that paper an example showing the existence of the projective limit when a. s. m. holds but violating the s. m. condition.]

12. Complete the following sketch of the proof of Proposition 4.1. If $\{x_\alpha, \alpha \in \mathcal{F}\} \in \Omega = \varprojlim(\mathcal{X}_\alpha, g_\alpha)$, then $x_\alpha \in \mathcal{X}_\alpha = \mathcal{X}/\alpha$ so $x_\alpha = x + \alpha$ is a coset of $x \in \mathcal{X}$. Since $(\mathcal{X}_\alpha)^* = \alpha^\perp$ (the annihilator of α), for each $x^* \in \alpha^\perp, \langle x_\alpha, x^*\rangle = \langle x, x^*\rangle$ and $x^* \mapsto \langle x_\alpha, x^*\rangle$ is well defined and if $\alpha < \beta$, then $\langle x_\alpha, x^*\rangle = \langle x_\beta, x^*\rangle = \langle x, x^*\rangle = \langle y, x^*\rangle$ where $x_\beta = y + \beta$. So the threads $\{x_\alpha, \alpha \in \mathcal{F}\}$ define consistently linear mappings on all finite dimensional subspaces of $\mathcal{X}^*$, and the correspondence $\{x_\alpha, \alpha \in \mathcal{F}\} \mapsto x$ is a linear mapping from Ω to $(\mathcal{X}^*)'$, the algebraic dual of $\mathcal{X}^*$. Conversely, if x is a linear map on $\mathcal{X}^*$, and $\mathcal{Y} \subset \mathcal{X}^*$ is a finite dimensional subspace so that $\mathcal{Y} = \alpha^\perp$, for some $\alpha \in \mathcal{F}$, consider $x_\mathcal{Y} = x|\mathcal{Y}$. Then $x_\mathcal{Y}$ is a linear mapping on $\alpha^\perp$, whence $x_\mathcal{Y} \in (\alpha^\perp)^* = (\mathcal{X}_\alpha)^{**} = \mathcal{X}_\alpha$. So $x_\mathcal{Y} = x_\alpha, \alpha \in \mathcal{F}$. If $\alpha < \beta$, then verify that $g_{\alpha\beta}(x_\beta) = x_\alpha$ in this formulation, that $\{x_\alpha, \alpha \in \mathcal{F}\} \in \Omega, x \mapsto \{x_\alpha, \alpha \in \mathcal{F}\}$ is a linear bijection of $(\mathcal{X}^*)'$ to Ω, and that the topologies are as given.

13. (a) This exercise illustrates the special character of an abstract Wiener space $(B^*, \mathcal{X}, B)$. Let $i : \mathcal{X} \to B$ and $j : B^* \to \mathcal{X}^* \cong \mathcal{X}$ be the injection maps where $\mathcal{X}$ is a Hilbert space. Hence the images of

both maps are norm dense in the respective spaces. If $T : B \to B^*$ is a bounded linear operator, and if we define $A = j \circ T \circ i : \mathcal{X} \to \mathcal{X}$, then $A_1 = \frac{1}{2}(A + A^*)$ is of trace class and $A_2 = \frac{1}{2}(A_1 - A^*)$ is Hilbert-Schmidt, i.e., $(A_2 A_2^*)$ is of trace class, where "*" denotes the adjoint of the operation.

(b) In the construction of an abstract Wiener space $(B^*, \mathcal{X}, B)$, with $i : \mathcal{X} \to B$ as a dense injection, we started with a cylindrical probability $\mathcal{P} = \{(\mathcal{X}_\alpha, \Sigma_\alpha, P_\alpha, g_{\alpha\beta})_{\alpha<\beta} : \alpha, \beta \text{ in } D\}$ where P_α on $\mathcal{X}_\alpha$ is Gaussian. If μ_α is the correspondent of P_α on $\mathbb{R}^\alpha$, then $\mu_\alpha(A) = \int_A dF_\alpha$ where $A \subset \mathbb{R}^\alpha$ is a Borel set and F_α is given by Equation (5) of Section 1. If we specialize the family P_α or μ_α by taking the mean to be zero and the covariance matrix to be $tI, t > 0$, so that

$$\mu_\alpha^t(A) = (2\pi t)^{-n/2} \int\limits_A \exp\left[-\sum_{i=1}^n x_i^2/2t\right] dx_1, \ldots, dx_n ,$$

where $n = \dim(\mathcal{X}_a) = \dim(\mathbb{R}^\alpha)$, then the resulting $(B^*, \mathcal{X}, B)$ does not depend on $t > 0$. If $P^t = \lim\limits_{\leftarrow} P_\alpha^t$, then P^t is a Gaussian measure on $(B, \mathcal{B})$, for each $t > 0$. Letting $P_t(b, A) = P^t(A - b)$ for any $A \in \mathcal{B}, b \in B$, show that $P_t(b, \cdot) : \mathcal{B} \to [0, 1]$ is a Gaussian probability and $P_t(\cdot, A)$ is $\mathcal{B}$-measurable for each $t > 0$. Show further that for any $b_i \in B, t_i > 0$, the measures $P_{t_i}(b_i, \cdot), i = 1, 2$, are mutually absolutely continuous iff $t_1 = t_2$ and $b_1 - b_2 \in \mathcal{X}$, and mutually singular otherwise. [These assertions are not simple. In particular, for the last part one has to use a known dichotomy theorem, due to Hájek [1] and Feldman [1] which states that a pair of Gaussian measures on a space such as $(B, \mathcal{B})$ are always either mutually absolutely continuous or singular. For further details, see Gross [2] and [3] where other related results are found.]

14. We now give an example of an extension of the notion of an abstract Wiener space to a locally convex vector space which is not a Banach space (but contains an infinite dimensional Banach space). Let $C(\mathbb{R}^+)$ be the space of real continuous functions on $\mathbb{R}^+ = [0, \infty)$. Its locally convex topology is given by the set of semi-norms $\{p_m(\cdot), m \geq 1\}$ where $p_m(f) = \sup\{|f(t)| : t \in T_m\}$, with T_m compact and $T_m \uparrow \mathbb{R}^+$. Let $\mathcal{X} \subset C$ be the space of absolutely continuous f for which $\|f\|_{\mathcal{X}}^2 = \int_0^\infty |f'(t)|^2 dt < \infty$ so that $\mathcal{X}$ becomes a Hilbert space. Let

$$q(f) = \sqrt{2/\pi} \sup \left\{ \left| \int\limits_0^t (f'(u)/\sqrt{1+u^2}) du \right| : t \in \mathbb{R}^+ \right\} .$$

Then verify that $q(f) < \infty$ for $f \in \mathcal{X}$. Consider a Gaussian cylindrical measure $\mathcal{P}$ on C. We assert that $q(\cdot)$ is an m. s. n. relative to $\mathcal{P}$ and if B is the closure of $\mathcal{X}$ under q, then $B \subset C$ and $(B^*, \mathcal{X}, B)$ is an abstract Wiener space. Hence $\mathcal{P}$ admits a projective limit $(B, \mathcal{B}, P)$, and if $C \backslash B$ is assigned P-measure zero, then P can be extended to $(C, \mathcal{B}(C))$ where $\mathcal{B}(C)$ is the Borel σ-algebra of C. (See Finlayson [3].)

15. This exercise gives some elementary facts on projective limits of spaces. (a) Let $\{\Omega_i, i \in I\}$ be an indexed family of spaces and D be the directed set (by inclusion) of all finite subsets of I. If $\Omega_\alpha = \times_{i \in \alpha} \Omega_i, \alpha \in D$, and for $\alpha < \beta, \pi_{\alpha\beta} : \Omega_\beta \to \Omega_\alpha$ is the canonical mapping, let $\Omega = \varprojlim(\Omega_\alpha, \pi_{\alpha\beta})$ and $\Omega_I = \times_{i \in I} \Omega_i$. Show that there is a bijective mapping $u : \Omega_I \to \Omega$ (so these spaces may be identified, $\Omega \cong \Omega_i$). If each Ω_i is a Hausdorff space, then u is a homeomorphism when Ω_i and Ω are given the product and projective limit topologies. [*Hints:* If $p_\alpha : \Omega_I \to \Omega_\alpha$ is the coordinate projection, and $\pi_\alpha : \Omega \to \Omega_\alpha$ is the canonical mapping, then for $\alpha < \beta, p_\alpha = \pi_{\alpha\beta} \circ p_\beta$. As in Theorem 3.9, this implies the existence of u such that $p_\alpha = \pi_\alpha \circ u$. Since $p_\alpha(\omega) = p_\alpha(\omega'), \alpha \in D, \Rightarrow \omega = \omega', u$ is one-to-one, and then observe that it is onto.]

(b) If $I = \mathbb{N}, \Omega_n \supset \Omega_{n+1}$, and $g_{n(n+1)} : \Omega_{n+1} \to \Omega_n$ is the inclusion, show that $\Omega = \varprojlim(\Omega_n, g_{nm}) \cong \bigcap_{n=1}^{\infty} \Omega_n$ in the sense of (a). (Hence even if all $\Omega_n \neq \emptyset$, it is possible that $\Omega = \emptyset$.)

(c) Let $\{(\Omega_\alpha, g_{\alpha\beta})_{\alpha<\beta} : \alpha, \beta \text{ in } I\}$ be a projective system of spaces. If $J \subset I$ show that $\{(\Omega_a, g_{\alpha\beta})_{\alpha<\beta} : \alpha, \beta \text{ in } J\}$ is also a projective system and if Ω and $\tilde{\Omega}$ are their projective limits, then there is a $u : \Omega \to \tilde{\Omega}$ such that $u(\omega) = \{g_\alpha(\omega), \alpha \in J\}$. If D is the directed set of finite subsets of $I, \Omega_J = \varprojlim(\Omega_\alpha, \alpha \in J), J \in D$, let $u_{JK} : \Omega_K \to \Omega_J, J < K$, as above. Show that $\{(\Omega_j, u_{JK})_{J<K} : J, K \text{ in } D\}$ is a projective system, and that $\Omega \cong \varprojlim \Omega_J = \tilde{\Omega}$.

Bibliographical remarks

The fundamental existence theorem of Kolmogorov (Theorem 2.1 or 2.2) appears in his Foundations [1]. Its abstrat formulation as a result on projective limits and the basic Theorem 3.4 are due to Bochner ([1], [2]), and the topological properties of the limit measure in this theorem to Choksi [1]. There are some accounts of a representation theorem in the literature, the earliest of which seems to be due to Segal [1]. The results in the present form, and particularly Theorems 3.6 and 3.7, have been worked out by the author. Theorem 3.6 given here is an improved version from that of the first edition. The present form and proof are obtained in collaboration with V. V. Sazonov. See also Schrieber, Sun, and Bharucha-Reid [1] for a related assertion. For an exposition of Prokhorov's result (Theorem 3.8) and related work, see Bourbaki [1] and Schwartz [1] and Rao [11].

Applications of projective limits to the construction of measures on linear spaces has its origins in the work of Gel'fand, and an account of his results appears in Gel'fand and Vilenkin [1]. Proposition 4.2, Theorem 4.4 and the associated discussion follow the work in Schwartz [1]. The theory of abstract Wiener spaces in general and the basic Theorem 4.5 in particular are due to Gross ([1], [2]). The connection between Prokhorov's condition (cf., Theorem 3.9) and Gross's postulation of the existence of a measurable semi-norm does not seem to have been explicitly noted in the literature. See the author [10] on this point.

Finally, Section 5 contains further developments complementing the text. Several important results on projective limit theory have been given by Sion and his students, of which 5.11 is a sample result. Many other contributions and references to these and related works may be found in the above cited papers, in addition to those specifically indicated in the text. See also Schwartz [2] for other applications, and Prokhorov [2] for a discussion on weak distributions and characteristic functionals.

Chapter II

Conditioning and martingales

The concept of conditioning, or equivalently conditional expectation, is one of the most fundamental notions in Probability. After some motivation for the concept, we discuss some immediate properties of conditional expectations (and probabilities). Then the regularity of conditional probabilities and some related results are discussed. Next the concept of a martingale, the basic inequalities, and the decompositions of Riesz, Doob (in the discrete case) and Jordan type [for (sub) martingales] are presented. The discrete parameter convergences of martingales, including the Andersen-Jessen theorems, are discussed. Related Complements with some details, are contained as exercises in the last section.

2.1 Definition and properties of conditioning

(a) *Introduction.* In the first chapter we defined the probability space as a triple (Ω, Σ, P), representing a mathematical model for the description of physical phenomena. To motivate a new concept, suppose that an event $A(\in \Sigma)$ is known to have occurred. How should one assign probabilities to other members of Σ, using the knowledge of A? Since A and most events from Σ should be related, one has to consider a new class $\{A \cap B : B \in \Sigma\} = \Sigma(A)$, the *trace* (or restriction) σ-algebra of events on A. If $P_A : \Sigma(A) \to \mathbb{R}^+$ is the restriction of P to $\Sigma(A)$, then $P_A(\Omega) = P(A) \leq 1$. Hence an obvious normalization (to make it a probability) is to define $P_A(B) = P(A \cap B)/P(A)$ provided $P(A) > 0$. Then $P_A = P$ when $A = \Omega$ and (Ω, Σ, P_A), or equivalently $\left(A, \Sigma(A), \frac{1}{P(A)}P\right)$, is the new probability space serving as a model for the changed problem. This may be extended if there is a finite or countable collection $\{A_n, n \geq 1\}$ of events, in place of one A in the above procedure.

To discuss the last situation, it is convenient to consider a random variable X on (Ω, Σ, P_A), when X is integrable on (Ω, Σ, P). Thus the expectation $E_A(X)$ of X relative to the new triple is given by:

$$E_A(X) = \int_\Omega X \, dP_A = \frac{1}{P(A)} \int_\Omega \chi_A X \, dP = \frac{1}{P(A)} \int_A X \, dP. \tag{1}$$

Here χ_A denotes the indicator of A. If $P(A^c) > 0$, we may similarly define $E_{A^c}(X)$ and if $f_X(\omega) = E_A(X)$, for $\omega \in A, = E_{A^c}(X)$ for $\omega \in A^c$, then $f_X : \Omega \to \mathbb{R}$ is a two valued function. If $\{A_n, n \geq 1\}$ is a partition of $\Omega, A_n \in \Sigma, P(A_n) > 0, n \geq 1$ then we can define $f_X(\cdot)$ as:

$$f_X = \sum_{i=1}^{\infty} E_{A_i}(X)\chi_{A_i}, \tag{2}$$

which is thus an elementary function on Ω, and it is measurable relative to $\mathcal{B} = \sigma(A_n, n \geq 1) \subset \Sigma$, the σ-algebra generated by the sets shown. Note that, if $P(A_n) = 0$ for some n, then $E_{A_n}(X)$ is undefined for that n. But $\chi_{A_n} = 0$, a.e. (P) for such n, and if one then assigns an $\alpha \in \mathbb{R}$ to $E_{A_n}(X)$, let $\tilde{f}_X$ be the corresponding function. Clearly $\tilde{f}_X$ is not uniquely defined, and any two such assignments can differ on a set of P-measure zero. If we identify functions which agree outside of P-null sets, then $\tilde{f}_X$ is uniquely defined by (2) up to a set of P-measure zero, and this equivalence class is indifferently denoted by $E^{\mathcal{B}}(X)$, called the conditional expectatation of X given the partition $\{A_n, n \geq 1\}$, or the generated algebra $\mathcal{B}$. Let us state the concept precisely as:

1. Definition. Let X be an integrable random variable on (Ω, Σ, P) and $\{A_n, n \geq 1\} \subset \Sigma$ be a partition of Ω. Then the *conditional expectation* of X given (or relative to) the partition, or $\mathcal{B} = \sigma(A_n, n \geq 1)$ is the elementary function defined, up to a P-null set, by

$$E^{\mathcal{B}}(X) = \sum_{i=1}^{\infty} \left(\frac{1}{P(A_i)} \int_{A_i} X \, dP \right) \chi_{A_i} . \tag{3}$$

If $X = \chi_B, B \in \Sigma$, the the equation $P^{\mathcal{B}}(B) = E^{\mathcal{B}}(\chi_B)$ defines, uniquely outside a P-null set, the *conditional probability* of B given $\mathcal{B}$. Thus (3) becomes

$$P^{\mathcal{B}}(B) = \sum_{i=1}^{\infty} P_{A_i}(B)\chi_{A_i}, \quad B \in \Sigma . \tag{4}$$

It should be stressed that both (3) and (4) define $E^{\mathcal{B}}(X)$ and $P^{\mathcal{B}}(B)$ uniquely only outside of a P-null set, and this qualification cannot be avoided. Also it is essential to note that both $E^{\mathcal{B}}(X)$ and $P^{\mathcal{B}}(B)$ are in general *functions* which are $\mathcal{B}$-measurable when the latter is completed (or they equal $\mathcal{B}$-measurable functions outside of P-null sets). But we still refer to them as $\mathcal{B}$-measurable, following custom. The other point is that $\mathcal{B}$ was countably generated by a partition. This was used, in (3) and (4), in order to piece together the simple results involving single sets. It is thus clear that if $\mathcal{B}$ is not countable partition generated, this procedure fails. Note also that the above definition uses only the fact that the ratios, such as $\frac{P(A\cap B)}{P(A)}$, are well defined (even the value that $P(\Omega) = 1$ is not crucial). These observations will be used in the considerations that follow in this and the succeeding chapters.

The above definition, being of a constructive nature, cannot be extended to more general σ-algebras which will appear in most of the work on the subject. To treat the general case, let us express Definition 1 in a different form. Let $B \in \mathcal{B}$ be arbitrary; let $\{A_i\}_{i\in\mathbb{N}}$ generate $\mathcal{B}(\mathbb{N} = \{1,2,3,\dots\})$. Since every element of $\mathcal{B}$ is a union of some subcollection of $\{A_i\}_{i\in\mathbb{N}}$, it follows that $B = \bigcup_{j\in J} A_j$ where J is a countable or finite subset of $\mathbb{N}$, and, if $B^c = \Omega - B$, we have $\Omega = B^c \cup \bigcup_{j\in J} A_j = \bigcup_{i'=1}^{\infty} A_{i'}$. Thus integrating (3) relative to P we get

$$\int_B E^{\mathcal{B}}(X)dP = \int_B \sum_{j\in J}\left[\frac{1}{P(A_j)}\left(\int_{A_j} X\, dP\right)\chi_{A_j}\right]dP + \int_{B\cap B^c} E^{\mathcal{B}}(X)dP$$
$$= \sum_{j\in J}\int_{A_j} X\, dP = \int_{\bigcup_{j\in J} A_j} X\, dP = \int_B X\, dP\,. \tag{5}$$

If $P_{\mathcal{B}} = P|\mathcal{B}$, then (5) can be written ($\mathcal{B}'$ is P-completion of $\mathcal{B}$)

$$\int_B E^{\mathcal{B}}(X)dP_{\mathcal{B}} = \int_B X\, dP, \quad B \in \mathcal{B}'\,. \tag{6}$$

Here X is not necessarily $\mathcal{B}'$-measurable, but $E^{\mathcal{B}}(X)(\cdot)$ is. If $\nu(B) = \int_B X\, dP$, then $\nu(\cdot) : \mathcal{B} \to \mathbb{R}$ is a σ-additive function which is absolutely continuous relative to $P_{\mathcal{B}}$. Hence ($P_{\mathcal{B}}$ being finite on $\mathcal{B}$), by the Radon-Nikodým theorem

$$\nu(B) = \int_B f_X\, dP_{\mathcal{B}}, \quad B \in \mathcal{B}\,, \tag{7}$$

for an essentially (i. e., outside of $P_{\mathcal{B}}$-null set) unique $\mathcal{B}$-measurable function f_X. So comparing (7) and (6) the essential uniqueness immediately yields again $f_X(\omega) = E^{\mathcal{B}}(X)(\omega)$, for almost all $\omega \in \Omega$. Since for the Radon-Nikodým theorem, no such restriction on $\mathcal{B}$ (that it be countable partition generated) is needed, the $f_X(\cdot)$ given by (7) can be taken as the definition of conditional expectation, as it coincides with (the constructive) Definition 1 when $\mathcal{B}$ is countable partition generated. This was proposed by A. N. Kolmogorov in 1933. *The general concept is one of the most fundamental, least intuitive, and most complicated notions in probability theory.* We now state it precisely and investigate its structure.

2. Definition. Let (Ω, Σ, P) be any probability space and $\mathcal{B} \subset \Sigma$ be a σ-algebra. If $X : \Omega \to \mathbb{R}$ is an integrable function, then the essentially unique $\mathcal{B}'$-measurable function $E^{\mathcal{B}}(X)(\cdot)$ satisfying the equation

$$\int_B E^{\mathcal{B}}(X) dP_{\mathcal{B}} = \int_B X \, dP, \quad B \in \mathcal{B}, \tag{8}$$

is called the *conditional expectation* of X relative to $\mathcal{B}$, and if we set $P^{\mathcal{B}}(A) = E^{\mathcal{B}}(\chi_A), A \in \Sigma$, then $P^{\mathcal{B}}(\cdot)$ is called the *conditional probability* on Σ relative to $\mathcal{B}$, or given $\mathcal{B}$.

This general definition and even the special case of (3) are clearly more involved than the ordinary concepts. In fact, even the existence of $E^{\mathcal{B}}, P^{\mathcal{B}}$ depend on the (nontrivial) Radon-Nikodým theorem as seen above. But, also in the simple case, there is the family of probability spaces $\{(\Omega, \Sigma, P_A), A \in \Sigma, P(A) > 0\}$. The corresponding general case is the class $\{(\Omega, \Sigma, P^{\mathcal{B}}), \mathcal{B} \subset \Sigma\}$. However, an immediate *warning* is in order. The function $P^{\mathcal{B}}(\cdot)$ is not uniquely defined, and $P^{\mathcal{B}} : \Sigma \times \Omega \to [0, 1]$, cannot be used as a measure on Σ unless somehow $P^{\mathcal{B}}(\cdot)(\omega), \omega \in \Omega$ is selected as a unique set function on Σ. For each $A \in \Sigma$, we can define $P^{\mathcal{B}}(A)$ uniquely outside of a null set N_A which depends on A, and if $\mathcal{B}$ is not partition generated this "unique selection," called a "version" of $P^{\mathcal{B}}$, may or may not be possible. In any case it is a nontrivial task to find such a version. We must know the associated problems and properties precisely for applications. Therefore a careful study of conditional expectations and probability functions is necessary, and it will be taken up in what follows. Let us record some immediate and useful properties of conditional expectations following from Definitions

1 and 2. These will present a basis for the general work in addition to giving an important lead to a deeper analysis.

To begin with, it will be useful to give an alternative form of the above definition before its properties are ascertained. Let (Ω, Σ, P) be a probability space and $(E, \mathcal{E})$ be a measurable space. If $h : \Omega \to E$ is a mapping which is measurable relative to $(\Sigma, \mathcal{E})$, let $Q = P \circ h^{-1}$ be the image probability on $\mathcal{E}$, i. e., $Q(A) = P(h^{-1}(A)), A \in \mathcal{E}$. (If h is not onto, we take $h^{-1}(A), A \subset E$ to be empty whenever no point of Ω is mapped into A. Such a procedure is called the "complete inverse image" of A.) Then $Q : \mathcal{E} \to [0,1]$ is a probability and $\mathcal{B}_n = h^{-1}(\mathcal{E})$ is a σ-subalgebra of Σ. It is clear from definition that $P_{\mathcal{B}_h} \circ h^{-1} = Q$ where $P_{\mathcal{B}_h} = P|\mathcal{B}_h$. If $X : \Omega \to \mathbb{R}$ is any measurable (Σ) and P-integrable function, then the set function μ_X defined by: $\mu_X(A) = \int_A X \, dP_{\mathcal{B}_h}, A \in \mathcal{B}_h$ is a finite σ-additive function on $\mathcal{B}_h$, and if $\nu_X = \nu_{h,X} = \mu_X \circ h^{-1}$, i. e., $\nu_X(B) = \mu_X(h^{-1}(B)) = \int_{h^{-1}(B)} X \, dP,\ B \in \mathcal{E}$, then $\nu_X : \mathcal{E} \to \mathbb{R}$ is σ-additive on $\mathcal{E}$ and is absolutely continuous relative to Q. Hence by the Radon-Nikodým theorem (Q is finite) there is an $\mathcal{E}$-measurable Q-unique integrable function g_X (determined by X and h) such that

$$\int_B g_X \, dQ = \nu_X(B) = \int_{h^{-1}(B)} X \, dP_{\mathcal{B}_h}, \quad B \in \mathcal{E} . \tag{9}$$

The function g_X on $(E, \mathcal{E}, Q)$ is called the *conditional expectation* of X *given* $\mathcal{B}_h$, or *simply given* h, and is denoted by

$$g_X(x) = E^{\mathcal{B}_h}(X)(x) = E(X|h = x), \quad x \in E . \tag{10}$$

If $E = \Omega, h =$ identity, and $\mathcal{E} = \mathcal{B} \subset \Sigma$ (a σ-algebra), then $E^{\mathcal{B}_h}(X) = E^{\mathcal{B}}(X)$, and (9) reduces to (8). The definition (9) is useful in some applications. For instance, let $E = \mathbb{N}, \mathcal{E} = \mathcal{N}$ (the power set), and define $h : \Omega \to \mathbb{N}$ as follows. If $\{A_i\}_{i \in \mathbb{N}} \subset \Sigma$ is a partition of Ω, let $h(A_n) = n$ so that $\mathcal{B} = h^{-1}(\mathcal{N})$ and $Q(\{n\}) = P(A_n)$. Then $(\mathbb{N}, \mathcal{N}, Q)$ is the image probability space, and (3) is obtained, i. e., if $X : \Omega \to \mathbb{R}$ is any integrable random variable, then (9) implies with $B = \{n\}$ there,

$$g_X(n) \cdot Q(\{n\}) = \int_B g_X dQ = \int_{h^{-1}(B) = A_n} X \, dP, \quad B \in \mathcal{N} , \tag{11}$$

and since $g_X(n) = E(X|h = n) = E^{\mathcal{B}_h}(X)(n) = E_{A_n}(X)$ in the old notation, and $Q(\{n\}) = P \circ h^{-1}(\{n\}) = P(A_n)$, (11) can be expressed, when $P(A_n) > 0$, as the familiar function (cf. (1)):

$$E_{A_n}(X) = \frac{1}{P(A_n)} \int_{A_n} X \, dP \, . \tag{12}$$

The equation (10) or (12) has the following interpretation. Suppose $\mathcal{B} \subset \Sigma$ is any σ-algebra. Then $E^{\mathcal{B}}(X)$ is $\mathcal{B}'$-measurable whenever it is defined. If $A \in \mathcal{B}$ is an atom, i. e., $A_1 \subset A, A_1 \in \mathcal{B}$ implies either $P(A - A_1) = 0$, or $P(A_1) = 0$, then the $\mathcal{B}'$-measurable function $E^{\mathcal{B}}(X)$ must be a constant a. e., on A. Hence by the defining equation (8), writing $(E^{\mathcal{B}}(X))(A) = E_A(X)$, we recover (1):

$$E_A(X)P(A) = \int_A E^{\mathcal{B}}(X) dP = \int_A X \, dP \, . \tag{13}$$

Consequently $E_A(X)$ is the average of X on A relative to P when $P(A) > 0$ and this is the value of $E^{\mathcal{B}}(X)$ on A. Equations (12) and (13) show that the function $E^{\mathcal{B}}(X)$ takes "fewer" values than X, and this fact is sometimes stated as a "smoothing property" of the conditional expectation operation. If $\mathcal{B} = \sigma(Y)$, the σ-algebra generated by the random variable $Y : \Omega \to \mathbb{R}$, then $E^{\mathcal{B}}(X)$ is also denoted by $E(X|Y)$ or $E^Y(X)$ (or even $E_Y(X)$) by some authors.

(b) *Properties.* The form of (8) yields the following basic properties: (i) Taking $B = \Omega$, we have $E(E^{\mathcal{B}}(X)) = E(X)$, (ii) from the essential uniqueness of the Radon-Nikodým derivative, $E^{\mathcal{B}}(X) = X$ a. e., whenever X is $\mathcal{B}$-measurable and $E^{\mathcal{B}}(X) = E(X)$ a. e., if $\mathcal{B} = \{\emptyset, \Omega\}$, (iii) if $X \geq 0$, a. e., so $\nu_X(\cdot) \geq 0, E^{\mathcal{B}}(X) \geq 0$ a. e., (iv) writing $X = X^+ - X^-$, where $X^+ = \max(X, 0), X^- = -\min(X, 0)$, the positive and negative parts, then $E^{\mathcal{B}}(X) = E^{\mathcal{B}}(X^+) - E^{\mathcal{B}}(X^-)$, a. e., (v) $X = a$, a. e., implies $E^{\mathcal{B}}(X) = a$, a. e., (by (ii)), and (vi) $E^{\mathcal{B}} : X \mapsto E^{\mathcal{B}}(X)$ is a linear operation, i.e., if X_1, X_2 are integrable, then $E^{\mathcal{B}}(a_1X_1 + a_2X_2) = a_1E^{\mathcal{B}}(X_1) + a_2E^{\mathcal{B}}(X_2)$, a. e. This last relation and (iii) imply that $E^{\mathcal{B}}(\cdot)$ is a positive linear operator on the space of all integrable functions, i. e., on $L^1(\Omega, \Sigma, P) = L^1(P)$. For, by (iii), $E^{\mathcal{B}}$ is positivity preserving so that using $-|X| \leq X \leq |X|$, we deduce that a. e., $-E^{\mathcal{B}}(|X|) \leq E^{\mathcal{B}}(X) \leq E^{\mathcal{B}}(|X|)$, or $|E^{\mathcal{B}}(X)| \leq E^{\mathcal{B}}(|X|)$ and hence:

$$\int_A |E^{\mathcal{B}}(X)|dP_{\mathcal{B}} \leq \int_A E^{\mathcal{B}}(|X|)d\mu = \int_A |X|dP < \infty\,, \quad A \in \mathcal{B}\,.$$

Thus $E^{\mathcal{B}}(X)$ is integrable and in fact, $\|E^{\mathcal{B}}(X)\|_1 \leq \|X\|_1$ where $\|\cdot\|_1$ is the L^1-norm. This shows that $E^{\mathcal{B}}$ is a contractive linear mapping on $L^1(P)$ and indicates that it should be analyzed as a linear transformation on such a class of function spaces. Before getting into a detailed discussion it is desirable to consider further properties involving (α_1) the continuity of $E^{\mathcal{B}}$ on function spaces, and (α_2) the "smoothing" nature of $E^{\mathcal{B}}$.

To discuss the continuity of $E^{\mathcal{B}}(\cdot)$, it will be useful to prove the conditional Jensen inequality which has considerable interest for applications. We record some more simple properties of $E^{\mathcal{B}}(\cdot)$ to use in the ensuing computations.

3. Proposition. *Let $\{X_n, n \geq 1, Y, Z\} \subset L^1(\Omega, \Sigma, P)$ and X be a random variable on (Ω, Σ). If $\mathcal{B} \subset \Sigma$ is a σ-algebra, then the following statements are true:*

(a) [Monotone Convergence] $Y \leq X_n \uparrow X$, a. e., *implies* $E^{\mathcal{B}}(X_n) \uparrow E^{\mathcal{B}}(X)$, a.e.

(b) [Fatou's Inequalities] $Y \leq X_n$, a.e., *all n implies* $E^{\mathcal{B}}(\liminf_n X_n) \leq \liminf_n E^{\mathcal{B}}(X_n)$, a.e., *and* $X_n \leq Z$, *all n implies* $E^{\mathcal{B}}(\limsup_n X_n^n) \geq \limsup_n E^{\mathcal{B}}(X_n)$, a.e.

(c) [Dominated Convergence] $Y \leq X_n \leq Z$, a.e., *all n and $X_n \to X$* a.e., *implies* $\lim_n E^{\mathcal{B}}(X_n) = E^{\mathcal{B}}(X)$, *in L^1* and a.e.

Proof. The proofs are simple modifications of the classical (unconditional) versions. To see what changes are needed let us prove (a) and the rest can be similarly verified. Clearly, in (8), $E^{\mathcal{B}}(X)$ *exists if only* X^+ or X^- *is integrable.*

If $A \in \mathcal{B}$, then by the definition of conditional expectation $E^{\mathcal{B}}$ (see (8)),

$$\int_A X_n dP = \int_A E^{\mathcal{B}}(X_n)dP_{\mathcal{B}}. \tag{14}$$

Since $Y \leq X_n \leq X_{n+1} \leq X$ a.e., implies $E^{\mathcal{B}}(Y) \leq E^{\mathcal{B}}(X_n) \leq E^{\mathcal{B}}(X_{n+1}) \leq E^{\mathcal{B}}(X)$ a.e., we deduce that $\lim_n E^{\mathcal{B}}(X_n) \leq E^{\mathcal{B}}(X)$, a.e.

Note that $Y \le X$, and Y integrable implies $\int_A X\,dP \ge \int_A Y\,dP > -\infty$ so that $E^{\mathcal{B}}(X)$ exists a.e. (may be $= +\infty$ but $\ne -\infty$ by the above remark on $X^{\pm}$) and $E^{\mathcal{B}}(Y) \in L^1(\Omega, \mathcal{B}, P_{\mathcal{B}})$. Since $E^{\mathcal{B}}(X)$ is $\mathcal{B}$-measurable, the classical Lebesgue Monotone Convergence theorem applies to both sides of (14) and yields for $A \in \mathcal{B}$:

$$\int_A \lim_n E^{\mathcal{B}}(X_n)dP_{\mathcal{B}} = \lim_n \int_A E^{\mathcal{B}}(X_n)dP_{\mathcal{B}} = \lim_n \int_A X_n\,dP$$
$$= \int_A X\,dP = \int_A E^{\mathcal{B}}(X)dP_{\mathcal{B}}\,. \tag{15}$$

Since the integrands of the extreme integrals are $\mathcal{B}'$-measurable in (15), and $A \in \mathcal{B}$ is arbitrary, we must have $\lim_n E^{\mathcal{B}}(X_n) = E^{\mathcal{B}}(X)$ a.e. This yields (a) and the rest are similar. □

Let us now turn to a very useful and important Jensen's inequality for conditional expectations. Since the result is also valid for σ-finite measure spaces, and problems arise later in this generality, the proof will be given with this point in mind. We then provide alternative proofs for illustration and variety.

It is first necessary to recall various characterizations of continuous (or equivalently measurable) real convex functions on the line $\mathbb{R}$ from classical analysis.

4. Proposition. *Let $\varphi : \mathbb{R} \mapsto \bar{\mathbb{R}}$ be a Borel function where $\bar{\mathbb{R}} = [-\infty, \infty]$. Then φ is convex (and then automatically continuous) iff one of the following holds:*

(1) there exists a nondecreasing function $g : \mathbb{R} \to \bar{\mathbb{R}}$, such that for any $a, b \in \mathbb{R}, a \le x \le b$,

$$\varphi(x) = \varphi(a) + \int_a^x g(t)dt, \tag{16}$$

and $g(\cdot)$ is strictly increasing iff $\varphi(\cdot)$ is strictly convex.

(2) φ is the upper envelope of a countable family of lines $f_n(\cdot)$ where $f_n(x) = a_n x + b_n$, each f_n touching φ at only one point if φ is strictly convex.

As an immediate consequence of the first part, i.e., of (16), we have the following useful property of continuous real convex functions:

5. Corollary. *If φ is a continuous convex function on $\mathbb{R}$ to $\mathbb{R}$, then either there exists a point $a \in \mathbb{R}$ such that φ is monotone decreasing on $(-\infty, a]$ and increasing on $[a, \infty)$, or is a monotone function everywhere on $\mathbb{R}$.*

The reader who is not familiar with the form (16) should consult the classical monographs of Hardy-Littlewood-Pólya ([1], p. 95), or Zygmund ([1]). For a proof of Jensen's theorem, as well as for much of the work on the subject, the following commutativity property of the conditional expectations is also needed.

6. Proposition. (a) *Let X, Y be random variables on (Ω, Σ, P) such that X and XY are integrable. If $\mathcal{B} \subset \Sigma$ is a σ-algebra and Y is $\mathcal{B}$-measurable, then (averaging property)*

$$E^{\mathcal{B}}(XY) = YE^{\mathcal{B}}(X), \text{ a.e.}$$

(b) *If $\mathcal{B}_1 \subset \mathcal{B}_2 \subset \Sigma$ are σ-algebras and X is integrable, then*

$$E^{\mathcal{B}_1}(E^{\mathcal{B}_2}(X)) = E^{\mathcal{B}_2}(E^{\mathcal{B}_1}(X)) = E^{\mathcal{B}_1}(X), \text{ a.e.}$$

In other words, the operators $E^{\mathcal{B}_1}$ and $E^{\mathcal{B}_2}$ commute on $L^1(\Omega, \Sigma, P)$.

(c) *If X, Y are integrable random variables such that XY is integrable, $\mathcal{B}_1 \subset \mathcal{B}_2 \subset \Sigma$ are σ-algebras and Y is $\mathcal{B}_2$-measurable, then*

$$E^{\mathcal{B}_1}(XY) = E^{\mathcal{B}_1}(YE^{\mathcal{B}_2}(X)), \text{ a.e.}$$

Proof. (a) We need to show for all $A \in \mathcal{B}$ that

$$\left(\int_A E^{\mathcal{B}}(XY)dP_{\mathcal{B}} = \right)\int_A XY\, dP = \int_A YE^{\mathcal{B}}(X)\, dP_{\mathcal{B}}\ . \tag{17}$$

However, if $Y = \chi_B, B \in \mathcal{B}$, then the desired result becomes

$$\int_A X\chi_B dP = \int_{A\cap B} X\, dP = \int_{A\cap B} E^{\mathcal{B}}(X) DP_{\mathcal{B}}$$
$$= \int_A \chi_B E^{\mathcal{B}}(X) dP_{\mathcal{B}}, \quad A \cap B \in \mathcal{B},$$

which is true by definition of the conditional expectation. Hence by the linearity of our integral, the desired result (17) holds for any simple function $Y_n = \sum_{i=1}^{n} a_i \chi_{A_i}, A_i \in \mathcal{B}, a_i \in \mathbb{R}$. Writing $X = X^+ - X^-$, it follows that

$$\int_A X^{\pm} Y \, dP = \int_A Y E^{\mathcal{B}}(X^{\pm}) dP_{\mathcal{B}} \tag{18}$$

holds for all $Y \geq 0$ a.e., since there exist $0 \leq Y_n \uparrow Y$ a.e., and (18) holds for each Y_n so that by the Dominated Convergence we can interchange the limit and integral. The general case is now obtained from a linear combination of (18), and (17) follows.

(b) Since $E^{\mathcal{B}_1}(X)$ is $\mathcal{B}_2'$-measurable and $E^{\mathcal{B}_2}(1) = 1$ a.e., it is obvious that $E^{\mathcal{B}_2} E^{\mathcal{B}_1}(X) = E^{\mathcal{B}_1}(X)$ a.e. On the other hand, by definition (cf. (8)), for all $A \in \mathcal{B}_1 \subset \mathcal{B}_2$.

$$\int_A E^{\mathcal{B}_1}(E^{\mathcal{B}_2}(X)) dP_{\mathcal{B}_1} = \int_A E^{\mathcal{B}_2}(X) dP_{\mathcal{B}_2} = \int_A X \, dP = \int_A E^{\mathcal{B}_1}(X) dP_{\mathcal{B}_1}$$

and since the integrands of the extreme integrals are $\mathcal{B}_1'$-measurable, and $A \in \mathcal{B}_1$ is arbitrary, it follows that

$$E^{\mathcal{B}_1}(E^{\mathcal{B}_2}(X)) = E^{\mathcal{B}_1}(X) \text{ a.e.}$$

(c) This is a consequence of (a) and (b) above. In fact, $Y E^{\mathcal{B}_2}(X) = E^{\mathcal{B}_2}(XY)$ a.e., by (a) and since $\mathcal{B}_1 \subset \mathcal{B}_2, E^{\mathcal{B}_1}(E^{\mathcal{B}_2}(XY)) = E^{\mathcal{B}_1}(XY)$ a.e., completing the proof. □

The commutativity relations of conditional expectations expressed in this proposition are often used in numerous computations. For instance, the following is a consequence.

7. Corollary. *Let $X_1, \ldots, X_n$ be n random variables on (Ω, Σ, P) and X_1 be integrable. Then*

$$\begin{aligned} E(X_1) &= E(E(X_1 | X_2, \ldots, X_n)) \\ &= E(E(\cdots(E(X_1 | X_2, \ldots, X_n) | X_3, \cdots X_n) \cdots) | X_n)) \text{a.e.} \end{aligned}$$

The above results enable us to prove some deep results for conditional expectations. Note that if P on $\mathcal{B}_1$ is σ-finite, then Proposition 3 is valid without any change. *In such a general case, we say that $\mathcal{B}_1$ is a rich σ-subalgebra for P, or P-rich.* Thus, if P is a probability, every σ-subalgebra of Σ is rich for each P. The following general form of the conditional Jensen inequality is substantially due to Chow [1].

8. Theorem. *Let $\mathcal{B} \subset \Sigma$ be a σ-algebra and (Ω, Σ, P) a measure space. Let X, Y be a pair of real random variables on (Ω, Σ) such that Y is $\mathcal{B}$-measurable. Suppose X, Y are such that the indefinite integrals $\int_A X\, dP$, $\int_B Y\, dP$, $A \in \Sigma$, $B \in \mathcal{B}$ exist as real numbers, or more generally, define σ-finite (on $\mathcal{B}$) set functions. Let $\varphi : \mathbb{R} \to \mathbb{R}$ be a continuous convex function such that $\nu(A) = \int_A \varphi^+(X)dP$, $A \in \mathcal{B}$ defines a σ-finite (on $\mathcal{B}$) measure. If $\mathcal{B}$ is P-rich and either (i) $Y = E^{\mathcal{B}}(X)$ a.e., or (ii) $Y \leq E^{\mathcal{B}}(X)$ a.e., and φ is also nondecreasing, then $\varphi(Y) \leq E^{\mathcal{B}}(\varphi(X))$, a.e. If, moreover, φ is strictly convex (and $\mathcal{B}$ is complete), then $\varphi(Y) = E^{\mathcal{B}}(\varphi(X))$ a.e., iff $X = Y$ a.e.*

Remark. If φ is a continuous concave functions in (i) and is also nonincreasing in (ii) where $Y \leq E^{\mathcal{B}}(X)$ is still in force, then $\varphi(Y) \geq E^{\mathcal{B}}(\varphi(X))$ a.e. holds. This follows from the fact that $-\varphi$ is convex and the theorem applies. We note that the proof and the result hold true if P and $\mathcal{B}$ are such that $P|\mathcal{B}(= P_{\mathcal{B}})$ has the Radon-Nikodým property. (This is equivalent to saying that $P_{\mathcal{B}}$ is "localizable." See e.g. Rao [11]; σ-finiteness is sufficient.) In any case we can conclude that $E^{\mathcal{B}}(1) = 1$ a.e. $(P_{\mathcal{B}})$. Thus if $\mathcal{B} = \{\emptyset, \Omega\}$ (so $E^{\mathcal{B}}$ = expectation), then the admissible function P on $\mathcal{B}$ (so that it is rich) is necessarily a probability measure. The generality therefore is possible only for nontrivial $\mathcal{B} \subset \Sigma$ in this theorem.

Proof. By (16) above, the convex function φ can be expressed as

$$\varphi(x) = \varphi(a) + \int_a^x g(t)dt, \quad -\infty < a, x < \infty,$$

where $g(\cdot) : \mathbb{R} \to \mathbb{R}$ is nondecreasing and in fact can be taken as the right (or the left) derivative of $\varphi(\cdot)$ which exists everywhere. It follows from this representation that

$$\varphi(x) - \varphi(a) \geq g(a)(x - a), \quad \text{all } a, x \text{ in } \mathbb{R}. \tag{19}$$

Since $g(\cdot)$ in particular is a Borel function on the line, the composite function $g \circ Y$ is $\mathcal{B}$-measurable. We prove the result with $P_{\mathcal{B}}\sigma$-finite.

Assume the hypothesis of (i). For any integer $m > 0$, the set $\bar{V}_m = \{\omega : |g \circ Y|(\omega) \leq m\} \in \mathcal{B}$. If $V \in \mathcal{B}$ is arbitrary, let $V_m = V \cap \bar{V}_m$. Then on V_m, $g(Y)(X - Y)$ is integrable since X, Y are integrable. Let $\Omega_n \in \mathcal{B}$

be a sequence of sets $\Omega_n \subset \Omega_{n+1}$, such that $\int_{\Omega_n} g^+(X)dP < \infty$. By σ-finiteness such sets exist and $\lim_n \Omega_n = \Omega$. Hence $V_m \cap \Omega_n \in \mathcal{B}$ and we have

$$\int_{\Omega_n \cap V_m} g(Y)(X-Y)dP = \int_{\Omega_n \cap V_m} E^{\mathcal{B}}(g(Y)(X-Y))dP_{\mathcal{B}}$$

$$= \int_{\Omega_n \cap V_m} g(Y)E^{\mathcal{B}}(X-Y)dP_{\mathcal{B}}, \text{ since } g(Y) \text{ is}$$

$\mathcal{B}$-measurable and Proposition 6(a) applies,

$$= 0, \text{ by hypothesis in (i)}.$$

Now (19) and the above equation imply

$$\int_{\Omega_n \cap V_m} (\varphi(X) - \varphi(Y))dP \geq 0\,, \tag{20}$$

for all m, n. Let $U_n = \{\varphi(Y) \geq 0\} \cap \Omega_n$ so that on $U_n, \varphi \circ Y = \varphi^+ \circ Y$, and (20) yields, since $U_n \in \mathcal{B}$,

$$\int_{\Omega_n \cap V_m \cap U_n} [\varphi^+(X) - \varphi^+(Y)]dP \geq \int_{\Omega_n \cap V_m \cap U_n} [\varphi(X) - \varphi(Y)]dP \geq 0. \tag{21}$$

It follows from (21) that

$$\int_{\Omega_n} \varphi^+(X)dP \geq \int_{\Omega_n \cap V_m \cap U_n} \varphi^+(X)dP \geq \int_{\Omega_n \cap V_m \cap U_n} \varphi^+(Y)dP \geq 0.$$

Letting $m \to \infty$, so that $V_m \uparrow V \cap \Omega_n$, and taking the arbitrary set V to be $\{\omega : \varphi(Y(\omega)) > 0\}$, we get

$$0 \leq \int_{\Omega_n} \varphi^+(Y)dP \leq \int_{\Omega_n} \varphi^+(X)dP < \infty\,. \tag{22}$$

Thus, with (20), for any $V \in \mathcal{B}$ we have, on noting that the integrals are finite and then letting $m \to \infty$ there, by Monotone Convergence:

$$\int_{\Omega_n \cap V} E^{\mathcal{B}}(\varphi(X))dP_{\mathcal{B}} = \int_{\Omega_n \cap V} \varphi(X)dP \geq \int_{\Omega_n \cap V} \varphi(Y)dP_{\mathcal{B}}\,. \tag{23}$$

Letting $n \to \infty$ in (23) and using the fact that the integrals in question define σ-finite set functions (cf., (22)), we can again apply the Monotone Convergence theorem to deduce

$$\int_V E^{\mathcal{B}}(\varphi(X))dP_{\mathcal{B}} \geq \int_V \varphi(Y)dP_{\mathcal{B}}\,, \quad V \in \mathcal{B}\,. \tag{24}$$

Since the integrands in (24) are $\mathcal{B}'$-measurable, and $V \in \mathcal{B}$ is arbitrary, we have:

$$E^{\mathcal{B}}(\varphi(X)) \geq \varphi(Y)(=\varphi(E^{\mathcal{B}}(X))) \text{ a.e.} \tag{25}$$

To prove (ii) note that under the strengthened hypothesis of this case with the result of (i) one has

$$\varphi(Y) \leq \varphi(E^{\mathcal{B}}(X)) \leq E^{\mathcal{B}}(\varphi(X)) \text{ a.e.} \tag{25$'$}$$

To prove the last part, suppose φ is strictly convex and $E^{\mathcal{B}}(\varphi(X)) = \varphi(E^{\mathcal{B}}(X))$, a.e. One may assume that $\varphi(X) \in L^1(\Omega, \Sigma, P)$ and $P_{\mathcal{B}}$ is σ-finite for both parts. In (19) take $a = E^{\mathcal{B}}(X)(\omega) = Y(\omega)$, $x = X(\omega), \omega \in \Omega$. Using an argument preceding (20), we can conclude that $X \cdot g(Y) \in L^1(\Omega, \Sigma, P)$. [In fact, for any X for which $\varphi(|X|)$ is integrable where $\varphi(0) = 0$, it is true that $X \cdot g(X)$ is also integrable. We leave a proof of this statement to the reader with the remark that it also follows from a standard result in Orlicz space theory.] Thus integrating (19) after this (pointwise true) substitution, it follows that

$$\begin{aligned}
\int_\Omega \varphi(X)dP &\geq \int_\Omega \varphi(Y)dP_{\mathcal{B}} + \int_\Omega g(Y)(X-Y)dP, \\
&= \int_\Omega E^{\mathcal{B}}(\varphi(X))dP_{\mathcal{B}} + \int_\Omega E^{\mathcal{B}}(g(Y)[X-Y])dP, \text{ by the} \\
&\qquad \text{equality hypothesis in (25) and the definition of } E^{\mathcal{B}}, \\
&= \int_\Omega E^{\mathcal{B}}(\varphi(X))dP_{\mathcal{B}}, \text{ since the last term is zero} \\
&\qquad \text{on using the averaging property of } E^{\mathcal{B}}, \\
&= \int_\Omega \varphi(X)dP.
\end{aligned}$$

Thus there is equality throughout. This fact and (19) yield:

$$0 \leq \int_\Omega [\varphi(X) - \varphi(Y) - g(Y)(X-Y)]dP = 0. \tag{25$''$}$$

But the integrand is non-negative a.e. so that by (25″) it must vanish a.e. (P). Since by Proposition 4(1), g is strictly increasing, on the (Σ)-measurable set $A = \{\omega : X(\omega) \neq Y(\omega)\}$ there is strict inequality in (19) and that if $P(A) > 0$ there must be strict inequality in (25) which is impossible. Hence $P(A) = 0$, and so $Y(\omega) = X(\omega)$ for $\omega \in \Omega - A$. Since $\mathcal{B}$ is complete, $A \in \mathcal{B}$ and hence X is $\mathcal{B}$-measurable. The converse implication is trivial when X is $\mathcal{B}$-measurable. This completes the proof of the theorem. □

Remark. It is interesting to note that the conditional Jensen inequality is true for a general class of nontrivial measures, while the non-conditional (or classical) Jensen inequality is only valid for probability measures. This is one of the instances where a greater generality and flexibility are obtained for conditional measures than the ordinary ones.

We now sketch another proof for probability measures.

Alternative proof of Theorem 8. In this proof one uses the characterization from Proposition 4(2) of a convex function. Thus φ is the upper envelope of a countable collection of affine functions $\ell_n(\cdot)$, where $\ell_n(x) = a_n x + b_n \leq \varphi(x)$, each $x \in \mathbb{R}$ and suitable a_n, b_n. From this, the positivity of the linear operator $E^{\mathcal{B}}(\cdot)$ and the finiteness of the measure P, we deduce that

$$E^{\mathcal{B}}(\varphi(X)) \geq E^{\mathcal{B}}(\ell_n(X)) = a_n E^{\mathcal{B}}(X) + b_n = \ell_n(E^{\mathcal{B}}(X)), \text{ a.e.} \qquad (26)$$

Taking the supremum of the countable collection of random variables on both sides of (26) it follows that

$$E^{\mathcal{B}}(\varphi(X)) \geq \varphi(E^{\mathcal{B}}(X)) \text{ a.e.}$$

which proves (i), and (ii) follows from (i) as before. The proof of the equality conditions is again the same as in the first method. A different proof of the Jensen inequality for probability measures is outlined in the Complements section. □

The continuity properties of the operator $E^{\mathcal{B}}$ on $L^p(\Sigma)$, $p \geq 1$, can now be established with the help of the above theorem, extending the simpler case of $p = 1$ considered earlier.

9. Theorem. *Let $\mathcal{B} \subset \Sigma$ be a σ-algebra and (Ω, Σ, P) a probability space. Then the conditional expectation operator $E^{\mathcal{B}} : L^p(\Omega, \Sigma, P) \to L^p(\Omega, \Sigma, P), 1 \leq p \leq \infty$, is a positive linear contractive (i.e., $\|E^{\mathcal{B}}(f)\|_p \leq \|f\|_p$) projection. Its range is $L^p(\Omega, \mathcal{B}, P_{\mathcal{B}})$.*

Proof. If $f \in L^p(\Omega, \Sigma, P) \subset L^1(\Omega, \Sigma, P)$ (since $P(\Omega) = 1$) then f is integrable and hence $E^{\mathcal{B}} : f \mapsto E^{\mathcal{B}}(f) \in L^1(\Omega, \mathcal{B}, P_{\mathcal{B}})$ exists. Moreo- ever, with Proposition 6(b), $E^{\mathcal{B}}$ is a positive linear operator such that $E^{\mathcal{B}} E^{\mathcal{B}} = E^{\mathcal{B}}$; so it is also a projection. Thus only a proof of the contraction property remains.

If $p = +\infty$, then $|f| \leq \|f\|_\infty$, a.e., so that $|E^{\mathcal{B}}(f)| \leq E^{\mathcal{B}}(|f|) \leq E^{\mathcal{B}}(\|f\|_\infty) = \|f\|_\infty$, a.e., by the order preserving property of $E^{\mathcal{B}}$. Hence $\|E^{\mathcal{B}}(f)\|_\infty \leq \|f\|_\infty$, and $E^{\mathcal{B}}$ is contractive in this case. This also shows that $E^{\mathcal{B}}(f) \in L^\infty(\Omega, \mathcal{B}, P_{\mathcal{B}})$. If $1 \leq p < \infty$, let $\varphi(x) = |x|^p$. Then $\varphi(\cdot)$ is evidently a convex function on $\mathbb{R}$ and $f, \varphi(f) \in L^1(\Omega, \Sigma, P)$. Consequently, by Theorem 8, $E^{\mathcal{B}}(\varphi(f)) \geq \varphi(E^{\mathcal{B}}(f))$ a.e., and by definition of $E^{\mathcal{B}}$

$$\|f\|_p^p = \int_\Omega \varphi(f) dP = \int_\Omega E^{\mathcal{B}}(\varphi(f)) dP_{\mathcal{B}} \geq \int_\Omega \varphi(E^{\mathcal{B}}(f)) dP_{\mathcal{B}} = \|E^{\mathcal{B}}(f)\|_p^p .$$

It follows that $E^{\mathcal{B}}(f) \in L^p(\Omega, \mathcal{B}, P_{\mathcal{B}})$ and $\|E^{\mathcal{B}}(f)\|_p \leq \|f\|_p$. Thus $E^{\mathcal{B}}$ is contractive in this case also with range in $L^p(\Omega, \mathcal{B}, P_{\mathcal{B}})$. Since clearly $E^{\mathcal{B}}$ is the identity on $L^p(\Omega, \mathcal{B}, P_{\mathcal{B}})$ its range is actually the whole space, in all cases, $1 \leq p \leq \infty$. □

The above result shows that $E^{\mathcal{B}}$ is a functional transformation on the L^p spaces, $1 \leq p \leq \infty$, and the proof indicates that only the convexity of $\varphi(x) = |x|^p$, $p \geq 1$ was used here. Some general cases are given in the Complements.

For the next important property, we need the following measure theoretical result due essentially to Doob [1], and Dynkin [1].

10. Lemma. *Let (Ω, Σ) and $(S, \mathcal{S})$ be a measurable spaces and $(\mathbb{R}, \mathcal{A})$ be the Borel line. Let $f : \Omega \to S$ be $(\Sigma, \mathcal{S})$ measurable (i.e., $f^{-1}(\mathcal{S}) = \Sigma(f) \subset \Sigma$) and $g : \Omega \to \mathbb{R}$ be a mapping. Then g is $(\Sigma(f), \mathcal{A})$-measurable (i.e., $g^{-1}(\mathcal{A}) \subset \Sigma(f)$) iff there is a measurable $h : S \to \mathbb{R}$ such that $g = h \circ f$.*

Proof. The "if" part is immediate. It follows from the identity

$$\begin{array}{ccc}\Omega & \xrightarrow{f} & S \\ & \searrow g & \downarrow h \\ & & \mathbb{R}\end{array} \quad g^{-1}(\mathcal{A}) = f^{-1}(h^{-1}(\mathcal{A})) \subset f^{-1}(\mathcal{S}) = \Sigma(f).$$

Conversely, let g be $(\Sigma(f), \mathcal{A})$-measurable. Then by the classical structure theory of measurable functions, there exists a sequence of simple $(\Sigma(f), \mathcal{A})$-measurable g_n such that $g_n(\omega) \to g(\omega)$ for all $\omega \in \Omega$. Suppose we can find $h_n : S \to \mathbb{R}$, which is $(\mathcal{S}, \mathcal{A})$-measurable, such that $g_n = h_n \circ f, n \geq 1$. Then the set of points S_0 at which $\{h_n, n \geq 1\}$ converges satisfies $f(\Omega) \subset S_0 \subset S$ and $S_0 \in \mathcal{S}$. If $h(s) = \lim_n h_n(s), s \in S_0$ and $h(s) = 0$ for $s \in S - S_0$, then $g = h \circ f$ is the desired representation.

To construct h for simple g, let $g = \sum_{i=1}^{k} a_i \chi_{A_i}$, $A_i \in \Sigma(f), i = 1, \ldots, k$ disjoint. Then there exist $B_i \in \mathcal{S}$ such that $A_i = f^{-1}(B_i)$ by definition of $\Sigma(f)$. Let $C_1 = B_1$, and for $i > 1, C_i = B_i - \bigcup_{j=1}^{i-1} B_j$ be a disjunctification. Then $C_i \in \mathcal{S}$, and $f^{-1}(C_i) = f^{-1}(B_i) - \bigcup_{j=1}^{i-1} f^{-1}(B_j) = A_i - \bigcup_{j=1}^{i-1} A_j = A_i$ since A_i are disjoint. Let $h = \sum_{i=1}^{k} a_i \chi_{C_i}$. Then $h \circ f(\omega) = \sum_{i=1}^{k} a_i \chi_{C_i}(f(\omega)) = \sum_{i=1}^{k} a_i \chi_{f^{-1}(C_i)}(\omega) = g(\omega), \omega \in \Omega$, and this satisfies all the requirements. □

11. Remarks. If $(S, \mathcal{S}) = (\mathbb{R}^n, \mathcal{A}^n)$ so that $f = (f_1, \ldots, f_n)' : \Omega \to \mathbb{R}^n$, the above result becomes $g = h \circ f = h(f_1, \ldots, f_n)$ for $h : \mathbb{R}^n - \mathbb{R}$. As another useful specialization, let $(S, \mathcal{S}) = (\mathbb{R}^T, \mathcal{B}_T)$ where T is an index set and $\mathcal{B}_T$ is the σ-algebra generated by the cylinder sets of $\mathbb{R}^T$, as in Theorem I.2.1. Then if $f : \Omega \to \mathbb{R}^T$ is $(\Sigma, \mathcal{B}_T)$ measurable, $\Sigma(f) = f^{-1}(\mathcal{B}_T)$, and if g is $(\Sigma(f), \mathcal{A})$-measurable, then there is an $h : \mathbb{R}^T \to \mathbb{R}, (\mathcal{B}_T, \mathcal{A})$-measurable such that $g = h \circ f$ by the above proposition. Since $\mathcal{A}$ is countably generated (by open intervals with rational end points), $h^{-1}(\mathcal{A}) \subset \mathcal{B}_T$ iff h depends at most on a countable set of coordinates (cf., I.5.9). So there are $\{t_i, i \geq 1\} \subset T$ such that $g(\omega) = h(f_{t_1}(\omega), \ldots, f_{t_i}(\omega), \ldots)$.

The announced property of $E(X|Y)$ is given by

12. Proposition. *Let X, Y be two random variables on (Ω, Σ, P) and*

X be P-integrable. If $\mathcal{B} = \sigma(Y) = Y^{-1}(\mathcal{A})$, the σ-algebra generated by Y, then there is a Borel function $h_X : \mathbb{R} \to \mathbb{R}$ such that $E^{\mathcal{B}}(X) = h_X \circ Y$, a.e., *where $(\mathbb{R}, \mathcal{A})$ is the Borel line.*

Proof. Since X is integrable, $E^{\mathcal{B}}(X)$ exists and is $\mathcal{B}'$-measurable. Taking $g = E^{\mathcal{B}}(X), f = Y$ in the Doob-Dynkin lemma, we deduce the existence of $h_X : \mathbb{R} \to \mathbb{R}$ such that $g = h_X \circ f$, and h_X is Borel measurable, as asserted. □

The above result is sometimes stated verbally as: "The conditional expectation of an integrable X, given Y, is a function of the function Y." (We again consider only probability measures P unless stated otherwise.)

Recall from Section I.1 that X, Y are said to be mutually independent random variables if for all real a, b

$$P[X < a, Y < b] = P[X < a]P[Y < b]. \tag{27}$$

Consequently if $\mathcal{B}_X, \mathcal{B}_Y$ are the σ-algebras generated by X and Y in Σ, then (27) is equivalent to:

$$P(A \cap B) = P(A)P(B), \text{ all } A \in \mathcal{B}_X,\ B \in \mathcal{B}_Y. \tag{27$'$}$$

If $\mathcal{B}_i \subset \Sigma, i = 1, 2$ are a pair of σ-algebras, then they are said to be *independent* if for all $A \in \mathcal{B}_1, B \in \mathcal{B}_2$ Equation (27′) holds. Let X be an integrable random variable, *then X and $\mathcal{B}$ are independent* if $\mathcal{B}_X$ and $\mathcal{B}$ are independent, where $\mathcal{B}_X$ is as in (27′). Now we observe that $E^{\mathcal{B}}(X) = E(X)$ a.e., whenever X and $\mathcal{B}$ are independent. In fact, for any $B \in \mathcal{B}, X$ and χ_B are independent random variables. Hence with (8)

$$\int_B E(X)dP_{\mathcal{B}} = E(X)P(B) = E(X)E(\chi_B) = E(X\chi_B = \int_B X\, dP$$
$$= \int_B E^{\mathcal{B}}(X)dP_{\mathcal{B}}\ . \tag{28}$$

Since $E(X)$ is a constant, so that it is $\mathcal{B}$-measurable, the extreme integrands can be identified, a.e. From this we can state the following: If X, Y are any two independent integrable random variables on (Ω, Σ, P), then with the notation of (10)

$$E(X|Y) = E(X), \text{ a.e.,} \qquad E(Y|X) = E(Y), \text{ a.e.} \tag{29}$$

For a convenient reference we state the general case as:

13. Proposition. *If X is an integrable random variable on (Ω, Σ, P) and $\mathcal{B} \subset \Sigma$ is a σ-algebra such that X and $\mathcal{B}$ are independent, then* $E^{\mathcal{B}}(X) = E(X)$ a.e.

We remark that the existence of independent σ-algebras $\mathcal{B}_i$ can be proved by considereing a Cartesian product of the given spaces and the $\mathcal{B}_i$ are identified cylindrically in the product algebras. This is the "adjunction procedure" (cf., Remark 3.6(ii) below).

The following property of $E^{\mathcal{B}}$ will be useful for some work in martingale theory and lies somewhat deeper.

14. Theorem. *If X, Y are two integrable random variables on (Ω, Σ, P) such that $E(X|Y) = Y$* a.e., *and $E(Y|X) = X$* a.e., *then $X = Y$* a.e.

Proof. If it is given that $X, Y \in L^p(\Omega, \Sigma, P), p \geq 2$, then the proof is very easy. Indeed, since $L^p \subset L^2$ for all $p \geq 2$ (because $P(\Omega) < \infty$), we may restrict to the L^2-case. The defining Equation (8), the hypothesis, and Proposition 6(a) imply (letting $\mathcal{B} = \sigma(Y)$)

$$\begin{aligned} 0 \leq \int_\Omega (X-Y)^2 dP &= \int_\Omega E^{\mathcal{B}}(X^2 - 2XY + Y^2) dP_{\mathcal{B}} \\ &= \int_\Omega E^{\mathcal{B}}(X^2) dP_{\mathcal{B}} - 2\int_\Omega Y E^{\mathcal{B}}(X) dP_{\mathcal{B}} + \int_\Omega Y^2 dP_{\mathcal{B}} \\ &= \int_\Omega X^2 dP - 2\int_\Omega Y^2 dP_{\mathcal{B}} + \int_\Omega Y^2 dP_{\mathcal{B}} \\ &= \int_\Omega (X^2 - Y^2) dP. \end{aligned} \tag{30}$$

Interchanging X and Y in this equation and noting that the hypothesis is symmetric in X and Y, it follows from (30) that

$$0 \leq \int_\Omega (X-Y)^2 dP = \int_\Omega (X^2 - Y^2) dP = \int_\Omega (Y^2 - X^2) dP, \tag{31}$$

so that $X = Y$ a.e. However, in the given (weaker) hypothesis, X, Y need not be in any L^p for $p > 1$ and a different argument is necessary.

For the general case, we will thus assume only that $X, Y \in L^1(\Omega, \Sigma, P)$. Then we claim that there exists a continuous *strictly* convex function $\varphi : \mathbb{R} \to \mathbb{R}^+$ such that $\varphi(0) = 0, \varphi'(x)$ is strictly increasing,

$\varphi(-x) = \varphi(x)$ and $E(\varphi(|X|)) < \infty, E(\varphi(|Y|)) < \infty$. [This is a consequence of an interesting classical theorem, due essentially to de la Vallée Poussin (1915), which states that a uniformly integrable part of $L^1(\Omega, \Sigma, P)$ is contained in a bounded set, or in a solid sphere, of an Orlicz space $L^\varphi(\Omega, \Sigma, P)$ where φ is as above. However, the proof of the present case is simple and it is included here.] Suppose, for a moment, this has been established. Then the proof is completed with Jensen's inequality (Theorem 8) as follows:

$$E(\varphi(X)|Y) \geq \varphi(E(X|Y)) = \varphi(Y), \text{a.e.},$$

with equality iff X is $\mathcal{B}_Y$-measurable (since φ is strictly convex), and then the hypothesis implies $X = Y$ a.e. Indeed,

$$\int_\Omega \varphi(X)dP = \int_\Omega E(\varphi(X)|Y)dP > \int_\Omega \varphi(Y)dP \tag{32}$$

unless $X = Y$, a.e. Interchanging X, Y here and remembering the symmetry in the hypothesis between X and Y, we get the opposite inequality in (32). This implies $X = Y$ a.e. as desired.

It remains to prove the existence of a φ of the above type. First note that $\varphi_0(X)$ and $\varphi_0(Y)$ are integrable where $\varphi_0(x) = |x|$. We show that there is a strictly convex φ such that $\varphi_0(x) \leq \varphi(x) \leq 2\varphi_0(x)$ by a very simple construction.

Let $\varphi_1'(x) = \varphi_0'(x)(2 - e^{-x})$ for $x \geq 0$, and set $\varphi(x) = \int_0^{|x|} \varphi_1'(t)dt$. Then $\varphi(\cdot)$ is strictly convex and satisfies the above inequalities with φ_0. Hence $E(\varphi(X)) \leq 2E(\varphi_0(X)) < \infty$, and similarly $E(\varphi(Y)) \leq 2E(\varphi_0(Y)) < \infty$. The thus obtained φ is clearly not unique. This, however, is immaterial here, and it completes the argument. □

2.2 Conditional expectations and projection operators

The basic theory of conditional expectations given in the above section admits an abstraction and an apparent simplification. This is achieved by considering $E^{\mathcal{B}}$ as a functional operation on spaces such as $L^p, p \geq 1$. We treat real L^p-spaces, except in a few indicated places.

1. Theorem. *Let (Ω, Σ, P) be a probability space and $\mathcal{B} \subset \Sigma$ be a σ-algebra. If $L^p(\Sigma)$ and $L^p(\mathcal{B}), 1 \leq p \leq \infty$ are the Lebesgue spaces* *on (Ω, Σ, P) and $(\Omega, \mathcal{B}, P_{\mathcal{B}})$ respectively, then there exists a projection $Q : L^p(\Sigma) \to L^p(\Sigma), Q(L^p(\Sigma)) = L^p(\mathcal{B})$, namely, $Q = E^{\mathcal{B}}$, which is positive, contractive, and $Q1 = 1$ a.e. If $\mathcal{B}_1 \subset \mathcal{B}_2 \subset \Sigma$ are σ-algebras and $Q_i = E^{\mathcal{B}_i}, i = 1, 2$ are the corresponding projections, then $Q_1Q_2 = Q_2Q_1 = Q_1$. Moreover, for each $1 < p \leq \infty, Q(L^p(\Sigma)) = L^p(\mathcal{B}), Q = E^{\mathcal{B}}$ has a unique extension to $\hat{Q}, \hat{Q}(L^1(\Sigma)) = L^1(\mathcal{B})$, where $\hat{Q}$ is an L^1-continuous conditional expectation.*

This is simply a restatement of Proposition 1.6 and Theorem 1.9 except perhaps for the last assertion. But this is immediate because $L^p(\Sigma) \subset L^1(\Sigma)$ so that Q is not only defined on a subspace of $L^1(\Sigma)$, but is also continuous in the L^1-norm by Theorem 1.9. Since the set $L^p(\Sigma)$ is norm dense in $L^1(\Sigma)$, the uniformly continuous linear function Q on a dense subset $L^1(\Sigma)$ has a unique extension to all of $L^1(\Sigma)$ by the so-called "principle of extension by continuity" on metric spaces. [The quoted result from classical analysis, which is very useful, is this: "If $\mathcal{X}, \mathcal{Y}$ are two metric spaces and $\mathcal{Y}$ is complete, if $\mathcal{D} \subset \mathcal{X}$ is a dense subset, then a uniformly continuous function $f : \mathcal{D} \to \mathcal{Y}$ has a unique extension $\hat{f} : \mathcal{X} \to \mathcal{Y}$ which is uniformly continuous on $\mathcal{X}$." Here neither the spaces nor f need be linear.]

The above theorem raises the following natural questions. (1) Are there other contractive projection operators on $L^p(\Sigma)$ onto $L^p(\mathcal{B}), p \geq 1$ which are not conditional expectations? (2) If $Q : L^p(\Sigma) \to L^p(\Sigma)$ is a contractive projection, under what conditions is it a conditional expectation? (I.e., characterize the subclass.) Both these questions are converses to the result of Theorem 1 in a certain sense and, as one might expect, are nontrivial. However, the answers to these questions provide consirable ease in the convergence theory of conditional expectations.

A complete solution to question (1) above can be presented as follows. It will facilitate the argument if we assume that all σ-algebras below are complete since otherwise, we have to identify functions separately if they differ on subsets of sets of measure zero. We assume this completeness property for convenience. Also our presentation proceeds step by step to the general case.

2. Theorem. *Let (Ω, Σ, P) be a probability space and $\mathcal{B} \subset \Sigma$ be a σ-algebra. If $L^p(\Sigma)$ and $L^p(\mathcal{B})(\subset L^p(\Sigma)), 1 \leq p < \infty$ are the Lebesgue spaces on (Ω, Σ, P) and $(\Omega, \mathcal{B}, P_{\mathcal{B}})$ respectively, then every linear contractive projection Q on $L^p(\Sigma)$, with range $L^p(\mathcal{B})$, coincides with the* *conditional expectation $E^{\mathcal{B}}(\cdot)$, i.e., $Q(f) = E^{\mathcal{B}}(f)$, all $f \in L^p(\Sigma)$.*

Proof. The proof is divided into two parts according as $1 < p < \infty$ and $p = 1$. Two arguments will be used for the two cases separately.

CASE 1. $1 < p < \infty$. Let $q = \frac{p}{p-1}$. If $f_0 \in L^p(\mathcal{B})$ is a bounded element, then it is clear that $f_0 \in L^q(\mathcal{B}) \subset L^q(\Sigma)$. The aim here (and later in Case 2) is to prove $\int_A Qf\, dP = \int_A f\, dP$, for all $f \in L^p(\Sigma)$ and $A \in \mathcal{B}$.

Recall that $(L^p(\Sigma))^*$, the adjoint space of $L^p(\Sigma)$, can be identified as $L^q(\Sigma)$ for $1 \leq p < \infty$ (cf., Royden [1], p. 246). The operator $Q^* : L^q \to L^q$, defined by the formula:

$$\langle Qf, g\rangle = \int_\Omega (Qf)\cdot g\, dP = \int_\Omega f\cdot (Q^* g) dP = \langle f, Q^* g\rangle,$$

for all $f \in L^p(\Sigma)$ and $g \in L^q(\Sigma)$ which exists, is called the *adjoint* of Q and is a linear operator. Moreover $\|Q\| = \|Q^*\|$, and $(Q^*)^2 = Q^*$ if $Q^2 = Q$; so Q^* is a contractive projection if Q is. In fact, if $Q^2 = Q$ and $g \in L^q(\Sigma)$, then

$$\langle f, Q^* g\rangle = \langle Qf, g\rangle = \langle Q^2 f, g\rangle = \langle Qf, Q^* g\rangle = \langle f, Q^{*2} g\rangle. \tag{1}$$

Since $f \in L^p(\Sigma)$ is arbitrary, $Q^* g = Q^{*2} g$ and so Q^* is a projection. As regards the norm equation, we have

$$\begin{aligned}
\|Q^*\| &= \sup\{\|Q^* g\|_q : \|g\|_q \leq 1\}, \text{ by definition,}\\
&= \sup\{\sup\{|\langle f, Q^* g\rangle| : \|f\|_p \leq 1\} : \|g\|_q \leq 1\},\\
&\quad \text{since } \|h\|_q = \sup\left\{\left|\int_\Omega hf\, dP\right| : \|f\|_p \leq 1\right\},\\
&= \sup\{\sup\{|\langle Qf, g\rangle| : \|f\|_p \leq 1\} : \|g\|_q \leq 1\},\\
&\quad \text{by definition of } Q \text{ and } Q^*,\\
&= \sup\{\sup\{|\langle Qf, g\rangle| : \|g\|_q \leq 1\} : \|f\|_p \leq 1\}\\
&= \sup\{\|Qf\|_p : \|f\|_p \leq 1\} = \|Q\|.
\end{aligned}$$

Hence $\|Q^*\| \leq 1$. [The above calculation clearly holds for all bounded operators in a Banach space if L^q is treated as $(L^p)^*$.]

Let $h_0 = Q^* f_0 \in L^q(\Sigma)$ (recall $f_0 \in L^q(\mathcal{B})$). We now show that h_0 is not merely in $L^q(\Sigma)$, but is in $L^q(\mathcal{B})$ itself. First note that

$$\|h_0\|_q = \|Q^* f_0\|_q \leq \|Q^*\| \|f_0\|_q \leq \|f_0\|_q. \tag{2}$$

On the other hand, if $A \in \mathcal{B}$ so that $\chi_A \in L^p(\mathcal{B}) = Q(L^p(\Sigma))$, and since $f_0 \in L^p(\Sigma)$ implies f_0 is integrable, we have

$$\begin{aligned}\int_A f_0 dP_{\mathcal{B}} &= \int_\Omega \chi_A f_0 dP_{\mathcal{B}} = \int_\Omega Q(\chi_A)\cdot f_0 dP_{\mathcal{B}}, \text{ since } Q\chi_A = \chi_A, \\ &= \int_\Omega \chi_A (Q^* f_0) dP = \int_\Omega \chi_A h_0 dP = \int_A h_0 dP \\ &= \int_A E^{\mathcal{B}}(h_0) dP_{\mathcal{B}}. \end{aligned} \tag{3}$$

But f_0 and $E^{\mathcal{B}}(h_0)$ are $\mathcal{B}$-measurable and $A \in \mathcal{B}$ is arbitrary, so it follows from (3) that $f_0 = E^{\mathcal{B}}(h_0)$ a.e. Now by Theorem 1 $E^{\mathcal{B}}$ is a contraction in L^q so that

$$\|f_0\|_q = \|E^{\mathcal{B}}(h_0)\|_q \leq \|h_0\|_q . \tag{4}$$

Equations (2) and (4) yield $\|f_0\|_q = \|h_0\|_q$. Since $1 < q < \infty$, the real function φ given by $\varphi(x) = |x|^q$ is strictly convex, so that by Theorem 1.8 (Jensen's inequality)

$$\varphi(f_0) = \varphi(E^{\mathcal{B}}(h_0)) \leq E^{\mathcal{B}}(\varphi(h_0)), \text{ a.e.,} \tag{5}$$

and inequality must hold on a set of positive measure unless h_0 is $\mathcal{B}$-measurable. The first possibility yields

$$\|f_0\|_q^q = \int_\Omega \varphi(f_0) dP < \int_\Omega E^{\mathcal{B}}(\varphi(h_0)) dP = \int_\Omega \varphi(h_0) dP = \|h_0\|_q^q \tag{6}$$

and contradicts the fact that $\|f_0\|_q = \|h_0\|_q$. Thus the second possibility must hold, i.e., h_0 is $\mathcal{B}$-measurable. So $f_0 = E^{\mathcal{B}}(h_0) = h_0$ a.e., and f_0 is a fixed point of Q^*. Then for any $A \in \mathcal{B}, \chi_A \in L^p(\mathcal{B}) \cap L^q(\mathcal{B})$, and for any $f \in L^p(\Sigma)$,

$$\begin{aligned}\int_A Qf\, dP_{\mathcal{B}} = \int_\Omega \chi_A \cdot Qf\, dP_{\mathcal{B}} &= \int_\Omega Q^*(\chi_A)\cdot f\, dP \\ &= \int_\Omega \chi_A f\, dP, \text{ since } Q^*\chi_A = \chi_A, \\ &= \int_A f\, dP = \int_A E^{\mathcal{B}}(f) dP_{\mathcal{B}} \end{aligned} \tag{7}$$

Here $L^p(\Sigma) \subset L^1(\Sigma)$ is used. From (7) it follows that, since $Qf \in L^p(\mathcal{B})$ implying Qf is $\mathcal{B}$-measurable, $Qf = E^{\mathcal{B}}(f)$ for all $f \in L^p(\Sigma)$. The result is proved in this case.

CASE 2. $p = 1$. Since Q is identity on $L^1(\mathcal{B})$ and constants are in $L^1(\mathcal{B})$, we deduce that $Q1 = 1$ a.e. Hence if $0 \leq f \leq 1$ a.e. is an arbitrary element of $L^1(\Sigma)$, then by the linearity and contractiveness of Q,

$$\begin{aligned} 1 - \int_\Omega f\, dP &= \int_\Omega (1-f) dP_{\mathcal{B}} \geq \int_\Omega |Q(1-f)| dP_{\mathcal{B}}, \text{ since } \|Q\| \leq 1, \\ &= \int_\Omega |(1 - Qf)| dP, \text{ since } Q1 = 1, \\ &\geq 1 - \int_\Omega Qf\, dP. \end{aligned}$$

Consequently,

$$\int_\Omega f\, dP \leq \int_\Omega Qf\, dP \leq \int_\Omega |Qf| dP \leq \int_\Omega f\, dP. \tag{8}$$

It follows that there is equality throughout and that $Qf \geq 0$, a.e. Hence $Qg \geq 0$, for any bounded and then by linearity for all $g \geq 0$ in $L^1(\Sigma)$. Moreover (8) also implies

$$\int_\Omega f\, dP = \int_\Omega Qf\, dP_{\mathcal{B}} \tag{9}$$

for bounded, and hence (by density, and linearity of Q) for all $f \in L^1(\Sigma)$. We assert that (9) holds if Ω is replaced by any $A \in \mathcal{B}$.

In fact, consider for $A \in \mathcal{B}, 0 \leq f \leq \chi_A$, and $f \in L^1(\Sigma)$, $h = Q(\chi_A f) - Q(\chi_A Qf)$. We claim that $h = 0$, a.e. Indeed, since Q is positive one has $0 \leq Qf \leq Q\chi_A = \chi_A$, and

$$\chi_A \pm h = Q[\chi_A(\chi_A \pm f \mp Qf)] \leq Q(\chi_A \pm f \mp Qf) = \chi_A. \tag{10}$$

Thus (10) implies that the pair of inequalities $\chi_A + h \leq \chi_A$ and $\chi_A - h \leq \chi_A$ are both true. Equivalently, $h \leq 0$ *and* $-h \leq 0$ simultaneously. But $h \in L^1$. So $h = 0$ a.e., and $Q(\chi_A f) = \chi_A Q(f)$ a.e. (since $\chi_A Q(f) \in$

$L^1(\mathcal{B})$). We next show that this last equation is true for all $0 \le f \le 1$ a.e., in $L^1(\Sigma)$ (and hence by linearity for all bounded and, by continuity of Q, for all $f \in L^1(\Sigma)$). If $A \in \mathcal{B}$ is arbitrary, $0 \le f\chi_A \le \chi_A$ and $0 \le f\chi_{A^c} \le \chi_{A^c}$ so that the special case implies (with $f\chi_A$ and $f\chi_{A^c}$),

$$Q(f) = Q(f\chi_A) + Q(f\chi_{A^c}) = \chi_A Q(f\chi_A) + \chi_{A^c} Q(f\chi_{A^c}), \text{ a.e.} \tag{11}$$

Hence, using the fact (cf. (10)) $Q(\chi_A g) = Q(\chi_A Qg), g \in L^1(\Sigma)$, one has

$$\begin{aligned}\int_A Q(f)dP_{\mathcal{B}} &= \int_A Q(f\chi_A)dP = \int_A \chi_A Q(f\chi_A)dP,\\ &= \int_\Omega Q(\chi_A Q(f\chi_A))dP = \int_\Omega Q(\chi_A f)dP,\\ &= \int_\Omega \chi_A f\, dP = \int_A E^{\mathcal{B}}(f)dP_{\mathcal{B}},\end{aligned} \tag{12}$$

where we used (9) in the second to last equality of (12). Since the extreme integrands in (12) are $\mathcal{B}$-measurable and $A \in \mathcal{B}$ is arbitrary, we can conclude that $E^{\mathcal{B}}(f) = f$, a.e., for bounded, and then for all $f \in L^1(\Sigma)$. This is desired conclusion. □

3. Definition. A contractive linear map $T : L^p(\Sigma) \to L^p(\Sigma), p \ge 1$, taking real elements into real elements, is said to be an *averaging operator* if the following two conditions hold:

(a) $T(fTg) = (Tf)(Tg), f, g \in L^p(\Sigma) \cap L^\infty(\Sigma)$,

(b) $Tf_0 = f_0$, for some $0 < f_0 \in L^p(\Sigma) \cap L^\infty(\Sigma)$.

(a) is called the *averaging identity.*

It is clear that, with $f_0 = 1$ a.e., Proposition 1.6 says: if $T = E^{\mathcal{B}}(\cdot)$ for a $\mathcal{B} \subset \Sigma$, then T satisfies (a) and (b) above. To see the class of operators admitted in Definition 3, we note the following simple result.

4. Lemma. *Every averaging operator $T : L^p(\Sigma) \to L^p(\Sigma), 1 \le p \le \infty$, as defined above, is a contractive projection on $L^p(\Sigma) \to L^p(\Sigma)$.*

Proof. Let $f \in L^p(\Sigma) \cap L^\infty(\Sigma)$ and $g = Tf \in L^p(\Sigma)$. Let $0 < f_0$ be as in (b). Then by (a) and (b),

$$T(f_0 g) = T(f_0 \cdot Tf) = (Tf_0) \cdot (Tf) = f_0 g. \tag{13}$$

On the other hand,

$$T(f_0 \cdot g) = T(g \cdot Tf_0) = (Tg) \cdot (Tf_0) = f_0(Tg). \tag{14}$$

Since $f_0 > 0$, a.e., (13) and (14) yield

$$T^2 f = T(Tf) = T(g) = g = T(f). \tag{15}$$

Since $f \in L^p \cap L^\infty$ is arbitrary $T^2 = T$ on this set, which is dense in L^p, and by continuity the result holds on all $L^p, 1 \leq p \leq \infty$, so that T is a contractive projection, since (by definition) an averaging operator is contractive. □

The above result shows that the averaging operators and conditional expectations are a subclass of contractive projections on $L^p(\Sigma), 1 \leq p \leq \infty$.

Suppose the range of a projection is not explicitly given, as in Theorem 2, but more information on the operator (such as being an averaging) is available. In many cases it is possible to identify the range of such an operator as the one given in Theorem 2. The following result plays a crucial role in this characterization. [Here {(a), (b), (c)}, {(a), (b′), (c′)} give *two* theorems.]

5. Theorem. *Let $L^p(\Sigma), 1 \leq p \leq \infty$ be the Lebesgue space on a measure space (Ω, Σ, μ). If $\mathcal{M} \subset L^p(\Sigma)$ is a closed subspace then it is of the form $L^p(\mathcal{B})$ for some rich σ-algebra $\mathcal{B}$ relative to μ iff the following conditions hold:*

(a) *$f \in \mathcal{M}$ implies $f^* \in \mathcal{M}$, (f^* = complex conjugate of f).*

(b) *The real functions of $\mathcal{M}$ form a lattice and there is an $0 < f_0 \in \mathcal{M}$ where f_0 is bounded and is determined by $\{\chi_{A_i}\}_1^\infty, \mu(A_i) < \infty$ (f_0 is called a* **weak unit**) *so that $f_0 \in \mathcal{M}, \{\chi_{A_i}\}_1^\infty \subset \mathcal{M}$, and $f_0 = \sum_{i=1}^{\infty} a_i \chi_{A_i}$ for some suitable scalars a_i.*

(c) *$0 \leq f_n \leq f_{n+1}, f_n \in \mathcal{M}, f_n \uparrow f, f \in L^p(\Sigma)$ implies $f \in \mathcal{M}$. The same result holds true if* (b) *and* (c) *are replaced by (b′) and (c′) below, but keeping (a):*

(b′) *The bounded functions of $\mathcal{M}$ form a dense algebra with a weak unit $f_0 \in \mathcal{M}$ (cf., (b) about the weak unit).*

(c′) *$f_n \in \mathcal{M}$, bounded, $f_n \to f$* a.e., *$f \in L^p(\Sigma)$*, and *$|f_n| \leq |f|$ implies $f \in \mathcal{M}$.*

Remark. If $\mu(\Omega) < \infty$, f_0 will usually be the constant function $f_0 = 1$. Note that conditions (c) and (c′) are automatic if $1 \le p < \infty$ by the Dominated Convergence theorem. We shall first present two applications in which Theorem 5 plays a key role, and then demonstrate the latter result.

The first statement is as follows:

6. Theorem. *Let $Q : L^p(\Sigma) \to L^p(\Sigma), 1 \le p < \infty$ be a positive contractive projection on the function space $L^p(\Sigma)$ on a probability space (Ω, Σ, P), such that $Q1 = 1$ where 1 is the identically one function. Then there exists a σ-algebra $\mathcal{B} \subset \Sigma$ such that $Q(f) = E^{\mathcal{B}}(f), f \in L^p(\Sigma)$.*

Proof. We shall prove the result for the real $L^p(\Sigma)$. The easy modification for the complex case is left to the reader. Thus let $\mathcal{M} = Q(L^p(\Sigma))$. Since Q is a projection, $\mathcal{M}$ is a closed subspace of $L^p(\Sigma)$, and $1 = Q1 \in \mathcal{M}$. The fact that $\|Q\| \le 1$ implies $\mathcal{M}$ is also a lattice. Indeed, let $f \in \mathcal{M}$. It suffices to show $|f| \in \mathcal{M}$ since $\max(f,g) = \frac{f+g+|f-g|}{2}$ implies that the linear space $\mathcal{M}$ is then closed under lattice operations. However,

$$|f| = |Qf| \le Q|f|. \tag{16}$$

Since Q is a positive operator, from $f \le |f|$ we deduce that $Qf \le Q|f|$ and hence $|Qf| \le Q|f|$. On the other hand, the contractivity property yields

$$\|Q(|f|)\|_p \le \||f|\|_p = \|f\|_p. \tag{17}$$

Equations (16) and (17) imply $\|f\|_p = \|Q(|f|)\|_p$. Now if in (16) there is inequality on a set of positive measure, then it must be true that $|f|^p < (Q|f|)^p$ on a set of positive measure so that

$$\|f\|_p^p = \int_\Omega |f|^p dP < \int_\Omega (Q|f|)^p dP = \|Q(|f|)\|_p^p. \tag{18}$$

This contradicts the equality resulting from (16) and (17). So $|f| = Q(|f|)$, a.e. Hence $|f| \in \mathcal{M}$.

It is clear that (a) and (b) (with $f_0 = 1$) of Theorem 5 hold for $\mathcal{M}$. To see that (c) also holds (it is here that $1 \le p < \infty$ of L^p is used), let $0 \le f_n \le f_{n+1} \uparrow f$ where $f \in L^p(\Sigma)$ and $f_n \in \mathcal{M}$. Then

$0 \le f - f_n \downarrow 0$, a.e. and $(f - f_n)^p \le f^p$. Since $f^p \in L^1(\Sigma)$, by the Dominated Convergence theorem,

$$\lim_{n\to\infty} \|f - f_n\|_p^p = \lim_{n\to\infty} \int_\Omega (f - f_n)^p dP = \int_\Omega \lim_{n\to\infty} (f - f_n)^p dP = 0. \quad (19)$$

Hence $f_n \to f$ in norm and since $\mathcal{M}$ is closed, $f \in \mathcal{M}$. Thus (c) holds. By Theorem 5, there is a σ-algebra $\mathcal{B} \subset \Sigma$ such that $\mathcal{M} = L^p(\mathcal{B})$ and $Q(L^p(\Sigma)) = L^p(\mathcal{B}), Q$ is contractive and $Q^2 = Q$. Consequently, by Theorem 2, $Q = E^{\mathcal{B}}$, as asserted. □

Remarks. 1. The same theorem holds for σ-finite measures if $Q1 = 1$ is replaced by $Qf_0 = f_0$ for a weak unit f_0, i.e., $Q\chi_{A_n} = \chi_{A_n}$ each n if f_0 is determined by $\{\chi_{A_n}\}_1^\infty$.

2. It must be noted that there is a redundancy in the hypothesis of the above theorem. If $p = 1$, for instance, then a contractive projection Q with $Q1 = 1$ (or $Qf_0 = f_0$) is automatically positive. This follows from the computations for (8).

The second part of Theorem 5 yields a similar characterization with averaging operators:

7. Theorem. *Let $T : L^p(\Sigma) \to L^p(\Sigma), 1 \le p < \infty$ be an averaging operator with $T1 = 1$ a.e., and (Ω, Σ, P) being a probability space. Then there is a σ-algebra $\mathcal{B} \subset \Sigma$ such that $T = E^{\mathcal{B}}$.*

Proof. If $\mathcal{M} = (L^p(\Sigma))$, then it is a closed subspace (since T is a projection by Lemma 4) and $1 \in \mathcal{M}$. If $\{f, g\} \subset \mathcal{M}$, f or g bounded, then $T(fg) = T(fTg) = (Tf)(Tg) = fg \in \mathcal{M}$. Hence bounded elements of $\mathcal{M}$ form an algebra. We assert that T maps bounded elements into bounded elements. In fact, let $f \in L^p(\Sigma), |f| \le 1$ a.e., and $g = Tf$. Then by the averaging identity and induction we get $T(g^2) = g^2$ and

$$T(fg^{n-1}) = (Tf)T(g^{n-1}) = g^n. \quad (20)$$

Hence, with $p^{-1} + q^{-1} = 1$, we have

$$\begin{aligned}\|g^n\|_p &= \sup\left\{\int_\Omega |g^n h|dP : \|h\|_q \leq 1\right\}\\ &= \sup\left\{\int_\Omega |T(fg^{n-1})h|dP : \|h\|_q \leq 1\right\}\\ &\leq \sup\left\{\int_\Omega |g^{n-1} h|dP : \|h\|_1 \leq 1\right\}, \quad \text{since } |f| \leq 1, \|T\| \leq 1,\\ &= \|g^{n-1}\|_p \leq \|g\|_p = \|Tf\|_p \leq \|f\|_p. \end{aligned} \tag{21}$$

Let $A = \{\omega : |g|(\omega) > 1\}$. Then $\|g^n \chi_A\|_p \leq \|g^n\|_p \leq \|f\|_p$ by (21) and since $|g|^n \uparrow \infty$ a.e. on A, we deduce, by Monotone Convergence, that $P(A) = 0$ and $|g| \leq 1$ a.e.

Let $f \in \mathcal{M}$ and $\epsilon > 0$. By the density of $L^\infty(\Sigma)$ in $L^p(\Sigma)$, there is an $f_\epsilon \in L^\infty(\Sigma), \|f - f_\epsilon\|_p < \epsilon$. If $g_\epsilon = Tf_\epsilon$, then $g_\epsilon \in \mathcal{M}$, bounded and $\|f - g_\epsilon\|_p = \|Tf - Tf_\epsilon\|_p \leq \|f - f_\epsilon\|_p < \epsilon$. So (a) and (b′), and (c′) of Theorem 5 are satisfied. Thus $\mathcal{M} = L^p(\mathcal{B})$ for a unique $\mathcal{B} \subset \Sigma$, and by Theorem 2 it follows that $T = E^{\mathcal{B}}$, as asserted. □

We now turn to Theorem 5, since its utility is shown in the above results.

Proof of Theorem 5. If $L^p_{\mathbb{R}}$ is the real space, $L^p(\Sigma) = L^p_{\mathbb{R}}(\Sigma) + iL^p_{\mathbb{R}}(\Sigma)$ (vector addition). If $\mathcal{M}_{\mathbb{R}} \subset L^p_{\mathbb{R}}(\Sigma)$ is the subspace satisfying the hypothesis (here (a) is vacuous for $\mathcal{M}_{\mathbb{R}}$), then in the complex case, with (a), we have $\mathcal{M} = \mathcal{M}_{\mathbb{R}} + i\mathcal{M}_{\mathbb{R}}$. Thus, if the real case is true so that there is a μ-rich σ-algebra $\mathcal{B} \subset \Sigma$ such that $\mathcal{M}_{\mathbb{R}} = L^p_{\mathbb{R}}(\mathcal{B})$, we then have $\mathcal{M} = L^p_{\mathbb{R}}(\mathcal{B}) + iL^p_{\mathbb{R}}(\mathcal{B}) = L^p(\mathcal{B}) \subset L^p(\Sigma)$. So we assume that $L^p(\Sigma)$ is real, and prove $\{(b'), (c')\} \Rightarrow \{(b), (c)\} \Rightarrow \{\mathcal{M} = L^p(\mathcal{B})\} \Rightarrow \{(b), (c)\}$ and $\{(b'), (c'\}$.

I. If $\mathcal{A} \subset \mathcal{M}$ is an algebra satisfying (b′) and (c′), then the L^p-closure $\bar{\mathcal{A}} = \mathcal{M}$ is a Banach lattice. [If $p = \infty$, it is in Loomis [1], p. 9.]

For, let $g \in \mathcal{A}$. It suffices to show that $|g| \in \bar{\mathcal{A}}$. We may take $|g| \leq 1$. Consider for $\epsilon > 0$, $(g^2 + \epsilon^2)^{1/2} = (1 + \epsilon^2)^{1/2}\left(1 + \frac{g^2 - 1}{1+\epsilon^2}\right)^{1/2} = (1 + \epsilon^2)^{1/2} \sum_{n=0}^{\infty} \binom{\frac{1}{2}}{n}([g^2 - 1)/(1 + \epsilon^2)]^n)$. If $f \in \mathcal{A}$, then the fact that

$[g^2/(1+\epsilon^2)]^n f \in \mathcal{A}$ for each n implies $(g^2+\epsilon^2)^{1/2} f \in \bar{\mathcal{A}}$ since the series converges in the norm of L^p. In fact,

$$\|(g^2+\epsilon^2)^{1/2} f\|_p \leq \sum_{n=0}^{\infty} \binom{\frac{1}{2}}{n} \|f\|_p (1+\epsilon^2)^{-n+2^{-1}} = \|f\|_p (2+\epsilon^2)^{1/2} < \infty. \tag{22}$$

Since $\bar{\mathcal{A}}$ is closed, we conclude that $(g^2+\epsilon^2)^{1/2} f \in \bar{\mathcal{A}}$, and then $\epsilon > 0$ being arbitrary, we deduce that $|g| f \in \bar{\mathcal{A}}$ for each $f \in \mathcal{A}$. However, by hypothesis $f_0 \in \mathcal{A}$ and so from the definition of weak unit, $\{\chi_{A_n}, n \geq 1\} \subset \mathcal{A}$. Thus taking $f = \chi_{A_n} \in \mathcal{A}$, one sees that $|g|\chi_{A_n} \in \bar{\mathcal{A}}$ for each $n \geq 1$. By (c$'$), $\sum_{k=1}^{n} |g|\chi_{A_k} \in \bar{\mathcal{A}}$ and this tends to $|g|$ since $\bigcup_{k=1}^{\infty} A_k = \Omega$, and hence $|g| \in \bar{\mathcal{A}}$. We remark that the same argument applies even if there is no weak unit. But then the analysis becomes somewhat involved. Since $\mathcal{M}$ is the norm closure of $\mathcal{A}$ $(= \bar{\mathcal{A}})$, by (b$'$) the assertion follows. Hence $\{(b'), (c')\} \Rightarrow \{(b), (c)\}$.

II. If (b), (c) hold for $\mathcal{M}$, then $\mathcal{B} = \{A : \chi_A f_0 \in \mathcal{M}\}$ is a σ-algebra.

For, since $f_0 = \sum_{i=1}^{\infty} a_i \chi_{A_i} \in \mathcal{M}, \Omega \in \mathcal{B}$. If A and B are in $\mathcal{B}$, then $\chi_{A\cap B} f_0 = \chi_A \cdot \chi_B \cdot f_0 = \min(\chi_A f_0, \chi_B f_0) \in \mathcal{M}$, since $\mathcal{M}$ is a lattice. So $A \cap B \in \mathcal{B}$. Consequently,

$$\chi_{A-B} f_0 = (\chi_A - \chi_{A\cap B}) f_0 \in \mathcal{M}, \qquad \chi_{A\cup B} f_0 = (\chi_A + \chi_{B-A}) f_0 \in \mathcal{M}. \tag{23}$$

Thus $A - B \in \mathcal{B}, A \cup B \in \mathcal{B}$, and hence, $\mathcal{B}$ is an algebra.

If $A_n \in \mathcal{B}, A_n \uparrow A$, then $\chi_{A_n} f_0 \uparrow \chi_A f_0$ a.e., and $\chi_{A_n} f_0 \in \mathcal{M} \subset L^p(\Sigma)$. But it is clear that $\chi_A f_0 \in L^p(\Sigma)$. Hence we deduce, by (c) that $\chi_A f_0 \in \mathcal{M}$, and thus $A \in \mathcal{B}$. Now $\mathcal{B}$ is an algebra, and it is closed under countable unions, by the above. Hence it is a σ-algebra. Since by hypothesis of (b) $\{\chi_{A_n}, n \geq 1\} \subset \mathcal{M}$, it follows that $A_n \in \mathcal{B}$ and hence f_0 is $\mathcal{B}$-measurable.

III. Under the conditions of (b) and (c), $\mathcal{M} = L^p(\mathcal{B})$.

For, it $B_i \in \mathcal{B}$ disjoint, then $f = \sum_{i=1}^{n} b_i \chi_{B_i} \in L^p(\mathcal{B})$. But $f f_0 \in \mathcal{M}$ and since $\chi_{A_n} \in \mathcal{M}$, for each n, by (b) we may deduce that $f \in \mathcal{M}$ itself. So every step function of $L^p(\mathcal{B})$ belongs to $\mathcal{M}$. If $1 \leq p < \infty$, then such functions are dense in $L^p(\mathcal{B})$. Consequently $L^p(\mathcal{B}) \subset \mathcal{M}$ because the latter is closed. If $p = \infty$, then $0 \leq f \in L^p(\mathcal{B}), A \in \mathcal{B}$ implies

$f\chi_A$ is $\mathcal{B}$-measurable so that it is the limit of an increasing sequence of simple functions. Hence by (c) we conclude that $f\chi_A \in \mathcal{M}$. Since $A \in \mathcal{B}$ is arbitrary, this implies $f \in \mathcal{M}$ and then $L^p(\mathcal{B}) \subset \mathcal{M}$, both being lattices.

To prove the opposite inclusion, we observe that $\mathcal{M}$ is a lattice by (b), and so it suffices to consider $0 \leq f \in \mathcal{M}$ and show that $f \in L^p(\mathcal{B})$. For any $a > 0, B \in \mathcal{B}$, and $n \geq 1$, consider $f_n = \min[n(f - a\chi_B)^+, \chi_B]$. Then $f_n \in \mathcal{M}$. But $f_n \uparrow \tilde{f}$ and $f_n \leq \chi_B$ where $\chi_B \in \mathcal{M}$ by Step II. Hence by condition (c), $\tilde{f} \in \mathcal{M}$. However, $\tilde{f} = \chi_{A_a}$ where $A_a = \{\omega : (f(\omega) - a)\chi_B(\omega) > 0\}$. But the weak unit f_0 being $\mathcal{B}$-measurable, we conclude that $\mathcal{B}$ is μ-rich and hence the support of $\tilde{f}$ is σ-finite (in Σ to begin with). Hence we can find a sequence $\{B_n\}_1^\infty \subset \mathcal{B}, B_n \uparrow \Omega$ (and $f\chi_{B_n}$ is $\mathcal{B}$-measurable by the result just proven). So $f\chi_{B_n} \uparrow f$ and f is $\mathcal{B}$-measurable. Since $f \in L^p(\Sigma)$, this implies $f \in L^p(\mathcal{B})$. Thus $\mathcal{M} \subset L^p(\mathcal{B})$. So $\mathcal{M} = L^p(\mathcal{B})$ with the earlier inclusion (whatever $1 \leq p \leq \infty$ is).

IV. We have now proved that if (a) – (c) or (a) – (c′) hold, then $\mathcal{M}$ is of the form $L^p(\mathcal{B})$. Conversely if $\mathcal{M} = L^p(\mathcal{B})$ where $\mathcal{B}$ is a μ-rich σ-algebra in Σ, then as noted before, there is a weak unit $f_0 \in L^p(\mathcal{B})$. It is then trivial that conditions (a) – (c) and (a) – (c′) hold for this space, for all $1 \leq p \leq \infty$. This completes the proof of the theorem. □

2.3 Conditional probability measures

(a) *Structure.* The concept of a conditonal probability function was introduced in Definition 1.2. Some of the consequences which are easily deduced from the properties of conditional expectations will be stated here for reference.

1. Proposition. *If (Ω, Σ, P) is a probability space, $\mathcal{B} \subset \Sigma$ is a σ-algebra, and $P^{\mathcal{B}}(\cdot) : \Sigma \to L^\infty(\mathcal{B})$, defined by $P^{\mathcal{B}}(A) = E^{\mathcal{B}}(\chi_A), A \in \Sigma$ where $E^{\mathcal{B}}(\cdot)$ is the conditional expectation relative to $\mathcal{B}$, then the conditional probability function $P^{\mathcal{B}}(\cdot)$ has the following properties:*

(i) $0 \leq P^{\mathcal{B}}(A) \leq 1$ a.e., $A \in \Sigma$, *(ii) $P(A) = 0$ or 1 implies $P^{\mathcal{B}}(A) = 0$ or 1* a.e., *respectively, (iii) if $\{A_n\}_1^\infty \subset \Sigma$ is a monotone sequence and $A = \lim_n A_n$, then $P^{\mathcal{B}}(A) = \lim_n P^{\mathcal{B}}(A_n)$* a.e., *and (iv) $\{A_n\}_1^\infty \subset \Sigma$*

disjoint, $A = \bigcup_{n=1}^{\infty} A_n$ *implies* $P^{\mathcal{B}}(A) = \sum_{n=1}^{\infty} P^{\mathcal{B}}(A_n)$ a.e.

Proof. All the statements are simple consequences of Proposition 1.3 and can also be proved directly from the definition. We outline the latter possibility for one of them, say (iv).

By definition of $P^{\mathcal{B}}(\cdot)$ one has the *functional equation:* if $B \in \mathcal{B}$, then

$$\int_B P^{\mathcal{B}}(A) dP_{\mathcal{B}} = \int_B E^{\mathcal{B}}(\chi_A) dP_{\mathcal{B}} = \int_B \chi_A \, dP = P(A \cap B), \quad A \in \Sigma. \quad (1)$$

Thus if $\{A_n\}_1^{\infty} \subset \Sigma$ are disjoint and $A = \bigcup_{n=1}^{\infty} A_n$, then by (1) and the σ-additivity of $P(\cdot)$ on Σ we get

$$\begin{aligned} P(A \cap B) = \sum_{n=1}^{\infty} P(A_n \cap B) &= \sum_{n=1}^{\infty} \int_B P^{\mathcal{B}}(A_n) dP_{\mathcal{B}} \\ &= \int_B \left(\sum_{n=1}^{\infty} P^{\mathcal{B}}(A_n) \right) dP_{\mathcal{B}}. \end{aligned} \quad (2)$$

Therefore, by (1) and (2)

$$\int_B P^{\mathcal{B}}(A) dP_{\mathcal{B}} = \int_B \left(\sum_{n=1}^{\infty} P^{\mathcal{B}}(A_n) \right) dP_{\mathcal{B}}, \quad B \in \mathcal{B}.$$

Thus (iv) follows, since the integrands are $\mathcal{B}$-measurable and hence can be identified outside a $P_{\mathcal{B}}$-null set. □

Since $P^{\mathcal{B}}(A)$ is a function and not a number, the above proposition says that it is a mapping of Σ into the positive part of the unit ball of $L^{\infty}(\mathcal{B})$ and is pointwise σ-additive, a.e. This implies that if we can select from the equivalence classes $P^{\mathcal{B}}(A)$ a single representative function (or a "version"), then $P^{\mathcal{B}}(\cdot)(\omega)$ may be a probability measure for each $\omega \in \Omega - N$ where $P(N) = 0$. It would then be possible to consider essentially all the results obtained with the measures $P(\cdot)$ for the functions $P^{\mathcal{B}}(\cdot)(\omega)$ for each $\omega \in \Omega - N$. However, this is not generally possible if Σ is not generated by a countable partition of sets, since each exceptional set N above, related to $P^{\mathcal{B}}(A)$, depends on A also.

A superficial examination of Proposition 1 says the following: (1) We have $\{P^{\mathcal{B}}(A) : A \in \Sigma\} \subset L^\infty(\mathcal{B}) \subset L^p(\mathcal{B}), 1 \leq p < \infty$ and, if $A_n \in \Sigma, A_n \uparrow A$, then $0 \leq P^{\mathcal{B}}(A_n) \uparrow P^{\mathcal{B}}(A) \leq 1$ a.e., and this implies the norm convergence in $L^p(\mathcal{B})$ for $p \in [1, \infty)$ (but not for $p = \infty$). Thus $P^{\mathcal{B}}(\cdot) : \Sigma \to L^p(\mathcal{B})$ is a function space (or vector space $L^p(\mathcal{B})$)-valued, *strongly* (i.e., in norm) σ-additive bounded positive set function, in the sense that $\left\| P^{\mathcal{B}}(A) - \sum_{n=1}^{m} P^{\mathcal{B}}(A_n) \right\|_p \to 0$ as $m \to \infty$ for any disjoint sequence $\{A_n\}_1^\infty \subset \Sigma$ with $A = \bigcup_{n=1}^{\infty} A_n$. Such a set function is called a *vector measure.* In what follows, we need a few elementary properties of these measures and a definition of the integral relative to them. [Standard references are Dunford and Schwartz ([1], IV.10), and N. Dinculeanu [1].] A brief account will be included to make the exposition intelligible. (2) The pointwise a.e. σ-additivity for $P^{\mathcal{B}}(\cdot)(\omega)$, which is meaningless for general vector measures, implies a special feature of these conditional probability functions. When does, in fact, $P^{\mathcal{B}}$ behave like a scalar measure? This will be discussed in this section.

To underscore the point that $P^{\mathcal{B}}(\cdot)(\omega)$ cannot always be treated as a scalar measure to use the abstract Lebesgue integration on (Ω, Σ), for almost all ω, we present a simple counterexample. Now $P^{\mathcal{B}}(\cdot)(\omega)$ may be taken as a scalar measure if there is a possibility of finding a version, $P(\cdot, \cdot)$ (of $P^{\mathcal{B}}(\cdot)$) of two variables such that $P(\cdot, \cdot) : \Sigma \times \Omega \to \mathbb{R}^+$, where (i) $P(\cdot, \omega) : \Sigma \to \mathbb{R}^+$ is a probability measure for each $\omega \in \Omega$ (or for each $\omega \in \Omega - N$ with $P(N) = 0$) and (ii) $P(A, \cdot) : \Omega \to \mathbb{R}^+$ is a $\mathcal{B}$-measurable function such that $P(A, \cdot) = P^{\mathcal{B}}(A)$ a.e., for each $A \in \Sigma$. This implies that the vector measure $P^{\mathcal{B}}(\cdot)$ should behave like a scalar measure when evaluated at each $\omega \in \Omega$. When such a measure exists, then the function $\{P^{\mathcal{B}}(A)\}_{A \in \Sigma}$ is called a *regular* conditional probability. First we give an example to show that this is not always true, and then proceed to the simple (but necessary) properties of integration of scalar functions relative to the vector measure $P^{\mathcal{B}}$, for any σ-algebra $\mathcal{B} \subset \Sigma$.

2. Counterexample. The following example shows that $P^{\mathcal{B}}(\cdot)(\omega)$ cannot be treated as a scalar measure.

Suppose $P^{\mathcal{B}}_\omega(\cdot)(= P(\cdot, \omega))$ is always a scalar measure so that we can employ the standard Lebesgue approach to integration. Thus for any $0 \leq f \in L^\infty(\Omega, \Sigma, P)$ and $\mathcal{B} \subset \Sigma$, we may define its integral in the following two ways for each $\omega \in \Omega$:

$$\int_\Omega f\, dP_\omega^{\mathcal{B}} = \sup\left\{\int_\Omega g\, dP_\omega^{\mathcal{B}} : 0 \leq g \leq f, g \text{ is } \Sigma\text{-simple}\right\}, \tag{3}$$

$$= \inf\left\{\int_\Omega h\, dP_\omega^{\mathcal{B}} : f \leq h, h \text{ is } \Sigma\text{-simple}\right\}. \tag{4}$$

The equality between (3) and (4) is the basic fact in the theory of abstract Lebesgue integration with finite positive measures. (See, e.g., Rao [11], Theorem 4.1.4(b) on p. 140.) Here $P_\omega^{\mathcal{B}}$ is such by assumption for each ω. To see that $P^{\mathcal{B}}(\cdot)$ need not behave as a scalar measure so that the equality between (3) and (4) need not hold, we consider the following specialization.

Let $\Omega = [0,1], \mathcal{B}$ the σ-algebra of Borel sets of Ω, and P be the Lebesgue measure. Let $A \subset \Omega$ be a Lebesgue nonmeasurable set, such that it has Lebesgue outer measure one, and inner measure zero, i.e., $\inf\{P(E) : E \supset A, E \in \Sigma\} = 1$ and $\sup\{P(F) : F \subset A, F \in \Sigma\} = 0$ where Σ is the Lebesgue σ-algebra of Ω. Let $\tilde{\Sigma}$ be the σ-algebra determined by Σ and A. That such an enlargement is possible follows from the classical theory. Let Q be an extension of P to $\tilde{\Sigma}$, with $Q(A \cap E) = P(E), E \in \Sigma$. (See, e.g., Rao [11], p. 73.) Since Q is a finite measure, the μ-completion $\mathcal{B}_\mu = \Sigma$, and the conditional probability function $Q^{\mathcal{B}}(\cdot) : \tilde{\Sigma} \to L^\infty(\mathcal{B})$ exists. We claim that $Q_\omega^{\mathcal{B}}, \omega \in \Omega$ does not behave like a measure. For, suppose on the contracry it does. Then (3) and (4) must hold for any $0 \leq f \in L^\infty(\tilde{\Sigma})$. So consider $f = \chi_A$. By definition of A and $\tilde{\Sigma}$, every set $F \in \Sigma, F \subset A$ satisfies $Q(F) = 0$. Observe that $\tilde{\Sigma}$ is generated by $\Sigma_0 = \Sigma_1 \cup \Sigma_2$ where

$$\Sigma_1 = \{A \cap E : E \in \Sigma\}, \quad \Sigma_2 = \{A^c \cap E : E \in \Sigma\}.$$

Hence, recalling that $Q^{\mathcal{B}}(E) = E^{\mathcal{B}}(\chi_E) = 0$ a.e., if $Q(E) = 0$ and 1 a.e., if $Q(E) = 1$ we see at once that (3) has value zero, and (4) has value one. thus the left integral in (3) cannot be defined for $Q^{\mathcal{B}}$ as a Lebesgue integral by *assuming* that $Q_\omega^{\mathcal{B}}$ behaves like a measure, $\omega \in \Omega$.

The above example shows that if we put no restrictions on $\tilde{\Sigma}$ (it is not sufficient that $\mathcal{B}$ be countably generated since in the above it is the Borel σ-algebra), then clearly $Q^{\mathcal{B}}(\cdot)$ must be treated differently (e.g., as a vector measure).

Unlike the scalar case, it is now necessary to distinguish the concepts of variation and semi-variation of a vector measure. [These are identical for scalar measures.] The latter is the most useful one in integration because it is always finite on Ω for a vector measure while the variation need not be. These two concepts are as follows. If $\nu : \Sigma \to \mathcal{X}$ (a Banach space) is a vector measure, its *variation* on $A \in \Sigma$, denoted $|\nu|(A)$, is defined as: ($\Sigma(A)$ is the trace of Σ on A)

$$|\nu|(A) = \sup\left\{\sum_{i=1}^{n} \|\nu(A_i)\|_{\mathcal{X}} : A_i \in \Sigma(A), \text{ disjoint}\right\}, \tag{5}$$

and $|\nu|(\cdot)$ is known to be σ-additive on Σ. If $|\nu|(\Omega) < \infty$, then ν is said to be of *bounded variation.* The *semi-variation* of ν on $A \in \Sigma$, denoted $\|\nu\|(A)$, is defined as: ($a_i \in \mathbb{R}$ or $\in \mathbb{C}$)

$$\|\nu\|(A) = \sup\left\{\left\|\sum_{i=1}^{n} a_i\nu(A_i)\right\|_{\mathcal{X}} : |a_i| \leq 1, A_i \in \Sigma(A), \text{ disjoint}\right\}. \tag{6}$$

Clearly $\|\nu(A)\| \leq \|\nu\|(A) \leq |\nu|(A), A \in \Sigma$. Moreover, $\|\nu\|(\cdot)$ is monotone increasing. Since ν is σ-additive, one verifies that $\|\nu\|(\cdot)$ is σ-subadditive. It is also clear that $|\nu|(A) = 0$ iff $\|\nu\|(A) = 0$, and that $|\nu|(A) = \|\nu\|(A)$ if $\mathcal{X}$ is the space of scalars. Further, one can show that there exists a probability measure λ on Σ such that $\lim_{\lambda(A)\to 0} \|\nu\|(A) = 0$. We see below that $\lambda = P$ serves the purpose if $\nu = P^{\mathcal{B}}$.

Let us compute (5) and (6) for $P^{\mathcal{B}} : \Sigma \to \mathcal{X} = L^1(\mathcal{B})$, to gain some facility with these concepts. Thus for $A \in \Sigma$,

$$\begin{aligned} |P^{\mathcal{B}}|(A) &= \sup\left\{\sum_{i=1}^{n} \|P^{\mathcal{B}}(A_i)\|_1 : A_i \in \Sigma(A), \text{ disjoint}\right\} \\ &= \sup\left\{\sum_{i=1}^{n} \int_{\Omega} E^{\mathcal{B}}(\chi_{A_i})dP : A_i \in \Sigma(A), \text{ disjoint}\right\} \\ &= P(A). \end{aligned} \tag{7}$$

The corresponding computation is more involved for $\mathcal{X} = L^p(\mathcal{B}), p > 1$.

But (6) is easy for this more general case. Thus for $A \in \Sigma$,

$$\|P^{\mathcal{B}}\|(A) = \sup\left\{\left\|\sum_{i=1}^{n} a_i P^{\mathcal{B}}(A_i)\right\|_p : A_i \in \Sigma(A), \text{ disjoint}, \ |a_i| \le 1\right\}$$

$$\le \sup\left\{\left\|\sum_{i=1}^{n} P^{\mathcal{B}}(A_i)\right\|_p : A_i \in \Sigma(A), \text{ disjoint}\right\}$$

$$= \sup \|E^{\mathcal{B}}(\chi_A)\|_p \le [P(A)]^{1/p}. \tag{8}$$

So $P^{\mathcal{B}}$ has a finite semi-variation in $L^p(\mathcal{B})$, for $1 \le p < \infty$, and when $p = 1$, we note that $\|P^{\mathcal{B}}\| = |P^{\mathcal{B}}| = P$. It is also clear that the dominating (or "control") measure λ can be taken as P by (7) and (8). This is used to state that $f_n \to f$ a.e. $[\nu$ or $P^{\mathcal{B}}]$ to mean $f_n \to f$ a.e. $[\lambda$ or $P]$.

Now to define an integral, let $f = \sum_{i=1}^{n} a_i \chi_{A_i}, A_i \in \Sigma$, disjoint. Then the vector integral of f relative to $P^{\mathcal{B}}$ can be given by the expression:

$$\int_A f \, dP^{\mathcal{B}} = \sum_{i=1}^{n} a_i P^{\mathcal{B}}(A \cap A_i), \quad A \in \Sigma. \tag{9}$$

The usual measure theory argument shows that the integral in (9) does not depend on the representation of the simple function f, and it is clearly an element of $L^p(\mathcal{B})$. If $f \in L^p(\Sigma)$, we define its integral relative to $P^{\mathcal{B}}$, provided the following two conditions hold:

(a) There exits a sequence $\{f_n, n \ge 1\}$ of simple functions such that $f_n \to f$ a.e. $[P]$ (or $P^{\mathcal{B}}$).

(b)$\{\int_A f_n dP^{\mathcal{B}}, n \ge 1\} \subset L^p(\mathcal{B})$, given by (9), is a norm Cauchy sequence for each $A \in \Sigma$. (Let f_A be this limit.) Then the function f is said to be integrable, for $P^{\mathcal{B}}$, on $A \in \Sigma$, and the integral is denoted by

$$f_A = \int_A f \, dP^{\mathcal{B}}, \quad A \in \Sigma. \tag{10}$$

The thus defined integral is termed the *Dunford-Schwartz* (or $D - S$) *integral.*

We record some consequences of the integral defined by (10). If f is simple, then it follows that ($q = p/(p-1)$)

$$\left\|\int_A f \, dP^{\mathcal{B}}\right\|_p \le \|f\|_\infty \|P^{\mathcal{B}}\|(A), \text{ and } \le \|f\|_p (P(A))^{1/q} \le \|f\|_p. \tag{11}$$

This inequality can be extended to all elements of $L^p(\Sigma), 1 \le p < \infty$, by the density of simple functions and the special cases (11), on noting that $f \mapsto \int_\Omega f \, dP^{\mathcal{B}}$ is a linear mapping. Thus all elements of $L^p(\Sigma)$ are $D - S$ integrable relative to $P^{\mathcal{B}}$. Fix $1 \le p < \infty$.

We use the following properties of this (vector) integral.

(1) The integral (10) is well defined; i.e., if $g_n \to f$ a.e. and satisfies (a) and (b), we still get (10) and it does not depend on the sequence used. Any two versions of (10) agree a.e. $[\lambda]$, the *control measure*.

(2) If f is $P^{\mathcal{B}}$-integrable, then $\lim_{P(A)\to 0} \int_A f \, dP^{\mathcal{B}} = 0$ as an element of $L^p(\mathcal{B})$.

(3) The mapping $f \mapsto \int_\Omega f \, dP^{\mathcal{B}}$ is linear and order preserving (hence continuous), $f \in L^p(\Sigma)$.

(4) The Monotone and Dominated convergence theorems hold where the convergence of the integrals is in the $L^p(\mathcal{B})$-norm.

These results are not trivial but (like in the classical case) all the details have to be written down. We omit the proofs, by referring the reader to the above noted excellent monographs of Dunford-Schwartz, or Dinculeanu. The monotone convergence is meaningful for vector measures ν only if their value space $\mathcal{X}$ is a Banach lattice, as here, and we sketch a proof of this special case. For convenience, set $\int_\Omega f \, dP^{\mathcal{B}} = +\infty$, if $f \ge 0$ and $\| \int_\Omega f \, dP^{\mathcal{B}} \|_p = \infty$.

3. Proposition. *Let* $\{0 \le f_n, n \ge 1\} \subset L^p(\Sigma)$ *be a nondecreasing sequence with the pointwise* a.e. *limit* f. *Then using the above convention for nonintegrability, we have*

$$\lim_{n\to\infty} \int_\Omega f_n \, dP^{\mathcal{B}} = \int_\Omega f \, dP^{\mathcal{B}}, \tag{12}$$

where the equality holds in the norm of $L^p(\Sigma), 1 \le p < \infty$.

Proof. Observe that the monotonicity of the norm $\| \cdot \|_p$ and the fact that when $p < \infty$, it is absolutely continuous (i.e., $f_n \downarrow 0, f_n \in L^p(\Sigma) \Rightarrow \|f_n\|_p \to 0$), together, show that the order convergence implies norm convergence in $L^p(\Sigma)$. [Here order convergence means, for any sequence $\{h_n, n \ge 1\} \subset L^p(\Sigma)$ with $\inf_k \sup_{n\ge k} h_n = \sup_k \inf_{n\ge k} h_n = h$ such that $h \in L^p(\Sigma)$, then the (order) limit of h_n is h.] Since $f \ge f_n$ for all n, and the integral is order preserving, we only need to prove (12) if $\sup_n \int_\Omega f_n dP^{\mathcal{B}}$ is in $L^p(\Sigma)$.

By definition of the $D-S$ integral, we can find a sequence of step functions $g_{nm} \uparrow f_n$ as $m \to \infty$ satisfying (a) and (b). Then define $h_{nk} = \max\{g_{jk}; 1 \le j \le n\}$. Clearly $h_{nn} \uparrow f$ a.e., and this sequence satisfies conditions (a) and (b), since $\sup_n \| \int_\Omega h_{nn}\, dP^{\mathcal{B}} \|_p < \infty$. Hence,

$$\int_\Omega f\, dP^{\mathcal{B}} = \lim_n \int_\Omega h_{nn}\, dP^{\mathcal{B}} \qquad (\in L^p(\mathcal{B})). \tag{13}$$

On the other hand $f \ge f_n \ge h_{nn}$, all n, so that $\int_\Omega f\, dP^{\mathcal{B}} \ge \int_\Omega f_n dP^{\mathcal{B}} \ge \int_\Omega h_{nn}\, dP^{\mathcal{B}}$. Taking limits on n and using (13), we get (12). □

Remark. This result is also true for $L^\infty(\Sigma)$, even though the norm $\|\cdot\|_\infty$ is not absolutely continuous. But then one uses the definition of the integral only in terms of order and the fact that the space is a boundedly σ-complete lattice, i.e., a Banach lattice such that each countable set with an upper bound has a least upper bound. Thus such integration theory, somewhat more general than the $D-S$ definition, can be developed for the $P^{\mathcal{B}}$-measures. In the context of $L^\infty(\Sigma)$ (or more precisely for the Banach space of real continuous functions on a compact Stone space), this integral was studied by Wright [1]. These ideas are also included in the more general theory of order preserving integration processes due to McShane [1].

As a consequence of Proposition 1.3, one has for any $0 \le f \in L^p(\Sigma)$, if $\{g_n, n \ge 1\}$ is a sequence of simple functions with $g_n \uparrow f$ a.e. (and hence in $L^p(\Sigma)$-norm), $E^{\mathcal{B}}(g_n \chi_A)(= \int_A g_n dP^{\mathcal{B}}) \uparrow E^{\mathcal{B}}(f\chi_A)$ a.e., and in norm, $1 \le p < \infty$. Consequently $\{\int_A g_n dP^{\mathcal{B}}, n \ge 1\}$ is Cauchy in $L^p(\mathcal{B})$ and we have by the above proposition (or even from definition)

$$E^{\mathcal{B}}(f\chi_A) = \int_A f\, dP^{\mathcal{B}}, \quad A \in \Sigma, \tag{14}$$

where the integral exists in the context of (10), *but not necessarily in the Lebesgue sense.* Thus we have proved the following conceptually important result.

4. Proposition. *If (Ω, Σ, P) is a probability space, and $\mathcal{B} \subset \Sigma$ is a σ-algebra, then the conditional probability function $P^{\mathcal{B}} : \Sigma \to L^p(\mathcal{B}), 1 \le p < \infty$, is a vector valued measure of finite semi-variation (and of variation P if $p = 1$), in terms of which we have:*

$$E^{\mathcal{B}}(f) = \int_\Omega f\, dP^{\mathcal{B}} \in L^p(\mathcal{B}), \quad f \in L^p(\Sigma), \tag{15}$$

where the integral exists in the Dunford-Schwartz sense.

The following specialization is worthy of note.

5. Corollary. *Let (Ω, Σ, P) be a probability space and $\mathcal{B}_1 \subset \mathcal{B}_2 \subset \Sigma$ be σ-algebras. If X, Y are random variables on Ω such that Y is $\mathcal{B}_2$-measurable and X, Y, XY are P-integrable, then*

$$\int_\Omega XY dP^{\mathcal{B}_1} = \int_\Omega Y \, dP^{\mathcal{B}_1} \int_\Omega X \, dP^{\mathcal{B}_2} \in L^1(\mathcal{B}_1), \tag{16}$$

where all these integrals are taken in the sense of (10).

Proof. Since $E^{\mathcal{B}_1}(XY) = E^{\mathcal{B}_1}(YE^{\mathcal{B}_2}(X))$, a.e., by Proposition 1.6(c), (16) is an immediate consequence of (15). □

6. Remarks. (i) We note for later use that the above formula can be combined with the abstract Lebesgue integral as follows. Let $X_1, \ldots, X_n$ be n random variables on (Ω, Σ, P) and Y be integrable. If $\mathcal{B}_k = \sigma(X_i, k \leq i \leq n)$ is the σ-algebra generated by $X_k, \ldots, X_n$, then $\mathcal{B}_n \subset \mathcal{B}_{n-1} \subset \cdots \subset \mathcal{B}_k \subset \Sigma$, and

$$\int_\Omega Y \, dP = \int_\Omega dP \int_\Omega dP^{\mathcal{B}_n} \cdots \int_\Omega Y \, dP^{\mathcal{B}_1}. \tag{17}$$

This is a reformulation of Corollary 1.7. Results of this kind are usually given for the restricted class of "regular" conditional probability functions so that all the integrals on the right side of (17) can be taken in the Lebesgue sense, instead of with the $D - S$ integration as here. But using the latter, it is clearly unnecessary to make such a restriction. However, evaluation of $D - S$ integrals is harder than the Lebesgue integrals.

(ii) It was seen in Proposition 1.11 that if Y and $X_k, 1 \leq k \leq n$, are mutually independent, then $E^{\mathcal{B}_k}(Y) = E(Y), 1 \leq k \leq n$. Hence $\int_\Omega Y \, dP^{\mathcal{B}_k} = \int_\Omega Y \, dP$ in (17), and the integral reduces to the Lebesgue case. Even though this is one of the simplest cases, we must still ask whether it is always possible to make this assumption, i.e., do there exist such classes of independent random variables on a given probability space? In general, this may not be the case. However, the problem is easily resolved by embedding Ω in a larger set by an *adjunction procedure.* The latter is as follows.

If $(\Omega_1, \Sigma_1, P_1)$ is the given probability triple, let $(\Omega_2, \Sigma_2, P_2)$ be any other probability space, and consider (Ω, Σ, P) the product of these two, i.e., $\Omega = \Omega_1 \times \Omega_2, \Sigma = \Sigma_1 \otimes \Sigma_2$, and $P = P_1 \otimes P_2$. We identify $A \in \Sigma_1$ with the cylinder set $A \times \Omega_2 \in \Sigma, P_1(A) = P(A \times \Omega_2)$. If X is a random variable on (Ω_1, Σ_1), it is identified with Y on (Ω, Σ) where $Y(\omega_1, \omega_2) = X(\omega_1)$ for all $(\omega_1, \omega_2) \in \Omega$, so that X and Y will have the same distribution functions since $P\{\omega : Y(\omega) < x\} = P_1\{\omega_1 : X(\omega_1) < x\}, x \in \mathbb{R}$. Thus the new space (Ω, Σ, P) will have a finer structure. If $(\Omega_1, \Sigma_1, P_1) = (\Omega_2, \Sigma_2, P_2)$ and Y is as above, $Z(\omega_1, \omega_2) = X(\omega_2), (\omega_1, \omega_2) \in \Omega$, then Y, Z are independent on Ω and are identically distributed.

In proceeding to the analysis of the structure of conditional probability functions, the following simple result is useful, particularly for the regularity of these function space valued measures.

7. Proposition. *Let (Ω, Σ, P) be a probability space and $(\tilde{\Omega}, \tilde{\Sigma})$ be a measurable space. If $X : \Omega \to \tilde{\Omega}$ is a $(\Sigma, \tilde{\Sigma})$-measurable mapping, i.e., $X^{-1}(\tilde{\Sigma}) \subset \Sigma$, and if $\mathcal{B} \subset \Sigma$ is a σ-algebra, let $P^{\mathcal{B}}(\cdot) : \Sigma \to L^\infty(\mathcal{B}) = L^\infty(\Omega, \mathcal{B}, P)$ be the conditional probability function and $\tilde{P}^{\mathcal{B}}(\cdot) = P^{\mathcal{B}} \circ X^{-1}(\cdot)$, defined by*

$$\tilde{P}^{\mathcal{B}}(\tilde{A}) = P^{\mathcal{B}}(X^{-1}(\tilde{A})), \qquad \tilde{A} \in \tilde{\Sigma}, \tag{18}$$

be the image of $P^{\mathcal{B}}(\cdot)$ so that $\tilde{P}^{\mathcal{B}} : \tilde{\Sigma} \to L^\infty(\mathcal{B})(\subset L^p(\mathcal{B}), 1 \leq p < \infty)$. Then $\tilde{P}^{\mathcal{B}}(\cdot)$ is a vector measure on $\tilde{\Sigma}$ into $L^p(\mathcal{B})$ such that for any $\tilde{P}^{\mathcal{B}}$ integrable real f on $\tilde{\Omega}$,

$$\int_{\tilde{\Omega}} f(\tilde{\omega})\tilde{P}^{\mathcal{B}}(d\tilde{\omega}) = \int_{\Omega} f \circ X(\omega) P^{\mathcal{B}}(d\omega) \tag{19}$$

where the integrals are taken in the Dunford-Schwartz sense.

Proof. Since $X^{-1} : \tilde{\Sigma} \to \Sigma$ is a σ-homomorphism, it is clear that (18) implies that $\tilde{P}^{\mathcal{B}}(\cdot)$ is a σ-additive set function with values in $L^\infty(\mathcal{B})$, having the same properties as $P^{\mathcal{B}}(\cdot)$. If $f = \chi_{\tilde{A}}$, then (19) is simply (18) and by linearity the integrals agree for all simple $\tilde{\Sigma}$-measurable functions f_n. If $f_n \to f$ pointwise and $\{\int_{\tilde{A}} f_n d\tilde{P}^{\mathcal{B}}, n \geq 1\} \subset L^p(\mathcal{B})$ is Cauchy for each $\tilde{A} \in \tilde{\Sigma}$, then the same must be true for the sequence $\{\int_A (f_n \circ X) dP^{\mathcal{B}}, n \geq 1\}$, all $A \in X^{-1}(\tilde{\Sigma})$. It follows that (19) holds for all $\tilde{P}^{\mathcal{B}}$ integrable f (in the sense of (10)). Observe that to define integrability relative to $\tilde{P}^{\mathcal{B}}$, we need a scalar measure $\tilde{P}$ (such as P for $P^{\mathcal{B}}$) associated with $\tilde{P}^{\mathcal{B}}$, as in (7) and (11). [Particularly this appears in property (2) following (11). That such a scalar measure λ always exists for any vector measure ν was established in the $D-S$ theory of integration. Such a λ is at times called a "control measure" of ν.] However, one can see immediately from (7) that $\tilde{P} = P \circ X^{-1}$ serves the purpose. This completes the proof. □

(b) *Regularity.* As observed before, the regularity of a conditional probability function is important for some parts of stochastic analysis, especially in the study of Markov processes. In this book, the latter processes are only touched on and regularity will not be invoked in much of our work. However, it will be very useful to understand the problem and recognize the dangerous twists in the application of conditional measures. This, moreover, gives an appreciation of the role of the vector integrals (10) in this theory, in lieu of the abstract Lebesgue measures. So we include a brief account of regularity here.

It was noted prior to the counterexample above that, if $P(\cdot,\cdot)$ is a version of $P^{\mathcal{B}}(\cdot)$ (so $P(A,\omega) = P^{\mathcal{B}}(A)(\omega)$) on (Ω, Σ, P), then it is called "regular" whenever $P(\cdot,\omega)$ is a scalar measure for almost all $\omega(\in \Omega)$ and $P(A,\cdot)$ is a Σ-measurable function. When this is possible, then the integrals in (19) above can be regarded as the Lebesgue integrals. We present the definition of regularity somewhat differently and then give a characterization as an *equivalent* concept with the above statement.

8. Definition. Let (Ω, Σ, P) be a probability triple and $(\tilde{\Omega}, \tilde{\Sigma})$ a topological measurable space where $\tilde{\Omega}$ is a Hausdorff space and $\tilde{\Sigma}$ is a σ-algebra containing the closed sets. If $X : \Omega \to \tilde{\Omega}$ is a measurable mapping, let $\psi = P^{\mathcal{B}} \circ X^{-1}$ be the image conditional measure, so that $\psi : \tilde{\Sigma} \to L^{\infty}(\mathcal{B})$ and $\mathcal{B} \subset \Sigma$ is a σ-algebra. (a) The function $\psi(\cdot)$, (and also $P^{\mathcal{B}}$), is called *inner regular* relative to X whenever the following condition is satisfied: For each relatively compact open set $A \subset \tilde{\Omega}$,

$$\psi(A) = \sup\{\psi(C) : C \subset A, C \text{ compact}\}. \tag{20}$$

[The terms *left-regular*, and *wide sense conditional distribution* are also used for the same concept.] (b) The function $\psi(\cdot)$, or $P^{\mathcal{B}}(\cdot)$, is termed *outer regular* (or *right-regular*, or *strict sense conditional distribution*) if for every relatively compact Borel set $A \subset \tilde{\Omega}$,

$$\psi(A) = \inf\{\psi(B) : B \supset A, B \text{ open}\}. \tag{21}$$

If $\psi(\cdot)$ is both inner and outer regular, then it is called *regular*.

Since $\psi(\cdot) : \tilde{\Sigma} \to L^\infty(\mathcal{B})$ is a positive uniformly bounded (by 1) mapping, and (Ω, Σ, P) is a finite measure space, equations (20) and (21) define $\psi(A)$ equivalent to a measurable function for each $A \in \tilde{\Sigma}$, and $0 \le \psi(A) \le 1$ a.e. This statement is true even if P is replaced by a σ-finite (or "localizable") measure on Σ, and one can select a measurable function from the equivalence class, as a consequence of an important result known as the lifting theorem, (cf. Section III.2 later).

If Ω itself is a topological space and Σ is the Borel algebra on Ω, then taking X = identity, the notion of regularity applies directly to $P^{\mathcal{B}}$ itself (instead of $\psi = P^{\mathcal{B}} \circ X^{-1}$). The next result shows that every (image) conditional probability ψ is inner regular relative to each random variable X with *values in a metric space* $\tilde{\Omega}$. However, if the image of $P^{\mathcal{B}}(\cdot)$ is regular relative to one X, it need not be regular relatie to other mappings, unless the original (Ω, Σ, P) itself is restricted. A good sufficient condition is that P should be a "perfect" probability as given in Definition 11 below.

9. Theorem. *Let (Ω, Σ, P) be a probability space, $\mathcal{B} \subset \Sigma$ a complete σ-algebra and $\tilde{\Omega}$ be an arbitrary metric space. Let $\tilde{\Sigma}$ be the Borel (= Baire here) σ-algebra of $\tilde{\Omega}$. If $X : \Omega \to \tilde{\Omega}$ is a $(\Sigma, \tilde{\Sigma})$-measurable mapping, then $P^{\mathcal{B}}$ or $\tilde{P}^{\mathcal{B}} = P^{\mathcal{B}} \circ X^{-1} : \tilde{\Sigma} \to L^\infty(\Omega, \mathcal{B}, P)$ is always inner regular (cf., (20)) relative to X. Moreover, the integral (19) relative to $P^{\mathcal{B}}$ on Ω can be defined in the Lebesgue sense iff $P^{\mathcal{B}}$ or ψ is regular relative to X. Thus the integral (10) given by the Dunford-Schwartz definition is also obtainable from the abstract Lebesgue theory iff ψ is regular on $\tilde{\Sigma}$ relative to X.* [Thus the regularity of a conditional probability stated prior to the counterexample is the same as that given by Definition 8.]

We shall not include a proof of this result here. It is not essential to the present exposition. (Details of the argument for this as well as the next few remarks and results may be found in the author's article [3].) We now give a companion result to the above.

If S is a topological space, $\mathcal{A}$ is a ring of subsets of $S, \nu : \mathcal{A} \to \mathcal{X}$ (a Banach space) is a vector measure, then it is *inner regular* relative to

$\mathcal{A}$, whenever, for each $A \in \mathcal{A}$ and $\varepsilon > 0$, there is a relatively compact $F \in \mathcal{A}$ such that $\bar{F} \subset A$ ($\bar{F}$ is the closure of F) and for any $E \subset A - F, E \in \mathcal{A}$, it follows that $\|\nu(E)\| < \varepsilon$. Similarly ν is *outer regular* if for each $A \in \mathcal{A}$ and $\varepsilon > 0$, there is a set $G \in \mathcal{A}$, such that $A \subset \mathrm{int}(G)$ and for any $H \subset G - A, \|\nu(H)\| < \varepsilon$. Finally ν is *regular* on $\mathcal{A}$ if it is both inner and outer regular there. The following abstract result is useful for us. It is not too difficult, though nontrivial. (For this and related results, see Dinculeanu ([1], p. 313ff), or the above reference.)

10. Proposition. *Let Ω be a locally compact space and Σ be the σ-ring generated by the compact (or compact G_δ)-subsets of Ω. If $\nu : \Sigma \to \mathcal{X}$ (a Banach space), is a vector measure then it is regular iff ν is inner or outer regular on Σ. Moreover, ν is always regular on Σ if its variation measure $|\nu|(\cdot)$ (cf., (5)) is finite on compact sets of Ω and is regular on Σ (or only inner [outer] regular on Σ).*

The last part is particularly useful for conditional probability measures $\nu = P^{\mathcal{B}} : \Sigma \to \mathcal{X} = L^1(\mathcal{B})$, because $|\nu| = P$ by (7). Using this we shall also strengthen the last part of Theorem 9 regarding the regularity of $P^{\mathcal{B}}$ (or ψ) relative to *all* random variables X, by a suitable restriction of P. The desired condition is "perfectness", introduced by B. V. Gnedenko and A. N. Kolmogorov (1949). Its analysis is due to V. V. Sazonov [1] where other references on the subject may be found. We indicate the concepts.

11. Definition. A probability space $(S, \mathcal{A}, \mu)$ is called *perfect* if for each random variable $X : S \to \mathbb{R}$, and any subset $A \subset \mathbb{R}$ (A is not necessarily Borel) such that $X^{-1}(A) \in \mathcal{A}$, there exists a Borel set $B \subset A$ such that $\mu(X^{-1}(A)) = \mu(X^{-1}(B))$.

If $\mu' = \mu \circ X^{-1}$, the image measure in $\mathbb{R}$, then we know from the classical Carathéodory measurability that the condition $X^{-1}(A)$ be μ-measurable forces A to be μ'-measurable. Thus what this definition demands is that the μ'-measurable sets should be in the μ'-completion of the Borel sets of $\mathbb{R}$, for each X. Hence *each* μ' is a Lebesgue-Stieltjes measure.

The following characterization facilitates it in applications:

12. Proposition. *A probability space $(S, \mathcal{A}, \mu)$ is perfect iff for each random variable $X : S \to \mathbb{R}$, there is a Borel set $B (= B_X)$ in the range of X such that $\mu(X^{-1}(B)) = 1$.*

Proof. If $X : S \to \mathbb{R}$ is an arbitrary random variable, let $\mathcal{R}_X = \{A \subset \mathbb{R} : X^{-1}(A) \in \mathcal{A}\}$. Clearly $\mathcal{R}_X$ is a σ-algebra containing the Borel algebra $\tilde{\mathcal{B}}$. Let $A \in \mathcal{R}_X$ and consider $S_1 = X^{-1}(A)$. If $X(S_1) \cap A = \emptyset$, then $S_1 = \emptyset$, by definition of the complete inverse image. In this case we take $B = \emptyset$ in $\tilde{\mathcal{B}}$, so that $\mu(X^{-1}(B)) = \mu(X^{-1}(A)) = \mu(S_1) = 0$. If $S_1 \neq \emptyset$ define $X_1(\omega) = X(\omega)$ for $\omega \in S_1$, and α_0 for $\omega \in S - S_1$, where α_0 is some point in $X(S_1) \subset \mathbb{R}$. Then $X_1(S) = X_1(S_1) \subset A$ and $X_1^{-1}(\tilde{\mathcal{B}}) = X_1^{-1}(\tilde{\mathcal{B}}(A)) \subset \mathcal{A}$, and hence X_1 is a random variable. (Here $\tilde{\mathcal{B}}(A)$ is the trace σ-algebra.) By hypothesis there exists a Borel set $B \subset X_1(S) = X_1(S_1) \subset A$ such that $\mu(X_1^{-1}(B)) = \mu(S) = 1$. On the other hand, if $\alpha_0 \in X(S_1) - B$ then $X_1^{-1}(B) = X^{-1}(B)$ while for $\alpha_0 \in B$ we have (the disjoint union of) $X_1^{-1}(B)$ as:

$$X_1^{-1}(B) = X^{-1}(B) \cup (S - X^{-1}(A)).$$

Hence, $\mu(X^{-1}(B)) = \mu(X_1^{-1}(B))$ in the first case, and if $\alpha_0 \in B, 1 = \mu(S) = \mu(X_1^{-1}(B)) = \mu(X^{-1}(B)) + 1 - \mu(X^{-1}(A))$.It follows again that $\mu(X^{-1}(A)) = \mu(X^{-1}(B))$, so that by Definition 11, μ is perfect.

Conversely, if μ is perfect, taking $A = \mathbb{R}$, we can find a $B \in \tilde{\mathcal{B}}$ (by definition), such that $\mu(X^{-1}(B)) = 1$ for each random variable, and the condition is satified. □

13. Remark. It is obvious that there exist nonperfect probability measures. Indeed, the measure Q on $\tilde{\Sigma}$ constructed in the counterexample above is such a measure. On the positive side, if S is a topological space, $\mathcal{A}$ is the σ-algebra generated by the class $B^b(S)$ of all real bounded functions on S (i.e., $\mathcal{A} = \sigma\{\cup f^{-1}(\tilde{\mathcal{B}}) : f \in B^b(S)\}$), then every inner regular (for the compact class $\mathcal{C} \subset \mathcal{A}$) probability is perfect. If S is a separable metric space and $\mathcal{A}$ is its Borel σ-algebra, then every Radon probability on $\mathcal{A}$ is perfect and conversely. If (Ω, Σ, P) is perfect, $X : \Omega \to \mathbb{R}$ is a random variable, then $(\mathbb{R}, \mathcal{R}_X, P \circ X^{-1})$ as well as $(\mathbb{R}, \tilde{\mathcal{B}}, P \circ X^{-1})$ are perfect where $\mathcal{R}_X \supset \tilde{\mathcal{B}}$ as defined in the above proof. Several other interesting results on perfect probabilities are given by Sazonov [1], where an amplification of these remarks may be found.

If $X : \Omega \to \mathbb{R}$ is a random variable, $\mathcal{B} \subset \Sigma$ is a σ-algebra and $P^{\mathcal{B}} : \Sigma \to L^\infty(\mathcal{B})$ is the conditional probability then we have seen in (7) that $|P^{\mathcal{B}}|(\cdot) = P(\cdot)$ and $|P^{\mathcal{B}} \circ X^{-1}|(\cdot) = P \circ X^{-1}(\cdot)$. Hence by the results 9, 10, and 12 above, we have the following useful consequence:

14. Corollary. *Let (Ω, Σ, P) be a probability space, $\mathcal{B} \subset \Sigma$ a σ-algebra, and $X : \Omega \to \mathbb{R}$ a random variable. Then the conditional probability $P^{\mathcal{B}}$ (or $P^{\mathcal{B}} \circ X^{-1}$) is regular for X, whenever there is a Borel set $B \subset X(\Omega)$ of full measure, i.e., $P(X^{-1}(B)) = 1$. Thus if P is a perfect probability, then $P^{\mathcal{B}} \circ X^{-1}$ is a regular conditional probability for each random variable X.*

This result was established by Doob in ([1], p. 31] by a different argument. In this case (of regularity) we may employ the Lebesgue integration and reduce the proof of the conditional Jensen inequality (Theorem 1.8) to the classical case of scalar measures.

15. Discussion. The problems related to the existence of regular conditional probability measures are doubtless caused due to the presence of a large collection of P-null sets in (Ω, Σ, P). It is therefore tempting to postulate a theory of conditional probability measures in the same way that Kolmogorov [1] has proposed his axiomatic theory for the abstract (= unconditional) probabilities which is the basis of all the current work. In fact this point of view has been pursued and developed by A. Rényi [1], so as to coincide with the Kolmogorov model when specialized. We include a few comments on this. Let (Ω, Σ) be a measurable space and $\mathcal{A} \subset \Sigma$ be a nonempty class, and suppose that a function $P(\cdot|\cdot) : \Sigma \times \mathcal{A} \to \mathbb{R}^+$ is given satisfying the following three axioms:

Axiom I. For any $A \in \Sigma, B \in \mathcal{A}, 0 \leq P(A|B) \leq 1$ and $P(B|B) = 1$.

Axiom II. For each $B \in \mathcal{A}, P(\cdot|B)$ is σ-additive.

Axiom III. (a) For each $A \in \Sigma, B \in \mathcal{A}, P(A|B) = P(A \cap B|B)$, and (b) for each $A \in \Sigma, \{B, C\} \subset \mathcal{A}$, with $A \subset B \subset C$,

$$P(A|B)P(B|C) = P(A|C).$$

If $\mathcal{A} = \{\Omega\}$, then $\tilde{P}(\cdot) = P(\cdot|\Omega)$ defines an ordinary probability function on the σ-algebra Σ. Axiom III distinguishes the Rényi approach, while Axiom II incorporates the regularity into the definition. Now $P(\cdot|B)$ being a finite measure, if $A \in \Sigma$ is such that $P(A|B) = 0$ for all $B \in \mathcal{A}$, then A cannot be in $\mathcal{A}$ for otherwise, by Axiom I, we must have $P(A|A) = 1$ (and $= 0$ at the same time). In particular $\emptyset \notin \mathcal{A}$. Thus the "$P(\cdot|\mathcal{A})$-null sets" are banished from $\mathcal{A}$. But this case automatically excludes problems involving random variables with continuous probability distributions. (One may give such excluded examples even with Gaussian processes.) To include those problems, we can think of adjoining the null sets to $\mathcal{A}$ by modifying Axiom I, and saying that $P(B|B) = 1$ if B is not null, and $P(A|A)$ is assigned any specific value for null sets A to agree with $P(B|A) = P(A \cap B)/P(A)$. However, this will bring back all the "exceptional null sets" of the earlier theory. Thus while Rényi's new axiomatic approach is of interest for some problems, as illustrated in his book [2], it is *not* a complete generalization of the conditional probability theory as expounded in this section. A detailed treatment of both these models is given in the author's recent book [12].

(c) *Characterization.* The preceding work clearly demonstrates the fact that, in general, the conditional probability functions are, and must be treated as, a distinguished subclass of vector (or function space) valued measures. Since this class plays a vital role in many classes of stochastic processes, it is helpful if we can characterize vector measures, on the real L^p-spaces, which coincide with conditional probabilities. One such result will be presented here.

16. Theorem. *Let (Ω, Σ, P) be a probability space, and $\nu : \Sigma \to L^p(\Sigma), 1 \leq p < \infty$, be a vector measure. Then ν is a conditional probability $P^{\mathcal{B}}$ for a unique σ-algebra $\mathcal{B} \subset \Sigma$, iff (i) $\nu(A) \geq 0, A \in \Sigma$, (ii) $\nu(\Omega) = 1$, and $|\nu|(A) \leq P(A)$ where $|\nu|(\cdot)$ is the variation measure of ν (cf. (5)), and (iii) for any A, B, C in Σ, one has the integral equation:*

$$\int_\Omega \chi_A(\nu(B) \vee \nu(C))dP = \int_\Omega \nu(A)(\nu(B) \vee \nu(C))dP, \tag{22}$$

where "$\vee$" denotes the maximum. When these conditions hold $\mathcal{B}$ is given by $\mathcal{B} = \sigma\{A \in \Sigma : \nu(A) = \chi_A$, a.e.$\}$.

Proof. If $\nu = P^{\mathcal{B}}$ for some σ-algebra $\mathcal{B} \subset \Sigma$, then by (7) and Proposition 1, (i) and (ii) are true. Since $P^{\mathcal{B}}(A) = E^{\mathcal{B}}(\chi_A)$, we also have, from the averaging property of $E^{\mathcal{B}}$, that

$$\int_\Omega \chi_A(\nu(B) \vee \nu(C))dP = \int_\Omega E^{\mathcal{B}}(\chi_A)(\nu(B) \vee \nu(C))dP_{\mathcal{B}} \tag{23}$$

which is (22). Only the converse is nontrivial.

Let ν be a vector measure satisfying the given conditions. Then the linear operator $T : L^p(\Sigma) \to L^p(\Sigma)$ given by the Dunford-Schwartz integral, $Tf = \int_\Omega f\, d\nu, f \in L^p(\Sigma)$, has the following properties. By (i) $Tf \geq 0$ for all simple functions $f \geq 0$, and then we can conclude that T is a positive operator if it is shown to be continuous.

The continuity of T is implied from its (stronger) property of contractivity. The latter follows from (ii). In fact, if $f = \sum_{i=1}^{n} a_i \chi_{A_i}, A_i \in \Sigma$, disjoint, then assuming that $\|f\|_p = \alpha > 0$, for nontriviality,

$$\|Tf\|_p = \left\| \int_\Omega f\, d\nu \right\|_p = \alpha \left\| \sum_{i=1}^{n} \frac{a_i}{\alpha} \nu(A_i) \right\|_p \leq \|f\|_p \|\nu\|(\Omega), \tag{24}$$

by (6) (with $\mathcal{X} = L^p(\Sigma)$ there). This is true and trivial for $\alpha = 0$. However $\|\nu\|(\Omega) \leq |\nu|(\Omega) \leq P(\Omega) = 1$ so that $\|Tf\|_p \leq \|f\|_p$ and T is a contraction. Also $T1 = 1$ by (ii) and T is a positive contractive linear operator with 1 as a fixed point.

Let T^* be the adjoint of T on $(L^p(\Sigma))^* = L^q(\Sigma), p^{-1} + q^{-1} = 1, 1 \leq p < \infty$. Now $T\chi_A = \nu(A)$. So we have, by (25) for $A \in \Sigma$,

$$\int_\Omega \chi_A (T\chi_B \vee T\chi_C) dP = \int_\Omega \chi_A T^*(T\chi_B \vee T\chi_C) dP. \tag{25}$$

It follows from (i) and (ii) that $0 \leq \nu(A) \leq 1$ a.e., and then T maps bounded functions into bounded functions. Since A is arbitrary in (25) we have the important equation:

$$T\chi_B \vee T\chi_C = T^*(T\chi_B \vee T\chi_C), \text{ a.e.} \tag{26}$$

On the other hand, taking $C = \emptyset$ in (25) and extending the resulting equation (first for simple functions) using the continuity of T, one gets for all bounded $\{f, g\} \subset L^p(\Sigma)$:

$$\int_\Omega fTg\, dP = \int_\Omega (Tf)(Tg) dP. \tag{27}$$

If $f = \chi_A, A \in \Sigma$, this yields

$$\int_A Tg dP = \int_\Omega \chi_A Tg dP = \int_\Omega (T\chi_A)(Tg) dP. \tag{28}$$

We also have, with (27), for all $A \in \Sigma$

$$\int_{\Omega} (T\chi_A)(Tg)dP = \int_{\Omega} gT\chi_A dP = \int_{\Omega} (T^*g)\chi_A dP = \int_{A} T^* g dP. \quad (29)$$

It follows from (28) and (29), and the arbitrariness of $A \in \Sigma$, that $Tg = T^*g$, a.e., for all $g \in L^\infty(\Sigma)$. This together with (26) implies the equation $T\chi_B \vee T\chi_B = T(T\chi_B \vee T\chi_C)$ a.e., $\{B, C\} \subset \Sigma$. Now taking $C = \emptyset$ and using the linearity of T, this yields that $T^2 f = Tf$ for all simple $f \in L^p(\Sigma)$. Since such functions are dense in $L^p(\Sigma)$, we deduce that $T : L^p(\Sigma) \to L^p(\Sigma)$ is a positive contractive projection with $T1 = 1$ a.e. Hence by Theorem 2.6 there is a unique σ-algebra $\mathcal{B} \subset \Sigma$ such that $T = E^{\mathcal{B}}$, and then $\mathcal{B}$ is as stated in the theorem. □

The above characterization of a conditional probability function shows the effectiveness of the Dunford-Schwartz integral in the present work. We now show how the same integral can be used with ease for a generalized Fubini theorem.

(d) *Products.* Let (Ω, Σ, P) be a probability space, $(E, \mathcal{E})$ be a measurable space and $h : \Omega \to E$ be a measurable mapping. Then $Q = P \circ h^{-1} : \mathcal{E} \to [0, 1]$ is a probability and if $\mathcal{B} = \mathcal{B}_h = h^{-1}(\mathcal{E})$, then, as shown in (11) of Section 1, for any P-integrable random variable X on Ω, the conditional expectation of X relative to $\mathcal{B}_h$ exists and is given by (cf., Equation (12) there):

$$g_X(e) = E(X|h(\omega) = e). \quad (30)$$

If $\psi(A, e) = g_A(e) = E(\chi_A|h(\omega) = e)$ $(= E^{\mathcal{B}}(\chi_A)(h(\omega) = e))$, then the conditional measure $\psi : \Sigma \times \mathbb{E} \to [0, 1]$ has the following description. $\psi : \Sigma \to L^p(\mathcal{E})$ is a vector measure $(1 \leq p < \infty)$ and $\psi(A) \in L^\infty(\mathcal{E})$. So (30) can be expressed by the functional equation:

$$\tilde{P}(A, B) = P(A \cap h^{-1}(B)) = \int_{B} \psi(A, e)dQ(e), \qquad A \in \Sigma, B \in \mathcal{E}. \quad (31)$$

Conversely, suppose a vector measure $\psi : \Sigma \to L^p(\mathcal{E})$, such that $0 \leq \psi(A) \leq 1$ a.e., is given. Then (31) defines a measure $\tilde{P}$ on the product σ-algebra $\Sigma \bigotimes \mathcal{E}$ such that the "marginal" measure $\tilde{P}(\Omega, \cdot)$ is $\{Q(B), B \in \mathcal{E}\}$. This discussion is made precise in the following result in which ψ need not be a conditional measure(even though its variation is a probability on Σ).

17. Proposition. *Let (Ω, Σ) and $(E, \mathcal{E})$ be measurable spaces and $Q : \mathcal{E} \to [0,1]$ be a probability measure. If $\psi : \Sigma \to L^p(\mathcal{E})$ is a vector measure $(1 \leq p < \infty)$ and $0 \leq \psi(A) \leq 1$ a.e., $\psi(\Omega) = 1$, a.e. $(Q), \psi$ with finite variation on $L^1(\mathcal{E})$, then there exists a probability P on $(\Omega \times E, \Sigma \bigotimes \mathcal{E})$ such that $P(\Omega, \cdot) = Q(\cdot)$, and P satisfies the equation:*

$$P(A \times B) = \int_B \psi(A)(e) dQ(e), \qquad A \in \Sigma, B \in \mathcal{E}, \tag{32}$$

Moreover, if $f : \Omega \times E \to \mathbb{R}$ is measurable and integrable relative to the product (scalar) measure $|\psi| \bigotimes Q$, where $|\psi|(\cdot)$ is the variation measure of ψ, then we have

$$\int_{\Omega \times E} f \, dP = \int_E \Big[\int_\Omega f(\omega, e) \psi(d\omega) \Big] Q(de), \tag{33}$$

where the inner integral on the right exists in the sense of (10) and the others are abstract Lebesgue integrals.

Proof. It suffices to prove (32) for the rectangles $\{A \times B : A \in \Sigma, B \in \mathcal{E}\}$, since they form a semi-algebra $\mathcal{S}$ which generates $\Sigma \bigotimes \mathcal{E}$. Thus for $A \in \Sigma, B \in \mathcal{E}$, consider $P(A \times B)$ given by the integral in (32). Then $P : \mathcal{S} \to [0,1]$ is well-defined. To see that P is σ-additive on the semi-algebra $\mathcal{S}$, let $\{A_i \times B_i, i \geq 1\}$ be a sequence of disjoint elements of $\mathcal{S}$ whose union is $A \times B \in \mathcal{S}$. Since $\chi_{A \times B} = \chi_A \chi_B$, it is clear that

$$\chi_A \chi_B = \sum_{i=1}^{\infty} \chi_{A_i} \chi_{B_i}, \tag{34}$$

the series converging a.e. In fact, for each $(\omega, e) \in A \times B$, it belongs only to one $A_i \times B_i$ so that only one term is non-zero on the right, and equality holds. Fix $e \in E$, and treat (34) as a series of functions in $\omega \in \Omega$. Then we may integrate both sides relative to the vector measure $\psi(\cdot)$ and use the Monotone Convergence theorem, given as Proposition 3 above to get:

$$\chi_B(e) \psi(A)(e) = \sum_{i=1}^{\infty} \psi(A_i)(e) \chi_{B_i}(e) \qquad \text{a.a. } (e). \tag{35}$$

Since e is arbitrary and all terms are nonnegative and measurable $(\mathcal{E})$, we may treat (35) as a series of functions in e and integrate relative to the measure Q to obtain (with the classical Lebesgue Monotone Convergence):

$$P(A \times B) = \sum_{i=1}^{\infty} P(A_i \times B_i). \tag{36}$$

Thus P is σ-additive on $\mathcal{S}$ and, by the Carathéodory extension procedure, the outer measure P^* generated by the pair $(\mathcal{S}, P)$ is σ-additive on $\mathcal{A}_{P^*}$, the P^*-measurable sets. Also $\mathcal{S} \subset \mathcal{A}_{P^*}$ and $P^*|\mathcal{S} = P$. Since $\mathcal{A}_{P^*}$ is the maximal (complete for P^*) σ-algebra, it follows that $\Sigma \bigotimes \mathcal{E} \subset \mathcal{A}_{P^*}$. $P^*|\Sigma \bigotimes \mathcal{E}$, which we denote again by P, is the desired measure. It is obvious that $P(\Omega \times B) = Q(B), B \in \mathcal{E}$ so that Q is the "marginal" measure of P.

Since f is $|\psi| \bigotimes Q$-integrable and $f(\cdot, e)$ is Σ-measurable, it follows from definition of the vector integral (10) that $g(e) = \int_\Omega f(\omega, e)\psi(d\omega)(e)$ defines a function $g(\cdot) \in L^1(\mathcal{E})$. Observe that if ψ is regular, then the above integral can be interpreted in the abstract Lebesgue sense by Theorem 9. In any case, (33) follows from the fact that one may use (32) for simple functions and then invoke Proposition 3. $\square$

Note that $\Phi(A) = P(A \times E), A \in \Sigma$ is also σ-additive since $\psi(\cdot)$ is bounded and strongly σ-additive. Hence $\Phi : \Sigma \to [0, 1]$ is a probability. This can be used to state the following:

18. Corollary. *Let $\psi : \Sigma \to L^\infty(\mathcal{E}), Q : \mathcal{E} \mapsto [0, 1]$ be a vector and a probability measure respectively, as in the proposition. Then $\Phi : \Sigma \to [0, 1]$, defined by $\Phi(A) = \int_E \psi(A)(e)dQ(e)$, is a probability on Σ, and for any Σ-measurable $f : \Omega \to \mathbb{R}$ which is $|\psi|(\cdot)$-integrable we have*

$$\int_\Omega f(\omega)d\Phi(\omega) = \int_E g(e)dQ(e) \tag{37}$$

where

$$g(e) = \int_\Omega f(\omega)\psi(d\omega)(e), \text{ a.e.}$$

The integral for g is in the sense of (10), and the first for f is a Lebesgue integral. [In case ψ is regular, then all are abstract Lebesgue integrals.]

If $\psi(A)(e)$ does not depend on e, or $L^\infty(\mathcal{E})$ is one dimensional, then Proposition 17 reduces to one form of the Fubini theorem. The preceding proposition admits an immediate extension if there are n-components. However, there is a nontrivial infinite dimensional extension for it corresponding to the classical Fubini-Jessen theorem to be given in Section 3.5.

2.4 Martingale concepts and inequalities

(a) *Introduction.* At the end of Section I.1, we have pointed out (but not considered) the concept of a (point) martingale, awaiting the analysis of conditional expectations. It is now possible, therefore, to introduce martingales and some of the immediate relatives. After the definition of this process, we shall explain its motivation and derive some easy consequences and then present the fundamental inequalities. Later we consider the decomposition and convergence results essential for an in-depth analysis of the subject.

1. Definition. Let $\{X_t, t \in T\} \subset L^1(\Omega, \Sigma, P)$ be a stochastic process where $T \subset \mathbb{R}$ and P is a probability. If $\mathcal{B}_n = \sigma(X_{t_i}, 1 \leq i \leq n)$ is the σ-algebra generated by the X_{t_i} shown, where $t_1 < t_2 < \cdots < t_{n+1}, t_i \in T$, are any t-points, then the process is called a *martingale* whenever we have

$$E^{\mathcal{B}_n}(X_{t_{n+1}}) = X_{t_n}, \text{ a.e.}, \qquad n \geq 1. \tag{1}$$

This is also suggestively written as:

$$E(X_{t_{n+1}} | X_{t_1}, \ldots, X_{t_n}) = X_{t_n}, \text{ a.e.}, \qquad n \geq 1. \tag{2}$$

The second form, which is by definition the same as (1), has the following interpretation of a game of chance. Namely, if X_{t_1} is the forture of a gambler at time t_1, and his fortunes in the next n games, played at times $t_2, \ldots, t_n$, are $X_{t_2}, \ldots, X_{t_n}$, which are thus known to him, then (2) says that his expected fortune at the $(n+1)$st game is the same as his present (i.e., the nth) one. Such a game can be considered fair. Equation (1) is simply a precise mathematical version of the verbal statement "that the first n outcomes are known to him". Thus we may abstract the concept as follows: If X_t is $\mathcal{F}_t$-measurable (or $\mathcal{F}_t$-*adapted*) where $\mathcal{F}_t \subset \Sigma, \mathcal{F}_t \subset \mathcal{F}_{t'}$ for $t < t'$, are σ-algebras, then $\{X_t, \mathcal{F}_t, t \in T\}$, above, is a martingale iff for each $s < t$ one has:

$$E^{\mathcal{F}_s}(X_t) = X_s, \text{ a.e.} \tag{3}$$

We make some observations on the statement leading to (3). First, the index T of the process represents the progress of a phenomenon (usually time) and thus should at least be a partially ordered set. Second, since $E^{\mathcal{F}}$ is a projection (1) or (3) says that $E^{\mathcal{F}_s}(X_t) \in L^p(\Omega, \mathcal{F}_s, P) = L^p(\mathcal{F}_s), p \geq 1$, if $X_t \in L^p(\Omega, \Sigma, P)$ and $t \geq s$. However, if $p = 2$, there are other projections (of norm 1) onto subspaces containing X_s. If $\mathcal{S}_n = \overline{\text{sp}}\{X_{t_1}, \ldots, X_{t_n}\}$ is the closed linear span of $X_{t_1}, \ldots, X_{t_n}$ in $L^2(\Sigma)$, then, on $L^2(\Sigma)$, there exists an orthogonal projection Q_n with $Q_n(L^2(\Sigma)) = \mathcal{S}_n$. If (1) holds when $E^{\mathcal{B}_n}$ is replaced by Q_n there, a new process called the *wide sense martingale* is obtained. Thus if $t_1 < \cdots < t_{n+1}, n \geq 1, t_i \in T$ and $\{X_t, t \in T\} \subset L^2(\Sigma)$, we have:

$$Q_n(X_{t_{n+1}}) = X_{t_n}, \text{ a.e.} \tag{4}$$

The preceding result does not have a natural analog in other L^p-spaces, $p \neq 2$. For instance, we may consider, following (3), $\mathcal{S}_t = \overline{\text{sp}}\{X_s, s \in T, s \leq t\}$ and ask whether there is a $Q_t, Q_t(L^p(\Sigma)) = \mathcal{S}_t$ such that $Q_t(X_{t'}) = X_t$ a.e., for $t < t', t, t'$ in T. But if $\mathcal{S}_t$ is infinite dimensional, there need not exist a bounded (much less contractive) projection onto $\mathcal{S}_t$. There are many other complications in this search. But the work in Sections 1-3 implies that $E^{\mathcal{B}}(L^p(\Sigma)) = L^p(\mathcal{B})$ and $E^{\mathcal{B}}$ is a contractive projection, and thus (1) is meaningful in all L^p-spaces while the wide sense concept is generally interesting only in L^2. In the case of Gaussian processes (in L^2), it can be verified without difficulty that both concepts coincide. Finally, since X_t's are real it makes sense to weaken (1) or (3) by replacing equality with an inequality. This leads to the following:

2. Definition. A (real) stochastic process $\{X_t, t \in T\} \subset L^1(\Omega, \Sigma, P)$, $T \subseteq \mathbb{R}$, is called a *submartingale* if for any $n \geq 1, t_1 < \cdots < t_{n+1}, t_i \in T$, it is true that

$$E(X_{t_{n+1}} | X_{t_1}, \ldots, X_{t_n}) = E^{\mathcal{B}_n}(X_{t_{n+1}}) \geq X_{t_n}, \text{ a.e.,} \tag{5}$$

and it is called a *supermartingale* if instead

$$E(X_{t_{n+1}} | X_{t_1}, \ldots, X_{t_n}) = E^{\mathcal{B}_n}(X_{t_{n+1}}) \leq X_{t_n}, \text{ a.e.} \tag{6}$$

In general, we may replace these, for a process $\{X_t, \mathcal{F}_t, t \in T\} \subset L^1(\Sigma)$ by

$$E^{\mathcal{F}_s}(X_t) \geq X_s, \text{ a.e.}, [E^{\mathcal{F}_s}(X_t) \leq X_s \text{ a.e.}]. \quad s \leq t \tag{7}$$

for a *sub-[super-]martingale* where $\mathcal{F}_s \subset \mathcal{F}_t$ for $s \leq t$ in T. [Occasionally one abbreviates $\{X_t, \mathcal{F}_t, t \in T\}$ to $\{X_t, t \in T\}$, if the meaning is clear.]

Thus a martingale is both a sub- and a supermartingale. In the gambling terminology a submartingale is a favorable game since the gambler's expected fortune in the next game is not less than his present fortune. The supermartingale by contrast is an unfavorable game (to the gambler). Clearly if $\{X_t, \mathcal{F}_t, t \in T\}$ is a supermartingale then $\{-X_t, \mathcal{F}_t, t \in T\}$ is a submartingale and conversely. Thus we need consider only one of these two processes in the development of a mathematical theory. In what follows (Ω, Σ, P) usually stands for the underlying probability space and other symbols, in place of P, will be used when (occasionally) nonfinite measures are also considered.

One must show the existence of such processes as martingales and submartingales. But in Section I.3 we have already seen that a martingale is a special projective system related to a stochastic base. It is thus a simple consequence of the Kolmogorov (Bochner) theorem that these processes exist. In fact if (Ω, Σ, P) is any probability space and $\nu : \Sigma \to \mathbb{R}$ is σ-additive and P-continuous, $\mathcal{F}_n \subset \Sigma, \mathcal{F}_n \uparrow$ are σ-algebras, then $\{X_n, \mathcal{F}_n, n \geq 1\}$ is such a process where $X_n = \frac{d\nu_n}{dP_n}$, $\nu_n = \nu|\mathcal{F}_n, P_n = P|\mathcal{F}_n$. The submartingale case then follows from this by the decomposition in Theorem 2.1 below and an enlargement of (Ω, Σ, P), as noted in Remark 3.6.

(b) *Simple properties.* The following elementary properties are consequences of the definitions, and will be used without comment. All processes are real valued unless the contrary is explicitly stated.

(i) If $\{X_t, \mathcal{F}_t, t \in T\}$ and $\{Y_t, \mathcal{F}_t, t \in T\}$ are two (sub-) martingales with the same index set $T \subseteq \mathbb{R}$ (the σ-algebras $\mathcal{F}_t \uparrow \subset \Sigma$), on (Ω, Σ, P), then $\{aX_t + bY_t, \mathcal{F}_t, t \in T\}$ is also a (sub-) martingale for any real (nonnegative) a, b.

This follows from the linearity and isotone property of $E^{\mathcal{F}_t}$. (Here if the X_t and Y_t processes are real martingales and we set $a = 1, b = i$, then the resulting process is *defined* as a complex martingale.)

(ii) If $\{X_t, \mathcal{F}_t, t \in T\}$ is a (sub-) martingale and $\varphi : \mathbb{R} \to \mathbb{R}$ is a (nondecreasing) convex function such that $E(|\varphi(X_{t_0})|) < \infty$, where E is the expectation and $t_0 \in T$, then $\{\varphi(X_t), \mathcal{F}_t, t \in T, t \leq t_0\}$ is a submartingale.

For, evidently $\varphi(X_t)$ is $\mathcal{F}_t$-adapted, since the hypothesis implies that φ is continuous, and hence we need only check the submartingale property. Now $E^{\mathcal{F}_t}(X_{t_0}) \geq X_t$, a.e., for $t < t_0, t \in T$, with equality in the martingale case. Hence by the conditional Jensen inequality (cf., Theorem 1.8) and monotonicity in the submartingale case,

$$\varphi(X_t) \leq \varphi(E^{\mathcal{F}_t}(X_{t_0})) \leq E^{\mathcal{F}_t}(\varphi(X_{t_0})), \text{ a.e.,} \tag{8}$$

since $\varphi(X_{t_0}) \in L^1(\Sigma)$ by hypothesis. Thus we only need to show that $\varphi(X_t) \in L^1(\Sigma)$ to conclude from (8) that $\{\varphi(X_t), \mathcal{F}_t, t \leq t_0, t \in T\}$ is a submartingale. Since $\varphi(\cdot)$ is convex, it has a support line at each point on the curve, i.e., $ax + b \leq \varphi(x), x \in \mathbb{R}$ for some real a, b. Thus

$$aX_t + b \leq \varphi(X_t) \leq E^{\mathcal{F}_t}(\varphi(X_{t_0})), \text{a.e.} \tag{9}$$

Since the measure space is finite and $X_t \in L^1(\Sigma)$, (9) implies $\varphi(X_t) \in L^1(\Sigma), t \leq t_0, t \in T$. Finally if $s < t < t_0, s \in T, t \in T$, then we can replace the pair (t, t_0) by (s, t) which thus shows that the new process is a submartingale .

Some useful examples of φ are the following:

(a) Let $\varphi(x) = e^{-x}, x \geq 0$, so $\varphi : \mathbb{R}^+ \to \mathbb{R}^+$ is a convex function. If $\{X_t, \mathcal{F}_t, t \in T\}$ is a positive martingale then $\{e^{-X_t}, \mathcal{F}_t, t \in T\}$ is a uniformly bounded positive submartingale.

(b) Let $\varphi(x) = \max(x, 0) = x^+$. If $\{X_t, \mathcal{F}_t, t \in T\}$ is a martingale so that $\{-X_t, \mathcal{F}_t, t \in T\}$ is also a martingale, then $\varphi(X_t) = X_t^+, \varphi(-X_t) = X_t^-$, and thus $\{X_t^\pm, \mathcal{F}_t, t \in T\}$ are submartingales. If $\varphi(x) = |x|^p, p \geq 1$, then $\{|X_t|^p, \mathcal{F}_t, t \in T\}$ is a submartingale whenever $E(|X_t|^p) < \infty$ for $t \in T$. This can be slightly generalized as follows: If $\{X_t^{(i)}, \mathcal{F}_t, t \in T\}, i = 1, 2$ are any two submartingales and $Y_t = \max(X_t^{(1)}, X_t^{(2)})$ then $\{Y_t, \mathcal{F}_t, t \in T\}$ is also a submartingale. In fact, let $t < t'$ in T and $A \in \mathcal{F}_t$ so that $A \cap [X_t^{(1)} < X_t^{(2)}]$ belongs to $\mathcal{F}_t$ also, and

$$\begin{aligned}\int_A Y_t \, dP &= \int_{A\cap[X_t^{(1)} < X_t^{(2)}]} X_t^{(2)} \, dP + \int_{A\cap[X_t^{(1)} \geq X_t^{(2)}]} X_t^{(1)} \, dP \\ &\leq \int_{A\cap[X_t^{(1)} < X_t^{(2)}]} X_{t'}^{(2)} \, dP + \int_{A\cap[X_t^{(1)} \geq X_t^{(2)}]} X_{t'}^{(1)} \, dP \leq \int_A Y_{t'} \, dP,\end{aligned}$$

where we used the submartingale property of the $X_t^{(i)}$-process in the second line above for $i = 1, 2$.
(c) If $\varphi(\lambda) = \lambda \log^+ \lambda, (\log^+ \lambda = 0$ for $\lambda \leq 1)$ and $\{X_t, t \in T\}$ is a submartingale, then so is $\{\varphi(X_t), \mathcal{F}_t, t \in T\}$ when $E(\varphi(X_t))$ exists, $t \in T$.

(iii) If $\{X_t, \mathcal{F}_t, t \in T\}$ is a submartingale, $T \subseteq \mathbb{R}$, then the function $f : T \to \mathbb{R}$ defined by $f(t) = E(X_t)$, is nondecreasing and is a constant iff the process is a martingale.

For, if $t_1 < t_2$ in T, since $X_{t_1} \leq E^{\mathcal{F}_{t_1}}(X_{t_2})$ a.e., by hypothesis, we have

$$f(t_1) = E(X_{t_1}) \leq E(E^{\mathcal{F}_{t_1}}(X_{t_2})) = E(X_{t_2}) = f(t_2), \tag{10}$$

and there is equality for every pair (t_1, t_2) in (10) iff $X_{t_1} = E^{\mathcal{F}_{t_1}}(X_{t_2})$ a.e., so that the process is a martingale. Thus in this case $f(t) = \alpha \in \mathbb{R}$ and f is a constant.

(iv) If $\{X_t, \mathcal{F}_t, t \in T\}$ is as in (iii), and $t \in [t_0, t_1] \cap T$ where t_0, t_1 are in T, then

$$E(|X_t|) \leq -E(X_{t_0}) + 2E(|X_{t_1}|).$$

For, since $|X_t| = 2X_t^+ - X_t$ and $E(X_t) \geq E(X_{t_0})$ by (10), we have

$$\begin{aligned} E(|X_t|) &\leq 2E(X_t^+) - E(X_{t_0}) = 2E(X_t \chi_{[X_t \geq 0]}) - E(X_{t_0}) \\ &\leq 2E(E^{\mathcal{F}_t}(X_{t_1}) \cdot \chi_{[X_t \geq 0]}) - E(X_{t_0}), \text{ since } X_t \text{ is a} \\ &\qquad \text{submartingale and } t \leq t_1, \\ &\leq 2E(E^{\mathcal{F}_t}(|X_{t_1}|) - E(X_{t_0}) = 2E(|X_{t_1}|) - E(X_{t_0}). \end{aligned} \tag{11}$$

(v) Let $\{X_t, \mathcal{B}_t, a \leq t \leq b\}$ be a martingale, $\mathcal{B}_t = \sigma(X_s, s \leq t)$, and let $\{Y_s, \tilde{\mathcal{B}}_s, a \leq s \leq b\}$ also be a martingale where $Y_s = X_{b-s+a}$ and $\tilde{\mathcal{B}}_t = \sigma(Y_s, s \leq t)$. This means the X_t process and its "reversed" process are both martingales. Then the process is trivial in that $X_a = X_t = X_b$ a.e., for $a \leq t \leq b$.

For, the hypothesis implies that for any $a \leq t_1 < t_2 \leq b$ we have $E(X_{t_2}|X_{t_1}) = X_{t_1}$ a.e., and $E(X_{t_1}|X_{t_2}) = X_{t_2}$ a.e. Hence by Theorem 1.14, $X_{t_1} = X_{t_2}$ a.e. Since t_1, t_2 are arbitrary, the result follows.

(vi) Let $X \in L^1(\Omega, \Sigma, P)$. Then the family $\{E^{\mathcal{B}}(X) : \mathcal{B} \subset \Sigma, \mathcal{B}$ a σ-algebra$\}$ is uniformly integrable.

In fact, if we write the σ-algebras as $\mathcal{B}_\alpha, \alpha \in I$ where I is an index set, and if $Y_\alpha = E^{\mathcal{B}_\alpha}(X)$, then Y_α is $\mathcal{B}_\alpha$-measurable and $\{Y_\alpha, \alpha \in I\} \subset L^1(\Omega, \Sigma, P)$. By definition of uniform integrability it must be shown that

$$\lim_{\lambda \to \infty} \int_{[|Y_\alpha| > \lambda]} |Y_\alpha| dP = 0, \text{ uniformly in } \alpha \in I. \tag{12}$$

[This condition is usually stated somewhat differently. All such forms are *equivalent* to (12) here.] Now by a simple application of the conditional Jensen inequality, we deduce that $|Y_\alpha| = |E^{\mathcal{B}_\alpha}(X)| \leq E^{\mathcal{B}_\alpha}(|X|)$ a.e., and since $[|Y_\alpha| > \lambda] \in \mathcal{B}_\alpha$,

$$\int_{[|Y_\alpha| > \lambda]} |Y_\alpha| dP \leq \int_{[|Y_\alpha| > \lambda]} E^{\mathcal{B}_\alpha}(|X|) dP = \int_{[|Y_\alpha| > \lambda]} |X| dP. \tag{13}$$

But $P[|Y_\alpha| > \lambda] \leq \frac{1}{\lambda} E(|Y_\alpha|) \leq \frac{1}{\lambda} E(E^{\mathcal{B}_\alpha}(|X|)) = \frac{1}{\lambda} E(|X|) \to 0$ as $\lambda \to \infty$, uniformly in α. Hence as $\lambda \to \infty$, the right side of (13) tends to zero uniformly in $\alpha \in I$ which implies (12). We remark that (12) is also a consequence of the Hölder inequality for Orlicz spaces and Corollary 1.15.

As a consequence of the above property, we deduce that a martingale or a positive submartingale of the form $\{X_t, \mathcal{F}_t, a \leq t \leq b\}$ is uniformly integrable. Indeed, for any $a \leq t \leq b$, we have $0 \leq |X_t| \leq E^{\mathcal{F}_t}(|X_b|)$ in either case by (ii). Now

$$\{E^{\mathcal{F}_t}(|X_b|) : a \leq t \leq b\} \subset \{E^{\mathcal{F}}(|X_b|) : \mathcal{F} \subset \Sigma, \mathcal{F} \text{ a } \sigma\text{-algebra}. \tag{14}$$

But the right side (larger) collection is uniformly integrable, by the above property. This implies the assertion.

In the preceding result, the positivity hypothesis on the submartigale may appear undesirable. However, some additional hypothesis is needed for the truth of the statement. The following result, due to Doob, adds to the above property if we consider *decreasing* indexed submartigales on a probability space.

(vii) Let $\{X_t, \mathcal{F}_t, t \in \tilde{T}\}$ be a decreasing indexed submartingale, i.e., if $t = -\tau$ and $Y_\tau = X_t$ then $\{Y_\tau, \mathcal{B}_\tau, \tau \in (-\tilde{T})\}$ is a submartingale where $\mathcal{B}_\tau = \sigma(Y_t, t \leq \tau)$. Then it is terminally uniformly integrable iff $\inf_t E(X_t) > -\infty$. The last term means: For any $\epsilon > 0$, there exists a $t_\epsilon \in \tilde{T}$ and a $\delta_\epsilon > 0$ such that $\int_{[|X_t|>\delta_\epsilon]} |X_t| dP < \epsilon$ uniformly in $t \leq t_\epsilon, t \in \tilde{T}$. This is ordinary uniform integrability of $\tilde{T}$ is countable and has a least element.

Proof. Let $f(t) = E(X_t)$. So $f : \tilde{T} \to \mathbb{R}$ is monotone and by hypothesis $\alpha_0 = \inf_t f(t) > -\infty$. Let $A_t^\lambda = \{\omega : |X_t(\omega)| > \lambda\}$. Then $P(A_t^\lambda) \to 0$ as $\lambda \uparrow \infty$ uniformly in $t \leq t_1, t, t_1$ in $\tilde{T}$. Indeed, if $t_0 \leq t \leq t_1$, with (11) (and t_1 is chosen later, t_0, t, t_1 in $\tilde{T}$) : $P(A_t^\lambda) = \int_{[|X_t|>\lambda]} dP \leq \frac{1}{\lambda} E(|X_t|) \leq \frac{1}{\lambda}[-E(X_{t_0}) + 2E(X_{t_1})] \leq [-\alpha_0 + 2E(|X_{t_1}|)]/\lambda \to 0$ as $0 < \lambda \uparrow \infty$, uniformly in $t \leq t_1$ for fixed $t_1 \in \tilde{T}$. Consider $B_t^\lambda = [X_t > \lambda]$ and $C_t^\lambda = [X_t < -\lambda], \lambda > 0$, so that $A_t^\lambda = B_t^\lambda \cup C_t^\lambda$, a disjoint union, all sets being in $\mathcal{F}_t$. Hence

$$\begin{aligned}\alpha_0 \leq E(X_t) &= \int_{B_t^\lambda} X_t dP + \int_{C_t^\lambda} X_t dP + \int_{[|X_t|\leq\lambda]} X_t dP \\ &= -\int_{A_t^\lambda} |X_t| dP + \int_{B_t^\lambda} X_t dP + \int_{[X_t \geq -\lambda]} X_t dP.\end{aligned}$$

Thus for any $t \leq t_1$, using the hypothesis $X_t \leq E^{\mathcal{F}_t}(X_{t_1})$ a.e., we have:

$$\int_{A_t^\lambda} |X_t| dP \leq \int_{B_t^\lambda} X_{t_1} dP + \int_{\Omega - C_t^\lambda} X_{t_1} dP - \alpha_0. \tag{15}$$

Now given $\epsilon > 0$, since $f(t) \downarrow \alpha_0$, we choose $t_1 \in \tilde{T}$ such that $f(t_1) - \alpha_0 < \epsilon/2$. Then select large enough $\lambda_1 (= \delta_\epsilon)$ such that $\lambda > \lambda_1, t \leq t_1 \Rightarrow \int_{A_t^\lambda} |X_{t_1}| dP < \epsilon/2$. This is possible since $P(A_t^\lambda) \to 0$ uniformly in $t \leq t_1$ as $\lambda \to \infty$. Hence with (15) we deduce that $\int_{A_t^\lambda} |X_t| dP < \epsilon$, for $\lambda > \lambda_1$ and $t \leq t_1$. The converse implication is immediate.

(c) *Inequalities.* Let us present some fundamental inequalities for martingales which generalize the classical Kolmogorov inequality. These results are due to Doob, Lévy, Hájek and Rényi, and are used in both theory and applications.

3. Theorem. *Let $\{X_k, \mathcal{F}_k, 1 \leq k \leq n\}$ be a submartingale. Then for each $\lambda \in \mathbb{R}$,*

$$\lambda P\left\{\omega : \max_{1\leq k\leq n} X_k(\omega) \geq \lambda\right\} \leq \int_{\left[\max_{1\leq k\leq n} X_k \geq \lambda\right]} X_n dP \leq E(|X_n|), \tag{16}$$

and

$$\lambda P\left\{\omega : \min_{1\le k\le n} X_k(\omega) \le \lambda\right\} \ge \int\limits_{\left[\min\limits_{1\le k\le n} X_k \le \lambda\right]} X_n \, dP - E(X_n - X_1)$$
$$\ge E(X_1) - E(|X_n|), \qquad (17)$$

Proof. Set $M = \left\{\omega : \max\limits_{1\le k\le n} X_k(\omega) \ge \lambda\right\}$. We first express the event M as a finite disjoint union of sets, each from $\mathcal{F}_k, 1 \le k \le n$ and then apply the submartingale hypothesis. The reader will find that this type of decomposition occurs frequently in calculating such probabilities. Thus let $M_1 = \{\omega : X_1(\omega) \ge \lambda\}$ and for $k > 1$, define M_k as:

$$M_k = \{\omega : X_k(\omega) \ge \lambda, X_j(\omega) < \lambda, 1 \le j \le k-1\},$$

so that M_k is the set such that X_k is the first random variable in this sequence which exceeds λ. Clearly $M_k \in \mathcal{F}_k$ and $M = \bigcup\limits_{k=1}^{n} M_k$, a disjoint union. Hence

$$\int\limits_M X_n \, dP = \sum_{k=1}^{n} \int\limits_{M_k} X_n \, dP$$
$$\ge \sum_{k=1}^{n} \int\limits_{M_k} X_k \, dP, \quad \text{since } \{X_k\}_1^n \text{ is a submartingale,}$$
$$\ge \lambda \sum_{k=1}^{n} P(M_k) = \lambda P\left(\bigcup_{k=1}^{n} M_k\right) = \lambda P(M).$$

This proves (16).

A similar decomposition will be used to prove (17). Let $N = \left\{\omega : \min\limits_{1\le k\le n} X_k(\omega) \le \lambda\right\}$. As before, let $N_1 = \{\omega : X_1(\omega) \le \lambda\}$ and N_k is the set on which X_k is the first random variable which does not exceed λ, so that for $k > 1$

$$N_k = \{\omega : X_k(\omega) \le \lambda, X_j(\omega) > \lambda, 1 \le j \le k-1\}.$$

Clearly $N = \bigcup_{k=1}^{n} N_k, N_k \in \mathcal{F}_k$ disjoint, and $N_k \subset \left(\bigcup_{i=1}^{k-1} N_i\right)^c$. Thus

$$E(X_1) = \int_{N_1} X_1 dP + \int_{N_1^c} X_1 dP \leq \lambda \int_{N_1} dP + \int_{N_1^c} X_2 dP,$$

since $N_1^c \in \mathcal{F}_1$ and $\{X_1, X_2\}$ is a submartingale,

$$= \lambda P(N_1) + \int_{N_2} X_2 dP + \int_{N_1^c \cap N_2^c} X_2 dP$$

$$\leq \lambda[P(N_1) + P(N_2)] + \int_{N_1^c \cap N_2^c} X_3 dP, \text{ since } N_1^c \cap N_2^c \in \mathcal{F}_2,$$

$$\cdots\cdots$$

$$\leq \lambda \sum_{i=1}^{n} P(N_i) + \int_{\bigcap_{i=1}^{n} N_i^c} X_n dP$$

$$= \lambda P(N) + \int_{\Omega} X_n dP - \int_{N} X_n dP.$$

Hence

$$\lambda P(N) \geq \int_{N} X_n dP - \int_{\Omega} (X_n - X_1) dP.$$

This establishes (17). □

We now present some important aspects of the above inequalities, of which these are extensions. They are useful for several applications. The following one is the classical *Kolmogorov inequality.*

4. Theorem. *Let $X_1, \ldots, X_n$ be independent random variables such that $\alpha_i = E(X_i)$ and $\sigma_i^2 = \mathrm{Var} X_i = E(X_i - \alpha_i)^2$ exist. Then for any $\epsilon > 0$,*

$$P\left[\max_{1 \leq k \leq n} \left|\sum_{i=1}^{k} (X_i - \alpha_i)\right| \geq \epsilon\right] \leq \frac{1}{\epsilon^2} \sum_{i=1}^{n} \sigma_i^2. \tag{18}$$

Proof. Let $Y_i = X_i - \alpha_i, S_k = \sum_{i=1}^{k} Y_i$ and $\mathcal{F}_k = \sigma\{X_i, 1 \leq i \leq k\}$. First we note that $\{S_k, \mathcal{F}_k, 1 \leq k \leq n\}$ is a martingale. In fact,

$$E^{\mathcal{F}_k}(S_{k+1}) = E^{\mathcal{F}_k}(S_k + Y_{k+1}) = S_k + E^{\mathcal{F}_k}(Y_{k+1}) = S_k, \text{ a.e.,}$$

since Y_{k+1} is independent of $\mathcal{F}_k, E^{\mathcal{F}_k}(Y_{k+1}) = E(Y_{k+1}) = 0$, and S_k is $\mathcal{F}_k$-adapted. Hence $\{S_k^2, \mathcal{F}_k, 1 \le k \le n\}$ is a submartingale by Property (ii) since $E(S_k^2) = \sum_{i=1}^{k} \sigma_i^2 < \infty$. Thus by (16)

$$\epsilon^2 P\left[\max_{1 \le k \le n} S_k^2 \ge \epsilon^2\right] \le E(S_n^2) = \sum_{i=1}^{n} \sigma_i^2,$$

which is (18). □

If $n = 1$, then (18) becomes:

$$P[|X_1 - \alpha_1| \ge \epsilon] \le \frac{\sigma_1^2}{\epsilon^2}, \tag{19}$$

and this is the classical *Čebyšev inequality*. [A simple direct proof of (19) also follows from definition of the integral. But the result shows the extent of the generalization involved in (18) and hence (16).]

The next two results are due to Rényi and Hájek, and they complement the above inequalities in some cases.

5. Proposition. *Let $a_n \ge a_{n+1} \searrow 0$ be a sequence and $\{X_n, \mathcal{F}_n, n \ge 1\}$ be a positive submartingale. If $\sum_{n=1}^{\infty} (a_n - a_{n+1})E(X_n) < \infty$, then for any $\lambda \ge 0$ we have*

$$\lambda P\left[\sup_{n \ge 1} a_n X_n \ge \lambda\right] \le \sum_{n=1}^{\infty} (a_n - a_{n+1})E(X_n). \tag{20}$$

Proof. The method of proof is similar to the above. Thus let $M = \left\{\omega : \sup_{n \ge 1} a_n X_n(\omega) \ge \lambda\right\}$ and $M_1 = \{\omega : a_1 X_1(\omega) \ge \lambda\}$. For $k > 1$ define:

$$M_k = \{\omega : a_i X_i(\omega) < \lambda, 1 \le i \le i-1, a_k X_k(\omega) \ge \lambda\}.$$

Then $M_k \in \mathcal{F}_k, k \ge 1$ and $M = \bigcup_{k \ge 1} M_k$, a disjoint union. If $S = \sum_{n=1}^{\infty} (a_n - a_{n+1})X_n$ then by hypothesis $0 \le E(S) < \infty$ so that $S < \infty$

a.e. As before consider

$$E(S) = \sum_{n=1}^{\infty}(a_n - a_{n+1}) \int_{\Omega} X_n dP \geq \sum_{n=1}^{\infty}(a_n - a_{n+1}) \int_{M} X_n dP,$$

since $X_n \geq 0$ a.e.,

$$= \sum_{n=1}^{\infty}(a_n - a_{n+1}) \sum_{k=1}^{\infty} \int_{M_k} X_n dP \geq \sum_{k=1}^{\infty} \sum_{n \geq k}(a_n - a_{n+1}) \int_{M_k} X_k dP,$$

by the submartingale property,

$$= \sum_{k=1}^{\infty} a_k \int_{M_k} X_k dP, \quad \text{since } a_k \geq a_{k+1} \text{ and } a_k \searrow 0,$$

$$\geq \lambda \sum_{n=1}^{\infty} \int_{M_n} dP = \lambda P\left(\bigcup_{n=1}^{\infty} M_n\right) = \lambda P(M).$$

This proves (20) . □

If $a_k = 1$ for $1 \leq k \leq n$, and $a_k = 0$ for $k > n$ then (20) reduces to (16) in this case. Also if $\{S_k, \mathcal{F}_k, 1 \leq k \leq n\}$ is considered as in Theorem 4, so that $\{S_n^2, \mathcal{F}_k, k \geq 1\}$ is a positive submartingale, the above choice of a_k's implies that (18) follows from (20). It should be noted that theorem 3 is *not* a special case of the above result since the submartingale there need not be positive.

Let $X_1, X_2, \ldots$ be a sequence of independent random variables such that $E(X_n) = 0$,and $E(X_n^2) = \sigma_n^2$. If $S_n = \sum_{k=1}^{n} X_k$, and $n_0 \geq 1$ is an integer let $Y_k = S_{n_0+k-1}^2$. Consider the (positive) submartingale $\{Y_k, \mathcal{F}_{n_0+k-1}, k \geq 1\}$ which is obtained by omitting the first $(n_0 - 1)$ terms in $\{S_n^2, \mathcal{F}_n, n \geq 1\}$. Suppose that, in the above proposition $a_k = (n_0 + k - 1)^{-2}$. It is then clear that

$$\begin{aligned} E\left(\sum_{k=1}^{\infty}(a_k - a_{k+1})Y_k\right) &= \sum_{k=1}^{\infty}(a_k - a_{k+1}) \sum_{i=1}^{n_0+k-1} \sigma_i^2 \\ &= a_1 \sum_{i=1}^{n_0} \sigma_i^2 + \sum_{i=2}^{\infty} a_i \sigma_{n_0+i-1}^2 \end{aligned} \tag{21}$$

With this specialization of the proposition one has the result:

6. Corollary. *Let $\{X_n, n \geq 1\}$ be a sequence of independent random variables such that $E(X_n) = 0, E(X_n^2) = \sigma_n^2$ and $\sum_{n=1}^{\infty} \frac{\sigma_n^2}{n^2} < \infty$. Then for any $\lambda > 0, S_n = \sum_{i=1}^{n} X_i$, and for $n_0 \geq 1$, we have*

$$P\left[\sup_{n \geq n_0} \left|\frac{S_n}{n}\right| \geq \lambda\right] \leq \frac{1}{\lambda^2}\left(\frac{1}{n_0^2}\sum_{i=1}^{n_0}\sigma_i^2 + \sum_{n \geq n_0+1}\frac{\sigma_n^2}{n^2}\right). \tag{22}$$

Hence one has a form of the Kolmogorov strong law of large numbers for such a sequence:

$$P\left[\lim_{n\to\infty}\frac{S_n}{n} = 0\right] = 1. \tag{23}$$

Proof. We only need to establish (23). Now the sets in (22) are decreasing as $n_0 \to \infty$. Consequently

$$\begin{aligned}\lim_{n_0\to\infty} P\left[\sup_{n\geq n_0}\left|\frac{S_n}{n}\right| \geq \lambda\right] &= P\left[\limsup_n \left|\frac{S_n}{n}\right| \geq \lambda\right] \\ &\leq \frac{1}{\lambda^2}\lim_{n_0\to\infty}\left(\frac{1}{n_0^2}\sum_{n=1}^{k}\sigma_n^2 + \sum_{n=k+1}^{\infty}\frac{\sigma_n^2}{n^2}\right)\end{aligned} \tag{24}$$

since clearly $n_0 \geq k \geq 1, \frac{1}{n_0^2}\sum_{i=1}^{n_0}\sigma_i^2 \leq \frac{1}{n_0^2}\sum_{i=1}^{k}\sigma_i^2 + \sum_{n=k+1}^{n_0}\frac{\sigma_n^2}{n^2}$. It follows that the right side of (24) tends to zero by first letting $n_0 \to \infty$, and then $k \to \infty$. But this implies (23), as asserted. □

The following inequality extends (16) when the random variables are nonnegative and it is useful in applications as well as in convergence theory.

7. Proposition. *Let $f, g \geq 0$ be two random variables on (Ω, Σ, P) such that for all $\lambda \geq 0$ we have:*

$$\lambda P\{\omega : f(\omega) \geq \lambda\} \leq \int_{\{\omega: f(\omega)\geq\lambda\}} g(\omega)dP(\omega). \tag{25}$$

Then for any nondecreasing function $\varphi : \mathbb{R}^+ \to \mathbb{R}^+$ with $\varphi(0) = 0$, the following inequality is true:

$$\int_\Omega \varphi(f(\omega))dP \leq \int_\Omega g(\omega)dP \int_0^{f(\omega)} \frac{d\varphi(t)}{t}. \tag{26}$$

Proof. Note that (25) is satisfied for positive submartingales by (16), if we set $f = \max\{X_k : 1 \le k \le n\}$ and $g = X_n$ there. This is in fact an important motivation. We now establish (26) in its generality.

Let $Q = P \circ f^{-1} : \mathcal{B} \to \mathbb{R}^+$ be the image measure by $f : \Omega \to \mathbb{R}^+$ of P, where $\mathcal{B}$ is the Borel algebra of $\mathbb{R}^+$. Then using the elementary relation between P and Q (see Equation (I.1.4)) we have

$$\int_\Omega \varphi(f(\omega))dP = \int_{\mathbb{R}^+} \varphi(t)dQ(t) = -\int_{\mathbb{R}} \varphi(t)dQ([t,\infty)),$$

which is true if φ is a simple function and then the general case follows by a standard approximation,

$$= \int_{\mathbb{R}^+} Q([t,\infty))d\varphi(t), \text{ since } \varphi(0) = 0 = Q(\emptyset),$$

$$\le \int_{\mathbb{R}^+} P(f^{-1}([t,\infty)))d\varphi(t)$$

$$\le \int_{\mathbb{R}^+} \frac{d\varphi(t)}{t} \int_{[f \ge t]} g(\omega)dP, \text{ by (25)},$$

$$= \int_\Omega g(\omega)dP \int_0^{f(\omega)} \frac{1}{t}d\varphi(t). \tag{27}$$

This implies the proposition. □

It should be noted that the proof did not use the finiteness of the measure P. This observation will be used in Theorem 9 below.

The result leads to certain powerful maximal inequalities. The first one has been obtained by Doob and the second ones is a (specialized) version of a result of Marcinkiewicz.

8. Theorem. *Let $\{X_k, \mathcal{F}_k, 1 \le k \le n\}$ be a positive submartingale. Then for $p \ge 1$, we have the maximal inequalities:*

$$\int_\Omega \left(\max_{1\le k\le n} X_k^p\right) dP \le \begin{cases} q^p \int_\Omega X_n^p dP, & \text{if } p > 1, q = p/p-1, \\ \frac{e}{e-1}[P(\Omega) + \int_\Omega X_n \log^+ X_n dP], & \text{if } p = 1. \end{cases} \tag{28}$$

Proof. If $Y = \max_{1 \le k \le n} X_k$, then the submartingale hypothesis implies by (16) that (25) is true, for Y and X_n as noted above. Taking $\varphi(x) = |x|^p$, we have by (26),

$$\begin{aligned} \|Y\|_p^p = \int_\Omega Y^p dP &\le \int_\Omega X_n(\omega) \frac{p}{p-1} (Y(\omega))^{p-1} \, dP, \\ &\le q \|X_n\|_p \|Y^{p-1}\|_q, \text{ by Hölder's inequality}, \\ &= q \|X_n\|_p (\|Y\|_p)^{p/q}. \end{aligned} \tag{29}$$

If $\|Y\|_p = 0$, then (28) is true and trivial, and $\|Y\|_p < \infty$ since L^p is a lattice, $n < \infty$. So let $0 < \|Y\|_p < \infty$. Then we may divide both sides of (29) by $(\|Y\|_p)^{p/q}$ and the result is (28). If $p = 1$, then let $\varphi(t) = t, |t| \ge 1; = 0$ for $0 \le |t| < 1$. Then (26) yields

$$\int_\Omega (Y(\omega) - 1) dP \le \int_\Omega \varphi(Y(\omega)) dP \le \int_{[Y \ge 1]} X_n(\omega) \log^+ Y(\omega) dP.$$

But for any $a \ge 0, b > 0$ we have the elementary inequality

$$a \log b = a \log^+ a + a \log \frac{b}{a} \le a \log^+ a + \frac{b}{e},$$

since $a \log \frac{b}{a}$ has its maximum for $a = \frac{b}{e}$ for each fixed $b > 0$. Hence

$$\int_\Omega Y \, dP - P(\Omega) \le \int_\Omega X_n \log^+ X_n dP + \frac{1}{e} \int_\Omega Y \, dP.$$

This is precisely (28) for $p = 1$, since Y is integrable. □

In the above result we considered the mapping between a submartingale and its maximum functional, i.e., $X_n \mapsto \max_{i \le n}(X_i) = Y = T(X_n)$. Note that $0 \le X_i \le E^{\mathcal{F}_i}(X_n)$ a.e., implies X_n determines Y uniquely. In fact, if we introduce a norm for the positive submartingale as $\|\{X_i\}_1^n\| = \|X_n\|_p$ and $\|Y\|_p$ is the usual L^p-norm of Y then $\|X_n\|_p = 0$ iff $X_i = 0$ a.e., $1 \le i \le n$ so that $Y = 0$ a.e. Thus $T : X_n \mapsto Y = T(X_n)$ is sublinear (see below) on the cone of positive submartingales $\{X_i\}_1^n$ to the cone of positive elements Y of the L^p-space. We abstract this idea to obtain with (26) a version of a classical result known as the Marcinkiewicz interpolation theorem (cf., Zygmund [1], XII.4.6):

9. Theorem. *Let $T : L^p(\Omega, \Sigma, \mu) \to L^p(\Omega, \Sigma, \mu)$ be a sublinear mapping, i.e., $|T(f + ag)| \leq |Tf| + |aTg|$ a.e., for f, g in $L^p(\Sigma)$. Suppose T satisfies the following two inequalities:*

(a) $\mu\{\omega : |Tf(\omega)| > \lambda\} \leq \frac{C}{\lambda}\|f\|_1$, *for $f \in L^1(\mu), \lambda > 0$, and a constant $C < \infty$.*

(b) $\|Tf\|_\infty \leq C\|f\|_\infty$, *for the same constant $C, f \in L^\infty(\Sigma)$.*

Then for any $1 < p < \infty$ and $f \in L^p(\Sigma), Tf \in L^p(\Sigma)$. Moreover,

$$\|Tf\|_p \leq C_p\|f\|_p, \qquad C_p = 2C \cdot \left[\frac{1}{p-1}\right]^{1/p}. \tag{30}$$

Proof. Taking $\varphi(x) = |x|^p, p > 1$, in Proposition 7, so that (27) becomes (with Tf for f there),

$$\int_\Omega |Tf|^p d\mu = p\int_0^\infty \mu[(Tf)| \geq t]t^{p-1}dt = p(2C)^p \int_0^\infty \mu[|(Tf)| > 2Ct]t^{p-1}dt. \tag{31}$$

To simplify this using (a) and (b), consider $f = g_t + h_t$, where for $t > 0$, $g_t = \min(f, t)$ and $h_t = f - g_t = (f - t)^+$. Thus $\|g_t\|_\infty \leq t$ and so by (b) $|Tg_t| \leq tC$, a.e. Hence by the sublinearity of T we have

$$\begin{aligned}
\mu[|Tf| > 2Ct] &\leq \mu[|Tg_t| > Ct] + \mu[|Th_t| > Ct] \\
&= \mu[|Th_t| > Ct] \leq \frac{C}{Ct}\|h_t\|_1, \text{ by (a) and (b)}, \\
&= \frac{1}{t}\int_\Omega |h_t|d\mu = \frac{1}{t}\int_0^\infty \mu\{\omega : h_t(\omega) > \lambda\}d\lambda, \text{ since } h_t \geq 0, \\
&= \frac{1}{t}\int_0^\infty \mu[f > \lambda + t]d\lambda.
\end{aligned} \tag{32}$$

Substituting (32) in (31) we get:

$$\int_\Omega |Tf|^p d\mu \le p(2C)^p \int_0^\infty t^{p-2} \left(\int_t^\infty \mu\{\omega : |f(\omega)| > \lambda\} d\lambda \right) dt$$

$$= p(2C)^p \int_0^\infty \int_0^\lambda t^{p-2} dt \mu[|f| > \lambda] d\lambda$$

$$= \frac{p}{p-1}(2C)^p \int_0^\infty \lambda^{p-1} \mu[|f| > \lambda] d\lambda$$

$$= \frac{1}{p-1}(2C)^p \int_\Omega |f|^p d\mu, \text{ by the first part of (31).}$$

This is (30), and hence the result follows. □

10. Remark. We note that, using the known and easily provable result stating $\|f\|_p \to \|f\|_\infty$ as $p \to \infty, f \in L^1(\mu)$, from (30) we get $\|Tf\|_\infty \le 2C\|f\|_\infty$ on letting $p \to \infty$. This is a bigger bound than (b) of the theorem. Conditions (a) and (b) on T are referred to as T being of *weak type* (1, 1) and of *strong type* (∞, ∞) and the conclusion (30) then is that T is of *strong type* (p, p) for each $1 < p < \infty$. Since, for $p = 1$, this inequality says nothing, the second half of (28) gives additional information while the first half is a consequence of (30). It can be shown that, for $p = 1$ without further conditions, the inequality (30) is actually false. The hypothesis on T being very weak we could not improve the bound in (30). It is of interest to note that if, in place of (a), we have the stronger inequality $\|Tf\|_1 \le C\|f\|_1$, so that $\mu[|Tf| > \lambda] \le \frac{1}{\lambda} \int_\Omega |Tf| d\mu \le \frac{C}{\lambda}\|f\|_1$ is implied, then we can conclude from (30) that $\|Tf\|_p \le 2C_p\|f\|_p$ for $1 \le p \le \infty$. Thus if T is a linear operator defined on $L^1 \to L^1$ and $L^\infty \to L^\infty$ satisfying the strengthened (a) and (b) (or even of weak type (1, 1) and strong type (∞, ∞)) then it is defined $(p > 1)$ and continuous on all L^p-spaces. Here we did not use the finiteness of the measure μ anywhere. This extension of the result is significant.

After introducing the concept and giving the properties of stopping time transformations, we shall see later (Chapter IV) that several new inequalities which complement the above on martingales can be proven and then new results can be established. We turn here to some decomposition results.

2.5 Decompositions of discrete parameter martingales

There are three basic decompositions which play a key role in martingale theory. Corresponding to the well-known and elementary decomposition of a real valued function into its positive and negative parts, a real martingale can be expressed as a difference of two positive martingales. There is an intimate relation between sub- (and super-) martingales and sub- (and super-) harmonic functions; and the classical F. Riesz decomposition (1930) of a superharmonic function into a harmonic function and a potential, translates into a useful result for supermartingales. Finally, another decomposition of a sub- (or super-) martingale into a martingale and a monotone function which is analogous to, but different from, the Riesz decomposition, is due to Doob (1953) for the discrete paramter case, to Meyer (1962) and Itô-Watanabe (1965) in the continuous parameter case. The discrete parameter results will be given here. But one needs sharper tools, involving the stopping time transformation theory, for the continuous parameter case and it will be postponed to Chapter V where other related results are detailed and used in stochastic integration.

We start with the simpler result in the discrete parameter case, and it will be called the *Doob decomposition*.

1. Theorem. *A sequence $\{X_n, \mathcal{F}_n, n \geq 1\} \subset L^1(\Omega, \Sigma, P)$ is a submartingale iff it admits a decomposition:*

$$X_n = X_n' + \sum_{j=1}^{n} A_j, \text{ a.e.}, \qquad n \geq 1 \tag{1}$$

where $\{X_n', \mathcal{F}_n, n \geq 1\}$ is a martingale, $A_j \geq 0$, a.e., and A_j is $\mathcal{F}_{j-1}$-adapted for $j \geq 2$. In fact, A_j is given for $j \geq 2$, by

$$A_j = E^{\mathcal{F}_{j-1}}(X_j) - X_{j-1}, \textit{ with } A_1 = 0, \text{ a.e.} \tag{2}$$

The decomposition (1) is unique outside of a P-null set.

Proof. Let A_j be defined by (2) for the given process. Then the submartingale hypothesis implies (cf., Equation (4.7)) $A_j \geq 0$ a.e., and A_j

is $\mathcal{F}_{j-1}$-adapted. Define X'_n by (1). To see that $\{X'_n, \mathcal{F}_n, n \geq 1\}$ is a martingale, since X'_n is clearly $\mathcal{F}_n$-adapted, consider, for $n \geq 2$,

$$\begin{aligned} E^{\mathcal{F}_{n-1}}(X'_n) &= E^{\mathcal{F}_{n-1}}\left(X_n - \sum_{j=1}^{n} A_j\right) = E^{\mathcal{F}_{n-1}}(X_n) - \sum_{j=1}^{n} A_j, \text{ a.e.,} \\ &= (E^{\mathcal{F}_{n-1}}(X_n) - A_n) - \sum_{j=1}^{n-1} A_j \\ &= X_{n-1} - \sum_{j=1}^{n-1} A_j = X'_{n-1}, \text{ a.e., by (2).} \end{aligned}$$

This shows that the X'_n-process is a martingale.

Conversely suppose that the decomposition (1) holds with $A_j \geq 0$ a.e., $\mathcal{F}_{j-1}$-adapted, and that the X'_n-process is a martingale. To see that the X_n-process is a submartingale, consider again

$$\begin{aligned} E^{\mathcal{F}_{n-1}}(X_n) &= E^{\mathcal{F}_{n-1}}(X'_n) + \sum_{j=1}^{n} A_j, \text{ since the last term is} \\ &\quad \mathcal{F}_{n-1}\text{-measurable,} \\ &= X'_{n-1} + \sum_{j=1}^{n-1} A_j + A_n = X_{n-1} + A_n \geq X_{n-1}, \quad \text{a.e.,} \end{aligned}$$

since $A_n \geq 0$, a.e. Thus the X_n-process is a submartingale.

To prove the uniqueness of the decomposition, let $X_n = Y'_n + \sum_{j=1}^{n} B_j$, be another decomposition with similar properties. Then, together with (1) we have

$$Z_n = Y'_n - X'_n = \sum_{j=1}^{n}(A_j - B_j), \qquad n \geq 1. \tag{3}$$

So $\{Z_n, \mathcal{F}_n, n \geq 1\}$ is a martingale, and since the right side of (3) is $\mathcal{F}_{n-1}$-adapted one sees that Z_n is $\mathcal{F}_{n-1}$-adapted as well as a martingale. Hence $Z_{n-1} = E^{\mathcal{F}_{n-1}}(Z_n) = Z_n$ a.e. Thus $Z_1 = Z_2 = \cdots$, a.e. But $Z_1 = A_1 - B_1 = 0$, a.e., and therefore $Z_n = 0$ a.e., for all $n \geq 1$. Hence by (3), $X'_n = Y'_n$ a.e., and $A_j = B_j, j \geq 1$, since n is arbitrary. This proves uniqueness. □

For supermartingales, the decomposition (1) takes the following form which we state for reference. If $X = \{X_n, \mathcal{F}_n, n \geq 1\} \subset L^1(\Omega, \Sigma, P)$, then it is a supermartingale iff X admits an a.e. unique decomposition:

$$X_n = X_n' - \sum_{i=1}^{n}(X_{i-1} - E^{\mathcal{F}_{i-1}}(X_i)) = X_n' - A_n, \qquad A_1 = 0, \text{a.e.} \quad (4)$$

where $A_j \geq 0$ a.e., $\mathcal{F}_{j-1}$-adapted for $j > 1$, and $\{X_n', \mathcal{F}_n, n \geq 1\}$ is a martingale.

It is obvious that the above argument cannot extend if the index set is an interval of $\mathbb{R}$ since then the sums have to be replaced by "stochastic integrals" and the appropriate concepts should be first introduced. This is done in Chapter V. There we shall present the corresponding solution. Let us turn to the other decompositions.

The following result is the *Jordan decomposition* for martingales. Its application in the pointwise convergence theory proves convenient, since we can restrict ourselves to nonnegative martingales. It should be noted that this decomposition is also valid for general indexes, though we shall consider only the discrete index at this time. The alternative proof of the converse part below is due to Meyer [2] and is an adaption of the argument of the Riesz decomposition theorem below.

2. Theorem. *Let $\{X_n, \mathcal{F}_n, n \geq 1\}$ be a martingale. Then the X_n-process admits a decomposition*

$$X_n = X_n^{(1)} - X_n^{(2)}, \text{ a.e.}, \quad n \geq 1 \quad (5)$$

where $\{X_n^{(j)}, \mathcal{F}_n, n \geq 1\}, j = 1, 2$ are positive martingales, iff $\sup_n E(|X_n|) < \infty$. Moreover, the $X_n^{(j)}$ processes can be chosen such that $\sup_n E(|X_n|) = E(X_n^{(1)}) + E(X_n^{(2)})$, and then the decomposition (5) is unique.

Proof. Suppose at first that (5) holds. Then $0 \leq X_n^{(j)} \in L^1(\Omega, \Sigma, P)$ and $K_j = E(X_n^{(j)}), j = 1, 2$ is independent of n by Property (iii) of Section 4(b). Since $|X_n| \leq X_n^{(1)} + X_n^{(2)}$ a.e., we have $\sup_n E(|X_n|) \leq K_1 + K_2 < \infty$. Thus only the converse is nontrivial.

We next observe that if $\lambda_n(A) = \int_A X_n dP$, then $\{\lambda_n, \mathcal{F}_n, n \geq 1\}$ is a set martingale such that $|\lambda_n|(\Omega) = \int_\Omega |X_n| dP \leq \sup_n E(|X_n|) < \infty$. Therefore $\lambda_n = \xi_n - \eta_n$ is the desired decomposition if we define ξ_n (so

$\eta_n = \xi_n - \lambda_n$) as follows: $0 \le \xi_n = \sup_{m \ge n} \lambda_m^+$ on $\mathcal{F}_n$, so that ξ_n is $P_{\mathcal{F}_n}$-continuous (Vitali-Hahn-Saks theorem) and $X_n^{(1)} = \frac{d\xi_n}{dP_{\mathcal{F}_n}}$. Similarly $X_n^{(2)} = \frac{d\eta_n}{dP_{\mathcal{F}_n}}$. This sketch gives (5). [See, e.g., Rao [11] p. 301 for details.] The argument holds for general indexes also.

To present an alternative argument, first note that $\{X_n^{\pm}, \mathcal{F}_n, n \ge 1\}$ are positive submartingales by Property (ii) (Example (b)) of Section 4(b) and moreover $\sup_n E(X_n^{\pm}) \le \sup_n E(|X_n|) < \infty$. For any fixed but arbitrary n, consider

$$X_{nm} = E^{\mathcal{F}_n}(X_{n+m}^+), \quad Y_{nm} = E^{\mathcal{F}_n}(X_{n+m}^-), \text{ a.e.,} \quad m \ge 1. \tag{6}$$

But

$$X_{nm} - Y_{nm} = E^{\mathcal{F}_n}(X_{n+m}^+ - X_{n+m}^-) = E^{\mathcal{F}_n}(X_{n+m}) = X_n, \text{ a.e.} \tag{7}$$

We claim that $X_{n(m+1)} \ge X_{nm}$ and similarly $\{Y_{nm}, m \ge 1\}$ is an increasing sequence. Indeed,

$$\begin{aligned} X_{n(m+1)} &= E^{\mathcal{F}_n}(X_{n+m+1}^+) = E^{\mathcal{F}_n} E^{\mathcal{F}_{n+m}}(X_{n+m+1}^+), \text{ by Proposition 1.6,} \\ &\ge E^{\mathcal{F}_n}(X_{n+m}^+), \text{ since } \{X_n^+, \mathcal{F}_n, n \ge 1\} \text{ is a submartingale} \\ &\qquad \text{by Property (ii(b)) of Section 4 above,} \\ &= X_{nm}, \text{ by (6).} \end{aligned}$$

Replacing X_n by $-X_n$ we get a similar result about the Y_{nm}-process. So $X_{nm} \uparrow X_n^{(1)}$, a.e., as $m \to \infty$ and similarly $Y_{nm} \uparrow X_n^{(2)}$, a.e. By (6) $X_n^{(j)} \in L^1(\Omega, \Sigma, P)$, because of conditional Monotone Convergence and the fact that $\sup_n E(X_n^{\pm}) < \infty$. Now (7) implies that $X_n = X_n^{(1)} - X_n^{(2)}$ a.e. Let us show that $\{X_n^{(j)}, \mathcal{F}_n, n \ge 1\}$ are martingales, $j = 1, 2$.

Since $X_n^{(j)}$ is $\mathcal{F}_n$-adapted, for the martingale property it suffices to consider one of them say the $X_n^{(1)}$-process. Thus

$$X_{n(m+1)} = E^{\mathcal{F}_n}(X_{n+m+1}^+) = E^{\mathcal{F}_n}(E^{\mathcal{F}_{n+1}}(X_{n+m+1}^+)) = E^{\mathcal{F}_n}(X_{(n+1)m}). \tag{8}$$

Letting $m \to \infty$ in (8) and using the conditional Monotone Convergence criterion we get $X_n^{(1)} = E^{\mathcal{F}_n}(X_{n+1}^{(1)})$, a.e. Thus $X_n = X_n^{(1)} - X_n^{(2)}$ is a decomposition. Also $E(|X_n|) \le E(X_n^{(1)}) + E(X_n^{(2)})$ and the right side is independent of n. Hence $\sup_n E(|X_n|) \le E(X_n^{(1)}) + E(X_n^{(2)})$.

To see that there is equality here, by (6) we have $E(X_{nm} + Y_{nm}) = E(E^{\mathcal{F}_n}(|X_{n+m}|)) \leq \sup_n E(|X_n|)$. Taking limits as $m \to \infty$ and using the Monotone Convergence again on the left we get $E(X_n^{(1)}) + E(X_n^{(2)}) \leq \sup_n E(|X_n|)$.

The uniqueness remains. If $X_n = Y_n^{(1)} - Y_n^{(2)}$ is a second such decomposition, then evidently $X_n^+ \leq Y_n^{(1)}$ a.e., and by (6)

$$X_n^{(1)} = \lim_{m\to\infty} X_{nm} = \lim_m E^{\mathcal{F}_n}(X_{n+m}^+) \leq \lim_m E^{\mathcal{F}_n}(Y_{n+m}^{(1)}) = Y_n^{(1)}, \quad \text{a.e.} \tag{9}$$

Replacing X_n by $(-X_n)$ here we get $X_n^{(2)} \leq Y_n^{(2)}$ and since $E(X_n^{(1)}) + E(X_n^{(2)}) = \sup_n E(|X_n|) = E(Y_n^{(1)}) + E(Y_n^{(2)})$ for any $n \geq 1$, we must have $X_n^{(1)} = Y_n^{(1)}, X_n^{(2)} = Y_n^{(2)}$, a.e. establishing uniqueness. □

Note that in the above proof in showing that $X_n^+ \leq X_n^{(1)}$, a.e., we only used the fact that X_n^+ is a positive submartingale. Thus if $\{X_n, \mathcal{F}_n, n \geq 1\}$ is any submartingale, by Property (ii) of Section 4, $\{X_n^+, \mathcal{F}_n, n \geq 1\}$ is a positive submartingale. Hence the above argument yields the following result. It is also obtainable directly by the first method indicated at the beginning of the proof.

3. Proposition. *Let $\{X_n, \mathcal{F}_n, n \geq 1\}$ be a submartingale such that $\sup_n E(|X_n|) < \infty$. Then there exists a positive martingale $\{Y_n, \mathcal{F}_n, n \geq 1\}$ such that $X_n^+ \leq Y_n$, a.e., $n \geq 1$, and $E(Y_n) = \sup_n E(X_n^+)$.*

We now consider the Riesz decomposition for the discrete indexed supermartingales. It is a translation of the classical result due to F. Riesz (1930) on the decomposition of a superharmonic function. Let us introduce the concept of a potential for this purpose.

4. Definition. A potential is a positive supermartingale $\{X_n, \mathcal{F}_n, n \geq 1\}$ such that $X_n \to 0$ in L^1-norm as $n \to \infty$, i.e., $E(X_n) \to 0$.

The main result on *Riesz's decomposition* is as follows:

5. Theorem. *Let $\{X_n, \mathcal{F}_n, n \geq 1\}$ be a supermartingale. Then the following two statements are equivalent:*

(i) *The X_n-process dominates a submartingale, i.e., there is a submartingale $\{Y_n, \mathcal{F}_n, n \geq 1\}$ such that $Y_n \leq X_n$, a.e. for $n \geq 1$.*

(ii) *there is a martingale* $\{X_n^{(1)}, \mathcal{F}_n, n \geq 1\}$ *and a potential* $\{X_n^{(2)}, \mathcal{F}_n, n \geq 1\}$ *such that we have an a.e. unique decomposition:*

$$X_n = X_n^{(1)} + X_n^{(2)}, \text{ a.e.,} \quad n \geq 1. \tag{10}$$

Proof. That (ii) implies (i) is immediate. In fact, $X_n \geq X_n^{(1)}$, a.e., for all n, since $X_n^{(2)} \geq 0$, a.e. However $\{X_n^{(1)}, \mathcal{F}_n, n \geq 1\}$ is a martingale (hence a submartingale) and the X_n-process dominates the $X^{(1)}$-process. Thus only that (i)⇒(ii) is nontrivial.

Now let (i) hold, so that the supermartingale $\{X_n, \mathcal{F}_n, n \geq 1\}$. dominates some submartingale $\{Y_n, \mathcal{F}_n, n \geq 1\}$. The construction of $X_n^{(1)}$ is that noted in the proof of Theorem 2 which was an abstraction of the present (earlier) result. Thus for each $n \geq 1$, and $k \geq 0$, define

$$X_{nk} = E^{\mathcal{F}_n}(X_{n+k}), \text{ a.e.} \tag{11}$$

Then $X_{nk} \geq X_{n(k+1)}$, a.e. In fact, since $\mathcal{F}_n \subset \mathcal{F}_{n+1}$,

$$\begin{aligned} X_{n(k+1)} &= E^{\mathcal{F}_n}(X_{n+k+1}) = E^{\mathcal{F}_n}(E^{\mathcal{F}_{n+k}}(X_{n+k+1})) \\ &\leq E^{\mathcal{F}_n}(X_{n+k}) = X_{nk}, \text{ a.e.,} \end{aligned} \tag{12}$$

since the X_n-process is a supermartingale. Thus $X_{nk} \downarrow X_n^{(1)}$, a.e., as $k \to \infty$. But $X_{nk} = E^{\mathcal{F}_n}(X_{n+k}) \geq E^{\mathcal{F}_n}(Y_{n+k}) \geq Y_n$, a.e., by the domination hypothesis. Hence $X_n^{(1)} \geq Y_n$, a.e., and $X_n^{(1)} \in L^1(\Omega, \mathcal{F}_n, P)$ because $X_n^{(1)}$ is $\mathcal{F}_n$-adapted.

To prove the validity of the decomposition (10) with this $X_n^{(1)}$, we show that $\{X_n^{(1)}, \mathcal{F}_n, n \geq 1\}$ is a martingale so that $X_n^{(2)} = X_n - X_n^{(1)}$ defines a supermartingale and then one sees that the $X_n^{(2)}$-process is a potential. Thus consider

$$\begin{aligned} E^{\mathcal{F}_n}(X_{n+1}^{(1)}) &= E^{\mathcal{F}_n}\left(\lim_{k\to\infty} X_{(n+1)k}\right) = \lim_{k\to\infty} E^{\mathcal{F}_n}(X_{(n+1)k}), \text{ by} \\ &\qquad \text{Proposition 1.3(c)}, \\ &= \lim_{k\to\infty} E^{\mathcal{F}_n}(E^{\mathcal{F}_{n+1}}(X_{n+1+k})) = \lim_{k\to\infty} E^{\mathcal{F}_n}(X_{n+1+k}) \\ &= \lim_{k\to\infty} X_{n(k+1)} = X_n^{(1)}, \text{ a.e.} \end{aligned} \tag{13}$$

This shows that the $X_n^{(1)}$-process is a martingale and so the $X_n^{(2)}$-process is a supermartingale. To see that the $\{X_n^{(2)}\}_1^\infty$ is a potential, note that

$$X_n^{(2)} = X_n - \lim_{k\to\infty} X_{nk} = X_n - \lim_{k\to\infty} E^{\mathcal{F}_n}(X_{n+k}) \geq X_n - X_n = 0, \text{ a.e.,}$$

by the supermartingale inequality for the X_n-process. Moreover,

$$E^{\mathcal{F}_n}(X^{(2)}_{n+k}) = E^{\mathcal{F}_n}(X_{n+k} - X^{(1)}_{n+k}) = X_{nk} - X^{(1)}_n, \text{ a.e.,} \tag{14}$$

by (11) and (13). It follows that $\lim_{k\to\infty} E^{\mathcal{F}_n}(X^{(2)}_{n+k}) = X^{(1)}_n - X^{(1)}_n = 0$, a.e., and

$$\begin{aligned}
\lim_{k\to\infty} E(X^{(2)}_{n+k}) &= \lim_{k\to\infty} E(E^{\mathcal{F}_n}(X^{(2)}_{n+k})) \\
&= \lim_{k\to\infty} E(X_{nk} - X^{(1)}_n), \text{ by (14),} \\
&= E\left(\lim_{k\to\infty} X_{nk} - X^{(1)}_n\right), \text{ by the Dominated Convergence,} \\
&= 0.
\end{aligned} \tag{15}$$

Since $X^{(2)}_n \geq 0$, a.e., this implies $\|X^{(2)}_{n+k}\|_1 \to 0$ for any n and hence $\{X^{(2)}_n\}_1^\infty$ is a potential. Thus $X_n = X^{(1)}_n + X^{(2)}_n$ satisfies (10).

Finally to prove uniqueness, let $X_n = Y^{(1)}_n + Y^{(2)}_n$ be another decomposition satisfying (ii). Then

$$X_{nk} = E^{\mathcal{F}_n}(X_{n+k}) = E^{\mathcal{F}_n}(Y^{(1)}_{n+k}) + E^{\mathcal{F}_n}(Y^{(2)}_{n+k}) = Y^{(1)}_n + E^{\mathcal{F}_n}(Y^{(2)}_{n+k}), \text{ a.e.} \tag{16}$$

Letting $k \to \infty$ and noting that $E(E^{\mathcal{F}_n}(Y^{(2)}_{n+k})) = E(Y^{(2)}_{n+k}) \to 0$, being a potential, we conclude that $0 \leq E^{\mathcal{F}_n}(Y^{(2)}_{n+k}) \to 0$ in measure and hence for a subsequence $E^{\mathcal{F}_n}(Y^{(2)}_{n+k_i}) \to 0$ a.e., as $k_i \to \infty$. With this (16) becomes

$$X^{(1)}_n = \lim_{k_i\to\infty} X_{nk_i} = Y^{(1)}_n, \text{ a.e.,} \quad n \geq 1.$$

Hence $X^{(2)}_n = X_n - X^{(1)}_n = X_n - X^{(1)}_n = Y^{(2)}_n$, a.e., giving uniqueness. □

The following consequence has some interest.

6. Corollary. *Any positive supermartingale* $\{X_n, \mathcal{F}_n, n \geq 1\}$ *admits an* a.e. *unique decomposition* (10).

The proof of the above theorem shows that we may strengthen (i) to: "X_n dominates a martingale". In fact this is the original form of the classical Riesz theorem. Namely, if u is a superharmonic function on a bounded domain in $\mathbb{R}^2$ and h^* is a harmonic minorant on each open subset and continuous on its boundary contained in the domain of u, then $u = p + h$ uniquely, where p is a potential and h is a harmonic function. (See Rado [1] for an account of the theory of, and references to, Riesz.) In view of this and since there is interest in this form in Chapter V, we state the analog of the classical version for reference.

7. Theorem. (Riesz decomposition, second form). *Let $\{X_n, \mathcal{F}_n, n \geq 1\}$ be a supermartingale. Then the following two statements are equivalent:*

(i) *the X_n-process dominates a martingale,*

(ii) *the X_n-process admits an a.e. unique decomposition $X_n = X_n^{(1)} + X_n^{(2)}, n \geq 1$, where $\{X_n^{(1)}, \mathcal{F}_n, n \geq 1\}$ is a martingale and $\{X_n^{(2)}, \mathcal{F}_n, n \geq 1\}$ is a potential.*

This holds for continuous parameter processes also, as seen later. The same proof applies to the general case after we show, in Chapter III, that measurability problems do not create difficulties in this context.

To understand the first condition in the above theorem, one recalls that by Theorem 1 every supermartingale is dominated by a martingale and (i) asks that it dominate a martingale. Thus if (i) is true, the supermartingale is bounded by a pair of martingales, one above and one below. In Chapter V we analyze the relations between the Riesz and Doob decompositions more closely.

Consider a supermartingale $\{X_n, \mathcal{F}_n, n \geq 1\}$ such that $\sup_n E(|X_n|) < \infty$. If $Y_n = -X_n$, then $\{Y_n, \mathcal{F}_n, n \geq 1\}$ is a submartingale such that $\sup_n E(|Y_n|) < \infty$ and hence by Proposition 3 above there is a (positive) martingale $\{Z_n, \mathcal{F}_n, n \geq 1\}$ which dominates the Y_n-process. hence $X_n = -Y_n \geq -Z_n$, a.e., $n \geq 1$. But $\{-Z_n, \mathcal{F}_n, n \geq 1\}$ is also a martingale. Thus X_n dominates the $\{-Z_n\}_1^\infty$ process. By the above theorem, we then have the following result which shows that L^1-bounded (i.e., $\sup_n E(|X_n|) < \infty$) supermartingales are always enclosed between two martingales, and admit the Riesz decomposition. Thus:

8. Corollary. *Let $\{X_n, \mathcal{F}_n, n \geq 1\}$ be a supermartingale such that $\sup_n E(|X_n|) < \infty$. Then we have an a.e. unique decomposition $X_n = X_n^{(1)} + X_n^{(2)}, n \geq 1$, where $\{X_n^{(1)}, \mathcal{F}_n, n \geq 1\}$ is a martingale and $\{X_n^{(2)}, \mathcal{F}_n, n \geq 1\}$ is a potential.*

The Riesz decomposition clearly shows that there is a close relation between the (super-) martingale and potential theory. However, before consideration of such specialization it is necessary to develop the convergence theory of martingales. We now turn to this aspect.

2.6 Convergence theorems

Using the decompositions of the last section, we prove the basic martingale and submartingale convergence theorems in the discrete indexed case. All the later developments depend on these results. Consequently, here and in the Complements section, we present some different proofs of the basic convergence results. This gives insight into the numerous and far-reaching extensions of the theory. We prove the main result for probability spaces here, but all these statements hold, with minor modifications, for infinite measures. The point of view followed is to build the theory step by step to a general form. Also some of the results are presented with more than one proof, when such are available, as they suggest generalizations.

The following is the basic convergence theorem and is due to Doob [1]. The proof here is essentially that of Isaac's [1]. Doob's earlier method will be outlined in the Complements section.

1. Theorem. *Let* $\{X_n, \mathcal{F}_n, n \geq 1\}$ *be a martingale on* (Ω, Σ, P) *such that* $\sup_n E(|X_n|) < \infty$. *Then* $\lim_{n\to\infty} X_n = X_\infty$ *exists* a.e., *and* $E(|X_\infty|) \leq \liminf_{n\to\infty} E(|X_n|) < \infty$.

Proof. The idea of the proof is to reduce it to a bounded positive submartingale case using the Jordan decomposition (Theorem 5.2) and the simple Property (ii) of Section 4; and then we obtain the conclusion with the simpler L^2-methods together with a maximal inequality. For clarity the argument is given in steps.

I. Since the martingale is L^1-bounded, by Theorem 5.2, we can express it uniquely as $X_n = X_n^{(1)} - X_n^{(2)}$, a.e., $n \geq 1$ where $\{X_n^{(j)}, \mathcal{F}_n, n \geq 1\}, j = 1, 2$ are positive martingales. Hence the result follows if we show that every positive martingale converges a.e. When this is established, it follows that $|X_n| \to |X_\infty|$ a.e., and then Fatou's lemma implies that $E(|X_\infty|) \leq \liminf_{n\to\infty} E(|X_n|) < \infty$.

II. Let $\{X_n, \mathcal{F}_n, n \geq 1\}$ be therefore a positive martingale. Then by Property (ii (a)) of Section 4, $\{e^{-X_n}, \mathcal{F}_n, n \geq 1\}$ is a positive and bounded (by 1) submartingale. Moreover, $X_n \to X_\infty$ a.e., iff $e^{-X_n} \to e^{-X_\infty}$ a.e. In fact, the "if" being evident, suppose that $e^{-X_n} \to Y$ a.e. We now show that $Y > 0$ a.e. If this is false, let $Y = 0$ on A with $P(A) > 0$. Then on $A, X_n \to \infty$. But $E(X_n) =$ a constant for all n, so that (by Fatou) $\underline{\lim}_n E(X_n) = \infty$, contradicting the hypothesis $\sup_m E(X_n) < \infty$. Thus $P(A) = 0$, and $Y > 0$ a.e. So $X_\infty = -\log Y \geq 0$ a.e. and $X_n \to X_\infty$ a.e. Hence it suffices to show that every bounded positive submartingale converges a.e.

III. We now assert, more generally, that any L^2-bounded positive submartingale $\{X_n, \mathcal{F}_n, n \geq 1\}$ converges in norm and a.e.

For, since $\varphi(x) = x^2$ defines an increasing and convex function on $\mathbb{R}^+$, by the simple Property (ii) of Section 4, $\{X_n^2, \mathcal{F}_n, n \geq 1\}$ is also a submartingale. By the L^2-boundedness of the X_n-process, and the fact that $E(X_n^2)$ is increasing, we have $\lim_{n\to\infty} E(X_n^2) \uparrow a < \infty$ as $n \to \infty$. Hence for $n > m$ consider the identity

$$0 \leq E(X_n^2 - X_m^2) = E((X_n - X_m)^2) + 2E(X_m(X_n - X_m)). \qquad (1)$$

Since the left side tends to zero as m (and n) tends to ∞, each of the terms on the right side must tend to zero provided we show that $E(X_m(X_n - X_m)) \geq 0$. But this follows from the submartingale property of X_n:

$$\begin{aligned} E(X_m(X_n - X_m)) &= E[E^{\mathcal{F}_m}(X_m(X_n - X_m))], \quad m < n, \\ &= E[X_m(E^{\mathcal{F}_m}(X_n) - X_m)] \geq 0. \end{aligned}$$

Hence the $\{X_n\}_1^\infty$ is a Cauchy sequence in L^2 and thus it converges in norm.

To prove the pointwise convergence, consider $\{X_k - X_m, \mathcal{F}_k, m < k \leq n\}$. This is a submartingale when $m \geq 1$ is fixed. Hence for any

$\epsilon > 0$ (since $\inf_{x\in A} x = -\sup_{x\in A}(-x), A \subset \mathbb{R}$):

$$P\left[\max_{m<k\le n} |X_k - X_m| \ge \epsilon\right] \le P\left[\max_{m<k\le n} (X_k - X_m) \ge \epsilon\right] + P\left[\min_{m<k\le n} (X_k - X_m) \le -\epsilon\right]$$
$$\le \frac{E(|X_n - X_m|)}{\epsilon} - \frac{E(X_{m+1} - X_m) - E(|X_n - X_m|)}{\epsilon}$$

by Theorem 4.3,

$$\le \frac{2E(|X_n - X_m|) + E(|X_{m+1} - X_m|)}{\epsilon} \tag{2}$$

Since $E(|X_m - X_n|) \le [E(X_n - X_m)^2 \cdot P(\Omega)]^{1/2} \to 0$, $E(|X_{m+1} - X_m|) \to 0$, as $n, m \to \infty$, by the last paragraph (because $P(\Omega) < \infty$) we see that for m large enough the left side of (2) can be made arbitrarily small. Thus from (2) we deduce that $\max_{k\ge m} |X_k - X_m|$ is a.e. finite so that $\sup_{k\ge 1} X_k$ and $\inf_{k\ge 1} X_k$ are finite a.e. Moreover,

$$P\left[\overline{\lim_n} X_n - \underline{\lim_n} X_n \ge 2\epsilon\right] \le 2 \lim_n P\left[\sup_{k\ge n} |X_k - X_n| \ge \epsilon\right] = 0, \tag{3}$$

by (2). Thus $\lim_n X_n = X_\infty$ exists a.e. (and in L^2), and this completes the demonstration. □

The argument of Step III above suggests that in some cases $\lim_n E(X_n) = E(X_\infty)$ also holds. In fact, a characterization of this equality and related matters for martingales can be given as follows.

2. Theorem. *Let $\{X_n, \mathcal{F}_n, n \ge 1\}$ be a martingale and $\mathcal{F}_\infty$ denote $\sigma\left(\bigcup_{n=1}^{\infty} \mathcal{F}_n\right)$. Then the following conditions are equivalent:*

(a) *$K = \lim_n E(|X_n|) < \infty$ and $\{X_n, \mathcal{F}_n, 1 \le n \le \infty\}$ is a martingale.*

(b) *$\{X_n, n \ge 1\}$ is (L^1-bounded and) uniformly integrable.*

(c) *$K < \infty$ and $K = E(|X_\infty|)$ where $X_\infty = \lim_n X_n$ a.e., (K as in (a)).*

(d) $\{X_n, n \geq 1\} \subset L^1$ *is a Cauchy sequence.*
(e) *There exists a symmetric convex function* $\varphi : \mathbb{R} \to \mathbb{R}^+, \varphi(0) = 0, \frac{\varphi(x)}{x} \uparrow \infty$ *as* $x \uparrow \infty$, *and* $\sup_n E(\varphi(|X_n|)) < \infty$, *[i.e., the martingale lies in a ball of the Orlicz space* $L^\infty(\Sigma)$*].*
(f) *There exists a convex* φ *as in* (e) *and an* X_∞ *with* $E(\varphi(X_\infty/\alpha_0)) < \infty$ *and* $E\left(\varphi\left(\frac{X_n - X_\infty}{\alpha_0}\right)\right) \to 0$ *as* $n, m \to \infty$ *for some* $\alpha_0 \geq 1$.

Proof. All parts are different formulations of uniform integrability in the context of martingales and the result illuminates the key role played by such a condition. It is to be shown that each statement is equivalent to (b).

(a) $\Rightarrow$ (b). Since $\{X_n, \mathcal{F}_n, 1 \leq n \leq \infty\}$ is a martingale, by Property (ii) of Section 4, $\{|X_n|, \mathcal{F}_n, 1 \leq n \leq \infty\}$ is a positive submartingale, and so $|X_n| \leq E^{\mathcal{F}_n}(|X_\infty|)$, a.e. Hence by Property (vi) (see Equation (14)) of Section 4, $\{|X_n|, n \geq 1\}$ is uniformly integrable. This clearly implies (b). Note that $\sup_n E(|X_n|) < \infty$ is automatic when $X_n = E^{\mathcal{F}_n}(X_\infty)$, a.e., $X_\infty \in L^1(P)$.

(b) $\Rightarrow$ (a). Uniform integrability of $\{X_n, n \geq 1\}$ implies it is L^1-bounded. Thus $\|X_n\|_1 \leq K_0 < \infty$. Hence by Theorem 1, $X_n \to X_\infty$, a.e. We show that $E^{\mathcal{F}_n}(X_\infty) = X_n$, a.e. By hypothesis, the sequence is a martingale. Thus for any $A \in \mathcal{F}_m$, and $n > m$,

$$\int_A X_m \, dP_{\mathcal{F}_m} = \int_A E^{\mathcal{F}_m}(X_n) dP_{\mathcal{F}_m} = \int_A X_n \, dP. \tag{4}$$

Letting $n \to \infty$ in (4), we see that the validity of our result follows if the interchange of the limit and integral is justified in the last term. However, $\{X_n, n \geq 1\}$ is uniformly integrable and so by Vitali's theorem we may interchange the limits for any $A \in \mathcal{F}_m \subset \mathcal{F}_\infty = \sigma\left(\bigcup_{n \geq 1} \mathcal{F}_n\right)$. Thus (4) implies, since $X_n \to X_\infty$, a.e., on letting $n \to \infty$ for fixed m:

$$\int_A X_m \, dP_{\mathcal{F}_m} = \int_A X_\infty \, dP = \int_A E^{\mathcal{F}_m}(X_\infty) dP_{\mathcal{F}_m}, \quad A \in \mathcal{F}_m. \tag{5}$$

Since the integrands are $\mathcal{F}_m$-measurable they can be identified a.e. Thus $X_m = E^{\mathcal{F}_m}(X_\infty), m \geq 1$. Hence $\{X_n, \mathcal{F}_n, 1 \leq n \leq \infty\}$ is a martingale proving (a).

(b) ⇒ (c). Since $\sup_n \|X_n\|_1 < \infty, X_n \to X_\infty$ a.e., and the uniform integrability of $\{X_n, n \geq 1\}$ implies the same of $\{|X_n|, n \geq 1\}$ (and hence by Vitali's theorem again), we deduce that $E(|X_n|) \to E(|X_\infty|) < \infty$ as $n \to \infty$.

(c) ⇒ (b). Since $E(|X_n|) \uparrow K < \infty, \sup_n E(|X_n|) = K < \infty$ and $X_n \to X_\infty$ a.e. as before. Hence $E(|X_n|) \uparrow E(|X_\infty|)$ implies that $\{|X_n|, n \geq 1\}$ is uniformly integrable. This result in fact does not depend on the martingale property at all. Let us add the simple proof.

Since clearly for any $A \in \Sigma$

$$\int_A |X_n| dP \leq \int_A ||X_n| - |X_\infty|| dP + \int_A |X_\infty| dP, \tag{6}$$

it suffices to show that the right side tends to zero as $P(A) \to 0$ uniformly in n. One has by the preceding paragraph and the identity

$$|X_n| + |X_\infty| = \min(|X_n|, |X_\infty|) + \max(|X_n|, |X_\infty|), \tag{7}$$

that $E(\min(|X_n|, |X_\infty|)) \to E(|X_\infty|)$ because of the fact that $\min(|X_n|, |X_\infty|) \to |X_\infty|$, a.e. (dominated by $|X_\infty|$ which is integrable). Then $E(\max[|X_n|, |X_\infty|]) \to E(|X_\infty|)$ by (7) and the hypothesis (c). However we also have

$$||X_n| - |X_\infty|| = \max(|X_n|, |X_\infty|) - \min(|X_n|, |X_\infty|). \tag{8}$$

Hence $E[||X_n| - |X_\infty||] \to 0$ as $n \to \infty$. Consequently, given an $\epsilon > 0$, there is an $n_0 (= n_0(\epsilon))$ such that $E[||X_n| - |X_\infty||] < \epsilon/2$ if $n \geq n_0$. Also there is a δ_ϵ such that for any $A \in \Sigma$ with $P(A) < \delta_\epsilon$ we have $\int_A h \, dP < \epsilon/2$ where $h \in \{|X_k|, |X_\infty|, 1 \leq k \leq n_0\}$, a finite set. Hence (6) becomes, for any $A \in \Sigma$ with $P(A) < \delta_\epsilon$,

$$\int_A |X_n| dP \leq E(||X_n| - |X_\infty||) + \int_A |X_\infty| dP < \epsilon/2 + \epsilon/2 = \epsilon, \qquad n \geq 1. \tag{9}$$

Thus $\{X_n, n \geq 1\}$ is uniformly integrable, proving (b).

(b) ⇔ (d). With (b), $X_n \to X_\infty$ a.e., and hence $|X_n - X_\infty| \to 0$ a.e. But the uniform integrability of (b) shows $\{|X_n - X_\infty|, n \geq 1\}$ is also uniformly integrable since $\{|X_n| + |X_\infty|, n \geq 1\}$ has that property.

Hence by Vitali's theorem $E(|X_n - X_\infty|) \to 0$ as $n \to \infty$ which is (d). Conversely, that every Cauchy sequence in $L^1(\Sigma)$ is uniformly integrable is a trivial consequence of the argument between (8) and (9) above so that (b) follows.

(a) $\Rightarrow$ (e). As noted before, (a) implies $X_n = E^{\mathcal{F}_n}(X_\infty)$ a.e., and hence $|X_n| \leq E^{\mathcal{F}_n}(|X_\infty|)$, a.e., since $\{|X_n|, \mathcal{F}_n, 1 \leq n \leq \infty\}$ is a submartingale. By Corollary 1.15 there is a symmetric convex function $\varphi : \mathbb{R} \to \mathbb{R}^+, \varphi(0) = 0, \frac{\varphi(x)}{x} \uparrow \infty$, and $\varphi(|X_\infty|) \in L^1(\Sigma)$. Moreover, by Property (ii) of Section 4, $\{\varphi(|X_n|), \mathcal{F}_n, n \geq 1\}$ is a submartingale since $\varphi(\cdot)$ must be increasing on $\mathbb{R}^+$. Hence for all $n \geq 1$,

$$E(\varphi(|X_n|)) \leq E(\varphi(E^{\mathcal{F}_n}(|X_\infty|))) \leq E(E^{\mathcal{F}_n}(\varphi(|X_\infty|))) = E(\varphi(|X_\infty|)) < \infty. \quad (10)$$

(e) $\Rightarrow$ (b). Observe that one may use the Hölder inequality for Orlicz spaces and deduce that $E(|X_n|\chi_A) \to 0$ uniformly in n as $P(A) \to 0$. An independent proof is as follows. Let $K_0 = \sup_n E(\varphi(|X_n|))$. Since $\frac{\varphi(t)}{t} \uparrow \infty$ we must have $\frac{t}{\varphi(t)} = \epsilon_t \searrow 0$, as $t \to \infty$, as for large enough $t, \varphi(t) > 0$. Hence

$$\int_{[|X_n|>t]} |X_n| dP = \int_{[|X_n|>t]} \frac{|X_n|}{\varphi(|X_n|)} \varphi(|X_n|) dP \leq \epsilon_t \int_{[|X_n|>t]} \varphi(|X_n|) dP \leq \epsilon_t \cdot K_0. \quad (11)$$

Since K_0 is an absolute constant, letting $t \to \infty$, we deduce that the integrals tend to zero uniformly in n. Hence (b) holds.

(e) $\Leftrightarrow$ (f). Since (e) implies (b) which implies (a), $X_n \to X_\infty$ a.e., and $\{|X_n|, \mathcal{F}_n, 1 \leq n \leq \infty\}$ is a submartingale. Then for the φ of (e), $\{\varphi(|X_n|), \mathcal{F}_n, 1 \leq n \leq \infty\}$ is also a submartingale, $\varphi(|X_n|) \to \varphi(|X_\infty|)$ a.e. So $E(\varphi(|X_\infty|)) \leq \lim_n E(\varphi(|X_n|)) \leq E(\varphi(X_\infty))$, and $\{\varphi(|X_n|)\}_1^\infty$ is uniformly integrable. Replacing X_n by $|X_n - X_\infty|$, where $\varphi(|X_\infty|)$ is integrable, we deduce that $\{\varphi(|X_n - X_\infty|/\alpha_0), n \geq 1\}$ is uniformly integrable. [If $\varphi(\cdot)$ did not satisfy a growth condition ($\varphi(2t) \leq C\varphi(t), t \geq 0$), then we need to put in an $\alpha_0 \geq 1$, as here, to insure integrability. But this changes nothing because one is only replacing X_n with

$\frac{1}{\alpha_0}X_n, 1 \leq n \leq \infty$.] Since $\varphi(|X_n - X_\infty|/\alpha_0) \to 0$ a.e., the result follows by the Vitali theorem again. So (f) is true. Conversely since $\varphi(|X_n|)$ and $\varphi(|X_\infty|)$ are integrable, we have by the convexity of φ,

$$\varphi\left(\frac{|X_n|}{2\alpha_0}\right) \leq \frac{1}{2}\varphi\left(\frac{|X_n - X_\infty|}{\alpha_0}\right) + \frac{1}{2}\varphi\left(\frac{|X_\infty|}{\alpha_0}\right).$$

Hence

$$\sup_n E\left(\varphi\left(\frac{|X_n|}{2\alpha_0}\right)\right) \leq \frac{1}{2}E\left(\varphi\left(\frac{|X_\infty|}{\alpha_0}\right)\right) < \infty, \text{ by hypothesis.} \quad (12)$$

If $\tilde{\varphi}(t) = \varphi\left(\frac{t}{2\alpha_0}\right)$, then $\tilde{\varphi}$ satisfies, together with (12), the requirements of (e). This proves all the assertions. □

Even though submartingales constantly intervened in the above proofs, we have not established a convergence assertion for them. Interestingly enough, such a general result can be deduced from Theorem 1.

3. Theorem. *Let $\{X_n, \mathcal{F}_n, n \geq 1\}$ be a submartingale such that $\sup_n E(|X_n|) < \infty$. Then $X_n \to X_\infty$ a.e., and $E(|X_\infty|) \leq \varliminf_n E(|X_n|) < \infty$.*

Proof. By Theorem 5.1, X_n can be uniquely decomposed as

$$X_n = X_n' + \sum_{k=1}^{n} A_k, \qquad A_k \geq 0, \text{ a.e.,} \quad (13)$$

where $\{X_n', \mathcal{F}_n, n \geq 1\}$ is a martingale, and A_k is $\mathcal{F}_{k-1}$-adapted. Hence

$$0 \leq E\left(\sum_{k=1}^{n} A_k\right) = \sum_{k=1}^{n} E(A_k) = E(X_n) - E(X_n') \leq \sup_n E(|X_n|) - E(X_1'). \quad (14)$$

Since the right side is independent of n, by the Monotone Convergence theorem, we get $\sum_{k=1}^{\infty} E(A_k) < \infty$ so that $\sum_{k=1}^{\infty} A_k < \infty$, a.e. On the other hand,

$$E(|X_n'|) \leq E(|X_n|) + E\left(\sum_{k=1}^{n} A_k\right) \leq \sup_n E(|X_n|) + \sum_{k=1}^{\infty} E(A_k) < \infty. \quad (15)$$

Hence $\sup_n E(|X'_n|) < \infty$ and $X'_n \to X'_\infty$, a.e., by Theorem 1. One therefore deduces that

$$\lim_n X_\infty = \lim_n \left(X'_n + \sum_{k=1}^{n} A_k \right) = X'_\infty + \sum_{k=1}^{\infty} A_k = X_\infty, \text{ a.e.}$$

The integral inequality follows by Fatou's lemma. □

It is now possible to present various conditions related to uniform integrability for submartingales, and we state the result leaving the proof to the reader since it is analogous to that of Theorem 2.

4. Theorem. *Suppose $\{X_n, \mathcal{F}_n, n \geq 1\}$ is a submartingale and $\mathcal{F}_\infty = \sigma\left(\bigcup_{n\geq 1} \mathcal{F}_n\right)$. Then the following statements are equivalent:*

(a) $K = \sup_n E(|X_n|) < \infty$, *$\{X_n, \mathcal{F}_n, 1 \leq n \leq \infty\}$ is a submartingale and $K = E(|X_\infty|)$ where $X_\infty = \lim_n X_n$ a.e., which exists.*
(b) *$\{X_n, n \geq 1\}$ is uniformly integrable.*
(c) $K = \lim_n E(|X_n|) < \infty$, *and $K = E(|X_\infty|)$, where $X_\infty = \lim_n X_n$, a.e., which exists by Theorem 3.*
(d) ,(e), (f) *of Theorem 2 verbatim here.*

We remark that in (a) above $\{X_n, \mathcal{F}_n, 1 \leq n \leq \infty\}$ being a submartingale does not by itself imply the equation $K = E(|X_\infty|)$, since the sequence is not necessarily uniformly integrable. This is in contrast to the martingale case. We also note that a submartingale is always dominated by a positive submartingale. In fact if $\{X_n, \mathcal{F}_n, n \geq 1\}$ is a submartingale, then with the decomposition in Theorem 5.1, we have

$$|X_n| \leq |X'_n| + \sum_{k=1}^{n} A_k = Y_n, \text{ a.e.,} \qquad n \geq 1. \tag{16}$$

Clearly $\{Y_n, \mathcal{F}_n, n \geq 1\}$ is a positive submartingale dominating the X_n-process [$\{X_n^+, \mathcal{F}_n, n \geq 1\}$ is another.] Also the hypothesis $\sup_n E(|X_n|) < \infty$ in Theorems 1 and 3 is equivalent to the apparently weaker condition that $\sup_n E(X_n^+) < \infty$. In fact, as in (11) of Section 4 above,

$$E(|X_n|) = -E(X_n) + 2E(X_n^+) \leq -E(X_1) + 2\sup_n E(X_n^+) < \infty, \tag{17}$$

since $E(X_1) \leq E(X_n)$. For martingales it is even sufficient if $\sup_n E(X_n^-) < \infty$.

The following basic result is due to Andersen and Jessen [2] and exhibits the close relations between martingales and projective limits. The proof leads to certain other extensions of interest.

5. Theorem. *Let* (Ω, Σ, P) *be a probability space and* $\nu : \Sigma \to \mathbb{R}$ *be* σ*-additive (hence bounded). If* $\mathcal{F}_n \subset \Sigma, \mathcal{F}_n \subset \mathcal{F}_{n+1}$ are σ-algebras, *let* $\mathcal{F}_\infty = \sigma\left(\bigcup_{n\geq 1} \mathcal{F}_n\right)$, *and* $\nu_n = \nu|\mathcal{F}_n, P_n = P|\mathcal{F}_n, 1 \leq n \leq \infty$ *be the restrictions. Suppose* $X_n = \frac{d\nu_n^c}{dP_n}$, *the Radon-Nikodým derivative of the absolutely continuous part* ν_n^c *of* ν_n *relative to* P_n *using the Lebesgue decomposition. Then* $X_n \to X_\infty$ a.e. [P], *and moreover,* $X_\infty = \frac{d\nu_\infty^c}{dP_\infty}$, a.e. [P]. *In particular, if* $\mathcal{F}_\infty = \Sigma$ *and* ν *is* P*-continuous, then* $X_\infty = \frac{d\nu}{dP}$, a.e. [P].

Proof. From the existence of Radon-Nikodým derivatives, we see that X_n is $\mathcal{F}_n$-adapted. Let $X_* = \liminf_n X_n, X^* = \limsup_n X_n$. These are $\mathcal{F}_\infty$-adapted and $X_* \leq X^*$, a.e. [P]. The main part follows if one shows that $C = \{\omega : X_*(\omega) < X^*(\omega(\}$ is P-null. We establish the result in steps.

I. Let $C_{ab} = \{\omega : X_*(\omega) \leq a < b \leq X^*(\omega)\}$. Then $C = \cup\{C_{ab} : a, b$ rational$\}$, a countable union. So it suffices to show that $P(C_{ab}) = 0$ for each $a < b$. Let $H_a = \{\omega : X_*(\omega) \leq a\}, K_b = \{\omega : X^*(\omega) \geq b\}$. Then H_a, K_b, and C_{ab} are in $\mathcal{F}_\infty$ and $C_{ab} = H_a \cap K_b$. The following estimates on the measure of H_a and K_b are the key to proof.

II. For all $A \in \mathcal{F}_\infty$ and *any* reals a, b it is true that $\nu(H_a \cap A) \leq aP(H_a \cap A), \nu(K_b \cap A) \geq bP(K_b \cap A)$.

In fact, consider the first inequality. Let $a_n \downarrow a$ be a sequence of numbers and define $H_n = \left\{\omega : \inf_{r\geq 1} X_{n+r}(\omega) < a_n\right\}$. So $H_n \in \mathcal{F}_\infty$. We express H_n as a disjoint union with the familiar procedure used for inequalities in Section 4. Thus let $H_{n1} = \{\omega : X_{n+1}(\omega) < a_n\}$ and for $r > 1$,

$$H_{nr} = \{\omega : X_{n+r}(\omega) < a_n, X_{n+j}(\omega) \geq a_n, 1 \leq j \leq r-1\}. \quad (18)$$

Clearly $H_n = \bigcup_{r=1}^{\infty} H_{nr}, H_{nr} \in \mathcal{F}_{n+r}$ and disjoint. Also $H_1 \supset H_2 \supset \cdots$

and $H_a = \bigcap H_n$ since $a_n \downarrow a$, and $H_{nr} \subset \{\omega : X_{n+r}(\omega) < a_n\}$. Hence if $A \in \bigcup_{n\geq 1} \mathcal{F}_n$, so that $A \in \mathcal{F}_{n_0}$, for some n_0 and for all $n \geq n_0, r \geq 1$, we have $H_{nr} \cap A \in \mathcal{F}_{n+r}$. Consequently

$$\nu(H_n \cap A) = \nu[\bigcup_{r\geq 1}(H_{nr} \cap A)] = \sum_{r=1}^{\infty} \nu_{n+r}(H_{nr} \cap A), \text{ since } \nu|\mathcal{F}_n = \nu_n,$$
$$= \sum_{r=1}^{\infty} \nu_{n+r}^c(H_{nr} \cap A) = \sum_{r=1}^{\infty} \int_{H_{nr} \cap A} X_{n+r}\, dP_{n+r},$$
$$\text{since } \nu_n^c = \nu_n \text{ on } H_n,$$
$$\leq \sum_{r=1}^{\infty} a_n P(H_{nr} \cap A) = a_n P(H_n \cap A). \tag{19}$$

Since $|\nu|(H_1) < \infty$ and $H_n \downarrow H_a$, we get on letting $n \to \infty$ in (19):

$$\nu(H_a \cap A) = \lim_n \nu(H_n \cap A) \leq \lim_n a_n P(H_n \cap A) = aP(H_a \cap A). \tag{20}$$

This implies that (20) is true for all $A \in \bigcup_{n\geq 1} \mathcal{F}_n$ and if ζ is defined as:

$$\zeta(A) = aP(H_a \cap A) - \nu(H_a \cap A) \geq 0, \qquad A \in \bigcup_{n\geq 1} \mathcal{F}_n, \tag{21}$$

then ζ is a σ-additive finite set function on the algebra $\bigcup_{n\geq 1} \mathcal{F}_n$ and hence by the Hahn extension theorem it has a unique σ-additive extension to $\mathcal{F}_\infty$. This gives the first inequality. The second inequality is similar. Alternatively, replacing X_n, a and ν by $-X_n, -b$ and $-\nu$ in the above we get the second inequality from the first one.

III. $\lim_{n\to\infty} X_n = X_\infty$, a.e., exists.

For, since $C_{ab} = H_a \cap K_b$ in the notation of Step I, let $A = C_{ab}$ in (21). Then

$$aP(C_{ab}) \geq \nu(C_{ab}) \geq bP(C_{ab}). \tag{22}$$

But $a < b$. Hence (22) is possible only if $P(C_{ab}) = 0$. In view of Step I, this proves the assertion.

IV. $X_\infty = \frac{d\nu_\infty^c}{dP_\infty}$, a.e. [P].

For, since $X_n \to X_\infty$ a.e. and $\nu_n^c(A) = \int_A X_n dP_n, n \geq 1$, so that $|X_n| \to |X_\infty|$ a.e., one has $\int_\Omega |X_\infty| dP \leq \underline{\lim}_n \int_\Omega |X_n| dP \leq |\nu^c|(\Omega) < \infty$.

The boundedness of (the signed measure) ν, implies that X_∞ is finite a.e. Let N_1 be the set of divergence of the X_n sequence and $N_2 = \{\omega : |X_\infty|(\omega) = \infty\}$. Then $P(N_i) = 0, i = 1, 2$ and let $\Omega_0 = \Omega - (N_1 \cup N_2)$. So $X_n(\omega) \to X_\infty(\omega), \omega \in \Omega_0, X_\infty(\omega)$ is finite and $\Omega_0 \in \mathcal{F}_\infty$.

For any $\delta > 0$, define an elementary function X_δ by

$$X_\delta = \sum_{n=-\infty}^{\infty} n\delta\chi_{D_{n,\delta}}, \qquad D_{n,\delta} = \{\omega : n\delta \le X_\infty(\omega) < (n+1)\delta\}.$$

Then $0 \le X_\infty - X_\delta \le \delta$ and $X_\delta \to X_\infty$ uniformly (in ω), as $\delta \to 0$ through a sequence. Thus on Ω_0 we have

$$X_\delta \le X_\infty < X_\delta + \delta. \tag{23}$$

Since on $\Omega_0, X_\infty = X_* = X^*$, the inequalities of Step II imply with $a = (n+1)\delta$ and $b = n\delta, H_n = \{\omega : X_\infty(\omega) \le (n+1)\delta\}, K_n = \{\omega : X_\infty(\omega) \ge n\delta\}$, so that $D_{n,\delta} \subset H_n \cap K_n \cap \Omega_0$,

$$n\delta P[D_{n,\delta} \cap \Omega_0 \cap A] \le \nu[D_{n,\delta} \cap \Omega_0 \cap A] \le (n+1)\delta P[D_{n,\delta} \cap \Omega_0 \cap A]. \tag{24}$$

Summing over n and using the definition of X_δ one has

$$\int_{\Omega_0 \cap A} X_\delta dP_\infty \le \nu(\Omega_0 \cap A) \le \delta P(\Omega) + \int_{\Omega_0 \cap A} X_\delta dP_\infty, \qquad A \in \mathcal{F}_\infty, \tag{25}$$

Hence (23) and (25) yield (since $P(\Omega) = 1$)

$$\nu(\Omega_0 \cap A) - \delta \le \int_{\Omega_0 \cap A} X_\infty dP_\infty \le \nu(\Omega_0 \cap A) + \delta. \tag{26}$$

Letting $\delta \to 0$ and since $\nu_\infty^c(\cdot) = \nu_\infty(\Omega_0 \cap \cdot)$, (26) implies

$$\int_A X_\infty dP_\infty = \int_{\Omega_0 \cap A} X_\infty dP_\infty = \nu_\infty^c(A), \qquad A \in \mathcal{F}_\infty. \tag{27}$$

Thus $N = N_1 \cup N_2$ is the singular set of ν_∞ relative to P_∞ and $X_\infty = \frac{d\nu_\infty^c}{dP_\infty}$, a.e. [P].

If $\Sigma = \mathcal{F}_\infty$, then it is clear that $\nu_\infty = \nu, P_\infty = P$, and $\nu(N) = 0$ when ν is P-continuous, as asserted. $\square$

An immediate question is a comparison of this result with Theorem 1 or 3. Since $\nu_n = \nu_{n+1}|\mathcal{F}_n = \nu|\mathcal{F}_n$, if N_n and N_{n+1} are the singular sets of ν_n and ν_{n+1} for P_n and P_{n+1}, we must have $P_n(N_n) = 0 = P_{n+1}(N_{n+1})$ and since $P_n = P_{n+1}|\mathcal{F}_n$, it follows that $N_n \subseteq N_{n+1}$. Thus by the Lebesgue-Radon-Nikodým theorem,

$$\int_A X_n dP_n + \nu_n(N_n \cap A) = \nu_n(A) = \nu_{n+1}(A)$$

$$= \int_A X_{n+1} dP_{n+1} + \nu_{n+1}(N_{n+1} \cap A), \tag{28}$$

$A \in \mathcal{F}_n$. Since $\nu = \nu^+ - \nu^-$ we may consider ν^+ and ν^- separately. Thus if $\nu \geq 0$, then (28) implies, since $\nu_n = \nu_{n+1}|\mathcal{F}_n, \nu_{n+1}(N_{n+1} \cap A) - \nu_n(N_n \cap A) = \nu_{n+1}[(N_{n+1} - N_n) \cap A] \geq 0$, that

$$\int_A X_n dP_n \geq \int_A X_{n+1} dP_{n+1} = \int_A E^{\mathcal{F}_n}(X_{n+1}) dP_n, \qquad A \in \mathcal{F}_n, \tag{29}$$

and $\{X_n, \mathcal{F}_n, n \geq 1\}$ is a supermartingale. So by Theorem 3, since $E(|X_n|)$ is bounded, $X_n \to X_\infty$ a.e. Hence Theorem 6 follows from the earlier theory, except for the identification of the limit. Since we already know (cf., Problem I.5.6) that projective limits need not always exist, i.e., if we started with $\{\nu_n, \mathcal{F}_n, n \geq 1\}$ such that $\nu_n(A) = \int_A X_n dP, A \in \mathcal{F}_n$, then $\nu = \lim_n \nu_n$ need not be σ-additive even if $\sup_n |\nu_n|(\Omega) < \infty$, the above result appears to be weaker than Theorem 1. However, one can demonstrate that Theorem 1 is obtained from Theorem 6 so that the two approaches are equivalent. For a proof of this equivalence and a related discussion, one may refer to the author's book (Rao [12], pp. 181-183).

It is clear that Step II is the key to the preceding proof. If we analyze those inequalities, the fact that ν_n is a restriction of ν was needed only to conclude Equation (19). More explicitly,

$$\nu(H_n \cap A) = \sum_{r=1}^{\infty} \nu_{n+r}(H_{nr} \cap A) \leq a_n P(H_n \cap A). \tag{30}$$

If $\{X_n, n \geq 1\}$ is a sequence of integrable random variables on (Ω, Σ, P), $\nu_n(A) = \int_A X_n dP, A \in \Sigma$, we may impose a suitable hypothesis so that (30) is true and then the proof goes through verbatim. So one gets a conceivably more general result. For instance, our $\mathcal{F}_\infty$ may be replaced by a σ-algebra such that X_* and $X^*, \sup_{n\geq r} X_n$ and $\inf_{n\geq k} X_n$ are measurable, i.e., a "tail" σ-algebra. Thus let $\mathcal{T}_n = \sigma(X_k : k \geq n)$ and $\mathcal{T} = \bigcap_{n\geq 1} \mathcal{T}_n$. Let $\mathcal{F}_n = \sigma(X_k, 1 \leq k \leq n)$ and $\mathcal{F}_\infty = \sigma\left(\bigcup_{n\geq 1} \mathcal{F}_n\right)$. Then a possible *hypothesis about* (30) is (since as $n \to \infty$ $\nu(H_n \cap A) \to \nu(H \cap A)$ is the key part) the following:

$$\lim_{n\to\infty}\left(\lim_{m\to\infty}\sum_{k=0}^{m}\nu_{n+k}(H_{nk}\cap A)\right) = \tilde{\nu}(H\cap A), \qquad A \in \mathcal{T}, \tag{31}$$

where $H = \lim_{n\to\infty} \bigcup_{k=0}^{\infty} H_{nk}$, $H_{nk} \in \sigma(X_n, \ldots, X_{n+k})$ are disjoint, and $\tilde{\nu}$ is a (necessarily additive) set function on $\mathcal{T}(H \in \mathcal{T}$ is then true).

With this, Theorem 5 takes the following form and the identical proof is omitted.

6. Theorem. *Let* $\{X_n, \mathcal{F}_n, n \geq 1\} \subset L^1(\Omega, \Sigma, P)$ *be a stochastic process satisfying the above condition* (31). *If* $|\nu|(\Omega) < \infty$, *then* $X_n \to X_\infty$ a.e., *and* $X_\infty \in L^1(\Omega, \mathcal{T}, P)$. *Moreover, if* ν *is* σ*-additive, then* $X_\infty = \frac{d\nu^c}{dP_{\mathcal{F}}}$ a.e.

As remarked already, the key to the proof is the pair of inequalities of Step II of Theorem 6 and with (31) they are now true in this case. Notice that with (31) we need no longer demand that ν be σ-additive for the convergence statement. This is evident from the Andersen-Jessen proof of the two inequalities. If $\{X_n, \mathcal{F}_n, n \geq 1\}$ is a martingale and $\nu_n(\cdot) = \int_{(\cdot)} X_n dP$, then we see easily that $\nu_n = \nu_{n+1}|\mathcal{F}_n$ and $\sup_n |\nu_n|(\Omega) < \infty$ imply that (31) is true on $\mathcal{F}_n(\supset \mathcal{T})$. Hence Theorem 1 follows also from the above extended form of the Anderson-Jessen theorem. This observation essentially is in Loève [1]. We present another useful extension of Theorem 6. It is due to Chow [1] and explains a related aspect of the convergence theory and identification of the limit.

To state the result, it is first necessary to recall an important fact about finitely additive bounded set functions–the Yosida-Hewitt decomposition–which could be more profitably used. Such a function ν on an algebra $\Sigma_0 \to \mathbb{R}$ can be uniquely decomposed as $\nu = \nu_1 + \nu_2$ where

$\nu_1(\nu_2)$ is a σ-additive (respectively purely finitely additive) bounded set function. The description of ν_2 means that there is no nonzero σ-additive set function μ such that $\mu^{\pm} \le \nu_2^{\pm}$. (See Dunford-Schwartz [1], p. 163, or Rao [11], p. 182 for proofs.) With this, the following extension of Theorem 5 (or 6) holds:

7. Proposition. *Let $\mathcal{F}_n$ be an increasing sequence of σ-algebras in Σ of the probability space (Ω, Σ, P) and let $\nu : \Sigma_0 \to \mathbb{R}$ be a bounded additive set function where $\Sigma_0 = \bigcup_{n\ge1} \mathcal{F}_n$. Let $P_n = P|\mathcal{F}_n, \nu_n = \nu|\mathcal{F}_n$; and $\nu_n = \nu_n^1 + \nu_n^2$ be the Yosida-Hewitt decomposition for each n. If $X_n = \frac{d((\nu_n)^1)^c}{dP_n}$ and similarly $X_\infty = \frac{d(\nu^1)^c}{dP_\infty}$* a.e. *where $(\nu_n)^1$ is the σ-additive part of ν_n, and where $(\nu_n^1)^c$ and $(\nu^1)^c$ are the P_n- and P-continuous parts of ν_n^1 and ν^1 given by the Lebesgue decomposition, then $X_n \to X_\infty$,* a.e. [P].

Proof. It is clear that this result is the same as Theorem 5 if ν is σ-additive so that $\nu^2 = 0$ and $\nu_n^2 = 0$ all $n \ge 1$. Since one can express $\nu = \nu^+ - \nu^-$ by the classical Jordan decomposition consider $\nu^{\pm}$ separately and get the general case by linearity, we may and do assume for this proof that $\nu \ge 0$.

Let then $\nu = \nu^1 + \nu^2$ be the Yosida-Hewitt decomposition. Consider the restriction $\nu_n = \nu_n^1 + \nu_n^2$ of ν to $\mathcal{F}_n$ where ν_n^1 is σ-additive. So by another application of the above decomposition we get

$$(\nu_n)^1 + (\nu_n)^2 = \nu_n^1 + (\nu_n^2)^1 + (\nu_n^2)^2 \tag{32}$$

where $(\nu_n)^2$ and $(\nu_n^2)^2$ are purely finitely additive. Now applying the Lebesgue decomposition relative to P_n to $(\nu_n)^1, \nu_n^1$, and $(\nu_n^2)^1$ and taking the Radon-Nikodým derivatives, we get

$$X_n = Y_n + Z_n, \qquad n \ge 1 \tag{33}$$

where $Y_n = \frac{d(\nu_n^1)^c}{dP_n}, Z_n = \frac{d((\nu_n^2)^1)^c}{dP_n}$, and $X_n = \frac{d((\nu_n)^1)^c}{dP_n}$ a.e. By Theorem 5, $Y_n \to X_\infty$ a.e. where $X_\infty = \frac{d(\nu^1)^c}{dP_\infty}$ a.e. So, to prove the theorem, it suffices to show that $Z_n \to 0$ a.e. Let us establish the result: If there is a density Z_n for the restriction to $\mathcal{F}_n$ of a purely finitely additive set function on Σ_0, then it converges to zero a.e. (this is precisely what happened in the counterexample I.5.6.) We deduce this by using Theorem 3. [However, Theorem 1 can be derived from Theorem 5, so that the present proof, which now used a mixture of these results, can be made independent of Theorem 3.]

Since $\nu^2 \geq 0$ by (29), $\{Z_n, \mathcal{F}_n, n \geq 1\}$ is a positive supermartingale. Hence $0 \leq \sup_n E(Z_n) \leq E(Z_1) < \infty$, and $Z_n \to Z_\infty$ a.e. [P] by Theorem 3. Also by (29), $\int_A Z_n dP \leq \int_Z Z_{n-1} dP \leq \nu_{n_1}^2(A)$ for all $n, A \in \mathcal{F}_n$. Hence by Fatou's lemma, $0 \leq \int_A Z_\infty dP \leq \nu^2(A)$, $A \in \bigcup_{n\geq 1} \mathcal{F}_n = \Sigma_0$. But ν^2 is purely finitely additive and $\int_{(\cdot)} Z_\infty dP$ is σ-additive so that it must vanish. Thus $Z_\infty = 0$, a.e. must hold. □

Remark. As a consequence of this proposition, the last statement of Theorem 6 is true without the additional assumption of σ-additivity if we replace ν^c there with $(\nu^1)^c$, the absolutely continuous part of $\nu^1(= \nu - \nu^c)$ relative to P, in the above Yosida-Hewitt decomposition. Thus $X_n \to X_\infty = \frac{d(\nu^1)^c}{dP_{\mathcal{F}}}$, a.e. [P].

Thus far only the increasing indexed processes are considered. The decreasing indexed martingales (and submartingales) are somewhat different (particularly in their extensions to σ-finite cases to be indicated in Complements, cf., Exercise 17) from the increasing ones. (They are useful, for instance, in martingale formulations of ergodic theorems.) In the decreasing case the direct martingale and the Andersen-Jessen methods of proof coincide. The following result is also due to the latter authors.

8. Theorem. *Let (Ω, Σ, P) be a probability space and $\mathcal{F}_n \subset \Sigma, \mathcal{F}_n \supset \mathcal{F}_{n+1}$. Let $\nu : \Sigma \to \mathbb{R}$ be a σ-additive (hence bounded) function and $\nu_n = \nu|\mathcal{F}_n, P_n = P|\mathcal{F}_n$. If $\mathcal{F}_\infty = \bigcap_{n\geq 1} \mathcal{F}_n$ and $X_n = \frac{d\nu_n^c}{dP_n}$, then $X_n \to X_\infty$ a.e., and moreover $X_\infty = \frac{d\nu_\infty^c}{dP_\infty}$, a.e. where ν_n^c, ν_∞^c are respectively the P_n, P-continuous parts of ν_n, ν_∞.*

Proof. As in the proof of Theorem 5, if $X_* = \liminf_n X_n, X^* = \limsup_n X_n$, then we show that $X_* = X^*$ a.e. Hence by Fatou's lemma it follows that the common value is integrable and so finite a.e. If $X_\infty = \frac{d\nu_\infty^c}{dP_\infty}$, then it is $\mathcal{F}_\infty$-measurable (and so are X_*, X^*); and, since we may decompose $\nu = \nu^+ - \nu^-$ and treat $\nu^\pm$, for this proof we again assume that $\nu \geq 0$. Then $\nu_n^c|\mathcal{F}_\infty \leq \nu_\infty$ by an argument used for (29). Hence

$$\int_A X^* dP_\infty \geq \nu^c(A) = \int_A X_\infty dP_\infty \geq \varliminf_n \int_A X_n dP \geq \int_A X_* dP, \ A \in \mathcal{F}_\infty, \tag{34}$$

because all random variables are nonnegative. This shows $X_* \leq X_\infty \leq X^*$, a.e. Hence $X_\infty = X_* = X^*$ a.e., and $\frac{d\nu^c}{dP_\infty} = X_\infty$ a.e. follows as soon as we establish that the set $\{\omega : X_*(\omega) < X^*(\omega)\}$ is P-null. We again present the proof in steps.

I. For $a < b$ rationals, $[X_* < X^*] = \bigcup_{a,b} C_{ab}, C_{ab} = [X_* < a < b < X^*]$, and the C_{ab} are in $\mathcal{F}_\infty$. Let $H_a = [X_* < a], K_b = [X^* > b]$. Then $C_{ab} = H_a \cap K_b$ and since the union is countable, it suffices to show that $P[C_{ab}] = 0$. For this the following key inequalities are needed.

II. $\nu(H_a \cap A) \leq aP(H_a \cap A), \nu(K_b \cap A) \geq bP(K_b \cap A), A \in \mathcal{F}_\infty, \{a, b\} \subset \mathbb{R}$.

For, let $H_n = \left[\inf_{k \leq n} X_k < a\right]$. Then $H_n \subset H_{n+1}$ and $H_a = \bigcup_{r \geq 1} \bigcap_{n \geq r} H_n$. We prove the stronger assertion that $\nu(H_n \cap A) \leq aP(H_n \cap A)$ for all n, and then the result follows on taking limits as $n \to \infty$.

Thus let $H_{nn} = \{\omega : X_n(\omega) < a\}$, and for $k < n$, $H_{kn} = \{\omega : X_k(\omega) < a, \ X_j(\omega) \geq a, k+1 \leq j \leq n\}$. Then $H_{kn} \in \mathcal{F}_k$ and $H_{kn} \subset [X_k < a], H_n = \bigcup_{k \leq n} H_{kn}$, a disjoint union. Hence for $A \in \mathcal{F}_\infty \subset \mathcal{F}_k \subset \mathcal{F}_n$, we have

$$\begin{aligned} \nu(H_n \cap A) &= \sum_{k \leq n} \nu(H_{kn} \cap A) = \sum_{k \leq n} \nu_k^c(H_{kn} \cap A), \text{ since } \nu = \nu_k^c \text{ on } H_{kn}, \\ &= \sum_{k \leq n} \int_{H_{kn} \cap A} X_k dP_k < a \sum_{k \leq n} P(H_{kn} \cap A) = aP(H_n \cap A). \end{aligned} \tag{35}$$

This proves the first inequality and the second one is similar, or is deduced from the first by replacing ν, X_n, a with $-\nu, -X_n, -b$ as before.

III. To complete the proof, since $C_{ab} \subset H_a, C_{ab} \subset K_b$ for a, b in $\mathbb{R}, a < b$, let us take $A = C_{ab}$ in Step II, to get

$$aP(C_{ab}) \geq \nu(C_{ab}) \geq bP(C_{ab}).$$

Since $a < b$, this can hold only if $P(C_{ab}) = 0$. Hence $X_* = X^*$ a.e., and the proof is complete. □

If $\{X_n, \mathcal{F}_n, n \geq 1\}$ is a decreasing martingale in $L^1(\Omega, \Sigma, P)$, then $\nu_n(\cdot) = \int_{(\cdot)} X_n \, dP_n$ defines a set function such that $\nu_1|\mathcal{F}_n = \nu_n, \nu_n = \nu_n^c$ and the hypothesis of the above theorem is satisfied. Thus the following is a consequence.

9. Proposition. *Let $\{X_n, \mathcal{F}_n, n \leq 0\}$ be a martingale in $L^1(\Omega, \Sigma, P)$, and $\mathcal{F}_\infty = \bigcap_{n \leq 0} \mathcal{F}_n$. Then $X_{-n} \to X_{-\infty}$ a.e. as $n \to \infty$ and $\{X_n, \mathcal{F}_n, -\infty \leq n \leq 0\}$ is a martingale and hence is uniformly integrable. So $E(|X_{-\infty}|) = \lim_{n \to \infty} E(|X_{-n}|)$ and $X_{-n} \to X_{-\infty}$ in L^1-norm also.*

If we consider martingales with two sided index sets, we may deduce the following result from the above one and Theorem 2.

10. Proposition. *Let $\mathcal{F}_{n-1} \subset \mathcal{F}_n \subset \mathcal{F}_{n+1}$ be a doubly infinite sequence of σ-algebras in (Ω, Σ, P) with $\mathcal{F}_{-\infty} = \bigcap_n \mathcal{F}_n, \mathcal{F}_\infty = \sigma\left(\bigcup_n \mathcal{F}_n\right)$. If $X \in L^1(\Sigma)$ and $X_n = E^{\mathcal{F}_n}(X)$ so that $\{X_n, -\infty < n < \infty\}$ is uniformly integrable, then,*

(a) $\lim_{n \to \infty} X_n = X_\infty = E^{\mathcal{F}_\infty}(X)$ a.e.,

(b) $\lim_{n \to -\infty} X_n = X_{-\infty} = E^{\mathcal{F}_{-\infty}}(X)$ a.e.

From these results and Theorems 3 and 4, one may easily deduce the corresponding results for submartingales.

We remark that the decreasing index case above may be treated with nonnegative suffixes. For instance, let $Y_n = X_{-n}, \mathcal{B}_n = \mathcal{F}_{-n}$. Then $\mathcal{B}_n \downarrow \subset \Sigma$ and $\{Y_n, \mathcal{B}_n, n \geq 1\}$ is a decreasing (sub-) martingale, $E^{\mathcal{B}_{n+1}}(Y_n) = (\leq) Y_{n+1}$ a.e., etc.

The proof of Theorem 8 and particularly the key inequality (35) shows how we should modify (31) to present an analog of Theorem 6 related to the "decreasing index case." The corresponding condition is ($\mathcal{T}$ being a tail σ-algebra)

$$\lim_{n \to \infty} \sum_{k=1}^{n} \nu_k(H_{kn} \cap A) = \tilde{\nu}(H \cap A), \qquad A \in \mathcal{T} \tag{36}$$

where $H = \lim_{n \to \infty} \bigcup_{k=1}^{n} H_{kn}, H_{kn} \in \sigma(X_k, \ldots, X_n)$, disjoint, and $\tilde{\nu}$ on $\mathcal{T}$ is a (necessarily additive) set function.

Then with (36) in place of (31), Theorem 6 has an exact analog and we leave the restatements to the reader. (See Exercise 12.) Note that it is not always easy to test conditions (31) and (36) in applications.

2.7 Complements and exercises

1. Prove the conditional Jensen inequality with the following alternative argument. Let (Ω, Σ, P) be a probability space, $X : \Omega \to \mathbb{R}$ a random variable such that $E(\varphi(X))$ exists where $\varphi : \mathbb{R} \to \mathbb{R}$ is a convex function. Then X is integrable and $E^{\mathcal{B}}(\varphi(X)) \geq \varphi(E^{\mathcal{B}}(X))$, a.e., for any σ-algebra $\mathcal{B} \subset \Sigma$. Show that this is true if X is simple and if one recalls $\varphi\left(\sum_{1}^{n} p_i a_i\right) \leq \sum_{i=1}^{n} \varphi(a_i)p_i, p_i \geq 0, \sum_{i=1}^{n} p_i = 1$. Next verify the general case with the conditional Monotone Convergence theorem.

2. Let (Ω, Σ, μ) be an arbitrary measure space and $\varphi : \mathbb{R} \to \mathbb{R}^+$ be a convex function such that $\varphi(-x) = \varphi(x)$, and $\varphi(x) = 0$ iff $x = 0$. If $f : \Omega \to \mathbb{R}$ such that $\int_\Omega \varphi(f)d\mu < \infty, \mathcal{B} \subset \Sigma$ is any σ-algebra, show that there exists an operator $E^{\mathcal{B}} : f \mapsto E^{\mathcal{B}}(f), E^{\mathcal{B}}(f)$ is $\mathcal{B}$-measurable and $\int_B f\, d\mu = \int_B E^{\mathcal{B}}(f)d\mu_{\mathcal{B}}$, for all $B \in \mathcal{B}_0 = \{A \in \mathcal{B} : \mu(A) < \infty\}$. Moreover, $E^{\mathcal{B}}(\varphi(f)) \geq \varphi(E^{\mathcal{B}}(f))$, a.e. (μ). Thus $E^{\mathcal{B}}$ is a *generalized conditional expectation*, and is useful in abstract analysis. [*Hints*: Let $\tilde{\mathcal{B}} \subset \mathcal{B}$ be the σ-ring generated by $\mathcal{B}_0$. Consider $f \geq 0$ and $\nu_f(A) = \int_A f\, d\mu,\ A \in \Sigma_0 = \{A \in \Sigma : \mu(A) < \infty\}$. Then $\nu_f : \Sigma_0 \to \bar{\mathbb{R}}^+$ is σ-additive and μ-continuous. The φ-bounded variation of ν_f on $A \in \Sigma$, *by definition*, is given by: $I_\varphi(\nu_f : A) = \sup\left\{\sum_{i=1}^{n} \varphi(\nu_f(A_i)/\mu(A_i))\mu(A_i) : A_i \in \Sigma(A)\right\}$. Verify that $I_\varphi(\nu_f : A) = \int_A \varphi(f)d\mu$–this is shown in Dunford-Schwartz ([1], III.2.20) if $\varphi(x) = |x|^p$ and the proof extends. Thus $I_\varphi(\nu_f) = I_\varphi(\nu : \Omega) < \infty$ by hypothesis. If $\tilde{\nu}_f = \nu_f|\mathcal{B}$, then $I_\varphi(\tilde{\nu}_f) \leq I_\varphi(\nu_f)$ and so the set $S = \text{supp}(\tilde{\nu}_f) \in \tilde{\mathcal{B}}$. There exist $S_n \in \mathcal{B}_0, S_n \uparrow S$. Let $\mathcal{B}_n = \mathcal{B}(S_n)$. Then $\tilde{\mathcal{B}}$ is generated by the ring $\bigcup_{n=1}^{\infty} \mathcal{B}_n$. Hence in the nontrivial case, $\int_A f\, d\mu = \int_A \tilde{f}_n d\mu_n, A \in \mathcal{B}_n$, for a unique $\mathcal{B}_n$-measurable $\tilde{f}_n$, by the Radon-Nikodým theorem. By the uniqueness, $\tilde{f}_{n+1} = \tilde{f}_n$ on S_n. Show that $\tilde{f}_n \to \tilde{f}$ a.e., and $\varphi(\tilde{f}) \in L^1(\mu)$. Define $E^{\mathcal{B}} : f \mapsto \tilde{f}$ if $\tilde{\mathcal{B}} \neq \{\emptyset\}$ and set $E^{\mathcal{B}}(f) = 0$ if $\tilde{\mathcal{B}} = \{\emptyset\}$. Now check that $E^{\mathcal{B}}(\cdot)$ has all the stated properties. The most general conditioning concept is in Dinculeanu [2].]

3. A somewhat different generalization, due to Dynkin [1], of conditional expectations which has applications in Markov Processes, can be given as follows. If (Ω, Σ, P) is a probability space, we define a conditional expectation on a subset $\Omega_0 \in \Sigma$ of Ω. Let $\tilde{\mathcal{B}}_0 = \mathcal{B}(\Omega_0)$ for

a σ-algebra $\mathcal{B} \subset \Sigma$. If $f : \Omega \to \mathbb{R}$ is a measurable function such that $\int_{\Omega_0} |f| dP < \infty$, the set function $\nu_0(A) = \int_A f dP, A \in \tilde{\mathcal{B}}_0$ is σ-additive and there is a unique $\tilde{\mathcal{B}}_0$ measurable $\tilde{f}$ with $\nu_0(A) = \int_A \tilde{f} dP, A \in \tilde{\mathcal{B}}_0$. Then the mapping $E^{\tilde{\mathcal{B}}_0} : f \mapsto \tilde{f}$ on Ω_0 is called a conditional expectation. Show that the following modified properties hold if $p \geq 1$: [$f \leq g$ in Ω_0 means $f\chi_{\Omega_0} \leq g\chi_{\Omega_0}$.]

(i) $E^{\tilde{\mathcal{B}}_0}$ is a positive linear operator on $L^p(\Omega, \Sigma(\Omega_0), P) = L^p(\Sigma(\Omega_0))$, where $\Sigma(\Omega_0)$ is again the trace of Σ on Ω_0.

(ii) $E^{\tilde{\mathcal{B}}_0}(\cdot)$ is a contraction on the $L^p(\Sigma(\Omega_0))$-spaces, and, if $\{f, g, fg\} \subset L^1(\Sigma(\Omega_0))$ and g is $\tilde{\mathcal{B}}_0$-measurable, then $E^{\tilde{\mathcal{B}}_0}(fg) = gE^{\tilde{\mathcal{B}}_0}(f)$, a.e., on Ω_0. Does $E^{\tilde{\mathcal{B}}_0}$ extend to $L^1(\Omega)$ itself?

4. The independence concept of Section I.1 has a generalization to "conditional independence." If (Ω, Σ, P) is a probability space and $\mathcal{B}_i \subset \Sigma, i = 1, 2, 3$ are σ-algebras, then $\mathcal{B}_1$ and $\mathcal{B}_3$ are *conditionally independent relative* to $\mathcal{B}_2$ (or given $\mathcal{B}_2$) if for any $A \in \mathcal{B}_1, B \in \mathcal{B}_3$ we have

$$P^{\mathcal{B}_2}(A \cap B) = P^{\mathcal{B}_2}(A) \cdot P^{\mathcal{B}_2}(B), \text{ a.e.} \tag{*}$$

If $\mathcal{B}_2 = \{\emptyset, \Omega\}$, is a probability space and $\mathcal{B}_i \subset \Sigma, i = 1, 2, 3$ are σ-algebras, then $\mathcal{B}_1$ and $\mathcal{B}_3$ are *conditionally independent relative* to $\mathcal{B}_2$ (or given $\mathcal{B}_2$) if for any $A \in \mathcal{B}_1, B \in \mathcal{B}_3$ we have

$$P^{\mathcal{B}_2}(A \cap B) = P^{\mathcal{B}_2}(A) \cdot P^{\mathcal{B}_2}(B), \text{a.e.} \tag{*}$$

If $\mathcal{B}_2 = \{\emptyset, \Omega\}$, this is the (mutual) independence considered in Chapter I. The concept extends to the conditional independence of the random variables X_1 and X_3 relative to X_2 if the generated σ-algebras $\mathcal{B}_i = \sigma(X_i), i = 1, 2, 3$ have the property defined above. Show that $\mathcal{B}_1$ and $\mathcal{B}_3$ are conditionally independent relative to $\mathcal{B}_2$ iff either (i) $P^{\mathcal{B}_{12}}(A) = P^{\mathcal{B}_2}(A)$, a.e., $A \in \mathcal{B}_3$, or (ii) $P^{\mathcal{B}_{23}}(B) = P^{\mathcal{B}_2}(B)$, a.e., $B \in \mathcal{B}_1$, where $\mathcal{B}_{12}(\mathcal{B}_{23})$ is the smallest σ-algebra containing $\mathcal{B}_1$ and $\mathcal{B}_2$ ($\mathcal{B}_2$ and $\mathcal{B}_3$). [*Hints:* If (i) holds, since $E^{\mathcal{B}_2} = E^{\mathcal{B}_2}E^{\mathcal{B}_{12}}$, then $E^{\mathcal{B}_2}(\chi_A E^{\mathcal{B}_{12}}(\chi_B)) = E^{\mathcal{B}_2}(\chi_A E^{\mathcal{B}_2}(\chi_B))$ for $A \in \mathcal{B}_1, B \in \mathcal{B}_3$, giving $(*)$. Conversely, if $(*)$ is true, this equation holds, and hence

$$\int_C \chi_A E^{\mathcal{B}_{12}}(\chi_B) dP_{\mathcal{B}_2} = \int_C \chi_A E^{\mathcal{B}_2}(\chi_B) dP_{\mathcal{B}_2}, \qquad \text{for all } C \in \mathcal{B}_2.$$

We can replace $P_{\mathcal{B}_2}$ by $P_{\mathcal{B}_{12}}$ here and see that this holds for all $C \in \mathcal{B}_2$ and all $A \in \mathcal{B}_1$ so that the integrals also are equal for the σ-algebra containing the collection of sets in $\{\mathcal{B}_1, \mathcal{B}_2\}$, which is $\mathcal{B}_{12}$. We can then identify the integrals to get (i); similarly for (ii).]

5. This problem gives some useful formulas for conditional distributions. Let (Ω, Σ, P) be a probability space and $(\mathbf{E}, \mathcal{E})$ a measurable space. Let $\Omega = \mathbb{R}$, and $\mathbf{E} = \mathbb{R}$.

(a) Let Σ and $\mathcal{E}$ contain the Borel sets. Let X be a random variable on Ω and $Y : \Omega \to \mathbf{E}$ be measurable for $(\Sigma, \mathcal{E})$. If $\mathcal{B}_Y = Y^{-1}(\mathcal{E}) \subset \Sigma$, suppose the conditional probability function $P^{\mathcal{B}_Y}$ on Σ is regular, as defined in Section 3. In particular this is always true if P itself is regular. Let

$$F_{X|Y}(x|y) = P^{\mathcal{B}_Y}(A_x)(y) = E(\chi_{A_x}|Y = y)$$

in the sense of Equation (10) of Section 1. Here $A_x = (-\infty, x) \in \Sigma$. By regularity of $P^{\mathcal{B}_Y}$, $F_{X|Y}(\cdot|y)$ is a distribution function and $F_{X|Y}(x|\cdot)$ is measurable with respect to $\mathcal{E}$. If $F_X(y) = P\{\omega : Y(\omega) < y\}$ is the distribution function of Y, show that the joint distribution $F_{X,Y}(x, y) = P[X < x, Y < y]$ is given by:

$$F_{X,Y}(x, y) = \int_{-\infty}^{y} F_{X|Y}(x|t)\,dF_Y(t) \qquad y \in \mathbb{R},$$

and that $F_{X|Y}(x|y) = F_X(x)$, all $x \in \mathbb{R}$ if X, Y are independent.

(b) Under the same hypotheses as in (a), show that

$$\int_A \int_B dF_{X,Y}(x, y) = \int_A \left[\int_B F_{X|Y}(dx|y) \right] dF_Y(y), \qquad A \in \Sigma, \quad B \in \mathcal{E},$$

using the following outline. By regularity, $F_{X|Y}(\cdot|y)$ is a distribution function, so let $\nu(A|y) = \int_A F_{X|Y}(dx|y)$. Let $\mathcal{C}$ be the collection of sets $C \in \mathcal{E}$ such that for all $A \in \Sigma$,

$$\int_A \int_C dF_{X,Y}(x, y) = \int_C \nu(A|y)\,dF_Y(y).$$

Show that $\mathcal{C} \subset \mathcal{E}$ is closed under monotone limits, unions, and differences of sets of Σ, and deduce finally that $\mathcal{C} = \mathcal{E}$.

6. (a) In this problem we illustrate the use of perfectness of a measure in relation to the independence concept. Recall that two real random variables X, Y on (Ω, Σ, P) are independent iff

$$P(X^{-1}(A) \cap Y^{-1}(B)) = P(X^{-1}(A)) \cdot P(Y^{-1}(B)) \qquad (*)$$

for all Borel sets A, B. This definition is due to Steinhaus. One can also define independence of X, Y iff $(*)$ holds for all $A, B \subset \mathbb{R}$ for which $\{X^{-1}(A), Y^{-1}(B)\} \subset \Sigma$. This is due to Kolmogorov. Call these two concepts (S)-independent and (K)-independent, respectively. It is clear that (K)-independence implies (S)-independence. To see that the converse is false we construct, following B. Jessen (1948), two random variables X, Y on Ω_i such that the algebra $\Sigma_1 = (X,Y)^{-1}(\mathcal{P}) \cap \Sigma \supsetneq (X,Y)^{-1}(\mathcal{B}) = \Sigma_2$ and that $(*)$ holds for Σ_2, but not for Σ_1. Here $\mathcal{P}$ is the power set and $\mathcal{B}$ the Borel σ-algebra in the plane in which the vector random variable $Z = (X, Y)$ takes its values. Let $\Omega_1 = \Omega_2 = [0, 1]$, and $\Omega = \Omega_1 \times \Omega_2$. If μ is the Lebesgue measure on Ω_1, and $Q \subset \Omega_1$ is a set of Lebesgue outer measure one and inner measure zero, then let $\mathcal{A}_0 = \mathcal{A}_1 \cup \mathcal{A}_2$ where $\mathcal{A}_1 = \{A \cap Q : A \subset \mathbb{R}, \text{Borel set}\}$, $\mathcal{A}_2 = \{B \cap Q^c : B \subset \mathbb{R}, \text{Borel set }\}$. Let $\mathcal{A}$ be the σ-algebra generated by $\mathcal{A}_0$ and, if $A \in \mathcal{A}_0$, it is of the form $A = (B_1 \cap Q) \cup (B_2 \cap Q^c)$ for Borel sets B_1, B_2. Define $\mu_i(A) = \mu(B_i), i = 1, 2$ so that $\mu_i = \mu$ on the Borel algebras of $\Omega_i, i = 1, 2$. Also $Q \in \mathcal{A}, \mu_1(Q) = 1, \mu_2(Q) = 0, \mu_2(Q^c) = 1, \mu_1(Q^c) = 0$. Let X, Y be the identity maps on Ω_1 and Ω_2 respectively, and show that with $\Sigma = \mathcal{A} \otimes \mathcal{A}, \Sigma_1 \supsetneq \Sigma_2$. Let P be the measure defined by $P = \frac{1}{2}[\mu_1 \otimes \mu_2 + \mu_2 \otimes \mu_1]$. Show that P coincides with $\mu \otimes \mu$ on Borel sets, and that X, Y above are (S)-independent, but not (K)-independent. [In the above definition of $(*)$, take $A = \Omega_1 \times Q, B = Q \times \Omega_2$ and note that $P(X^{-1}(A) \cap Y^{-1}(B)) = 0 \neq P(X^{-1}(A) \cdot P(Y^{-1}(B)) = \frac{1}{4}$.]

(b) Show, however, if (Ω, Σ, P) is a perfect measure space then (S)- and (K)-independence concepts coincide. [Observe that $P \circ Z^{-1}$ is a Lebesgue-Stieltjes measure for every real random variable Z, and hence the measurable sets in both definitions are the same. This discrepancy in the general case is one of the motivations for the introduction of the concept of perfectness, and various related notions.]

7. This problem illustrates the relations between contractive projections and conditional expectation operators on $L^p(\Sigma)$ spaces with an alternative argument. The spaces are based on (Ω, Σ, μ).

(a) If $T : L^p(\Sigma) \to L^p(\Sigma), p \geq 1$ is a positive contractive projection where $\mu(\Omega) < \infty$, let $T1 = 1$, a.e. If $\mathcal{M} = T(L^p(\Sigma))$ and $1 \leq p < \infty$, show directly that $\mathcal{B} = \{B \in \Sigma : \chi_B \in \mathcal{M}\}$ is a σ-algebra such that $\mathcal{M} = L^p(\mathcal{B})$, and hence $T = E^{\mathcal{B}}$. [*Hints:* With $A, B \in \mathcal{B}, \chi_{A\cap B} = \chi_A - |\chi_A - \chi_B| = \chi_B - |\chi_A - \chi_B|$ is in the vector lattice $\mathcal{M}$ and so $A \cap B \in \mathcal{B}$. Then $\chi_{A\cup B} = \chi_A + \chi_B - \chi_{A\cap B} \in \mathcal{M}$ and $A \cup B \in \mathcal{B}$. Finally since $p < \infty$, show that $\chi_{\bigcup_{i=1}^{n} A_i} \in \mathcal{M}$ implies $\chi_{\bigcup_{i=1}^{\infty} A-i} \in \mathcal{M}$ and $\bigcup_{i=1}^{\infty} A_i \in \mathcal{B}$. Deduce that $\mathcal{M} = L^p(\mathcal{B})$.]

(b) If $p = \infty$, then also $\mathcal{M}$ is a complete vector lattice in (a). Show that, using the last part of Theorem 2.5 $\mathcal{M} = L^\infty(\mathcal{B})$ for a σ-algebra $\mathcal{B} \subset \Sigma$, and conclude that $Tf = E^{\mathcal{B}}(fh), f \in L^\infty(\Sigma)$ for a unique $h \in L^1(\Sigma)$ with $E^{\mathcal{B}}(h) = 1$ a.e. [*Hints*: Only that $\mathcal{M}$ is a vector lattice is to be established. If $f \in \mathcal{M}$, it suffices to show the maximum of f and 0 exists in $\mathcal{M}$. Since $f^+ \geq f$ and T is order preserving, $Tf^+ \geq f$ and $Tf^+ \geq 0$ so that $Tf^+ \geq f^+$. If $0 \leq g \in \mathcal{M}$ is any element with $g \geq f$ then $g = Tg \geq Tf^+ \geq f^+$ and the element $Tf^+(\in \mathcal{M})$ is the maximum of f and 0. So $\mathcal{M}$ is a vector lattice. Note that $\mathcal{M}$ is also closed under a norm given by order, i.e., $f \in \mathcal{M}$ implies $-\|f\|_\infty \leq f \leq \|f\|_\infty$, a.e., and if $-\alpha \leq f \leq \alpha$ a.e., then $\alpha \geq \|f\|_\infty$. Thus $\|f\|_\infty =$ ess $\inf\{\alpha > 0 : -\alpha \leq f \leq \alpha\}$. The latter is the order norm. Hence the order norm and the L^∞-norm agree. So $\mathcal{M} = L^\infty(\mathcal{B})$ holds, as in the last part of Theorem 2.5.]

8. Let $T : L^p(\Sigma) \to L^p(\Sigma), 1 \leq p < \infty$ be a bounded linear operator such that $T(L^\infty(\Sigma)) \subset L^\infty(\Sigma)$, and T satisfies the averaging identity: $T(fTg) = (Tf)\cdot(Tg), f, g \in L^p(\Sigma) \cap L^\infty(\Sigma)$. If either (i) $\mu(\Omega) < \infty$, or (ii) $\mu(\Omega) = \infty$ and there is a weak unit f_0 in $L^p(\Sigma)$ such that $Tf_0 = f_0$, show that the set $\mathcal{C} = \{f : T(fg) = fT(g)$, for all g in $L^p(\Sigma) \cap L^\infty(\Sigma)\}$ is a closed subspace of $L^p(\Sigma)$ such that $T(L^p(\Sigma)) \subset \mathcal{M} \subset \mathcal{C}$ where $\mathcal{M} = L^p(\mathcal{B})$ for a σ-algebra $\mathcal{B} \subset \Sigma$. Deduce that $Tf = E^{\mathcal{B}}(fg), f \in L^p(\Sigma)$ for a $g \in L^{p'}(\Sigma)$ where $p' = p/(p-1)$, and $E^{\mathcal{B}}(g)$ is bounded. In case (ii) $\mathcal{M}$ is the range of T and $E^{\mathcal{B}}(g) = 1$, a.e. (*Hint:* Follow the arguments of Section 2 (use Theorem 2.5). If T is positive, then the second case can be reduced to the preceding problem. Cf., e.g., Moy [1], Dinculeanu and Rao [1].)

9. (a) The result of the above problem admits an extension to a class of Orlicz spaces. Let Φ be a continuous convex function such that

$\Phi(0) = 0, \Phi(x) = \Phi(-x)$, and $\Phi(x) > 0$ if $x > 0$. It is called a Young function. Let $L^{\Phi}(\Sigma)$ be an Orlicz space on a probability triple (Ω, Σ, P). (See Krasnoselskii-Rutickii [1] or Rao-Ren [1] about these spaces.) If $T : L^{\Phi}(\Sigma) \to L^{\Phi}(\Sigma)$ is an averaging operator with $T1 = 1$, show that there exist a σ-algebra $\mathcal{B} \subset \Sigma$ and a nonnegative (Σ)-measurable function g such that for $f \in L^{\Phi}(\Sigma)$,

$$Tf = E^{\mathcal{B}}(fg), \qquad E^{\mathcal{B}}(g) = 1, \text{ a.e.,} \tag{*}$$

The same characterization $(*)$ holds true if T satisfies instead: (i) $T((Tf) \vee (Tg)) = (Tf) \vee (Tg)$, (ii) $T1 = 1$, and (iii) T is a contraction. These results have natural analogs if μ is σ-finite when (ii) is replaced by $Tf_0 = f_0$ for a weak unit f_0. Here $g = 1$ a.e., if $\Phi(2x) \leq C\Phi(x)$, for $x \geq 0, 0 < C < \infty$, but not necessarily so otherwise. [*Hints:* This is somewhat more involved than the L^p-case. First show that $\mathcal{M} = T(L^{\Phi}(\Sigma))$ is a measurable subspace, $\mathcal{M} = L^{\Phi}(\mathcal{B})$, where $\mathcal{B} = \{A \in \Sigma : \chi_A \in \mathcal{M}\}$. The result of Theorem 2.5 applies. Then after some work one can show that $(*)$ obtains.] (b) Let Φ be as above and Ψ be the complementary Young function, i.e. $\Psi(y) = \sup\{x|y| - \Phi(x) : x \geq 0\}$. It is then known that $L^{\Psi}(\Sigma)$ can be identified with a subspace of $(L^{\Phi}(\Sigma))^*$. Suppose $T : L^{\Phi}(\Sigma) \to L^{\Phi}(\Sigma)$ is a contractive operator such that its adjoint T^* restricted to $L^{\Psi}(\Sigma)$ maps $L^{\Psi}(\Sigma) \to L^{\Psi}(\Sigma)$ and is an averaging there as defined above. Using the result of (a) [and with some further work], show that T^* is positive, $\int_{\Omega} Tf\, dP = \int_{\Omega} f\, dP$, and $Tf = T^*f$ for all $f \in L^{\infty}(\Sigma) \cap L^{\Phi}(\Sigma) \cap L^{\Psi}(\Sigma)$. Deduce that T can be represented in the form $(*)$, and so it is also an averaging. (In connection with this problem, see the author's papers [1] and especially [2_{IV}].)

10. Let $\{X_n, \mathcal{F}_n, n \geq 1\} \subset L^2(\Omega, \Sigma, P)$ be a martingale and let $Y_n = X_n - X_{n-1}$ with $X_0 = 0$ a.e., $n \geq 1$, where P, as usual, denotes a probability on Σ. Show that $\{Y_n, n \geq 1\}$ is an orthogonal sequence which satisfies $E^{\mathcal{F}_m}(Y_n) = 0$ a.e., for $1 \leq m < n$. Conversely if $\{Y_n, n \geq 1\}$ is any integrable sequence on (Ω, Σ, P) and $E^{\mathcal{F}_m}(Y_n) = 0$ for each $1 \leq m < n$ where $\mathcal{F}_n = \sigma(Y_i, 1 \leq i \leq n)$, then $\{X_n, \mathcal{F}_n, n \geq 1\}$ is a martingale where $X_n = \sum_{i=1}^{n} Y_i$.

11. Let $\{X_n, n \geq 1\} \subset L^2(\Omega, \Sigma, P)$ be a sequence of independent random variables with means zero and variances $\{\sigma_n^2, n \geq 1\}$. If $S_n =$

$\sum_{i=1}^{n} X_i$ and $\mathcal{F}_n = \sigma(X_i, 1 \leq i \leq n)$, show that $\{S_n, \mathcal{F}_n, n \geq 1\}$ is a martingale and that if $\sum_{n=1}^{\infty} \sigma_n^2 < \infty$ then $\sum_{i=1}^{\infty} X_i = \lim_n S_n$ exists a.e. and in L^2-norm. [A direct proof with Kolmogorov's inequality is also possible.]

12. Complete the details of proof of the generalized Andersen-Jessen Theorem, i.e., of Theorem 6.6. Similarly state and prove the corresponding decreasing parameter version (with inequality (36) of Section 6).

13. (a) Let $\{X_n, \mathcal{F}_n, n \geq 1\}$ be a submartingale on a probability space (Ω, Σ, P) and $c_n \downarrow 0$. Then, for each $\epsilon > 0$, establish the "maximal inequality"

$$\epsilon P\left[\max_{1 \leq k \leq n} c_k X_k \geq \epsilon\right] \leq c_1 E(X_1^+) + \sum_{k=2}^{n} c_k E(X_k^+ - X_{k-1}^+).$$

[*Hint:* If $A_i = [X_j < \epsilon/c_j, 1 \leq j \leq i-1,\ c_i X_i \geq \epsilon] (\in \mathcal{F}_n)$, then $\int_{\bigcup_{i=1}^{n} A_i} (X_{n+1}^+ - X_n^+) dP \geq 0$, and follow the proof of Proposition 4.5.]

(b) If in (a), $X_n \geq 0$, the submartingale lies in $L^p(\Sigma)$ for some $p \geq 1$, and for $c_k \downarrow 0$, $\sum_{k=1}^{\infty} c_k^p E(X_k^p - X_{k-1}^p) < \infty$, show that $\lim_{k \to \infty} c_k X_k = 0$, a.e. [*Hint:* Since $\{X_k^p, \mathcal{F}_k, k \geq 1\}$ is a submartingale, apply (a) and the following elementary but useful *Kronecker lemma:* If $0 \leq a_n \uparrow \infty$, and $b_n \in \mathbb{R}$ such that $\lim_n \sum_{k=1}^{n} \frac{b_k}{a_k}$ is convergent, then $\lim_n \frac{1}{a_n} \sum_{i=1}^{n} b_i = 0$.

Proof. Let $c = a_n - a_{n-1}$, $a_0 = 0$, and $d_n = \sum_{k=1}^{n} \frac{b_k}{a_k}, d_0 = 0$. So $\sum_{k=1}^{n} b_k = \sum_{k=1}^{n} a_k(d_k - d_{k-1}) = a_n d_n - \sum_{k=1}^{n} c_k d_{k-1}$, $\frac{1}{a_n} \sum_{k=1}^{n} b_k \to d - d = 0$ since $d_n \to d$ implies $\frac{1}{a_n} \sum_{k=1}^{n} c_k d_{k-1} = d + \sum_{k=1}^{n} \frac{c_k}{a_n}(d_k - d) \to d$. Let $b_n = (X_n^p - X_{n-1}^p)(\omega)$ and $c_k = a_k^{-p}$ in the problem. In this connection, see Chow [2].]

14. We present an alternative proof of Theorem 6.1 using an upcrossings lemma due to Doob [1]. Let $\{X_n, \mathcal{F}_n, n \geq 1\}$ be an adapted stochastic process in $L^1(\Omega, \Sigma, P)$. If $-\infty < a < b < \infty$, for the sequence of real numbers $\{X_n(\omega), n \geq 1\}, \omega \in \Omega$, define the "upcrossings" by this sequence of $[a, b]$ as follows: Let $k_1(\omega) = \min\{i : X_i(\omega) \leq a\}, k_2(\omega) =$

$\min\{i > k_1(\omega) : X_i(\omega) \geq b\}$; and by induction, let $k_{2j+1}(\omega) = \min\{i > k_{2j}(\omega) : X_i(\omega) \leq a\}$, $k_{2j+2}(\omega) = \min\{i > k_{2j+1}(\omega) : X_i(\omega) \geq b\}$. As usual we set $\min\{\emptyset\} = +\infty$. Let $\beta_n(\omega) = \max\{j : k_{2j}(\omega) \leq n\}$ with $\max\{\emptyset\} = 0$. Then β_n is interger-valued and is a measurable function of $X_1, \ldots, X_n$, and for each $\omega, \beta_n(\omega)$ is called the number of upcrossings of $[a, b]$ by the sequence $\{X_1(\omega), \ldots, X_n(\omega)\}$. Let us define the counting sequence: $u_i = \chi_{\bigcup_{j=1}^{i-1} A_j^i}$ where $A_j^i = \{\omega : k_{2j}(\omega) < i \leq k_{2j+1}(\omega)\}$, so that $u_i(\omega) = 1$ iff the sequence $\{X_1(\omega), \ldots, X_{i-1}(\omega)\}$ completes an upcrossing of $[a, b]$, and $u_i(\cdot)$ is $\sigma(X_1, \ldots, X_{i-1})$-measurable, $i \geq 2$.

(a) If $\{X_i, \mathcal{F}_i, 1 \leq i \leq n\}$ is a submartingale in $L^1(\Omega, \Sigma, P)$ and β_n is the upcrossings function of the process for the interval $[a, b]$, prove the key *upcrossings inequality*:

$$E(\beta_n) \leq \frac{1}{b-a} \int_{\{\omega: X_n(\omega) \geq a\}} (X_n - a)dP \leq \frac{E(|X_n|) + |a|}{b-a}.$$

[*Hints:* Let $X = \sum_{i=3}^{n} u_i(X_i - X_{i-1})$. By the submartingale property $E(X) \geq 0$. We may assume $\beta_n(\omega) > 0$ for some $\omega \in \Omega$, so that $k_{2\beta_n}(\omega) \leq n, k_{2\beta_n+1}(\omega) \leq n$ but $k_{2\beta_n+2}(\omega) > n$. Then $X(\omega) = (X_{k_3}(\omega) - X_{k_2}(\omega)) + \cdots + (X_{k_{2\beta_n-1}}(\omega) - X_{k_{2\beta_n-2}}(\omega)) + (X_n(\omega) - X_{k_{2\beta_n}}(\omega)) \leq (a-b)(\beta_n(\omega) - 1) + (X_n(\omega) - X_{2\beta_n}(\omega)) \leq (X_n(\omega) - a) + \beta_n(\omega)(a-b)$ because $X_{k_{2\beta_n}}(\omega) \geq b$ and $X_n(\omega) > a$ since $k_{2\beta_n+2}(\omega) > n$. Hence we deduce that $0 \leq E(X) \leq (a-b)E(\beta_n) + E((X_n - a)^+$.]

(b) If $\{X_n, \mathcal{F}_n, n \geq 1\}$ is a submartingale such that $K_0 = \sup_n E(|X_n|) < \infty$, then $X_n \to X_\infty$ a.e. and $E(|X_\infty|) \leq \underline{\lim}_n E(|X_n|) = K_0$. [*Hints:* Let $X^* = \limsup_n X_n$, and $X_* = \liminf_n X_n$. Then $A = \{\omega : X_*(\omega) < X^*(\omega)\} = \bigcup_{r_1 < r_2} A_{r_1 r_2}$, where $A_{r_1 r_2} = \{\omega : X_*(\omega) < r_1 < r_2 < X^*(\omega)\}$, r_i being rational. If β_n is the upcrossings function of $[r_1, r_2]$ by $\{X_1, \ldots, X_n\}$, then $\beta_n(\omega) \uparrow \infty$ for each $\omega \in A_{r_1 r_2}$. But by (a), $E(\beta_n) \leq \frac{K_0 + |r_1|}{r_2 - r_1} < \infty$, all n. By Monotone Convergence, this will be possible only if $A_{r_1 r_2}$ is P-null.]

15. The following special case of the martingale convergence is due to P. Lévy (1937). In view of Proposition 6.9 this can be used in obtaining the full result as given in the preceding problem. *Lévy's Theorem.* For any $A \in \mathcal{F}_\infty$ where $\mathcal{F}_n \uparrow \mathcal{F}_\infty \subset \Sigma$ on (Ω, Σ, P), $P^{\mathcal{F}_n}(A) = E^{\mathcal{F}_n}(\chi_A) \to$

$E^{\mathcal{F}_\infty}(\chi_A) = \chi_A = P^{\mathcal{F}_\infty}(A)$, a.e. [*Hints:* Let $0 < \epsilon, \delta < 1$. Now $A \in \mathcal{F}_\infty = \sigma\left(\bigcup_n \mathcal{F}_n\right) \Rightarrow$ there is an $n_0, B \in \mathcal{F}_{n_0}$, such that $P(A\Delta B) < \frac{\epsilon\delta}{2}$. If $C_n = \{\omega : 1 - P^{\mathcal{F}_n}(A)(\omega) \geq \epsilon\}$ so that $C_n \in \mathcal{F}_n, n \geq 1$, let $D_1 = B \cap C_1, \ldots, D_n = B \cap C_n - \bigcup_{i=1}^{n-1} D_i$. Thus $D_n \in \mathcal{F}_n$ and the D_n are disjoint. For $n \geq n_0$,

$$P(D_n - A) = P(D_n) - P(D_n \cap A) = \int_{D_n} (1 - P^{\mathcal{F}_n}(A))dP \geq \epsilon P(D_n),$$

by definition of $P^{\mathcal{F}_n}$. Since $D = \bigcup_{n \geq n_0} D_n \subset B$, we have

$$\frac{\epsilon\delta}{2} \geq P(B - A) \geq P(D - A) \geq \epsilon P(D).$$

So $P(D) < \delta/2$, and for $\omega \in B - D, 1 - P^{\mathcal{F}_n}(A)(\omega) < \epsilon$ if $n \geq n_0$. Let $N = (A\Delta B) \cup D$. Then $P(N) < \delta$ and for $n \geq n_0, \omega \in A - N$ we have $P^{\mathcal{F}_n}(A) \geq 1 - \epsilon$. Letting $\epsilon \to 0$ and then $\delta \to 0$ we get $\lim_n P^{\mathcal{F}_n}(A) = 1$ for almost all $\omega \in A$; similarly with $\Omega - A$. Combining, one funds $\lim_n P^{\mathcal{F}_n}(A) = \chi_A$, for all $A \in \mathcal{F}_\infty$. *Extension:* Now $E^{\mathcal{F}_n}(f) \to f$ a.e., if f is an $\mathcal{F}_\infty$ simple function. Since such functions are dense in $L^1(\mathcal{F}_\infty)$, and since $\sup_n |E^{\mathcal{F}_n}(f)| < \infty$ a.e., by Theorem 4.3, it follows that $E^{\mathcal{F}_n}(f) \to f$ for all $f \in L^1(\mathcal{F}_\infty)$ by a classical theorem of Banach's (cf. Dunford-Schwartz [1], IV.11.2). The latter is a deep result, and shows that the pointwise convergence of (sub-) martingales cannot be expected to emerge easily.]

16. Following Definition 4.1 we noted that a martingale can be thought of intuitively as a fair game. This may be "generalized" as follows: A process $\{X_n, \mathcal{F}_n, n \geq 1\} \subset L^1(\Omega, \Sigma, P)$, where $\mathcal{F}_n \uparrow \subset \Sigma$ are σ-algebras, is *fairer with time*, if for each $\epsilon > 0$

$$\lim_{\substack{m,n\to\infty \\ n \geq m}} P[|E^{\mathcal{F}_m}(X_n) - X_m| > \epsilon] = 0. \qquad (*)$$

Thus a martingale, being fair, remains so with time and satisfies $(*)$. The new process may thus be called an *approximate martingale.* Some properties of such a process are presented in this problem.

(a) Let $\{X_n, \mathcal{F}_n, n \geq 1\} \subset L^1(\Sigma)$ be a uniformly integrable approximate martingale. Then $X_n \to X_\infty$ in $L^1(\Sigma)$. [*Hints:* Write $Y_{mn} =$

$E^{\mathcal{F}_m}(X_n)$ for $n \geq m$. If $\mathcal{L} = \{X_n, n \geq 1\}$ and $\mathcal{T} = \{Y_{mn}, n \geq m \geq 1\}$, then the uniform integrability of $\mathcal{L}$ implies that of $\mathcal{T}$. In fact by the classical de la Vallée Poussin criterion (cf. Section 1 for a case), there is a convex $\varphi : \mathbb{R}^+ \to \mathbb{R}^+$ such that $\frac{\varphi(t)}{t} \uparrow \infty$ as $t \uparrow \infty, \varphi(0) = 0$ and $E(\varphi(X_n)) \leq K_0 < \infty$ all n. We may also assume that $\varphi(2t) \leq C\varphi(t), t \geq 0, 0 < C < \infty$, for convenience (or use $\tilde{\varphi}(\cdot) = \varphi(\frac{1}{\alpha_0}\cdot)$, some $\alpha_0 \geq 1$). By Theorem 1.8, $E(\varphi(Y_{mn})) = E(\varphi(E^{\mathcal{F}_m}(X_n))) \leq E(E^{\mathcal{F}_m}(\varphi(X_n))) = E(\varphi(X_n)) \leq K_0 < \infty$; the collection $\mathcal{T}$ is thus uniformly integrable. Hence $\mathcal{L} \cup \mathcal{T}$ is also uniformly integrable, and $\lim_{P(A)\to 0} \int_A |Y_{nm}| dP = 0$ uniformly in n, m. Since the process is an approximate martingale for each $\epsilon > 0, \delta > 0$ there exist $m_0 (= m_0(\epsilon, \delta))$ and n_0 $(= n_0(\epsilon, \delta)), n_0 \geq m_0$, such that $P\left[|X_{m_0} - Y_{m_0,n_0}| > \frac{\epsilon}{2}\right] < \delta$ and, by the uniform integrability of $\mathcal{L} \cup \mathcal{T}$, we may choose a $\delta_1 > 0$ such that $P(A) < \delta_1$ implies $\int_A |X_m - Y_{mn}| dP < \frac{\epsilon}{2}$, uniformly in m, n. Let $\delta \leq \delta_1$ above and using the corresponding m_0, n_0, we have for $n \geq m \geq n_0$,

$$(+) E(|X_m - Y_{mn}|) = \int\limits_{[|X_m - Y_{mn}| \leq \frac{\epsilon}{2}]} |X_m - Y_{mn}| dP + \int\limits_{[|X_m - Y_{mn}| > \frac{\epsilon}{2}]} |X_m - Y_{mn}| dP < \epsilon.$$

Next note that $\{Y_{mn}\}_{n \geq m}$ is a Cauchy sequence in $L^1(\Sigma)$ for $m \geq m_0$. In fact, let $m \leq n < n'$. So $E(|Y_{mn} - Y_{mn'}|) = E(|E^{\mathcal{F}_m}(X_n - X_{n'})|) = E(|E^{\mathcal{F}_m} E^{\mathcal{F}_n}(X_n - X_{n'})|) = E(|E^{\mathcal{F}_m}(X_n - Y_{nn'})|) \leq E(E^{\mathcal{F}_m}(|X_n - Y_{nn'}|)) = E(|X_n - Y_{nn'}|) < \epsilon$, by (+) and Jensen's inequality. Hence $Z_m = \lim_n Y_{mn}$ exists in the mean of $L^1(\Sigma), Z_m \in L^1(\Sigma)$. However, each Y_{mn} is $\mathcal{F}_m$- measurable so that Z_m is also. We now assert that $\{Z_m, \mathcal{F}_m, m \geq m_0\}$ is a uniformly integrable martingale. Since $E^{\mathcal{F}_m}(Z_{m+1}) = E^{\mathcal{F}_m}\left(\lim_n Y_{(m+1)n}\right) = \lim_n E^{\mathcal{F}_m}(Y_{(m+1)n})$ (by the uniform integrability of $\mathcal{T}$) $= \lim_n E^{\mathcal{F}_m}(E^{\mathcal{F}_{m+1}}(X_n)) = \lim_n E^{\mathcal{F}_m}(X_n) = \lim_n Y_{mn} = Z_m$, a.e., it follows that $\{Z_m, \mathcal{F}_m, m \geq m_0\}$ is a martingale. But each Z_m is the L^1-limit of $\{Y_{mn}; n \geq m \geq m_0\}$. Hence by Fatou's inequality $E(\varphi(|Z_m|)) \leq \liminf_n E(\varphi(|Y_{mn}|)) \leq K_0 < \infty$, and by the above it holds for $m \geq m_0$. Thus by the de la Vallée Poussin criterion, the martingale is uniformly integrable and $Z_m \to Z_\infty$ a.e., and in L^1 (cf. Theorem 6.2). So $Z_m = E^{\mathcal{F}_m}(Z_\infty)$ a.e. Finally, if $m \geq n_0, E(|X_m - Z_m|) \leq \lim_n [E(|X_m - Y_{mn}|) + E(|Y_{mn} - Z_m|)] \leq \epsilon$, by (+) and the fact that $Y_{mn} \to Z_m$ in L^1. Since $\epsilon > 0$ is arbitrary we deduce from $Z_m \to Z_\infty$ in $L^1(\Sigma)$ that $X_m \to X_\infty$ ($= Z_\infty$ a.e.) in $L^1(\Sigma)$. Notice the similarity of proofs here and in Theorem 5.2.]

(b) An arbitrary sequence of random variables $\{X_n, n \geq 1\} \subset L^1(\Sigma)$ is strongly convergent (i.e. is a Cauchy sequence) iff it is a uniformly integrable approximate martingale relative to some increasing sequence of σ-algebras $\mathcal{F}_n \subset \Sigma$. [*Hint:* If the given sequence is Cauchy then it is uniformly integrable and if we take $\mathcal{F}_n = \mathcal{F}_\infty = \sigma(X_i, i \geq 1)$ for all n so that the process trivially satisfies (*), it is an approximate martingale. The converse is a consequence of (a). This shows that an arbitrary uniformly integrable approximate martingale does not admit a *pointwise* a.e. convergence, and the notion of "fairer with time" is too weak a concept for such a conclusion.]

(c) If $\{X_n, \mathcal{F}_n, n \geq 1\} \subset L^1(\Sigma)$ is an approximate martingale, $\Delta_1 = 0$, and for $j > 1, \Delta_j = E^{\mathcal{F}_{j-1}}(X_j) - X_{j-1}$, let $Y_n = X_n - \sum_{j=1}^{n} \Delta_j = X_n - Z_n$, as in the Doob decomposition. Then $\{Y_n, \mathcal{F}_n, n \geq 1\}$ is a martingale but the Δ_j may have either sign. If $\{X_n\}_1^\infty, \{Y_n\}_1^\infty$ are both uniformly integrable, then $Z_n \to Z_\infty$ in $L^1(\Sigma)$, and $X_n \to X_\infty$ a.e., iff $Z_n \to Z_\infty$ a.e. [Regarding this problem, see Blake [1] and Subrahmanian [1].]

17. Let (Ω, Σ, μ) be a measure space, $\mathcal{F}_{n+1} \subset \mathcal{F}_n \subset \Sigma$ be σ-algebras such that each $\mathcal{F}_n$ is μ-rich. If $\nu : \Sigma \to \bar{\mathbb{R}}$ is σ-additive, $|\nu|(\Omega) < \infty, \mu_n = \mu|\mathcal{F}_n$, and $\nu_n = \nu|\mathcal{F}_n$, let $f_n = \frac{d\nu_n}{d\mu_n}$. Show that $f_n \to f_\infty$ a.e., even though $\frac{d\nu_\infty}{d\mu_\infty}$ need *not* exist, where $\nu_\infty = \nu|\mathcal{F}_\infty \left(\mathcal{F}_\infty = \bigcap_n \mathcal{F}_n\right)$. If $\mu(\Omega) < \infty$, then show that $\{f_n, \mathcal{F}_n, 1 \leq n \leq \infty\}$ is a uniformly integrable martingale. [*Hint:* Analyze and apply the procedure of proof of Theorem 6.8.]

18. Let $X = \{X_n, n \in \mathbb{N}\}$ be a (discrete) stochastic process and $f : \mathbb{N} \to \mathbb{R}$ be a mapping. Let P and P_f be the canonical measures of the X and $X + f$ processes in its Kolmogorov representation (cf., I.2.3(3))). Then f is called an *admissible translate* of X if P_f is P-continuous. There exist processes with admissible f but for which αf is not so, for some $\alpha \in \mathbb{R}$ (i.e., $P_{\alpha f}$ is not P-continuous). Let P_n be the n-dimensional measure of $(X_1, \ldots, X_n)$, and assume P_n has a density $p_n > 0$ a.e. (Lebesgue), $n \geq 1$. (Such processes exist.) If ${}_\alpha Y_n = \frac{p_n(X_1 - \alpha f(1), \ldots, X_n - \alpha f(n))}{p_n(X_1, \ldots, X_n)}$, and $\mathcal{B}_n = \sigma(X_1, \ldots, X_n)$, show that

$\{_\alpha Y_n, \mathcal{B}_n, n \geq 1\}$ is a positive martingale for P, and if $U_n^2 = {}_\alpha Y_n$, then $\{U_n, \mathcal{B}_n, n \geq 1\}$ is a uniformly integrable supermartingale, but $\{\varphi(U_n), n \geq 1\} \subset L^1(P)$, with $\varphi(x) = x^2$, is not uniformly integrable. [Admissible translates are important in Statistical Inference Theory, cf., Pitcher [1].]

19. Let $\{\mu_n, n \geq 1\}$ be a sequence of probability measures on the Borelian space $(\mathbb{R}^k, \mathcal{B})$. Using the projective limit theory show that there is a probability space (Ω, Σ, P) and a martingale $\{X_n, n \geq 1\}$ on it, with values in $\mathbb{R}^k$, such that $\mu_n = P \circ X_n^{-1}$, iff (i) $\int_{\mathbb{R}^k} |x| d\mu_n(x) < \infty, n \geq 1$, and (ii) for any continuous real concave function ξ on $\mathbb{R}^k$, the sequence $\{\int_{\mathbb{R}^k} \xi(x) d\mu_n(x), n \geq 1\}$ is nonincreasing. If $k = 1$ and ξ is moreover nonincreasing, then the same characterization holds if "martingale" is replaced by "submartingale" in the above. [Observe that, given the "marginals" μ_n, we have to show the existence of a probability measure on $\mathbb{R}^k \times \mathbb{R}^k$ (and thus on finitely many products) with the given μ_n, μ_m as its projections or marginals, so that I.2.1 applies. This is the key point of the existence proof and it depends on a variational inequality. For this and related results, see Strassen ([1], Theorem 7).]

Bibliographical remarks

The basic concept of conditional expectation relative to any σ-subalgebra of a probability space, together with its functional properties, has been given by Kolmogorov in his Foundations [1]. The conditional Jensen inequality, in its general form (Theorem 1.8), is due to Chow [1] and the simple equality proof is due to N. Dinculeanu (private communication). The result of theorem 1.14, in the case $p = 2$, goes back to Girshick and Savage [1], and the proof of the general case is taken from the author's paper [5]. Conditional expectations on very general measure spaces are discussed in Dinculeanu [2].

The first detailed attempt on the study of conditional expectations and averaging operators on L^p-spaces, from a probabilistic point of view, appears to be due to Moy [1] which was based in part on the pioneering work of J. Kampé de Fériet (see [1], [2] and references there to his earlier work). Later many researchers worked on these problems, in more general spaces, with nonfinite measures. The regularity of conditional probabilities was considered by Doob [1], and there is still active research going on in this area. The presentation in the text follows the author's papers ([2], [3], [4]) and an extensive bibliography for the earlier work may be found in [3] where the vector measure point of view for the conditional probability functions was advocated, apparently for the first time (see also Dubins [1] and Olson [1] for an anticipation of the idea). A comprehensive treatment of conditioning is given in the recent book by the author (Rao [13]).

The concept of a martingale seems to have been introduced by P. Lévy and J. Ville in the late 1930's, but their importance in probability theory and in general analysis has been recognized by Doob who proved the key convergence theorems and extended the work to submartingales. An account of all this appears in his monograph [1], which will also be referred to concerning his earlier work. The treatment in the text, however, incorporated many later developments and ideas, as noted there. The equivalence of Andersen-Jessen theory and that of Doob's seems to have been given explicitly only in the author's papers ([5], [7]). See also the work of Johansen and Karush [1].

Several important results are discussed in the Complements section. We shall not repeat the textual references which have already been indicated. The reader would note that we have not touched the continuous parameter processes in this chapter. One needs the "stopping times" transformations for this purpose, and some technical problems have to be settled before that. This and the preceding chapter contain the set of results that are obtainable with the basic ideas of clasical analysis.

Chapter III

Stochastic function theory

The main aim of this chapter is to study in some detail certain technical problems arising in the treatment of continuous parameter stochastic processes. The concepts of separability and measurability are introduced and analyzed for general classes of processes. This can be done abstractly and more rapidly through the use of lifting theory and a brief discussion of the lifting theorem is included. The existence of separable and measurable modifications under various conditions is established. We illustrate these results by proving some stochastic function theoretical results, including Kolmogorov's criterion for sample path continuity. Then we present some convergence theorems for continuous parameter martingales, under certain Vitali conditions. As an adjunct we include a general result on the existence of projective limits of projective systems of conditional probability measures, generalizing the classical case of Tulcea's theorem. This work prepares for many refinements of martingale theory, with stopping times to be treated in the next chapter.

3.1 Separability and related hypotheses: standard case

If $\{X_t, t \in T\}$ is a real stochastic process on a probability space (Ω, Σ, P), then in many studies, such as stochastic differential and integral calculus or statistical inference problems on such processes, it will be necessary to consider events involving expressions of the form $\sup\{X_t : t \in T\}$, $\inf\{X_t : t \in T\}$, $\lim_{\pi} \sum_{\pi} X_{t_i}$, $(\pi = \{t_1, \dots, t_n\} \subset T$, π's directed by refinement) or $\liminf_{t \to t_0} \frac{X_t - X_{t_0}}{t - t_0}$. However, if $T = (a, b)$, $a < b$, then the above functions need not be measurable since T is uncountable, and indeed one may construct simple examples resulting in nonmeasurability.

Therefore it is necessary to find conditions on the process in order that the above quantities are measurable or at least differ from measurable objects only on subsets of a P-null set. Notice that if the probability space is replaced by its canonical representation (Kolmogorov's theorem) $(\mathbb{R}^T, \mathcal{B}_T, P)$ then, as noted in Exercise I.5.9, a set $A \in \mathcal{B}_T$ iff it is determined by a countable collection of indices $J_A \subset T$. Since the above noted quantities are determined by uncountable sets of T, it is clear that there is a nontrivial measurability problem in the treatment of uncountable or "continuous" indexed processes. We thus introduce the desired concept, called separability, at two levels of generality for convenience.

1. Definition. Let $\{X_t, t \in T\}$ be a real stochastic process on a probability space (Ω, Σ, P) where $T \subset \mathbb{R}$. If $\mathcal{A}$ is a class of Borel sets of $\mathbb{R}$, let $V(I, A)$ be the ω-set defined for $I \subset T$, $A \in \mathcal{A}$ by $V(I, A) = \{\omega : X_t(\omega) \in A \text{ for every } t \in I\} = \bigcap_{t \in I} \{\omega : X_t(\omega) \in A\}$ (i.e., $V(I, A)$ is the set of "paths" ω that remain in A for the duration of the "time" I). Then the X_t-process is said to be *separable relative* to $\mathcal{A}$ if for each open $I \subset T$ and $A \in \mathcal{A}$ we have: (i) $V(I, A) \in \Sigma$ and (ii) $P(V(I, A)) = \inf\{P(V(J, A)) : J \subset I, \text{ finite}\}$. If $\mathcal{A}$ is the class of closed sets, we often omit the qualification "relative to $\mathcal{A}$."

Since the sups, infs, etc., can be expressed in terms of union and intersection operations, the above definition includes all the results of interest in this work. Because generally $V(I, A)$ is nonmeasurable for Σ, the separability condition restricts the class of continuous parameter processes admitted in such operations preserving measurability. But (i) and (ii) are automatic if T is countable, so that the conditions are suitably abstracted for an uncountable index set.

Two processes $\{X_t, t \in T\}$ and $\{Y_t, t \in T\}$ on (Ω, Σ, P) are said to be *equivalent* if $X_t = Y_t$ a.e., $t \in T$. A set of functions $\{Z_t, t \in T\}$, $Z_t : \Omega \to \mathbb{R}$, can satisfy $X_t(\omega) = Z_t(\omega)$ for all $\omega \in \Omega - N_t, P(N_t) = 0$. In this case also "$Z_t$ is equivalent to X_t" a.e., but it will be a stochastic process iff each Z_t is measurable for (Σ). Clearly $X_t = Z_t$ a.e. implies that Z_t is measurable for (Σ) if every subset of a P-null set is in Σ. Thus the problem will not arise if (Ω, Σ, P) is complete. Any equivalent stochastic process $\{Y_t, t \in T\}$ to $\{X_t, t \in T\}$ is also called a *modification* of $\{X_t, t \in T\}$. The following extremely simple example is instructive in amplifying the differences between these concepts.

Let (Ω, Σ, P) be the Lebesgue unit interval and consider $\{X_t, t \in T\}$ and $\{Y_t, t \in T\}$, two processes where $Y_t = 0$ for all $t \in T = [0,1]$, and $X_t(\omega) = 1$ if $t = \omega$, $\omega \in A$, and $= 0$ otherwise, $A \subset \Omega$ being a set to be specified. It is clear that both are measurable and hence stochastic processes for every $A \subset \Omega$, and that $X_t = Y_t$, a.e. So Y_t is a modification of X_t, but the Y_t-process has continuous sample paths everywhere, while the X_t-process has discontinuous sample paths a.e. on A (and a.e. on Ω if $A = \Omega$). If A is a Lebesgue nonmeasurable set, and $C = [0, a]$, $0 < a < 1$, $I = T$, then $V(I, C) = C \cap A$, $A \notin \Sigma$ (using the notation of Definition 1) so that the X_t-process is not separable. But it is equivalent to a separable process; namely, $\{Y_t, t \in T\}$. Thus replacing a nonseparable one by a separable modification, when it exists, allows us to continue further analysis. This indicates the importance of knowing the separability of a process or at least the existence of a separable modification. In particular, if we consider the canonical representation of the process ($\Omega = \mathbb{R}^T$), and if the process is right (or left) continuous, i.e., $X_t(\omega) = \omega(t) = \omega(t+0) = X_{t+0}(\omega)$ (or $X_{t-0}(\omega) = X_t(\omega)$) for each $t \in T$, then $V(I, C) = \bigcap_{t \in I} X_t^{-1}(C) \in \Sigma$, for each $I \subset T$ and closed $C \in \mathcal{A}$; so (i) holds, and condition (ii) of Definition 1 is immediate. Thus such a process is separable. But, as the preceding example indicates there may be many different modifications. Also we usually replace $\mathbb{R}$ by $\bar{\mathbb{R}}$ in applying this concept. Hence one can take $\mathcal{A}$ as compact sets.

It is evident that, in Definition 1, we have not used any special properties of X_t being real valued or T being a subset of $\mathbb{R}$. Since in applications X_t can take values in a more general space and the index set can be multidimensional, we first restate the concept for later use in the following more abstract form.

2. Definition. Let (Ω, Σ, P) be a probability space and $(S, \mathcal{B})$ be a measurable space where S is completely regular (i.e., a topological space in which points are closed and a point and a closed set not containing it can be separated by a real bounded continuous function) and $\mathcal{B}$ is the Baire σ-algebra of S (i.e., the smallest σ- algebra relative to which every bounded continuous function on S is measurable). Let

$X_t : \Omega \to S$ be $(\Sigma, \mathcal{B})$- measurable and $X_t(\Omega)$ be relatively compact in S. Then $\{X_t, t \in T\}$ is a stochastic process with T as an index set. If $\mathcal{F}$ is a class of subsets of T such that $\cup_{F \in \mathcal{F}} F = T$ and $\mathcal{A}$ is a class of compact Baire sets, then the process is *separable* (for $\mathcal{A}$, but again this is often omitted) if for each $I \in \mathcal{F}$ and $C \in \mathcal{A}$, we have (i) $V(I, C) = \cap_{t \in I} X_t^{-1}(C) \in \Sigma$, and (ii) $P(V(I,C)) = \inf\{P(V(J,C)) : J \subset I, \text{ finite}\}$. If moreover, T is a topological space and $D \subset T$ is dense denumerable, then it is called a *universal separating set* of the X_t-process if for each $I \in \mathcal{F}$, $C \in \mathcal{A}$, $V(I \cap D, C) - V(I, C)$ is a P-null set (depending on I and C), so that $P(V(I \cap D, C)) = P(V(I, C))$. Here $\mathcal{F}$ is taken as a class of open sets forming a base for the topology of T.

In the remainder of this section we prove that if $S = \mathbb{R}$ and $T \subset \mathbb{R}$ or slightly more generally S and T are compact separable spaces, then a stochastic process $\{X_t, t \in T\}$ on a given probability space always admits a separable modification with a dense set $D \subset T$ as a universal separating set. However, in the general case we need to use the lifting theorem, to be given in the next section. The present result will be sufficient for many applications. In the case that $S = \mathbb{R}$, $T \subset \mathbb{R}$, it is essentially due to Doob [1] and Ambrose [1].

3. Theorem. *Let $\{X_t, t \in T\}$ be a real stochastic process on a probability space (Ω, Σ, P) where $T \subset \mathbb{R}$. Then there exists an extended real valued stochastic process $\{\tilde{X}_t,\ t \in T\}$ on the same probability space such that it is separable (for the closed sets of the compactified line $\bar{\mathbb{R}}$) and $P[X_t = \tilde{X}_t] = 1, t \in T$, i.e., $\tilde{X}_t$ is a modification of $X_t, t \in T$. Moreover, there exists a dense denumerable set D which is a universal separating set for the X_t- and hence for the $\tilde{X}_t$-process.*

Proof. We arrange the proof such that it is also valid if the range $\bar{\mathbb{R}}$ is replaced by a compact metric space S. Let us first establish the last part on the existence of a universal separating set $D \subset T$ without mention of separability.

Let C be a compact subset of $\bar{\mathbb{R}}$ and $I \subset T$ be any set. If $F \subset I$ is a finite set, then $V(F, C) \in \Sigma$; and if $\alpha = \inf\{P(V(J, C)) : J \subset I, \text{ finite}\}$ then for each n, there is a finite set $H_n \subset I$ such that $P(V(H_n, C)) <$

$\alpha+1/n$. Letting $D_{I,C} = \cup_{n=1}^{\infty} H_n \subset I$, one may take $D_{I,C}$ as a universal separating set of the process for the pair (I, C), since one can (and does) assume its density in I by adding a countable set of points if necessary. Now let $\mathcal{T}$ be the collection of all open intervals of $\mathbb{R}$ with rational end points, and $\mathcal{C}$ be a countable class of compact sets of $\bar{\mathbb{R}}$ whose complements form a neighborhood base of $\bar{\mathbb{R}}$. For each pair $(I \cap T, C)$, $I \in \mathcal{T}$, $C \in \mathcal{C}$, we have a countable dense set $D_{I,C} \subset I \cap T$ which is a universal separating set. If $D = \cup\{D_{I,C} : I \in \mathcal{T}, C \in \mathcal{C}\}$, then D is a dense subset of T and we claim that it is a universal separating set for the process (relative to $\mathcal{C}$). Clearly D is countable.

If $A \subset T$ is an open set then, by the well-known properties of the real line, it is the union of a countable collection of disjoint open intervals so that we may also write it as $A = \lim_n I_n$ where $I_n \subset I_{n+1}$ are open intervals, and $I_n - I_{n-1} \in \mathcal{T}$. If $B \subset \bar{\mathbb{R}}$ is any compact set, then it is expressible as $B = \cap_n B_n$, $B_n \subset B_{n-1}$ for some $B_n \in \mathcal{C}$ since $\mathcal{U} = \{C^c : C \in \mathcal{C}\}$ is a base of the topology of $\bar{\mathbb{R}}$. Hence we need only establish that

$$P[V(D \cap A, B)] = \inf\{P[V(J, B)] : J \subset A, \text{ finite}\}. \tag{1}$$

But each finite $J \subset I_{n_0}$ for large enough n_0. Hence

$$\begin{aligned} P[V(D \cap A, B)] &= P\left[\bigcap_n V(D \cap A, B_n)\right] \\ &= \lim_n P[V(D \cap A, B_n)] \\ &\le \lim_n P[V(D \cap I_{n_0}, B_n)] \\ &\le \lim_n P[V(D \cap J, B_n)] \\ &\le P[V(J, B)], \qquad \text{since } \lim_n B_n = B. \end{aligned} \tag{2}$$

Hence taking infimum as $J \subset A$ varies we get (1).

From (1) we conclude that for each compact $B \subset \bar{\mathbb{R}}$, and each $t \in T$,

$$P[V(D, C)] = P[V(D \cup \{t\}, C)], \quad C \in \mathcal{C}, \tag{3}$$

since D is a universal separating set. Thus $N_{t,C} = V(D, C) - V(D \cup \{t\}, C)$ is a P-null set for every $C \in \mathcal{C}$, and also $P(N_{t,B}) = 0$ for any compact $B \subset \bar{\mathbb{R}}$. Moreover if ($\mathcal{C}$ being countable) $N_t = \cup\{N_{t,C} :$

$C \in \mathcal{C}\}$, then $P(N_t) = 0$ and since $B = \cap_n B_n$, $B_n \in \mathcal{C}$, we see that $N_{t,B} \subset N_t$ holds. It follows that D is a universal separating set for each modification of the X_t-process, if the latter exists.

We now establish the existence of a separable modification $\tilde{X}_t$ of the X_t-process. For each $I \in \mathcal{T}$, $\omega \in \Omega$, let $A_0(I,\omega) = \{X_t(\omega) \in \mathbb{R} : t \in I\}$ and $A(I,\omega) = \overline{A}_0(I,\omega)$, the closure in $\bar{\mathbb{R}}$. Then $A(I,\omega)$ is compact in $\bar{\mathbb{R}}$ and $I_1 \subset I_2$ implies $A(I_1,\omega) \subset A(I_2,\omega)$. Moreover $A(D \cap I,\omega)$ is nonempty for each $I \in \mathcal{T}$ so that for each $t \in T$, $A(t,\omega) = \cap\{A(D \cap I,\omega) : t \in I, I \in \mathcal{T}\}$ is nonempty. Next define the process $\{\tilde{X}_t, t \in T\}$, on noting that $X_t \in A(t,\omega)$ for each $t \in T$ and $\omega \in N_t^c$, as:

$$\tilde{X}_t(\omega) = X_t(\omega), \text{ if } t \in D, \omega \in \Omega, \text{ or if } t \in T - D \text{ and } \omega \in N_t^c. \tag{4}$$

If $\omega \in N_t$ and $t \in T$ then by density of D in T, there exists a sequence $t_n \in D$ such that $t_n \to t$ and $\{X_{t_n}(\omega)\}_{n=1}^{\infty} \subset \bar{\mathbb{R}}$. Since $\bar{\mathbb{R}}$ is a compact space, there exists a convergent subsequence $\{X_{t_{n_j}}(\omega)\}_{j=1}^{\infty}$ with limit $\bar{X}_t(\omega) \in \bar{\mathbb{R}}$. We define $\tilde{X}_t(\omega) = \bar{X}_t(\omega)$ in this case ($\omega \in N_t$). Since on N_t^c, $\tilde{X}_t = X_t$ it is measurable for the trace σ-algebra $\Sigma(N_t^c) \subset \Sigma$. But by definition of $\bar{X}_t$, it is also clear that $\bar{X}_t$ is measurable for $\Sigma(N_t) \subset \Sigma$. Thus $\{\tilde{X}_t,\, t \in T\}$ is a stochastic process and since $P(N_t) = 0$, $P[X_t = \tilde{X}_t] = 1$ for each $t \in T$ with D as a separating set.

It remains to show that the $\tilde{X}_t$-process is separable for $\mathcal{C}$. This will follow if we prove that the sets $\tilde{V}(T \cap I, C)$, $\tilde{V}(D \cap I, C)$ defined for the $\tilde{X}_t$-process are the same for each $I \in \mathcal{T}$, $C \in \mathcal{C}$ (cf., Definition 1). Since $\tilde{V}(D \cap I, C) \supset \tilde{V}(T \cap I, C)$, let ω be a point in the first set so that $\tilde{X}_t(\omega) \in C$ for all $t \in D \cap I$. By definition of the $\tilde{X}_t$-process, when $\tilde{X}_t(\omega) \in C$ for all $t \in D \cap I$, one has $\tilde{X}_t(\omega) = X_t(\omega) \in A(t,\omega)$. This is also true if $t \in (T - D) \cap I$ and $\omega \in N_t^c$. If $\omega \in N_t$ then $\tilde{X}_t(\omega) = \lim_{n_j} X_{t_{n_j}}(\omega)$, for $t_{n_j} \in D \cap I$ since $D \cap I$ is dense in $T \cap I$. But by definition $X_{t_{n_j}}(\omega) = \tilde{X}_{t_{n_j}}(\omega) \in C$, for each $t_{n_j} \in D \cap I$, and since C is closed the limit $\tilde{X}_t(\omega) \in C$ as well. Hence $\tilde{V}(D \cap I, C) \subset \tilde{V}(T \cap I, C)$ also holds. This proves the separability of the $\tilde{X}_t$-process, and thus all the assertions are established. □

4. Remark. A more general situation needs different methods when the range is not necessarily a separable metric space (e.g. a completely regular space) and T is some index set. Such a problem will be solved below using a lifting theorem. It will clarify the special construction involved here. We also note in passing that if we define $\tilde{X}_t(\omega)$ for $\omega \in N_t$ as any element of $A(t, \omega)$ in the above proof, then the resulting function while agreeing with the X_t on N_t^c need not be measurable (since the subsets of N_t may not be in Σ) unless Σ is completed. Thus the separability of $\tilde{X}_t$ (with $X_t = \tilde{X}_t$ a.e.) is valid for non-complete Σ, but in the general case this will no longer be possible. Also the separability of a process does not imply anything on the regularity of its sample functions, since if $\Omega = \{\omega\}$, a singleton, every process is separable.

In many problems such as stochastic integration, to be treated later, it will be necessary to know the joint measurability of a stochastic process $\{X_t, t \in T\}$ considered as a function $X(\cdot,\cdot) : T \times \Omega \to S$. Thus if $\mathcal{T}$ is a σ-algebra of subsets of T and (Ω, Σ, P) is a probability space, S being completely regular with $\mathcal{B}$ as its Baire σ-algebra, then we say that $\{X_t, t \in T\}$ is a *measurable process* if the function $X(\cdot,\cdot)$ is measurable relative to $(\mathcal{T} \otimes \Sigma)$, i.e., $X^{-1}(\mathcal{B}) \subset \mathcal{T} \otimes \Sigma$. Since $X(t, \cdot) = X_t(\cdot)$ by definition, and for each t, X_t is $(\Sigma, \mathcal{B})$-measurable by virtue of the fact that $\{X_t, t \in T\}$ is a stochastic process, we would like to know the conditions under which it is a measurable process. Obviously not every stochastic process is a measurable process. For instance, if $\Omega = \{\omega\}$ is again a singleton, $T = [0, 1]$ is the Lebesgue interval, and $X(t, \omega) = f(t)$, where $f(\cdot) : T \to \mathbb{R}$ is a Lebesgue nonmeasurable function, then $\{X_t, t \in T\}$ is not a measurable process. However, in a sense, such an example is exceptional. The next result, in the case of the Kolmogorov representation, gives some insight regarding the preceding comment, and is an extension, due to Nelson [1], of Doob's work [1]. In what follows we take $\mathcal{T}$ as the Baire (= Borel) σ-algebra of the separable T.

5. Theorem. *Let T, S be compact Hausdorff spaces each with a countable base. Let $\Omega = S^T$, $\Sigma = \mathcal{B}_T$ where $\mathcal{B}$ is the Borel σ-algebra of S, and $\{X_t, t \in T\}$ be the canonical representation of a stochastic process on (Ω, Σ) so that $\omega \in \Omega$ implies $X_t(\omega) = \omega(t) \in S$ and $X_t^{-1}(\mathcal{B}) \subset \Sigma$, $\mathcal{B}_T$ being the smallest σ-algebra relative to which each X_t is measurable. Then the (stochastic) function $X(\cdot,\cdot) : T \times \Omega \to S$ has the following properties: (i) the set $A = \{(t, \omega) : \omega(\cdot)$ is discontinuous at $t\} \in \mathcal{T} \otimes \tilde{\Sigma}$, and (ii) $X(\cdot,\cdot)$ is Borel measurable on the set A^c, i.e., $X : A^c \to S$ is*

$((\mathcal{T} \otimes \tilde{\Sigma})(A^c), \mathcal{B})$-measurable, $\tilde{\Sigma}$ being the Borel σ-algebra of Ω.

Remark. It is possible that a t-section of A, A_t has positive probability. On A^c the sample paths $X_{(\cdot)}(\omega) = \omega(\cdot)$ are all continuous and the process is measurable there. Note that P played no part in the hypotheses or conclusions thus far, but we are considering the larger $\tilde{\Sigma}$ ($\supset \Sigma$) here.

Proof. The hypothesis implies that T and S are metrizable. Let d_1, d_2 be the corresponding metrics on them. Thus the Baire and Borel sets of S are also the same. Let (by compactness) $\{G_i^n\}_{i=1}^{k_n}$ be an open covering of T of diameter at most $1/n$ for each $n \geq 1$. If $f : S \to \mathbb{R}$ is a continuous function, consider $\{f(X(t,\omega)),\ t \in T,\ \omega \in \Omega\}$. Let $\{H_i^n\}_{i=1}^{k_n}$ be a disjunctification of $\{G_i^n\}_{i=1}^{k_n}$, and define (omitting f from Y_n^f notation):

$$Y_n^+(t,\omega) = \sum_{i=1}^{k_n} \chi_{H_i^n}(t) \sup\{f(X(s,\omega)) : s \in H_i^n\}, \tag{5}$$

and similarly $Y_n^-(t,\omega)$, replacing 'sup' by 'inf' in (5). Let us show that Y_n^+ and Y_n^- are $(\mathcal{T} \otimes \tilde{\Sigma})$-measurable and that $\lim_{n\to\infty} Y_n^{\pm}(t,\omega) = Y^{\pm}(t,\omega)$ exists. This will essentially complete the argument.

Since $G_i^n \in \mathcal{T}$, it is clear that $\chi_{H_i^n}$ is $\mathcal{T}$-measurable for each i and n. To see the second factor in (5) is $\tilde{\Sigma}$-measurable, we observe that for any subset $H \subset T$ and $x \in \mathbb{R}$,

$$\begin{aligned}\left\{\omega : \sup_{t\in H} f(X(t,\omega)) \leq x\right\} &= \bigcap_{t\in H} \{\omega : f(X(t,\omega)) \leq x\} \\ &= \bigcap_{t\in H} \{\omega : f(\omega(t)) \leq x\},\end{aligned} \tag{6}$$

because $X(t,\omega) = \omega(t)$. Since f is continuous, the right side set of (6) is a closed cylinder and hence is in $\tilde{\Sigma}$. (If H is countable then it is in Σ itself.) Similarly,

$$\left\{\omega : \inf_{t\in H} f(X(t,\omega)) < x\right\} = \bigcup_{t\in H} \{\omega : f(\omega(t)) < x\} \tag{7}$$

is an open cylinder and so is in $\tilde{\Sigma}$. It follows that Y_n^+, Y_n^- are $\mathcal{T} \otimes \tilde{\Sigma}$-measurable. Next define

$$Y^+(t,\omega) = \limsup_{s \to t} f(X(s,\omega))(= \inf_{\varepsilon > 0} \sup_{t,s \in G} \{f(X(s,\omega)) : \mathrm{diam}(G) \le \varepsilon\}). \tag{8}$$

If $H_{i_0}^n$ is a set containing $t \in T$ (and there is exactly one such set) then

$$Y^+(t,\omega) \le Y_n^+(t,\omega) \le \sup_{s \in H_{i_0}^n} f(X(s,\omega)). \tag{9}$$

If now we let $n \to \infty$, so that $\mathrm{diam}(H_{i_0}^n) \le \frac{1}{n} \to 0$, then the right side of (9) tends to $Y^+(t,\omega)$ by definition (see (8)). Since Y_n^+ is $\mathcal{T} \otimes \tilde{\Sigma}$-measurable, so is Y^+ ($= \lim_n Y_n^+$ pointwise). Similarly $Y^- = \lim_n Y_n^-$ is $\mathcal{T} \otimes \tilde{\Sigma}$-measurable. In general however $f(X(t,\omega)) \ne Y^\pm(t,\omega)$ unless $f(X(\cdot,\omega)) = f(\omega(\cdot))$ is upper (or lower) semicontinuous at t.

We now define the set A_f of discontinuity points of $f \circ X$:

$$A_f = \{(t,\omega) : Y^+(t,\omega) > Y^-(t,\omega)\}. \tag{10}$$

Then A_f is in $\mathcal{T} \otimes \tilde{\Sigma}$ for each $f \in C(S)$, the space of real continuous functions on the given compact space S. Hence there exists a dense denumerable set $\{f_n\}_{n=1}^\infty \subset C(S)$, and if $A = \cup_{n=1}^\infty A_{f_n}$, then $A \in \mathcal{T} \otimes \tilde{\Sigma}$ and it is the set of discontinuity points of $X(\cdot,\cdot)$ since f_n is continuous on S. Thus $X(\cdot,\cdot)$ is continuous on A^c and therefore is $((\mathcal{T} \otimes \tilde{\Sigma})(A^c), \mathcal{B})$-measurable. □

We now present another result analyzing the discontinuity set $A = \cup_n A_{f_n}$, and in particular investigate the measurability of X on A. Let T, S be compact metric spaces, as in the above theorem, and let $\{X_t, t \in T\}$ be a stochastic process with the canonical representation on (Ω, Σ, P) with values in S. A point $t_0 \in T$ is said to be a *fixed point of discontinuity* of the process if $t_n \in T$, $t_n \to t_0$ in the topology of T, then it is false that $X_{t_n} \to X_{t_0}$ with probability one, i.e., almost all sample paths are not continuous at t_0, so that there is positive probability for a sample path discontinuity at t_0. This can be stated alternatively as follows. Let $U(t,r) = \{s \in T : d_1(s,t) < r\}$ be the ball of radius r and center at $t \in T$, where $d_1(\cdot,\cdot)$ is the distance function of T (and let d_2 be that of S). If for each $\varepsilon > 0$, we define the cylinder set of Ω by the equations $\Delta(t_1,t_2,\varepsilon) = \{\omega : d_2(\omega(t_1),\omega(t_2)) \le \varepsilon\}$

where $X_{t_i}(\omega) = \omega(t_i)$ is used here, then the class of all "ε-continuous functions" of Ω at $t \in T$ is

$$\Delta_t^\varepsilon = \bigcup_{n=1}^{\infty} \bigcap \left\{ \Delta(t_1, t_2, \varepsilon) : t_1, t_2 \in U\left(t, \frac{1}{n}\right) \right\}. \tag{11}$$

The *continuity set* of ω's at t is then $\Delta_t = \cap_{n=1}^{\infty} \Delta_t^{1/n}$. Since $\Delta(t_1, t_2, \varepsilon)$ is closed, so that the set in braces in (11) is closed, we deduce that $\Delta_t^\varepsilon \in \Sigma$ for each $\varepsilon > 0$. Hence $\Delta_t \in \Sigma$ whatever be the probability measure P on Σ. Thus $t_0 \in T$ is a fixed point of discontinuity of the process iff $P(\Delta_{t_0}) < 1$. Let $\Theta_P \subset T$ be the set of all fixed discontinuity points of the process. Let Σ_P be the P-completion of Σ $(\tilde{\Sigma} \subset \Sigma_P,\ P$ being a regular measure).

The measurability and discontinuity of such a process are connected as follows:

6. Theorem. *Let S, T be compact metric spaces and (Ω, Σ, P) be the canonical representation of a stochastic process $\{X_t, t \in T\}$ with values in $(S, \mathcal{B})$ where $\Sigma = \mathcal{B}_T$, $\Omega = S^T$. Suppose μ is a Radon measure on the Borel σ-algebra $\mathcal{T}$ of T. Let $\Delta_\mu = \{\omega \in \Omega : \omega(\cdot)$ is continuous at μ-almost all $t \in T\}$. Then $\Delta_\mu \in \Sigma_P$, and $P(\Delta_\mu) = 1$ iff ($\Theta_P \in \mathcal{T}$ and) $\mu(\Theta_P) = 0$. When the last condition holds and when the product measure space $(T \times \Omega, \mathcal{T} \otimes \tilde{\Sigma}, \mu \otimes P)$ is complete, then $X(\cdot, \cdot)$ is measurable, i.e., the family $\{X_t, t \in T\}$ is a measurable stochastic process.*

Proof. Let A be the discontinuity set of ω's at some t, as in Theorem 5. Then $A \in \mathcal{T} \otimes \tilde{\Sigma}$ by that result. But $A_t = \{\omega \in \Omega : (t, \omega) \in A\} = \Omega - \Delta_t$ and the ω-section $A_\omega = \{t \in T : (t, \omega) \in A\} = \Theta_P$ by definition. Hence $\Theta_P \in \mathcal{T}$ and $A_t \in \tilde{\Sigma}$. Consequently by Fubini's theorem we have:

$$\int_\Omega \mu(A_\omega) dP = \int_T P(A_t) d\mu = \int_\Omega \int_T \chi_A \, d\mu dP. \tag{12}$$

Hence $P(A_t) = 1 - P(\Delta_t) = 0$ for almost all t iff $\mu(A_\omega) = 0$ for almost all ω. When this holds (12) implies $(\mu \otimes P)(A) = 0$. If now $(T \times \Omega, \mathcal{T} \otimes \tilde{\Sigma}, \mu \otimes P)$ is complete, then Theorem 5 (ii) yields the measurability of the process. It remains to show that $\Delta_\mu \in \Sigma_P$ and $P(\Delta_\mu) = 1$.

Clearly $\Delta_\mu \subset \Delta_t$ for almost all t. Let $T_0 \subset T$ be the exceptional set; $\mu(T_0) = 0$. Then $\Delta_\mu \subset \Delta_t$, $t \in T_1 = T - T_0$ and $\Delta_\mu = \cap\{\Delta_t : t \in T_1\}$. Since $\{\Delta_t : t \in T_1\} \subset \Sigma_P$, by a standard computation using the fact that Σ_P is complete, we deduce that $\Delta_\mu \in \Sigma_P$. If D is the class of finite subsets of T_1, it is directed under inclusion. Let $\tau \in D$. Then letting $\Delta_\tau = \cap_{t\in\tau} \Delta_t \in \Sigma$, it is clear that $\Delta_\mu = \cap\{\Delta_\tau : \tau \in D\}$ and that the class $\{\Delta_\tau, \tau \in D\}$ is filtering to the left. Since $P(\Delta_t) = 1$, $t \in T_1$, we have $1 = P(\Delta_{t_1} \cup \Delta_{t_2}) = P(\Delta_{t_1}) + P(\Delta_{t_2}) - P(\Delta_{t_1} \cap \Delta_{t_2})$; so $P(\Delta_{t_1} \cap \Delta_{t_2}) = 1$. Thus $P(\Delta_\tau) = 1, \tau \in D$. From this we may conclude (cf. also Theorem 2.4 below) that $P(\Delta_\mu) = \lim_\tau P(\Delta_\tau) = 1$. □

It is clear that in the above results we may allow T and S to be σ-compact metric spaces. Thus $T \subseteq \mathbb{R}$ and $S = \mathbb{R}$ may be admitted. The last part has the consequence that, if the process has no fixed points of discontinuities and $\mu \otimes P$ is completed (μ, P being Radon), the canonical representation of the process $\{X_t, t \in T\}$ is automatically measurable. In particular if μ is the Lebesgue measure, and $T \subset \mathbb{R}$ and $S \subset \mathbb{R}$ are compact intervals then almost all $\omega \in \Omega$ are Riemann integrable on T iff almost no $t \in T$ (Lebesgue) is a fixed discontinuity of the process. This is a consequence of the classical fact that a bounded real function on a compact interval is Riemann integrable iff its discontinuity points form a set of Lebesgue measure zero.

The absence of fixed discontinuities does not imply that a process has almost all continuous sample functions. Each ω may have some discontinuity (e.g. a jump) at some $t \in T$, the latter depending on ω. If $T \subseteq \mathbb{R}$ and the process is separable then a discontinuity point of any sample function of $\{X_t, t \in T\}$ which is not a fixed discontinuity is called a *moving point of discontinuity*.

The preceding two theorems show that the existence of a separable and measurable process can be deduced somewhat more quickly if the measures are complete. When we complete the measures, however, the range space of the process may be allowed to be more general. In particular, the insistence that they be separable can be dropped. These improvements are possible with the lifting theory, *for the given process itself.* We turn to a brief discussion of the latter point as needed for the present work.

3.2 Remarks on a lifting operation and applications

The concept of lifting is, generally speaking, a continuous selection from equivalence classes in a topological space. This operation is important for our study when the separability of the range space of a stochastic process is not assumed. Let us motivate the concept, state it precisely, and then use it in what follows.

Let G be a topological group and H a closed subgroup. Then the factor space $F = G/H$ of all (left) cosets $\{xH : x \in G\}$ is made into a topological space by declaring that a set $A \subset F$ is open iff $\pi^{-1}(A) \subset G$ is open where $\pi : G \to F$ is the (onto) quotient map ($\pi : x \mapsto xH$). One defines a local cross-section of H in G as a transformation $h : F \to G$ which takes a neighborhood V of the identity of F continuously into G such that $\pi \circ h$ is the identity mapping on V. If $V = F$, then one says that h is a cross-section of H in G. Now the existence of a local-cross section for topological groups, in general, is known only for a small class. In our case the problem and its solution can be given as follows.

If $G = \mathcal{L}_p(\Sigma)$, an abelian vector group, of all real pth power integrable functions on (Ω, Σ, P) with the usual semi-norm $\|\cdot\|_p$, and if $H = \{f \in G : \|f\|_p = 0\}$, then $F = L_p(\Sigma) = G/H$. The question here reduces to: Does H admit a cross-section? Note that G is generally infinite dimensional. There exists, in such a situation, an example (due to Hanner) without a local cross-section, with H as a zero dimensional closed subgroup of G. In fact one can show that if P is non-atomic and $1 \leq p < \infty$, then H does not admit a cross-section. However, if $p = \infty$, even when P is σ-finite, we have that the subgroup H (fortunately) admitting a cross-section. This is a deep result and is required for our purposes. It is used to show the existence of separable and measurable modifications of a stochastic process with values even in a completely regular space. This is also useful for the projective limits of product conditional probability measures. With this in mind we shall now state the concept of cross-section in our context, called a *lifting map*, precisely and present an existence theorem for various measures and σ-algebras Σ. The result has been established in this generality by Tulcea and Tulcea [1], and a somewhat simpler proof is in Sion [2]. A further simplification, due to Traynor [1], is given with complete details by the author elsewhere (cf. Rao [11], Chapter 8). So we omit the proof and refer the reader to the latter source.

Let (Ω, Σ, P) be a complete space and $\mathcal{N} = \{A \in \Sigma : \mu(A) = 0\}$, the class ($\sigma$-ideal) of μ-null sets. Let $\tilde{\Sigma} = \Sigma/\mathcal{N}$, the quotient algebra; then $\tilde{A} \in \tilde{\Sigma}$ iff $\tilde{A}$ is the residue class of $A (\operatorname{mod} \mathcal{N})$ or the coset of A in Σ. Similarly let M^∞ be the linear space of all real measurable f on Ω for which $\|f\|_\infty < \infty$ where the latter is the $\mathcal{L}^\infty$-norm. If $N^\infty = \{f \in M^\infty : \|f\|_\infty = 0\}$, then $L^\infty(\Sigma) = M^\infty / N^\infty$ is the usual Lebesgue space of equivalence classes $\tilde{f}$ of essentially bounded measurable functions. Let $\tilde{\rho} : L^\infty(\Sigma) \to M^\infty$ be a mapping satisfying the conditions: (1) if $\tilde{f} = f + N^\infty$, then $\tilde{\rho}(\tilde{f}) = f \in M^\infty$, (2) $\tilde{f} = \tilde{g}$ implies $\tilde{\rho}(\tilde{f}) = \tilde{\rho}(\tilde{g})$, (3) $\tilde{\rho}(\tilde{1}) = 1$, (4) $\tilde{f} \geq \tilde{0} \Rightarrow \tilde{\rho}(\tilde{f}) \geq 0$, (5) $\tilde{\rho}(a\tilde{f} + b\tilde{g}) = a\tilde{\rho}(\tilde{f}) + b\tilde{\rho}(\tilde{g})$, $a, b \in \mathbb{R}$, and (6) $\tilde{\rho}(\tilde{f}\tilde{g}) = \tilde{\rho}(\tilde{f})\tilde{\rho}(\tilde{g})$.

The mapping $\tilde{\rho}$ is a cross-section of N^∞ in the earlier terminology. Thus if $\pi : M^\infty \to L^\infty(\Sigma)$ is the canonical mapping, then $\tilde{\rho}(\pi(f)) = f$ for all $f \in M^\infty$. We would like to establish the existence of such a mapping $\tilde{\rho}$. However, it turns out to be convenient to transfer the problem to M^∞ itself. For this we regard $L^\infty(\Sigma) = \{\tilde{f} : \tilde{f} = f + N^\infty, f \in M^\infty\}$ as the space of cosets and identify it as a set with M^∞. Thus these cosets partition M^∞ into a disjoint union, and we can define a mapping $\rho : M^\infty \to M^\infty$ such that for each $h \in \tilde{f}$, we let $\rho(h) = f = \tilde{\rho}(\tilde{f})$. Then ρ is well-defined and is a constant on each coset (or equivalence class) $\tilde{f}$. Thus ρ and $\tilde{\rho}$ determine each other uniquely. So it suffices to establish the existence of ρ on M^∞. Now the above conditions (1)–(6) can be restated for ρ as:

(i) $\rho(f) = f$, a.e., (ii) $f = g$, a.e.$\Rightarrow \rho(f) = \rho(g)$, (iii) $\rho(1) = 1$, (iv) $f \geq 0$, a.e.$\Rightarrow \rho(f) \geq 0$, (v) $\rho(af + bg) = a\rho(f) + b\rho(g)$, $a, b \in \mathbb{R}$, and (vi) $\rho(fg) = \rho(f)\rho(g)$.

While $\tilde{\rho}$ is an operator between the distinct linear spaces $L^\infty(\Sigma)$ and M^∞, the operator ρ maps M^∞ into itself. If only the first five ((1)–(5) or (i)–(v)) properties are assumed it is called a *linear lifting*. If (iv) or (4) is strengthened to (vi) or (6), then the mapping ρ or $\tilde{\rho}$ is termed a *lifting*. We note that in the second formulation the properties of ρ are very similar to those of a conditional expectation operator as discussed in Section II.1. However the latter are true only a.e., and not everywhere as demanded in the definition of a lifting. Hence the present result needs a much more detailed treatment.

At present there are three different methods of proof of the existence of ρ. We discuss briefly the ideas for an appreciation of the technical details necessary for the proofs. The first method is somewhat similar to that of establishing the Hahn-Banach theorem. For, if $\Sigma_0 \subset \Sigma$ is the algebra of null sets and their complements, then $M^\infty(\Sigma_0)$ consists of functions which are constants a.e. So if $f = a$ a.e. is an element then we define $\rho_0(f) = a$. One easily checks that ρ_0 is a lifting. Now assume the existence of ρ on $M^\infty(\Sigma_1)$ for an algebra $\Sigma_1 \subset \Sigma$ containing all null sets, and extend it to $M^\infty(\Sigma_2)$, where Σ_2 is the generated and completed σ-algebra of Σ_1 and a set $A \in \Sigma - \Sigma_1$. Next consider a chain of such σ-subalgebras and show that the corresponding liftings have a maximal element, using Zorn's lemma. All of this can be done for any linear lifting. The maximality property then implies the existence of a lifting on M^∞. This is the original proof of Tulcea and Tulcea [1]. The second method, due also to the same authors, is to use some properties of the normed algebra $L^\infty(\Sigma)$ and its Stone-Gel'fand representation (cf., e.g., Dunford-Schwartz [1], p. 312) as well as some facts about the "Lebesgue lower density". The third proof, due to Sion [2], is somewhat of a different nature. It relates to differentiation and was included in the first edition of this book. A simplification of the latter idea is a key point of Traynor's [1] work, and it can be found in Rao [11].

For the third method we need to translate the conditions (i)–(iv) of ρ on M^∞ to those on the sets in Σ. Recall that for each $\tilde{A} \in \tilde{\Sigma} = \Sigma/\mathcal{N}$, sets A, B are in $\tilde{A}$ iff the symmetric difference $A \Delta B$ is in $\mathcal{N}$. Thus if $\rho : M^\infty \to M^\infty$ is a lifting and $A \in \Sigma$ so that $\chi_A \in M^\infty$, then by (vi) $\rho(\chi_A) = (\rho(\chi_A))^2$. Hence $\rho(\chi_A)$ takes only the values 0 or 1. So $\rho(\chi_A) = \chi_B$ where $B = \{\omega : \rho(\chi_A)(\omega) = 1\} \in \Sigma$. The mapping $\lambda : A \mapsto B$ in this correspondence is well-defined and $\rho(\chi_A) = \rho(\chi_{\lambda(A)})$. If $A \Delta B \in \mathcal{N}$ then $\chi_A, \chi_B \in \tilde{\chi}_A$ so that by (ii) $\rho(\chi_A) = \rho(\chi_B)$, and hence we deduce that $\lambda(A) = \lambda(B)$, and if Σ is complete then $\chi_A \in M^\infty$ implies $\chi_{\lambda(A)} \in M^\infty$, for all $A \in \Sigma$. Also $A \Delta \lambda(A) \in \mathcal{N}$. Thus conditions (i)–(vi) can be restated, when (Ω, Σ, P) is complete, as follows: (I) $\lambda(A) = A$ a.e., (II) $A = B$ a.e. $\Rightarrow \lambda(A) = \lambda(B)$, (III) $\lambda(A \cap B) = \lambda(A) \cap \lambda(B)$, and (IV) $\lambda(A^c) = (\lambda(A))^c$.

From (III) and (IV) we deduce that $\lambda(\emptyset) = \emptyset$, $\lambda(\Omega) = \Omega$, and $\lambda(A \cup B) = \lambda(A) \cup \lambda(B)$. Thus if $\lambda : \Sigma \to \Sigma$ satisfying (I)–(IV) is given then we may define $\rho : M^\infty \to M^\infty$ by the equation $\rho(\chi_A) = \chi_{\lambda(A)}$, and then extend for all simple functions

$$f = \sum_{i=1}^{n} a_i \chi_{A_i} \in M^\infty,$$

with $\rho(f) = \sum_{i=1}^{n} a_i \chi_{\lambda(A_i)}$. It is clear that ρ does not depend on the representation of f and that $\|\rho(f)\|_\infty = \|f\|_\infty$. The thus defined ρ is a lifting on the dense subspace of all simple functions of M^∞, and has a unique extension satisfying (i)–(vi) to all of M^∞. Hence it is sufficient to have the existence of λ on Σ satisfying (I)–(IV), and the latter is termed a (set) lifting. Hereafter we assume that μ is not a zero measure and that Σ is $\sigma(\Sigma_0)$ completed for μ, where Σ_0 is the δ-ring of sets of finite μ-measure.

We recall that for (Ω, Σ, μ), the measure has the *finite subset property* if for any $A \in \Sigma, \mu(A) > 0$ implies the existence of a $B \subset A$, $B \in \Sigma, 0 < \mu(B) < \infty$, and it has the *direct sum property* if there is $\{A_i, i \in I\} \subset \Sigma$, $A_i \cap A_j$, $i \neq j$, $\mu(A_i) < \infty$, $\Omega - \cup_{i \in I} A_i \in \mathcal{N}$, and each $A \in \Sigma_0$ intersects at most countably many A_i's in positive measure. Thus σ-finiteness concept is subsumed; see also Exercise 7.8 below. But it was shown by Ryan [1] that the existence of a lifting implies the direct sum property of μ. We thus present the general result.

1. Theorem. *Let (Ω, Σ, μ) be a Carathéodory generated measure space where μ has the finite subset property and is nonzero. Then there exists a lifting on Σ iff μ has the direct sum property. [In particular, if Ω is a locally compact space and μ is a Radon measure then it has the latter property and so a Haar measure on a locally compact group always admits a lifting.]*

The direct sum property of a measure is also called *strict localizability*. We now recall the associated notion termed *localizability* and discuss its exact relation to the lifting property.

2. Definition. (a) Let (Ω, Σ, μ) be a complete (Carathéodory generated) measure space such that μ has the finite subset property. Then μ is said to be *localizable* iff every collection $\mathcal{C} \subset \Sigma$ has a supremum in Σ, i.e., there exists a $C \in \Sigma$ such that (i) $A \in \mathcal{C} \Rightarrow A - C \in \mathcal{N}$, and (ii) if $\tilde{C} \in \Sigma$ satisfies (i), then $C - \tilde{C} \in \mathcal{N}$.

(b) If for the measure space of (a), $\mathcal{M}(\Sigma)$ denotes the class of all (Σ)-measurable $\bar{\mathbb{R}}$-valued functions on Ω, then we say that a collection $\mathcal{A} \subset \mathcal{M}(\Sigma)$ has a *supremum* $f_0 \in \mathcal{M}(\Sigma)$ iff (i) $f_0 \geq f$, a.e., for all $f \in \mathcal{A}$, and (ii) if $\bar{f}_0 \in \mathcal{M}(\Sigma)$ is any other function with property (i), then $\bar{f}_0 \geq f_0$, a.e.

We have already noted that every σ-finite measure has the direct sum property. However, localizability is not more stringent than the latter. It is not hard to show that every measure with the direct sum property is localizable. The converse is not true however. It was an open question for some years, but in 1978 D. H. Fremlin gave a counter example to this effect (cf., Rao [11], p. 79, Ex. 5 on the implication).

The next result illustrates the usefulness of localizability concept and it will be used in applications later.

3. Theorem. *Let (Ω, Σ, μ) be a localizable measure space and $\mathcal{M}(\Sigma)$ be the set of all extended real valued measurable functions on Ω, i.e., the space introduced in Definition 2(b). If $\mathcal{A} \subset \mathcal{M}(\Sigma)$ is any nonempty set, then it has a supremum f_0 (of course in $\mathcal{M}(\Sigma)$). If also the elements of $\mathcal{A}$ are nonnegative, and the set is directed upwards, then*

$$\int_\Omega f_0 d\mu = \int_\Omega \sup\{f : f \in \mathcal{A}\} d\mu = \sup\left\{\int_\Omega f\, d\mu : f \in \mathcal{A}\right\}. \qquad (1)$$

If $\mathcal{A}$ is bounded above and there is a lifting ρ on $M^\infty(\Sigma)$ (or equivalently μ has the direct sum property by Theorem 1) such that $\rho(f) \geq f$ for each $f \in \mathcal{A}$, then $\rho(f_0) \geq f_0$ also holds.

Proof. Let $C_r = \{A_f^r : f \in \mathcal{A}\}$ where $A_f^r = \{\omega : f(\omega) \geq r\}$, so that $C_r \subset \Sigma$, and let B_r be the supremum of C_r. This supremum exists since μ is localizable. If $r' > r$ (so $A_f^r \supset A_f^{r'}$) and if $B_{r'}$ is the supremum of $C_{r'}$ then $B_{r'} - B_r \in \mathcal{N}$ and hence $C_r = \cup\{B_{r'} : r' \text{ rational}\} \in \Sigma$, $C_r - B_r \in \mathcal{N}$. Also $C_r \supset C_{r'}$ for $r < r'$. Define $g_r = r$ on C_r, and $= -\infty$ on $\Omega - C_r$, so that $g_r \in \mathcal{M}(\Omega)$. Let $f_0 = \sup\{g_r : r \text{ rational}\}$. Then $f_0 \in \mathcal{M}(\Sigma)$, and we show that f_0 is a supremum of $\mathcal{A}$, by verifying the conditions of Definition 2(b).

Thus let $f \in \mathcal{A}$, and r be a rational number. The set $D_r = \{\omega : f(\omega) > r > f_0(\omega)\}$ is measurable and $A_f^r - B_r \supset D_r - B_r$. Since the former is a null set and Σ is complete we see that $D_r - B_r$ is null. Also $D_r - C_r \subset D_r - B_r$ and so $D_r - D_r \cap C_r = D_r - C_r$ is null. But it is clear from definition that $D_r \cap C_r = \emptyset$. Thus D_r is null so that $f \leq f_0$ a.e. for each $f \in \mathcal{A}$.

To see that f_0 is the least upper bound, let $\bar{f}$ be any other upper bound of $\mathcal{A}$, so that $f \leq \bar{f}$ a.e. for all $f \in \mathcal{A}$. If $\bar{B}_r = \{\omega : \bar{f}(\omega) \geq r\} \in \Sigma$, and $f \in \mathcal{A}$, then $A_f^r \subset \bar{B}_r$ a.e. So $\bar{B}_r$ is also an upper bound of C_r and hence $B_r - \bar{B}_r$ and then $C_r - \bar{B}_r$ are null sets. Thus $N = \cup\{C_r - \bar{B}_r : r \text{ rational}\}$ is a μ-null set. If $\omega \in \Omega - N$ then

$$\begin{aligned} f_0(\omega) &= \sup\{r : \omega \in C_r, r \text{ rational}\}, \quad \text{by definition,} \\ &\leq \sup\{r : \omega \in \bar{B}_r, r \text{ rational}\}, \quad \text{since } C_r \subset \bar{B}_r \text{ a.e.,} \\ &= \bar{f}(\omega). \end{aligned} \tag{2}$$

Hence f_0 is a supremum of $\mathcal{A}$.

Now let us first dispose of the last part and then we come back to establish (1). Thus let $\mathcal{A} \subset (M^\infty)^+$ be bounded by k_0, say. If μ is strictly localizable so that there is a lifting ρ on $M^\infty(\Sigma)$ and since $k_0 \geq f_0 \geq f \geq 0$ we have $\rho(f_0) \geq \rho(f) \geq f$ a.e. for each $f \in \mathcal{A}$. Hence $\rho(f_0)$ is also an upper bound of $\mathcal{A}$, and so $\rho(f_0) \geq f_0$, a.e., where we used the fact that ρ preserves order.

To prove (1), we now only assume that μ is localizable but that $\mathcal{A}$ is filtering to the right, with nonnegative elements. As before let f_0 be the supremum of this set. Let $\beta = \sup\{\|f\|_1 : f \in \mathcal{A}\}$. Clearly $\|f_0\|_1 \geq \beta$. To see that there is equality here, we may assume that $\beta < \infty$. Then there is a sequence $f_n \in \mathcal{A}$, such that $\beta = \lim_n \|f_n\|_1$. But $\mathcal{A}$ is right filtering; so we may take $f_n \leq f_{n+1}$, a.e., and if $g = \lim_n f_n$ then $\|g\|_1 = \beta$, by the monotone convergence theorem. Also g is an upper bound of $\mathcal{A}$. In fact, if there is an element $h \in \mathcal{A}$ such that $\{g < h\}$ has positive measure then by directedness, $h \geq g$ a.e., and we get

$$\beta = \|g\|_1 < \|h\|_1 \leq \sup\{\|f\|_1 : f \in \mathcal{A}\} = \beta, \tag{3}$$

which is a contradiction. Thus $g \geq f_0$ a.e., since f_0 is a least upper bound. Hence $\beta = \|g\|_1 \geq \|f_0\|_1$ and with the earlier inequality we must have equality. This proves (1). $\square$

We now present a situation where the two concepts of Definition 2 coincide.

Recall that $\mathcal{P} = \{\{A_i\}_{i\in I} : 0 < \mu(A_i) < \infty$, A_i's have pairwise μ-null intersection$\}$ is a nonempty collection if (Ω, Σ, μ) has the finite subset property. Ordering $\mathcal{P}$ by inclusion, we note by Zorn's lemma that there exists a maximal collection $\mathcal{C}$, in $\mathcal{P}$. In the following $\mathcal{C}$ stands for such a collection.

4. Theorem. *Let (Ω, Σ, μ) be a localizable measure space and $\mathcal{C} = \{B_i\}_{i\in I_0}$, be a maximal family in $\mathcal{P} \subset \Sigma$, of a.e. disjoint positive μ-finite sets. If the cardinality of I_0 is at most of the continuum, then μ has the direct sum property so that both concepts coincide.*

This result is given for general information and will not be used below. The details may be found, e.g., in Rao [11], p. 431, and are not reproduced here.

For the separability problem, however, we need the following adjunct of Theorem 1.

5. Proposition. *Let (Ω, Σ, μ) be a strictly localizable space and $(S, \mathcal{B})$ be a measurable space, where S is completely regular and $\mathcal{B}$ its Baire σ-algebra. Let $M^\infty(\Sigma, S)$ be the set of $f : \Omega \to S$ which are $(\Sigma, \mathcal{B})$-measurable and each $f(\Omega)$ is relatively compact. If ρ is a lifting of $M^\infty(\Sigma, \mathbb{R})$, then it determines a unique $\rho' : M^\infty(\Sigma, S) \to M^\infty(\Sigma, S)$ such that (i) $\rho'(f) = f$ a.e., (ii) $f = g$ a.e. $\Rightarrow \rho'(f) = \rho'(g)$, and (iii) $\rho(h \circ f) = h \circ \rho'(f)$ for all $f \in M^\infty(\Sigma, S)$ and all real bounded continuous functions h on S.*

Proof. Let $f \in M^\infty(\Sigma, S)$ and $K = \overline{f(\Omega)}$, the closure of $f(\Omega)$ in S. Then K is compact by hypothesis. For any $\omega \in \Omega$, $h \in C(K)$ the space of real continuous functions on K, define ${}_f\ell_\omega(h) = \ell_\omega(h) = \rho(h(f))(\omega)$. (We do not display f hereafter as it is fixed throughout.) Since ρ is a lifting on $M^\infty(\Sigma, \mathbb{R})$, it is multiplicative. Hence

$$\begin{aligned}\ell_\omega(h_1 h_2) &= \rho((h_1 \cdot h_2)(f))(\omega) = \rho(h_1(f)h_2(f))(\omega) \\ &= \rho(h_1(f))(\omega)\rho(h_2(f))(\omega) = \ell_\omega(h_1)\ell_\omega(h_2). \qquad (4)\end{aligned}$$

Since ρ is also linear and $\rho(1) = 1$, we conclude that ℓ_ω is a multiplicative linear functional (with $\ell_\omega(1) = 1$) on $C(K)$, of unit norm. However, every such functional on $C(K)$ is an evaluation, i.e., $\ell_\omega(h) =$

$h(b)$, $h \in C(K)$ for a unique $b \in K$ depending on ω (and f). Here is a quick proof of this statement.

If Δ is the class of all multiplicative linear functionals on $C(K)$ and $T = \{\tau_a : \tau_a(h) = h(a), a \in K, h \in C(K)\}$, the set of evaluation functionals, then $T \subset \Delta$, and the inclusion is one-to-one. In fact if $\{a_1, a_2\} \subset K$, $a_1 \neq a_2$, let $B \subset K$ be a closed set such that $a_1 \notin B$, $a_2 \in B$. By Urysohn's lemma there is $f \in C(K)$ with $f(a_1) = 1$ and $f(B) = 0$. So $\tau_{a_1}(f) = f(a_1) = 1 \neq 0 = \tau_{a_2}(f)$ and hence $a \mapsto \tau_a$ is one-to-one. If Δ is given the weak*-topology, i.e., the neighborhood basis is $\{x^* : |x^*(h_i) - x_0^*(h_i)| < \varepsilon, i = 1, \ldots, n\}$, $h_i \in C(K)$, $\{x_0^*, x^*\} \subset (C(K))^*$, then this topology and that of K can be clearly identified and so K and T can be identified. One shows now, under this identification, that $K = \Delta$. If this is false (it is evident Δ is closed), there exists an $0 \neq x_0^* \in \Delta - K$ and the neighborhood of x_0^* which is disjoint from K. Thus there is an $\varepsilon > 0$, $h_0 \in C(K)$ such that $\{x^* : |x^*(h_0) - x_0^*(h_0)| < \varepsilon\} \cap K = \emptyset$. If $h_1 = h_0 - x_0^*(h_0) \cdot 1$, then $|x^*(h_1)| < \varepsilon$ for all x^* in the above neighborhood, and so $|\tau_a(h_1)| = |h_1(a)| \geq \varepsilon > 0$ for all $a \in K$. Since K is compact, and $h_1 \in C(K)$, this implies $h_1^{-1} \in C(K)$. Thus $1 = x_0^*(1) = x_0^*(h_1 h_1^{-1}) = x_0^*(h_1)x_0^*(h_1^{-1}) = 0$ since $x_0^*(h_1) = 0$. This contradiction shows $K = \Delta\,(= T)$.

Define a mapping $\rho'(f) : \Omega \to S$ by the equation $\rho'(f)(\omega) = b\,(= b_f)$. Since b_f is uniquely determined by ω, by the above proof, $\rho'(f)$ is well-defined and $\rho'(f)(\Omega) \subset K$. Also for $h \in C(K)$, $h(\rho'(f))(\omega) = h(b) = \tau_b(h) = {}_f\ell_\omega(h) = \rho(h(f))(\omega)$ by definition and the preceding paragraph. Thus $h(\rho'(f)) = \rho(h(f))$ for all $h \in C(K)$, $f \in M^\infty(\Sigma, S)$, and ρ' satisfies (iii). Since $\rho(h(f)) = h(f)$ a.e., for any $h \in C(K)$, it follows that $\rho'(f) = f$ a.e., by the above equation so that (i) is true. (However this does not imply that $\{\omega : \rho'(f)(\omega) \neq f(\omega)\}$ is μ-null without some extra condition, e.g., the separability of S.) For (ii), if $f = g$ a.e., then $h(\rho'(f)) = \rho(h(f)) = \rho(h(g)) = h(\rho'(f))$, for all $h \in C_b(S)$, the space of bounded real continuous functions on S. The middle equalities imply that $h(f) = h(g)$ a.e., $h \in C_b(S)$. So (ii) is also true.

Finally, to see that ρ' is uniquely determined by ρ, let ρ'' be another such lifting. Then by (iii) $h(\rho'(f)) = \rho(h(f)) = h(\rho''(f))$, for all $h \in C_b(S)$. Since this means $\rho'(f) = \rho''(f)$, $f \in M^\infty(\Sigma, S)$, we conclude that $\rho' = \rho''$. $\square$

Observe that, in the above proof, all functions of $C_b(S)$ (or $C(K)$) are not utilized. In each case any subalgebra of these spaces separating the points of S (or K) would have been sufficient. Using this fact we may extract a useful consequence of this result on the existence of a linear lifting on $M^\infty(\Sigma, S)$ when S is a Banach space or an adjoint space, where $f \in M^\infty(\Sigma, S)$ iff for each $x^* \in S^*$ (or $x \in \mathcal{X}$ when $S = \mathcal{X}^*$) $x^*(f) \in M^\infty(\Sigma, \mathbb{R})$ ($f \circ x = \hat{x}(f) \in M^\infty(\Sigma, \mathbb{R})$) and $f(\Omega)$ is relatively weakly (or weak$^*-$) compact in S. This is because S is locally convex (so S^* separates the points of S), and it is completely regular. We only consider the case of weak measurability, leaving the modifications for the weak* case to the reader.

6. Corollary. *If there is a lifting ρ on $M^\infty(\Sigma, \mathbb{R})$, then there exists uniquely a (linear) lifting ρ' on $M^\infty(\Sigma, S)$, S being a Banach space and functions are weakly measurable as above, in that (i) $\rho'(f) = f$, a.e., (ii) $f = g$, a.e. $\Rightarrow \rho'(f) = \rho'(g)$, and (iii) $\rho(x^*(f)) = x^*(\rho'(f))$, for all $x^* \in S^*, f \in M^\infty(\Sigma, S)$.*

Proof. If we temporarily denote the weakly measurable space as $M_S^\infty(\Sigma)$ and the space when S (with weak topology $\mathcal{T}$) is completely regular as $M^\infty(\Sigma, S)$, then the result follows from the preceding proposition as soon as we show that $M_S^\infty(\Sigma) = M^\infty(\Sigma, S)$.

Define $h_{x^*} : S \to \mathbb{R}$ by $h_{x^*}(s) = x^*(s)$, for each $x^* \in S^*$. So $h_{x^*} \in C_b(S)$. If $\mathcal{H} = \{h_{x^*} : x^* \in S^*\}$, then the Hahn-Banach theorem implies that $\mathcal{H}$ separates points of the space $(S, \mathcal{T})$. Let $f \in M^\infty(\Sigma, S)$. Then by definition $f(\Omega)$ is relatively compact in $(S, \mathcal{T})$, and $h(f)$ is μ-measurable for all $h \in \mathcal{H} \subset C_b(S)$. Hence $h_{x^*}(f) = x^*(f) \in M^\infty(\Sigma)$. So $f \in M_S^\infty(\Sigma)$. Conversely, let $f \in M_S^\infty(\Sigma)$. By definition $f(\Omega)$ is again relatively compact and if $h_i \in \mathcal{H}, i = 1, \ldots, n$ is any finite collection then $(h_1 \cdots h_n)(f) = h_1(f) \cdots h_n(f)$ is μ-measurable. But all such collections separate points of S and hence (by the Stone-Weierstrass theorem) $h(f)$ is μ-measurable for all $h \in C_b(S)$. So $f \in M^\infty(\Sigma, S)$. Thus these two sets define the same space as asserted. □

The following consequence of Theorem 3 and Proposition 5 will be useful in some applications.

7. Remark. If ρ is a lifting on $M^\infty(\Sigma, \mathbb{R})$ and $f \in M^\infty(\Sigma, S)$ where S is completely regular, then the hypothesis that $\rho'(f) = f$ implies for each closed (not necessarily Baire) set $K \subset S$, $B = f^{-1}(K) \in \Sigma$ and $\rho(\chi_B) \leq \chi_B$.

In fact, let $\mathcal{A} = \{h \in C_b(S) : h = 1 \text{ on } K, 0 \leq h \leq 1\}$. Then for each $a \in K, (h \circ f)(a) = 1$ and hence $\inf\{h \circ f : h \in \mathcal{A}\} = \chi_B$. Since $h \circ f \in M^\infty(\Sigma, \mathbb{R})$, and there is a lifting on this space (so μ is strictly localizable), we conclude that χ_B is measurable by Theorem 3. Since $\rho'(f) = f$, by Proposition 5, $h \circ f = h(\rho'(f)) = \rho(h \circ f), h \in \mathcal{A}$. Now $h_1 \geq h_2$ clearly orders $\mathcal{A}$ (downward). We may apply the last part of Theorem 3 and conclude that (since $\chi_B \leq h \circ f$) $\chi_B \leq \inf_h(h \circ f) = \rho(\inf_h h \circ f) = \rho(\chi_B)$. This establishes the remark. The point is that not only $f^{-1}(\mathcal{B}) \subset \Sigma$, but the localizability implies $f^{-1}(\bar{\mathcal{B}}) \subset \Sigma$, where $\bar{\mathcal{B}}(\supset \mathcal{B})$ is the Borel σ-algebra of S. These results will be used in the stochastic function theory in the next section; and other applications are included as Complements and Exercises.

3.3 Separability and measurability: general case

As noted in Remark 1.4, we are in a position to establish the separability of a stochastic process with values in a completely regular space when it is defined on an arbitrary complete (Carathódory generated) probability space. The result of Theorem 1.3 did not require the completeness of the measure space, but then it treats only a special case. The following theorem, from Tulcea and Tulcea [1], is the desired generalization. We use Definition 1.2 without further comment.

1. Theorem. *Let (Ω, Σ, P) be a complete probability space and $\{X_t, t \in T\}$ be a stochastic process, $X_t : \Omega \to S$, where $(S, \mathcal{B})$ is a completely regular measurable space, X_t being $(\Sigma, \mathcal{B})$-measurable. Then there exists a process $\{Y_t, t \in T\}$ on the same measurable space, which is a measurable modification of the X_t-process, i.e., $X_t = Y_t$, a.e., $t \in T$. In fact, if ρ' is a lifting on $M^\infty(\Sigma, S)$ (cf., Proposition 2.5), then $Y_t = \rho'(X_t)$, $t \in T$, gives a measurable modification.*

Proof. By definition of measurability, for each $h \in C_b(S)$, the space of real bounded continuous functions on S, $h \circ X_t$ is μ-measurable and $X_t(\Omega)$ is relatively compact in S. To see that the latter condition is not more restrictive than that of Theorem 1.3, we recall that every completely regular space S can be embedded (continuously) as a dense subset of a compact Hausdorff space $\check{S}$ in such a way that each bounded continuous function on S has a unique continuous extension to $\check{S}$, by the Stone-Čech compactification theorem and $S = \mathbb{R}$ is covered. Hence $C_b(S) \subset C(\check{S})$ densely and $X_t(\Omega) \subset \check{S}$ is automatically relatively compact in the latter space. Thus X_t may be regarded as an element of $M^\infty(\Sigma, \check{S})$. So for all complete probability spaces, Theorem 1.3 is properly extended by the present result. With this understanding we now proceed with the proof.

By hypothesis and Proposition 2.5, there is a lifting ρ on $M^\infty(\Sigma, S)$. If $C \subset S$ is any compact set then $X_t^{-1}(C) \in \Sigma$ by Remark 2.7. Moreover, if $Y_t = \rho'(X_t)$, then the fact that $Y_t = X_t$ a.e., implies $\rho'(Y_t) = Y_t$. So by the same remark $B_t = Y_t^{-1}(C)$ satisfies $\rho(\chi_{B_t}) \leq \chi_{B_t}$ for each $t \in T$. Writing λ for the set lifting induced by ρ, we have for any finite set $J \subset T$ (ρ and ρ' are related by Proposition 2.5 and $\rho(\chi_A) = \chi_{\lambda(A)}$),

$$\begin{aligned} \lambda(V(J,C)) &= \lambda\left(\bigcap_{t\in J} B_t\right) \\ &= \bigcap_{t\in J} \lambda(B_t) \\ &\subset \bigcap_{t\in J} B_t, \quad \text{since } \rho(\chi_{B_t}) \leq \chi_{B_t}, \\ &= V(J,C), \end{aligned} \tag{1}$$

where we are using the notation of Definition 1.2, i.e., $V(J,C) = \{\omega : Y_t(\omega) \in C, \text{ for all } t \in J\} = \bigcap_{t\in J} Y_t^{-1}(C)$. Since the measure space is finite (hence strictly localizable), $V(I,C) = \cap\{V(J,C) : J \subset I, \text{ finite}\}$ is measurable by Theorem 2.3 (and its last part) for any $I \subset T$, because the collection $\{V(J,C), J \subset I, \text{ finite}\}$ is ordered by inclusion and that $\tilde{\rho}(V(I,C)) \supset V(I,C)$ the difference being a P-null set. So $V(I,C) \in \Sigma$. The directedness of the collection implies that we can apply Eq. (1) of Theorem 2.3. Consequently,

$$P(V(I,C)) = \inf\{P(V(J,C)) : J \subset I, \ \text{finite}\}. \tag{2}$$

Thus both conditions of Definition 1.2 are satisfied and $\{Y_t, t \in T\}$ is a separable modification. □

This is a short proof of a more inclusive result than that of Theorem 1.3 when completeness of (Ω, Σ, P) is assumed or if the cardinality of Σ is at most of the continuum. However, the general theory of lifting is decidedly more advanced; but such a study reveals what is involved in the structure of separable modifications. We may now proceed to an analysis of such separable processes. Here, if the original process is not separable then one replaces it by a modification provided by the above theorem.

If the index set is suitably restricted, it is easy to obtain a universal separating set of the process. The following is a possibility and its proof is almost identical to that of Theorem 1.3, and is left to the reader.

2. Theorem. *Let $\{X_t, t \in T\}$ be a stochastic process on a probability space (Ω, Σ, P) to a completely regular space $(S, \mathcal{B})$. Suppose S has a countable base and T is also a topological space with a countable base. Then there exists a separable modification $\{Y_t, t \in T\}$ on the same measure space of the given process and a dense denumerable set $D \subset T$ as a universal separating set for both versions.*

For many applications it will be useful to know what classes of dense subsets $D \subset T$ are universally separating for the process. The following result contains one such condition.

3. Theorem. *Let $\{X_t, t \in T\}$ be a stochastic process on (Ω, Σ, P) to $(S, \mathcal{B})$ and suppose that T and S satisfy the conditions given in Theorem 2. Suppose also that for each real bounded continuous f on S and each $t_0 \in T, \varepsilon > 0$ we have $P\{\omega : |f \circ X_t - f \circ X_{t_0}|(\omega) \geq \varepsilon\} \to 0$ as $t \to t_0$ in T (i.e., $f \circ X_t \to f \circ X_{t_0}$ in probability). Then every dense denumerable subset of T is a universal separating set for the given process.*

Proof. We first note that the separability of the completely regular S implies its metrizability (again by the classical Urysohn metrization theorem). So, let $d(\cdot, \cdot)$ be the metric function in this identification. Then the condition on the process is equivalent (as a simple computation reveals) to:

$$\lim_{t \to t_0} P\{\omega : d(X_t(\omega), X_{t_0}(\omega)) \geq \varepsilon\} = 0, \tag{3}$$

and this is the *stochastic continuity* of the process at $t_0 \in T$. Since $t_0 \in T$ is arbitrary the process is *stochastically continuous* on T.

The hypothesis of Theorem 2 is included here. So there is a separable modification of the process with the universally separating set $D_0 \subset T$. Let $D \subset T$ be any other dense denumerable set. Replacing X_t by Y_t of Theorem 2 if necessary, we assume that the given process is separable and show that D is also its universal separating set when the process is stochastically continuous.

Suppose $\mathcal{G}$ is a (countable) basis for the topology of T. If $I \in \mathcal{G}$, let $A(I \cap D, \omega)$ be the closure of $\cup\{X_t(\omega) : t \in I \cap D\}$ for each $\omega \in \Omega$. If $t_0 \in D_0 \cap I$, and $N_{t_0}(I) = \{\omega : X_{t_0}(\omega) \in S - A(I \cap D, \omega)\}$, then we assert that $N_{t_0}(I)$ is a subset of a P-null set. In fact, since $t_0 \in I$ and $D \cap I$ is dense in I, there is a sequence $t_n \in I \cap D$ with $t_n \to t_0$. Then $X_{t_n} \to X_{t_0}$ in probability by hypothesis. Hence if P^* is the outer measure generated by (Σ, P) we have

$$\begin{aligned} P^*(N_{t_0}(I)) &= P^*\{\omega : X_{t_0} \notin A(I \cap D, \omega)\} \\ &\leq P\left\{\omega : \lim_{n \to \infty} d(X_{t_n}(\omega), X_{t_0}(\omega) > 0\right\} \\ &= \lim_{k \to \infty} P\left\{\omega : \lim_{n \to \infty} d(X_{t_n}(\omega), X_{t_0}(\omega)) \geq \frac{1}{k}\right\} \\ &\leq \lim_{k \to \infty} \lim_{n \to \infty} P\left\{\omega : d(X_{t_n}(\omega), X_{t_0}(\omega)) \geq \frac{1}{k}\right\} = 0, \quad \text{by (3).} \end{aligned} \tag{4}$$

Thus if $N_1 = \cup\{N_{t_0}(I), t_0 \in D_0, I \in \mathcal{G}\}$, then since $\mathcal{G}$ and D_0 are countable, we deduce that $P^*(N_t) = 0$. Hence by separability (of the process) for every compact $C \subset S$, the sets $V(I \cap D, C), V(I \cap D_0, C)$ and $V(I, C)$ differ by a P-null set, where $V(I, C) = \cap_{t \in I} X_t^{-1}(C)$. Since $I \in \mathcal{G}$ is arbitrary, D is a universal separating set, as asserted. □

We note that, if P itself is a Carathéodory generated measure so that it is complete, $P^* = P$ and in place of Theorem 2, Theorem 1 can be invoked. In particular, if the process has continuous sample paths or if T is an interval and $X_t = X_{t+}$ (i.e., the paths are right continuous) then the above result is applicable.

For many problems, especially in stochastic integration, it will be useful to know the relations between separable and measurable processes. The latter are defined and analyzed in Section 1 in some cases. The next result contains further information.

Recall that a process $\{X_t, t \in T\}$ on a probability space with values in $(S, \mathcal{B})$ is measurable if, considered as a function, $X(\cdot,\cdot) : T \times \Omega \to S$ is (jointly) measurable for $(\mathcal{T} \otimes \Sigma, \mathcal{B})$ where $\mathcal{T}$ is a σ-algebra of T and $\mathcal{B}$ is the Baire σ-algebra of the completely regular space S. This can be stated in an equivalent form as: for each $f \in C_b(S)$, $f \circ X_t$ is a real measurable process relative to $\mathcal{T} \otimes \Sigma$, for each such f. We now have the following:

4. Theorem. *Let $\{X_t, t \in T\}$ be a stochastic process on a complete probability space (Ω, Σ, P) into a completely regular $(S, \mathcal{B})$, where S need not be separable but T is a locally compact metric space with a countable base. Let $(T, \mathcal{T}, \mu)$ be a Radon measure space and suppose that X_t is stochastically continuous at μ-almost all t in T. Then there exists a separable modification of X_t which is also measurable relative to $(\mathcal{T} \otimes \Sigma, \mathcal{B})$ when the product σ-algebra is completed for $\mu \otimes P$.*

Proof. Let ρ and ρ' be the liftings on $M^\infty(\Sigma, \mathbb{R})$ and $M^\infty(\Sigma, S)$ given by Proposition 2.5. Then by Theorem 1, if $Y_t = \rho'(X_t)$, we have $\{Y_t, t \in T\}$ to be a separable modification of the given process. Hence for each fixed but arbitrary $f \in C_b(S), Z_t = f \circ Y_t = f \circ \rho'(X_t) = \rho(f \circ X_t)$ is a real separable (for closed sets) process for $t \in T$. By the relations between the measurablities noted above, it suffices to show that $\{Z_t, t \in T\}$ is a measurable process under the hypothesis of its stochastic continuity. We now establish this.

Let $D \subset T$ be a dense denumerable set. By Theorem 3 it is a universal separating set for the process $\{Z_t, t \in T\}$. For this proof we may assume that T is compact also, since a real function on $T_0 \times \Sigma$ is $\mu \otimes P$-measurable for each compact $T_0 \subset T$ implies the measurability on $T \times \Omega$, by the classical Luzin theorem (cf., e.g., Rao [11], p. 147) with respect to $(\mathcal{T} \otimes \Sigma, \mathcal{B})$ as given. So let T be compact from now on.

For each n, let $\{G_i^n\}_{i=1}^{k_n}$ be an open covering of T such that the diameter of each G_i^n is atmost $\frac{1}{n}$, as in the proof of Theorem 1.5, and let $\{H_i^n\}_{i=1}^{k_n}$ be its disjunctification. We may assume that H_i^n is nonempty for each i and by the density of $D \subset T$, one can take $t_i^n \in D \cap H_i^n$. (Thus the open covering may be taken to have these properties as well.) Define

$$V_n(t,\omega) = \sum_{i=1}^{k_n} Z_{t_i^n}(\omega)\chi_{H_i^n}(t).$$

Then V_n is $\mu \otimes P$-measurable.

Since clearly H_i^n is in $\mathcal{T}$, it follows that $V_n(t,\cdot) \to Z_t$ in measure as $n \to \infty$ for almost all t. In fact, for any $\varepsilon > 0$,

$$P\{\omega : |V_n(t,\omega) - Z_t(\omega)| \geq \varepsilon\} = P\{\omega : |Z_{t_i^n}(\omega) - Z_t(\omega)| \geq \varepsilon\} \to 0, \tag{5}$$

as $n \to \infty$ by the stochastic continuity of Z_t, and the fact that each $t \in T$ belongs to exactly one H_i^n for each n. Hence $\{V_n(t,\cdot)\}_{n=1}^{\infty}$ is a Cauchy sequence in probability (i.e., for the metric induced by "in probability"). Since Z_t and V_n are bounded random variables, by Fubini's theorem we deduce that

$$\int_{T\times\Omega} |V_n(t,\omega) - V_m(t,\omega)| d\mu \otimes P = \int_T \left[\int_\Omega |V_n(t,\omega) - V_m(t,\omega)| dP\right] d\mu, \tag{6}$$

and the right side tends to zero as $n, m \to \infty$, by (5) and the bounded convergence theorem. Thus $\{V_n, n \geq 1\}$ is Cauchy in $L^1(\mathcal{T} \otimes \Sigma)$ and hence it converges in $\mu \otimes P$-measure. So there is a subsequence $V_{n_i} \to V$ pointwise on N_1^c where $N_1 \subset T \times \Omega$, $\mu \otimes P(N_1) = 0$. Also the quantity [ ] on the right side of (6) tends to zero for $t \in M_1^c$ where $\mu(M_1) = 0$. Thus $V_{n_i}(t,\cdot) \to V(t,\cdot)$ for all $\omega \in (N_1^c)_t$ and $t \in M_1^c$ as $i \to \infty$. Then on $N_1^c \cap [M_1^c \times \Omega]$, $V(t,\omega) = Z_t(\omega)$. It follows that Z is equivalent to a $\mu \otimes P$-measurable process since each subset of N_1 is in $\mathcal{T} \otimes \Sigma$ by hypothesis. On each P-null section $(N_1)_t$, one may take $V(t,\cdot) = Z_t$ for each $t \in M_1$, and on M_1^c we can find a sequence $t_n \in D$ such that $t_n \to t$ so that $Z_{t_n} \to V(t,\cdot)$. Thus D is also a universal separating set of the process $\{V(t,\cdot), t \in T\}$. Since $P\{\omega : V(t,\omega) = Z_t(\omega)\} = 1$ for $t \in M_1^c$ and on M_1, $V(t,\cdot) = Z_t(\cdot)$, we conclude that $\{V(t,\cdot), t \in T\}$ is a separable and measurable modification of the Z_t-process. This means $\{f \circ Y_t, t \in T\}$ is separable and measurable, for each $f \in C_b(S)$, or that $\{X_t, t \in T\}$ has a separable and measurable modification. □

If S is also separable (i.e., is a separable metric space) and locally compact, we may apply Theorem 1.3 in place of Theorem 1 so that (Ω, Σ, P) need not be complete. This result was proved by Doob [1], and the above proof is a modification of his, which however uses lifting theory, to take the nonseparability of S into account. If $T \subset \mathbb{R}$, then the reader may verify that the left or right continuity of the sample paths ($X(\cdot, \omega)$ is left or right continuous for a.a. (ω)) implies the statements (4) and (5). So the result is true for those cases also.

We state the above classical separable version for comparison.

5. Corollary. *Let T be a locally compact separable space (e.g., $\mathbb{R}^n$ or a subset of $\mathbb{R}^n$) and S a separable metric space. If $\{X_t, t \in T\}$ is a stochastic process on a probability space (Ω, Σ, P), $(T, \mathcal{T}, \mu)$ is a Radon measure space, and X_t is stochastically continuous on $T - T_0$, with $\mu(T_0) = 0$, then there exists a separable and measurable modification $\{Y_t, t \in T\}$ of the given process. Also any dense denumerable subset of T is a universal separating set of the Y_t- (and X_t-) process.*

6. Discussion. It will be of interest to analyze the role played by the stochastic continuity, in the measurablity problem, of a process presented in Theorem 4. We consider the case that $S = \mathbb{R}$ for simplicity. Since (Ω, Σ, P) is a complete (Carathéodory generated) probability space, there is a (set) lifting λ on Σ by Theorem 2.1 so that λ satisfies conditions (I)–(IV) of that section. The sets $\{\lambda(A), A \in \Sigma\}$ provide a basis for a (not necessarily Hausdorff) topology on Ω, and one can prove that each continuous function on Ω is P-measurable, and conversely, for each measurable function f on Ω there is a unique continuous function $\bar{f} : \Omega \to \bar{\mathbb{R}}$ such that $f = \bar{f}$ a.e. (The mapping $f \mapsto \bar{f}$ preserves the algebraic operations whenever they are defined.) We discuss this topology in the exercises section. Since P is Carathéodory generated and μ is Radon, the product measure $\mu \otimes P$ becomes outer regular on the topological measurable space $(T \times \Omega, \mathcal{T} \otimes \Sigma)$, with the product topology. Under these conditions the classical theorem of Luzin, noted above, implies that a function $X : T \times \Omega \to \mathbb{R}$ is $\mu \otimes P$-measurable iff for each $\varepsilon > 0$, there is a closed set $C_\varepsilon \subset T \times \Omega$ such that $\mu \otimes P(C_\varepsilon^c) < \varepsilon$ and on C_ε, X coincides with a continuous function. But in our problem the process satisfies (i) $X(t, \cdot) : \Omega \to \mathbb{R}$ is P-measurable (hence in the λ-topology is equivalent to a continuous function), and (ii) $X(\cdot, \omega)$ on

$T - T_\varepsilon^\omega$, is continuous ($\mu(T_\varepsilon^\omega) < \varepsilon$) for each ω, (these two by themselves do not imply the joint continuity of $X(\cdot,\cdot)$) and (iii) the (additional) stronger condition of stochastic continuity. It is this last property that allows us to extend the separate continuity of sections to joint continuity. Thus in the earlier terminology, we have the joint measurability. Also note that, since the sample function continuity of a process implies trivially the stochastic continuity, Theorem 4 applies to all such processes. For canonical representations, Theorems 1.5 and 1.6 complement the present result.

We shall see in applications that the generality of T (and S) is useful. So we continue to consider general parameter indexes for processes. Another specialized concept in modifications will be needed for later work.

7. Definition. Let (Ω, Σ, P) be a probability space and $\{\Sigma_t, t \in T\}$ be a filtering (to the right) sequence or net of σ-subalgebras of Σ where T is an ordered index set (thus $t < t'$ in T implies $\Sigma_t \subset \Sigma_{t'}$). An adapted stochastic process $\{X_t, \Sigma_t, t \in T\}$ on (Ω, Σ, P) with range $(S, \mathcal{B})$, a completely regular (Baire) measurable space, is said to be *progressively measurable* relative to the net $\{\Sigma_t, t \in T\}$, if $\mathcal{T}$ is a σ-algebra of T such that its points are measurable and for each $A \in \mathcal{T}$, we have $X(\cdot,\cdot)\chi_{A\times\Omega} : A \times \Omega \to S$ to be measurable for $(\mathcal{T}(A) \otimes \Sigma_A, \mathcal{B})$ where $\mathcal{T}(A)$ is the trace of $\mathcal{T}$ on A and $\Sigma_A = \sigma(\bigcup_{t\in A} \Sigma_t)$.

This concept has the following interpretation. If $T = \mathbb{R}$, the trajectory $X(\cdot,\omega)$ can be thought of as a description of an experiment ω, progressing in time and that $X(t,\cdot)$ depends only on the past, i.e., on $\{X(s,\cdot), s \leq t\}$. Thus for any $A = (-\infty, t]$, in $\mathcal{T}$, Σ_A represents all the "information" of the experiment until the present instant t, and $\Sigma_t = \sigma\{X_s, s \leq t\}$. We consider it here since the next result shows that a measurable process $\{X_t, t \in T\}$ admits a progressively measurable modification relative to the (natural) net of σ-algebras Σ_t, defined above. It finds applications in the theory of martingales, "optional stopping" problems, and stochastic integration.

8. Theorem. *Let T be an ordered subset of a locally compact separable (ordered) metric space. Let (Ω, Σ, P) be a (Carathéodory generated) complete probability space and $(T, \mathcal{T}, \mu)$ be a Radon measure space. If*

$\{X_t, \Sigma_t, t \in T\}$ is an adapted $(\mathcal{T} \otimes \Sigma, \mathcal{B})$-measurable process, where $X_t : \Omega \to S$ with $(S, \mathcal{B})$ as a completely regular (Baire) measurable space, then there exists a progressively measurable and separable (relative to the compact sets of S) process which is a modification of the X_t-process.

Proof. Since the hypothesis of Theorem 4 is included here we can take (its conclusion) that X_t is a separable process. Equivalently, if $f \in C_b(S)$ and $Y_t = f \circ X_t$, then $\{Y_t, \Sigma_t, t \in T\}$ is a real (bounded) separable adapted process for each fixed but arbitrary f; and it is jointly measurable (for $\mathcal{T} \otimes \Sigma$) by hypothesis. We now show that there is a progressively measurable modification.

As in the proof of Theorem 4, we may (and do) assume that T is compact. Since Y is bounded and measurable $Y(\cdot,\cdot)$ is $\mu \otimes P$-integrable. But $\mu \otimes P(T \times \Omega) < \infty$; so there exists a sequence of simple functions $\{Y_n(\cdot,\cdot)\}_{n=1}^{\infty}$ such that

$$\int_{T\times\Omega} |Y_n - Y| d\mu \otimes P \to 0, \qquad \text{as } n \to \infty, \tag{7}$$

where we have a representation for Y_n as:

$$Y_n(t,\omega) = \sum_{i=1}^{k_n} \chi_{A_i^n}(t)\chi_{B_i^n}(\omega), \qquad a_i^n \in \mathbb{R}, \tag{8}$$

$A_i^n \in \mathcal{T}, B_i^n \in \Sigma$ and $\{A_i^n \times B_i^n\}_{i=1}^{k_n}$ is a partition of $T \times \Omega$. This follows from the standard facts: (a) simple functions $\left\{\sum_{i=1}^{k_n} c_i^n \chi_{C_i^n}, C_i^n \in \mathcal{T} \otimes \Sigma\right\}_{n\geq 1}$ are dense in $L^1(\mathcal{T} \otimes \Sigma)$, (b) $\mu \otimes P(\cup_{i=1}^{k_n} C_i^n) < \infty$ implies the existence of a measurable rectangle $A_i^n \times B_i^n \in \mathcal{T} \times \Sigma$, such that $\mu \otimes P(C_i^n \Delta (A_i^n \times B_i^n))$ is arbitrarily small, and (c) adjusting the coefficients c_i^n, the form (8) results. Leaving the easy verification of these statements, we sketch the rest of the argument.

From (7) it follows that $Y_n \to Y$ in $\mu \otimes P$-measure and so for a subsequence $Y_{n'} \to Y$, a.e. In particular, $Y_{n'}(t,\cdot) \to Y(t,\cdot)$ a.e. [P], for a.a.(t). So for any $A \in \mathcal{T}$, $Y_{n'}(t,\omega)\chi_A(t) \to Y(t,\omega)\chi_A(t)$, for a.a.$(t,\omega)$. But for large enough n', $Y_{n'}(\cdot,\cdot)\chi_A$ is $\mathcal{T}(A) \otimes \Sigma_t$-measurable since Y_t is Σ_t-adapted. If we set $Z(t,\omega)$ to be the limit of $Y_{n'}(t,\omega)$ whenever this limit exists, and $= 0$ otherwise, then $Z(\cdot,\cdot)\chi_A$ is $\mathcal{T}(A) \otimes \Sigma_A$-measurable. It is now easily verified that $\{Z(t,\cdot), t \in T\}$ is a progressively measurable modification of $\{Y(t,\cdot), t \in T\}$. Hence (by Theorem

1) the Y_t-process (and so also the X_t-process) admits a separable and progressively measurable modification. □

The above result when $S = \mathbb{R}$ and $T \subset \mathbb{R}$ has been given by Chung and Doob [1], who have introduced the term "progressively measurable". Since the sample left or right continuity in this case is well-defined, and either implies the measurability of the X_t-process, it follows that all such processes admit progressively measurable modifications.

As applications of these results, we consider some sample function properties of stochastic processes, both when they are canonically represented and when they are not. A treatment of these problems, given in the next section, will be needed for stochastic integration and elsewhere.

3.4 Stochastic functions: regularity properties

The regularity properties of sample functions, of a separable (and measurable) stochastic process, require a further nontrivial analysis. In this section some results on the discontinuities of a canonical process, the Kolmogorov criterion for its continuity, and related questions are treated. It will be seen that the existence of processes with special properties usually needs extensions of the fundamental Kolmogorov-Bochner theorems.

Our first result is a refinement of Theorem 1.6 when the hypothesis is strengthened, and this leads to some useful specializations. To motivate and relate the stochastic result, we recall certain concepts from classical function theory.

Let (M, d), or M, be a metric space and $f : \mathbb{R} \to M$ be a mapping. Then f has a *discontinuity of the second kind* at $t_0 \in \mathbb{R}$ iff there is a sequence $\{t_n, n \geq 1\}$, the t_n tending to t_0 monotonely, such that for some $\varepsilon > 0$ and any n_0 it is true that $d(f(t_n), f(t_0)) \geq \varepsilon$ for infinitely many $n > n_0$. In case there are only finitely many $n > n_0$ in this statement, then f is said to have *a point discontinuity* (or be *free of oscillatory discontinuity*, or has a *a jump discontinuity* when $M = \mathbb{R}$) at t_0. The points of the latter kind are isolated and hence the set of point discontinuities is at most countable and nowhere dense in $\mathbb{R}$. The converse statement is also true and so it can be taken as a definition of point discontinuity. This was first proposed by H. Hankel (cf.,*Math. Annalen* **20**(1882), page 90). The definition of Hankel admits the following stochastic extension. Let T be a topological space, and M a metric space as above. If $f \in M^T$, let $\Delta(f)$ be the set of discontinuity points of f, so that if $\mathcal{U}(t)$ is a neighborhood base at $t \in T$, then $t_0 \in \Delta(f)$ iff $F(t_0) > 0$ where

$$F(t) = \inf\{\sup\{d(f(u), f(v)) : u, v \in U\} : U \in \mathcal{U}(t)\}. \qquad (1)$$

If $\Delta_n(f) = \{t : F(t) \geq \frac{1}{n}\}$, then $\Delta(f) = \cup_{n=1}^{\infty} \Delta_n(f)$. For each n, $\Delta_n(f)$ is closed. In fact, $\{t : F(t) < a\} = \emptyset$ for $a \leq 0$, and if $a > 0$, for each t in this set and $0 < \varepsilon < a - F(t)$, there is an open set U containing t such that in the right side of (1), the sup{} is less than $F(t) + \varepsilon$. So $U \subset \{F(t) < a\}$ and $\Delta(f)$ is closed. Thus $\Delta(f)$ is an F_σ-set. We use this information below. In this general form $f \in M^T$ is said to have only *point discontinuities* iff $\Delta(f)$ is of the first category in T. We translate this to the stochastic case as follows.

Let $\{X_t, t \in T\}$ be a stochastic process on (Ω, Σ, P) with values in a metric space M, and T be a topological space. Suppose it is already represented in canonical form, so that $\Omega = M^T, \Sigma = \mathcal{B}_T = \otimes_{t \in T} \mathcal{B}_t$ where $\mathcal{B}_t = \mathcal{B}$, the Borel σ-algebra of M, P is the (regular) Baire probability measure on Σ, and $X_t(\omega) = \omega(t)$, $\omega \in \Omega$. Let D_p be the class of all sample functions of the process with only point discontinuities. Thus $D_p = \{\omega \in \Omega : \Delta(\omega) \subset T, \text{of first category}\}$. Clearly D_p contains all continuous sample functions. One has the following description of D_p, without considering Σ_P, but with some restrictions on T and M. We let $\tilde{\Sigma}$ be the Borel σ-algebra generated by Σ, and denote the unique (regular) extension of P onto it by the same letter, where M is compact.

1. Theorem. *Let T and M be compact metric spaces and D_p be the set of sample functions of $\{X_t, t \in T\}$ with only point discontinuities where the process has the canonical representation. Then (i) $D_p \in \tilde{\Sigma}$, and (ii) if the process has no fixed discontinuities, $P(D_p) = 1$.*

Proof. Since $\tilde{\Sigma}$ is not complete, the result of (i) does not follow from Theorem 1.6; and the present hypothesis and conclusion are both stronger than the former. We first give a proof of (i).

Let δ and d be the distance functions of T and M respectively. If $\varepsilon > 0, S \subset T$, let $D(\varepsilon, S)$ be the set of ε-continuous sample paths ω at each point of S, i.e.,

$$D(\varepsilon, S) = \bigcup_{n=1}^{\infty} \bigcap \left\{ \left\{ \omega : d(\omega(t), \omega(s)) \leq \varepsilon,\ \delta(s,t) < \frac{1}{n} \right\} : s,t \in S \right\}. \tag{2}$$

Since each set in the union is closed, one deduces that $D(\varepsilon, S)$ is an F_σ- set. By the separability of T, we take a denumerable dense set $T_0 \subset T$ and if $\mathcal{U}_0$ is a countable base of open sets of T, $\mathcal{U}_0$ admits a locally finite refinement $\mathcal{U}$. So every point $t \in T$ has a neighborhood which meets with only finitely many members of $\mathcal{U}$ and we may also assume the refinement to have the property that each of its members has diameter at most $\frac{1}{r}, r \geq 1$. Let $T_{mk} \in \mathcal{U}$ be the kth set (in some enumeration) so that it is within a distance of $\frac{1}{m}$ from T_0. Thus each point of T is within a distance of $\frac{1}{m}$ from some point of T_{mk} for a k. From this we deduce that $\omega \in D_p$ iff for each n, $\Delta_n(\omega)$ is at a positive distance from some T_{mk} for each m and some k (i.e., $\Delta_n(\omega)$ is nowhere dense), or in symbols

$$\omega \in \bigcap_{j=1}^{\infty} \bigcap_{m=1}^{\infty} \bigcup_{k=1}^{\infty} D\left(\frac{1}{j}, T_{mk}\right). \tag{3}$$

Since from (2) and (3), the union being an F_σ-set, D_p is the above displayed quantity, we conclude that D_p is an $F_{\sigma\delta}$-set and hence is a Borel subset of Ω, i.e., $D_p \in \tilde{\Sigma}$ (but is not necessarily in Σ). Thus (i) is true.

To prove (ii), let $f \in C(M)$, the space of real continuous functions which is separable. Since the present hypothesis implies that of Theorem 1.5, we can use some of the computations of its proof. Given that the process has no fixed discontinuity points, the set A of that theorem is empty, and by (5) and (8) there, one concludes that $Y^+(t,\omega) = (f \circ X)(t,\omega)$ for each $t \in T$ and a.a. (ω), because $f(X(\cdot,\omega)) = f(\omega(\cdot))$ is upper semicontinuous for a.a. (ω) since in fact it is continuous. We now use a simple extension of the result (I.5.9) to conclude, from the fact that D_p is an $F_{\sigma\delta}$, the existence of a *countable* set $J \subset T$ such that $D_p = D_{pJ}$ a.e., where D_{pJ} is determined by its values on J (i.e., $\omega \in \Omega$ and $\omega(t) = \tilde{\omega}(t)$ for $t \in J, \tilde{\omega} \in D_{pJ}$ then

$\omega \in D_{pJ}$). If $\Sigma = \tilde{\Sigma}$, then this is precisely the result given there stating that $D_p = D_{pJ}$ a.e. We have to show this equality in the present (slightly larger) class $\tilde{\Sigma}$ (i.e., for the Borel, and not merely the Baire, class). In general such an extension need not obtain, but it holds for regular Baire measure spaces (Ω, Σ, P). The latter property is available for the canonical representations by the last part of Theorem I.3.4. Then a standared measure theory result says that each Borel set B of finite measure can be expressed as $B = A \Delta N$ where $A \in \Sigma$ (i.e., a Baire set) and N is a Borel set of measure zero (cf., e.g., Royden [1], p. 314, or Rao [11], p. 453). Thus $B = A$ a.e. for the Borel extension of P. This means in our case $D_p = D'_p$ a.e. for some Borel set D'_p and then by (I.5.9) noted above there is a countable set $J \subset T$ such that $D_p = D_{pJ}$ a.e., where $D_{pJ} = D'_p$.

We can now complete the argument as follows. Since $Y^+(j, \omega) = (f \circ X)(j, \omega)$ for $j \in J$ and a.a. (ω), by Theorem 1.5 $Y^+(j, \omega) = \lim_{n \to \infty} Y_n^+(j, \omega)$ and each Y_n^+ has a finite number of point discontinuities. So $Y^+(\cdot, \omega)$ can have only point discontinuities. But $C(M)$ is a separable space, and so for a dense denumerable set $\{f_n, n \geq 1\}$ the function $(f_n \circ X)$ has only point discontinuities for a.a. ω, and all $n \geq 1, j \in J$. Thus $X(\cdot, \omega)$ is equal to a function on J having only point discontinuities for a.a. (ω). But this means it is true for all $\omega \in D_{pJ}$, and so $P(D_{pJ}) = 1$. Since $D_p = D_{pJ}$ a.e.$[P]$, (ii) follows. □

Remark. The above result admits an extension if T and M are replaced by $T \subset \mathbb{R}, M = \mathbb{R}^n$ by reducing it to the compact case and observing that the Kolmogorov (Bochner) Theorem I.2.2 (I.3.4) applies to give the regularity of P. Indeed T, M can be σ-compact metric spaces.

Although all continuous sample functions are in D_p, the above result does not tell enough about this particular subset. For instance, if it is known that almost all sample paths of a process are continuous (we show the existence of such processes below), does D_p coincide with this set? The answer is not entirely simple. We prove this result within a somewhat more general context. The relevant changes in the measure space to make such a set measurable are useful. (With $T \subset \mathbb{R}, M = \mathbb{R}$, it is due to Doob.)

2. Theorem. *Let $\{X_t, t \in T\}$ be a canonically represented process* *on the space* (Ω, Σ, P) *with values in a compact (or* σ*-compact) metric space* M *(having at least two points), and* T *be a (non-finite) compact metric space. If the process has almost all sample functions continuous, then the set of continuous functions* $C \subset \Omega(= M^T)$ *is nonmeasurable (for* Σ*), and has outer measure one, i.e.,* C *is a thick set. In this case the triple* (Ω, Σ, P) *can be replaced by* $(C, \tilde{\Sigma}, \tilde{P})$ *where* $\tilde{\Sigma}$ *is the trace* σ*-algebra* $\Sigma(C) = \{A \cap C : A \in \Sigma\}$ *and* $\tilde{P}(A \cap C) = P(A)$*. If* $\alpha \subset T$ *is a finite set and* $A_i \subset M$ *is a Borel set so that* $P_\alpha(\times_{i\in\alpha} A_i) = P \circ \pi_\alpha^{-1}(\times_{i\in\alpha} A_i)$ *is the finite dimensional distribution of the given process, then* $P \circ \pi_\alpha^{-1} = \tilde{P} \circ \pi_\alpha^{-1}$ *is true and the modified stochastic process* $\{\tilde{X}_t, t \in T\}$ *has* $(C, \tilde{\Sigma}, \tilde{P})$ *as its probability space. Thus* $\tilde{X}_t$ *is "indistinguishable" from* X_t *in the sense that both processes have the same finite dimensional distributions.*

Proof. To see that C is nonmeasurable, suppose the contrary. Then by the same remark used in the preceding, there is a countable $J \subset T$ such that every element of C is determined on J; i.e., if $\omega \in \Omega = M^T$ and $\tilde{\omega} \in C$ such that $\omega(t) = \tilde{\omega}(t)$ for all $t \in J$ then $\omega \in C$. Now, if $t_0 \in T - J$, then for each $\omega' \in C$ we have $\omega'(t_0) = \lim_{t_i \to t_0} \omega'(t_i), t_i \in J$ by continuity. Since the spaces are Hausdorff the limits are unique. Choose a function $\omega_0 \in \Omega$ such that $\omega_0(t) = \tilde{\omega}(t)$ for all $t \in J$, but ω_0 is not continuous at t_0. (Such functions clearly exist.) By our assumption then $\omega_0 \in C$ and hence must be continuous. This contradiction shows that $C \notin \Sigma$. Note that we may even assume that J is dense in T by adding a countable set of points from T, if necessary, and the situation is unaltered.

To see that C is a thick set, we show that every measurable $B \subset \Omega$ such that $B \cap C = \emptyset$ satisfies $P(B) = 0$. Then $P^*(C) = \inf\{P(B^c) : B \text{ as above}\} = 1$ follows. Since $B \in \Sigma$, there exists a countable $J_1 \subset T$ such that B (and B^c) are determined by J_1; i.e., $\omega \in \Omega, \tilde{\omega} \in B$ ($\tilde{\tilde{\omega}} \in B^c$) and $\omega(t) = \tilde{\omega}(t)(\tilde{\tilde{\omega}}(t) = \omega(t))$ all $t \in J_1$, implies $\omega \in B(\omega \in B^c)$. Since almost all sample paths are continuous, by hypothesis, for each $t \in J_1$ there is a P-null set N_t such that each $\omega \in N_t^c$ is continuous at t, so each ω determines an $\tilde{\omega} \in C$ satisfying $\omega(t) = \tilde{\omega}(t)$. Hence if $N = \cup\{N_t : t \in J_1\}$ then $N \in \Sigma, P(N) = 0$ and each $\omega \in N^c$ coincides with an $\tilde{\omega} \in C$ on J_1. But $C \cap B = \emptyset$ or that $C \subset B^c$. Hence $\omega \in N^c$ implies the existence of an $\tilde{\omega} \in C \subset B^c$ such that $\omega(t) = \tilde{\omega}(t)$, $t \in J_1$, and so $\omega \in B^c$. This means $N^c \subset B^c$ or $B \subset N$, and $P(B) = 0$ so that $P^*(C) = 1$.

Let us define a new measure space. Set $\tilde{\Omega} = C, \tilde{\Sigma} = \{A \cap C : A \in \Sigma\}$, and $\tilde{P}(A \cap C) = P(A)$. If $A = \Omega$, then $C \in \tilde{\Sigma}, \tilde{P}(C) = 1$, and we need only show that $\tilde{P}$ is well-defined and σ-additive. The rest will be immediate. Suppose $A_1 \cap C = A_2 \cap C, A_i \in \Sigma$. We claim that $P(A_1) = P(A_2)$. Let $A = A_1 - A_1 \cap A_2 (\in \Sigma)$. Then $A \cap C = A_1 \cap C - A_1 \cap (A_2 \cap C) = \emptyset$, since $A_1 \cap C = A_2 \cap C$. Hence $A^c \supset C$ and then, if P^* is the outer measure generated by (Σ, P),

$$1 = P^*(C) \leq P(A^c) = 1 - P(A) \leq 1, \quad C \text{ being thick.} \tag{4}$$

It follows from (4) that

$$0 = P(A) = P(A_1 - A_1 \cap A_2) = P(A_1) - P(A_1 \cap A_2), \tag{5}$$

and $P(A_1) = P(A_1 \cap A_2) (= P(A_2)$ by symmetry). So $\tilde{P}$ is well-defined on $\tilde{\Sigma}$. For the σ-additivity of $\tilde{P}$, let $\{A_n\}_{n=1}^{\infty} \subset \tilde{\Sigma}$ be a disjoint sequence, and let $A_n = B_n \cap C, B_n \in \Sigma$, be a representation. Since the B_n need not be disjoint, let $E_1 = B_1$ and for $n > 1, E_n = B_n - \cup_{i=1}^{n-1} B_i$ be a disjunctification. Then $(B_n - E_n) \cap C = \emptyset$ so that $P(B_n - E_n) = \tilde{P}((B_n - E_n) \cap C) = 0$. But $E_n \subset B_n$, so that $P(B_n) = P(E_n)$, and we have

$$\begin{aligned} \sum_{n=1}^{\infty} \tilde{P}(A_n) &= \sum_{n=1}^{\infty} P(B_n) = \sum_{n=1}^{\infty} P(E_n) \\ &= P\left(\cup_{n=1}^{\infty} E_n\right) = P\left(\cup_{n=1}^{\infty} B_n\right) = \tilde{P}\left(\cup_{n=1}^{\infty} A_n\right). \end{aligned} \tag{6}$$

We may now define a process $\{\tilde{X}_t, t \in T\}$ by the equation $\tilde{X}_t(\omega) = \omega(t)$, for $\omega \in C$. Then one has since $C \subset \Omega = M^T$,

$$\begin{aligned} \tilde{P}\left[\bigcap_{i=1}^{n} \{\omega : \tilde{X}_{t_i}(\omega) \in A_i\}\right] &= \tilde{P}[C \cap \pi_{\alpha}^{-1}(A)], \quad \text{with} \\ & A = \times_{i \geq 1} A_i,\ A_k = M \text{ for } k > n, \\ &= P[\pi_{\alpha}^{-1}(A)] = P\left[\bigcap_{i=1}^{n} \{\omega : X_{t_i}(\omega) \in A_i\}\right], \end{aligned}$$

where $\pi_\alpha : \Omega \to \times_{i=1}^n M^i (M^i = M, \alpha = \{t_1, \ldots, t_n\})$ is the coordinate projection of the Kolmogorov existence theorem. Here the A_i are Borel sets of M. □

Regarding non-measurability of C above, we have the following result.

3. Proposition. *Let T, M be two compact (or σ-compact) metric spaces, and $\Omega = M^T$. Let $C \subset \Omega$ be the set of all continuous functions. Then C is an $F_{\sigma\delta}(F_{\sigma\delta\sigma})$ set, so that it is a Borel subset of Ω, and in fact we have:*

$$C = \bigcap_{m=1}^{\infty} \bigcup_{n=1}^{\infty} \left\{ \omega : d(\omega(s), \omega(t)) \leq \frac{1}{m}, \text{ for all } \delta(s,t) \leq \frac{1}{n}; s, t \in T \right\}, \tag{7}$$

where d and δ are metrics on M and T ($C = \bigcup_{k=1}^{\infty} C_k$, where C_k is as in (7) with T_k for T, $T = \bigcup_{k=1}^{\infty} T_k$, and $T_k \nearrow$, compact).

Proof. Since continuity is a local property, the σ-compact case follows from the compact case, and it suffices to prove the result with the latter assumption. Note that the set $\{\omega : d(\omega(s), \omega(t)) \leq \frac{1}{m}\} \subset \Omega$ is closed for each (s,t) verifying $\delta(s,t) \leq \frac{1}{n}$ and hence the set in $\{\}$ of (7) is an $F_{\sigma\delta}$ and so is a Borel set. Clearly $C \supset$ the right side. For the opposite inequality, observe that the set in $\{\}$ is the class of all $\omega \in \Omega$ which satisfy the given conditions for all $m \geq 1, n \geq 1$. Thus the right side of (7) says that it is the set of all uniformly continuous $\omega \in \Omega$ on T. Since T is compact this accounts for all continuous functions on T to M. Thus it contains C and hence equality holds. The σ-compactness case then follows by considering each T_k and applying the above result.
□

The preceding two results are significant only when we show the existence of processes having almost all continuous sample paths. For this we present in the following theorem a useful criterion (due to Kolmogorov) in order that a process may have continuous sample paths a.e., and then deduce that the Brownian motion process has this property. More importantly, this result gives conditions on the *finite* (in fact *two)dimensional* distributions, and we verify them for the Brownian family (cf., I.5.4). By Theorem 2 the existence of such a process follows at once.

4. Theorem. *Let $T \subset \mathbb{R}$ be a compact interval and (M, d) a complete separable metric space (also called a Polish space). Suppose $\{X_t, t \in T\}$ is a stochastic process on a probability space (Ω, Σ, P) with values in $(M, \mathcal{B})$ where $\mathcal{B}$ is the Borel σ-algebra of M. If there exist constants $\alpha > 0, \beta > 0$ and $K > 0$ such that for any pair of points t_1, t_2 of T and $\varepsilon > 0$,*

$$P\{\omega : d(X_{t_1}, X_{t_2})(\omega) > 0\} \leq K\varepsilon^{-\beta} |t_1 - t_2|^{1+\alpha}, \tag{8}$$

then the process may be taken to be separable and almost all of its sample functions are continuous.

Before proving this result, we illustrate its use on the Brownian motion $\{X_t, t \in T\}$ which is a real Gaussian process with mean zero and with independent increments, such that $E(|X_{t_1} - X_{t_2}|^2) = \sigma^2 |t_1 - t_2|$ for some constant σ^2 which we take it to be unity. Thus we have the following:

5. Corollary. *The sample functions of a Brownian motion process $\{X_t, t \in T\}$ on (Ω, Σ, P) are almost all continuous where $T \subset \mathbb{R}$ is any compact interval.*

Proof. By definition, for t_1, t_2 in T with $(t_1 < t_2)$ and $x \in \mathbb{R}$, we have

$$P\{\omega : X_{t_2}(\omega) - X_{t_1}(\omega) < x\} = [2\pi(t_2 - t_1)]^{-1/2} \int_{-\infty}^{x} \exp[-\frac{1}{2}u^2/(t_2 - t_1)]du. \tag{9}$$

But an elementary computation shows that

$$E(|X_{t_2} - X_{t_1}|^4) = 3|t_1 - t_2|^2, \tag{10}$$

so that by Čebyšev's inequality using fourth moments,

$$P\{\omega : |X_{t_2}(\omega) - X_{t_1}(\omega)| > \varepsilon\} \leq E(|X_{t_2} - X_{t_1}|^4)/\varepsilon^4 = 3(t_2 - t_1)^2/\varepsilon^4. \tag{11}$$

Thus (8) is true with $\alpha = 1, \beta = 4$, and $K = 3$ and the result follows. □

As we remarked above, from this the existence of Brownian motion is deduced when we complete the proof of the above theorem. However, we obtain that result from the following slightly more general statement which is adapted from Loève [1] and Neveu [1].

6. Theorem. *Let $\{X_t, t \in T\}$ be a separable stochastic process on (Ω, Σ, P) where $T \subset \mathbb{R}$ is a compact interval and X_t takes its values in a complete separable metric space M. Suppose there exist two positive non-decreasing functions f, g on the interval $(0, h)$ such that*

$$(i) \int_0^h \frac{f(x)}{x} dx < \infty, \quad \text{and} \quad (ii) \int_0^h \frac{g(x)}{x^2} dx < \infty, \quad h > 0. \tag{12}$$

If the process satisfies the inequality

$$P\{\omega : d(X_{t+h}(\omega), X_t(\omega)) \geq f(h)\} \leq g(h), t \in T, \tag{13}$$

then almost all sample functions of the process are continuous.

We first deduce Theorem 4 from this result before giving its proof.

Proof of Theorem 4. Since condition (8) clearly implies stochastic continuity by Theorem 3.3, the process is separable with any dense denumerable subset of T as a universal separating set. To see that (13) follows from (8) for suitable f and g satisfying (12), note that $\frac{\beta}{\alpha} > 0$ and choose $0 < \gamma < \frac{\beta}{\alpha}$. If $K > 0$ is given and $\varepsilon > 0$ is arbitrary, set $f(x) = x^\gamma$ and $g(x) = Kx^{1+\delta}$ where $\delta = \alpha - \beta\gamma > 0$. Then on the interval $(0, h)$,f, g are positive and increasing, and (i), (ii) of (12) are true. By (8) we have

$$P\{\omega : d(X_{t+h}(\omega), X_t(\omega)) \geq f(h)\} \leq K(f(h))^{-\beta} |h|^{1+\alpha} = g(h), \quad t \in T.$$

Since this is (13), the result follows. □

Proof of Theorem 6. Since the integrals in (12) are finite, it is clear that $\lim_{x\to 0} g(x) = 0 = \lim_{x\to 0} f(x)$, and by (13) then $X_{t+h} \to X_t$ in probability as $h \to 0$. Hence by Theorem 3.3, any dense denumerable subset of T is a universal separating set for the separable process. Let $S \subset T$ be the set of all dyadic numbers which is thus a separating set. Hence if

$s = k2^{-n}, n \geq 0$ and k are integers, so $s \in S$ are numbers dense in T, then by the compactness of M,

$$\sup_{T\cap[|t-\tau|\leq h]} d(X_t(\omega), X_\tau(\omega)) = \sup_{S\cap[|s-s'|\leq h]} d(X_s(\omega), X_{s'}(\omega)), a.a.(\omega). \tag{14}$$

We therefore may restrict our attension to S in computations below.

Since $T \subset \mathbb{R}$ is an interval, say $T = [a,b], -\infty < a < b < \infty$, for each n, define the random variables

$$Y_n = \sup\{d(X_{(k+1)2^{-n}}, X_{k2-n}) : a2^n \leq k < b2^n\}. \tag{15}$$

Then we have (with k starting at the smallest integer $\geq a2^n$ and taking all integers $< b2^n$)

$$\begin{aligned} P[Y_n \geq f(2^{-n})] &\leq P[\bigcup_{k\geq a2^n}^{b2^n-1} [d(X_{(k+1)2^{-n}}, X_{k2^{-n}}) \geq f(2^{-n})]] \\ &\leq \sum_{k\geq a2^n}^{b2^n-1} P[d(X_{(k+1)2^{-n}}, X_{k2^{-n}}) \leq f(2^{-n})] \\ &\leq (b-a)2^n g(2^{-n}), \quad \text{by (13)}. \end{aligned} \tag{16}$$

Considering the interval $(0,h)$ of (12) as $(0, 2^{-m}) = \cup_{n\geq m}[2^{-n-1}, 2^{-n})$ we deduce that $f(2^{-n})$ and $2^n g(2^{-n})$ are general terms of convergent series. In fact,

$$\log 2 \sum_{n>m} f(2^{-n}) \leq \sum_{n=m}^{\infty} \int_{2^{-n}}^{2^{-n+1}} \frac{f(x)}{x} dx \leq \int_0^{2^{-m}} \frac{f(x)}{x} dx < \infty,$$

and

$$\sum_{n>m} 2^n g(2^{-n}) \leq \sum_{n=m}^{\infty} \int_{2^{-n}}^{2^{-n+1}} \frac{g(x)}{x^2} dx \leq \int_0^{2^{-m}} \frac{g(x)}{x^2} dx < \infty.$$

Hence, if $A_n = \{\omega : Y_n(\omega) \leq f(2^{-n})\}$, and $A = \limsup_n A_n$, then

$$\begin{aligned} P(A) &= P(\bigcap_{k=1}^{\infty} \bigcup_{n\geq k} A_n) \\ &\leq P(\bigcup_{n\leq k} A_n) \leq \sum_{n\geq k} P(A_n) \\ &\leq \sum_{n\geq k} (b-a)2^n g(2^{-n}) < \infty. \end{aligned}$$

Since this is true for all k, we must have $P(A) = 0$, i.e., $P(A^c) = 1$. But $\omega \in A^c (= \liminf_n A_n^c)$ implies $Y_n(\omega) < f(2^{-n})$ for all $n \geq k(\omega)$. This means

$$\sum_{n\geq k(\omega)} Y_n(\omega) < \sum_{n\geq k(\omega)} f(2^{-n})$$

$$\leq \frac{1}{\log 2} \int_0^{2^{-n}} \frac{f(x)}{x} dx < \infty, \quad k(\omega) \geq m. \tag{17}$$

It follows that $\sum_{n=1}^{\infty} Y_n(\omega) < \infty$ for each $\omega \in A^c$, i.e., for a.a. (ω).

Let $t \in T$ and $k2^{-n} \in S$ be such that $|t - k2^{-n}| < 2^{-n}$. Since $a \leq t \leq b$ we have, on using (15) and the triangle inequality ($t = k2^{-n} + \sum_{m\geq(n+1)} \theta_m 2^{-m}, \theta_m = 0 \, or \, 1$),

$$d(X_t, X_{2^{-n}}) \leq \sum_{m>n} Y_m, \qquad \text{a.e.} \tag{18}$$

In a similar way for s such that $|t - s| < 2^{-n}$ (density of S),

$$d(X_t, X_s) \leq d(X_t, X_{k2^{-n}}) + d(X_s, X_{k2^{-n}}) \leq 2 \sum_{m\geq n} Y_m, \quad \text{a.e.}$$

From this it follows that

$$\sup_{|t-s|<2^{-n}} d(X_t, X_s) \leq 2 \sum_{m\geq n} Y_m, \quad \text{a.e. by (18).} \tag{19}$$

Letting $n \to \infty$ and using (17) we get (19) tending to 0, a.e. Hence $\{X_t, t \in T\}$ has almost all continuoous sample paths. □

Note that (17) and (19) contain somewhat more information than the final conclusion of the theorem. In fact in (17), $k(\cdot)$ is an integer valued random variable and if we set $N(\omega) = 2^{-k(\omega)}$, then $N(\cdot)$ is a positive bounded random variable and (17) and (19) can be expressed, for $\frac{h}{2} \leq 2^{-n} < h$, as

$$\sup \left\{ d(X_t, X_s) : |t - s| < \frac{h}{2} \right\} \leq \frac{2}{\log 2} \int_0^h \frac{f(x)}{x} dx, \qquad \text{if } h \leq N(\cdot), \text{ a.e.} \tag{20}$$

Choosing f and g as in the proof of Kolmogorov's theorem above we can present the "modulus" of continuity for processes satisfying the hypothesis of Theorem 4, and then a sharpening of Corollary 5.

7. Proposition. *Let $\{X_t, t \in T\}$ be a process satisfying the hypothesis of Theorem 4. Then for every $0 < \gamma < \frac{\alpha}{\beta}$ it follows that*

$$h^{-\gamma} \sup_{|t-s|<h} d(X_t, X_s) = o(1), \quad \text{a.e., as } h \to 0. \tag{21}$$

In particular, if $\{X_t, t \in T\}$ is a Brownian motion, as in Corollary 5, then for each $0 < \varepsilon < \frac{1}{2}$,

$$h^{(-1/2)+\varepsilon} \sup_{|t-s|<h} |X_t - X_s| \to 0, \text{a.e., as } h \to 0^+. \tag{22}$$

Thus almost all sample paths of a Brownian motion process are Hölder continuous of order $0 < \varepsilon < \frac{1}{2}$.

Proof. By (20),

$$\sup_{|t-s|<h} d(X_t, X_s) \leq 2 \sup_{|t-s|<h/2} d(X_t, X_s) < \frac{4}{\log 2} \int_0^h \frac{f(x)}{x} dx, \text{ a.e.}$$

Now take $f(x) = \varepsilon_1 x^\gamma$ and $g(x) = K\varepsilon_1 x^{1+\delta}$ where $\delta = \alpha - \beta\gamma > 0$ as in the proof of Theorem 4 for any constant $\varepsilon_1 > 0$. Then again we have

$$\int_0^h \frac{f(x)}{x} dx = \varepsilon_1 \frac{h^\gamma}{\gamma} < \infty,$$

and using this in the first line we get

$$\sup_{|t-s|<h} d(X_t, X_s) \leq \frac{4}{\log 2} \varepsilon_1 \frac{h^\gamma}{\gamma}, \quad \text{a.e.} \tag{23}$$

Since $0 < \gamma < \alpha/\beta$ is fixed and $\varepsilon_1 > 0$ is arbitrary, (23) implies (21).

Let $0 < \varepsilon < 1/2$ be given. Since in the Brownian case $X_t - X_s$ is Gaussian with mean zero and variance $= E((X_t - X_s)^2) = \alpha_0 |t - s| = \sigma^2$, we note that for any $p \geq 0$,

$$\begin{aligned} E(|X_t - X_s|^p) &= \sqrt{\frac{2}{\pi\sigma^2}} \int_0^\infty u^p e^{-u^2/2\sigma^2} du \\ &= \frac{\Gamma(\frac{p+1}{2})}{\sqrt{\pi}} (2\sigma^2)^{p/2} = K_p |t-s|^{[(p-2)/2]+1} . \end{aligned} \tag{24}$$

Let $p = \frac{1}{\varepsilon} > 2$ so that $\alpha = \frac{p-2}{2} > 0$, $\beta = p$ and let $0 < \gamma < \frac{\alpha}{\beta} = \frac{p-2}{2p} = \frac{1}{2} - \varepsilon$. Thus for any $0 < \varepsilon < \varepsilon' < \frac{1}{2}, \gamma = \frac{1}{2} - \varepsilon'$ we have from (21)

$$h^{-(1/2)+\varepsilon} \sup_{|t-s|<h} |X_t - X_s| \to 0 \text{ a.e. as } h \to 0^+.$$

Since ε and ε' are arbitrary in the ranges, this is the same as (22). □

It is of interest to note that the set C_α of the class of all Hölder continuous functions of order α in $\Omega = M^T$ is a thick set for the Brownian motion probability P on the Baire σ-algebra of Ω if $0 \leq \alpha < 1/2$. We have treated the continuous function case C $(\alpha = 0)$ earlier. It is clear that $C_\alpha \subset C_{\alpha'}$ if $\alpha > \alpha'$. It is easy to show that for $\alpha > 1/2, P(C_\alpha) = 0$ (for the Brownian case again). This is deduced from the fact that $Y = (X_t - X_s)/(t-s)^{1/2}$ is a Gaussian random variable with mean zero and variance one, independent of $t - s (> 0)$. Thus $\omega \in C_\alpha$ iff $|X_t - X_s|(\omega) \leq k|t-s|^\alpha$ and we have

$$P\left\{\omega : \frac{|X_t - X_s|(\omega)}{|t-s|} \leq k|t-s|^{\alpha-(1/2)}\right\} = \frac{1}{2\pi} \int_0^{k|t-s|^{\alpha-(1/2)}} e^{-x^2/2}\, dx,$$

and since the left side is independent of $|t-s|$, and the pair (t,s) is arbitrary, we conclude that the set is P-null. We leave the details to the reader. If $\alpha = 1/2$, then $C_{1/2}$ is neither a thick nor a null set. Thus the special processes demand a nontrivial special sample function analysis. For instance, the Poisson process (mentioned in Chapter I) in its canonical form can be shown to have almost all of its sample paths right continuous and monotone, and this is a thick set for the Poisson measure. Then the classical Kolmogorov theorem (I.2.2), as modified in the last part of Theorem 2 above, should be employed to show the existence of such processes. A general projective limit theorem, applicable to these situations and including (I.2.2), will be given in the last section of this chapter. It could be readily employed to establish several existence results on processes.

3.5 Continuous and directed parameter martingale convergence

In a number of problems (e.g., inferences on processes) one finds a need for a treatment of martingale convergence when the indexing is either uncountable or not linearly ordered. The measurability and separability of such a process intervene in the analysis. Since these concepts and some of their implications are discussed in the preceding sections, we can now treat the above problems. As noted in Chapter II, some results admit immediate extensions. These include the Riesz and Jordan decompositions and the basic inequalities. Some others, such as the Doob decomposition, are deeper and have to wait for new tools. We consider those results that are obtainable with the preceding work.

(a) *Linearly ordered index set.* When the index set is part of the real line, so that it may be identified with time, the convergence theory of separable martingales proceeds as in the discrete case. We state their precise versions.

1. Theorem. *Let $\{X_t, t \in T\}$, $T \subset \mathbb{R}$, be a real separable submartingle on a probability space (Ω, Σ, P). Then the following statements are true:*

(i) if the process has no fixed points of discontinuities, the only discontinuities (if any) of almost all sample functions are jumps;

(ii) almost all sample functions are bounded on compact sets of T; and

(iii) for almost all $\omega \in \Omega, X_{(\cdot)}(\omega)$ has a left (or a right) limit at each $t \in T$ which is a left (or right) limit point of T.

Proof. This is an easy consequence of the earlier theory, and we include an outline. If $t_0 \in T$ is not a fixed discontinuity, then there exist sequences $\{t_n, n \geq 1\}$ from T such that $t_n \uparrow t_0$(or $t_n \downarrow t_0$) and the process $\{X_{t_n}, \mathcal{F}_n, 1 \leq n \leq \infty\}$ is an integrable submartingle sequence where $\mathcal{F}_n = \sigma(X_{t_i}, 1 \leq i \leq n)$, with X_{t_0} closing on the right. Since t_0 is not a fixed discontinuity, $X_{t_n} \to X_{t_0-}$ a.e. By a similar argument, for $t_n \downarrow t_0$, we get $X_{t_n} \to X_{t_0+}$ a.e. Hence when $X_{t_0+} \neq X_{t_0} \neq X_{t_0-}$ there can be a jump discontinuity only. [If these limits are not a.e., then there will be an oscillatory discontinuity and t_0 will be a fixed point of discontinuity.] Thus (i) follows and (iii) is similar.

Regarding (ii), we may represent the compact set as $[a,b] \cap T$ with $\{a,b\} \subset T$. The process has a universal separating set $S \subset T$ and we may assume that $\{a,b\} \subset S$. Hence by separability and Theorem II.4.3, for any $\lambda > 0$,

$$P\{\omega : \sup_{t \in [a,b] \cap T} X_t(\omega) \geq \lambda\} = P\{\omega : \sup_{t \in [a,b] \cap S} X_t(\omega) \geq \lambda\} \leq \frac{1}{\lambda} E(|X_b|), \tag{1}$$

where S can be approximated by $S_n \uparrow S$ (S_n is finite and $\{a,b\} \subset S_n$) and then the limits can be interchanged in the middle term. As $\lambda \to \infty$, the right side tends to zero. So a.a. paths are bounded above. A similar reasoning shows that a.a. paths are bounded below, and this gives (ii). □

Since the jump discontinuities can be atmost countable, say T_0, the above result may be stated without the separability hypothesis as one which coincides a.e. with a process (its separable version) on $T - T_0$ where the new one satisfies the hypothesis of the above result. Just as a monotone function can be replaced by one which is right (or left) continuous by a redefinition or a "regularization" at its countable set of jumps, the above theorem indicates that a similar procedure may be used for submartingales also. Let us introduce the necessary concepts and show how this can be formulated in the present context.

Let (Ω, Σ) be a measurable space and $\mathcal{F}_i \subset \Sigma$ be a σ-algebra. If I is a directed set, let $\{\mathcal{F}_i, i \in I\}$ be a filtering (to the right) family of σ-subalgebras, so that $i < i'$ implies $\mathcal{F}_i \subset \mathcal{F}_{i'}$. Define $\mathcal{F}_{i+} = \bigcap_{j>i} \mathcal{F}_i$ and $\mathcal{F}_{i-} = \sigma\left(\bigcup_{j<i} \mathcal{F}_j\right)$ where '<' (or '>') is the ordering of I. The family $\{\mathcal{F}_i, i \in I\}$ is *right (left) order continuous* at i iff $\mathcal{F}_i = \mathcal{F}_{i-}$ ($\mathcal{F}_i = \mathcal{F}_{i-}$) and *order continuous* at i iff both the right and left concepts hold simultaneously. If these conditions hold at each i of I, then the qualification "at i" will be omitted from the above. There are other continuity definitions for these σ-algebras to be right continuoous using the L^p-norms or the Fréchet metric $P(A \Delta B)$. Since the latter concepts are rarely used in our work, we can omit the qualification "order" in the above definition.

We discuss some properties of this continuity, when $I = T \subset \mathbb{R}$, in the context of martingales where if $t_n \downarrow t$ (or $s_n \uparrow t$), and set $X_{t+} =$

$\lim_n X_{t_n}$ (and $X_{t-} = \lim_n X_{s_n}$) a.e., whenever these limits exist. (They exist at each $t \in T$ which is not a fixed discontinuity by the above theorem.)

2. Theorem. *Let $\{X_t, \mathcal{F}_t, t \in T\}$, $T \subset \mathbb{R}$, be a real separable submartingale on (Ω, Σ, P). Then the following assertions are true if there are no fixed points of discontinuities:*

(i) $\{X_{t-}, \mathcal{F}_{t-}, t \in T\}$ is a submartingale whenever $X_t \geq 0$,a.e., (and without the last additional condition, it is a martingale if the original one is).

(ii) $E^{\mathcal{F}_t}(X_{t+}) \geq X_t$, a.e., $E^{\mathcal{F}_{t-}}(X_t) \geq X_{t-}$, a.e., and $\{X_{t+}, \mathcal{F}_{t+}, t \in T\}$ is a submartingale (and a martingale if the original one is).

(iii) If $\{\mathcal{F}_t, t \in T\}$ is right [left] continuous and $\varphi(t) = E(X_t)$, then $X_t = X_{t+}$ ($X_t = X_{T-}$), a.e. for each $t \in T$ iff $\varphi(t+) = \varphi(t)$ ($\varphi(t-) = \varphi(t)$); i.e., X_t admits a right [left] continuous modification iff $\varphi(\cdot)$ does [when $X_t \geq 0$, a.e. also].

Proof. As in the preceding case, this result is an easy consequence of the earlier theory.

First note that by separability, X_{t+}, X_{t-} are random variables so that (i) and (ii) are meaningful. For (i), let $S \subset T$ be a universal separating set. If $t \in T$, there exist $t_n \in S$, $t_n \uparrow t$, and $\{X_{t_n}, \mathcal{F}_{t_n}, n \geq 1\}$ is a submartingale and $X_{t-} = \lim_n X_{t_n}$ a.e., by Theorem 1. So X_{t-} is $\mathcal{F}_{t-}$ adapted. Since $X_t \geq 0$ a.e., and the process is closed on the right by X_t, $\{X_{t_n}, n \geq 1\}$ is uniformly integrable. So $X_{t_n} \to X_{t-}$ in $L^1(P)$ also. Let $s > t$ and $s_n \in S, s_n \uparrow s$. Then $\{X_{s_n}, \mathcal{F}_{s_n}, n \geq 0\}$ is also uniformly integrable, and $X_{s_n} \to X_{s-}$ a.e., and in $L^1(P)$. Thus $s > s_n > t > t_n$, $\{s_n, t_n, n \geq 1\}$ increasing as above, imply that

$$E^{\mathcal{F}_{t_n}}(X_{s_m}) \geq X_{t_n} \text{ a.e.,} \tag{2}$$

and by the uniform integrability of $\{E^{\mathcal{F}_{t_n}}(X_{s_m}), n \geq 1\}$ we have

$$\lim_{n\to\infty} E^{\mathcal{F}_{t_n}}(X_{s_m}) = E^{\mathcal{F}_{t-}}(X_{s_m}) \geq X_{t-} \text{ a.e.} \tag{3}$$

The L^1-convergence of $\{X_{s_m}, m \geq 1\}$ implies (cf., Proposition II.1.3)

$$E^{\mathcal{F}_{t-}}(X_{s-}) \geq X_{t-} \text{ a.e.} \tag{4}$$

In the martingale case, these are equalities. So (i) follows from (4).

For (ii), let $t \in T$. There exist $t_n \in S$, $t_n \downarrow t$ so that $\{X_{t_n}, \mathcal{F}_{t_n}, n \geq 1\}$ is a uniformly integrable decreasing martingale and $E(X_{t_n}) \geq E(X_t))$. Hence $X_{t_n} \to X_{t+}$ a.e., as well as in $L^1(P)$ and $E^{\mathcal{F}_t}(X_{t_n}) \geq X_t$ a.e. This implies $E^{\mathcal{F}_t}(X_{t+}) \geq X_t$ a.e., which is the first half. Regarding the second part, let $t \in T$ and so there exist $t_n \in S$, $t_n \uparrow t$. Then $\{X_{t_n}, \mathcal{F}_{t_n}, n \geq 1\}$ is an L^1-bounded submartingale so that $X_{t_n} \to X_{t-}$ a.e. Also $E^{\mathcal{F}_{t_n}}(X_t) \geq X_{t_n}$ a.e., and by the uniform integrability of $\{E^{\mathcal{F}_{t_n}}(X_t), n \geq 1\}$ we deduce that $E^{\mathcal{F}_{t_n}}(X_t) \to E^{\mathcal{F}_{t-}}(X_t)$ a.e., and in $L^1(P)$. Hence $E^{\mathcal{F}_{t-}}(X_t) \geq X_{t-}$ a.e. The absence of fixed points of discontinuities is used in all these limits.

Regarding the X_{t+}-process, it is clear that X_{t+} is $\mathcal{F}_{t+}$-adapted. Choose monotone decreasing sequences s_n, t_n from S such that $s < s_n < t < t_n$ with limits s, t respectively. Then by earlier arguments $X_{s_n} \to X_{s+}$ and $X_{t_n} \to X_{t+}$ a.e., and in $L^1(P)$ (uniform integrability). Since $s_m < t < t_n$ so that $E^{\mathcal{F}_{s_m}}(X_{t_n}) \geq X_{s+}$ a.e., we deduce after taking limits on n, m that $E^{\mathcal{F}_{s+}}(X_{t_n}) \geq X_{s+}$ a.e. and finally that $E^{\mathcal{F}_{s+}}(X_{t+}) \geq X_{s+}$ a.e. Hence $\{X_{t+}, \mathcal{F}_{t+}, t \in T\}$ is a submartingale, with equalities in the case of martingales.

Lastly by the right continuity hypothesis we have $\mathcal{F}_t = \mathcal{F}_{t+}$. Then by (ii), since X_{t+} is $\mathcal{F}_t$-adapted, $X_{t+} \geq X_t$ a.e. and $E(X_{t+} - X_t) \geq 0$. There is equality here iff $X_{t+} = X_t$ a.e., i.e., iff the process is right continuous. But this means $E(X_{t+}) = \varphi(t)$. If $t_n \searrow t$, with $t_n \in S$, then by a similar argument $E(X_{t+}) = \lim_n E(X_{t_n}) = \lim_n \varphi(t_n) = \varphi(t+)$ and $\varphi(t) = \varphi(t+)$. Thus φ is right continuous. Conversely, if the latter condition holds then $\lim_n E(X_{t_n}) = E(X_{t+})$ and since $X_{t_n} \to \tilde{X}_\infty$ a.e., and in $L^1(P)$ by uniform integrability, $E^{\mathcal{F}_{t+}}(\tilde{X}_\infty) = E^{\mathcal{F}_t}(\tilde{X}_\infty) = \tilde{X}_\infty \geq X_t$ a.e., we must have

$$\lim_n E(X_{t_n}) = E(\tilde{X}_\infty) \geq E(X_t) = \varphi(t) = \lim_n \varphi(t_n) = \lim_n E(X_{t_n}). \tag{5}$$

So there is equality throughout. Thus if $Y_t = X_{t+}$, then $\{Y_t, \mathcal{F}_t, t \in T\}$ is a right continuous submartingale and is a right continuous modification of the X_t-process. The left continuity case is similar. □

The preceding results show the importance of uniform integrability in the continuous parameter case. Its usefulness will be seen even more strikingly when we consider the stopping time transformations in the next chapter.

(b) *Directed index sets.* In contrast to the preceding work where T is a linearly ordered index set, the corresponding statements when T is merely a directed set do not hold in general. But martingales with a directed index arise, e.g., in inference theory, differentiation, and elsewhre. Motivated by such applications, we present a few results following for the most part Chow [1]. The results depend on certain restrictions, called *Vitali conditions* which we now introduce.

3. Definition. (a) Let (Ω, Σ, μ) be a measure space and $\{\mathcal{F}_i, i \in I\}$ be a right filtering family of σ-subalgebras of Σ. Let $A \in \Sigma$ and $K_i \in \mathcal{F}_i, i \in I$. Then $\{K_i, i \in K\}$ is called a *fine covering* of A iff for each $i' \in I, A \subset \bigcup_{i \geq i'} K_i$; it is a μ- *essential fine covering* if the last inclusion holds outside of a μ-null set. (b) The family $\{\mathcal{F}_i, i \in I\}$ is said to satisfy the *Vitali condition* V_0 if for each $A \in \Sigma$ with $\mu(A) < \infty$, each essential fine covering $\{K_i, i \in I\}$ of A and any $\varepsilon > 0$, there exists a finite set $\{i_1, \ldots, i_{n_\varepsilon}\} \subset I$ and a.e. disjoint $L_j \in \mathcal{F}_{i_j}, L_j \subset K_{i_j}$ a.e., such that $\mu\Big(A - \bigcup_{j=1}^{n_\varepsilon} L_j\Big) < \varepsilon$, (or equivalently $\mu\Big(A - \bigcup_{j=1}^{\infty}\Big) = 0$).

The dependence of V_0 on μ is "nominal" in that if I is countable and linearly ordered, then it holds for all μ. We establish this fact for finite measures, although the result is true for localizable $\mu_{\mathcal{F}_i}$.

4. Proposition. *Let μ and μ' be two finite measures on (Ω, Σ) which are equivalent, i.e., they have the same class of null sets. Then a family $\{\mathcal{F}_i, i \in I\}$ satisfies V_0 relative to μ iff it satisfies V_0 relative to μ'. Moreover, if I is linearly ordered and countable, then V_0 is satisfied for any finite μ.*

Proof. Let $A \in \Sigma$, and $\{K_i, i \in I\}$ be an essential fine covering of A relative to μ. Then by V_0, for each $\varepsilon_1 > 0$ there exist n_{ε_1} and $L_j \subset K_{i_j}, j = 1, \ldots, n_{\varepsilon_1}, L_j \in \mathcal{F}_{i_j}$ such that $L_j \cap L_{j'}$ is μ-null for $j \neq j'$, and $\mu\Big(A - \bigcup_{j=1}^{n_{\varepsilon_1}} L_j\Big) < \varepsilon_1$. If μ' is any equivalent finite measure (to μ), we need to show that V_0 holds for it also. Now by the Radon-Nikodým theorem, $f = \frac{d\mu'}{d\mu}$ exists and is finite a.e.$[\mu]$. If $B_n = \{\omega : \frac{1}{n} <$

$f(\omega) < n\}$ and $S = \{\omega : 0 < f(\omega) < \infty\}$, then $B_n \uparrow S$ so that for any $\varepsilon > 0$, there exists $n_0(= n_0(\varepsilon))$ and $C = B_{n_0}$ such that $\mu'(S - C) < \varepsilon/2$. But $\omega \in C \Rightarrow \frac{1}{n_0} < f(\omega) < n_0$, and f is bounded there.

To prove the V_0 property for μ', we note that the essential cover $\{K_i, i \in I\}$ of A is also one for each subset of A and , in particular, for $A \cap S$ and $A \cap C$. Let us choose the $\{L_J, j \geq 1\}$ sequence and n_{ε_1} to go with $\varepsilon_1 = \frac{\varepsilon}{2n_0}$. Thus the condition of V_0 for μ implies $\mu\Big(C \cap A - \bigcup_{j=1}^{n_\varepsilon} L_j\Big) < \frac{\varepsilon}{2n_0}$. Hence

$$\begin{aligned}
\mu'\left(A - \bigcup_{j=1}^{n_\varepsilon} L_j\right) &= \int_{A - \cup_{j=1}^{n_\varepsilon} L_j} f d\mu \\
&= \int_{(A-S)\cup((S-C)\cap A)} f d\mu + \int_{(C - \cup_{j=1}^{n_\varepsilon} L_j)\cap A} f d\mu \\
&< \mu'(S - C) + n_0 \mu\Big(A \cap C - \cup_{j=1}^{n_\varepsilon} L_j\Big), \\
&\qquad \text{since } \mu'(S^c) = 0, \\
&< \frac{\varepsilon}{2} + n_0 . \frac{\varepsilon}{2n_0} = \varepsilon.
\end{aligned}$$

Thus $\{\mathcal{F}_i, i \in I\}$ also satisfies V_0 relative to μ'. The converse follows by symmetry.

Suppose now that $I = N$, the countable linearly ordered set taken as the natural numbers for convenience. Since $A \subset \cup_{i \geq i'} K_i$ a.e., by the V_0 hypothesis, for $\varepsilon > 0, A \in \Sigma$, one may find a finite sequence $\{K_{i_j}, j = 1, \ldots, n_\varepsilon\}$ from the above countable union such that $\mu\Big(A - \cup_{j=1}^{n_\varepsilon} K_{i_j}\Big) < \varepsilon$. We can also assume that $K_{i_j} \subset K_{i_{j'}}$ for $i_j \leq i_{j'}$ by the linear ordering. Let $L_1 = K_{i_1}$, and $L_j = K_{i_j} - K_{i_{j-1}}$. Then $L_j \in \mathcal{F}_{i_j}$, the latter algebras increasing, L_j being disjoint, $\cup_j K_{i_j} = \cup_j L_j$, and $\mu\Big(A - \bigcup_{j=1}^{n_\varepsilon} L_j\Big) < \varepsilon$. $\square$

The following simple example shows that the last part can be false if I is uncountable.

Let (Ω, Σ, P) be the Lebesgue unit interval, $\mathcal{F}_i = \mathcal{F} = \Sigma, i \in \mathbb{R}^+ = I$. Let $K_i = \{i - [i]\}$, the one point set where $[i]$ is the integral part of i. So $P(K_i) = 0, i \in I$. If $A = \Omega$, then for any i', $A \subset \cup_{i \geq i'} K_i$ a.e., and $\{K_i, i \in I\}$ is an essential fine covering of A. But there is no countable subcovering since $P(A) = 1$ and $P\Big(\bigcup_{j=1}^{\infty} K_{i_j}\Big) = 0$ for countable subsequences.

Let us now prove a mean and a pointwise convergence theorem for martingales with a directed index set. It is essentially a specialization of a result due to Eberlein [1].

5. Theorem. *Let $\{X_i, \mathcal{F}_i, i \in I\}$ be a martingale, with a directed index, on a probability space (Ω, Σ, P). If the martingale is terminally uniformly integrable (i.e., $\|X_i\|_1 \leq K_0 < \infty$ and for any $\varepsilon > 0$, there are $i_\varepsilon \in I, \delta_\varepsilon > 0$ such that for all $i \geq i_\varepsilon, A \in \Sigma, P(A) < \delta_\varepsilon$ one has $\int_A |X_i| dP < \varepsilon$), then there is a unique $X_\infty \in L^1(\Omega, \mathcal{F}_\infty, P)$, where $\mathcal{F}_\infty = \sigma(\bigcup_i \mathcal{F}_i)$ such that $X_i \to X_\infty$ in $L^1(P)$ and $X_i = E^{\mathcal{F}_i}(X_\infty)$, a.e. Conversely, if there is an $X \in L^1(\Omega, \Sigma, P)$ such that $X_i = E^{\mathcal{F}_i}(X)$, a.e. $i \in I$ then the stated uniformity condition holds and $X_\infty = E^{\mathcal{F}_\infty}(X)$, a.e.*

Proof. If $\mu_i(A) = \int_A X_i dP, A \in \mathcal{F}_i$, then $\{\Omega, \mathcal{F}_i, \mu_i), i \in I\}$ is a set martingale. Let $\mu : \bigcup_i \mathcal{F}_i \to \mathbb{R}$ be defined as $\mu(A) = \mu_i(A)$, for $A \in \mathcal{F}_i$. Clearly μ is well defined and additive. The terminal uniform integrability implies that μ is P-continuous and hence it is σ-additive on $\bigcup_i \mathcal{F}_i$. Since $|\mu|(\Omega) < \infty$ by the uniform boundedness of μ_i, μ has a unique σ-additive extension to $\mathcal{F}_\infty$. So by the Radon-Nikodým theorem, there is a P-unique $X_0 = \frac{d\mu}{dP}$ belonging to $L^1(\Omega, \Sigma, P)$, such that for $B \in \mathcal{F}_i$, we have $\int_B X_i dP = \mu_i(B) = \mu(B) = \int_B X_0 dP$. Hence $X_i = E_i^{\mathcal{F}}(X_0)$,a.e. We claim that $\|X_i - X_0\|_1 \to 0$, so that $X_\infty = X_0$, a.e., must hold.

Since $\{X_i, \mathcal{F}_i, i \in I\} \subset L^1(\Omega, \mathcal{F}_\infty, P)$, we use the well-known remark that a net in a complete metric space converges if every countable subnet is Cauchy. (If the net did not converge, we can exhibit a subnet using the directedness of I, which fails to be Cauchy. For an application of this idea, see Dunford-Schwartz[1], p. 125.) Hence if $\{X_{i_n}, \mathcal{F}_{i_n}, n \geq 1\}$ is a subnet where $\{i_n, n \geq 1\}$ is an increasing sequence, then $X_{i_n} = E^{\mathcal{F}_{i_n}}(X_0)$ a.e., and so it converges (a.e. and) in $L^1(P)$ by Theorem II.6.2. Since this holds for every subnet, $X_i \to X_\infty$ in $L^1(P)$ by the remark and $X_\infty \in L^1(\Omega, \mathcal{F}_\infty, P)$, because the martingale is in this space. Hence $\|X_i\|_1 \to \|X_\infty\|_1$ so that $E(X_i \chi_B) \to E(X_\infty \chi_B)$ for each $B \in \mathcal{F}_\infty$. This and the last paragraph yield:

$$\lim_i \int_A X_i dP = \int_A X_\infty dP = \int_A X_0 dP = \nu(A), \qquad A \in \bigcup_i \mathcal{F}_i.$$

But ν is σ-additive and agrees with μ on $\bigcup_i \mathcal{F}_i$. So $\nu = \mu$ on $\mathcal{F}_\infty$. It follows that $X_\infty = X_0$ a.e., since both are $\mathcal{F}_\infty$-adapted.

Conversely, if $X_i = E^{\mathcal{F}_i}(X), X \in L^1(P)$, then the martingale is uniformly integrable since $\{E^{\mathcal{B}}(X), \mathcal{B} \subset \Sigma\}$ is uniformly integrable. The rest is immediate from the preceding work. □

Here is an interesting consequence of the above proof:

6. Corollary. *If $\{X_i, \mathcal{F}_i, i \in I\}$ is a submartingale, I directed, and $X_i = E^{\mathcal{F}_i}(X), i \in I$, for some $X \in L^1(P)$, then $X_i \to X_\infty$ in mean, and $X_\infty \in L^1(\Omega, \mathcal{F}_\infty, P), \mathcal{F}_\infty = \sigma\big(\bigcup_i \mathcal{F}_i\big) \subset \Sigma$, as above.*

Proof. Since $|X_i| \leq E^{\mathcal{F}_i}(|X|)$ a.e., by Jensen's inequality, we see that $\{X_i, \mathcal{F}_i, i \in I\}$ is uniformly integrable. Then by Theorem II.6.4, and the mean convergence above implies $X_i \to X_\infty$ in the same sense, one has $X_\infty \leq E^{\mathcal{F}_\infty}(X)$ a.e. Note that mean convergence and convergence in probability are the same under the present hypothesis (cf., Dunford-Schwartz [1], p. 125). □

The following result will be generalized in the next chapter, but this special case is sufficient for many applications.

7. Theorem. *Let (Ω, Σ, P) be a complete probability space and $\{\mathcal{F}_i, i \in I\}$ be a right filtering completed family of σ-subalgebras of Σ satisfying the condition V_0. If $\{f_i, \mathcal{F}_i, i \in I\}$ is a martingale lying in a ball of $L^1(P)$, then $f_i \to f_\infty$ a.e. as $i \uparrow$. [The same result holds if P is nonfinite but each $\mathcal{F}_i$ is P-rich.]*

Proof. The proof will be given in steps for convenience.

I. We noted that Jordan decomposition of a martingale is valid for directed sets (cf., Theorem II.5.2). Consequently f_i is expressible as $f_i = f_i^{(1)} - f_i^{(2)}$, where $\{f_i^{(j)}, \mathcal{F}_i, i \in I\}, j = 1, 2$ are nonnegative martingales since by hypothesis the given process is $L^1(P)$-bounded. Hence it suffices to prove the result under the assumption that $f_i \geq 0$, a.e., which we now do. [Below $\mathcal{F}'$ denotes the P-completion of $\mathcal{F}$.]

II. Suppose that the conclusion is false, to derive a contradiction. Let $f^* = \limsup_i f_i$, and $f_* = \liminf_i f_i$. Since P- is complete, it follows by Theorem 2.3 that f^* and f_* are P-measurable, and then $f^* > f_* \geq 0$ with the strict inequality holding on a set of positive measure. Hence there exist $0 < a < b$ such that $A = \{\omega : f^*(\omega) > b > a > f_*(\omega)\} \in \mathcal{F}'_\infty = \sigma\Big(\bigcup_{i \in I} \mathcal{F}_i\Big)'$ and $P(A) > 0$. If now $K_i = \{\omega : f_i(\omega) \leq a\}$, then $A \subset \bigcup_{i \geq i'} K_i$ a.e., for each $i' \in I$. We assert that for each $0 < \varepsilon < 1/2$ there exist $i_0 = i_\varepsilon \in I$ and a $B \in \mathcal{F}_{i_0}$ such that $B \subset K_{i_0}$, a.e., and

$$P(B \cap A) > (1 - \varepsilon)P(B). \tag{6}$$

If this is granted, then the proof is completed as follows.

III. By definition if $\omega \in K_{i_0}, 0 \leq f_{i_0}(\omega) \leq a$ a.e., and hence this is also true for $\omega \in B$. Let us define a fine covering of B. For $i \geq i_0$ let $L_i = B \cap [f_i \geq b]$ and $L_i = \Omega$ if $i \not\geq i_0$. Then $B \subset \bigcup_{i \geq i'} L_i$ for each $i' \in I$ and hence $\{L_i, i \in I\}$ is also such a covering for $B \cap A$. By the martingale property both B, L_i are in $\mathcal{F}_i$. Now by the V_0 condition and Step II, for the given ε there, we have a finite number of indices $i_1, \ldots, i_{n_\varepsilon}$ and a.e. disjoint $B_j \subset L_j, B_j \in \mathcal{F}_{i_j}, j = 1, \ldots, n_\varepsilon$ such that

$$P\Big[B \cap A - \bigcup_{j=1}^{n_\varepsilon} B_j\Big] < \varepsilon P(B \cap A). \tag{7}$$

By (6) and (7)we have

$$\begin{aligned} \varepsilon P(B) > \varepsilon P(B \cap A) &> P(B \cap A) - P\Big(\bigcup_{j=1}^{n_\varepsilon} B_j \cap B \cap A\Big) \\ &> (1 - \varepsilon)P(B) - P\Big(\bigcup_{j=1}^{n_\varepsilon} B_j \cap B \cap A\Big), \end{aligned}$$

so that

$$\sum_{j=1}^{n_\varepsilon} P(B \cap B_j) \geq P\Big(\bigcup_{j=1}^{n_\varepsilon} A \cap B_j \cap B\Big) > (1 - 2\varepsilon)P(B). \tag{8}$$

But $f_{i_j} \geq b$ on $B \cap B_j$. So (8) implies

$$\sum_{j=1}^{n_\varepsilon} \int_{B \cap B_j} f_{i_j} \, dP \geq P\Big(\bigcup_{j=1}^{n_\varepsilon} B_j \cap B\Big) \geq b(1 - 2\varepsilon)P(B). \tag{9}$$

By the directedness of I, there is a $k_0 \in I, k_0 \geq i_j, j = 1, \ldots, n_\varepsilon$ such that (martingale property) we have

$$\int_{B \cap B_j} f_{i_j} dP = \int_{B \cap B_j} f_{k_0} dP, \qquad B \cap B_j \in \mathcal{F}_{i_j}, \tag{10}$$

and if $C = \bigcup_{j=1}^{n_\varepsilon} B \cap B_j \subset B$, we futher have ($f_{k_0} \geq 0$ a.e.) with the a.e. disjointness of B_j,

$$\begin{aligned} \int_B f_{k_0} dP \geq \int_C f_{k_0} dP &= \sum_{j=1}^{n_\varepsilon} \int_{B \cap B_j} f_{k_0} dP, \\ &= \sum_{j=1}^{n\varepsilon} \int_{B \cap B_j} f_{i_j} dP \geq b(1 - 2\varepsilon)P(B), \end{aligned} \tag{11}$$

by (9) and (10). Since $i_0 \leq i_j \leq k_0$, and $B \subset K_{i_0}$, we deduce that

$$\int_B f_{k_0} dP = \int_B f_{i_0} dP \leq aP(B). \tag{12}$$

It follows from (11) and (12) that $a \geq (1 - 2\varepsilon)b$ because $P(B) > 0$. Since $a < b$ and $0 < \varepsilon < 1/2$, is arbitrary, this yields the desired contradiction, and hence we must have $f_i \to f_\infty$ a.e. So it remains to prove (6).

IV. To prove (6), suppose the contrary. Then we must have for some $0 < \varepsilon < 1/2$, an $i' \in I$ and $i \geq i', B \subset K_i, B \in \mathcal{F}_i$, the opposite of (6):

$$P(B \cap A) \leq (1 - \varepsilon)P(B). \tag{13}$$

Let $P_A(\cdot) = P(A \cap \cdot)$, the restriction of P to the trace $\Sigma(A)$. If $P_i = P|\mathcal{F}_i$, and $h_{A,i} = E^{\mathcal{F}_i}(\chi_A)$, then $P_A(B) = \int_B h_{A,i} dP_i \leq (1 - \varepsilon)P(B)$ by (13). Since $B \subset K_i$ is arbitrary ($B \in \mathcal{F}_i$), this implies $h_{A,i} \leq (1 - \varepsilon)$ a.e., on K_i and $\{h_{A,i}, \mathcal{F}_i, i \geq i'\}$ is a martingale. Since $\{K_i, i \in I\}$ is an essential fine covering of A, we can also deduce that $h_{A,i} \leq (1 - \varepsilon)$ a.e. on A. But by definition of $h_{A,i}$ it defines a terminally uniformly integrble martingale; and it follows by Theorem 5 that $h_{A,i} \to h_{A,\infty}$ in $L^1(P)$ and that $h_{A,\infty} = E^{\mathcal{F}_\infty}(\chi_A)$ a.e. Now $A \in \mathcal{F}_\infty$ so that $h_{A,\infty} = \chi_A$ a.e. Since $h_{A,\infty} \leq (1 - \varepsilon)$ a.e. on $A, P(A) > 0$, this gives the desired contradiction. Thus (6) must be true, and the P-rich case is now easy. □

With this we may now state the full continuous parameter analog of Theorem II.6.2.

8. Theorem. *Let (Ω, Σ, P) be a complete probability space and $\{\mathcal{F}_i, i \in I\}$ be a right filtering family of σ-subalgebras satisfying the Vitali condition V_0. For a martingale $\{f_i, \mathcal{F}_i, i \in I\}$ in $L^1(P)$ the following are equivalent statements:*

(i) the martingale is terminally uniformly integrable and $L^1(P)$-bounded,

(ii) $f_i \to f_\infty$ a.e., measurable, and $E(|f_i - f_\infty|) \to 0$,

(iii) $f_i \to f_\infty$ a.e., measurable, and $\lim_i E(|f_i|) \to E(|f_\infty|)$,

(iv) $\{f_i, \mathcal{F}_i, i \in I_1\}$ is a martingale where $I_1 = I \cup \{\infty\}$ and for each $i \in I$ we define $i \leq \infty$ so that I_1 is directed and $\mathcal{F}_\infty = \sigma\left(\bigcup_{i\in I} \mathcal{F}_i\right)'$.

The proof is now an extension of the discrete parameter case and is left to the reader. Other parts of the earlier result can be similarly formulated. Let us also remark that using the $\mathcal{F}$'s to be P-rich and the sets A, B to be of finite (positive) measure in the above proof one can easily give a version of the above theorem to the case that the P-measure is localizable. On the other hand, new techniques are needed for the corresponding result with submartingales since an analog of the Doob decomposition for the continuous parameter case has not yet been established. This is a nontrivial problem and will be considered in the following chapter when the index is the real line. (The directed index case is not completely resolved.)

3.6 A limit theorem for regular projective systems

To motivate the main result here, we restate the conditioning concept of Section II.1 in a suggestive form. Suppose that $(\Omega_i, \Sigma_i), i = 1, 2$ are measurable spaces and $\mu_1 : \Sigma_1 \to \mathbb{R}^+$ is a probability measure. If $X : \Omega_1 \to \Omega_2$ is a measurable mapping (i.e., an abstract random variable), then $\mu_2 = \mu_1 \circ X^{-1}$ is the image measure on Σ_2 and $\mu_1(A \cap X^{-1}(\cdot))$ is absolutely continuous relative to μ_2 on Σ_2 for each $A \in \Sigma_1$. Hence by the Radon-Nikodým theorem we have :

$$\mu_1(A \cap X^{-1}(B)) = \int_B Q(A, \omega_2) d\mu_2(\omega_2), \qquad A \in \Sigma_1, \qquad B \in \Sigma_2. \quad (1)$$

Since $\mu_1(\cdot \cap X^{-1}(B))$ is σ-additive for each $B \in \Sigma_2$, we see that $Q(\cdot, \omega_2)$ is σ-additive for almost all ω_2 (the exceptional set depending on the sequence used), and $Q(A, \cdot)$ is μ_2-measurable. In general, however, $Q(\cdot, \omega_2)$ is not a measure for all ω_2 although Q can be regarded as a vector measure in $M^\infty(\Sigma_2, \mu_2)$. In terms of Chapter II, $Q(\cdot, \cdot)$ is a conditional probability function and it is *regular* if there is a fixed μ_2-null set N such that $Q(\cdot, \omega)$ is a probability for *each* $\omega \in (\Omega_2 - N)$. If the regularity can be assumed then we can integrate measurable functions on (Ω_1, Σ_1) relative to $Q(\cdot, \omega)$ using the Lebesgue integration theory. Note that even then $Q(\cdot, \cdot)$ can only be considered as a mapping of Ω_2 into the set of all probability measures $\mathcal{M}^1(\Sigma_1)$, which need not be a separable subset of the positive part of the Banach space of scalar σ-additive set functions on (Ω_1, Σ_1) under total variation norm. The question now is whether we can select a regular version from $\{Q(\cdot, \omega_2), \omega_2 \in \Omega_2\}$ to be used in (1), and then combine them to get new (scalar) measures to use in our work. Here we present conditions for a family of (not necessarily topological) measurable spaces $(\Omega_i, \Sigma_i), i = 1, 2$ whenever such a selection is possible, and establish a product (limit) theorem for a wide class of regular systems of conditional measures.

To clarify further and to set down the desired terminology, suppose $(\Omega_i, \Sigma_i), i = 1, 2$ are Borelian measurable spaces, the Ω_i being locally compact. If $X : \Omega_1 \to \Omega_2$ is a continuous mapping and μ_1 is a Radon probability on Σ_1, then $\mu_2 = \mu_1 \circ X^{-1}$ on Σ_2 is μ_1-continuous, and in (1) $Q(\cdot, \cdot)$ can be chosen to satisfy $\rho'(Q(\cdot, \omega_2)) = Q(\cdot, \omega_2), \omega_2 \in \Omega_2$, where $\rho : M^\infty(\Sigma_1) \to M^\infty(\Sigma_2)$ is a lifting which exists by Theorem 2.1, and ρ' is induced by ρ as in Corollary 2.6 since $Q(\cdot, \cdot) \in M^\infty(\Sigma_1, S)$ with $S = \mathcal{M}^1(\Sigma_1)$. This fact is not obvious, but has been established by Tulcea and Tulcea ([1], p. 150).

It is also possible to present other conditions on the measure spaces without topology in order that Q is a regular conditional measure in (1). The following is one such result (cf., Sazonov [1]). In (1) let $\Omega_2 = \mathbb{R}$ and $X : \Omega_1 \to \Omega_2$ be a random variable such that $X(\Omega_1)$ is a universally measurable set, i.e., it is an element of $\tilde{\mathcal{B}}$, the Borel σ-algebra of $\mathbb{R}$ completted for all Lebesgue-Stieltjes probability measures. Then Q of

(1) has a version which is a regular conditional probability and which is even "perfect". There is a detailed treatment of regular conditioning in Rao ([13], Chapter 5). Here we present a general (not-necessarily topological) projective limit theorem for regular systems.

Let $\mathcal{F} = \{(\Omega_i, \Sigma_i, \mu_i, g_{ij}) : i < j \text{ in } D\}$ be a projective system of probability spaces, with D as a directed set (cf., Definition I.2.4). We say that:

(a) the system $\mathcal{F}$ is *almost separable* if for each $i \in D$, there exist a μ_i-null set N_i and a sequence $\{E_{ik}, k = 1, 2, \ldots\} \subset \Sigma_i$ such that for each pair of distinct points ω', ω'' of Ω_i with $\omega' \in (\Omega_i - N_i)$, we have $\omega' \in E_{ik}$ but $\omega'' \notin E_{ik}$ for some k. Thus the $\{E_{ik}\}_{k=1}^{\infty}$ sequence essentially separates points of Ω_i for each i.

(b) the system $\mathcal{F}$ is *regular* if for $i < j$ in D there is a regular conditional measure $\mu_{ij}(\cdot, \cdot) : \Sigma_j \times \Omega_i \to \mathbb{R}^+$, and (1) holds in the following form:

$$\mu_j(A \cap g_{ij}^{-1}(B)) = \int_B \mu_{ij}(A, \omega_i) d\mu_i, \qquad A \in \Sigma_j,\, B \in \Sigma_i. \tag{2}$$

With this background we have the following result, taken from Rao and Sazonov [1]:

1. Theorem. *If the projective system $\mathcal{F}$ is almost separable, regular, and satisfies the s.m. condition (cf., Definition I.3.2), then there exists a unique probability μ on Σ such that $\mu_i(A) = \mu(\pi_i^{-1}(A))$ for all $A \in \Sigma_i$ and $i \in D$ where $\pi : \Omega(= \varprojlim(\Omega_i, g_{ij})) \to \Omega_i$ is the canonical projection and Σ is the cylinder σ-algebra of Ω.*

This result which is, in a sense, more general than the corresponding work of Chapter I, will be demonstrated after some preliminaries. Let us first restate the definition of Σ, π and the additive ν given in Proposition I.3.1. If $\tilde{\Omega} = \times_{i \in I} \Omega_i$, the cartesian product, then $\Omega \subset \tilde{\Omega}$ and the inclusion can be proper. If $\tilde{\pi}_i : \tilde{\Omega} \to \Omega_i$ is the coordinate projection let $\pi_i = \tilde{\pi}_i | \Omega$, the restriction, and consider the cylinder σ-algebra $\tilde{\Sigma} = \sigma\Big(\bigcup_{i \in D} \tilde{\pi}_i^{-1}(\Sigma_i)\Big)$. Then Σ (of the theorem) is the trace of $\tilde{\Sigma}$ on Ω so that $\Sigma = \{\Omega \cap A : A \in \tilde{\Sigma}\}$. We next define $\nu(\cdot)$, on the algebra of the cylinders, i.e., on the algebra of sets $\tilde{A} = \cap\{\tilde{\pi}_i^{-1}(A_i), A_i \in \Sigma_i \text{for finitely many} i, \text{ and } A_i = \Omega_i \text{ otherwise}\}$.

Let $r(\tilde{A}) = \{i \in D : A_i \neq \Omega_i\}$. Then $r(\tilde{A})$ is a finite subset of D, and $r(\cdot)$ is defined for all measurable cylinders. We can now introduce ν as:

$$\nu(\tilde{A}) = \mu_i(\cap\{g_{ij}^{-1}(A_i) : i \in r(\tilde{A})\}), \qquad j > r(\tilde{A}). \tag{3}$$

To see that ν is well-defined, we can follow the argument of Proposition I.3.1. In fact, from the compatibility relations of g_{ij}'s, ($g_{ik} = g_{ij} \circ g_{ik}$ for $i < j < k$, and $\mu_i \circ g_{ij}^{-1} = \mu_j$) we have

$$\begin{aligned}\mu_j\Big(\bigcap_{i\in r(\tilde{A})} g_{ij}^{-1}(A_i)\Big) &= \mu_k\Big(g_{jk}^{-1}\Big(\bigcap_{i\in r(\tilde{A})} g_{ij}^{-1}(A_i)\Big)\Big)\\ &= \mu_k\Big(\bigcap_{i\in r(\tilde{A})} g_{jk}^{-1}\circ g_{ij}^{-1}(A_i)\Big)\\ &= \mu_k\Big(\bigcap_{i\in r(\tilde{A})} g_{ik}^{-1}(A_i)\Big)\\ &= \nu(\tilde{A}).\end{aligned} \tag{4}$$

Thus ν does not depend on j in (3). Applying the earlier argument (of Prop. I.3.1), it is easily seen that ν is uniquelly extendable to a finitely additive function on the algebra Σ_0 of cylinder sets, and it can (and will) be denoted by the same symbol. Since no topology is given for the Ω_i's one needs to use the regularity of the system $\mathcal{F}$ now. Here we modify and extend the arguments of Choksi [1] and Mallory-Sion [1] which are based on Tulcea's [1] work.

Thus to continue, let ν^* be the (Carathéodory) generated outer measure by the pair (Σ_0, ν). Then the classical measure theory results imply that $(\nu^*)^* = \nu^*$ and the ν^*-measurable sets form a complete σ-algebra, Σ', on which ν^* is σ-additive, $\Sigma_0 \subset \Sigma'$, and $\nu^*|\Sigma = \nu$, (cf., e.g., Rao [11], pp.41-49). In this case ν^* also satisfies:

$$\nu^*(A) = \inf\{\nu^*(B) : A \subset B, B \in \Sigma\}, \; A \subset \Omega. \tag{5}$$

We now establish some properties of ν^* in the following proposition and use them in the proof of the theorem.

2. Proposition. *Let the projective system $\mathcal{F}$ be almost separable and $D_1 \subset D$ be a directed set.*

(i) If $\Omega_1 = \{\tilde{\omega} \in \tilde{\Omega} : g_{ij}(\omega_j) = \omega_i, i < j$ *in* $D_1\}$*, then*

$$\nu^*(\tilde{\Omega} - \Omega_1) = 0, \tag{6}$$

where ν^* *is the outer measure as in (5).*

(ii) If $\mathcal{S}_1$ *is the algebra of cylinder sets of* $\tilde{\Omega}$*,* $\mathcal{F}$ *also satisfies the s.m. condition (of Bochner's), and* ν *on* $\mathcal{S}_1$ *is* σ*-additive, then we have*

$$\nu(\tilde{A}) = \nu_1(\tilde{A}) = \nu_1^*(\Omega \cap \tilde{A}), \quad \tilde{A} \in \mathcal{S}_1, \tag{7}$$

where $\nu_1 = \nu|\mathcal{S}_1$ *and* ν_1^* *is the outer measure generated by* $(\mathcal{S}_1, \nu_1)$*.*

(iii) If ν_1 *is* σ*-additive as in (ii), then* Ω *is a thick set relative to the (unique)* σ*-additive extension of* ν_1 *on to* $\sigma(\mathcal{S}_1) = \Sigma_1$*, and* $\nu_1(B) = \nu_1^*(\Omega \cap B)$*,* $B \in \Sigma_1$*.*

Proof. (i) Let $B \in \Sigma_i, i < j$ in D, and consider

$$\tilde{B}_{ij} = \tilde{\pi}_i^{-1}(B) \cap \pi_j^{-1}(g_{ij}^{-1}(B^c)).$$

Then the definition of ν (cf.,(3)) implies

$$\nu(\tilde{B}_{ij}) = \mu_j(g_{ij}^{-1}(B) \cap g_{ij}^{-1}(B^c)) = 0, \quad \text{as in (4)}. \tag{8}$$

Define the class $\tilde{\mathcal{E}} = \{\tilde{B}_{ij} : B \in \Sigma_i, i < j \text{ in } D_1\}$. By the almost separability condition we have

$$\Omega - \Omega_1 \subset \left(\bigcap\{\tilde{B} : \tilde{B} \in \tilde{\mathcal{E}}\}\right) \cup \left(\bigcup\{\pi_i^{-1}(N_i) : i \in D_1\}\right). \tag{9}$$

Since $\tilde{\mathcal{E}}$ and D_1 are countable, and the generated outer measure ν^* is σ-subadditive, we get from (8) and (9) that $\nu^*(\Omega - \Omega_1) = 0$, since $\nu(\tilde{\pi}_i^{-1}(N_i)) = 0$ for all $i \in D_1$. Thus (6) holds.

(ii) To see the validity of (7), given $\varepsilon > 0$ choose $\{B_n, n \geq 1\}$ from $\mathcal{S}_1$ such that $\Omega \cap \tilde{A}$ is covered by this sequence and that

$$\sum_{n=1}^{\infty} \nu(\tilde{B}_n) < \nu_1^*(\Omega \cap \tilde{A}) + \varepsilon. \tag{10}$$

This is possible since ν^* (hence ν_1^*) is a generated outer measure. Let $\tilde{A} = \bigcup_{k=1}^{m} \tilde{A}_k$, $\tilde{A}_k \in \mathcal{S}_1$ and use this $\tilde{A}$ in (10). Let $D_2 \subset D_1$ be a directed set such that

$$D_2 \supset \bigcup_{k=1}^{m} r(\tilde{A}_k) \cup \bigcup_{n=1}^{\infty} r(B_n).$$

Then D_2 is countable and by the s.m. condition we have

$$\tilde{A} \cap \Omega_1 \subset \bigcup_{n=1}^{\infty} \tilde{B}_n,$$

where Ω_1 is as in (i) with D_2 in place of D_1 there. Using (6) and the σ-additivity of ν_1 on $\mathcal{S}_1$ which implies $\nu = \nu_1 = \nu^*|\mathcal{S}_1$ (cf., Rao [11], p. 41) we have

$$\begin{aligned}\nu(\tilde{A}) &\leq \nu^*\Big(\bigcup_{n=1}^{\infty} \tilde{B}_n\Big) + \nu^*(\Omega - \Omega_1) \\ &\leq \sum_{n=1}^{\infty} \nu^*(\tilde{B}_n) + 0 < \nu_1^*(\Omega \cap \tilde{A}) + \varepsilon, \quad \text{by (10).}\end{aligned}$$

Hence $\nu(\tilde{A}) \leq \nu_1^*(\Omega \cap \tilde{A}) \leq \nu_1^*(\tilde{A}) = \nu(\tilde{A})$, yielding (7).

(iii) To prove that Ω is ν_1-thick, suppose the contrary. Then there exists an $\tilde{E} \in \Sigma_1, \tilde{E} \subset \Omega - \Omega_1, \nu_1(\tilde{E}) > 0$, where Ω_1 is as in (i). If $0 < \varepsilon < \nu_1(\tilde{E})/2$, then there is an $\tilde{A} \in \mathcal{S}(\Sigma_1 = \sigma(\mathcal{S}_1) = \sigma(\mathcal{S}))$ with $\mathcal{S}$ as the ring generated by $\mathcal{S}_1$, such that $\nu_1(\tilde{E}\Delta\tilde{A}) < \varepsilon$, (cf., e.g., Rao [11], Prop.2.2.16(iii)). Also we may express $\tilde{A} = \bigcup_{k=1}^{m} \tilde{A}_k, \tilde{A}_k \in \mathcal{S}$, disjoint, for some $m \geq 1$. Then

$$\begin{aligned}\sum_{k=1}^{m} \nu_1^*(\Omega \cap \tilde{A}_k) &\leq \sum_{k=1}^{m} \nu_1(\tilde{A}_k \cap \tilde{E}^c), \text{ since } \quad \tilde{E}^c \supset \Omega, \\ &= \nu_1(\tilde{A} \cap \tilde{E}^c) < \varepsilon. \end{aligned} \tag{11}$$

However, by (ii) we have

$$\begin{aligned}\sum_{k=1}^{m} \nu_1^*(\Omega \cap \tilde{A}_k) &= \sum_{k=1}^{m} \nu_1(\tilde{A}_k) \\ &= \nu(\tilde{A}) > \nu_1(\tilde{E}) - \varepsilon > \varepsilon. \end{aligned} \tag{12}$$

The contradiction between (11) and (12) shows that our initial supposition does not hold, and that Ω is a thick set.

Since ν_1^* is a generated outer measure and is Carathéodory regular, so that $\nu_1^*(\tilde{G}) = \inf\{\nu_1(\tilde{F}) : \tilde{G} \subset \tilde{F} \in \Sigma_1\}$, we get $\nu_1(\tilde{F}) = \nu_1^*(\Omega \cap \tilde{F}), \tilde{F} \in \Sigma_1$ as asserted. □

With the above auxiliary work we can establish the main result.

Proof of Theorem 1. First note that the regularity of $\mathcal{F}$ implies the following formula for any $i_1 < i_2 < \cdots < i_m, i_k \in D, 1 \leq k \leq m$:

$$\nu(\tilde{A}) = \int_{A_1} \mu_{i_1}(d\omega_{i_1}) \int_{A_2} \mu_{i_1 i_2}(d\omega_{i_2}, \omega_{i_1}) \cdots \int_{A_m} \mu_{i_{m-1} i_m}(d\omega_{i_m}, \omega_{i_{m-1}}), \tag{13}$$

where $\tilde{A} = \{\tilde{\omega} : \tilde{\pi}_{i_k}(\tilde{\omega}) \in A_k \in \Sigma_{i_k}, i_k \in D, 1 \leq k \leq m\}$ is a cylinder. This was verified in Choksi ([1], p. 333), but we include a short argument since our conditions are slightly different. By the regularity of $\mathcal{F}$, it is clear that (13) holds for $m = 1, 2$ and we use induction for $m \geq 3$. Thus suppose (13) holds for $(m-1)$, and then we have

$$\mu_{i_{m-1}}\left(\left[\bigcap_{k=1}^{m-2} g_{i_k i_{m-1}}(A_k)\right] \cap A_{m-1}\right)$$
$$= \int_{A_1} \mu_{i_1}(d\omega_{i_1}) \cdots \int_{A_{m-1}} \mu_{i_{m-2} i_{m-1}}(d\omega_{i_{m-1}}, \omega_{i_{m-2}}). \tag{14}$$

Since $\mu_{i_{m-1}}$ is a measure on $\Sigma_{i_{m-1}}$, and by hypothesis all the integrals in (14) are in the sense of Lebesgue, we may extend the identity first for simple and then for all nonnegative $\Sigma_{i_{m-1}}$-measurable functions f on $\Omega_{i_{m-1}}$ to obtain:

$$\int_{\cap_{k=1}^{m-1} g_{i_k i_{m-1}}(A_k)} f(\omega_{i_{m-1}}) \mu_{i_{m-1}}(d\omega_{i_{m-1}})$$
$$= \int_{A_1} \mu_{i_1}(d\omega_{i_1}) \cdots \int_{A_{m-2}} \mu_{i_{m-3} i_{m-2}}(d\omega_{i_{m-2}}, \omega_{i_{m-1}}) \times$$
$$\int_{A_{m-1}} f(\omega_{i_{m-1}}) \mu_{i_{m-2} i_{m-1}}(d\omega_{i_{m-1}}, \omega_{i_{m-2}}). \tag{15}$$

Letting $f(\omega_{i_{m-1}}) = \mu_{i_{m-1} i_m}(A_m, \omega_{i_{m-1}})$ in (15) and employing the regularity of $\mathcal{F}$ again we can simplify it to get:

$$\begin{aligned} LHS\, of\, (15) &= \mu_{i_m}\left(g_{i_{m-1} i_m}^{-1}(\cap_{k=1}^{m-1} g_{i_k i_{m-1}}^{-1}(A_k)) \cap A_m\right) \\ &= \mu_{i_m}\left([\cap_{k=1}^{m-1} g_{i_k i_m}^{-1}(A_k)] \cap A_m\right) \\ &= \mu_{i_m}\left(\cap_{k=1}^{m} g_{i_k i_m}^{-1}(A_k)\right), \quad \text{since} \quad g_{i_m i_m} = id, \\ &= \nu(\tilde{A}). \end{aligned} \tag{16}$$

Thus (13) holds for all $m \geq 1$.

Now the formula (13) enables us to apply the original Kolmogorov argument of Theorem I.2.1 to this case and deduce that ν is σ-additive on the cylinder algebra generated by any countable directed subset D_1 of D, as shown by C. Ionescu Tulcea [1] (cf., also Neveu [1], p. 162, and Rao [11], p. 367). The result is extended to the general index D, using Proposition 2, as follows.

Thus let $\{\tilde{A}_k, k \geq 1\}$ be a disjoint sequence of measurable rectangles whose union $\tilde{A}_0$, say, is also a similar rectangle. Let $D_1 = \{i_k \in D : j < i_k < i_{k+1}, \text{for all } j \in r(\tilde{A}_k), k \geq 0\}$ where $r(\tilde{A})$ is defined above (3). If $\mathcal{S}_1$ denotes the semiring of measurable rectangles $\tilde{A}$ such that $r(\tilde{A}) \subset D_1$, and $\nu_1 = \nu|\mathcal{S}_1$, then by the Ionescu Tulcea theorem recalled in the preceding paragraph ν_1 is σ-additive on $\mathcal{S}_1$. Let $\tilde{A}_k = \bigcap\{\tilde{\pi}_i^{-1}(A_{ki}) : i \in r(\tilde{A}_k)\}$ be a cylinder with proper bases $A_{ki} \in \Sigma_i, i \in r(\tilde{A}_k)$, and $\tilde{B}_k = \pi_{i_k}^{-1}(\bigcap\{g_{ii_k}^{-1}(A_{ki}) : i \in r(\tilde{A}_k)\})$. Then $\Omega \cap \tilde{A}_k = \Omega \cap \tilde{B}_k$ (using the composition rules of these mappings), and by Proposition 2(ii), we get

$$\sum_{k=1}^{\infty} \nu(\tilde{A}_k) = \sum_{k=1}^{\infty} \nu_1^*(\Omega \cap \tilde{A}_k) = \sum_{k=1}^{\infty} \nu_1^*(\Omega \cap \tilde{B}_k). \tag{17}$$

Let $\mathcal{S}$ be the algebra generated by $\mathcal{S}_1$ so that $\Sigma = \sigma(\mathcal{S})$ and let ν also denote the unique σ-additive extension of the corresponding set function on $\mathcal{S}_1$. If $\{\tilde{C}_k, k \geq 1\}$ is a disjunctification of $\{\tilde{B}_k, k \geq 1\}$, then $\Omega \cap \tilde{B}_k = \Omega \cap \tilde{C}_k, k \geq 1$, and $\tilde{C}_k \in \mathcal{S}$. Hence by the last two parts of Proposition 2, we get

$$\begin{aligned}
\sum_{k=1}^{\infty} \nu_1^*(\Omega \cap \tilde{B}_k) &= \sum_{k=1}^{\infty} \nu_1^*(\Omega \cap \tilde{C}_k) \\
&= \sum_{k=1}^{\infty} \nu_1(\tilde{C}_k), \qquad \text{(cf., (7))} \\
&= \nu_1\Big(\bigcup_{k=1}^{\infty} \tilde{C}_k\Big), \quad \text{since } \nu_1 \text{ is } \sigma\text{-additive}, \\
&= \nu_1\Big(\Omega \cap \bigcup_{k=1}^{\infty} \tilde{C}_k\Big), \quad \text{by (7)}, \\
&= \nu_1^*(\Omega \cap \tilde{A}_0) = \nu(\tilde{A}_0).
\end{aligned} \tag{18}$$

Thus (17) and (18) imply that ν is σ-additive on $\tilde{\Sigma}$.

We now apply Proposition 2(iii) to ν with $D_1 = D$ to conclude that Ω is a thick set. For each $\tilde{A} \in \tilde{\Sigma}$, if $A = \Omega \cap \tilde{A}$, then we let $\mu(A) = \nu(\tilde{A})$, so that $\mu : \Sigma \to \mathbb{R}^+$ is a well-defined σ-additive set function, and is a probability measure (cf., e.g., Rao [11], p. 75). Moreover, if $A_i \in \Sigma_i$, then

$$\mu(\pi_i^{-1}(A_i)) = \nu(\tilde{\pi}_i^{-1}(A_i)) = \mu_i(A_i), \quad i \in D. \tag{19}$$

Hence $\mu = \underleftarrow{\lim}(\mu_i, \pi_i)$, $\pi_i = g_{ij} \circ \pi_j$ for $i < j$, and μ is the desired probability measure. Its uniqueness is quickly established as follows.

Suppose μ' is another probability measure on Σ with μ_i as its marginal so that for each $i_k \in D, A_k \in \Sigma_{i_k}, 1 \le k \le n$, we have for all $j > i, i \in r(A)$, A being a rectangle,

$$A = \bigcap_{k=1}^{n} \pi_k^{-1}(A_k) = \pi_j^{-1} \bigcap_{k=1}^{n} g_{i_k j}^{-1}(A_k)). \tag{20}$$

Consequently,

$$\begin{aligned}
\mu(A) &= \mu\Big(\pi_j^{-1}\Big(\bigcap_{k=1}^{n} g_{i_k j}^{-1}(A_k)\Big)\Big) \\
&= \mu_j\Big(\bigcap_{k=1}^{n} g_{i_k j}^{-1}(A_k)\Big) \\
&= \mu'\Big(\pi_j^{-1}\Big(\bigcap_{k=1}^{n} g_{i_k j}^{-1}(A_k)\Big)\Big) = \mu'(A).
\end{aligned}$$

So $\mu = \mu'$ on all rectangles (20) of $\mathcal{S}$ and then on the algebra generated by them. Hence the same holds on $\Sigma(= \sigma(\mathcal{S}))$ as desired. □

Remark. If $(\Omega_i, \Sigma_i) = \big(\times_{j \in i} T_j, \bigotimes_{j \in i} \mathcal{T}_j\big)$, with $i \in D$, where D is the family of all finite subsets of a set J directed by inclusion and $\{(T_j, \mathcal{T}_j), j \in J\}$ is a collection of measurable spaces, we have $\Omega = \tilde{\Omega}$ and no separability or s.m. condition is needed for the above proof. However, regularity hypothesis is still necessary. We reformulate this case to state Ionescu Tulcea's [1] theorem, which is extended in the preceding result.

Thus let $\{(R_n, \mathcal{R}_n, n \ge 1\}$ be a family of measurable spaces, $\Omega_n = \times_{i=1}^{n} R_i, \Sigma_n = \bigotimes_{i=1}^{n} \mathcal{R}_i$, and $p_{n+1} : \mathcal{R}_{n+1} \times \Omega_n \to [0,1]$ be functions such that $p_{n+1}(\cdot,\omega_n)$ is a probability on $\mathcal{R}_{n+1}$ for each $\omega_n \in \Omega_n$ and $p_{n+1}(A,\cdot)$ is measurable relative to Σ_n, whatever be $A \in \mathcal{R}_{n+1}$. For $n > m \geq 1$ define p_{nm} by the formula below where we set $A^{(n)} = A_1 \times \cdots A_n \in \Sigma_n$, and $\omega_k = (x_1, \ldots, x_k), k \geq 1$:

$$p_{mn}(A^{(n)}, \omega_m) = \chi_{A^{(m)}} \int_{A_{m+1}} p_{m+1}(dx_{m+1}, \omega_m) \cdots \int_{A_n} p_n(dx_n, \omega_{n-1}). \tag{21}$$

If $P_1 : \mathcal{R}_1 \to [0,1]$ is any (initial) probability, for $n > 1$, we define $P_m(A) = \int_{\Omega_1} p_{1m}(A, \omega_1) P(d\omega_1), A \in \Sigma_m$, and similarly $P_n(B) = \int_{\Omega_m} p_{mn} P_m(d\omega_m), B \in \Sigma_n$, then (21) implies that $\{(\Omega_n, \Sigma_n, P_n, \pi_{mn}) : n \geq m \geq 1\}$ is a regular projective system of probability spaces. Hence Theorem 1 reduces to the following result due to Ionescu Tulcea [1]:

3. Theorem. *Let $\{(R_n, \mathcal{R}_n), n \geq 1\}$ be measurable spaces, $\{p_n(\cdot,\cdot), n \geq 1\}$ be regular conditional probability measures from which one defines the family $\{p_{mn}, n \geq m \geq 1\}$ by (21). Then they determine a projective system $\{(\Omega_n, \Sigma_n, P_n, \pi_{mn}) : n \geq m \geq 1\}$ of probability spaces relative to an initial probability P_1 on $\mathcal{R}_1$. Moreover, the system admits a limit (Ω, Σ, P) such that for each cylinder $\pi_n^{-1}(A^{(n)}), A^{(n)} \in \Sigma_n$, one has:*

$$\begin{aligned} P(\pi_n^{-1}(A^{(n)})) &= \int_{\Omega_m} p_{mn}(A^{(n)}, \omega_m) P_m(d\omega_m) \\ &= \int_{A_1} P_1(d\omega_1) \int_{A_2} p_2(dx_2, \omega_1) \cdots \int_{A_m} p_n(dx_n, \omega_n). \end{aligned} \tag{22}$$

Remark. Observe that, if all the p_n are independent of the parameters ω_n, then Theorem 3 is precisely the classical Fubini-Jessen theorem (cf., Dunford-Schwartz [1], III.11.20, Halmos [1], p. 157, or Rao [11], p. 346) of which it is an extension (and not of Theorem I.3.2 of Kolmogorov). This distinction reveals the difference between the conditions in these two results. Theorem 1 may be regarded as a generalization of these statements.

Formula (22) is of interest in proving the existence of Markov processes with the $p_n(\cdot,\cdot)$ as transition probabilities of a particle moving from states prior to the "time" $(n-1)$ to the next state at time n

and P_1 as the initial probability of entering state 1, and this description will be made precise later. Formula (22) is also useful in statistical estimation theory where $x_0 = \theta$ will be an unknown parameter, $\theta \in \Theta, p_n(A, x_1, \ldots, x_{n-1}, \theta)$ is the "posterior" probability of A after observing $x_1, \ldots, x_{n-1}$ and P_1 is a "prior" probability of θ.

3.7 Complements and exercises

1. Let $\{X_t, t \in [a,b]\}$ be a stochastic process on a probability space (Ω, Σ, P) into a complete separable metric space M. If the process is stochastically continuous on $[a,b]$ and has no fixed discontinuities, show that there exists a modification $\{Y_t, t \in [a,b]\}$ of the process such that almost all sample sample functions of the Y_t-process are right or left continuous at each $t \in [a,b]$.

2. Let $T \subset \mathbb{R}$ and $\{X_t, t \in T\}$ be a real separable process on (Ω, Σ, P). If $\overline{T}$ is the closure of T, suppose that for each $t \in \overline{T}$ either $\lim_{s \uparrow t} X_s = X_{t-}$ or $\lim_{s \downarrow t} X_s = X_{t+}$ exists stochastically. Then there is atmost a countable set $T_0 \subset \overline{T}$ such that $X_{t-}(\omega) = X_{t+}(\omega)$ for every $t \in \overline{T} - T_0$ and this is $X_t(\omega)$ if $t \in T$, for almost all ω.

3. Let (Ω, Σ, P) be a complete separable probability space and T an uncountable index set. Then there does not exist a real separable stochastic process $\{X_t, t \in T\}$ on it such that the X_t are mutually independent nontrivial (i.e. nonconstant) random variables.

4. The concept of separability can be extended slightly. Let (Ω, Σ, P) be a complete probability space, and $(\mathbb{E}, \mathcal{E})$ be a measurable space. If $\{X_t, t \in T\}$ is an $\mathbb{E}$-valued process, we say that it is separable whenever (i) $V(I, K) = \cap_{t \in I}\{\omega : X_t(\omega) \in K\} \in \Sigma$ for each $I \subset T, K \in \mathcal{E}$, and (ii) $P(V(I,K)) = \inf\{P(V(\alpha, K)) : \alpha \subset I, \text{finite}\}$. Show that there is a separable modification of the process relative to a given collection $\mathcal{K} \subset \mathcal{E}$ such that each countable subcollection with the finite intersection property has a nonempty intersection, and $\emptyset \in \mathcal{K}$. [Note that we have no lifting now, but the proof proceeds on the same lines as that of Theorem 1.3. This extension is due to Meyer [1].]

5. Let (Ω, Σ, P) be a Carathéodory generated probability space. Let $\mathcal{N}$ be the σ-ideal of null sets of Σ and $\tilde{\lambda} : \Sigma/\mathcal{N} \to \Sigma$ be the set lifting as in Section 2. (a) Show that $\{\tilde{\lambda}(\tilde{A}) : \tilde{A} \in \Sigma/\mathcal{N}\}$ forms a basis for a

topology $\mathcal{T}$ on Ω, called the "lifting topology". [$\mathcal{T}$ is uniformizable but not necessarily Hausdorff.] (b) In the topology $\mathcal{T}$, each real continuous function is measurable for (Σ), and for each (Σ)-measurable function $f : \Omega \to \mathbb{R}$, there exists a unique $\mathcal{T}$-continuous function $\overline{f}$ on the same space such that $f = \overline{f}$,a.e. Moreover the mapping $f \mapsto \overline{f}$ preserves algebraic operations (where defined). [*Hints*: For the first statement verify that each open set for $\mathcal{T}$ is in Σ by showing that (with localizability) an arbitrary union $\cup_\alpha \tilde{\lambda}(A_\alpha)$ has a (measurable) supremum in Σ. If f is measurable for (Σ) define $\overline{f}(\omega) = \sup\{r \in \mathbb{R} : \omega \notin \tilde{\lambda}((f^{-1}(-\infty, r))^{\sim})\}$ for each $\omega \in \Omega$. Now verify that $\overline{f}$ is continuous and that $f = \overline{f}$ a.e.] (c) Define a mapping F on Ω as: $F(\omega) = \{\tilde{A} \in \Sigma/\mathcal{N} : \omega \in \tilde{\lambda}(\tilde{A})\}$. Let $\tilde{\Omega}$ be the representation space of $\Sigma/\mathcal{N}$ as in I.3.6, or the points of $\tilde{\Omega}$ are the ultra filters $\mathcal{F}$ of the complete Boolean algebra $\Sigma/\mathcal{N}$. Then $\tilde{\Omega}$ is an extremally disconnected compact Hausdorff space whose basic open sets are $\Gamma(\tilde{A}) = \{\mathcal{F} : \tilde{A} \in \mathcal{F}\}$. Show that $F : \Omega \to \tilde{\Omega}$ is onto and if S is any compact Hausdorff space and $f : \Omega \to S$ is continuous, then there is a unique continuous $g : \tilde{\Omega} \to S$ such that $f = g \circ F$. Moreover $\mathcal{T}$ is the weakest topology making F continuous and that $\mathcal{T}$ is Hausdorff iff F is one-to-one. [Only the last statement needs a separate proof. For this verify that $\tilde{\lambda} = F^{-1} \circ \Gamma$. In connection with this result, see Fillmore [1].]

6. (Lifting operators need not always exist.) Let (Ω, Σ, P) be a probability space with a set of positive diffuse measure, and let $\mathcal{L}^p(P)$ be the usual Lebesgue space of pth power integrable functions on it. Then there does not exist a (even linear) lifting on $\mathcal{L}^p(P)$ if $1 \leq p < \infty$. [*Hints*: If there is a linear lifting ρ then $\rho(\tilde{f}) = \rho(f)$ for $f \in \tilde{f}$, where $\tilde{f} \in L^p(P) = \mathcal{L}^p(P)/(\text{null functions})$. Consider the mapping $\tilde{f} \mapsto \rho(\tilde{f})(\omega)$. This is a positivs (hence continuous) linear functional on the Banach lattice $L^p(P)$, and hence $\rho(\tilde{f})(\omega) = \int_\Omega f(x) g_\omega(x) dP(x)$ for a unique $\|g_\omega\|_q \leq 1, q = p/(p-1)$. If A is the diffuse set and $a = P(A) > 0$, there exist for each n a decomposition $A_1^{(n)}, \ldots, A_n^{(n)}$ of A such that $P(A_j^{(n)}) = a/n$. If $A^{(n)} = \bigcup_{j=1}^n \{\omega : \rho(\chi_{A_j^{(n)}})(\omega) = 1\}$, then $A = A^{(n)}$ a.e., and $P(\bigcap_{n=1}^\infty A^{(n)}) = a$. If $\omega_0 \in A \cap \bigcap_{n=1}^\infty A^{(n)}$ then $\rho(\chi_{A_j^{(n)}})(\omega_0) = 1 \leq \|\chi_{A_j^{(n)}}\|_p = (a/n)^{(1/p)} \to 0$ as $n \to \infty$, giving the desired contradiction. This argument is due to J. von Neumann.]

7. We present the classical Lebesgue differentiation theorem from the point of view of martingales with directed index sets. Let $(\mathbb{R}^n, \mathcal{B}, \mu)$

be the n-dimensional Lebesgue measure space and $I_k^\delta \subset \mathbb{R}^n, k \geq 1$, be a sequence of nondegenerate disjoint rectangles of finite measure, and $\bigcup_{k=1}^\infty I_k^\delta = \mathbb{R}^n$. Thus $\delta = \{I_k^\delta, k \geq 1\}$ is a partition of $\mathbb{R}^n$. Let $\mathcal{D}$ denote the class of all such partitions ordered by refinement. Let $\mathcal{F}^\delta = \sigma(I_k^\delta, k \geq 1) \subset \mathcal{B}$. Consider a subdirected set $\mathcal{D}^\alpha \subset \mathcal{D}$ such that for a subsequence $\{E_n, n \geq 1\} \subset \mathcal{B}$ we have $(i) \mu(E_k) > 0$ but $\mathrm{diam}(E_k) \to 0$ as $k \to \infty$, and (ii) there is a $\delta \in \mathcal{D}$ with $\mu(E_k) \leq \alpha\mu(I_k^\delta)$, for all $k \geq 1$ where $\alpha > 0$ is fixed. Then the following statements hold:

(a) $\{\mathcal{F}^\delta, \delta \in \mathcal{D}_\alpha\}$ satisfies the Vitali condition V_0 of Definition 5.3 for a given $\alpha > 0$.

(b) If $\nu : \mathcal{B} \to \mathbb{R}$ is finitely additive, and $x_\delta = \sum_{k=1}^{\infty} \frac{\nu(I_k^\delta)}{\mu(I_k^\delta)} \chi_{I_k^\delta}$, then $\{x_\delta, \mathcal{F}_\delta, \delta \in \mathcal{D}\}$ is a martingale on $(\mathbb{R}^n, \mathcal{B}, \mu)$. If also $\sup_{\delta \geq \delta_0 \in \mathcal{D}} \int_{\mathbb{R}^n} |x_\delta| d\mu < \infty$, which is true if ν is of bounded variation or if it is σ-additive, then the above martingale converges a.e. when $\mathcal{D}$ is restricted to $\mathcal{D}_\alpha, \alpha > 0$, and $x_\infty = \lim_{\delta \in \mathcal{D}_\alpha} x_\delta$ satisfies $\int_A |x_\infty| d\mu \leq k_A < \infty$ for each $A \in \mathcal{F}_{\delta_0}, \mu(A) < \infty$.

(c) Under the hypothesis of (b), the Dini derivative $D\nu$ of ν exists a.e., $(= x_\infty)$ for any $\alpha \geq 1$. Here $D\nu = \sup_{\{E_k, k \geq 1\} \subset \mathcal{B}} \limsup_{k \to \infty} \nu(E_k)/\mu(E_k)$, with $E_k \subset I_k^\delta, \delta \in \mathcal{D}^\alpha$.

(d) If $\nu_f(A) = \int_A f d\mu, A \in \mathcal{B}, \mu(A) < \infty$, and f is locally integrable, then the martingale $\{x_\delta, \mathcal{F}_\delta, \delta \in \mathcal{D}_\alpha\}$, defined as in (b) for this ν_f, converges a.e., and $x_\delta^f \to x_\infty^f = f$ a.e., and $f = D\nu_f$ a.e., in the notation of (c). [This result uses Theorem 5.7 and the computations are nontrivial. See in this connectionn Chow [1] and Hayes-Pauc [1].]

8. Show that every complete σ-finite measure space is (even strictly) localizable.

9. Using the ideas and methods of Section 3 stochastic extensions of the classical extension theorems such as Hahn-Banach, Tietze and the like can be given. We illustrate this fact for the well-known Dugundji generalization of Tietze's theorem.

Let T be a metric space and S a locally convex vector space. If $(T, \mathcal{T}, \mu)$ is a Radon measure space and (Ω, Σ, P) a probability space, let $X = \{X_t, t \in A\}$ be an S-valued stochastic process on Ω with almost all continuous sample paths and $A \subset T$ be closed and nonempty. Then there exists an S-vslued process $Y = \{Y_t, t \in T\}$ such that (i) it has almost all continuous sample paths, (ii) $f \circ Y = \{f \circ Y_t, t \in T\}$

is $\mu \otimes P$-measurable for each bounded continuous real f on S, (iii) $X(t,\omega) = Y(t,\omega), t \in A$ and almost all $\omega \in \Omega$, and (iv) $Y(T,\Omega)$ is a subset of the convex hull of $X(A,\Omega)$. For this work we take $\Omega = S^T$. Then one has the canonical representation of the processes, and one can identify S^A as a subset of S^T. [This result uses the ideas and proofs of Theorems 3.3 and 3.4, and can be established with a careful extension of the latter. For details and related results and references, one may consult the author's paper [6].]

10. Here we explain the use of right continuity of a net $\{\mathcal{F}_t, t \in T\}, T \subset \mathbb{R}$, of a probability space (Ω, Σ, P) for the work of modifications (cf.,Theorem 5.2(iii)). The latter essentially implies: if the right continuous net above of σ-subalgebras is complete and $\{X_t, \mathcal{F}_t, t \in T\}$ is a right continuous supermartingale, then it has a modification which is right continuous with left limits, called a *càdlàg*(= continue à droite et limite à gauche). This admits an extension for general processes: if the filtration of the net is as above, and $\{X_t, t \in T\}$ is any càdlàg process satisfying $|X_t| \leq Z, Z \in L^1(P), t \in T$, then there is a càdlàg modification of $\{Y_t, \mathcal{F}_t, t \in T\}$ where $Y_t = E^{\mathcal{F}_t}(X_t)$. This is due to Mertens [1] and Meyer [1], (see also K. M. Rao [3]). Thus in our treatment with supermartingales it is convenient to assume that these are càdlàg processes, as otherwise one can replace them with such modifications when the filtration is right continuous and complete.

Bibliographical remarks

The concept of separability of a continuous parameter stochastic process has been introduced by Doob in the late 1930's and its key role in the stochastic function theory has been recognized ever since. The basic Theorem 1.3 and some related results are due to him, as they are presented in final form in his monograph [1]. The separability of the range space of a process is essentially used in all the earlier work. The relevance of the lifting operator in these problems has been noted by Tulcea and Tulcea [1] who proved the most general form of the result, which we presented in Theorem 3.1. The general lifting theorem and the main results of Section 2 are due to the latter authors, and a different proof of the lifting theorem is in Sion [2]. Theorems 1.5 and

3.4, as well as 4.1 are adapted from Nelson [1]; and Theorem 4.6, from Neveu [1]. Theorem 2.1 is due to Ryan [1], and the general Theorems 2.3 and 2.4 are taken from McShane [2]. The very useful concept of localizability was introduced by Segal [2], and its significance in analysis has been clarified and emphasized by McShane [2] and Zaanen [1] independently.

The convergence theory of martingales with directed index sets is essentially due to Chow [1] (see also Hayes-Pauc [1]), on whose work much of Section 5(b) is based.

The fundamental result on the product (regular) conditional measures is due to Ionescu Tulcea (1949) and its extension to Choksi [1]. The generalization given in Theorem 6.1 is adapted from Rao and Sazonov [1]. Further developments on the disintegration problem with applications to martingales of measure valued functions can be found in the long memoir by Schwartz [3].

There is extensive literature on second order processes that can be presented with Hilbert space techniques. For some of this, see Gikhman-Skorokhod [2]. But a more streamlined treatment in this case, with the reproducing kernel theory of Aronszajn, can be given extending most of the earlier work. Some of this in the context of Gaussian processes is included in Neveu [2]. Other references and complements given in the text should be consulted.

Chapter IV

Refinements in martingale analysis

In order to obtain finer and specialized results from the basic theory of martingales, we introduce a new tool called the stopping time or optional transformation and investigate various properties of martingales under the effect of these mappings. A number of technical (measure theoretical) problems arise when such families are considered, and we present a detailed analysis of these processes together with their structure and limit theory. Both the directed and linearly ordered index sets of the (sub-) martingales are considered. A consequence here is the culmination of a proof of the existence of projective limits of certain systems and the associated class (D) martingales. The study leads to several decompositions of processes that are useful in applications.

We then specialize the results when the index sets are integers or subsets of the line, and consider classes of processes obtained from martingale differences or increments. Some extensions of the maximal martingale inequalities are proved. Finally, we treat the $\mathcal{H}^p$ theory, the space of BMO functions and their dualities. The Doob-Meyer decomposition of (sub or) supermartingales, when the index set is an arbitrary part of the line and the consequent stochastic integration will be taken up in the next chapter. A number of results complementing this work are given as graded exercises in the last section.

4.1 Stopping times as a new tool

(a) *Introduction.* In discussing the decomposition of a submartingale, we noted in Chapter II that new tools are needed for a solution of the corresponding continuous parameter problem. Regarding a process $\{X_t, t \in I\}$ as a function, $X(\cdot,\cdot) : I \times \Omega \to \mathbb{R}$, of two variables, new techniques are available involving a certain class of mapping(s) which enables the transformed problems solvable. To motivate the definition of these mappings, consider the following example from classical analysis: Let $f : \mathbb{C}\times\mathbb{C} \to \mathbb{C}$ be a complex function such that $f(\cdot, z_2) : \mathbb{C} \to \mathbb{C}$ and $f(z_1, \cdot) : \mathbb{C} \to \mathbb{C}$ are holomorphic (i.e. have power series expansions at each point in the domain). Then by the well-known Hartog's Theorem f is (jointly) holomorphic in its domain so that information on partial mappings implies a global property when the class of functions is suitably restricted. Since the stochastic process $\{X_t, t \in I\}$ is identifiable with the "new" function $X(t, \cdot) : \Omega \to \mathbb{R}$, our intended mappings on each such partial domains (i.e. I and Ω) separately should be useful for a general study. The availability of measures but lack of topologies in I and Ω shows that we have a different set of problems here (of measure theoretic nature) in considering compositions in contrast to the preceding example. To restrict our processes (or stochastic functions) suitably, we start with the following comment. If $I \subset \mathbb{R}$, then $X_t(\omega)$ is the value of the stochastic phenomenon observed at "time" t, and if $\tau(\omega)$ is the duration of the observation, then $[\tau(\cdot) \leq a]$ should reflect the characteristics of the process until the instant $a \in T$ (i.e. the event $\{\omega : \tau(\omega) \leq a\}$ should belong to $\mathcal{F}_t = \sigma\{X_s, s \leq t\}$), and the restricted class of processes should be such that $X \circ \tau = X(\tau(\cdot), \cdot) : \Omega \to \mathbb{R}$ is a random variable. We now make this precise with the concepts of a stopping time transformation, and of a stopping time process.

1. Definition. Let $\{\mathcal{F}_i, i \in I\}$ be a given family of σ-subalgebras, filtering to the right from a measurable space (Ω, Σ), where I is a directed set. Then a mapping $T : \Omega \to I$ is called a *stopping time, or optional, transformation* relative to the family $\{\mathcal{F}_i, i \in I\}$ if, for each $i \in I$, it is true that $\{\omega : T(\omega) \leq i\} \in \mathcal{F}_i$, and $\{\omega : T(\omega) \geq i\} \in \mathcal{F}_i$. A filtering family $\{T_j, j \in J\}$ of such T's is a *stopping time (or optional) process* if J is directed and $j_1 \leq j_2$ implies $T_{j_1}(\omega) \leq T_{j_2}(\omega), \omega \in \Omega$. [We use both terms "stopping time" and "optional" interchangeably according to convenience.]

Note that if $T(\omega)$ and i are not compatible, then $\{\omega : T(\omega) \leq i\}$ or $\{\omega : T(\omega) \geq i\}$ will be empty. If we have a measure P on Σ, then all the above relations are to hold a.e. Clearly for a linearly ordered and countable I it is sufficient to assume, in the definition of a stopping time T, that $\{\omega : T(\omega) \leq i\} \in \mathcal{F}_i$, or $\{\omega : T(\omega) < i\} \in \mathcal{F}_i$ or

$\{\omega : T(\omega) > i\} \in \mathcal{F}_i$ for each i since $\{\omega : T(\omega) < i\} = \bigcup_{n=1}^{\infty} \{\omega : T(\omega) \leq \delta_n\} \in \mathcal{F}_i (\delta_n < i)$ when the first of the conditions holds with $I = \{\delta_n\}_1^\infty$. In applications, we assume the range of T to be a countable set in I or if $I \subseteq \mathbb{R}$, then the range contains a countable dense set $\{\delta_n\}_1^\infty$. To understand this concept well, we first prove some general consequences. We again denote the sets, such as $\{\omega : T(\omega) < i\}, \{\omega : X(\omega) \leq a\}$, simply as $[T < i], [X \leq a]$ when convenient.

If $\{X_i, \mathcal{F}_i, i \in I\}$ is an adapted real process on (Ω, Σ) and $T : \Omega \to I$ is a stopping time relative to $\{\mathcal{F}_i, i \in I\}$, then a new function $Y = X_T (Y(\omega) = X_T(\omega) = X_{T(\omega)}(\omega) = X(T(\omega), \omega), \omega \in \Omega)$ may be defined, which is sometimes termed a "superposition" of the functions $X(\cdot, \cdot)$ and $T(\cdot)$. Since Y can be expressed as a composition of the mappings $\Omega \xrightarrow{T} I \xrightarrow{j} I \times \Omega \xrightarrow{X} \mathbb{R}$, where j is the embedding of I in $I \times \Omega$ $(j(T(\omega)) = (T(\omega), \omega))$, we see that Y is also a random variable. (This is a consequence of the countability of the range of T, and a proof in a general case is given in (5) below.) The following result indicates the power of these transformations.

2. Theorem. *Let $\{X_i, i \in I\}$ be an adapted stochastic process on (Ω, Σ) where I is countable and directed. Let $\{T_j, j \in J\}$ be a stopping time process where I and J are perhaps unrelated. Then we have:*

(i) (a) $\sigma\{A \cap [T_j \geq i] : A \in \mathcal{F}_i, i \in I\} = \sigma\{A \cap [T_j = i] : A \in \mathcal{F}_i, i \in I\}$ *and if this common σ-algebra is denoted by $\mathcal{G}_j$, then $\{Y_j, \mathcal{G}_j, j \in J\}$ is an adapted process where $Y_j(\cdot) = X(T_j(\cdot), \cdot)$.*

(b) *If I has a last element "∞", then*

$$\mathcal{C}_j = \{A \subset \Omega : A \cap [T_j \leq i] \in \mathcal{F}_i, i \in I\}, \tag{1}$$

is a σ-algebra, $\mathcal{C}_j \subset \mathcal{C}_{j'}$, for $j \leq j'$ and $\mathcal{G}_j \subset \mathcal{C}_j \subset \Sigma$.

(ii) *If $\{X_i, \mathcal{F}_i, i \in I\}$ is a submartingale on (Ω, Σ, P), I is linearly ordered and finite, then $\{Y_j, \mathcal{C}_j, j \in J\}$ as well as $\{Y_j, \mathcal{G}_j, j \in J\}$ are submartingales and are martingales if the original X_j-process is such. Let α, β be the first and last elements of I. Then $\mathcal{F}_\alpha \subset \mathcal{G}_j \subset \mathcal{C}_j \subset \mathcal{F}_\beta$ for all $j \in J$ and we have*

$$\int_A X_\alpha dP \leq \int_A Y_j dP \leq \int_A X_\beta dP, \qquad A \in \mathcal{F}_\alpha, \tag{2}$$

with equality in the martingale case.

Proof. We consider (i)(a). To prove the equality of the σ-algebras, let $A \in \mathcal{F}_i$ and call the two algebras of (i)(a) as $\mathcal{A}_1$ and $\mathcal{A}_2$. Now I is countable, and so for $i \in I, A \cap \{\omega : T_j(\omega) = i\} \in \mathcal{F}_i$. Then

$$A \cap [T_j \geq i] = \bigcup_{\delta_n \geq i} \{A \cap [T_j = \delta_n]\} \in \mathcal{A}_2. \tag{3}$$

Since A and i are arbitrary, we deduce that $\mathcal{A}_1 \subset \mathcal{A}_2$. On the other hand,

$$B = A \cap [T_j = i] = A \cap [T_j = i] \cap [T_j \geq i] = B \cap [T_j \geq i] \in \mathcal{A}_1, \tag{4}$$

whence $B \in \mathcal{F}_i$. Again the arbitrariness of A and i (and hence of B and i) implies $\mathcal{A}_2 \subset \mathcal{A}_1$ by (4), and so $\mathcal{A}_1 = \mathcal{A}_2 = \mathcal{G}_j$. We show that Y_j is $\mathcal{G}_j$-adapted, and that $\mathcal{G}_j$ is filtering to the right.

For any $a \in \mathbb{R}$, consider

$$\begin{aligned}[Y_j < a] &= \bigcup_{i \in I} [Y_j < a] \cap [T_j = i] \\ &= \bigcup_{i \in I} [X_i < a] \cap [T_j = i], \quad \text{by definition.}\end{aligned} \tag{5}$$

Since this union is countable, it follows that the right side set of (5) is in $\mathcal{A}_2$ so that Y_j is $\mathcal{A}_2$ $(= \mathcal{G}_j)$-measurable. (Note that I can be countable and not finite in (5).) Regarding monotonicity, let $A \in \mathcal{F}_i$, so that $A \cap [T_j = i] \in \mathcal{F}_i$, and let $j_1 \leq j_2$. Then $T_{j_1}(\omega) \leq T_{j_2}(\omega), \omega \in \Omega$. So $[T_{j_1} = i] \subset [T_{j_2} \geq i]$ and $B = A \cap [T_{j_1} = i] = B \cap [T_{j_2} \geq i] \in \mathcal{A}_1$, since $B \in \mathcal{F}_i$. Hence $\mathcal{G}_{j_1} \subset \mathcal{G}_{j_2}$, which proves (i)(a), since $\mathcal{G}_j$ is generated by the union of the trace algebras $\mathcal{F}_i \cap [T_j = i]$, as i ranges in I.

(i)(b) Suppose that I has a last element "∞" so that $i \in I$ implies $i \leq \infty$. Then for any $j \in J$ we assert that $\mathcal{C}_j$ of (1) is a σ-algebra. In fact, since $\emptyset \in \mathcal{C}_j$, let $A \in \mathcal{C}_j$. Then $A^c \cap [T_j \leq i] = [T_j \leq i] - A \cap [T_j \leq i] \in \mathcal{F}_i, i \in I$, and so $A^c \in \mathcal{C}_j$. Let $\{A_n\}_1^\infty \subset \mathcal{C}_j$. Then

$$\left(\bigcup_{n=1}^{\infty} A_n\right) \cap [T_j \leq i] = \bigcup_{n=1}^{\infty} (A_n \cap [T_j \leq i]) \in \mathcal{F}_i, \qquad i \in I.$$

Hence $\bigcup_{n=1}^{\infty} A_n \in \mathcal{C}_j$ and $\mathcal{C}_j$ is a σ-algebra even if I has no last element. However, $\mathcal{C}_j \subset \Sigma$ may be false in this generality. If "∞" is the last element of I, then for any $A \in \mathcal{C}_j$ we have

$$A = A \cap \bigcup_{i \in I} [T_j = i] = A \cap [T_j \leq \infty] \in \mathcal{F}_\infty \subset \Sigma. \tag{6}$$

Thus $\mathcal{C}_j \subset \Sigma$. it is clear from definitions that $\mathcal{G}_j \subset \mathcal{C}_j$. The fact that $\mathcal{C}_j \subset \mathcal{C}_{j'}$ for $j \leq j'$ follows as in (i)(a). This establishes (i).

(ii) Suppose that $\{X_i, \mathcal{F}_i, i \in I\}$ is a submartingale and I is finite and linearly ordered. Let $I = \{a_1 < a_2 < \cdots < a_n\}$, and $A \in \mathcal{G}_{j_1}, j_1 \leq j_2$ in J. If $A_0 = A \cap [T_{j_1} = a_k]$, then we assert that $A_0 \in \mathcal{F}_{a_k}$. In fact, let

$$\mathcal{U} = \{A \in \mathcal{G}_{j_1} : A \cap [T_{j_1} = a_k] \in \mathcal{F}_{a_k}, \text{ for some } 1 \leq k \leq n\}. \tag{7}$$

Then clearly $A \in \mathcal{U}$ implies $A^c \in \mathcal{U}$ and $\emptyset \in \mathcal{U}$. If $B_i \in \mathcal{U}, i = 1, 2$, then $[B_1 \cup B_2] \cap [T_{j_1} = a_k] = \bigcup_{i=1}^{2} \{B_i \cap [T_{j_1} = a_k]\} \in \mathcal{F}_{a_k}$ so $B_1 \cup B_2 \in \mathcal{U}$. It is also closed under monotone limits so that $\mathcal{U}$ is a σ-subalgebra of $\mathcal{G}_{j_1}$. But by (i)(a) and (7) we see that these sets generate $\mathcal{G}_{j_1}$, so $\mathcal{U} = \mathcal{G}_{j_1}$. Since $A = A \cap \bigcup_{k=1}^{n} [T_{j_1} = a_k]$, (7) also implies that $A \cap [T_{j_1} = a_k] \in \mathcal{F}_{a_k} \subset \mathcal{G}_{j_1}$, for $1 \leq k \leq n$. With this we now prove that $\{Y_j, \mathcal{G}_j, j \in J\}$ is a (sub) martingale since Y_j is $\mathcal{G}_j$-adapted. For this it is sufficient to prove that, if $A_0 = A \cap [T_{j_1} = a_{k_0}]$ for any fixed $1 \leq k_0 \leq n$, then

$$\int_{A_0} Y_{j_1} \, dP \leq \int_{A_0} Y_{j_2} dP. \tag{8}$$

Let $A_\alpha = A_0 \cap [T_{j_2} = \alpha]$, and $A^\alpha = A_0 \cap [T_{j_2} > \alpha]$ for $\alpha \in I$. It is clear that $A_{a_{k+1}} \cup A^{a_{k+1}} = A^{a_k}$ is a disjoint union, for $1 \leq k \leq n$, since $T_{j_1} \leq T_{j_2}$. Also $A_0 = A_{a_{k_0}} \cup A^{a_{k_0}}$. Thus

$$\begin{aligned}
\int_{A_0} Y_{j_1} \, dP &= \int_{[T_{j_1} = a_{k_0}] \cap A} Y_{j_1} \, dP = \int_{A_0} X_{a_{k_0}} dP, \quad \text{since } T_{j_1} = a_{k_0} \text{ on } A_0, \\
&= \int_{A_{k_{k_0}}} X_{a_{k_0}} dP + \int_{A^{a_{k_0}}} X_{a_{k_0}} dP, \quad \text{since the sets are disjoint,} \\
&= \int_{A_{a_{k_0}}} Y_{j_2} dP + \int_{A^{a_{k_0}}} X_{a_{k_0}} dP, \quad \text{since } T_{j_2} \text{ is constant on } A_{a_{k_0}},
\end{aligned}$$

$$\leq \int_{A_{a_{k_0}}} Y_{j_2} dP + \int_{A^{a_{k_0}}} X_{a_{k_0+1}} dP, \text{ by the submartingale property of } X_i\text{'s},$$

$$= \int_{A_{a_{k_0}}} Y_{j_2} dP + \int_{A_{a_{k_0+1}}} Y_{j_2} dP + \int_{a_{k_0+1}} X_{a_{k_0+1}} dP, \text{ by definition of } A^{\alpha}\text{'s},$$

$$\cdots\cdots$$

$$\leq \int_{\bigcup_{k \geq k_0} A_{a_k}} Y_{j_2} dP = \int_{A_0} Y_{j_2} dP, \tag{9}$$

since $X_{a_n} = Y_{j_2}$ on A^{a_n} and the decomposition stops there. This shows that the Y_j-process is a submartingale. Since there is equality throughout in (9) if the X_i's form a martingale, we conclude that Y_j's then form a martingale in the case.

Finally, if $\alpha = a_1, \beta = a_n$ in I, let $J' = J \cup \{\alpha, \beta\}$ where we define the ordering as: $\alpha \leq j \leq \beta$ for $j \in J$. Let $T_\alpha = \alpha, T_\beta = \beta$ be the constant stopping times. Then $Y_\alpha = X_\alpha, Y_\beta = X_\beta$ a.e., so that by (i)(a) we get $\mathcal{G}_\alpha = \mathcal{F}_\alpha, \mathcal{G}_\beta = \mathcal{C}_\beta = \mathcal{F}_\beta$ and $\mathcal{G}_\alpha \subset \mathcal{G}_j \subset \mathcal{C}_j \subset \mathcal{G}_\beta, j \in J$ and hence (8) reduces to (2). Since $\int_\Omega Y_j^+ dP \leq \int_\Omega X_{a_n}^+ dP < \infty$, this proves (ii) and the result follows, since in this case $\mathcal{G}_j = \mathcal{C}_j$ actually holds. □

Remark. If $\mathcal{F}_i$ are considered as P-rich σ-algebras, then the above theorem extends to non-finite (σ-finite or localizable) measures. This modification also applies in the following, but will not be mentioned. We shall extend the martingale part of the second half of the above result for countable directed index sets I in Theorem 13 below. However, the submartingale case is not valid in this generality as seen from the next example.

3. Counterexample. Let (Ω, Σ, P) be the Lebesgue unit interval, and $I = \{a, b, c, d\}$ with the partial ordering $a < b < d, a < c < d$ but b and c are not comparable. Let $\mathcal{F}_a = \{\emptyset, \Omega\}, \mathcal{F}_b = \mathcal{F}_c = \mathcal{F}_d = \{A = [0, \frac{1}{2}), B = [\frac{1}{2}, 1], \Omega, \emptyset\}$. Let $X_a = 0, X_b = \chi_A - \chi_B, X_c = -X_b, X_d = 1$. Then $\{X_i, \mathcal{F}_i, i \in I\}$ is a bounded submartingale. However, if $T_a = a, T_b = b$ on B and $= c$ on A, then $T_a \leq T_b$ are stopping times for $\{\mathcal{F}_i, i \in I\}$, and $Y_a = X_a = 0,\ Y_b = X \circ T_b = X_b\chi_B + X_c\chi_A = -(\chi_B -$

$\chi_A)^2$. Hence $0 = E(Y_a) = \int_\Omega Y_a dP > \int_\Omega Y_b dP = -\int_\Omega(\chi_A^2 + \chi_B^2)dP = -1$. So the Y_j-process cannot be a submartingale. Thus, for the validity of Theorem 2(ii), I must be linearly ordered, generally.

The stopping time transformations for the directed index sets have been primarily investigated by Chow [1], to whom many of the results of this sections are due. Note that Theorem 1(ii) is the first of a series of results below and if I is regarded as "time", then the Y–process is the X–process after a random sampling is made on the basis of "past and present" only, so that the Y_j may be called an *optional sampling* process obtained from the X–process with the stopping times $T_j, j \in J$. Several assertions below can be similarly interpreted.

(b) *General properties.* The above theorem shows that new processes can be generated from the old ones under stopping time transformations. To gain insight, we first show how the fundamental inequalities (cf., e.g. Theorem II.4.4) can be proved quickly as a consequence of Theorem 2(ii) above, and then present other useful results with the new tools.

4. Proposition. *Let $\{X_i, \mathcal{F}_i, i \in I\}$ be an integrable real submartingale on (Ω, Σ, P), and suppose that I is a linearly ordered countable index set. If $\alpha_0 = \inf\ I_0, \beta_0 = \sup I$, then one has:*

(i) *when $\beta_0 \in I$,*

$$\lambda P\left[\sup_{\alpha\in I} X_\alpha > \lambda\right] = \lambda P[A_\lambda] \leq \int_{A_\lambda} X_{\beta_0} dP, \qquad \lambda \in \mathbb{R}, \tag{10}$$

(ii) *when $\{\alpha_0, \beta_0\} \subset I$, then*

$$\lambda P\left[\inf_{\alpha\in I} X_\alpha < \lambda\right] = \lambda P[A^\lambda] \geq \int_\Omega X_{\alpha_0} dP - \int_{\Omega - A^\lambda} X_{\beta_0} dP, \qquad \lambda \in \mathbb{R}, \tag{11}$$

Proof. We essentially translate the earlier result (II.4.4) into the stopping time terminology in this proof. Since I is countable, there exist finite sets $I_n \uparrow I$ and hence, if we prove the result for $I_n = \{\alpha_1, \ldots, \alpha_n, \beta_0\}$, the general case follows by the Monotone Convergence. So assume that I is finite and linearly ordered, $I = \{\alpha_1, \ldots, \alpha_n, \alpha_{n+1}\}$ where $\alpha_{n+1} = \beta_0$.

Thus consider $\{X_{\alpha_j}, \mathcal{F}_{\alpha_j}, j = 1, \ldots, n+1\}$ and define the stopping times T_1 and T_2 as: $T_2(\omega) = \beta_0, \omega \in \Omega$, and T_1 as the first α_j such that $X_{\alpha_j} > \lambda$, i.e. (using the ordering in I)

$$T_1(\cdot) = \inf\{\alpha_j : X_{\alpha_j}(\cdot) > \lambda\} \qquad (\inf(\emptyset) = \beta_0). \tag{12}$$

If $A_j = \{\omega : X_{\alpha_k}(\omega) \le \lambda, 1 \le k \le j-1, X_{\alpha_j}(\omega) > \lambda\} \in \mathcal{F}_{\alpha_j}$ so that $A_\lambda = \bigcup_{j=1}^{n} A_j$, with $A_1 = \{\omega : X_{\alpha_1}(\omega) \ge \lambda\}$, then (12) means $T_1 = \alpha_j$ on A_j, and $= \beta_0$ on $\Omega - A_\lambda$, and so $T_1 \le T_2$. Now $[T_1 \le \alpha] = \bigcup_{\alpha_j \le \alpha} A_j$ if $\alpha < \beta_0$, and $= \Omega$ if $\alpha = \beta_0$. So $[T_1 \le \alpha] \in \mathcal{F}_\alpha$ and T_1, T_2 are stopping times of $\{\mathcal{F}_{\alpha_j}, 1 \le j \le n+1\}$. If $Y_1 = X \circ T_1$ and $Y_2 = X \circ T_2$, then $\mathcal{F}_{\alpha_1} \subset \mathcal{G}_1 \subset \mathcal{G}_2 = \mathcal{F}_{\beta_0}$ and $\{Y_j, \mathcal{G}_j\}_{j=1}^2$ is a submartingale by Theorem 2(ii) whose notation is used here. Hence $E^{\mathcal{G}_1}(Y_2) \ge Y_1$ a.e. and by (2) with $A = A_\lambda \in \mathcal{G}_1$, one has

$$\int_{A_\lambda} Y_1 dP \le \int_{A_\lambda} Y_2 dP = \int_{A_\lambda} X_{\beta_0} dP. \tag{13}$$

But $A_\lambda = \bigcup_{j=1}^{n} A_j$ is a disjoint union, and so the left side of (13) simplifies to

$$\int_{A_\lambda} Y_1 dP = \sum_{j=1}^{n} \int_{A_j} Y_1 dP = \sum_{j=1}^{n} \int_{A_\lambda} X_{\alpha_j} dP \ge \lambda \int_{\bigcup_{j=1}^{n} A_j} dP = \lambda P(A_\lambda). \tag{14}$$

Clearly (10) follows from (13) and (14).

The second inequality is similar. Thus let $T_1 = \alpha_0$ and $T_2 = \alpha_j$ on B_j, and $= \beta_0$ on $\Omega - A^\lambda$, where $A^\lambda = \bigcup_{j=1}^{n} B_j$, and $B_j = \{\omega : X_{\alpha_k}(\omega) \ge \lambda, 1 \le k \le j-1, X_{\alpha_j}(\omega) < \lambda\}, B_1 = \{\omega : X_{\alpha_1}(\omega) \le \lambda\}$, all disjoint. Thus $T_1 \le T_2$ are stopping times, as above, and let $Y_1 = X \circ T_1 = X_{\alpha_0}, Y_2 = X \circ T_2$, then $\mathcal{G}_1 = \mathcal{F}_{\alpha_1} \subset \mathcal{G}_2 \subset \mathcal{F}_{\beta_0}$ and $\{Y_j, \mathcal{G}_j\}_{j=1}^2$ is a submartingale by Theorem 2(ii). So $E(X_{\alpha_0}) = E(Y_1) \leq E(Y_2)$. Also

$$\begin{aligned}\lambda P(A^\lambda) + \int\limits_{\Omega - A^\lambda} X_{\beta_0} dP &= \sum_{j=1}^{N} \int\limits_{B_j} \lambda dP + \int\limits_{\Omega - A^\lambda} X_{\beta_0} dP \\ &\geq \sum_{j=1}^{n} \int\limits_{B_j} X_{\alpha_j} dP + \int\limits_{\Omega - A^\lambda} X_{\beta_0} dP \\ &= \int\limits_{\Omega} Y_2 dP \geq \int\limits_{\Omega} Y_1 dP = \int\limits_{\Omega} X_{\alpha_0} dP. \end{aligned} \tag{15}$$

Since X_{β_0} is integrable, (15) implies (11). In view of the initial remark, this establishes the result. □

In case α_0 and β_0 are not in I, we can obtain a slightly weaker set of inequalities using the above results (10) and (11), which yield the following:

$$\lambda P(A_\lambda) \leq \int\limits_{A_\lambda} X_{\beta_0} dP \leq \int\limits_{A_\lambda} X_{\beta_0}^{+} dP, \qquad \beta_0 \in I, \quad \lambda \in \mathbb{R}, \tag{10$'$}$$

$$\lambda P(A^\lambda) \geq \int\limits_{\Omega} X_{\alpha_0} dP - \int\limits_{\Omega - A^\lambda} X_{\beta_0}^{+} dP, \quad \{\alpha_0, \beta_0\} \subset I, \quad \lambda \in \mathbb{R}. \tag{11$'$}$$

If now $\{\alpha_0, \beta_0\} \not\subset I$, let $\tilde{I} = I \cup \{\alpha_0, \beta_0\}$ and consider the submartingale $\{X_\alpha^+, \mathcal{F}_\alpha, \alpha \in I\}$. Then (10′) and (11′) imply the following extension.

5. Corollary. *Let $\{X_\alpha, \mathcal{F}_\alpha, \alpha \in I\}$ be an integrable real submartingale where I is a linearly ordered countable index set. If $A_\lambda = \left[\sup_\alpha X_\alpha > \lambda\right]$ and $A^\lambda = \left[\inf_\alpha X_\alpha < -\lambda\right]$ for $\lambda \geq 0$, and $\alpha_0 = \inf\ I, \beta_0 = \sup\ I$, then one has:*

$$\lambda P(A_\lambda) \leq \liminf_{\alpha \uparrow \beta_0} \int\limits_{A_\lambda} X_\alpha^+ dP \leq \lim_{\alpha \uparrow \beta_0} \int\limits_{\Omega} X_\alpha^+ dP, \tag{16}$$

and

$$\lambda P(A^\lambda) \leq \lim_{\alpha \uparrow \beta_0} \int\limits_{\Omega} X_\alpha^+ dP - \lim_{\alpha \downarrow \alpha_0} \int\limits_{\Omega} X_\alpha dP. \tag{17}$$

We again observe that these inequalities are not valid if I is not linearly ordered. This is shown by a modification of the earlier example:

6. Counterexample. Let (Ω, Σ, P) be the Lebesgue unit interval and $I = \{a_1, a_2, a_3, a_4\}$ with a partial ordering $a_1 < a_2 < a_4, a_1 < a_3 < a_4$, but a_2 and a_3 are not comparable. Let $\mathcal{F}_1, \mathcal{F}_2$ be as in Example 3 and $\mathcal{F}_3, \mathcal{F}_4$ be defined as:

$$\mathcal{F}_3 = \{[0, \frac{1}{3}), [\frac{1}{3}, \frac{5}{6}), [\frac{5}{6}, 1), [0, \frac{5}{6}), [\frac{1}{3}, 1), [0, \frac{1}{3}) \cup [\frac{5}{6}, 1), \Omega, \emptyset\},$$
$$\mathcal{F}_4 = \sigma(\mathcal{F}_2 \cup \mathcal{F}_3).$$

Let $X_1 = \frac{1}{2}, X_2 = \chi_{[0,1/2)}, X_3 = \chi_{[1/3,5/6)}$ and $X_4 = 3\chi_{[1/3,1/2)}$. Then $\{X_i, \mathcal{F}_i, i \in I\}$ is a real (bounded) martingale. However if $\lambda = \frac{9}{10}$, then one finds that $\lambda P(A_\lambda) = \lambda P\left[\max_i X_i > \lambda\right] = \frac{3}{4} > \int_{A_\lambda} X_4 dP = \frac{1}{2}$, which contradicts (10). Thus the linear ordering hypothesis cannot be dropped from the statement of Proposition 4.

In Theorem 2(i) we saw that $\mathcal{C} \subset \Sigma$ if I has a largest element "∞". In general let $\mathcal{B} = \mathcal{C} \cap \Sigma$ so that $\mathcal{B}$ is a σ-subalgebra of Σ, where $\mathcal{C}$ is determined by the mapping T. Then $\mathcal{B}$ is called the σ-algebra of *events that occur prior to the instant* T, and denoted $\mathcal{B}_T$ or $\mathcal{B}(T)$. Thus, since $[T \le i] \in \mathcal{F}_i \subset \Sigma$ and $A \cap [T \le i] \in \mathcal{F}_i$ with $A \in \mathcal{C}, A \in \mathcal{B}_T$ iff $A \in \Sigma$. This may be stated as:

$$\mathcal{B}_T = \{A \in \Sigma : A \cap [T \le i] \in \mathcal{F}_i, i \in I\}. \tag{18}$$

Since $\mathcal{G}_T \subset \mathcal{C} \cap \Sigma = \mathcal{B}_T$, our earlier results hold if $\mathcal{G}_i$ is replaced by $\mathcal{B}_j (= \mathcal{B}(T_j))$. Taking $A_a = [T \le a] \in \mathcal{F}_a \subset \Sigma$, it follows that $A_a \cap \{\omega : T(\omega) \le b\} \in \mathcal{F}_b, b \in I$ so that $A_a \in \mathcal{B}_T$. Thus if $I = \mathbb{R}$ (since rationals are dense in I), we see that a stopping time T is always measurable for $\mathcal{B}_T$. Note however that T may not be measurable for $\mathcal{G}_T$ when it is strictly smaller than $\mathcal{B}_T$. If I is *countable*, then it is easily seen that $\mathcal{G}_T = \mathcal{B}_T$ can be taken. We thus use $\mathcal{B}_T$ in lieu of $\mathcal{G}_T$ hereafter. Regarding the family $\{\mathcal{B}_j, j \in J\}$ of $\{T_j, j \in J\}$ we establish a useful result, following Hunt [1].

7. Theorem. *Let $\{\mathcal{F}_i, i \in I\}$ be a directed family of filtering (to the right) σ-algebras of (Ω, Σ, P). Let I be countable and linearly ordered. If $\{X_i, \mathcal{F}_i, i \in I\}$ is an integrable submartingale, and $\{T_j, j \in J\}$ is a stopping time process relative to $\{\mathcal{F}_i, i \in I\}$, and $Y_j = X \circ T_j, \mathcal{B}_j = \mathcal{B}(T_j)$, then one has:*

(i) *$T_1 \le T_2$ implies $\mathcal{B}_1 \subset \mathcal{B}_2$.*

(ii) *If each* T_j *is bounded by an element* $\alpha \in I(\alpha = \alpha_j)$, *then* $\{Y_j, \mathcal{B}_j, j \in J\}$ *is a submartingale, and a martingale if the original* X*-process is such. Here* $E(Y_j^+) < \infty$ *for* $j \in J$, *but* $E(|Y_j|) = \infty$ *is possible, where as usual* $E(\cdot)$ *is the expectation symbol.*

Proof. Since Theorem 2(i) implies $\mathcal{C}_1 \subset \mathcal{C}_2$ for $T_1 \le T_2$, (i) is immediate and we only need to prove (ii) for which it suffices to take $J = \{j_1, j_2\}$, and consider T_1, T_2.

We can and do assume that $T_1 \le T_2 \le \beta$ for some $\beta \in I$, and show

$$\int_A Y_1 dP \le \int_A Y_2 dP, \qquad A \in \mathcal{B}_1, \tag{19}$$

with equality in the martingale case where $Y_i = X \circ T_i, i = 1, 2$. In fact, since I is countable, one has for $i \in I$,

$$\begin{aligned}\int_A Y_1 dP &= \sum_{i \le \beta} \int_{A \cap [T_1 = i]} Y_1 dP = \sum_{i \le \beta} \int_{A \cap [T_1 = i]} X_i dP \\ &\le \sum_{i \le \beta} \int_{A \cap [T_1 = i]} X_\beta dP = \int_A X_\beta dP, \end{aligned}\tag{20}$$

since $A \cap [T = i] \in \mathcal{C}_i$ by (18) and the X_i-process is a submartingale.

It is evident that (20) implies the integrability of Y_1^+. As in the proof of Proposition 4, there exist $I_n \uparrow I_\beta$ where $I_n = \{\alpha_1, \dots, \alpha_n\}$ is a finite linearly ordered set and where $I_\beta = \{i \in I : i \le \beta\}$. If $T_1^n(\omega) = T_1(\omega)$ when $T_1(\omega) \in I_n$, and $= \beta$ if not, then T_1^n is finitely valued and for each $\alpha \in I$,

$$[T_1^n \le \alpha] = \begin{cases} [T_1 \le \alpha] \in \mathcal{F}_\alpha, & \text{if } \alpha \in I_n \\ \emptyset \in \mathcal{F}_\alpha, & \text{if } \alpha \notin I_n \end{cases},$$

so that T_1^n is a stopping time of $\{\mathcal{F}_i, i \in I\}$. Since $T_1^n \le T_1$ we deduce that $\mathcal{B}(T_1^n) \subset \mathcal{B}_1$ by (i). Also for $A \in \mathcal{B}_1$, with $Y_1^n = X \circ T_1^n$,

$$\begin{aligned}\int_A Y_1^n dP &= \sum_{\alpha \in I_n} \int_{A \cap [T_1^n = \alpha]} X_\alpha dP \\ &= \sum_{\alpha \in I_n} \int_{A \cap [T_1 = \alpha]} X \circ T_1 dP = \int_{\bigcup_{\alpha \in I_n} A \cap [T_1 = \alpha]} Y_1 dP. \end{aligned}\tag{21}$$

Since Y_1^+ is integrable, we may let $n \to \infty$ to obtain from (21)

$$\lim_{n\to\infty} \int_A Y_1^n dP = \int_A Y_1 dP, \quad A \in \mathcal{B}_1. \tag{22}$$

Note that (22) is also true if Y_1 is replaced by Y_2 because $T_2 \leq \beta$.

Let $T_i^n(\omega) = T_i(\omega)$ for $T_i(\omega) \in I_n$, and $= \beta$ if not, $i = 1, 2$, where the I_n are as in the above computation. Then $T_1^n \leq T_2^n \leq \beta$ and $T_i^n, i = 1, 2$ are stopping times of $\{\mathcal{F}_i, i \in I\}$. By Theorem 2(ii), we deduce that (on comparing with sets of (18)) the process $\{Y_i^n, \mathcal{B}(T_i^n)\}_{i=1}^2$ is a submartingale. Hence

$$\int_A Y_1^n dP \leq \int_A Y_2^n dP, \quad A \in \mathcal{B}(T_1^n) \subset \mathcal{B}(T_2^n), \mathcal{B}(T_1^n) \uparrow \mathcal{B}_1. \tag{23}$$

Now letting $n \to \infty$ in (23) and using (22) we get (19). Since all the inequalities are equalities for the martingale case, it follows that $\{Y_j, \mathcal{B}_j, j \in J\}$ is a submartingale or a martingale according as the X-process. □

The following consequence is used in applications.

8. Corollary. *Let $\{X_i, \mathcal{F}_i, i \in I\}$ be a submartingale on (Ω, Σ, P) as in the theorem. If T is any stopping time of $\{\mathcal{F}_i, i \in I\}$ and $T_n = T \wedge i_n$ where $i_n \in I$ and $i_n < i_{n+1}, n \geq 1$, let $Y_n = X \circ T_n$. Then $\{Y_n, \mathcal{B}(T_n), n \geq 1\}$ is a submartingale, or a martingale if the X-process is such. In particular if $I = \mathbb{N} = \{1, 2, \ldots, \}, i_n = n$, we can even assert that $\{Y_n, \mathcal{F}_n, n \geq 1\}$ is a (sub-)martingale and moreover*

$$\int_\Omega X_1 dP \leq \int_\Omega Y_n dP \leq \int_\Omega X_n dP, \qquad n \geq 1, \tag{24}$$

with equality in the martingale case.

Proof. Only the last (special) case needs a proof. Since $T_1 = 1$ here $X_1 = Y_1$ a.e., and the first inequality of (24) is true. Regarding the second,

$$Y_n = X \circ T_n = \sum_{i=1}^{n-1} X_i \chi_{[T=i]} + X_n \chi_{[T \geq n]} = \sum_{i=1}^{n} (X_i - X_{i-1}) \chi_{[T \geq i]}.$$

Since $[T \geq i] \in \mathcal{F}_i$, this gives

$$\int_\Omega Y_n dP = \sum_{i=1}^{n-1} \int_{[T=i]} X_i dP + \int_{[T \geq n]} X_n dP \leq \int_\Omega X_n dP, \tag{25}$$

by the submartingale property of X_i's. So (24) is true. That Y_n is $\mathcal{F}_n$-adapted is clear. Also,

$$E^{\mathcal{F}_n}(Y_{n+1}) = \sum_{i=1}^{n}(X_i - X_{i-1})\chi_{[T \geq n]} + E^{\mathcal{F}_n}((X_n - X_{n-1})\chi_{[T \geq n]}) \geq Y_n, \text{ a.e.,}$$

since $E^{\mathcal{F}_n}[(X_n - X_{n-1})\chi_{(T \geq n)}] = \chi_{[T \geq n]} E^{\mathcal{F}_n}(X_n - X_{n-1}) \geq 0$ a.e., with equality in the martingale case both here and in (24). □

We remark that the hypothesis that T_j's are bounded by elements of I is needed for the integrability of various functions in Theorem 7. Otherwise, for instance, the random variables Y_j^+ need not be integrable,as seen from:

9. Example. Let $\{X_n, \mathcal{F}_n, n \geq 1\}$ be a submartingale on (Ω, Σ, P) and $T : \Omega \to \mathbb{N}$ be a stopping time of $\{\mathcal{F}_n, n \in \mathbb{N}\}$. If $Y = X \circ T$, then

$$\int_\Omega Y \, dP = \sum_{n=1}^{\infty} \int_{[T=n]} X_n dP, \tag{26}$$

by definition. If $X_n = n$, a.e., then $\{X_n, \mathcal{F}_n, n \geq 1\}$ is clearly a submartingale. Let $T : \Omega \to \mathbb{N}$ be defined by $P[T = n] = \frac{c}{n^2}, (c = \frac{6}{\pi^2})$. Then $\int_\Omega Y \, dP = \sum_{n=1}^{\infty} n \cdot \frac{c}{n^2} = \infty$ in (26). Here $Y \geq 0$, and is not integrable even though each X_n is bounded. The situation is not improved for martingales. In fact, let (Ω, Σ, P) be the Lebesgue unit interval, $A_n = \left[\frac{1}{2} - \frac{1}{2^{n+1}}, \frac{1}{2}\right)$ and $\mathcal{F}_n = \sigma\left\{\left(\frac{j}{2^{n+1}}, \frac{j+1}{2^{n+1}}\right], j = 1, \ldots, 2^{n+1} - 2, \left[0, \frac{1}{2^{n+1}}\right], A_n\right\}$. If $X_n = 2^{n+1}\chi_{A_n}$ and $T = \sum_{n=1}^{\infty} n\chi_{B_n}$, where the B_n are disjoint intervals such that $B_n \supset A_n$, $\sum_{n=1}^{\infty} P(B_n) = 1$, and $\sum_{n=1}^{\infty} nP(B_n) = \infty$. [For instance, set $B_n = \left[\frac{1}{2} - \frac{\alpha}{n^2}, \frac{1}{2}\right)$ and adjust α.] Then $\{X_n, \mathcal{F}_n, n \geq 1\}$ is a nonnegative martingale, but $\int_\Omega Y \, dP = \infty$.

The above example shows that first one should try to obtain results for locally integrable (sub-) martingales (the above two are such processes), and second one should consider a subclass of these processes $\{X_n, n \geq 1\}$ so that the transformed ones are integrable for any stopping time process. The following result contains some useful information on the second point.

10. Proposition. *Let $\{X_i, \mathcal{F}_i, i \in I\}$ be an integrable submartingale on (Ω, Σ, P) where I is a countable linearly ordered index set with $a_0 = \inf\{i : i \in I\} \in I$. If $\int_\Omega |X_i| dP \leq K_0 < \infty, i \in I$, and $T : \Omega \to I$ is any stopping time of $\{\mathcal{F}_i, i \in I\}$, then $\int_\Omega |X \circ T| dP \leq 3K_0 < \infty$, and $3K_0$ can be replaced by K_0 if either $X_i \geq 0$ a.e., or the process is a martingale.*

Proof. Let $\{a, b\} \subset I$ be any elements with $a < b$. Define a "bounded" stopping time as follows. If $A_{ab} = \{\omega : a \leq T(\omega) \leq b\}, B_a = \{\omega : T(\omega) < a\}$ and $C_b = \{\omega : T(\omega) > b\}$, let $T_{ab} = T$ on $A_{ab}, = a$ on B_α, and $= b$ on C_b. Then $T_{ab} : \Omega \to [a, b]$ (an "interval"), and it is clear that T_{ab} is a stopping time of $\{\mathcal{F}_i, i \in I\}$ bounded by elements of I. Let $Y_{ab} = X \circ T_{ab}$ be the transformed random variable (see (5)) for each a and b. By Theorem 7, $\{Y_{a\alpha}, \mathcal{F}_\alpha, \alpha \in \{i : i \geq a\}\}$ is a submartingale. Hence $\{X_i^+, \mathcal{F}_i, i \in I\}$ and $\{Y_{a\alpha}^+, \mathcal{F}_\alpha, \alpha \in \{i : i \geq a\}\}$ are also submartingales by the Property II.4(ii). Since $Y_{a\alpha}^+ = X^+ \circ T_{a\alpha}$, one has (on writing $Y_{aa}^+ = X_a^+$ and $Y_{bb}^+ = X_b^+$),

$$\int_A X_a^+ dP \leq \int_A Y_{a\alpha}^+ dP \leq \int_A X_b^+ dP, \qquad A \in \mathcal{F}_a, \text{ by (2).} \tag{27}$$

Taking $A = \Omega$, we get by hypothesis that

$$\int_\Omega Y_{a\alpha}^+ dP \leq \int_\Omega X_b^+ dP \leq \int_\Omega |X_b| dP \leq K_0. \tag{28}$$

Since $|Y_{a\alpha}| = 2Y_{a\alpha}^+ - Y_{aa}$, and $\int_\Omega Y_{aa} dP \geq \int_\Omega X_a dP$ by the first half of (2) applied to the original processes, it follows that

$$\int_\Omega |Y_{a\alpha}| dP \leq 2K_0 - \int_\Omega X_a dP \leq 3K_0 < \infty. \tag{29}$$

The linearly ordered countable index set I presents no new difficulty in invoking the pointwise convergence (Theorem II.6.4) for the

Y-process because of (29). Thus if $b_0 = \sup I$, then $Y_{a\alpha} \to (X \circ T)\chi_{[T \geq a]} + X_a \chi_{B_a}$ a.e. as $\alpha \to b_0$, and this tends to $X \circ T$ a.e., as $a \to a_0$. Hence the result follows by Fatou's lemma and (29). The last statement is immediate. $\square$

Let us extend the above result for a stopping time process of a general nature. This helps in obtaining an analog of Theorem 2(ii) for directed sets.

11. Proposition. *Let $\{X_i, \mathcal{F}_i, i \in I\}$ be an integrable submartingale [supermartingale] on (Ω, Σ, P) with I as a countable linearly ordered index set. If $\{T_j, j \in J\}$ is a stopping time process for $\{\mathcal{F}_i, i \in I\}$ let $Y_j = X \circ T_j$ be integrable and*

$$\liminf_{k \uparrow b_0} \int_{[T_j > k]} X_k^+ dP = 0 \left[or \ \liminf_{k \uparrow b_0} \int_{[T_j > k]} X_k^- dP = 0 \right], \ for\ each\ j \in J, \tag{30}$$

where $b_0 = \sup I$. Then $\{Y_j, \mathcal{B}_j, j \in J\}$ is a submartingale [supermartingale], and a martingale if the original X_i-process is such and both the conditions of (30) *hold.*

Proof. First observe that $\int_\Omega |X_i| dP \leq K_0 < \infty$ implies the integrability of Y_j for each $j \in J$ by Proposition 10; and here we assumed the conclusion without the L^1-boundedness of X_i's. With (30) let us establish the result.

Since Y_j is $\mathcal{G}_j$-measurable (cf. Theorem 2 for $\mathcal{G}_j$ and other details), let $A \in \mathcal{G}_j$ and $j < j'$. It is sufficient to show that

$$\int_A Y_j dP \leq \int_A Y_{j'} dP, \qquad A \in \mathcal{G}_j = \mathcal{B}_j. \tag{31}$$

(See (9) and (18).) For each $\alpha \in I$, define $A^\alpha = A \cap [T_i > \alpha]$, and $B_\alpha = A \cap [T_{j'} = \alpha]$. Since $T_j \leq T_{j'}$ we see from $A = \bigcup_{\alpha \in I} A_\alpha$, where $A_\alpha = A \cap [T_j = \alpha]$, that T_j and $T_{j'}$ are constant on A_α. Now using a computation similar to that of (9) one has:

$$\int_A Y_j dP \leq \int_{\bigcup_{\alpha_0 \leq \alpha \leq \beta} A_\alpha} Y_{j'} dP + \int_{A \cap [T_j > \beta]} X_\beta^+ dP. \tag{32}$$

By the σ-additivity of the integral, the first term on the right tends to the right term of (31) as $\beta \uparrow b_0$. Hence taking "$\liminf_{\beta \uparrow b_0}$" on both sides of (32) and using (30), we get (31). The remaining statements are similar. We note that if only one of the conditions of (30) holds and the X_i-process is a martingale, then by (32) the Y_j-process will only be a sub- [super-] martingale. □

The following sufficient condition has some interest in applications.

12. Corollary. *Let $\{X_i, \mathcal{F}_i, i \in I\}$ be a submartingale on (Ω, Σ, P) and I, a linearly ordered countable index set. Suppose that there is an integrable random variable Z such that $X_i \le E^{\mathcal{F}_i}(Z)$* a.e., *$i \in I$. Then for any stopping time process $\{T_j, j \in J\}$ of $\{\mathcal{F}_i, i \in I\}$, with $Y_j = X \circ T_j$, the process $\{Y_j, \mathcal{B}_j, j \in J\}$ is a submartingale (and a supermartingale if the inequality is reversed).*

Proof. It suffices to check that condition (30) is implied. Let $i_0 \in I$. First we conclude, by the submartingale property and the hypothesis, that the X_i-process is closed on the right by Z. If $\mathcal{F}_\infty = \sigma\left(\bigcup_{i \in I} \mathcal{F}\right)$, then $\{X_i, \mathcal{F}_i, i \in I, E^{\mathcal{F}_\infty}(Z), \mathcal{F}_\infty\}$ is a submartingale. Hence also $\{X_i^+, \mathcal{F}_i, i \in I, E^{\mathcal{F}_\infty}(Z^+), \mathcal{F}_\infty\}$ is a submartingale by Property II.4(ii). Moreover for $i_0 \in I$ and $i \ge i_0$,

$$E(|X_i|) \le 2E(X_i^+) - E(X_i) \le 2E(Z^+) - E(X_{i_0}) = K_0 < \infty. \quad (33)$$

Hence by Proposition 10, Y_j is integrable for each $j \in J$, and if $b_0 = \sup I$, then $\lim_{i \uparrow b_0} P[T_j > i] = 0$ (because T_j takes values in I). Thus

$$\liminf_{i \uparrow b_0} \int_{[T_j > i]} X_i^+ dP \le \liminf_{i \uparrow b_0} \int_{[T_j > i]} Z^+ dP = 0.$$

for each $j \in J$. Hence (30) holds as desired. □

With the properties of stopping times obtained thus far, we can extend the result of Theorem 2(ii) to directed index sets for martingales in the following form. The method is a model for similar extensions.

13. Theorem. *Let $\{X_i, \mathcal{F}_i, i \in I\}$ be a martingale on (Ω, Σ, P) where I is a countable directed index set. If $\{T_j, j \in J\}$ is a stopping time* *process of* $\{\mathcal{F}_i, i \in I\}$*, suppose that for each linearly ordered subset* $I_1 = \{\delta_n\}_1^\infty$ *of* I *and any stopping time* $T_0 : \Omega \to I_1$ *of* $\{\mathcal{F}_i, i \in I_1\}$ *the following conditions hold:*

$$\limsup_{n\to\infty} \int_{[T_0>\delta_n]} |X_{\delta_n}| dP < \infty, \quad \liminf_{n\to\infty} \int_{[T_0>\delta_n]} |X_{\delta_n}| dP = 0. \tag{34}$$

If $Y_j = X \circ T_j$ *is integrable for each* j*, then* $\{Y_j, \mathcal{B}_j, j \in J\}$ *is a martingale. In particular the* Y_j*-process is always a martingale under* (34) *when* $\int_\Omega |X_j| dP \le K_0 < \infty, i \in I$.

Proof. Since each $Y_j = X \circ T_j$ is integrable, we have to show that

$$\int_A Y_j dP = \int_A Y_{j'} dP, \qquad A \in \mathcal{B}_j, \ \ j < j'.$$

But $A = \bigcup_{\alpha \in I} A \cap [T_j = \alpha]$ is a disjoint countable union, and $B = A \cap [T_j = \alpha] \in \mathcal{F}_\alpha$ so that it suffices to establish this if A is replaced by B for any fixed but arbitrary $\alpha \in I$.

Let $I_\alpha = \{i \in I : i \ge \alpha\}$. Then I_α is countable. Since on $B, Y_j = X_\alpha$, we may express the above integrals as

$$\int_B X_\alpha dP = \int_B Y_{j'} dP = \sum_{i \in I_\alpha} \int_{B\cap[T_{j'}=i]} X_i dP. \tag{35}$$

To prove (35), let us label I_α as $\{\alpha_n : \alpha_n \ge \alpha, n \ge 1\}$ since it is countable. Choose $\beta_1 \in I$ such that $\beta_1 \ge \alpha_1$, and then, by induction if $\beta_1 < \beta_2 \cdots < \beta_k$ are selected such that $\beta_j \ge \alpha_j$, let $\beta_{k+1} \in I$ be an element, if any, such that $\beta_{k+1} > \{\beta_k, \alpha_{k+1}\}$ and we stop the selection if there is no such β_{k+1}. Thus we have $I'_\alpha = \{\alpha < \beta_1 < \beta_2 < \cdots\} \subset I$, and I'_α is a linearly ordered set with $\alpha_n \le \beta_n$ for each $n \ge 1$.

Define a mapping $T_0 : \Omega \to I'_\alpha$ as follows. If $E_0 = \{\omega : T_{j'}(\omega) \ne \alpha\}$ and $E_k = \{\omega : T_{j'}(\omega) = \alpha_j, j \ge 1 \text{ and } \alpha_j < \beta_k\}$, then let

$$T_0(\omega) = \begin{cases} \alpha & \text{for } \omega \in E_0, \\ \beta_k & \text{for } \omega \in E_k, k \ge 1. \end{cases}$$

It is clear that T_0 is a stopping time of $\{\mathcal{F}_{\beta_k}, k \ge 1\}$ and $T_0 \ge \alpha$. If $Y_0 = X \circ T_0$ then it is integrable. In fact for any $\beta \in I'_\alpha$, since the

$|X_i|$-process is a submartingale, we have:

$$\begin{aligned}\int_{\Omega} |Y_0|dP &= \int_{[\alpha<T_0\leq\beta]} |Y_0|dP + \int_{[T_0>\beta]} |Y_0|dP, \\ &= \sum_{\beta_k\leq\beta} \int_{[T_0=\beta_k]} |X_{\beta_k}|dP + \int_{[T_0>\beta]} |Y_0|dP, \\ &\leq \sum_{\beta_k\leq\beta} \int_{[T_0=\beta_k]} |X_{\beta}|dP + \int_{[T_0>\beta]} |Y_0|dP \\ &= \int_{[\alpha<T_0\leq\beta]} |X_{\beta}|dP + \int_{[T_0>\beta]} |Y_0|dP.\end{aligned}$$

This is finite if the last term is such. To prove this, suppose the contrary. Then there exists an $\epsilon > 0$, and a stopping time $\tau_0 : \Omega \to I'_\alpha$ of $\{\mathcal{F}_{\beta_k}, k \geq 1\}$ and a sequence $\{\delta_k\}_1^\infty \subset I'_\alpha$ such that $\int_{[\delta_k<\tau_0\leq\delta_{k+1}]} |Y_0|dP \geq k\epsilon$. Now for any $\delta < \delta'$ in I'_α we have

$$\begin{aligned}\int_{[\delta\leq\tau_0<\delta']} |Y_0|dP &= \sum_{\delta\leq\beta_k<\delta'} \int_{[\tau_0=\beta_k]} |X_{\beta_k}|dP \leq \sum_{\delta\leq\beta_k<\delta'} \int_{[\tau_0=\beta_k]} |X_{\delta'}|dP, \\ &= \int_{[\delta\leq\tau_0<\delta']} |X_{\delta'}|dP, \quad \text{by the submartingale property} \\ &\qquad\qquad \text{of } |X_i|\text{'s.} \end{aligned} \tag{36}$$

Taking $\delta = \delta_k$ and $\delta' = \delta_{k+1}$, we thus get the last term of (36) to be $\geq \epsilon k > 0$. But, if $\tau' : \Omega \to I'_\alpha$ is defined as

$$\tau'(\omega) = \begin{cases} \delta_1, & \text{if } \omega \in [\tau_0 < \delta_1] \\ \delta_{k+1}, & \text{if } \delta \in [\delta_k \leq \tau_0 < \delta_{k+1}], \end{cases} \quad k \geq 1,$$

then it is a stopping time of $\{\mathcal{F}_{\beta_k}, k \geq 1\}$, and (36) implies

$$0 < k\epsilon \leq \int_{[\delta_k\leq\tau_0<\delta_{k+1}]} |X_{\delta'}|dP = \int_{[\tau'=\delta_{k+1}]} |X_{\delta_{k+1}}|dP \leq \int_{[\tau'\geq\delta_{k+1}]} |X_{\delta_{k+1}}|dP.$$

Since k is arbitrary, the "lim sup" of the right side must be infinite and this contradicts the first condition of (34). Hence our supposition is false and $\int_\Omega |Y_0|dP < \infty$. Since $T_0 > \alpha$, considering α as a constant stopping time, we can apply Proposition 11, using the conditions in

(34), to conclude that the pair $\{X_\alpha, Y_0\}$ is a martingale for $\{\mathcal{F}_\alpha, \mathcal{B}_0\}$. Thus

$$\int_B X_\alpha dP = \int_B Y_0 dP. \tag{37}$$

On the other hand, using the martingale property of X_j's,

$$\begin{aligned}\int_B Y_{j'} dP &= \sum_{n=1}^{\infty} \int_{B\cap[T_{j'}=\alpha_n]} X_{\alpha_n} dP = \sum_{n=1}^{\infty} \int_{B\cap[T_{j'}=\alpha_n]} X_{\beta_n} dP, \ (\alpha_n < \beta_n),\\ &= \sum_{n=1}^{\infty} \int_{B\cap[T_0=\beta_n]} X_{\beta_n} dP, \quad \text{since } T_0 = \beta_n \text{ on } E_n,\\ &= \sum_{k=1}^{\infty} \int_{B\cap[T_0=\beta_k]} (X\circ T_0) dP = \int_B Y_0 dP = \int_B X_\alpha dP, \quad \text{by (37)}.\end{aligned}$$

This gives (35) as desired. □

Remark. As we noted in Example 3 above, this result need not be true for submartingales when I is only directed. Thus there is a sharp division between martingales and submartingales in the case of a general index (in addition to the distinctions when the range space is a vector lattice and $I = \mathbb{N}$, cf., Theorem II.6.4 for a simple case).

The following consequence of Proposition 11 and the above result is useful.

14. Corollary. *Let $\{X_i, \mathcal{F}_i, i \in I\}$ be a martingale on (Ω, Σ, P) where I is a countable directed index set. Let $\{X_i\}_{i\in I} \subset L^1(\Omega, \Sigma, P)$ be in a ball. Suppose for each linearly ordered set $I_1 = \{\delta_n\}_1^\infty \subset I$ and any stopping time process $\{T_j, j \in J\}$ of $\{\mathcal{F}_i, i \in I_1\}$ we have*

$$\liminf_{n\to\infty} \int_{[T_j > \delta_n]} X_{\delta_n}^+ dP = 0, \qquad j \in J. \tag{38}$$

Then $\{Y_j, \mathcal{B}_j, j \in J\}$ is a martingale where $Y_j = X \circ T_j$. In particular if each T_j is bounded (above) by an element of I, or the X-process is uniformly integrable, then the Y_j-process is a martingale without further conditions.

It is clear that the stopping time transformations allow us to produce new processes from the old ones, and sometimes to refine the theory by considering a subclass of the processes in question. To use these mappings effectively, we specialize I to be a subset of $\mathbb{R}$, and consider several properties in the next section. A more detailed analysis of processes with directed index sets, using stopping times effectively, is found in Edgar and Sucheston [2].

4.2 A calculus of real valued stopping times

To obtain finer results with a stopping time process $\{T_j, j \in J\}$ of a family $\{\mathcal{F}_i, i \in I\}$ of σ-algebras and to relax the countability hypothesis on I, it is desirable to have some topology and order for I. If $J \subset I$ then we may also discuss the continuity of T_j in $j \in J$. Hence *we shall take $I \subset \bar{\mathbb{R}}$ for the work below.*

Recall that a net $\{\mathcal{F}_t, t \in I\}$ of σ-subalgebras of (Ω, Σ) is *right (left) continuous* iff for each $t, \mathcal{F}_t = \mathcal{F}_{t+} \left(= \bigcap_{s>t} \mathcal{F}_s\right)$ (respectively, $\mathcal{F}_t = \mathcal{F}_{t-} = \sigma\left(\bigcup_{s<t} \mathcal{F}_s\right)$). In general $\mathcal{F}_{t-} \subset \mathcal{F}_t \subset \mathcal{F}_{t+}$ with a possibility of strict inclusions. When I is uncountable, one has to distinguish, in the definition of a stopping time $T: \Omega \to I$, whether one needs $[T < t] \in \mathcal{F}_t$ or $[T \leq t] \in \mathcal{F}_t$ for $t \in I$ because of the following simple observation.

1. Lemma. *Let $\{\mathcal{F}_t, t \in I\}$ be a net of σ-subalgebras of (Ω, Σ), a measurable space, and $I \subset \bar{\mathbb{R}}$ an interval. Then $[T < t] \in \mathcal{F}_t$ and $[T \leq t] \in \mathcal{F}_t$ for each $t \in I$, in the definition of a stopping time $T: \Omega \to I$ of the net, are equivalent iff the net $\{\mathcal{F}_t, t \in I\}$ is right continuous.*

Proof. Since $[T < t] = \bigcup_{n=1}^{\infty} [T \leq t - 1/n] \in \sigma\left(\mathcal{F}_{t-1/n}\right) = \mathcal{F}_{t-} \subset \mathcal{F}_t$ if the second definition is taken, it is seen that the first one follows from the second without further hypothesis. On the other hand, the first definition implies:

$$[T \leq t] = \bigcap_{n=1}^{\infty} [T < t + 1/n] \in \mathcal{F}_{t+} \supset \mathcal{F}_t,$$

and this gives the second definition iff $\mathcal{F}_{t+} = \mathcal{F}_t$. □

In view of the above lemma, if we replace the given net $\{\mathcal{F}_t, t \in I\}$ by its right continuous modification $\tilde{\mathcal{F}}_t = \mathcal{F}_{t+}$, then a stopping time T relative to $\{\tilde{\mathcal{F}}_t, t \in I\}$ is unchanged for either definition. For convenience, *hereafter we assume that the net of σ-algebras $\{\mathcal{F}_t, t \in I\}$ of a problem is a right continuous system.*

2. Proposition. *Let (Ω, Σ) be a measurable space and $\{\mathcal{F}_t, t \in I\}$ be a net of σ-subalgebras of Σ, as above. If $T : \Omega \to I$ is a stopping time of this net, $(I \subset \bar{\mathbb{R}})$ and $\sigma(T)$ is the σ-algebra generated by T, then $\sigma(T) \subset \mathcal{B}_T$ (the σ-algebra of events 'prior' to T). Moreoever the following statements are true:*

(a) *If $\alpha : \Omega \to I$ is a $\mathcal{B}_T$-measurable mapping such that $\alpha \geq T$, then α is also a stopping time of $\{\mathcal{F}_t, t \in I\}$.*

(b) *If $\{T_n, n \geq 1\}$ is a sequence of stopping times of the net $\{\mathcal{F}_t, t \in I\}$, then $\sup_n T_n, \inf_n T_n, \limsup_n T_n, \liminf_n T_n$ are all stopping times of the same net, and for all $A \in \mathcal{B}_1 (= \mathcal{B}(T_1)), A \cap [T_1 \leq T_2] \in \mathcal{B}_2 (= \mathcal{B}(T_2))$. In particular, $[T_1 < T_2], [T_2 < T_1]$ and $[T_1 = T_2]$ are in $\mathcal{B}_1 \cap \mathcal{B}_2$.*

(c) *If $I = \mathbb{R}^+$ and T_1, T_2 are stopping times of $\{\mathcal{F}_t, t \in I\}$, then $T_1 + T_2$ is also a stopping time. [Neither the difference nor scalar multiplies of stopping times need be stopping times.]*

(d) *If $T_n \downarrow T$ pointwise, and $\{T_n, n \geq 1\}$ are stopping times of $\{\mathcal{F}_t, t \in I\}, (I \subset \bar{\mathbb{R}})$, then $\mathcal{B}_T = \bigcap_{n=1}^{\infty} \mathcal{B}_n$. [$T$ is a stopping time by* (b).*]*

Proof. To see that $\sigma(T) \subset \mathcal{B}_T$, we recall that $A \cap [T \leq t] \in \mathcal{F}_t, t \in I$, for $A \in \mathcal{B}_T$. Let $A_0 = [T \leq t_0] \in \sigma(T), t_0 \in I$. Then $A_0 \in \Sigma$ and

$$A_0 \cap [T \leq t] = \begin{cases} [T \leq t_0], & \text{if } t \geq t_0, (\in \mathcal{F}_{t_0} \subset \mathcal{F}_t) \\ [T \leq t], & \text{if } t < t_0, (\in \mathcal{F}_t), \end{cases} \tag{1}$$

so that in either case this set is in $\mathcal{F}_t$. By definition of $\mathcal{B}_T$, then $A_0 \in \mathcal{B}_T$. Since $t_0 \in I$ is arbitrary and A_0 is a generator of $\sigma(T)$, we conclude that $\sigma(T) \subset \mathcal{B}_T$ and so T is $\mathcal{B}_T$-measurable.

(a) By hypothesis $[\alpha \leq t] \in \mathcal{B}_T$, for all $t \in I$. Since

$$\alpha \geq T \text{ one has } [\alpha \leq t] = [\alpha \leq t] \cap [T \leq t] \in \mathcal{F}_t, \qquad t \in I. \tag{2}$$

Hence α is a stopping time of $\{\mathcal{F}_t, t \in I\}$.

(b) The assertions follow from the relations:

$$\left[\sup_n T_n \leq t\right] = \bigcap_{i=1}^{\infty}[T_i \leq t] \in \mathcal{F}_t, \qquad t \in I.$$

$$\left[\inf_n T_n < t\right] = \bigcup_{n=1}^{\infty}[T_n < t] \in \mathcal{F}_t, \qquad t \in I, \text{ so } \left[\inf_n T_n \leq t\right] \in \mathcal{F}_t,$$

by Lemma 1. [$\mathcal{F}_t = \mathcal{F}_{t+}$ is used here.]

Since $\limsup_n T_n = \inf_k \sup_{n\geq k} T_n$ and $\liminf_n T_n = \sup_k \inf_{n\geq k} T_n$, it is clear that these are stopping times because of the above two relations.

Let $A \in \mathcal{B}_1$. Since $A_0 = A \cap [T_1 \leq T_2] \in \Sigma$, we have to show that $A_0 \cap [T_2 \leq t] \in \mathcal{F}_t$, for $t \in I$. However, writing $a \wedge b$ for $\inf(a, b)$,

$$A_0 \cap [T_2 \leq t] = (A \cap [T_1 \leq t]) \cap [T_2 \leq t] \cap [T_1 \wedge t] \in \mathcal{F}_t,$$

$t \in I$, since $[T_i \wedge t \leq t] \in \mathcal{F}_t$ and each set on the right is in $\mathcal{F}_t$. Hence $A_0 \in \mathcal{B}_2$. Taking $A = \Omega$, we get $[T_1 \leq T_2] \in \mathcal{B}_2$ and hence $[T_2 < T_1] = [T_1 \leq T_2]^c \in \mathcal{B}_2$ also. If $T = T_1 \wedge T_2$ then T is a stopping time of $\{\mathcal{F}_t, t \in I\}$ and $\mathcal{B}_T \subset \mathcal{B}_i, i = 1, 2$ by Theorem 1.7(i). Hence T, T_2 are $\mathcal{B}_2$-measurable by the first paragraph of this proof. Thus $[T_1 = T_2] = [T_2 = T] \in \mathcal{B}_2$. So $[T_1 < T_2] = [T_1 \leq T_2] - [T_1 = T_2] \in \mathcal{B}_2$. By symmetry these statements also hold if the subscripts 1 and 2 are interchanged. Hence the sets $[T_1 < T_2], [T_1 = T_2]$, and $[T_1 > T_2]$ belong to $\mathcal{B}_1 \cap \mathcal{B}_2$.

(c) Since $T_i \geq 0, i = 1, 2$, are stopping times of $\{\mathcal{F}_t, t \in I\}$,

$$[T_1 + T_2 \leq t] = \bigcap_{\substack{0\leq r\leq t \\ r-\text{rational}}} ([T_1 \leq r] \cap [T_2 \leq t - r]) \in \mathcal{F}_t, \quad t \in I, \tag{3}$$

so $T_1 + T_2$ is a stopping time of the same net $\{\mathcal{F}_t, t \in I\}$. Note that the above computation holds if $I \subset \mathbb{R}$ is a certain semigroup under addition.

(d) If $T_n \downarrow T$, then by Theorem 1.7, $\mathcal{B}_T \subset \bigcap_{n=1}^{\infty} \mathcal{B}_n$. For the oppositive inclusion, let $A \in \mathcal{B}_n$ for all n. Then

$$\bigcup_{n=1}^{\infty}(A \cap [T_n < t]) = A \cap \left[\inf_n T_n < t\right] = A \cap [T < t] \in \mathcal{B}_T,$$

by (b), and the following relation:

$$A \cap [T \leq t] = \bigcap_{n=1}^{\infty} (A \cap [T < t + 1/n]) \in \bigcap_{n=1}^{\infty} \mathcal{F}_{t+1/n} = \mathcal{F}_t, \quad t \in I, \tag{4}$$

by the right continuity of the filtration. Hence $A \in \mathcal{B}_T$. □

Remark. By the above proof, it is clear that if $\mathcal{F}_t = \mathcal{F}_{t+}$ we may define

$$\mathcal{B}_T = \{A \in \Sigma : A \cap [T < t] \in \mathcal{F}_t, t \in I\}. \tag{5}$$

Also, for any stopping time T and $\delta > 0, T + \delta$ is again a stopping time. Thus we may define $\mathcal{B}_{T+} = \bigcap_{\delta > 0} \mathcal{B}_{T+\delta} \supset \mathcal{B}_T$. Similarly $\mathcal{B}_{T-}$ and its completion $\mathcal{B}'_T$, when there is a measure, are meaningful. Some of these will be given as exercises.

Thus far measures did not enter into the discussion. With their introduction, we may consider the problem of (progressively measurable) modifications. If $\{X_t, \mathcal{F}_t, t \in I\}$ is a measurable process and if $\{T_j, j \in J\}$ is a stopping time process, we need to know, for some problems, the existence of (progressively) measurable modifications of the transformed process $\{Y_j, \mathcal{B}_j, j \in J\}$ where $Y_j = X \circ T_j$. We answer this question here. (See Definition III.3.7 regarding progressive measurability.)

3. Proposition. *Let (Ω, Σ) be a measurable space, $I \subset \mathbb{R}$ be an interval and $\{\mathcal{F}_t = \mathcal{F}_{t+}, t \in I\}$ be a net of σ-subalgebras of Σ. If $\{X_t, \mathcal{F}_t, t \in I\}$ is an adapted real stochastic process which is progressively measurable relative to $\{\mathcal{F}_t, t \in I\}$ and $\mathcal{I} \otimes \Sigma$, where $\mathcal{I}$ is the Borel σ-algebra of I, and if $\{T_j, j \in J\}$ is a stopping time process of $\{\mathcal{F}_t, t \in I\}, J \subset \mathbb{R}$, with $\lim_{j \downarrow j_0} T_j = T_{j_0}$ (or $T_j = T_{j+}, j \in J$) for each $j_0 \in J$, then $\{Y_j = X \circ T_j, \mathcal{B}'_j, j \in J\}$ is progressively measurable for $\mathcal{J}(J) \bigotimes \mathcal{B}_j$ where $\mathcal{B}_j = \mathcal{B}(T_j)$ and $\mathcal{J}(J)$ is the Borel σ-algebra of J. Moreover, if P is a probability on Σ and μ is the Lebesgue measure on $\mathbb{R}$ and the σ-algebras are completed for the respective measures P, μ and $\mu \bigotimes P$, then $\{Y_j, \mathcal{B}_j, j \in J\}$ also admits a (progressively measurable and) separable modification for the family $\{\mathcal{B}_j, j \in J\}$, even if $\{X_t, \mathcal{F}_t, t \in I\}$ is only an adapted process. In particular, if the X_t-process is a uniformly integrable martingale, then the Y_j-process has also this additional property. [This is an optional sampling property.]*

Proof. Since I is separable, it follows by Theorem 1.2 that $\{Y_j, \mathcal{B}_j, j \in J\}$ is an adapted process on (Ω, Σ). Also by the preceding proposition, T_j is $\mathcal{B}_j$-measurable and if μ and P are measures as in the statment, then $\{Y_j, \mathcal{B}_j, j \in J\}$ satisfies the conditions of Theorem III.3.8, and hence it admits a progressivley measurable and separable (relative to closed sets of $\mathbb{R}$) modification. Thus the result is true in this case.

We now consider the problem without measures, and show that $\{Y_j, j \in J\}$ is progressively measurable for $\{\mathcal{B}_j, j \in J\}$ if $\{X_t, t \in I\}$ is such for the family $\{\mathcal{F}_t = \mathcal{F}_{t+}, t \in I\}$. Let T_{j_0} be a fixed but arbitrary stopping time of $\{\mathcal{F}_t, t \in I\}$ and let $S_t = t \wedge T_{j_0}, t \in I$. Then S_t is a stopping time of $\{\mathcal{F}_t, t \in I\}$ (Proposition 2) and $\tilde{Y}_t = X \circ S_t = X_{t \wedge T_{j_0}}$ is $\mathcal{B}(S_t)$-measurable. But $S_t \leq T_{j_0}$ so that $\mathcal{B}(S_t) \subset \mathcal{B}_{j_0} = \mathcal{B}(T_{j_0})$. We now establish that (i) $\{\tilde{Y}_t, t \in I\}$ is $\mathcal{I} \bigotimes \mathcal{B}_{j_0}$-adapted, and with this (ii) $\{Y_j, j \in A\}$ is $\mathcal{I}(A) \bigotimes \mathcal{B}_A$-measurable for any (Borel) $A \subset J$, where $\mathcal{B}_A = \sigma\left(\bigcup_{j \in A} \mathcal{B}_j\right)$. This will prove the result.

Consider the stochastic function $\tilde{Y} : I \times \Omega \to \mathbb{R}$. For any Borel set $B \subset \mathbb{R}$, we have the decomposition:

$$\begin{aligned}\{(t, \omega) : \tilde{Y}(t, \omega) &= X_{t \wedge T_{j_0}(\omega)}(\omega) \in B\} \\ &= \{(t, \omega) : t < T_{j_0}(\omega), X_t(\omega) \in B\} \\ &\cup \{(t, \omega) : t \geq T_{j_0}(\omega), X_{T_{j_0}(\omega)}(\omega) \in B\}. \qquad (6)\end{aligned}$$

Each of the right side sets in (6) can be expressed as follows:

$$\begin{aligned}&\{(t, \omega) : t < T_{j_0}(\omega), X_t(\omega) \in B\} \\ &= \bigcup_{\substack{t < s \in I \\ s-\text{rational}}} \{(t, \omega) : s < T_{j_0}(\omega), X_t(\omega) \in B\} \\ &= \bigcup_{\substack{t < s \in I \\ s-\text{rational}}} [I \times ([T_{j_0} > s] \cap \{\omega : X_t(\omega) \in B\})].\end{aligned}$$

Since $s > t$ implies $\mathcal{F}_t \subset \mathcal{F}_s$ and $[T_{j_0} > s] \in \mathcal{F}_s$, the right side set is in $\mathcal{I} \bigotimes \mathcal{B}_{j_0}$. Similarly we have

$$\{(t, \omega) : T_{j_0}(\omega) \leq t, (X \circ T_{j_0})(\omega) \in B\} = \{t\} \times \{[T_{j_0} \leq t] \cap [X \circ T_{j_0} \in B]\},$$

and since $\mathcal{F}_t \subset \mathcal{B}_{j_0}$ and $Y_{j_0} = X \circ T_{j_0}$ is $\mathcal{B}_{j_0}$-measurable, the above set is in $\mathcal{I} \bigotimes \mathcal{B}_{j_0}$. Hence the right side of (6) is in $\mathcal{I} \bigotimes \mathcal{B}_{j_0}$ and this proves (i).

To prove (ii), we first claim that, by the right continuity of $\{T_j, j \in J\}$, it is a progressively measurable process on $J \times \Omega$. In fact, let $j_0 \in J$ and $B \subset I$ be a Borel set. Then

$$\begin{aligned}
&\{(j,\omega) \in J \times \Omega : j < j_0, T_j(\omega) \in B\} \\
&= \bigcup_{\substack{r<j_0 \\ r,j\in J \\ r-\text{rational}}} \{(j,\omega) : j < r, T_r(\omega) \in B\} \\
&= \bigcup_{\substack{r,j\in J \\ r-\text{rational}}} ([j < r] \cap J) \times [(T_r \wedge T_{j_0}) \in B]. \qquad (7)
\end{aligned}$$

Since $T_r \leq T_{j_0}$, and $T_r \wedge T_{j_0}$ is a stopping time of $\{\mathcal{F}_t, t \in B\}$, the last set on the right of (7) is in $\mathcal{F}_B = \sigma\left(\bigcup_{t\in B} \mathcal{F}_t\right) \subset \mathcal{B}_{j_0+}$. If $J_r = \{j \in J : j \leq r\}$, then the set of (7) is in $\mathcal{J}(J_{j_0}) \bigotimes \mathcal{B}_{j_0}$ since $\mathcal{B}_{j+} = \mathcal{B}_j$ by the right continuity of the $\{T_j, j \in J\}$-process. Hence the T_j-process is progressively measurable for $\{\mathcal{J}(J) \bigotimes \mathcal{B}_j, j \in J\}$. Thus the above progressive measurability implies that the mapping $(j,\omega) \mapsto (T_j(\omega), \omega)$ which is from $J_{j_0} \times \Omega$ to $I \times \Omega$ is $(\mathcal{J}(J_{j_0}) \bigotimes \mathcal{B}_{j_0}, \mathcal{I} \bigotimes \mathcal{B}_{j_0})$-measurable. By (i) the mapping $(s,\omega) \mapsto X_{s\wedge T_{j_0}(\omega)}(\omega)$ from $I \times \Omega$ to $\mathbb{R}$ is $(\mathcal{I} \bigotimes \mathcal{B}_{j_0}, \mathcal{R})$-measurable where $\mathcal{R}$ is the Borel σ-algebra of $\mathbb{R}$. Hence the composition $(j,\omega) \mapsto X_{T\wedge T_{j_0}(\omega)}(\omega) = Y(j,\omega)$, from $J_{j_0} \times \Omega$ to $\mathbb{R}$ is $(\mathcal{I}(J_{j_0}) \bigotimes \mathcal{B}_{j_0}, \mathcal{R})$-measurable. Thus $\{Y_j, \mathcal{B}_j, j \in J\}$ is a progressively measurable process, since $j_0 \in J$ is arbitrary. The last statement is an immediate consequence of Corollary 1.4. □

4. Remarks. 1. Since no modification of a process is possible in the absence of a measure, one needs the stronger right continuity in the hypothesis. Without this, however, we have to replace $\mathcal{B}_j$ by $\mathcal{B}_{j+}$ (and $\mathcal{F}_t = \mathcal{F}_{t+}$ is still in forece). If the σ-algebras are defined differently, then suitable forms of the above results can be obtained under a weaker hypothesis, and they are discussed in Chung and Doob [1], and Courrège and Priouret [1]. The preceding result is adapted from these sources.

2. In the above proposition, the family $\{T_j, j \in J\}$, was not restricted. The following specializations are of interest: (a) Let $T_t = T \wedge t$ for a fixed stopping time T of $\{\mathcal{F}_i, i \in I\}$ and $t \in J = I, T_t$ is called a truncation. Then $Y_t = X \circ T_t$. (b) Let $T_t = T + t$, if $I = J \subset \mathbb{R}^+$, and then T_t is called a translation of T. Thus the process $Y_t = X_{t+T}$ is

a "random shift" of the X_t-process. These two operations are also denoted by $\alpha_t(\cdot)$ and $\theta_t(\cdot)$ so that $\alpha_s(X_t) = X_t \circ \alpha_s = X_{t\wedge s}$ and $\theta_s(X_t) = X_t \circ \theta_s = X_{t+s}$. The above definitions are obtained if we consider α_T and θ_T and the $\alpha_T(X_t) - (\theta_T(X_t)-)$ process is said to be *stopped* at (or *translated by*) the random instant T. Here α_T, θ_T are used both for the point mappings (on Ω into Ω) and the induced mappings on the space of measurable functions on (Ω, Σ). Thus $X_t \circ \alpha_s(\omega) = X_{t\wedge s}(\omega)$ and $\alpha_s : X_t \mapsto X_{t\wedge s}$ are identified following a custom in classical analysis. In particular the point mappings α_s and θ_s are $(\mathcal{F}_{s\wedge t}, \mathcal{F}_t)$-and $(\mathcal{F}_{s+t}, \mathcal{F}_t)$-measurable respectively.

The preceding proposition shows that a major problem in this work is the measurability of the transformed processes under various stopping times. For a class of processes, we can present a result whose hypothesis may be verified with relative ease. If I and J are replaced by directed sets, then the corresponding statements present many difficulties. So we only consider the case that (I, J) are subsets of $(\mathbb{R}^+, \mathbb{R})$ and then obtain more detailed results. The following concept will be useful.

Let $A \subset I \times \Omega$ and $\{\mathcal{F}_t, t \in I\}$ be an increasing family of σ-algebras of Σ, as before. Then A is called a *progressively measurable set* iff (i) $A \in \mathcal{I} \bigotimes \Sigma$, and (ii) for each $B \in \mathcal{I}, \chi_B \chi_A : B \times \Omega \to \mathbb{R}$ is $(\mathcal{I}(B) \bigotimes \mathcal{F}_B, \mathcal{R})$-measurable where $\mathcal{F}_B = \sigma\left(\bigcup_{t\in B} \mathcal{F}_t\right)$ and $\mathcal{I}(B)$ is the trace of $\mathcal{I}$ on B (and $\mathcal{R}$ is the Borel σ-algebra of $\mathbb{R}$). If $A \subset I \times \Omega$ is a set then the function $D_A : \Omega \to \tilde{I} \cup \{\infty\}$ is defined on Ω as:

$$D_A(\omega) = \begin{cases} \inf\{t : (t, \omega) \in A\} & \\ +\infty, \text{ otherwise} & (\infty \geq t \in \bar{I}). \end{cases} \tag{8}$$

Then D_A is called the *debut* of A, with $\bar{I}$ denoting the closure of I. Note that, if T is a stopping time of $\{\mathcal{F}_t, t \in I\}$ and $\tilde{A} = \{(t, \omega) : T(\omega) \leq t\}$, then $\tilde{A}$ is progressively measurable, and $[D_{\tilde{A}} \leq t] = [T \leq t]$. Thus every stopping time of $\{\mathcal{F}_t, t \in I\}$ can be regarded as the (trivial) debut of a progressively measurable set, namely $\tilde{A}$. Here is a nontrivial example.

Let $I = \bar{\mathbb{R}}^+$, and $\{X_t, t \in I\}$ be a stochastic process on (Ω, Σ) with values in $(S, \mathcal{B})$, a completely regular (Baire) measurable space. If $B \in \mathcal{B}$, define $\delta_B : \Omega \to I$ and $\tau_B : \Omega \to I$, called respectively the *entry time* and the *hitting time* of B by the process, as:

$$\begin{aligned} \delta_B(\omega) &= \inf\{t \geq 0 : X_t(\omega) \in B\}, \qquad \omega \in \Omega, \\ \tau_B(\omega) &= \inf\{t > 0 : X_t(\omega) \in B\}, \qquad \omega \in \Omega. \end{aligned} \tag{9}$$

To justify the name "time" , we need to check that δ_B, τ_B are stopping times of the (natural) family $\{\mathcal{F}_t, t \in I\}$ where $\mathcal{F}_t = \sigma\{X_s, s \le t\}$. If $A = \{(t, \omega) : X_t(\omega) \in B\}$, then $\tau_B = D_A$ (or $= \delta_B$) if I is left open (closed). Thus the measurability of D_A provides a solution to the problem. Note that the mapping θ_s gives:

$$\begin{aligned} s + \delta_B \circ \theta_s(\omega) &= \inf\{t \ge s : X_t(\omega) \in B\}, \\ s + \tau_B \circ \theta_s(\omega) &= \inf\{t > s : X_t(\omega) \in B\}, \end{aligned} \tag{10}$$

so that we have

$$\lim_{s \downarrow 0}(s + \delta_B \circ \theta_s) = \lim_{s \downarrow 0}(s + \tau_B \circ \theta_s) = \tau_B, \tag{11}$$

and it suffices to show the measurability of δ_B or τ_B for each $B \in \mathcal{B}$. The measurability of D_A is deduced from the next general statement since A is a progressively measurable set in $I \times \Omega$.

In this connection we have the following result due to Blackwell and Freedman, as noted in Meyer([1], p. 248).

5. Theorem. *Let (Ω, Σ, P) be a complete probability space, and $\{\mathcal{F}_t, t \in I\}$ a filtering right continuous family of complete σ-algebras of Σ. If $A \subset I \times \Omega$ is any progressively measurable set, then the debut function D_A is a stopping time of $\{\mathcal{F}_t, t \in I\}$.*

The proof of this result depends on a number of properties of analytic sets and the Souslin operation, and the detailes are rather technical. We outline the argument in the Complements, (cf. Exercise 6.10). Here we illustrate its use on some examples of independent interest.

Let $\{X_t, t \ge 0\}$ be a real process on (Ω, Σ, P) without fixed discontinuities. For any $\varepsilon > 0$, define $T_0 = 0$ and, inductively, for each $\omega \in \Omega$, let

$$T_{n+1}(\omega) = \begin{cases} \inf\{t : t > T_n(\omega), |X_t(\omega) - X_{t-}(\omega)| > \varepsilon\} \\ +\infty, \text{ otherwise.} \end{cases} \tag{12}$$

We claim that $\{T_n, n \ge 0\}$ is a stopping time process of $\{\mathcal{F}'_t, t \in I\}$ where $\mathcal{F}_t = \sigma(X_s, s \le t)$ and $\mathcal{F}'_t$ is a completion of $\mathcal{F}_t$. This can be verified by a direct (but tedious) computation. We deduce it from the above theorem. Since the process has no fixed discontinuities, using a slight alteration (which we omit) of Theorem III.3.8, it may be assumed that the X_t-process is progressively measurable (or has such a modification). By induction, T_0 being a stopping time, suppose T_n is also. Then $A_1 = \{(t,\omega) : T_n(\omega) < t\}$ is progressively measurable (since $\mathcal{F}_{t+} = \mathcal{F}_t$ here). Also $A_2 = \{(t,\omega) : |X_t(\omega) - X_{t-}(\omega)| > \epsilon\}$ has the same property. Thus $A = A_1 \cap A_2$ is progressively measurable and $T_{n+1} = D_A$, so that T_{n+1} is a stopping time of $\{\mathcal{F}_t, t \geq 0\}$ by Theorem 5.

We also include the following result.

6. Proposition. *Let $\{\mathcal{F}_t, t \in I\}$ be a net of right continuous completed σ-algebras of (Ω, Σ, P). Let $\{T_n^i, n \geq 0\}, i = 1, 2$, be two stopping time processes of $\{\mathcal{F}_t, t \in I\}$ where $I = [a, b] \subset \bar{\mathbb{R}}^+$. Suppose further that $\lim_{n\to\infty} P[T_n^i < b] = 0, i = 1, 2$. Let $\{S_n, n \geq 0\}$ be a superposition (i.e. the upper envelope) of the two families $\{T_n^1, T_n^2, n \geq 0\}$ such that $S_n \leq S_{n+1}$. Then $\{S_n, n \geq 0\}$ is a stopping time process of $\{\mathcal{F}_t, t \in I\}$.*

Proof. By Proposition 2, $S_0 = T_0^1 \wedge T_0^2$ is a stopping time of $\{\mathcal{F}_t, t \in I\}$. Suppose that $a \leq S_0 \leq \cdots \leq S_{k-1} \leq b$ are stopping times, and use induction. Let $A_n^i = \{(t,\omega) : T_n^i(\omega) \leq t\}, i = 1, 2$. Then A_n^i is a progressively measurable set, as noted after (8). If $B = \{(t,\omega) : S_{k-1}(\omega) \leq t\}$, let $A = \bigcup_{i=1}^{2} \bigcup_{n=1}^{m} A_n^i$ where m corresponds to the used T_n^1, T_n^2 (if fewer T_n^1 than T_n^2 are used, take the remaining A_n^1's to be empty in the above symmetric union). Then both A and B are progressively measurable, and by the definition of S_k one has $[S_k < t] = [D_{A\cap B} < t]$. Since $\{\mathcal{F}_t, t \in I\}$ is complete and right continuous, it follows from Theorem 5, that the debut $D_{A\cap B}$ is measurable for $\{\mathcal{F}_t, t \in I\}$ and hence S_k is a stopping time of the same family. This completes the induction. Note that the same result holds if we consider the superposition of a finite number of stopping time processes of $\{\mathcal{F}_t, t \in I\}$. □

We proceed to an analysis of the structure of transformed martingales under stopping time transformations in a more detailed manner than that of Section 1, for the index $I \subset \mathbb{R}$.

4.3 Regularity properties of martingales

The work of the preceding two sections shows that a stochastic process may change its character under stopping time transformations and thus a martingale may not go into a martingale. However, those that remain martingales under classes of stopping time transformations must have more "regularity". We shall explore and classify such processes here. Uniform integrability plays a key role, and the work leads to the Doob decomposition (later) of continuous parameter submartingales. The main results of this section are Theorems 12 and 14 below. All the other results are needed for them, but some have independent interest.

Recall that a set $\mathcal{H} \subset L^1(\Omega, \Sigma, P)$ is uniformly integrable if it is bounded and $\lim_{P(A)\to 0} \int_A |h| dP = 0$ uniformly in $h \in \mathcal{H}$, where $P(\Omega) = 1$. We start with the case that $T : \Omega \to \mathbb{N}$, a stopping time with range in $\{1, 2, \dots\}$.

1. Proposition. *Let $X = \{X_n, \mathcal{F}_n, n \geq 1\}$ be a uniformly integrable submartingale on (Ω, Σ, P), and $\{T_j, j \in J\}$ be a collection of stopping times of $\{\mathcal{F}_n, n \in \mathbb{N}\}$. Then $\{Y_j = X \circ T_j, j \in J\}$ is a uniformly integrable family of random variables.*

Proof. The uniform integrability of the submartingale implies, by Theorem II.6.4, that $X_n \to X_\infty$, a.e., and in L^1-norm. If $U_n = E^{\mathcal{F}_n}(X_\infty)$, then letting $\mathcal{F}_\infty = \sigma\left(\bigcup_{n\geq 1} \mathcal{F}_n\right)$ we note that $\{U_n, \mathcal{F}_n, 1 \leq n \leq \infty\}$ is a martingale and $U_n \to U_\infty = X_\infty$, a.e., and in L^1-norm. Let $V_n = X_n - U_n$. Then $E^{\mathcal{F}_n}(V_{n+1}) \geq X_n - U_n = V_n$ so that $\{V_n, \mathcal{F}_n, n \geq 1\}$ is a submartingale, and $V_n \to 0$, a.e., as $n \to \infty$. Since both the X_n- and U_n-processes are uniformly integrable so is their difference, the V_n-process.

Consider $Y_j = X \circ T_j = U \circ T_j + V \circ T_j = \tilde{U}_j + \tilde{V}_j$ (say). To show that $\{Y_j, j \in J\}$ is uniformly integrable, it is clearly sufficient to show that both $\{\tilde{U}_j, j \in J\}$ and $\{\tilde{V}_j, j \in J)$ are uniformly integrable. But $E(|U_j|) \leq K_1 < \infty$ and $E(|V_j|) \leq K_2 < \infty, j \in J$. Hence by Proposition 1.10 we deduce that $E(|\tilde{U}_j|) \leq 3K_1 < \infty, E(|\tilde{V}_j|) \leq 3K_2 < \infty$, and we need only show that the indefinite integrals of $\tilde{U}_j$ and $\tilde{V}_j$ are uniformly absolutely small on P-small sets.

Let $\mathcal{B}_j = \mathcal{B}(T_j)$, the σ-algebra of events "prior to T_j". Define the constant mapping $T_\infty : \Omega \to \{\infty\}$ so that it is trivially a stopping time of $\{\mathcal{F}_n, n \leq \infty\}$ and $T_\infty \geq T_j$ for all $j \in J$. Thus $U \circ T_\infty = U_\infty = X_\infty$ a.e., by definition and the above two paragraphs. Letting $\tilde{\mathbb{N}} = \{\mathbb{N}, \infty\}$, we deduce that $\{\tilde{U}_j, \mathcal{B}_j, j \in \tilde{\mathbb{N}}\}$ is a martingale, by Theorem 1.7. Hence for each $j \in J, E^{\mathcal{B}_j}(X_\infty) = \tilde{U}_j$ a.e. so that $\{\tilde{U}_j, j \in J\}$ is uniformly integrable (by Theorem II.6.2). It remains to prove the same property for the $\tilde{V}_j$-process.

Since $V_n \to 0$ a.e. and in L^1-norm, by its uniform integrability, we can find, for any $\varepsilon > 0$, and $n_0 = n_0(\varepsilon)$, such that $|E(V_n)| < \varepsilon$ if $n \geq n_0$. Also $V_n = X_n - E^{\mathcal{F}_n}(X_\infty) \leq E^{\mathcal{F}_n}(X_\infty) - E^{\mathcal{F}_n}(X_\infty) = 0$, a.e., by the submartingale property of X_n's. Hence for any $j \in J, a < 0$, we have

$$\begin{aligned}
\left| - \int_{[\tilde{V}_j < a]} \tilde{V}_j dP \right| &= \sum_{i=1}^{n_0} \int_{[T_j = i] \cap [\tilde{V}_j < a]} -V_i dP + \int_{[T_j > n_0] \cap [\tilde{V}_j < a]} -\tilde{V}_j dP \\
&\leq \sum_{i=1}^{n_0} \int_{[T_j = i] \cap [V_i < a]} -V_i dP + \left| \int_\Omega \tilde{V}_j dP \right|, \\
&\qquad \text{since } -V_i \geq 0 \text{ a.e., } \tilde{V}_j = V_i \text{ on } [T_j = i], \text{ and} \\
&\qquad \tilde{V}_j = (V \circ T_j) \cdot \chi_{[T_j > n_0]}, \\
&\leq \sum_{i=1}^{n_0} \int_{[V_i < a]} -V_i dP + 3\varepsilon, \quad \text{since } E(-V_n) < \varepsilon, \text{ and} \\
&\qquad \text{Proposition 1.10 applies.}
\end{aligned} \tag{1}$$

Now letting $a \to -\infty$ (and then $n_0 \to \infty$), it follows that from the arbitrariness of ε, the right side is uniformly (in j) small. So $\{\tilde{V}_j, j \in J\}$ is uniformly integrable, since $P[V_i < a] \to 0$ as $a \to -\infty$. □

In the above proof we used the fact that $\mathbb{N}$ is the countable linear index set crucially in the decomposition $\tilde{V}_j = \sum_{i=1}^{n_0} V_i \chi_{[T_j = i]} + V_j \chi_{[T_j > n_0]}$. Such a procedure fails if $I \neq \mathbb{N}$. It can be shown that the conclusion of the proposition is false if $I \subseteq \mathbb{R}$ is uncountable. We now analyze this general case in more detail.

It will be useful to isolate a class in this analysis.

2. Definition. Let $\{X_t, \mathcal{F}_t, t \in I\}$ be a supermartingale on (Ω, Σ, P) and $\{T_j, j \in J\}$ be a collection of stopping times of $\{\mathcal{F}_t, t \in I\}, I \subset \mathbb{R}$. Then the X_t-process is said to be of *class* (D) if the family $\{Y_j = X \circ T_j, j \in J\}$ is uniformly integrable. The process is said to be of *class* (DL) if for each compact set $K \subset I$, the process $\{X_t, \mathcal{F}_t, t \in K\}$ is of class (D) relative to the family of all stopping times $\{\tilde{T}_j : \Omega \to K\}_{j \in J}$

of $\{\mathcal{F}_t, t \in K\}$, i.e. if $\{X \circ \tilde{T}_j\}_{j\in J}$ is uniformly integrable. Thus the class is "locally in (D)".

If $I = \mathbb{R}^+$, then $K = [0, a]$ can be taken in the above definition of (DL). By Proposition 1, if $I = \mathbb{N}$, every uniformly integrable supermartingale on (Ω, Σ, P) belongs to class (D). The following result gives a characterization of class (D) in the continuous parameter case. It is due to Johnson and Helms [1]. The class (D) was introduced by Doob motivated by the solutions of Dirichlet's problem in potential theory.

3. Proposition. *Let $\{X_t, \mathcal{F}_t, t \in I\}$ be a nonnegative supermartingale on $(\Omega, \Sigma, P), I \subset \mathbb{R}$, almost all of whose sample functions are continuous. Then the process is of class (D) iff $\lim_{n\to\infty} nP[X^* > n] = 0$, where X^* is equivalent to $\sup_{t\in I} X_t$. [The last supermum is equivalent to a measurable function by the localizability of a finite measure space (cf. Section III.2) and X^* is defined to be this random variable.]*

Proof. Suppose the process is in class (D). Then we can prove the desired condition even if the process is only right continuous and $\mathcal{F}_t = \mathcal{F}_{t+}$, or replacing $\mathcal{F}_t$ by $\mathcal{F}_{t+}$. This generality is useful.

Let X^* be as in the statement. Let $T_n(\omega) = \inf\{t : X_t(\omega) \geq n\}$, and $= \infty$ if this set is empty. Let $\mathcal{F}_\infty = \sigma\left(\bigcup_{t\in I} \mathcal{F}_t\right)$ and $\tilde{I} = I \cup \{\infty\}$. Then $\{T_n, n \geq 1\}$ is a stopping time process of the family $\{\mathcal{F}_t, t \in \tilde{I}\}$. In fact, $[T_n < t] = [D_A < t]$ for a suitable progressively measurable set $A \subset \tilde{I} \times \Omega$, where D_A is the debut of A. So by Theorem 2.5, each T_n is a stopping time of the above family and it is clear that $T_n \leq T_{n+1}$. Let $B_\lambda^n = \{\omega : (X \circ T_n)(\omega) > \lambda\}$. Since $\{X \circ T_j : j \in J\}$ is uniformly integrable, by the membership in (D), one has for $n \geq \lambda$, (since $\{X \circ T_n, \mathcal{B}(T_n), n \geq 1\}$ is a supermartingale by Proposition 1.11)

$$nP[X^* > n] \leq nP[B_n^n] = n\int_{B_n^n} dP$$

$$\leq \int_{B_\lambda^n} (X \circ T_n)dP \to 0, \qquad \text{as } \lambda \to \infty, \text{ since } P(B_\lambda^n) \to 0, \tag{2}$$

uniformly in n. This implies the direct part.

For the converse we use the full hypothesis on $\{\mathcal{F}_t, t \in I\}$. Let $\{T_n, T\}$ be a pair of stopping times of $\{\mathcal{F}_t, t \in I\}$, with T_n as defined above. If $\mathcal{B}(T)$ is the σ-algebra of events prior to T, as usual, and $S_n = T \wedge T_n \leq T$, then S_n is a stopping time and $\mathcal{B}(S_n) \subset \mathcal{B}(T)$ by Proposition 2.2. Also $\{X \circ S_n, X \circ T\}$ forms a supermartingale relative to $\{\mathcal{B}(S_n), \mathcal{B}(T)\}$ provided (i) these random variables are integrable and (ii) the condition $\liminf_{t\to\infty} \int_{[\tau>t]} X_t^- dP = 0$ for $\tau \in \{S_n, T\}$. (See Prop. 1.11.) For this we shall establish, using the sample function continuity, that the T_n exceeds any $T_j, j \in J$ for large enough n, and then deduce, after some computations, the membership of the X_t-process in (D). Here are the details.

Since the $X_t \geq 0$ (so $X_t^- = 0$) (ii) is trivial. For (i), it suffices to show that $\int_\Omega |X_t| dP \leq K_0 < \infty, t_0 \leq t \in I$, by Proposition 1.10. But, the process being a positive supermartingale, for any $t_0 \in I$ we have $\int_\Omega X_t dP \leq \int_\Omega X_{t_0} dP = K_0 < \infty$ for all $t \geq t_0, t \in I$. So (i) is also true. Hence the above process $\{X \circ S_n, X \circ T\}$ is a supermartingale for $\{\mathcal{B}(S_n), \mathcal{B}(T)\}$, so that

$$\int_{[X\circ S_n > n-1]} X \circ S_n dP \geq \int_{[X\circ S_n > n-1]} X \circ T\, dP, \qquad T \in \{T_j, j \in J\}. \tag{3}$$

Note that only right continuity of $X_t(\mathcal{F}_t)$ is used thus far.

Let us simplify (3). Since $T_n(\omega) > T(\omega)$ implies $S_n(\omega) = (T \wedge T_n)(\omega) = T(\omega)$ and then $(X \circ T)(\omega) = (X \circ S_n)(\omega) < n$ (by the definition of $T_n, X_t(\omega) > n$), so that for (3)

$$\begin{aligned}\int_{[X\circ S_n > n-1]} X \circ T dP &\leq \int_{[X\circ S_n > n-1]\cap[T<T_n]} X \circ S_n dP \\ &\quad + \int_{[X\circ S_n > n-1]\cap[T\geq T_n]} X \circ S_n dP \\ &= nP[X \circ T > n-1] + \int_{[X\circ T_n > n-1]} X \circ T_n dP,\end{aligned} \tag{4}$$

where we used the facts that $S_n = T$ on the first set and $S_n = T_n$ on the second and that $X_t \geq 0$ a.e. However, one also has

$$\begin{aligned}[X \circ S_n > n-1] &= ([X \circ T > n-1] \cap [T \leq T_n]) \\ &\quad \cup ([X \circ T_n > n-1] \cap [T > T_n]) \supset ([X \circ T > n-1] \\ &\quad \cap [T \leq T_n]) \cup ([T > T_n] \cap [X \circ T > n-1]), \\ &= [X \circ T > n],\end{aligned} \tag{5}$$

because $(X \circ T_n)(\omega) > n-1$ implies $(X \circ T)(\omega) \geq (X \circ T_n)(\omega) \geq n > n-1$ on $[T > T_n]$. Thus (4) and (5) yield, for (3), with the nonnegativity of X_t's,

$$\int_{[X \circ T > n-1]} X \circ T dP \leq nP[X^* > n-1] + \int_{[X \circ T_n > n-1]} X \circ T_n dP, \quad (6)$$

since $[X \circ T > n-1] \subseteq \bigcup_{\substack{t \in I \\ t-\text{rational}}} [X_t > n-1] \subseteq [X^* > n-1]$. By the supermartingale property of the X_t's and the work of Section III.5(a) specialized to the index set $I \subset \mathbb{R}$ (with separability), the now classical theory of discrete indexed (i.e. $I \subset \mathbb{N}$) martingale convergence results of Section II.6 apply. Thus $X \to X_\infty$ a.e. and $X_\infty \geq 0, \int_A X_t dP \geq \int_A X_\infty dP$, $A \in \mathcal{F}_t$. Hence on $[T = \infty], X \circ T = X_\infty$ so that (6) yields, if $B_n = [X \circ T_n > n-1]$,

$$\int_{[X \circ T > n-1]} X \circ T dP \leq nP[X^* > n-1] + \int_{[T_n = t_0] \cap B_n} X \circ T_n dP$$
$$+ \int_{[B_n \cap [T_n \in I - t_0] \cap [T_n < \infty]} X \circ T_n dP + \int_{[X_\infty > n-1]} X_\infty dP. \quad (7)$$

But the sample functions $X(\cdot, \omega)$ are (right) continuous for a.a. (ω). So $t_0 < T_n(\omega) < \infty$ implies that $\inf\{t : X(t,\omega) \geq n\}$ is attained, i.e. $(X \circ T_n)(\omega) = n$. Thus noting that $[X = n] \subset [X^* \geq n]$, we have the crucial inequality from (7) as

$$\int_{[X \circ T > n-1]} X \circ T dP \leq 2nP[X^* > n-1] + \int_{[X_{t_0} > n-1]} X_{t_0} dP$$
$$+ \int_{[X_\infty > n-1]} X_\infty dP. \quad (8)$$

Since the right side is independent of T and that it tends to zero as $n \to \infty$ by the integrability of X_{t_0}, X_∞ (and the hypothesis on the first term) (8) implies that $\{X \circ T_j; j \in J\}$ is a uniformly integrable family. Thus the process $\{X_t, t \in I\}$ belongs to (D). □

The above characterization of class (D) supermartingales can be stated in a more convenient form as follows in which $I = \mathbb{R}^+$.

4. Proposition. *Let $\{X_t, \mathcal{F}_t, t \in I\}$ be a nonnegative* a.e. *continuous supermartingale on a probability space (Ω, Σ, P). Let $X_t \to X_\infty$* a.e. *(This limit clearly exists.) Let $\mathcal{T} = \{T_j, j \in J\}$ be the class of all stopping times of $\{\mathcal{F}_t, t \in I\}$. Let $T_n = \inf\{t : X_t \geq n\} \in \mathcal{T}$. Then the following statements are equivalent:* (i) *$\{X_t, \mathcal{F}_t, t \in I\}$ belongs to (D).* (ii) *For any increasing sequence $\tau_n \in \mathcal{T}$, with $\tau_n \to \infty$, as $n \to \infty$, one has $\lim_n \int_{[\tau_n<\infty]}(X \circ \tau_n)dP = \int_\Omega X_\infty dP$, and* (iii) *$\lim_n \int_{[T_n<\infty]} X \circ T_n dP = \int_\Omega X_\infty dP$ for the particular $\{T_n, n \geq 1\}$ process defined above.*

Proof. (i) $\Leftrightarrow$ (iii) is established in the above proposition. To see that (i) $\Rightarrow$ (ii), note that $X_t \to X_\infty$ a.e. and $\{X \circ \tau_n, \mathcal{B}(\tau_n), n \geq 1\}$ is a nonnegative supermartingale. By (i) $\{X \circ \tau_n, n \geq 1]$ is uniformly integrable. So $X \circ \tau_n \to X_\infty$ in L^1-norm and a.e. So (ii) holds. It is clear that, on taking $\tau_n = T_n$, (ii) $\Rightarrow$ (iii). Since clearly (iii) $\Rightarrow$ (i), the result follows. □

Remark. The nonnegativity was used crucially in the above proofs. However, if this is dropped the implication (i) $\Rightarrow$ (iii) is false, as shown by an exercise. While every member of class (D) is uniformly integrable (following from definition) the exercise also shows that the converse is not true.

The following is a useful consequence for potentials. (Cf. Definition II.5.4.)

5. Corollary. *Let $\{X_t, \mathcal{F}_t, t \in \mathbb{R}^+\}$ be a right continuous potential, i.e.,* $0 \leq X_t \to 0$ a.e. *and in L^1, the X_t-process being a supermartingale. Let $T_n = \inf\{t : X_t \geq n\}$. (Thus T_n is a stopping time and $T_n < \infty$* a.e. *by Theorem 2.5 and the fundamental maximal inequalities.) Then the potential is of class (D) iff $\lim_{n\to\infty} \int_\Omega X \circ T_n dP = 0$.*

Proof. If $\{X_t, t \in \mathbb{R}^+\}$ is a class (D) potential, so that $X_t \to 0$ a.e., then $X \circ \tau_n \to 0$ a.e. for any stopping time process $\{\tau_n\}_1^\infty, \tau_n \to \infty$, of $\{\mathcal{F}_t, t \geq 0\}$. Also the membership in (D) implies the uniform integrability of $\{X \circ T, T \in \mathcal{T}\}$ where $\mathcal{T}$ is the class of all stopping times of $\{\mathcal{F}_t, t \in \mathbb{R}^+\}$. Hence by Vitali's theorem $\lim_n \int_\Omega X \circ \tau_n dP = \int_\Omega \lim_n X \circ \tau_n dP = 0$. In particular, this is true for $\{T_n, n \geq 1\}$.

Conversely, let $T \in \mathcal{T}$, and define $S_n = T$ on $[X \circ T \geq n]$, and $= \infty$ elsewhere. Then the given $T_n \leq S_n$ and hence $\{X \circ T_n, X \circ S_n\}$ is a supermartingale for $\{\mathcal{B}(T_n), \mathcal{B}(S_n)\}$ for each n, by Theorem 1.7. (We may use any countable dense subset of $\mathbb{R}^+$ in this application.) Hence $E(X \circ T_n) \geq E(X \circ S_n)$, and invoking the hypothesis

$$\int_{[X \circ T > n]} X \circ T dP \leq \int_{\Omega} X \circ S_n dP \leq E(X \circ T_n) \to 0, \qquad \text{as } n \to \infty.$$

Since the right side is independent of T, and $P[X \circ T > n] \to 0$ as $n \to \infty$, we conclude that $\{X \circ T, T \in \mathcal{T}\}$ is uniformly integrable. Hence the potential $\{X_t, t \geq 0\}$ belongs to class (D). $\square$

Remark. It follows from some later work that, if the parameter set is a compact interval, the positivity of the process can be weakened in the above proposition. Then the condition becomes: $nP\left[\sup_{t \in T} |X_t| > n\right] \to 0$ as $n \to \infty$.

It is clear that class $(D) \subset$ class (DL), from Definition 2. However, we should have some idea of the "size" of these classes. The following observations, due to Meyer, give an insight into this question.

6. Proposition. *Let (Ω, Σ, P) be a probability space and $\{\mathcal{F}_t, t \in I\}$ be a filtering right continuous family of σ-algebras, and $I \subset \mathbb{R}$. Then (i) every right continuous martingale $\{X_t, \mathcal{F}_t, t \in I\}$ belongs to class (DL), (ii) if the above martingale is also uniformly integrable then it is in class (D); (iii) every nonnegative right continuous submartingale is in class (DL), and (iv) every uniformly integrable right continuous supermartingale of class (DL) with $I = \mathbb{R}^+$, belongs to class (D).*

Proof. (i) [Let $\mathcal{T}$ be the class of all stopping times of $\{\mathcal{F}_t, t \in I\}$, where I is an interval in all cases.] Let $\{a, t_0\} \subset I$. Consider the compact set $I_a = \{t \in I : t_0 \leq t \leq a\}$. For any $T \in \mathcal{T}$ with $T \leq a, \{X \circ T, X_a\}$ is a martingale relative to $\{\mathcal{B}(T), \mathcal{B}_a = \mathcal{F}_a\}$ so that $E^{\mathcal{B}(T)}(X_a) = X \circ T$ a.e. But $\{E^{\mathcal{B}(T)}(X_a) : T \in \mathcal{T}, T \leq a\}$ is a uniformly integrable set of random variables. Since $a \in I$ and I_a are arbitrary (and compact) we deduce that the X_t-process belongs to class (DL).

(ii) If $\{X_t, t \in I\}$ is also a uniformly integrable martingale, then $X_t \to X_\infty$ a.e., and in L^1-mean. Hence we can apply the argument of (i) with $a = \infty$, in $\tilde{I} = I \cup \{\infty\}$. So $\{X_T = E^{\mathcal{B}(T)}(X_\infty) : T \in \mathcal{T}\}$ is uniformly integrable. Since this set is $\{X \circ T, T \in \mathcal{T}\}$, the process belongs to class (D) by definition.

(iii) Let $a \in I, T \in \mathcal{T}$ with $T \leq a$. Then again $\{X \circ T, X_a\}$ forms a submartingale for $\{\mathcal{B}(T), \mathcal{F}_a\}$ so that, $[X \circ T \geq n] \in \mathcal{B}(T)$ and

$$0 \leq \int_{[X \circ T \geq n]} X \circ T dP \leq \int_{[X \circ T \geq n]} X_a dP \to 0, \qquad \text{as } n \to \infty. \tag{9}$$

But $P[X \circ T \geq n] \to 0$ uniformly in T. Indeed by (9),

$$P[X \circ T \geq n] \leq \frac{1}{n} \int_{[X \circ T \geq n]} X_a dP \to 0, \qquad \text{as} \quad n \to \infty, \tag{10}$$

uniformly in $T \in \mathcal{T}$. Since $a \in I$ is arbitrary, (9) and (10) show that the X_t-process is of class (DL).

(iv) We first note that the hypothesis implies the Riesz decomposition $X_t = Y_t + Z_t$, where $\{Y_t, \mathcal{F}_t, t \in I\}$ is a martingale and $\{Z_t, \mathcal{F}_t, t \in I\}$ is a potential. In fact $X_t \to X_\infty$ a.e. and in L^1, by the uniform integrability, so that $Y_t = E^{\mathcal{F}_t}(X_\infty)$ and $Z_t = X_t - Y_t$ gives the desired decomposition since $Z_t = X_t - E^{\mathcal{F}_t}(X_\infty) \geq E^{\mathcal{F}_t}(X_\infty) = 0$ a.e., by the supermartingale property of the process. By (ii) $\{Y_t, t \in I\} \in (D)$, hence to (DL), and $\{X_t, t \in I\}$ is in (DL) by hypothesis. Now the linearity of these classes implies that the Z_t-process is in (DL). Since $\{Y_t\}_{t \in I}$ is uniformly integrable, it suffices to show that the Z_t-process is in class (D) which implies the same for the X_t-process.

Since $0 \leq Z_t \to 0$ a.e., and in L^1, as $t \to \infty$, given $\varepsilon > 0$, choose $a = t_0(\varepsilon) \in I$ such that $t \geq a$ implies $\int_\Omega Z_t dP < \varepsilon$. Let $T_a = T \wedge a \in \mathcal{T}$, for $T \in \mathcal{T}$. Then $\{Z \circ T_a, Z_a\}$ forms a supermartingale for $\{\mathcal{B}(T_a), \mathcal{F}_a\}$ as before. Hence

$$\begin{aligned} \int_{[Z \circ T > n]} Z \circ T dP &= \int_{[Z \circ T > n] \cap [T \leq a]} Z \circ T\, dP + \int_{[Z \circ T > n] \cap [T > a]} Z \circ T\, dP \\ &\leq \int_{[Z \circ T_a > n]} Z \circ T_a dP + \int_{[T > a]} Z \circ T\, dP, \text{ since } Z_t \geq 0. \end{aligned} \tag{11}$$

But $[T > a] \in \mathcal{B}(T)$ and $\int_\Omega Z_t dP \leq K_0 < \infty$, for $t \geq t_0(\varepsilon)$, by the positive supermartingale property. Hence $\{Z_a, Z \circ T\}$ is a supermartingale for any $T \in \mathcal{T}$ such that $T > a$ (cf. Theorem 1.7), i.e., take $T = \sup(\tilde{T}, a)$, for $\tilde{T} \in \mathcal{T}$, which by Proposition 2.2 belongs to $\mathcal{T}$. The first term on the right of (11) is bounded by $\int_{[Z \circ T_a > n]} Z_a dP$ which goes to zero, as $n \to \infty$. So it suffices to show that the second term tends to zero also. Now

$$\int_{[T>a]} Z \circ T \, dP \leq \int_{[T>a]} Z_a dP \leq \int_{\Omega} Z_a dP \leq \varepsilon \tag{12}$$

by the choice of a. Hence as $n \to \infty$, the right side of (11) is at most ε. By the arbitrariness of $\varepsilon > 0$, and of $T \in \mathcal{T}$, this shows that the Z_t-process is in (D). Hence the X_t-process is in class class (D). □

Remark. The above proof of (iv) actually shows that a right continuous potential of class (DL) already belongs to class (D). The preceding four propositions show that for a finer analysis of the continuous parameter martingales, class (D) is a well-behaved family of processes. This will be studied further.

There is another class, called "regular martingales," introduced by Snell [1] and considered further by Chow [1]. These are related to the projective limit theory of Chapter I, and we analyze this class here, allowing the index to be a linearly ordered set.

7. Definition. Let $\{\mathcal{F}_i, i \in I\}$ be an increasing family of σ-subalgebras of a probability space (Ω, Σ, P) with I as a linearly ordered set. A submartingale $\{X_t, \mathcal{F}_t, t \in I\}$ is said to be *regular* if for every stopping time process $\{T_j, j \in J\}$ of $\{\mathcal{F}_t, t \in I\}$ the transformed process $\{Y_j = X \circ T_j, \mathcal{B}(T_j) = \mathcal{B}_j, j \in J\}$ is also a submartingale. The submartingale is *semiregular* if for any stopping time T of $\{\mathcal{F}_t, t \in I\}$ and $A_t = \{\omega : T(\omega) \geq t\}, t \in I$, it is true that $X \circ T$ is integrable and $E^{\mathcal{F}_t}(\chi_{A_t}(X \circ T)) \geq \chi_{A_t} X_t$,, a.e. A martingale is (semi)-regular if it and its negative (each regarded as submartingales) are *both (semi-) regular*. (These concepts are meaningful even if only X_t^+ is summable. Also if I is countable or $I \subset \mathbb{R}$ and the submartingale is right continuous, we can have J countable.)

It is clear that the regularity hypothesis is more stringent than semiregularity. To see that the former actually implies the latter let $\{X_t, \mathcal{F}_t, t \in I\}$ be a regular submartingale, and T any stopping time of $\{\mathcal{F}_t, t \in I\}$. To verify the condition of semi-regularity, define, for any fixed but arbitrary $t_0 \in I$, a family $\{T_t, t \in I\}$ as

$$T_t = \begin{cases} t & \text{if } t < t_0, t \in I \\ (T \vee t_0) = \tilde{T}, & \text{if } t \geq t_0, t \in I. \end{cases} \tag{13}$$

Then $[T_t \leq a] = \emptyset$ if $a < t \leq t_0, = \Omega$ if $t \leq a < t_0$, and $= [\tilde{T} \leq a]$ if $t \geq t_0$, where $a \in I$. Since $\tilde{T}$ is a stopping time ($[\tilde{T} \leq a] = [T \leq a] \cap [t_0 \leq a] = \emptyset$, or $= [T \leq a]$ according to $t_0 > a$ or $t_0 \leq a$), it follows that $\{T_t, t \in I\}$ is a stopping time process of $\{\mathcal{F}_t, t \in I\}$ since I is linearly ordered. The regularity hypothesis implies (on noting that $\mathcal{B}_{t_0} = \mathcal{F}_{t_0}$ and $A_{t_0} \in \mathcal{F}_{t_0}$),

$$\begin{aligned} E^{\mathcal{F}_{t_0}}(\chi_{A_{t_0}}(X \circ T)) &= E^{\mathcal{F}_{t_0}}(\chi_{A_{t_0}}(X \circ T_t)), \quad \text{since } T = \tilde{T} = T_t \\ &\qquad \text{on } A_{t_0} \text{ for } t \geq t_0, \\ &= \chi_{A_{t_0}} E^{\mathcal{B}_{t_0}}(X \circ T_t), \ \text{ since } \ A_{t_0} \in \mathcal{B}_{t_0}, \\ &\geq \chi_{A_{t_0}}(X \circ T_{t_0}) = \chi_{A_{t_0}} X_{t_0} \ \text{ a.e.}, \end{aligned} \tag{14}$$

where in the last inequality the regularity hypothesis of the process is used. Thus the arbitrariness of $t_0 \in I$ implies in (14) that the process $\{X_t, t \in I\}$ is semi-regular.

On the other hand, if I is countable, the converse implication of the above is also true. More precisely:

8. Proposition. *Let $\{X_t, \mathcal{F}_t, t \in I\}$ be a submartingale where I is linearly ordered. Then the regularity of X_t implies its semi-regularity, and if moreover I is countable then the semi-regularity is equivalent to regularity. In particular, when I is countable any semi-regular submartingale is transformed into a submartingale by every stopping time process of $\{\mathcal{F}_t, t \in I\}$.*

Proof. We only need to show that, for countable I, semi-regularity implies regularity, since other statements are already proved above. Let $\{T_j, j \in J\}$ be a stopping time process of $\{\mathcal{F}_t, t \in I\}$ and $\{X_t, \mathcal{F}_t, t \in I\}$ be a semi-regular submartingale. Then $Y_j = X \circ T_j$ is integrable and we need to show that $\{Y_j, \mathcal{B}_j, j \in J\}$ is a submartingale, i.e., for j_1, j_2 we have to verify

$$\int_A Y_{j_1} dP \leq \int_A Y_{j_2} dP, \qquad A \in \mathcal{B}_{j_1}. \tag{15}$$

Now the definition of $\mathcal{B}_{j_1}$ (see (18) of Section 1) implies $A \in \mathcal{B}_{j_1}$ iff $A \cap [T_{j_1} \leq i] \in \mathcal{F}_i, i \in I$. Since i is countable by Theorem 1.2(2), this holds iff $A \cap [T_{j_1} = i] \in \mathcal{F}_i, i \in I$. Hence

$$\begin{aligned}
\int_{A\cap[T_{j_1}=i]} Y_{j_2} dP &= \int_{A\cap[T_{j_1}=i]} E^{\mathcal{F}_i}(Y_{j_2}) dP \\
&= \int_{A\cap A_i} E^{\mathcal{F}_i}(\chi_{A_i} Y_{j_2}) dP, \text{ with } A_i = [T_{j_1} = i] (\in \mathcal{F}_i), \\
&= \int_{A\cap A_i} E^{\mathcal{F}_i}(\chi_{B_i} Y_{j_2}) dP, \text{ with } B_i = [T_{j_1} \geq i] \supset A_i, \\
&\geq \int_{A\cap A_i} \chi_{B_i} X_i dP, \text{ by the semi-regularity of } X_t\text{'s}, \\
&= \int_{A\cap A_i} X_i dP = \int_{A\cap A_i} X \circ T_{j_1} dP. \qquad (16)
\end{aligned}$$

Summing both sides of (16) over $i \in I$, and noting that $A = \bigcup_{i\in I} (A_i, \cap A)$, we get (15). □

The following result adds information to the preceding one by relaxing the integrability hypothesis on the process.

9. Proposition. *Let $\{X_t, \mathcal{F}_t, t \in I\}$ be a submartingale on (Ω, Σ, P), with I countable and linearly ordered. If $\int_\Omega X_t^+ dP < \infty$, for $t \in I$, $\limsup_{t\uparrow b_0} \int_{[T>t]} X_t^+ dP < \infty$, and $\liminf_{t\uparrow b_0} \int_{[T>t]} X_t^+ dP = 0$, for each stopping time T of $\{\mathcal{F}_t, t \in I\}$ where $b_0 = \sup I$, then $\{X_t, \mathcal{F}_t, t \in I\}$ is a regular submartingale and (*) $E(X^+ \circ T) < \infty$ for each such T. On the other hand, if $\{X_t, \mathcal{F}_t, t \in I\}$ is regular, and (*) holds, then one has (**) $\lim_{t\uparrow b_0} \int_{[T>t]} X_t^+ dP = 0$.*

Proof. If we assume that $\int_\Omega |X_t| dP < \infty$ and $\int_\Omega |X \circ T_j| dP < \infty$, then this result is an immediate consequence of Proposition 1.11. However, the present hypothesis is weaker and so a different argument is needed to reduce the result to the above stated proposition.

Since $\{X_t, \mathcal{F}_t, t \in I\}$ is a submartingale, by Section II.4 (simple Property II(b) essentially), $\{X_t^+, \mathcal{F}_t, t \in I\}$ is an integrable submartingale. The result is reduced to the above stated proposition if we show that $E(X^+ \circ T) < \infty$ for any stopping time T of $\{\mathcal{F}_t, t \in I\}$. Thus for any $t < b_0$, one has by the submartingale property,

$$0 \leq \int_{[T\leq t]} X^+ \circ T dP = \sum_{\substack{i\in I \\ i\leq t}} \int_{[T=i]} X^+ \circ T\, dP \leq \sum_{\substack{i\in I \\ i\leq t}} \int_{[T=i]} X_t^+ dP$$
$$\leq \int_\Omega X_t^+ dP < \infty. \tag{17}$$

However we also have, by the hypothesis on $[T > t]$,

$$\int_\Omega X^+ \circ T dP = \int_{[T\leq t]} X^+ \circ T dP + \int_{[T>t]} X^+ \circ T\, dP \leq \int_{[T\leq t]} X_t^+ dP$$
$$+ \int_{[T>t]} X^+ \circ T\, dP, \tag{18}$$

and it suffices to show that the last term if finite. Since it is non-increasing for large enough t it should be shown to be finite. We assert that, as $t \uparrow b_0$, the last term actually tends to zero. If this is false, then there exists an $\varepsilon > 0$, and a sequence $\{\delta_n \uparrow\}_1^\infty \subset I$ such that $\int_{[\delta_n < T < \delta_{n-1}]=A_n} X^+ \circ T\, dP > n\varepsilon$. But then

$$n\varepsilon < \int_{A_n} X^+ \circ T\, dP = \sum_{\substack{i\in I \\ \delta_n < i \leq \delta_{n-1}}} \int_{[T=i]} X^+ \circ T\, dP \leq \int_{A_n} X_{\delta_{n+1}}^+ dP. \tag{19}$$

In particular, if $T_1 = \delta_{n+1}$ on A_n, and $= \delta_1$ on $B_1 = [T \leq \delta_1]$, then it is clear that T_1 is a stopping time of $\{\mathcal{F}_t, t \in I\}$. Also (19) becomes

$$n\varepsilon < \int_A X^+ \circ T\, dP \leq \int_{A_n} X_{\delta_{n+1}}^+ = \int_{[T_1=\delta_{n+1}]} X_{\delta_{n+1}}^+ dP. \tag{20}$$

Hence $\limsup_{\delta_n} \int_{[T_a \geq \delta_n]} X_{\delta_n}^+ dP = +\infty$, contradicting the hypothesis. Thus (18) must be finite which proves (*) and the direct part follows from Proposition 1.11.

For the last part, if (*) holds and $\{X_t^+, \mathcal{F}_t, t \in I\}$ is an integrable submartingale, then for any T, as above, by the first half of (18),

$$\limsup_{t\uparrow b_0} \int_{[T>t]} X^+ \circ T\, dP \leq \int_\Omega X^+ \circ T\, dP < \infty, \tag{21}$$

Since (18) is finite, clearly $\lim_{t\uparrow b_0} \int_{[T>t]} X^+ \circ T\, dP = 0$ holds if $\int_\Omega X_t^+ dP < \infty$. The latter follows from $(*)$ with $T \equiv t$ as a constant stopping time. □

In the above proof (see (18)-(20)) we actually established the following useful fact.

10. Proposition. *Let $\{X_t, \mathcal{F}_t, t \in I\}$ be a nonnegative (integrable) submartingale on (Ω, Σ, P) with I as a linearly ordered countable set. If for any stopping time T of $\{\mathcal{F}_t, t \in I\}$, $\limsup_{t\uparrow b_0} \int_{[T>t]} X_t dP < \infty$, ($b_0 = \sup I$), then $\lim_{t\uparrow b_0} \int_{[T>t]} X \circ T\, dP = 0$. Moreover, $\int_\Omega X \circ T\, dP < \infty$.*

Remark. The existence of a counterexample (cf., I.5.6) of a nonnegative nonuniformly integrable martingale shows that the above conclusion is strictly weaker than the condition $\lim_{t\uparrow b_0} \int_{[T>t]} X_t dP = 0$.

We shall now present a characterization of certain semi-regular submartingales when the index I is not necessarily countable.

11. Proposition. *Let $\{X_t, \mathcal{F}_t, t \in I\}$ be a submartingale on (Ω, Σ, P) with I linearly ordered and $\mathcal{F}_t = \mathcal{F}_{t+}, t \in I$. Then the process is semi-regular iff for each stopping time T of $\{\mathcal{F}_t, t \in I\}$, letting $T_t = T \wedge t$ and $Y_t = X \circ T_t$, which is $\mathcal{B}_t = \mathcal{B}(T_t)$ (hence $\mathcal{F}_t$)-adapted, the process $\{Y_t, \mathcal{F}_t, t \in I\}$ is a submartingale satisfying $E^{\mathcal{F}_t}(X \circ T) \geq Y_t$* a.e.

Proof. First observe that the right continuity of $\{\mathcal{F}_t, t \in I\}$ is needed to conclude that T_t's are stopping times, and since $T_t \leq t$ we have $\mathcal{B}_t = \mathcal{B}(T_t) \subset \mathcal{B}(t) = \mathcal{F}_t$ so that Y_t is $\mathcal{F}_t$-adapted.

The direct part is immediate. In fact, if $\{Y_t, \mathcal{F}_t, t \in I\}$ is a submartingale and $Y = X \circ T$ with $E^{\mathcal{F}_t}(Y) \geq Y_t$ a.e., then the set $A_t = [T \geq t] \in \mathcal{F}_t$. Further

$$\chi_{A_t} E^{\mathcal{F}_t}(Y) \geq \chi_{A_t} Y_t = \chi_{A_t} X \circ (T \wedge t) = \chi_{A_t} X_t, \text{ a.e.}$$

Hence $\{X_t, \mathcal{F}_t, t \in I\}$ is semi-regular, by Definition 7.

For the converse, let the process be semi-regular and $t_1 < t_2$. Then $A_{t_1}^c = [T < t_1] \in \mathcal{F}_{t_1}$ and on this set $T_2 = T \wedge t_2 = T \wedge t_1 = T$, so that

$$\chi_{A_{t_1}^c} E^{\mathcal{F}_{t_1}}(Y_{t_2}) = E^{\mathcal{F}_{t_1}}(\chi_{A_{t_1}^c} X \circ T) = E^{\mathcal{F}_{t_1}}(\chi_{A_{t_1}^c} X \circ T_{t_1}) = \chi_{A_{t_1}^c} Y_{t_1} \text{ a.e.} \tag{22}$$

On the other hand by the semi-regularity of the X_t-process.

$$\begin{aligned}\chi_{A_{t_1}} E^{\mathcal{F}_{t_1}}(Y_{t_2}) &\geq \chi_{A_{t_1}} X_{t_1} \text{ a.e.} \\ &= \chi_{A_{t_1}} (X \circ T_{t_1}) = \chi_{A_{t_1}} Y_{t_1} \text{ a.e.}\end{aligned} \tag{23}$$

Adding (22) and (23), we deduce that $\{Y_t, \mathcal{F}_t, t \in I\}$ is a submartingale.

To show that $Y = X \circ T$ closes the submartingale, since Y is integrable by the semi-regularity hypothesis, let us compute $E^{\mathcal{F}_t}(Y)$ on A_t as in the above paragraph. Thus $A_t \in \mathcal{F}_t$ so that on $A_t^c, T = T_t$ and then

$$\chi_{A_{tc}} E^{\mathcal{F}_t}(Y) = E^{\mathcal{F}_t}(\chi_{A_t^c} Y_t) = \chi_{A_t^c} Y_t \quad a.e.$$

Also

$$\begin{aligned}\chi_{A_t} E^{\mathcal{F}_t}(Y) &\geq \chi_{A_t} \text{ a.e., by the semi-regularity,} \\ &= \chi_{A_t} (X \circ T_t) = \chi_{A_t} Y_t \text{ a.e.}\end{aligned} \tag{24}$$

Hence, by addition, $E^{\mathcal{F}_t}(Y) \geq Y_t$, a.e., $t \in I$, as desired. □

Remark. In view of the conditions given in Corollary 1.8, it is not enough to demand only that $E^{\mathcal{F}_t}(Y_{t'}) \geq Y_t$ a.e., $t' > t$, for the sufficiency part of the above result, and in fact without the closure element $X \circ T$, the result is false.

All the preceding propositions are needed to prove the next two theorems which are the key results connecting the projective limits with the regularity of martingales. They also illuminate the role of the stopping time transformations in the existence of the limit of a projective system, in the absence of topological conditions.

Let $\{X_t, \mathcal{F}_t, t \in I\}$ be a submartingale on (Ω, Σ, P) and suppose that $\sup_t \int_\Omega |X_t| dP < \infty$, and I is linearly ordered. If $\mathcal{F}_0 = \bigcup_{t \in I} \mathcal{F}_t$, define

$$\nu(A) = \lim_{t \uparrow b_0} \int_A X_t dP, \qquad A \in \mathcal{F}_0, \quad b_0 = \sup I. \tag{25}$$

Then $\nu : \mathcal{F}_0 \to \mathbb{R}$ is clearly a well-defined additive bounded function. As seen in the proof of Theorem II.6.2 (or verified directly), if $\nu_{t_0}(A) = \lim_{t \uparrow b_0} \int_A X_t dP, A \in \mathcal{F}_{t_0}$, then $\nu_{t_0} = \nu | \mathcal{F}_{t_0}$ and is σ-additive. However, ν is only finitely additive on $\mathcal{F}_0$. We first consider the case that I is countable. Note that $\{\Omega, \mathcal{F}_t, \nu_t, t \in I\}$ is a projective system of signed measures, with connecting maps $g_{tt'} : \Omega \to \Omega$ as identities. We obtain conditions for the σ-additivity of ν, or for the existence of projective limits of such systems (with $I = \mathbb{N}$ for convenience). The following theorem is a key step for the general result.

12. Theorem. *Let $\{X_n, \mathcal{F}_n, n \geq 1\}$ be a submartingale on (Ω, Σ, P) and let $\sup_n \int_\Omega |X_n| dP < \infty$. Let ν be the associated set function through (25). If ν is σ-additive, then the process is regular. On the other hand, if the process is regular and $\limsup_n \int_{[T>n]} X_n^- dP = 0$ for each stopping time T of $\{\mathcal{F}_n, n \geq 1\}$, then ν is σ-additive. In particular, if $\{X_n, \mathcal{F}_n, n \geq 1\}$ is a martingale (with $\sup_n \int_\Omega |X_n| dP < \infty$), then ν is σ-additive iff the process is regular.*

Proof. We express ν as a sum of two additive set functions, corresponding to the decomposition of a submartingale into a martingale and an increasing process (cf. II.5.1), and then prove the result. Thus let $A_1 = 0$ and for $j \geq 1, A_j = E^{\mathcal{F}_{j-1}}(X_j) - X_{j-1} \geq 0$, a.e. Let $X'_j = X_j - \sum_{i=1}^{j} A_i$ so that A_j is $\mathcal{F}_{j-1}$-adapted, $\{X'_j, \mathcal{F}_j, j \geq 1\}$ is a martingale, and $B_j = \sum_{i=1}^{j} A_i$ is an increasing process. Also

$$\int_\Omega B_n dP = \int_\Omega X_n dP - \int_\Omega X'_n dP \leq \sup_n \int_\Omega |X_n| dP + \left| \int_\Omega X'_1 dP \right| < \infty,$$

so that $\lim_n B_n = \tilde{B} < \infty$, a.e. and is integrable. Hence

$$\nu_1(M) = \lim_n \int_M X'_n dP, \quad \nu_2(M) = \lim_n \int_M B_n dP = \int_M \tilde{B} dP, \quad M \in \mathcal{F}_0 \tag{26}$$

are well-defined, ν_2 is σ-additive, and $\nu = \nu_1 + \nu_2$. Now, if ν is σ-additive then so must ν_1 be (and conversely). We *assert* that this implies the regularity of the martingale $\{X'_n, \mathcal{F}_n, n \geq 1\}$. From this the regularity of the X_n-process is deduced as follows. Since $X_n = X'_n - B_n$, the (trivial) submartingale $\{B_n, \mathcal{F}_{n-1}, n \geq 1\}$ satisfies the hypothesis of Proposition 10 because $\sup_n \int_\Omega B_n dP = \int_\Omega \tilde{B} dP < \infty$. So $\int_\Omega B \circ T dP < \infty$ for any stopping time T of $\{\mathcal{F}_n, n \geq 1\}$, and for any stopping time process $\{T_j, j \in J\}$, we get $\{B \circ T_j, \mathcal{B}_j, j \in J\}$ to be a submartingale. Hence $\{B_n, n \geq 1\}$ is regular and $\{X_n, n \geq 1\}$ will be regular.

To prove the assertion, by Proposition 8, it suffices to show that $\{X'_n, \mathcal{F}_n, n \geq 1\}$ is semi-regular. Let $N_1 = [T \geq n]$ where T is any stopping time of $\{\mathcal{F}_n, n \geq 1\}$, and $N_2 = \{E^{\mathcal{F}_n}(X' \circ T) < X'_n\} \cap N_1$, where n is a fixed but arbitrary integer; then it is to be shown that $P(N_2) = 0$. Replacing the martingale $\{X'_n, n \geq 1\}$ by $\{-X'_n, n \geq 1\}$, we get the opposite inequality to be true on a null set, i.e., one has the desired result that $E^{\mathcal{F}_n}(X' \circ T) = X'_n$ a.e. on N_1.

To prove that $P(N_2) = 0$, consider the sets $M_i = N_2 \cap [T = i]$. Then $N_2 = \bigcup_{i \geq n} M_i$ is a disjoint union and since $N_2 \in \mathcal{F}_n$, we have $M_k \in \mathcal{F}_k$ for $k \geq n$. The σ-additivity of ν_1 and the martingale property of the $\{X'_n\}_1^\infty$ imply:

$$\begin{aligned}
\sum_{k \geq n} \nu_1(M_k) = \nu_1(N_2) &= \lim_n \int_{N_2} X'_n dP = \int_{N_2} X'_n dP \geq \int_{N_2} E^{\mathcal{F}_n}(X' \circ T) dP \\
&= \int_{N_2} (X' \circ T) dP = \sum_{k \geq n} \int_{M_k} (X' \circ T) dP \\
&= \sum_{k \geq n} \int_{M_k} X'_k dP = \sum_{k \geq n} \lim_m \int_{M_k} X'_m dP = \sum_{k \geq n} \nu_1(M_k),
\end{aligned} \tag{27}$$

because $M_k \in \mathcal{F}_k$ and $\int_{M_k} X'_m dP = \int_{M_k} X'_k dP$ for $m \geq k$ so that $\lim_m \int_{M_k} X'_m dP = \int_{M_k} X'_k dP$. Hence there is equality throughout in (27) and it follows that

$$\int_{N_2} X'_n dP = \int_{N_2} X' \circ T dP = \int_{N_2} E^{\mathcal{F}_n}(X' \circ T) dP. \tag{28}$$

The first and last terms on N_2 can be equal iff $P(N_2) = 0$. Thus $\{X'_n, \mathcal{F}_n, n \geq 1\}$ is regular as asserted.

For the opposite implication, assume the given hypothesis, and let T be any stopping time of $\{\mathcal{F}_n, n \geq 1\}$. Since ν_2 of (26) is always σ-additive, it suffices to show that ν_1 is σ-additive. We deduce this after showing that $\{X'_n, \mathcal{F}_n, n \geq 1\}$ is a regular martingale under the given conditions. Now $\{(X'_n)^+, \mathcal{F}_n, n \geq 1\}$ is a submartingale and $\sup_n \int_\Omega |X'_n| dP < \infty$. Hence $E((X'_n)^+ \circ T) < \infty$, by Proposition 10.

Since the X_t-process is regular and

$$E(X^+ \circ T) \le E(|X| \circ T) \le 3 \sup_n E(|X_n|) < \infty$$

(cf., Proposition 1.10), it follows from the last part of Proposition 9 that $\lim_n \int_{[T>n]} X_n^+ dP = 0$. This (and the hypothesis) implies that $\limsup_n \int_{[T>n]} |X_n| dP = 0$. But $|X_n'| \le |X_n| + B_n$ (and $\{B_n, n \ge 1\}$ is regular as noted above). Hence

$$\limsup_n \int_{[T>n]} |X_n'| dP \le \limsup_n \int_{[T>n]} |X_n| dP + \limsup_n \int_{[T>n]} B_n dP = 0. \tag{29}$$

Now (29) trivially implies the hypothesis of Proposition 1.11 so that for any stopping time process $\{T_j, j \in J\}$ of $\{\mathcal{F}_n, n \ge 1\}$, it follows that $\{X' \circ T_j, \mathcal{B}_j, j \in J\}$ is a martingale. Hence $\{X_n', \mathcal{F}_n, n \ge 1\}$ is regular by Definition 7. Note that in (20) one may replace 'lim sup' by 'lim' now.

We next deduce the σ-additivity of ν_1 with this fact by choosing a suitable stopping time T. Let $\{H_n\}_1^\infty \subset \mathcal{F}_0 = \bigcup_n \mathcal{F}_n$ be a disjoint sequence such that $H = \bigcup_{n=1}^\infty H_n \in \mathcal{F}_0$. Since $\mathcal{F}_n \uparrow\subset \mathcal{F}_0$, we may assume for convenience that $H_n \in \mathcal{F}_n$ (and $H \in \mathcal{F}_0$ so $H \in \mathcal{F}_{n_0}$ for some n_0). For $n \ge 1$, set $\tilde{T} = n$ on H_n, and $= +\infty$ on H^c. Then $\tilde{T}$ is a stopping time of $\{\mathcal{F}_n, n \ge 1\}$. With the finite aditivity of ν_1 one has for $n > n_0$, $\left(\text{so } H - \bigcup_{k=1}^n H_k \in \mathcal{F}_n\right)$,

$$\begin{aligned}
\nu_1(H) - \sum_{k=1}^n \nu_1(H_k) &= \nu_1\left(H - \bigcup_{k=1}^n H_k\right) = \nu_1\left(\bigcup_{k>n} H_k\right) \\
&= \int_{\bigcup_{k>n} H_k} X_n' dP, \quad \text{by the martingale property,} \\
&= \int_{[\tilde{T}>n]} X_n' dP.
\end{aligned} \tag{30}$$

Taking $T = \tilde{T}$ in (29), it follows from (29) and (30) that $\left|\nu_1(H) - \sum_{k=1}^n \nu_1(H_k)\right| \to 0$ as $n \to \infty$. Hence ν_1 is σ-additive on $\mathcal{F}_0$ so that ν is also.

Finally, to prove the last part, suppose that $\{X_n, \mathcal{F}_n, n \geq 1\}$ is a regular martingale with $\int_\Omega |X_n| dP \leq K_0 < \infty$. Then by Proposition 10 applied to $\{X_n^+, n \geq 1\}$, we deduce that $\int_\Omega X^+ \circ T\, dP < \infty$ and $\limsup_n \int_{[T>n]} X^+ \circ T dP = 0$ for any stopping time T of $\{\mathcal{F}_n, n \geq 1\}$. So by Proposition 9, using regularity, $\lim_{n\to\infty} \int_{[T>n]} X_n^+ dP = 0$. Considering $\{-X_n, \mathcal{F}_n, n \geq 1\}$, one gets by the same argument that $\lim_{n\to\infty} \int_{[T>n]} X_n^- dP = 0$. Thus the additional hypothesis of the first part of the theorem is automatic for regular martingales. Hence ν is σ-additive iff we have the regularity of the martingales by the first part. □

Remark. The main result is that ν_1 is σ-additive iff ν is and this is so iff $\{X'_n, \mathcal{F}_n, n \geq 1\}$ is regular. However, for the σ-additivity of ν some additional condition such as $\limsup_n \int_{[T>n]} X_n^- dP = 0$ is indispensable. A regular submartingale need not give a σ-additive ν by (25), as the following example shows.

13. Example. Let (Ω, Σ, P) be the Lebesgue unit interval and $\mathcal{F}_n = \sigma((0, 2^{-k}], 0 \leq k \leq n), X_n = -2^{-n}\chi_{(0,2^{-n}]}$. Then $\{X_n, \mathcal{F}_n, n \geq 1\}$ is a martingale (as one may easily verify with the generating sets). Since $X_n < 0$ a.e., the condition $\lim_n \int_{[T>n]} X_n^+ dP = 0$ is trivial. Thus $\{X_n, \mathcal{F}_n, n \geq 1\}$ is a regular submartingale by Proposition 2. But ν defined by (25) is such that $\nu(A) = -1$ or 0 for each $A \in \mathcal{F}_0 (\nu(A) = -1$ if $A = (0, \frac{1}{2}])$ so that it cannot be σ-additive. In the contrary case, since $X_n \to 0$ a.e., $\nu(A) \equiv 0$ all A. However, the martingale itself is not regular (as required by the last part of Theorem 12). In fact, let T be a stopping time given by: $T = n$ on $(2^{-n}, 2^{-n+1}], n \geq 1$. Then $X \circ T = 0$ a.e., and if $T_1 = T \wedge 1$, then $X \circ T_1 = X_1$ a.e. and $E^{\mathcal{F}_1}(X \circ T) = 0 > X_1$, a.e., instead of equality demanded by Definition 7. Note that $\liminf_n \int_{[T>n]} X_n^- dP > 0$. Thus regularity is a more stringent condition for martingales.

We can now present a general result linking the martingale and projective limit theories mentioned before. Recall that, by Definition I.3.3, a martingale $\{X_t, \mathcal{F}_t, t \in I\}$ on (Ω, Σ, P) determines a set martingale which is a projective system. In fact if $\nu_1(\cdot) = \int_{(\cdot)} X_t dP$, then $\nu_1 : \mathcal{F}_t \to \mathbb{R}$ is a bounded σ-additive set function such that for $t' < t, \nu_{t'} | \mathcal{F}_t = \nu_t$. If $g_{tt'} : \Omega_{t'} \to \Omega_t$ is identity (where $\Omega_t = \Omega = \Omega_{t'}$), then $\{(\Omega_\alpha, \mathcal{F}_\alpha, \nu_\alpha, g_{\alpha\beta})_{\alpha\leq\beta} : \alpha, \beta \text{ in } I\}$ is a projective system and $\nu(A) = \lim_\alpha \nu_\alpha(A), A \in \mathcal{F}_0 = \bigcup_{\alpha\in I} \mathcal{F}_\alpha$, is additive and uniquely defined. This ν is clearly the same as that of (25). We say that the above family is the *associated projective system* (of the martingale, $\{\mathcal{F}_t, t \in I\}$ being the base). We then have the following characterization.

14. Theorem. *Let $\{X_t, \mathcal{F}_t, t \in I, I \subset \mathbb{R}^+\}$ be a right continuous martingale where $\mathcal{F}_t = \mathcal{F}_{t+}$ and $\sup_t \int_\Omega |X_t| dP < \infty$. Then the following statements are equivalent.*

(i) *The martingale is regular.*

(ii) *The associated projective system $K_0 = \{(\Omega_\alpha, \mathcal{F}_\alpha, \nu_\alpha, g_{\alpha\beta})_{\alpha\leq\beta} : \alpha, \beta \text{ in } I\}$ admits a limit $(\Omega, \mathcal{F}, \nu)$, where $\mathcal{F} = \sigma(\mathcal{F}_0)$.*

(iii) *The set $\{X_t, t \in I\} \subset L^1(\Omega, \Sigma, P)$ is uniformly integrable.*

(iv) *The martingale is of class* (D).

(v) *There exists a bounded set $\{\tilde{\nu}_\alpha, \alpha \in I\}$ of signed measures on $\mathcal{F}$ such that* (a) *$\tilde{\nu}_\alpha | \mathcal{F}_\alpha = \nu_\alpha, \alpha \in I$, and* (b) *for each sequence $\alpha_1 < \alpha_2 < \cdots$ of indexes of I, the family $\{\tilde{\nu}_{\alpha_n}, n \geq 1\}$ is uniformly σ-additive (or equivalently, uniformly P-continuous) on $\mathcal{F}$.*

Proof. Observe that by the classical abstract analysis one has the implications (ii) $\Rightarrow$ (iii) $\Rightarrow$ (v) $\Rightarrow$ (ii). These will not be proved (one uses Dunford-Schwartz [1], IV.9.2). We establish: (i) $\Rightarrow$ (ii) and (iii) $\Rightarrow$ (iv) $\Rightarrow$ (i), using the above stated result (ii) $\Leftrightarrow$ (iii), and its right continuous modification.

Suppose (i) is true. Let $\alpha_1 < \alpha_2 < \cdots$ be an increasing sequence in I, such that $\alpha_n \uparrow b_0 = \sup I$. If $\{X_{\alpha_n}, \mathcal{F}_{\alpha_n}, n \geq 1\}$ is the corresponding process, then it follows from Definition 7 that this is a regular martingale. Hence by Theorem 12, if $\tilde{\mu}(A) = \lim_n \nu_{\alpha_n}(A), A \in \mathcal{F}_0 = \bigcup_n \mathcal{F}_{\alpha_n}$ (same $\mathcal{F}_0$ as before since the α_n sequence may be taken cofinal in the choice), then $\tilde{\mu}$ is σ-additive. it is clear that $\tilde{\mu} = \nu$ by the martingale property. Also $\tilde{\mu}$ is P-continuous, and

$$\nu_{\alpha_n}(A) = \int_A X_{\alpha_n} dP = \nu(A) = \int_A X_\infty dP, \qquad A \in \mathcal{F}_{\alpha_n}, \tag{31}$$

where $X_\infty = \frac{d\nu}{dP}$ a.e., (and is $\mathcal{F}(= \mathcal{F}_\infty)$-measurable). Thus $\{X_{\alpha_n}, \mathcal{F}_{\alpha_n}, 1 \leq n \leq \infty\}$ is a uniformly integrable martingale. So Theorem II.6.2(b) implies that K_0 admits the projective limit $(\Omega, \mathcal{F}, \nu)$, and (ii) holds.

Now (ii) $\Rightarrow$ (iii) (noted above), so $\{X_t, t \in I\}$ is uniformly integrable. Since it is right continuous and $\mathcal{F}_t = \mathcal{F}_{t+}$ (used for the first time), we may apply Proposition 6(ii) to conclude that the process belongs to class (D). Thus (iii) $\Rightarrow$ (iv).

Let us show that (iv) $\Rightarrow$ (i). Thus $\{X_t, \mathcal{F}_t, t \in I\}$ is a martingale of class (D). It follows that (take the constant stopping times) the set is uniformly integrable. By right continuity of the process, we may also assume for this proof that it is separable. Let $X_\infty = \lim_{t \uparrow b_0} X_t$ which then exists a.e., and $\int_\Omega |X_\infty| dP \leq \int_\Omega |X_t| dP \leq K < \infty$. Because of the right continuity it is sufficient to prove the result with I replaced by a dense subset, say the rationals I_0 (cf. the discussion of Section III.5(a)). Thus for any stopping time T of $\{\mathcal{F}_t, t \in I_0\}$, $\int_\Omega |X \circ T| dP \leq 3K$ by Proposition 1.10. Hence for any stopping time process $\{T_j, j \in J\}$ of $\{\mathcal{F}_t, t \in I_0\}$, the transformed process $\{Y_j = X \circ T_j, \mathcal{B}_j, j \in I_0\}$ is a martingale if we establish:

$$\limsup_{t \in I_0} \int_{[T>t]} |X_t| dP < \infty, \quad \text{and} \quad \liminf_{t \in I_0} \int_{[T>t]} |X_t| dP = 0, \tag{32}$$

by Theorem 1.13. (We may also apply Proposition 9 above in this situation.) Since $\{|X_t|, \mathcal{F}_t, t \in I_0\}$ is a positive uniformly integrable submartingale, it suffices to show for (32) that

$$(+) \lim_n \int_{[T>t]} |X \circ T| dP = \lim_t \int_{[T>t]} |X| \circ T \, dP = 0,$$

is true. But by the uniform integrability, $\int_\Omega |X_t| dP \leq K < \infty$ and $\limsup_n \int_{[T>t]} |X_t| dP \leq K$. Hence the condition (+) follows from Proposition 10 above. Thus (iv) $\Rightarrow$ (i). $\square$

The following consequence of this theorem is recorded for reference.

15. Corollary. *Let $\{X_t, \mathcal{F}_t, t \in I, I \subset \mathbb{R}\}$ be a right continuous martingale on (Ω, Σ, P) with $\mathcal{F}_t = \mathcal{F}_{t+}, t \in I$. If $\mu(A) = \int_A X_t dP, A \in \mathcal{F}_0 = \bigcup_{t \in I} \mathcal{F}_t$, and $\sup_t \int_\Omega |X_t| dP < \infty$, then $\mu : \mathcal{F}_0 \to \mathbb{R}$ is σ-additive iff the martingale is regular.*

16. Note. Motivated by the above theorem, we may formulate a *sub-* (or *super-*) *projective system* as followss. Let $\{(\Omega_\alpha, g_{\alpha\beta})_{\alpha\leq\beta} : \alpha, \beta \text{ in } D\}$ be a projective system of spaces and if $\{(\Omega_\alpha, \Sigma_\alpha, \mu_\alpha)\}$ are measure spaces, then it is a sub-(or super-) projective system if for $\alpha \leq \beta$ we have $g_{\alpha\beta}^{-1}(\Sigma_\alpha) \subset \Sigma_\beta$ and $\mu_\beta \circ g_{\alpha\beta}^{-1} \leq \mu_\alpha$ (or $\mu_\beta \circ g_{\alpha\beta}^{-1} \geq \mu_\alpha$). A projective system is thus a sub- and super- projective system simultaneously. If $g_\alpha : \Omega = \lim_{\beta} \Omega_\alpha \to \Omega_\alpha$ is the canonical mapping such that $g_{\alpha\beta} \circ g_\beta = g_\alpha$, then we may define an additive μ on $\Sigma_0 = \bigcup_{\alpha\in G} g_\alpha^{-1}(\Sigma_\alpha)$, when $|\mu_\alpha|(\Omega_\alpha) \leq K_0 < \infty, \alpha \in D$, by (recalling $\mu_\alpha = \mu_\alpha^* \circ g_\alpha^{-1}$ for a unique measure μ_α^*):

$$\mu(A) = \sup\{\mu_\alpha^*(A) : \alpha \in D\}, \qquad A \in \Sigma_0, \quad \mu_\alpha^* \circ g_\alpha^{-1} = \mu_\alpha, \tag{33}$$

(or 'sup' replaced by 'inf' in the superprojective case). In general μ is only additive, and $\mu \circ g_\alpha^{-1} \geq \mu_\alpha$ (or $\mu \circ g_\alpha^{-1} \leq \mu_\alpha$). By an analog of Prokhorov's theorem (I.3.8) it can be shown that one has: "If each Ω_α is a Hausdorff space and $(\Sigma_\alpha, \mu_\alpha)$ is a Radon measure space, $\sup_\alpha |\mu_\alpha|(\Omega_\alpha) < \infty, \{(\Omega_\alpha, \Sigma_\alpha, \mu_\alpha, g_{\alpha\beta})_{\alpha\leq\beta} : \alpha, \beta \text{ in } D\}$ is a sub- (or super-) projective system, then there exists a unique Radon measure μ on $\sigma(\Sigma_0)$ such that $\mu \circ g_\alpha^{-1} \geq \mu_\alpha$ (or $\mu \circ g_\alpha^{-1} \leq \mu_\alpha$), $\alpha \in D$". (For a proof, see Bourbaki [1] or Schwartz [1].) In the non-topological case Theorem 12 provides conditions if $D \subset \mathbb{R}$. We have to leave the discussions here.

4.4 More on convergence theory

The preceding work implies that, when a stochastic function under a stopping time process is transformed into a martingale, we can consider the convergence theory both for the given and the transformed families. If the new process is a martingale and the conditions of Sections II.6 and III.5 are satisfied, then the same convergence theory applies. However, if we can find conditions on the transformed one which imply certain convergence statements for the given process, then new results can be obtained. We present here some theorems by extending a few of the earlier propositions with the help of stopping times.

The first result is a generalization of Theorem III.5.7. It is due to Chow [1], and will be used to prove several other statements. Like the former, it uses the Vitali condition and the underlying ideas of proof are similar although a more detailed analysis is needed.

1. Theorem. *Let $\{X_i, \mathcal{F}_i, i \in I\}$ be a real martingale on a complete probability space (Ω, Σ, P) with I as a directed set. Let the family $\{\mathcal{F}_i, i \in I\}$ satisfy the Vitali condition V_0 (of III.5.3). Suppose that for each $i_0 \in I$, every subsequence $J = \{i_{n+1} \geq i_n \geq i_0 : n \geq 1\}$, and every stopping time T of $\{\mathcal{F}_j, j \in J\}$ we have*

$$\int_\Omega X^+ \circ T dP < \infty. \tag{1}$$

Then $X_i \to X_\infty$ a.e. and X_∞ is P-measurable.

Remark. The difference between the earlier result and this one is that the integrability condition (1) is much weaker than $\sup_i \int_\Omega |X_i| dP < \infty$ (cf. Proposition 1.10). This supremum may be infinite under (1) and X_∞ may be infinite on a set of positive P-measure as we show by an example. For completeness, we shall present the details of proof, since the technique is useful for the directed index processes.

Proof. As in the earlier proofs (cf. III.5.7 and 1.13) we assume that the result is false, and produce a countable set J, a stopping time T of $\{\mathcal{F}_j, j \in J\}$ for which (1) does not hold, and this will prove the theorem. We present the argument in steps. Set $\mathcal{F}_\infty = \sigma\left(\bigcup_{i \in I} \mathcal{F}_i\right)$ and denote by $\mathcal{F}'_\infty$ the P-completion of $\mathcal{F}_\infty$.

I. Suppose that the limit does not exist a.e. Then there are numbers $a, b, -\infty < a < b < \infty$, such that the set $A = [X^* > b > a > X_*]$ has positive measure where $X^* = \limsup_i X_i, X_* = \liminf_i X_i$, which are P-measurable (completeness of measure) by Theorem III.2.3, so that $A \in \mathcal{F}'_\infty$. Subtracting a and dividing by $b-a$ if necessary, we take $a = 0$ and $b = 1$ here, for convenience. It is asserted that for any $0 < \varepsilon < 1$ there exist an $i_0 = i_0(\varepsilon) \in I$ and a set $V_\varepsilon \in \mathcal{F}_{i_0}$ of positive measure such that $B_\varepsilon \subset [X_{i_0} > 1]$ and (writing B for B_ε)

$$P(A \cap B) > (1 - \varepsilon) P(B). \tag{2}$$

The proof of this statement is identical with that given in Step IV of the proof of Theorem III.5.7, and need not be repeated.

To use the hypothesis of V_0, we select an essential fine covering of B and a finite collection of almost disjoint sets approximating B and then define a stopping time T. The details are technical and the motivation for the following computations can be seen only after reading the two earlier results noted above. We therefore present them in the ensuing three steps to ease the burden on the reader.

II. Let $0 < d \leq P(B)$ and $0 < \varepsilon_1 < \frac{1}{3}\min(d, \varepsilon)$ so that $0 < 3\varepsilon_1 < d \leq 1$. Thus both d and ε_1 depend on ε. Define a covering: $K_i = \{\omega \in B : X_i(\omega) > 1\}$ if $i \geq i_0$ and $K_i = \Omega$ otherwise. Then $K_i \in \mathcal{F}_i$, and $\{K_i, i \geq i_o\}$ is an essential fine covering of B, and hence also of $A \cap B$. By V_0, there exist $n_1 = n_1(\varepsilon_1)$, and sets $L_j \subset K_{i_j}, L_j \in \mathcal{F}_{i_j}$ disjoint a.e., for a subsequence $i_j \geq i_0$ such that if $A_1 = \bigcup_{i=1}^{n_1} L_i$, then $P(B - A_1) < \varepsilon_1 P(B)$. Thus $P(B) - P(B \cap A_1) < \varepsilon_1 P(B)$ and with (2)

$$P(A_1) \geq P(B \cap A_1) > (1 - \varepsilon_1)P(B) > d(1 - \varepsilon_1) > d - \varepsilon_1. \quad (3)$$

We similarly find another fine covering, with the other inequality, for $B \cap A_1$. Indeed let $K_i' = \{\omega \in A_1 : X_i(\omega) < 0\}$, and $K_i' = \Omega$ if $X_i(\omega) \geq 0$ for some i. Let $j_1 \in I$ be an index such that $j_1 \geq i_j, j = 1, \ldots, n_1$. Then $\{K_i', i \geq j_1\}$ is an essential fine covering of $A_1 \in \mathcal{F}_{j_i}$ and also of $B \cap A_1$ since $B \in \mathcal{F}_{i_0} \subset \mathcal{F}_{j_1}$. Now by V_0, there exist $n_1' = n_1'(\varepsilon_1), L_j' \subset K_{\ell_j}'$, disjoint a.e., $L_j' \in \mathcal{F}_{\ell_j}, j = 1, \ldots, n_1', \ell_j \geq j_1$, such that, if $A_1' = \bigcup_{j=1}^{n_i'} L_j'$, we have $P(B \cap A_1 - A_1') < \varepsilon_1(1 - d)$. Since $L_j' \subset K_{\ell_j}' \subset A_1, (A_1' \subset A_1)$ we have

$$\begin{aligned} P(A_1') \geq P(B \cap A_1') &> P(B \cap A_1) - \varepsilon_1(1 - d) \\ &> d(1 - \varepsilon_1) - \varepsilon_1(1 - d) = d - \varepsilon_1 > d - 2\varepsilon_1 > 0, \text{ by (3)}. \end{aligned} \quad (4)$$

We now define two finite stopping times and transform the given martingale into another one (using Theorem 1.2(ii)), which will produce the desired estimates to contradict (1).

III. Let $j_1' \in I$ be such that $j_1' \geq \ell_j \geq j_1, j = 1, \ldots, n_1'$(by directedness of I). Define finite stopping times as:

$$\begin{aligned} T_1 &= i_0 \text{ on } \Omega - A, = j_1 \text{ on } A - A_1, = i_j \text{ on } L_j, 1 \leq j \leq n_1, \\ T_2 &= i_0 \text{ on } \Omega - A, = j_1 \text{ on } A - A_1, = j_1' \text{ on } A_1 - A_1', \\ &\qquad = \ell_j \text{ on } L_j', 1 \leq j \leq n_1'. \end{aligned}$$

Then it is clear that $T_1 \le T_2$ and both are stopping times of the appropriate families $\{\mathcal{F}_{i_j}, j = 0, 1, \ldots, n_1 + 1\}$ and a second one with $(j_1 = i_{n+1}, j_1' = i_{n'+1})$. Since T_1, T_2 take finitely many values, by Theorem 1.2, $\{X \circ T_1, X \circ T_2\}$ is a martingale. Also $L_j = [T_1 = i_j] \in \mathcal{B}(T_1)$ so $A_1 \in \mathcal{B}(T_1)$ and then the martingale property for $\{\mathcal{B}(T_1), \mathcal{B}(T_2)\}$ implies:

$$\int_{A_1} Y_1 dP = \int_{A_2} Y_2 dP, \qquad Y_r = X \circ T_r, \quad r = 1, 2. \tag{5}$$

Let us estimate the integral of $X_{j_1'}$ on $A - A_j'$ with (5) in evaluating the integral of (1). Thus on noting that $T_2 \le j_1'$, we have

$$\begin{aligned}
\int_{A_1 - A_1'} X_{j_1} dP &\ge \int_{A_1 - A_1'} X_{j'} dP + \int_{A_1'} X_{\ell_j} dP, \quad \text{since } X_{\ell_j} < 0, 1 \le j \le n_1', \\
&= \int_{(A_1 - A_1') \cap [T_2 = j']} X \circ T_2 dP + \sum_{j=1}^{n_1'} \int_{[T_2 = \ell_j]} X \circ T_2 dP \\
&= \int_{A_1} X \circ T_2 dP = \int_{A_1} Y_2 dP = \int_{A_1} Y_1 dP, \quad \text{by (5)}, \\
&= \sum_{j=1}^{n_1+1} \int_{A_1 \cap [T_1 = i_j]} X_{i_j} dP \ge \int_{A_1} dP, \quad \text{since } X_{i_j} > 1 \text{ on } L_{i_j}, \\
&\qquad\qquad 1 \le j \le n_1 + 1, \\
&= P(A_1) > d - \varepsilon_1, \quad \text{by (3)}.
\end{aligned} \tag{6}$$

We now iterate the procedure replacing Ω by A_1' and by its appropriate measurable subsets.

IV. Let $B_1 = B \cap A_1'$. Then by (4), $P(B_1) > d - \varepsilon_1 = d_1 > d - 2\varepsilon > 0$. Consider the process $\{Z_i, \mathcal{F}_i, i \in I, i \ge j_1'\}$ where $Z_i = X_i \chi_{A_1'}$. Since $A_1' \in \mathcal{F}_{j_1'}$, this new process is again a martingale satisfying the hypothesis of the theorem. So by Step III applied to this Z-process, we can find indices j_2, j_2' in I and sets A_2, A_2' in $\mathcal{F}_{j_2}$ and $\mathcal{F}_{j_2'}$ respectively such that $i_0 \le j_1 \le j_1' \le j_2 \le j_2'$ and $A_1 \supset A_1' \supset A_2 \supset A_2'$ and (6) is true if we replace d by d_1, ε_1 by $\frac{\varepsilon_1}{2}$, and (A_1, A_1') by (A_2, A_2'). Thus

$$\int_{A_2 - A_2'} Z_{j_2'} dP = \int_{A_2 - A_2'} X_{j_2'} \chi_{A_1'} dP \geq P(A_1' \cap A_2)$$
$$= P(A_2) \geq d_1 - \frac{\varepsilon_1}{2} = d - \varepsilon_1 - \frac{\varepsilon_1}{2}, \text{ since } A_2' \subset A_2 \subset A_1'. \tag{7}$$

Now let $B_2 = B_1 \cap A_2'$. Then the analog of (4) yields $P(B_2) > d_1 - \varepsilon_1 - \frac{\varepsilon_1}{2} = d_2 > d - 2\varepsilon_1 > 0$. So we may consider the new martingale $\{X_i \chi_{A_2'}, \mathcal{F}_i, i \in I, i \geq j_2'\}$, and by induction produce sequences $i_0 \leq j_1 \leq j_1' \leq \cdots \leq j_n \leq j_n' \leq \cdots$, and $A_1 \supset A_1' \supset \cdots \supset A_n \supset A_n' \supset \cdots$ with $A_n \in \mathcal{F}_{j_n}, A_n' \in \mathcal{F}_{j_n'}$, satisfying

$$\int_{A_n - A_n'} X_{j_n'} dP \geq d - \varepsilon_1 - \frac{\varepsilon_1}{2} - \cdots - \frac{\varepsilon_1}{2^{n-1}} > d - 2\varepsilon_1 > 0, \qquad n \geq 1. \tag{8}$$

V. Finally define a stopping time T as:

$$T = j_1' \text{ on } \Omega - A_1', = j_n' \text{ on } A_{n-1}' - A_n', \text{ for } n > 1.$$

If $J = \{j_n' \geq j_{n-1} \geq i_0, n > 1\}$, then T is clearly a stopping time of $\{\mathcal{F}_j, j \in J\}$, where J is linearly ordered. Then

$$\int_\Omega X^+ \circ T dP = \sum_{n=1}^{\infty} \int_{[T=j_n']} X_{j_n'}^+ dP \geq \sum_{n=1}^{\infty} \int_{A_n - A_n'} X_{j_n'}^+ dP$$
$$\geq \sum_{n=1}^{\infty} \left(d - 2\varepsilon_1 - \frac{\varepsilon_1}{2^{n-1}}\right) = \infty, \tag{9}$$

since $d - 3\varepsilon_1 > 0$. This contradicts (1), as desired. □

We now record some useful consequences of this result. In the above proof, the fact that the martingale property of $\{X_i, \mathcal{F}_i, i \in I\}$ implied the same for the $\{Y_1, Y_2\}$ with any *finite stopping times* T_1, T_2 of $\{\mathcal{F}_i, i \in I\}$ was used in Step III from which (6) followed. For submartingales (5) is no longer true, as we saw in Section 1 if I is only directed. However, if (5) is given as the hypothesis for submartingales, and the remaining conditions hold, then the result is again true in this generality, since the rest of the proof is unchanged. We state this precisely. The last part is in Chow [4].

2. Theorem. *Let $\{X_i, \mathcal{F}_i, i \in I\}$ be a submartingale on a complete proability space (Ω, Σ, P) with I directed and V_0 holding for $\{\mathcal{F}_i, i \in I\}$. Suppose that for any pair of finite stopping times $T_1 \leq T_2$, of $\{\mathcal{F}_i, i \in I\}$, $\{Y_j = X \circ T_j, \mathcal{B}(T_j)\}_1^2$ is a submartingale. If (1) is again assumed, then $X_i \to X_\infty$* a.e., *and X_∞ is P-measurable. In particular, if I is linearly ordered, countable, and* (1) *is true, then the conclusion obtains.*

The proof of Theorem 1 shows that the condition V_0 for $\{\mathcal{F}_i, i \in I\}$ may be slightly weakened to every countable subset J. Thus we also have the following result.

3. Proposition. *Let $\{X_i, \mathcal{F}_i, i \in I\}$ be a martingale on a complete (Ω, Σ, P) such that I is directed and for each countable (directed) subset $J \subset I, \{\mathcal{F}_j, j \in J\}$ satisfies V_0. If (1) holds, then $X_i \to X_\infty$* a.e., *and X_∞ is P-measurable.*

This and the fact that rationals are dense in $\mathbb{R}$ implies the following consequence where we also use (for the last part) that $\{X_t^+, \mathcal{F}_t, t \in \mathbb{R}\}$ is a submartingale if $\{X_t, \mathcal{F}_t, t \in \mathbb{R}\}$ is.

4. Corollary. *let $\{X_t, \mathcal{F}_t, t \in I, I \subseteq \mathbb{R}\}$ be a separable submartingale on (Ω, Σ, P). If I is countable or V_0 holds for each countable subset of I (if it is merely directed), and* (1) *is true, then$X_t \to X_\infty$* a.e., *as $t \uparrow b_0 =$* sup *I. In particular the same conclusion obtains if $\sup\limits_{t \in I} \int_\Omega X_t^+ dP < \infty$ instead of* (1).

We shall present an example to show that the condition (1) is strictly weaker than that of Theorem III.5.7.

5. Example. Let (Ω, Σ, P) be the Lebesgue unit interval and let $I_n = \left(0, \frac{1}{2} + \frac{1}{2^n}\right], J_n = \left(\frac{1}{2} + \frac{1}{2^n}, 1\right]$, and $\mathcal{F}_n = \sigma(I_k, 1 \leq k \leq n)$. Then $J_n \in \mathcal{F}_n, \mathcal{F}_n \subset \mathcal{F}_{n+1}$. Define a process as: $X_1 = 0$ and if $n > 1$, let

$$X_n = \alpha_n \chi_{I_n} - 2^n \chi_{I_{n-1} I_n} + X_{n-1} \chi_{J_{n-1}}, \qquad \alpha_n = (n-1)/P(I_n). \quad (10)$$

Then $\{X_n, \mathcal{F}_n, n \geq 1\}$ is an integrable adapted process. It is a martigale. To see this it suffices to check $\int_I X_n dP = \int_I X_{n-1} dP$ for each generator $I \in \mathcal{F}_{n-1}$. Thus

$$\begin{aligned}\int_I X_n dP &= (n-1)\frac{P(I_n \cap I)}{P(I_n)} - 2^n P[(I_{n-1} - I_n) \cap I] + \int_{I \cap J_{n-1}} X_{n-1} dP \\ &= n-2, \text{ if } I = I_{n-1} (\text{so } I \cap J_{n-1} = \emptyset, I \cap I_n = I), \\ &= \int_I X_{n-1} dP. \end{aligned} \tag{11}$$

If I is any other generator, a similar computation shows that Equation (11) holds and the process is a martingale. On the other hand

$$\int_\Omega X_n^+ dP = \alpha_n \int_{I_n} dP + \int_{J_n} X_{n-1}^+ dP = (n-1) + \int_{J_n} X_n^+ dP \geq n-1.$$

Thus $\sup_n \int_\Omega X_n^+ dP = \infty$, and hence Theorem III.5.7 is not applicable. Now let T be any stopping time of $\{\mathcal{F}_n, n \geq 1\}$. Then (10) implies

$$\int_\Omega X^+ \circ T \, dP = \sum_{i=1}^{\infty} \int_{[T=i]} X_i^+ dP = \sum_{i=1}^{\infty} [\alpha_i P(A_i \cap I_i) + \int_{A_0 \cap J_{i-1}} X_{i-1}^+ dP], \tag{12}$$

where $A_i = [T = i] \in \mathcal{F}_i$. But for each i, I_i is an atom of $\mathcal{F}_i$ so that either $A_i = I_i$ or $A_i \subset J_i = I_i^c$. In the latter case A_i can be one of a finite number of sets of $\mathcal{F}_i$. Since A_i are disjoint, if $A_i = I_i$ for some i_0, then the first term on the right of (12) has just one non-zero term $= \alpha_{i_0} P(I_0)$; and since $I_n \supset I_{n+1}$ and $J_n \subset J_{n+1}, \int_{J_n} X_n^+ dP = \int_{J_n} X_{n-1}^+ dP = \int_{J_n} X_1^+ dP = 0$, the right side sum of (12) is finite. Hence the hypothesis of Theorem 1 is satisfied and $X_n \to X_\infty$ a.e. (X_∞ may take infinite values on a set of positive measure, however).

The above example is not isolated. We present now a decomposition of non-L^1-bounded martingales. Several other useful decompositions can then be obtained. Note that given a martingale $\{X_n, \mathcal{F}_n, n \geq 1\}$ there always exists some stopping time T of $\{\mathcal{F}_n, n \geq 1\}$ such that $\int_\Omega X^+ \circ T dP < \infty$. (Take any constant T, for instance.) On the other hand, if $\mathcal{T}$ is the class of all stopping times of $\{\mathcal{F}_n, n \geq 1\}$, and $\alpha_T = \int_\Omega X^+ \circ T \, dP$, then by Example 5, the set $\{\alpha_T : T \in \mathcal{T}\} \subset \mathbb{R}^+$ is not bounded. It is bounded iff $\sup_n \int_\Omega X_n^+ dP < \infty$ or equivalently $\sup_n \int_\Omega |X_n| dP < \infty$, i.e. $\{X_n, n \geq 1\}$ is L^1-bounded, by the computation in Equation (17) following Theorem II.6.4. In fact, by Proposition 1.10, if $\sup_n \int_\Omega X_n^+ dP < \infty$, then $\sup_{T\in\mathcal{T}} \int_\Omega X^+ \circ T dP = K_0 < \infty$. Conversely, if $K_0 < \infty$, then taking $T = n$, the constant time, we get $\int_\Omega X_n^+ dP \leq K_0 < \infty$ and hence $\sup_n \int_\Omega X_n^+ dP \leq K_0 < \infty$. We now give a decomposition if only $\alpha_T < \infty$ for each $T \in \mathcal{T}$.

6. Theorem. *Let $\{X_n, \mathcal{F}_n, n \geq 1\}$ be a martingale on (Ω, Σ, P). For each stopping time $T \in \mathcal{T}$, assume that: $(*)$ $\int_\Omega X^+ \circ T dP = \alpha_T < \infty$. Then for each $\lambda > 0$ and $T \in \mathcal{T}$, there exist martingales $\{Y_n^{(i)}, \mathcal{F}_n, n \geq 1\}, 1 \leq i \leq 4$ (depending on T and λ), such that $X_n = \sum_{i=1}^4 Y_n^{(i)}$ with the following properties ($Y_n^{(1)}$ may be chosen independent of λ, but not the others):*

(i) *$\{Y_n^{(j)}, n \geq 1\}, j = 2,3,4$ are L^1-bounded, hence converge* a.e., *and $\{Y_n^{(1)}, n \geq 1\}$ also converges* a.e. *(even though it is not necessarily L^1-bounded).*

(ii) *For any $n_1 \geq 1$, we have*

$$\lambda P\left[\sup_{n\leq n_1} Y_n^{(1)} > \lambda\right] \leq 2\alpha_T - \int_\Omega X_1 dP + \int_\Omega X_{n_1}^+ dP$$
$$= \beta_T + \int_\Omega X_{n_1}^+ dP \ (say).$$

(iii) $\lambda P\left[\sup_n |Y_n^{(2)}| > 0\right] \leq 4\beta_T.$

(iv) $\left\|\sum_{n=1} |Y_n^{(3)} - Y_{n-1}^{(3)}|\right\|_1 \leq 4\beta_T,$

(v) $\|Y_n^{(4)}\|_\infty \leq 4\lambda, \|Y_n^{(4)}\|_2^2 \leq 12\lambda\beta_t,$ *and*

(vi) $\|Y_n^{(i)}\|_1 \leq 2\beta_T, i = 2,3,4.$

Remark. The detailed decomposition is stated for some applications. The result is an extension of the one due to Gundy [1]. In fact $\{X_n - Y_n^{(1)} = Z_n', n \geq 1\}$ is L^1-bounded and the corresponding decomposition is due to him. If the given martingale itself is L^1-bounded, then (and only then) we get a nontrivial inequality in (ii) if $n_1 \to \infty$, and in this case one may take, e.g., $Y_n^{(1)} \equiv 0$. Also this result is of interest only when the components $Y_n^{(i)}$ are (all) nontrivial. For instance, $X_n = Y_n^{(1)}$ and $Y_n^{(i)} = 0$ for $2 \leq i \leq 4$ is a decomposition satisfying (i)-(vi), a trivial one. This shows that the decomposition is nonunique. However, if $\{Y_n^{(1)} \neq X_n, n \geq 1\}$, then all the $\{Y_n^{(i)}, 2 < i \leq 4$ must be nontrivial, as a computation shows. Also using Theorem 2, this result admits an extension to submartingales. We omit the proof as it is not essential for the following work. (The details may be found in Rao [14].)

We considered a non-L^1-bounded martingale in the above theorem. It will be useful to know on what subsets of Ω the martingale converges in this case. Such results are of interest in structural analysis. The following theorem, due to Lamb [1], throws some light on this aspect of the convergence theory.

For each $0 < \delta \leq 1/2, \mathcal{F}_n \subset \mathcal{F}_{n+1} \subset \Sigma$, there exists an essentially unique set $S_n(\delta) \in \mathcal{F}_n$, called a *$\delta$-splitting* of $\mathcal{F}_n$ relative to $\mathcal{F}_{n+1}$, determined as follows: if $S(B, \delta) = \{\omega : 0 < P^{\mathcal{F}_n}(B)(\omega) \leq \delta\}, B \in \mathcal{F}_{n+1}$, where $P^{\mathcal{F}_n}(B)$ is a version of the conditional probability (or one chosen with a lifting operator of Section III.2), then one can show the existence of a set $A \in \mathcal{F}_{n+1}$ such that $P^{\mathcal{F}_n}(A) \leq \delta$ a.e., and $S(B, \delta) \subset S(A, \delta)$ a.e., for every $B \in \mathcal{F}_{n+1}$. Thus $S_n(\delta) = S(A, \delta)$ gives the δ-splitting. Let $\delta_k \downarrow 0$ and for each $n \geq 1$, let $S_n(\delta)$ be a δ-splitting of $\mathcal{F}_n$ relative to $\mathcal{F}_{n+1}$, and $S = \bigcap_{k \geq 1} \limsup_n S_n(\delta_k)$. Then the *regular part* of Ω relative to a filtering sequence $\{\mathcal{F}_n, n \geq 1\}$ of σ-algebras of Σ is defined as $\Omega^{(r)} = \Omega - S$. By the monotonicity of $S_n(\delta)$ in δ, it follows that S [and hence $\Omega^{(r)}$] is well-defined and does not depend on the particular $\{\delta_n\}_1^\infty$ sequence. We now prove the following result. (It holds for submartingales also, with simple changes.)

7. Theorem. *Let $\{X_n, \mathcal{F}_n, n \geq 1\}$ be a martingale and $\Omega^{(r)}$ be the regular part of Ω relative to $\{\mathcal{F}_n, n \geq 1\}$. Then $X_n(\omega) \to X_\infty(\omega)$ finite, for almost all ω in the set $\Omega^{(r)} \cap \left(\left[\sup_n X_n < \infty\right] \cap \left[\inf_n X_n > -\infty\right]\right)$. (It is possible that $\Omega^{(r)}$ has measure zero.)*

Proof. For each $m \geq 1$ and each $\lambda \in \mathbb{R}$ define a stopping time of $\{\mathcal{F}_n, n \geq 1\}$ by the following device:

$$T_m^\lambda = \inf\{i \geq m : P^{\mathcal{F}_i}[X_{i+1} > \lambda] > 0\}. \tag{13}$$

(We again leave it to the reader to check that T_m^λ is a stopping time, by Theorem 2.5.) Let $T_{mn}^\lambda = T_m^\lambda \wedge n$. Then by Corollary 1.8, $\{Y_{mn} = X \circ T_{mn}^\lambda, \mathcal{F}_n, n \geq 1\}$ is a martingale for each m. From (13) one can deduce that $Y_{mn} \leq \max(X_m, \lambda)$ if $n \geq m$. Since the last term is integrable, the given martingale is closed on the right, and hence $\lim_{n\to\infty} Y_{mn}$ exists a.e. and is finite a.e. We claim that for each $\omega_0 \in \Omega^{(r)} \cap \left[\sup_n X_n < \infty\right]$ there is a pair (m, λ), depending on ω_0, such that $T_m^\lambda(\omega_0) = +\infty$ (and similarly if "sup" is replaced by "inf" $> -\infty$). The preceding result then shows that $\lim_n Y_{mn}(\omega_0)$ exists and is finite for a.a. (ω_0) in the stated set.

Thus to prove the claim, if (i) ω_0 is in the above set, then $\sup_n X_n(\omega_0) < \infty$ and this implies clearly the existence of a $\lambda(=\lambda_{\omega_0})$ in $\mathbb{R}$ such that $\sup_n X_n(\omega_0) \leq \lambda$, and if (ii) $\omega_0 \in \Omega^{(r)} = S^c = \bigcup_{k\geq 1}\left(\limsup_n S_n(\delta_k)\right)^c = \bigcup_{k\geq 1}\left(\bigcup_{j\geq 1}\bigcap_{n\geq j} S_n^c(\delta_k)\right)$, then there is a $\delta(=\delta_{k_0}) > 0$, and a $j_0 \geq 1$ such that for all $n \geq j_0, \omega_0 \in S_n^c(\delta)$. Hence

$$P^{\mathcal{F}_i}[X_{i+1} > \lambda](\omega_0) > \delta, \quad \text{or} \quad = 0, \quad i \geq j_0, \tag{14}$$

by definition of $S_i(\delta)$, the δ-splitting set, excluding a P-null set. However, one has as $i \to \infty$

$$0 \leq P^{\mathcal{F}_i}[X_{i+1} > \lambda](\omega_0) \leq P^{\mathcal{F}_i}\left[\sup_n X_n > \lambda\right](\omega_0)$$
$$\to \begin{cases} 1 & \text{if } \sup_n X_n(\omega_0) > \lambda, \\ 0 & \text{otherwise,} \end{cases}$$

by the martingale convergence theorem (since $\{Z_i = P^{\mathcal{F}_i}[\sup X_n > \lambda], \mathcal{F}_i, i \geq 1\}$ is a bounded martingale). This, (i), and (14) imply that $P^{\mathcal{F}_i}[X_{i+1} > \lambda](\omega_0) = 0$ for some $i \geq m \geq j_0$, so that $T_m^\lambda(\omega_0) = +\infty$, as desired. $\square$

Remark. In this proof we only used that $0 < \delta < 1$ (not $\leq 1/2$). However, if the δ-splitting sets (with $\delta \leq 1/2$) are "fine enough" (i.e., if $S_{mn}(1/2)$ is the half splitting of $\mathcal{F}_m \subset \mathcal{F}_n, 1 \leq m < n < \infty$, then $\cup\{S_{mn}(1/2) : n > m\} = \Omega$ a.e. for each $m \geq 1$, and then $\{\mathcal{F}_n\}_1^\infty$ is called a *d-sequence* so that every $S_{mn}(\delta), 0 < \delta \leq 1/2$ also has the same property), then it can be shown that $\Omega^{(r)}$ is the best behaved set for the convergence theory. Thus there exist martingales which oscillate in any prescribed manner on $S = (\Omega^{(r)})^c$ while converging on $\Omega^{(r)}$. We shall present a related result later on as an exercise explaining this phenomenon. Unfortunately a better description of $S_n(\delta)$ is not available. These points should be compared with Theorem III.5.1.

The following result is an interesting application of Theorem 6, to some Paley-Littlewood inequalities, and is adapted from Stein [1].

8. Theorem. *Let (Ω, Σ, P) be a probability space, $\{\mathcal{F}_n, n \geq 1\}$ a filtering (to the right) sequence of σ-algebras, and $E^{\mathcal{F}_n}$ the conditional expectation. Let $T_a : L^2(P) \to L^2(P)$ be defined, with $a = (a_1, a_2, \dots), |a_i| \leq 1, a_i \in \mathbb{R}$, as:*

$$T_a(f) = \sum_{k=1}^{\infty} a_k [E^{\mathcal{F}_k}(f) - E^{\mathcal{F}_{k-1}}(f)], \quad f \in L^2(P). \tag{15}$$

Then T_a is linear, well-defined, and the following assertions hold:

(i) $\|T_a f\|_p \leq c_p \|f\|_p, f \in L^p(P), c_p > 0$ *depends only on* $p, 1 < p < \infty$.

(ii) $\lambda P[|T_a f| > \lambda] \leq c_1' \|f\|_1, \lambda > 0, f \in L^1 \cap L^2, c_1' > 0$, *an absolute constant.*

(iii) *If* $s^2(f) = \sum_{k=1}^{\infty} [E^{\mathcal{F}_k}(f) - E^{\mathcal{F}_{k-1}}(f)]^2$, $\mathcal{F}_\infty = \sigma\left(\bigcup_{n=1}^{\infty} \mathcal{F}_n\right)$, *then for* $1 < p < \infty$ *there exist constants* $c_{ip} > 0, i = 1, 2$ *depending only on* p, *such that*

$$c_{1p} \|E^{\mathcal{F}_\infty}(f)\|_p \leq \|s(f)\|_p \leq c_{2p} \|f\|_p, \qquad f \in L^2 \cap L^p. \tag{16}$$

Remark. In the above theorem, all the stated inequalities also hold if the measure space is nonfinite but the σ-subalgebras are P-rich. For this reason $f \in L^1 \cap L^2$ and $f \in L^2 \cap L^p$ were written. We must note that (16) is false if $p = 1$, but Theorem 5.3 below has a correct version. The proof depends on Theorem 6 and some inequalities of trignometric series (cf. Zygmund [1], V.8.4). We shall omit its detail here referring to the original sources. It is stated here for comparison purposes.

A key feature of both Theorems 6 and 8 is that the martingale differences φ_n (also called increments in the continuous parameter case) play a special role. Pursuing this point, we shall present a result in the spirit of Theorem 7 on the pointwise convergence on subsets of Ω, due to Doob [1] (see also Neveu [1]).

9. Theorem. *Let (Ω, Σ, P) be a probability space and $\{X_n, \mathcal{F}_n, n \geq 1\}$ be a real martingale on it. If $\{\varphi_n = X_n - X_{n-1}\}_1^\infty$ is the difference process ($X_0 = 0$, a.e.), then the following statements are true:*

(i) *If $A_p = \left\{\omega : \sum_{n=1}^{\infty} [E^{\mathcal{F}_n}(|X_{n+1}|^p](\omega) < \infty\right\}$, then for all $1 \leq p < \infty$, $X_n(\omega) \to X_\infty(\omega)$ (finite) for almost all $\omega \in A_p$. If $p = 1$ or 2 we may take for A_p the set $B = \left\{\omega : \sum_{n=1}^{\infty} E^{\mathcal{F}_n}(|\varphi_{n+1}|^p)(\omega) < \infty\right\}$.*

(ii) *If $A_0 = \left\{\omega : \limsup_n X_n(\omega) < \infty\right\}, \sup_n \varphi_n \in L^1(P)$, then $X_n(\omega) \to X_\infty(\omega)$ (finite) for almost all $\omega \in A_0$.*

(iii) *Let $\sup_n |\varphi_n| \in L^2(P)$. Then $X_n(\omega) \to X_\infty(\omega)$ (finite a.a. (ω)) iff $\sum_{n=1}^{\infty} E^{\mathcal{F}_n}(|\varphi_{n+1}|^2)(\omega) < \infty$, for a.a. (ω).*

(iv) *If $\{Z_n, \mathcal{F}_n, n \geq 0\}$ is an adapted process, $0 \leq Z_n \leq Z_{n+1}$ a.e., $B_p = \left\{\omega : Z_n(\omega) \to \infty, \text{ and } \sum_{n=1}^{\infty} [Z_{n-1}^{-1} E^{\mathcal{F}_n}(|\varphi_{n+1}|^p)](\omega) < \infty\right\}$, for any $1 \leq p < \infty$, then we have $\lim_{n\to\infty} Z_{n-1}^{-1}(\omega) \sum_{k=1}^{n} \varphi_k(\omega) = 0$ for a.a. $\omega \in B_p$.*

Remark. Several facts are collected at one place as they are all related to the martingale differences, since then useful consequences can be drawn from various specializations of the result.

Proof. It is convenient to note the following formula for use in all the parts. Thus let $\{Z'_n, \mathcal{F}_n, n \geq 1\}$ be an adapted integrable process. Let for any $\lambda > 0, T^\lambda = \inf\left\{n > 0 : \sum_{k=1}^{n} [E^{\mathcal{F}_k}(Z'_{k+1}) - Z'_k] > \lambda\right\}$. Then T^λ is a stopping time of $\{\mathcal{F}_n, n \geq 1\}$, and if $T_n^\lambda = T^\lambda \wedge n, Y_n = Z' \circ T_n^\lambda$,

$$Y_n = \sum_{k=1}^{n-1} Z'_k \chi_{[T^\lambda = k]} + Z'_n \chi_{[T^\lambda \geq n]} = \sum_{k=1}^{n-1} (Z'_{k+1} - Z'_k)\chi_{[T^\lambda > k]}. \tag{17}$$

However, $[T^\lambda > k] \in \mathcal{F}_k$. Hence

$$\begin{aligned} E(Y_n) &= \sum_{k=1}^{n-1} E[(E^{\mathcal{F}_k}(A'_{k+1} - Z'_k))\chi_{[T^\lambda > k]}], \\ &= E\left(\sum_{k < T_n^\lambda} E^{\mathcal{F}_k}(Z'_{k+1} - Z'_k)\right), \end{aligned} \tag{18}$$

since for $T^\lambda(\omega) \leq k$ the terms in the preceding sum vanish. Notice that the formula (18) is valid for *any* stopping time T of $\{\mathcal{F}_n, n \geq 1\}$ bounded by an integer, and the particular T^λ has not been used.

To prove (i), let $Z'_n = |X_n|^p$ so that the Z'-process is now a submartingale. Hence (19) implies, with the particular T^λ for this process,

$$E(Y_n) = E(|X \circ T_n^\lambda|^p) = E\left(\sum_{k<T_n^\lambda} [E^{\mathcal{F}_k}(|X_{k+1}|^p) - |X_k|^p]\right) \leq \lambda. \quad (19)$$

But by Corollary 1.8, $\{Y_n, \mathcal{F}_n, n \geq 1\}$ and $\{|X \circ T_n^\lambda|, \mathcal{F}_n, n \geq 1\}$ are submartingales so that by (19) and Theorem II.6.4 we deduce that Y_n converges a.e. as well as $X \circ T_n^\lambda \to X \circ T^\lambda$ a.e., as $n \to \infty$, on $A_p^\lambda = \left\{\omega : \sum_{k=1}^{\infty} [E^{\mathcal{F}_k}(|X_{k+1}|^p) - |X_k|^p](\omega) \leq \lambda\right\}$. Since $\lambda > 0$ is arbitrary and $A_p = \bigcup\{A_p^\lambda : \lambda > 0, \text{rational}\}$, the result of (i) holds on A_p.

Regarding A_1, A_2 it suffices to establish the result on A_1^λ, A_2^λ for any $\lambda > 0$. So if $p = 1$, then T^λ can be taken as: $T^\lambda = \inf\left\{n > 0 : \sum_{k=1}^{n} E^{\mathcal{F}_k}(|X_{k+1} - X_k|) > \lambda\right\}$. With this (19) becomes

$$\begin{aligned} E(Y_n) &= E\left(\sum_{k<T_n^\lambda} E^{\mathcal{F}_k}(|X_{k+1} - X_k + X_k| - |X_k|)\right) \\ &\leq E\left(\sum_{k<T_n^\lambda} E^{\mathcal{F}_k}(|X_{k+1} - X_k|)\right) \leq \lambda. \end{aligned} \quad (20)$$

The rest of the argument is unchanged. Hence A_1^λ may be replaced by the set B which may be properly contained in A_1^λ. In case $p = 2, A_2$ is precisely the set as given in statement since $E^{\mathcal{F}_k}(\varphi_{k+1} X_k) = 0$ a.e. Thus (i) is true in all respects.

To prove (ii) define the stopping time $\tau^\lambda = \inf\{n > 0 : X_n > \lambda\}$. Then with $Y_n = X \circ \tau_n^\lambda$ where $\tau_n^\lambda = \tau^\lambda \wedge n$, we deduce that (Corollary 1.8 again) $\{Y_n, \mathcal{F}_n, n \geq 1\}$ is a martingale. Moreover,

$$\begin{aligned} Y_n = X \circ \tau_n^\lambda &= \sum_{k=1}^{n-1} X_k \chi_{[\tau^\lambda = k]} + X_n \chi_{[\tau^\lambda \geq n]} \\ &\leq \lambda \chi_{[\tau^\lambda \leq n-1]} + \varphi_n \chi_{[\tau^\lambda \geq n]} + X_{n-1} \chi_{[\tau^\lambda \geq n]} \\ &\leq \lambda [\chi_{[\tau^\lambda \leq n-1]} + \chi_{[\tau^\lambda \geq n]}] + \sup_n \varphi_n \\ &\leq \lambda + \sup_n \varphi_n = V \text{ (say)}; \end{aligned} \quad (21)$$

but $V \in L^1(P)$, and thus $\sup_n E(Y_n^+) \leq E(|V|) < \infty$. Hence $Y_n \to Y_\infty$ a.e., and $Y_\infty = X \circ \tau^\lambda$ on $[\sup X_n \leq \lambda]$ for each $\lambda > 0$. This implies that on $A_0, X_n \to X_\infty$ a.e. (finite), since $Y_n = X_n$ on $\{\omega : \tau_n^\lambda(\omega) \leq \lambda\}$. So (ii) follows.

For (iii), note that $\sum_{n=1}^{\infty} E^{\mathcal{F}_n}(|\varphi_{n+1}|^2)(\omega) < \infty$ implies the existence of $\lim_n X_n(\omega) < \infty$ a.e. by (i), even without the additional hypothesis that $\sup_n |\varphi_n|^2$ is integrable. For the converse we need all the conditions. Given a $\lambda > 0$, let $\tau^\lambda = \inf\{n > 0 : |X_n| > \lambda\}, \tau_n^\lambda = \tau^\lambda \wedge n$, and $Y_n = X \circ \tau_n^\lambda$. Then the computation leading to (21) yields this time

$$|Y_n| \leq \lambda + \sup_n |\varphi_n|, \text{ a.e.} \tag{22}$$

Hence $E(Y_n^2) \leq 2\left[\lambda^2 + E\left(\sup_n |\varphi_n|^2\right)\right] < \infty$, so that $Y_n \to X \circ \tau^\lambda$ a.e. and in L^2, as $n \to \infty$. Hence by (i) (the case $p = 2$, cf. (19))

$$\begin{aligned}\lambda^2 \geq E(Y_\infty^2) &= \lim_n E(|X \circ \tau_n^\lambda|^2)\\ &= \lim_n E\left(\sum_{k<\tau_n^\lambda} E^{\mathcal{F}_k}(\varphi_{k+1}^2)\right) = E\left(\sum_{k<\tau^\lambda} E^{\mathcal{F}_k}(\varphi_{k+1}^2)\right)\end{aligned}$$

This means

$$\sum_{k=1}^{\infty} E^{\mathcal{F}_k}(\varphi_{k+1}^2)(\omega) < \infty \qquad \text{for a.a. } \omega \in B_\lambda = [\sup_n |X_n| \leq \lambda].$$

Since $\lambda > 0$ is arbitrary, the result follows as before and (iii) is true.

Finally to prove (iv), let $Z_n > 0$ a.e. If $V_n = \sum_{k=1}^{n} \varphi_k Z_{k-1}^{-1}$, then $\{V_n, \mathcal{F}_n, n \geq 1\}$ is clearly a martingale. Hence by (i) $\lim_n V_n = V_\infty < \infty$ a.e. on the set $\left[\sum_{k=0}^{\infty} Z_k^{-2} E^{\mathcal{F}_k}(\varphi_{k+1}^2) < \infty\right]$. On B_p, $Z_n \uparrow \infty$, so we have $\sum_{k=1}^{\infty} \varphi_k(\omega) Z_{k-1}^{-1}(\omega) < \infty$ and hence $\lim_n \frac{1}{Z_{n-1}} \sum_{k=1}^{n} \varphi_k = 0$ a.e. by Kronecker's lemma (cf. II.8.13(b)). This is (iv). □

Remark. Part (iii) of the above result can be improved by the following observation due to Austin [1]; and this opened up a whole sequence of investigations in martingale theory of a discrete parameter.

10. Extension of (iii). *If $\{X_n, \mathcal{F}_n, n \geq 1\}$ is an L^1-bounded martingale, then $\sum_{n=1}^{\infty} \varphi_n^2 < \infty$ a.e., where $\{\varphi_k\}_1^\infty$ is the martingale difference process.*

Proof. The condition implies by Theorem II.6.1 that $X_n \to X_\infty$ a.e. and is finite. So, letting $X_0 = 0$, consider

$$\sum_{k=1}^{n} \varphi_k^2 = \sum_{k=1}^{n} (X_k - X_{k-1})^2 = X_n^2 - 2\sum_{k=1}^{n} X_{k-1}X_k + 2\sum_{k=1}^{n-1} X_k^2$$
$$= X_n^2 - 2\sum_{k=1}^{n-1} \varphi_{k+1}X_k. \tag{23}$$

Since $X_k^2 \to X_\infty^2$ a.e. (finite) and the left side always has a limit, it is finite a.e. iff $\left|\sum_{k=1}^{n-1} \varphi_{k+1}X_k\right|(\omega)$ is bounded by a finite number (depending on ω) for a.a. (ω).

Let $\lambda > 0$ be arbitrry and set $E_n^\lambda = \{\omega : |X_k(\omega)| < \lambda, 1 \leq k \leq n\} \in \mathcal{F}_n$. Then on $A^\lambda \left(= \lim_n E_n^\lambda\right)$ each term of the martingale is bounded by λ and if $E_0 = \lim_n A^{\lambda_n} \left(= \bigcup_{n=1}^{\infty} A^{\lambda_n}\right)$ so that $E_0 = \left\{\omega : \sup_n |X_n(\omega)| < \infty\right\}$, we must have $P(E_0) = 1$. Indeed,

$$P[E_0^c] = \lim_{\lambda\to\infty} P\left[\sup_n |X_n| \geq \lambda\right] \leq \lim_\lambda \left[\frac{1}{\lambda} \sup_n E(|X_n|)\right] = 0,$$

by Corollary 1.5 and the fact that the martingale is L^1-bounded. Since $E_n^\lambda \in \mathcal{F}_n$ for any $\lambda > 0$, define $\widehat{X}_n = X_n \chi_{E_n^\lambda}$ and $\widehat{Y}_n = \sum_{k=1}^{n} \hat{\varphi}_k \widehat{X}_{k-1}$ where $\hat{\varphi}_k = \hat{X}_k - \hat{X}_{k-1}$. Then $\{\hat{Y}_n, \mathcal{F}_n, n \geq 1\}$ is a martingale since $\hat{\varphi}_k \hat{X}_{k-1} = (X_k \chi_{E_k^\lambda} - X_{k-1}\chi_{E_{k-1}^\lambda})X_{k-1}\chi_{E_{k-1}^\lambda} = (X_k - X_{k-1})X_{k-1}\chi_{E_{k-1}^\lambda}$ (because $E_k^\lambda \subset E_{k-1}^\lambda$) $= \varphi_k X_{k-1}\chi_{E_{k-1}^\lambda}$, so that $E^{\mathcal{F}_n}(\hat{\varphi}_{n+1}\widehat{X}_n) = X_n \chi_{E_n^\lambda} E^{\mathcal{F}_n}(\varphi_{n+1}) = 0$ a.e. On the other hand,

$$\int_\Omega \sup_n |\hat{Y}_{n+1} - \hat{Y}_n|^2 dP = \int_\Omega \sup_n |\hat{\varphi}_n X_{n-1}|^2 dP$$
$$\leq \lambda^2 \int_\Omega \sup_n |\hat{\varphi}_n|^2 dP \leq 4\lambda^4 < \infty. \tag{24}$$

Moreover, for any $n \geq 1$,

$$E\left(\sum_{k=1}^{n} E^{\mathcal{F}_k}(\hat{Y}_{k+1} - \hat{Y}_k)^2\right) = \sum_{k=1}^{n} E((\hat{\varphi}_{k+1}\hat{X}_k)^2) \leq \lambda^2 \sum_{k=1}^{n} E(\hat{\varphi}_{k+1}^2)$$
$$= \lambda^2 E(\hat{X}_n^2) \leq \lambda^4 < \infty. \tag{25}$$

It follows that $\sum_{n=1}^{\infty} E^{\mathcal{F}_n}(\hat{Y}_{n+1} - \hat{Y}_n)^2 = \sum_{n=1}^{\infty} E^{\mathcal{F}_n}(|\hat{\varphi}_{n+1}|^2) < \infty$, a.e. Hence (iii) of the theorem is applicable because of (24) and (25), so that $\hat{Y}_n \to \hat{Y}_\infty$ a.e. ($\leq \lambda$, and in L^2). This implies that $\hat{Y}_n \chi_{A^\lambda} \to \hat{Y}_\infty \chi_{A^\lambda}$ a.e. since $A^\lambda \subset E_n^\lambda$ for all n. However on $A^\lambda, \hat{Y}_n = \left(\sum_{k=1}^{n} \varphi_k X_{k-1}\right) \chi_{A^\lambda}$ so that in (23), $\left(X_n^2 - \sum_{k=1}^{n} \varphi_k^2\right) \chi_{A^\lambda} \to \hat{Y}_\infty \chi_{A^\lambda}$, a.e., for each $\lambda > 0$. The arbitrariness of $\lambda > 0$ and the finiteness of X_∞^2 a.e. implies $\sum_{k=1}^{\infty} \varphi_k^2 < \infty$ a.e. $\square$

The significance of this extension is that (even though $\sum_{n=1}^{\infty} \varphi_n^2 = s^2(X)$ need not be integrable) the convergence theory can be extended by considering the behavior of the differences process. Further this explains also why the inequalities of the type (16) are interesting in such a study. We discuss some useful consequences of the above results.

In Theorem 9(i), we have used the martingale property only in asserting that $\{|X_n|^p, \mathcal{F}_n, n \geq 1\}$ is a submartingale. Hence if $\{X_n, \mathcal{F}_n, n \geq 1\}$ is a positive submartingale, then that part holds true. Thus (ii) and (iii) take the following form for submartingales $\{X_n, \mathcal{F}_n, n \geq 1\}$:

(ii′) *Let* $\tilde{\varphi}_n = X_n - E^{\mathcal{F}_{n-1}}(X_n)$ *and* $\sup_n \tilde{\varphi}_n \in L^1(P)$, *then* $X_n(\omega) \to X_\infty(\omega)$ *exists for almost all* $\omega \in B_1 = \left\{\omega : \limsup_n X_n(\omega) < \infty\right\}$.

(iii′) *For* $\tilde{\varphi}_n$ *as above, if* $\sup_n |\tilde{\varphi}_n| \in L^2(P)$, *then* $X_n(\omega) \to X_\infty(\omega)$ *iff almost all* $\omega \in B_2 = \left\{\omega : \sum_{n=1}^{\infty} [E^{\mathcal{F}_n}(\tilde{\varphi}_n^2)](\omega) < \infty\right.$ *and* $\left.\sum_{n=1}^{\infty} [E^{\mathcal{F}_n}(X_{n+1}) - X_n](\omega) < \infty\right\}$.

These conditions clearly reduce to the preceding case for martingales, since then $\tilde{\varphi}_n = \varphi_n$ and $E^{\mathcal{F}_n}(\varphi_{n+1}) = 0$ a.e. The submartingale proofs are reduced to the martingale case by the Decomposition II.5.1. Thus

$$X_n = X_n' + \sum_{k=1}^{n} A_k, \qquad A_k \geq 0, \quad A_1 = 0 \text{ a.e.},$$

A_k is $\mathcal{F}_{k-1}$-adapted, $\{X'_n, \mathcal{F}_n, n \geq 1\}$ a martingale. Hence $\tilde{\varphi}_n = X_n - E^{\mathcal{F}_{n-1}}(X_n) = X'_n - X'_{n-1} = \varphi'_n$ (say) so that $\tilde{\varphi}_n$ agrees with a martingale difference, and $E^{\mathcal{F}_n}(X_{n+1}) - X_n = A_{n+1}$ a.e. Thus the result follows from the theorem.

(iv′) *Using (ii′), (iii′) in place of (ii), (iii), we get this part as: If* $\{\tilde{\varphi}_n, n \geq 1\}$ *is as above,* $0 \leq Z_n \uparrow, \mathcal{F}_n$*-adapted, then* $\lim_n \frac{1}{Z_n(\omega)} \sum_{k=1}^{n} \tilde{\varphi}_k(\omega) = 0$ a.a. (ω) *in* $B_n \cap \{\omega : Z_n(\omega) \to \infty$ *as* $n \to \infty\}$.

The following consequence, due to Paul Lévy, is frequently used in applications:

11. Corollary. *Let* $A_1, A_2, \ldots$ *be any sequence of measurable sets in* (Ω, Σ, P). *If* $\mathcal{F}_n = \sigma(A_1, \ldots, A_n)$ *is the* σ*-algebra generated by the sets shown, and* $X_{n+1} = E^{\mathcal{F}_n}(\chi_{A_{n+1}})$, *then* $A = \limsup_n A_n$ *and* $B = \left\{\omega : \sum_{n=1}^{\infty} X_n(\omega) = \infty\right\}$ *are* a.e. *equal; i.e.,* $A \Delta B$ *is a* P*-null set. [Intuitively this says that* $P\left[\limsup_n A_n\right] = P[A_n$ *occur infinitely often]* > 0 *if* $\sum_{n=1}^{\infty} E^{\mathcal{F}_n}(\chi_{A_{n+1}}) = \infty$ a.e., *and the probability is zero, if the series converges* a.e.*]*

Proof. Let $\varphi_n = \chi_{A_n} - X_n$ so that $E^{\mathcal{F}_n}(\varphi_{n+1}) = 0$ a.e., and let $Y_n = \sum_{k=1}^{n} \varphi_k$. Then $\{Y_n, \mathcal{F}_k, n \geq 1\}$ is a martingale. Since $\omega \in A\left(= \bigcap_{k=1}^{\infty} \bigcup_{n \geq k} A_n\right)$ iff $\sum_{n=1}^{\infty} \chi_{A_n}(\omega) = \infty$, and $Y_n \leq \sum_{i=1}^{n} \chi_{A_i}$, we deduce that $\sum_{n=1}^{\infty} \chi_{A_n}(\omega) < \infty$ implies $\limsup_n Y_n(\omega) < \infty$. Hence by Theorem 9(ii),

$$\lim_n Y_n(\omega) = Y_\infty(\omega) \quad \text{exists when} \quad \omega \notin A; \quad \text{or if} \quad \sum_{n=1}^{\infty} \chi_{A_n}(\omega) < \infty,$$

Thus $\sum_{n=1}^{\infty} X_n(\omega) = \sum_{n=1}^{\infty} \chi_{A_n}(\omega) - \lim_n Y_n(\omega)$ exists and is finite implies $\omega \notin A$ so that $\omega \notin B$. Multiplying by -1 we get

$$\sum_{k=1}^{\infty} \chi_{A_k}(\omega) = \sum_{k=1}^{\infty} X_k(\omega) - \lim_n(-Y_n(\omega)), \qquad \omega \notin B,$$

exists and is finite, so that $\omega \notin A$. Hence $\omega \in A$ iff $\omega \in B$ except possibly for a P-null set. This is precisely the stated result. $\square$

Note that $P(A) > 0$ if $\sum_{n=1}^{\infty} E^{\mathcal{F}_n}(\chi_{A_{n+1}}) = \infty$ a.e. However, if $\{A_n\}_1^\infty$ are *mutually independent*, we can strengthen this to say that $P(A) = 1$, because then $E^{\mathcal{F}_n}(\chi_{A_{n+1}}) = E(\chi_{A_{n+1}}) = P(A_{n+1})$ is no longer a random variable; it is a constant. To verify this strengthened conclusion, with $\sum_{n=1}^{\infty} E^{\mathcal{F}_n}(A_{n+1}) = \sum_{n=1}^{\infty} P(A_{n+1}) = \infty$, consider:

$$P(A) = 1 - P(A^c) = 1 - P\left(\bigcup_{k=1}^{\infty} \bigcap_{n \geq k} A_n^c\right) = 1 - \lim_{k\to\infty} \prod_{n\geq k} P(A_n^c)$$
$$= 1 - \lim_{k\to\infty} \prod_{n\geq k}(1 - P(A_n)) = 1,$$

since $\sum_{n\geq k} P(A_n) = +\infty$ implies $\prod_{n=k}^{\infty}(1 - P(A_n)) = 0$ for each k. We state this for reference as follows after noting that $\sum_{n=1}^{\infty} P(A_n) < \infty$ implies $\sum_{n=1}^{\infty} E^{\mathcal{F}_n}(\chi_{A_{n+1}}) < \infty$ a.e.

12. Corollary. (Borel-Cantelli lemma). *Let $A_1, A_2, \ldots$ be a sequence of measurable sets in (Ω, Σ, P). Then*

(i) $\sum_{n=1}^{\infty} P(A_n) < \infty$ *implies* $P\left(\limsup_n A_n\right) = 0$. *[A direct proof of this very simple.]*

(ii) $\sum_{n=1}^{\infty} P(A_n) = \infty$ *and* $\{A_n\}_1^\infty$ *are mutually independent implies* $P\left(\limsup_n A_n\right) = 1$. *[This is true if A_n's are only pairwise independent!]*

In many of the above results, a series such as $\sum_{k=1}^{n} \varphi_k \chi_{[T\geq k]}$ plays a key role. For instance the a.e. convergence of $\sum_{k=1}^{\infty} \varphi_k X_{k-1}$ was a main step in Result 10 above. This leads to the observation that in each case the martingale differences are multiplied by a "past" or a "predictable" element. We may state this precisely as follows: Let $\{V_{n+1}, \mathcal{F}_n, n \geq 1\}$ be an adapted process. For any other adapted process $\{X_n, \mathcal{F}_n, n \geq 1\}$ (or the family $\{\mathcal{F}_n, n \geq 1\}$), the V_n-process is said to be *predictable* for, and the process $\{(V\cdot X)_n, \mathcal{F}_n, n \geq 1\}$ a *predictable transform* of, the X-process where $(V\cdot X)_n = \sum_{k=1}^{n} \varphi_k V_k = \sum_{k=1}^{n} \varphi_k V_k = \sum_{k=1}^{n} (X_k - X_{k-1})V_k$ with $X_0 = 0$. In the first example $V_k = \chi_{[T\geq k]}$, and $V_k = X_{k-1}$ in the second case. It is interesting to note that every stopping time T determines a two valued decreasing predictable process and conversely the latter property determines a stopping time T of $\{\mathcal{F}_n, n \geq 1\}$. To see the (possibly non-obvious) converse, let $\{V_{k+1}, \mathcal{F}_k, k \geq 1\}$ be a two valued decreasing process. Define $T = \inf\{n > 0 : V_{n+1} = 0\}$ where $\inf\{\emptyset\} = \infty$. Then it is clear that T is a stopping time of $\{\mathcal{F}_n, n \geq 1\}$.

In applications involving "optimal stopping rules"–problems of Sequential Analysis and of certain optimal games–the following formulation is of interest. Let $\mathcal{T}$ be the collection of all stopping times of $\{\mathcal{F}_n, n \geq 1\}$ in (Ω, Σ, P). If $\{X_n, \mathcal{F}_n, n \geq 1\}$ is an adapted process lying in a ball of $L^1(P)$, find a $\tau \in \mathcal{T}$ such that

$$E(X \circ \tau) = \sup\{E(X \circ T) : T \in \mathcal{T}\}. \tag{26}$$

If a τ satisfying (26) exists, it is called *optimal*. More generally, we may ask: Find a process V^* in a class $\mathcal{V}_X$ of predictable processes for $\{X_n, \mathcal{F}_n, n \geq 1\}$, such that $E((V\cdot X)_\infty)$ exists and

$$E((V^* \cdot X)_\infty) = \sup\{E((V \cdot X)_\infty) : V \in \mathcal{V}_X\}. \tag{27}$$

To indicate the flavor of the problem we state a general result due to Alloin [1] who studied the problem extending some earlier work.

If $-\infty < a \leq b < \infty$, define $\mathcal{V}_1 = \{V \in \mathcal{V}_X : V = \{V_{n+1}, \mathcal{F}_n, n \geq 1\}, a \leq V_n \leq b\}$. Let $V^* = \{V_{0n}, \mathcal{F}_{n-1}, n \geq 1\}$ where $V_{0n} = b$ on $[E^{\mathcal{F}_{n-1}}(X_n) > X_{n-1}]$, and $= a$ otherwise.

Then one has the following result:

13. Proposition. *Let $\{X_n, \mathcal{F}_n, n \geq 1\}$ be a submartingale on (Ω, Σ, P) such that $\sup_n |X_n| \in L^1(P)$. Let $\mathcal{V}_2 = \{V \in \mathcal{V}_1 : V = \{V_n, \mathcal{F}_{n-1}, n \geq 1\}, V_n = V_n^+ - V_n^-$ (and $\{V_n^{\pm}\}_1^\infty$ are monotone in the same sense)$\}$. Then V^* defined above is in $\mathcal{V}$, and is an optimal predictable process for the given martingale in that (27) holds for this V^* with $\mathcal{V}_2$ in place of $\mathcal{V}_X$ there.*

We shall not prove this result here, since it will not be used below, and refer the reader to the original paper. Instead, we shall consider some other aspects of martingale differences and the spaces of such process in the next section.

4.5 Martingale differences and $\mathcal{H}^p$-spaces

The importance of the martingale difference process became evident in the preceding section. However, the final inequality in Theorem 4.8 was given only for $1 < p < \infty$. The case $p = 1$ needs a special and nontrivial treatment, involving some new ideas which however are intimately related to the classical notions of bounded mean oscillation of functions and the $\mathcal{H}^p$-spaces of Hardy. In the case $p = \infty$, the inequality is generally false, and the result for $p = 1$ is due to Davis [1]. We follow, for the most part, Garsia [2]. One notes the close relations between martingales and the theory of $\mathcal{H}^p$-spaces in this work.

Let $X = \{X_n, \mathcal{F}_n, n \geq 1\}$ be a martingale, $X_0 = 0$ a.e., and $\varphi_n = X_n - X_{n-1}$. As before let $X^* = \sup_n |X_n|, s_n(X) = \left[\sum_{k=1}^{n} \varphi_k^2\right]^{1/2}$ and $s(X) = \lim_n s_n(X)$. A space of *bounded mean oscillation, or* BMO, relative to a stochastic base $\{\mathcal{F}_n, n \geq 1\}$, is the class of all uniformly integrable martingales X defined as follows:

$$BMO = \left\{X : \sup_n \|E^{\mathcal{F}_n}(|X_\infty - X_{n-1}|^2)\|_\infty < \infty\right\}, \tag{1}$$

where $X_\infty = \lim_n X_n$ which exists a.e. and in $L^1(P)$, by Theorem II.6.2. Note that each L^∞-bounded martingale is in BMO. In fact, if $k_0 = \|X_\infty\|_\infty$ then one has

$$E^{\mathcal{F}_n}(|X_\infty - X_{n-1}|^2) \leq E^{\mathcal{F}_n}(|X_\infty|)^2 + E^{\mathcal{F}_n}(|X_{n-1}|^2) + 2E^{\mathcal{F}_n}(|X_{n-1}X_\infty|) \leq 4k_0^2 < \infty.$$

Thus BMO has sufficiently many elements. Define a functional on this space as:

$$\|X\|_B = \sup_n \|[E^{\mathcal{F}_n}(X_\infty - X_{n-1})^2]^{1/2}\|_\infty. \tag{2}$$

It is clear that $\|\cdot\|_B$ is a semi-norm and that BMO is a linear space. To see that $\|\cdot\|_B$ is a norm, suppose $\|X\|_B = 0$. Then $E^{\mathcal{F}_n}(|X_\infty - X_{n-1}|^2) = 0$, a.e., for all $n \geq 1$. On the other hand, for $k \geq n, E^{\mathcal{F}_n}$

$(\varphi_k\varphi_n) = 0$ and so $\sum_{k\geq n} \varphi_k^2 = |X_\infty - X_{n-1}|^2$ a.e., and is finite by (4.10). Hence $E^{\mathcal{F}_n}(|X_\infty - X_{n-1}|^2) = \sum_{k\geq n} E^{\mathcal{F}_n}(\varphi_k^2) = 0$ implying $E^{\mathcal{F}_n}(\varphi_k^2) = 0$, or $|\varphi_k| = 0$, a.e., $k \geq n \geq 1$ ($E^{\mathcal{F}_n}$ being faithful). Since $|\varphi_1| = |X_1|$, this shows that $X = 0$ a.e.

It is evident that iBMO,$\|\cdot\|_B\}$ is a normed linear space. That the BMO is also complete is not too hard to note. However, this follows as a by-product when we show that it is the adjoint space of another normed linear space, namely $\mathcal{H}^1$, which we now introduce. The *space* $\mathcal{H}^p, p \geq 1$, is the class of all martingales X, defined as:

$$\mathcal{H}^p = \{X : E(s(X)^p) < \infty\}, \tag{3}$$

and we take the norm as $\|X\|_p = \|s(X)\|_p$. If we represent X by its difference sequence $\{\varphi_1, \varphi_2, \dots\}$, then $\mathcal{H}^p$ is seen to be a subspace of $L^p(\ell^2)$ where $L^p(\mathcal{X})$ is the Banach space of $\mathcal{X}$-valued functions f, on Ω, such that $\|f\|^p$ is integrable on (Ω, Σ, P), with $\mathcal{X} = \ell^2$, the sequence Hilbert space. Thus it is clear that $\{\mathcal{H}^p, \|\cdot\|_p\}$ is a normed linear space. The completeness can be proved along standard lines. Since this is not hard and not needed below, we shall leave it to the reader.

The following basic inequality which implies that $\langle \mathcal{H}_1, BMO\rangle$ is paired in duality has been established (for an important special case) by Fefferman [1], and more generally by Fefferman and Stein [1], using nonprobabilistic methods.

1. Theorem. *Let $X = \{X_n, \mathcal{F}_n, n \geq 1\}$ and $Y = \{Y_n, \mathcal{F}_n, n \geq 1\}$ be two martingales such that $X \in \mathcal{H}^1$ and $Y \in BMO$, for the same base $\{\mathcal{F}_n, n \geq 1\}$. Then we have*

$$|E(X_nY_n)| \leq \sqrt{2}\|X\|_1\|Y\|_B. \tag{4}$$

Proof. Let $f_n = X_n - X_{n-1}, g_n = Y_n - Y_{n-1}, n \geq 1$. Since $Y \in BMO$, one has

$$g_n = E^{\mathcal{F}_n}\Big(\sum_{k\geq n} g_k\Big) = E^{\mathcal{F}_n}(Y_\infty - Y_{n-1}),$$

because $E^{\mathcal{F}_n}(g_k) = 0$ for $k > n$. Hence $g_n^2 \leq E^{\mathcal{F}_n}(Y_\infty - Y_{n-1})^2 \leq \|Y\|_B^2$, a.e., using the conditional Jensen inequality and (2). This result and the fact that $g_1 = Y_1$ a.e., together imply that $Y_n = \sum_{i=1}^{n} g_i \in L^\infty(P)$ so that $X_nY_n \in L^1(P)$ for each $n \geq 1$ and moreover

$$\begin{aligned}\int_\Omega X_nY_n dP &= \sum_{i=1}^{n}\int_\Omega f_ig_i dP + 2\sum_{1\leq i<j\leq n}\int_\Omega f_ig_i dP\\ &= \sum_{i=1}^{n}\int_\Omega f_ig_i dP, \qquad (5)\end{aligned}$$

since for $i < j, E(f_ig_i) = E[E^{\mathcal{F}_i}(f_ig_i)] = E[f_iE^{\mathcal{F}_i}(g_j)] = 0$. (The representation (5) is clearly true also for $\langle L^p, L^q\rangle$ instead of $\langle L^1, L^\infty\rangle$.) To simplify further, we multiply and divide the integrand by the root of $s_i = s_i(X)$ and use the CBS inequality twice to get

$$\begin{aligned}\left|\int_\Omega X_nY_n dP\right| &= \left|\sum_{i=1}^{n}\int_\Omega \frac{f_i}{\sqrt{s_i}}\cdot(\sqrt{s_i}g_i)dP\right|\\ &\leq \sum_{i=1}^{n}\left[\int_\Omega \frac{f_i^2}{s_i}dP\right]^{1/2}\left[\int_\Omega s_ig_i^2 dP\right]^{1/2}\\ &\leq \left[\sum_{i=1}^{n}\int_\Omega \frac{f_i^2}{s_i}dP\right]^{1/2}\left[\sum_{i=1}^{n}\int_\Omega s_ig_i^2 dP\right]^{1/2} \qquad (6)\end{aligned}$$

Regarding the right side of (6), we consider (with $s_0 = 0$)

$$\begin{aligned}\sum_{i=1}^{n}\int_\Omega \frac{s_i^2 - s_{i-1}^2}{s_i}dP &\leq 2\sum_{i=1}^{n}\int_\Omega (s_i - s_{i-1})dP, \text{ since } 2s_is_{i-1} \leq s_i^2 + s_{i-1}^2,\\ &= 2\int_\Omega s_n dP = 2\|s_n(X)\|_1 \leq 2\|X\|_1. \qquad (7)\end{aligned}$$

Then the second factor (with $s_0 = 0$) becomes:

$$\begin{aligned}\sum_{i=1}^{n}\int_\Omega g_i^2 s_i dP &= \sum_{k=1}^{n}\int_\Omega g_k^2\sum_{i=1}^{k}(s_i - s_{i-1})dP\\ &= \sum_{i=1}^{n}\int_\Omega (s_i - s_{i-1})\sum_{k=i}^{n} g_k^2 dP\end{aligned}$$

$$
\begin{aligned}
&= \sum_{i=1}^{n} \int_{\Omega} E^{\mathcal{F}_i}\left[(s_i - s_{i-1}) \sum_{k=1}^{n} g_k^2\right] dP \\
&= \sum_{i=1}^{n} \int_{\Omega} (s_i - s_{i-1}) E^{\mathcal{F}_i}\left(\sum_{k=i}^{n} g_k\right)^2 dP \\
&= \sum_{i=1}^{n} \int_{\Omega} (s_i - s_{i-1} E^{\mathcal{F}_i}(Y_n - Y_{i-1})^2 dP \\
&= \sum_{i=1}^{n} \int_{\Omega} (s_i - s_{i-1}) E^{\mathcal{F}_i}(E^{\mathcal{F}_n}(Y_\infty - Y_{i-1}))^2 dP \\
&\leq \sum_{i=1}^{n} \int_{\Omega} (s_i - s_{i-1}) E^{\mathcal{F}_i}(Y_\infty - Y_{i-1})^2 dP \\
&\leq \|Y\|_B^2 \cdot \sum_{i=1}^{n} \int_{\Omega} (s_i - s_{i-1}) dP \\
&= \|Y\|_B^2 \cdot \|s_n(X)\|_1 \leq \|Y\|_B^2 \cdot \|X\|_1. \qquad (8)
\end{aligned}
$$

Substituting (7) and (8) in (6) we get (4). □

This result admits an extension, in the manner of Hölder's inequality, to other $p \geq 1$ (cf. Theorem 6 below), and then it will be used in obtaining an integral representation of $(\mathcal{H}^p)^*$. First we use (4) in completing the inequalities for $p = 1$ in Theorem 4.8. We now present a simple construction of elements of BMO to employ it later.

2. Lemma. (i) *Let $\{Z_n, n \geq 1\}$ be any sequence of random variables on (Ω, Σ, P) such that $\sum_{n=1}^{\infty} |Z_n| \leq 1$,* a.e. *For each stochastic base $\{\mathcal{F}_n, n \geq 1\}$ if we define $X = \sum_{n=1}^{\infty} E^{\mathcal{F}_n}(Z_n)$, then the martingale $\{X_n = E^{\mathcal{F}_n}(X), n \geq 1\}$ is an element of* BMO *and satisfies $E^{\mathcal{F}_n}(X - X_{n-1})^2 \leq 5$* a.e., *for all $n > 1$.* (ii) *If on the other hand $\{Z_n, \mathcal{F}_n, n \geq 0\}$ is an adapted sequence, and $Z^* = \sup_n |Z_n|$, then the function $Y = \sum_{n=1}^{\infty} Z_{n-1} \left\{E^{\mathcal{F}_n}\left(\frac{1}{Z^*}\right) - E^{\mathcal{F}_{n-1}}\left(\frac{1}{Z^*}\right)\right\}$ defines a* BMO *element $\{Y_n = E^{\mathcal{F}_n}(Y), n \geq 1\}$ and moreover, $E^{\mathcal{F}_n}(Y - Y_{n-1})^2 \leq 2$* a.e., *for $n \geq 1$, so that $\|Y\|_B \leq \sqrt{2}$.*

Proof. (i) Observe that $E(|X|) \leq \sum_{n=1}^{\infty} E(E^{\mathcal{F}_n}(|Z_n|)) = \sum_{n=1}^{\infty} E(|Z_n|) \leq$

1. If we set $U_n = \sum_{k\geq n} E^{\mathcal{F}_n}(Z_k)$, then (for $n > 1$)

$$X_{n-1} = \sum_{k=1}^{n-1} E^{\mathcal{F}_k}(Z_k) + E^{\mathcal{F}_{n-1}}\left(\sum_{k\geq n} E^{\mathcal{F}_n}(Z_k)\right)$$
$$= \sum_{k=1}^{n-1} E^{\mathcal{F}_k}(Z_k) + E^{\mathcal{F}_{n-1}}(U_n),$$

and hence

$$X - X_{n-1} = U_n - E^{\mathcal{F}_{n-1}}(U_n). \tag{9}$$

To complete the proof we estimate the norm as follows.

$$\begin{aligned} E^{\mathcal{F}_n}(X - X_{n-1})^2 &= E^{\mathcal{F}_n}[U_n^2 + (E^{\mathcal{F}_{n-1}}(U_n))^2 - 2U_n E^{\mathcal{F}_{n-1}}(U_n)] \\ &\leq E^{\mathcal{F}_n}(U_n^2) + 2(E^{\mathcal{F}_n}|U_n|) \cdot E^{\mathcal{F}_{n-1}}(|U_n|) \\ &\quad + (E^{\mathcal{F}_{n-1}}(U_n))^2 \leq E^{\mathcal{F}_n}(U_n^2) + 3, \quad \text{a.e.,} \end{aligned} \tag{10}$$

since

$$E^{\mathcal{F}_j}(|U_n|) \leq \sum_{k\geq n} E^{\mathcal{F}_j} E^{\mathcal{F}_k}(|Z_k|))$$
$$= \sum_{k\geq n} E^{\mathcal{F}_j}(|Z_k|) = E^{\mathcal{F}_j}\left(\sum_{k\geq n} |Z_k|\right) \leq 1 \quad \text{a.e.,}$$

for $j \leq n$. To simplify the first term of (10), consider

$$\begin{aligned} E^{\mathcal{F}_n}(U_n^2) &= E^{\mathcal{F}_n}\left(\sum_{k\geq n}(E^{\mathcal{F}_k}(Z_k))^2\right) \\ &\quad + 2E^{\mathcal{F}_n}\left(\sum_{n\leq i<j<\infty} (E^{\mathcal{F}_i}(Z_i)E^{\mathcal{F}_j}(Z_j))\right) \\ &\leq 2 \sum_{n\leq i\leq j<\infty} E^{\mathcal{F}_n}[E^{\mathcal{F}_i}(|Z_i|) \cdot E^{\mathcal{F}_i}(|Z_j|)] \\ &= 2\sum_{i\geq n} E^{\mathcal{F}_n}\left(E^{\mathcal{F}_i}(|Z_i|)E^{\mathcal{F}_i}\left(\sum_{j\geq i}|Z_j|\right)\right) \\ &\leq 2\sum_{i\geq n} E^{\mathcal{F}_n}(|Z_i|) \cdot 1 \leq 2, \text{ a.e.,} \end{aligned} \tag{11}$$

since $\sum_{i\geq 1} |Z_i| \leq 1$. From (10) and (11) we deduce that $\sup_n E^{\mathcal{F}_n}(X - X_{n-1})^2 \leq 5$, a.e., establishing (i).

(ii) We may assume that $|Z_n| > 0$, for at least one n, so that $Z^* > 0$ on a set of positive measure. If we let $Z_n^* = \sup_{1\leq k\leq n} |Z_k|$, then $Z_n^* \leq Z^*$, a.e. Since $\mathcal{F}_n \subset \mathcal{F}_{n+1}$ it is clear that $E^{\mathcal{F}_n} - E^{\mathcal{F}_{n-1}}$ is a projection on $L^1(P)$ and is defined on all positive measurable functions mapping them into the space of measurable functions. Thus $h_n = (E^{\mathcal{F}_n} - E^{\mathcal{F}_{n-1}})\left(\frac{1}{Z^*}\right)$ is a (finite) measurable function, (set $h_n = 0$ on $A = [Z^* = 0]$). Now, for each n, if we let $W_n = \sum_{k=1}^{n} Z_{k-1}h_n$, then $E^{\mathcal{F}_n}(W_{n+1}) = W_n + Z_n E^{\mathcal{F}_n}(E^{\mathcal{F}_{n+1}} - E^{\mathcal{F}_n})\left(\frac{1}{Z^*}\right) = W_n$ a.e., so that $\{W_n, \mathcal{F}_n, n \geq 1\}$ is (as yet not necessarily inegrable) martingale, and W_n is finite a.e. Because of the special definition, we see that $E^{\mathcal{F}_n}((Z_m h_{m+1}) \cdot Z_k h_{k+1}) = 0$ a.e. for $n \leq k < m$. Thus the differences are "orthogonal". Hence for any $N \geq n$ we have

$$
\begin{aligned}
E^{\mathcal{F}_n}(W_N - W_n)^2 &= E^{\mathcal{F}_n}\left(\sum_{k=n}^{N} Z_{k-1}h_k\right)^2 \\
&= \sum_{k=n}^{N} E^{\mathcal{F}_n}(Z_{k-1}h_k)^2, \text{ a.e.,} \\
&\leq \sum_{k=n}^{N} E^{\mathcal{F}_n}(Z_{k-1}^* h_k)^2. \qquad (12)
\end{aligned}
$$

Since $Z_n^* \leq Z_{n+1}^*$ a.e., and $E^{\mathcal{F}_n} - E^{\mathcal{F}_{n-1}}$ is a projection, we have

$$
\begin{aligned}
-1 \leq -E^{\mathcal{F}_{n-1}}\left(\frac{Z_{n-1}^*}{Z^*}\right) &= Z_{n-1}^* E^{\mathcal{F}_{n-1}}\left(\frac{-1}{Z^*}\right) \\
&\leq Z_{n-1}^*[E^{\mathcal{F}_n} - E^{\mathcal{F}_{n-1}}]\left(\frac{1}{Z^*}\right) \\
&\leq Z_{n-1}^* E^{\mathcal{F}_n}\left(\frac{1}{Z^*}\right) \leq 1. \qquad (13)
\end{aligned}
$$

It follows that $(Z_{n-1}^* h_n)^2 \leq 1$ a.e. and (12) is finite for each W. Consequently (12) reduces to:

$$
E^{\mathcal{F}_n}(W_N - W_n)^2 \leq \sum_{k=n}^{N} E^{\mathcal{F}_n}[(Z_{k-1}^*)^2 h_k^2]
$$

$$
\begin{aligned}
&= \sum_{k=n}^{N} E^{\mathcal{F}_n}[Z_{k-1}^{*2} E^{\mathcal{F}_{k-1}}(h_k^2)] \\
&= \sum_{k=n}^{N} E^{\mathcal{F}_n}\left[Z_{k-1}^{*2} E^{\mathcal{F}_{k-1}}\left(\left(\frac{1}{Z^*}\right)\right)^2\right. \\
&\quad \left. - \left(E^{\mathcal{F}_{k-1}}\left(\frac{1}{Z^*}\right)\right)^2\right] \\
&= \sum_{k=1}^{N} E^{\mathcal{F}_n}\left[Z_{k-1}^{*2}\left(\left(E^{\mathcal{F}_k}\left(\frac{1}{Z^*}\right)\right)^2\right)\right. \\
&\quad \left. - \left(E^{\mathcal{F}_{k-1}}\left(\frac{1}{Z^*}\right)\right)^2\right)\Bigg] \\
&\le \sum_{k=n}^{N} E^{\mathcal{F}_n}\left(Z_k^{*2}\left(E^{\mathcal{F}_k}\left(\frac{1}{Z^*}\right)\right)^2\right) \\
&\quad - \sum_{k=n}^{N} E^{\mathcal{F}_n}\left(Z_{k-1}^{*2}\left(E^{\mathcal{F}_{k-1}}\left(\frac{1}{Z^*}\right)\right)^2\right) \\
&\le E^{\mathcal{F}_n}\left(Z_N^{*2}\left(E^{\mathcal{F}_N}\left(\frac{1}{Z^*}\right)\right)^2\right) \\
&\quad + E^{\mathcal{F}_n}\left(Z_{n-1}^{*2}\left(E^{\mathcal{F}_{n-1}}\left(\frac{1}{Z^*}\right)\right)^2\right) \le 2.
\end{aligned} \tag{14}
$$

Now taking $n = 0$, and noting that $W_0 = 0$ a.e. (by convention) and $\mathcal{F}_0 = \{\emptyset, \Omega\}$, (14) shows that $\{W_N, N \ge 1\}$ is $L^2(P)$-bounded, so $W_N \to W$ a.e. and in $L^2(P)$. Clearly $W = Y$ of the statement. Now letting $N \to \infty$ in (14) we get $E^{\mathcal{F}_n}(Y - Y_n)^2 \le 2$. This gives (ii) and the lemma. □

The asserted inequality can now be given as follows:

3. Theorem. *Let $X = \{X_n, \mathcal{F}_n, n \ge 1\}$ be a martingale in $L^1(P)$ and suppose that $X^* = \sup_n |X_n| \in L^1(P)$. If $s(X)$ is defined as before, then*

$$10^{-1/2}\|X^*\|_1 \le \|s(X)\|_1 \le (2 + \sqrt{5})\|X^*\|_1, \tag{15}$$

where we take $X_0 = 0$, a.e., in the definition of $s(X)$.

Proof. To establish the first half, we use a decomposition of X_n^* $\left(= \max_{k \le n} |X_k|\right)$ analogous to that of Theorem II.4.3. Thus for each

$1 \leq k \leq n$, define the disjoint sequence

$$A_k = \{\omega : X^*_{k-1}(\omega) < X^*_n(\omega), X^*_k(\omega) = X^*_n(\omega)\} \in \mathcal{F}_n. \tag{16}$$

Since by assumption $X_0 = 0$, let $Z_k = \chi_{A_k} \operatorname{sgn} X_k$, and consider

$$\begin{aligned}
\int_\Omega X^*_n dP &= \sum_{k=1}^n \int_{A_k} X^*_k dP = \sum_{k=1}^n \int_{A_k} |X_k| dP \\
&= \sum_{k=1}^n \int_\Omega Z_k X_k dP \\
&= \sum_{k=1}^n \int_\Omega E^{\mathcal{F}_k}(X_n) Z_k dP, \text{ since } \quad E^{\mathcal{F}_k}(X_n) = X_k \text{ a.e.}, \\
&= \sum_{k=1}^n \int_\Omega E^{\mathcal{F}_k}[X_n E^{\mathcal{F}_k}(Z_k)] dP, \quad \text{by the properties of } E^{\mathcal{F}}, \\
&= \sum_{k=1}^n \int_\Omega X_n E^{\mathcal{F}_k}(Z_k) dP = \int_\Omega X_n \left(\sum_{k=1}^n E^{\mathcal{F}_k}(Z_k) \right) dP \\
&= \int_\Omega X_n Y_n dP \quad \text{(say)}, \\
&\leq \sqrt{2} \|s(X)\|_1 \|Y\|_B, \quad \text{by Theorem 1,}
\end{aligned} \tag{17}$$

where $\{Y_m = E^{\mathcal{F}_m}(Y_n), \mathcal{F}_m, n \geq m \geq 1\}$ is a martingale. Recalling that $\sum_{k \geq 1} |Z_k| = 1$, we now apply Lemma 2(i). Thus (17) reduces to:

$$\int_\Omega X^*_n dP \leq \sqrt{2} \|s(X)\|_1 \cdot \sqrt{5} = \sqrt{10} \|s(X)\|_1. \tag{18}$$

Letting $n \to \infty$ and noting that $X^*_n \uparrow X^*$, we deduce the first half of (15) from (18).

To prove the second half of (15) one uses Lemma 2(ii) and Theorem 1 after some computations. Thus consider, as in (6) with the CBS inequality,

$$\int_\Omega s_n(X) dP = \int_\Omega \sqrt{X^*} \frac{s_n(X)}{\sqrt{X^*}} dP \leq \left[\|X^*\|_1 \left\| \frac{s_n^2(X)}{X^*} \right\|_1 \right]^{1/2}. \tag{19}$$

But writing $\varphi_k = X_k - X_{k-1}$, and using (23) of Section 4 above, we have $s_n^2(X) = X_n^2 - 2\sum_{k=1}^{n-1} \varphi_{k+1} X_k$ and so

$$\int_\Omega \frac{s_n^2(X)}{X^*} dP \leq \int_\Omega (X_n^2/X^*) dP + 2\left|\sum_{k=1}^{n} \int_\Omega (\varphi_k X_{k-1}/X^*) dP\right|. \tag{20}$$

Let us estimate the size of the quantity $Q = \sum_{k=1}^{n} E(\varphi_k X_{k-1}/X^*)$. Since $E^{\mathcal{F}_{k-1}}(\varphi_k) = 0$ a.e. and $E^{\mathcal{F}_k}(\varphi_k) = \varphi_k$ a.e., it follows that $E\left(\varphi_k E^{\mathcal{F}_{k-1}}\left(\frac{X_{k-1}}{X^*}\right)\right) = E\left(E^{\mathcal{F}_{k-1}}\left(\varphi_k E^{\mathcal{F}_{k-1}}\left(\frac{X_{k-1}}{X^*}\right)\right)\right) = 0$. Hence we may express Q as:

$$\begin{aligned}
Q &= \sum_{k=1}^{n} E\left(\varphi_k \left\{E^{\mathcal{F}_k}\left(\frac{X_{k-1}}{X^*}\right) - E^{\mathcal{F}_{k-1}}\left(\frac{X_{k-1}}{X^*}\right)\right\}\right) \\
&= \sum_{k=1}^{n} E\left(E^{\mathcal{F}_k}(X_n - X_{k-1})\left[E^{\mathcal{F}_k}\left(\frac{X_{k-1}}{X^*}\right) - E^{\mathcal{F}_{k-1}}\left(\frac{X_{k-1}}{X^*}\right)\right]\right), \\
&\qquad\qquad \text{since } E^{\mathcal{F}_k}(X_n) = X_k \text{ a.e.,} \\
&= \sum_{k=1}^{n} E\left(X_n \left[E^{\mathcal{F}_k}\left(\frac{X_{k-1}}{X^*}\right) - E^{\mathcal{F}_{k-1}}\left(\frac{X_{k-1}}{X^*}\right)\right]\right) \\
&\quad - \sum_{k=1}^{n} \left[E\left(\frac{X_{k-1}^2}{X^*}\right) - E\left(\frac{X_{k-1}^2}{X^*}\right)\right] \\
&= E\left(X_n \left[\sum_{k=1}^{n} E^{\mathcal{F}_k}\left(\frac{X_{k-1}}{X^*}\right) - E^{\mathcal{F}_{k-1}}\left(\frac{X_{k-1}}{X^*}\right)\right]\right) \\
&= E(X_n g_n),
\end{aligned} \tag{21}$$

where $g_n = \sum_{k=1}^{n} X_{k-1}(E^{\mathcal{F}_k} - E^{\mathcal{F}_{k-1}}\left(\frac{1}{X^*}\right)$. But this sequence satisfies the hypothesis of Lemma 2(ii) with $Z_n = X_n$. If $Y = \lim_n g_n$ a.e., (shown to exist there) then (21) reduces to:

$$|Q| = |E(X_n g_n)| \leq 2\|s_n(X)\|_1, \quad \text{by Theorem 1.} \tag{22}$$

Since $\frac{X_n^2}{X^*} \leq X^*$ a.e., we get from (19), (20), and (22):

$$\begin{aligned}
a = \int_\Omega s_n(X) dP &\leq \left[\|X^*\|_1 \left\{\|X^*\|_1 + 2|Q|\right\}\right]^{1/2} \\
&\leq \left[\|X^*\|_1 \left(\|X^*\|_1 + 4\|s_n(X)\|_1\right)\right]^{1/2}
\end{aligned}$$

If $b = \|X^*\|_1$, then we have

$$a^2 \leq b(b+4a) \quad \text{or} \quad (a-2b)^2 \leq 5b^2. \tag{23}$$

Hence $0 \leq a \leq (2+\sqrt{5})b$ which is the right half of (15). Thus (18) and (23) establish (15). □

Combining the results of Theorem 4.8 and the above one, we may state:

4. Theorem. *Let $X = \{X_n, \mathcal{F}_n, n \geq 1\}$ be a martingale on (Ω, Σ, P), $X^* = \sup_n |X_n| \in L^p(P)$. If $s(X)$ is the square function defined above, then there exist absolute constants $c_{ip} > 0, i = 1, 2$, such that*

$$c_{1p}\|s(X)\|_p \leq \|X^*\|_p \leq c_{2p}\|s(X)\|_p, \quad 1 \leq p < \infty. \tag{24}$$

Proof. For $p = 1$, (24) is a restatement of (15). So let $1 < p < \infty$. Then the hypothesis implies that the martingale is L^p-bounded and hence is uniformly integrable. Thus Theorem II.6.2 implies that $X_n \to X_\infty$ a.e. and in L^p. Moreover, $X_n = E^{\mathcal{F}_n}(X_\infty)$ a.e. By Theorem 4.8, we get for $X = \{X_n, \mathcal{F}_n, n \geq 1\}$, (which is first applied to the truncated martingale $E(X_\infty \wedge k)$ and then letting $k \to \infty$)

$$A_{1p}\|X_\infty\|_p \leq \|s(X)\|_p \leq A_{2p}\|X_\infty\|_p, \quad 1 < p < \infty, \tag{25}$$

since $E^{\mathcal{F}_\infty}(X_\infty) = X_\infty$ a.e. However, by Theorem II.4.8, $\|X^*\|_p \leq \frac{p}{p-1}\|X_\infty\|_p$, and since $|X_\infty| \leq X^*$, we have $\|X_\infty\|_p \leq \|X^*\|_p$. Hence (25) becomes

$$\|s(X)\|_p \leq A_{2p}\|X^*\|_p \leq \frac{p}{p-1}A_{2p}\|X_\infty\|_p \leq \frac{pA_{2p}}{A_{1p}(p-1)}\|s(X)\|_p. \tag{26}$$

Letting $c_{1p} = A_{2p}^{-1}$ and $c_{2p} = \frac{pA_{1p}^{-1}}{p-1}$, (24) follows from (26). □

Let us analyze the content of the preceding result. If $X = \{X_n, \mathcal{F}_n, n \geq 1\}$ is a martingale relative to a fixed stochastic base $\{\mathcal{F}_n, n \geq 1\}$ of Σ, and if $\varphi_n = X_n - X_{n-1} (X_0 = 0)$ so that $X_n = \sum_{k=1}^{n} \varphi_k$ (and conversely every martingale admits such a representation by II.7.10), then the following observation can be made whenever $X^* = \sup_n |X_n| \in L^p(\Sigma), p \geq 1$.

(*) If $\mathcal{M}_p = \mathcal{M}(\Omega, \Sigma, P; \mathcal{F}_n, n \geq 1)$ is the (vector) space of all martingales with $X^* \in L^p(\Sigma)$, then $\mathcal{M}_p$ may be regarded as a closed subspace of $L^p(\ell^2) = L^p(\Omega, \Sigma, P; \ell^2)$, the L^p-space of vector-valued (ℓ^2-valued) functions on $(\Omega, \Sigma, P), 1 \leq p < \infty$ where $X \cong (\varphi_1, \varphi_2, \dots)$ i.e., X is identified by its difference sequence.

This space $\mathcal{M}_p$ is another form of $\mathcal{H}^p$ of (3), but the present emphasis has some advantages. For instance, the classical results in abstract analysis suggest that perhaps a similar study may reveal further properties in looking at this space as a submanifold of $L^p(\ell^r)$, or even of the Orlicz spaces $L^\varphi(\ell^M)$ where φ and M are (convex) Young functions. If $r = 2$ (or $M(t) = t^2$) then such a general study, which is related to the "square function" $s(X)$ is clearly possible. For instance, one has the following result.

5. Theorem. *Let φ be a nonnegative convex function, $\varphi(0) = 0$ and $\varphi(2x) \leq C\varphi(x), x \geq 0, 0 < C < \infty$. Let $L^\varphi(P)$ be the Orlicz space over (Ω, Σ, P) (as noted in II.7.9). If $X = \{X_n, \mathcal{F}_n, n \geq 1\}$ is a martingale such that $X^* = \sup_n |X_n| \in L^\varphi(P)$, then there exist two absolute constants $c_{i\varphi} > 0, i = 1, 2$ such that*

$$c_{1\varphi}\|s(X)\|_\varphi \leq \|X^*\|_\varphi \leq c_{2\varphi}\|s(X)\|_\varphi \tag{27}$$

where $\|f\|_\varphi = \inf\left\{k > 0 : \int_\Omega \varphi\left(\frac{|f|}{k}\right) dP \leq 1\right\}$ is the norm of $L^\varphi(P)$.

It is clear that (27) reduces to (24) when $\varphi(x) = |x|^p, p \geq 1$, and (27) shows that the corresponding space $\mathcal{M}^\varphi$ of martingales can be identified as a closed subspace of $L^\varphi(\ell^2)$. The case of ℓ^M (for ℓ^2) here needs new techniques. We shall omit the proof of the above result which may be found in Burkholder-Davis-Gundy [1] or in a slightly simplified form in Garsia [3].

It is time to give a proof of the announced result: that BMO is the adjoint space of $\mathcal{H}^1$. With only a small amount of additional work we are able to characterize $(\mathcal{H}^p)^*, 1 \leq p \leq 2$. Since this reveals the connections with the corresponding results for $L^p(\mathcal{X})$, and has some possibility of generalizations, we consider the $\mathcal{H}^p$-case in this range. It will be seen that the result for $p > 2$ is obtainable from that for $p \leq 2$.

To understand the problem better, let us recall that $\mathcal{H}^p \subset L^p(\ell^2)$ with an isometric embedding. But from the classical analysis (cf. e.g., Dinculeanu [1], §13) it follows rather easily that its adjoint space $(L^p(\ell^2))^*$ is $L^q(\ell^2), p^{-1} + q^{-1} = 1, 1 \leq p < \infty, ((L^p(\mathcal{X}))^*$ is equivalent to $L^q(\mathcal{X}^*)$ if $\mathcal{X}$ is reflexive). Since each element of $(\mathcal{H}^p)^*$ extends, preserving the norm (by the Hahn-Banach theorem), to an element of $(L^p(\ell^2))^*$, it follows that $(\mathcal{H}^p)^* \subset L^q(\ell^2)$ where the embedding is a topological isomorphism into. However, $\mathcal{H}^p$ is not a measurable subspace of $L^p(\ell^2)$ in terms of the theory of Chapter II, and so the abstract result noted above is not sharp enough to describe the subspace $(\mathcal{H}^p)^*$ of $L^q(\ell^2)$ precisely. The essence of the present work is that we are able to characterize this space if $1 \leq p \leq 2$ with the martingale theory playing a key role. It would be unnecessary to do the same for $p > 2$ as the work follows from a duality argument of $L^p(\ell^2)$. One introduces a class of spaces $\mathcal{K}^q, q \geq 2$, and shows that $(\mathcal{H}^p)^* = \mathcal{K}^q$ (topolgical equivalence) and then it will imply that the norm introduced in $\mathcal{K}^q$ is *equivalent* to that of $L^q(\ell^2)$-norm, whence to the $\mathcal{H}^q$-norm. Here are the details.

Consider an L^2-bounded (hence uniformly integrable) martingale $X = \{X_n, \mathcal{F}_n, n \geq 1\}$ on (Ω, Σ, P) so that $X_n \to X_\infty$ a.e. and in $L^2(P), X_n = E^{\mathcal{F}_n}(X_\infty)$. Define a class $G_X \subset L^2(\Omega, \mathcal{F}_\infty, P)$ as:

$$G_X = \{g : E^{\mathcal{F}_n}(|X_\infty - X_{n-1}|^2) \leq E^{\mathcal{F}_n}(g^2), n \geq 1, g \in L^2(\mathcal{F}_\infty, P)\}. \tag{28}$$

If we set $\|X\|_p^* = \inf\{\|g\|_p : g \in G_X\}$, then $\|\cdot\|_p^*$ is a norm $(p \geq 1)$ and the class $\{\mathcal{K}^q, \|\cdot\|_q^*\}$ of all martingales $X = \{X_n, \mathcal{F}_n, n \geq 1\}$, $\sup_n \|X_n\|_2 < \infty$, for which $\|X\|_q^* < \infty (q \geq 2)$, is a normed linear space. Here and below the stochastic basis $\{\mathcal{F}_n, n \geq 1\}$ is fixed and will not be mentioned again. Since for any $f \in L^1(P)$, $\lim_{q\to\infty} \|f\|_q = \|f\|_\infty$, it follows easily that $\|\cdot\|_\infty^* = \|\cdot\|_B$ and that $\mathcal{K}^\infty = BMO$. The completeness of the $\mathcal{K}^q$-spaces is deduced from the representation theorem. To gain an understanding of the norm of (28), we include a brief discussion, in terms of martingale theory, which explains the inequality used there.

Let $\varphi_n = X_n - X_{n-1}, X_0 = 0$. Then by the uniform integrability, $X_n = \sum_{k=1}^{n} \varphi_k, X_\infty = \sum_{k=1}^{\infty} \varphi_k$. Let $Z = (\varphi_1, \varphi_2, \dots)$ so that $Z(\omega) \in \ell^2, \omega \in \Omega$. If $Z_n = (\varphi_1, \dots, \varphi_n, 0, 0, \dots)$, then $|X_\infty - X_{n-1}|^2(\omega) =$

$\|Z(\omega) - Z_{n-1}(\omega)\|^2 = \left(\sum_{k\geq n} \varphi_k^2\right)(\omega)$ (with ℓ^2-norm). Let $U_n = E^{\mathcal{F}_n}(|X_\infty - X_{n-1}|^2) = E^{\mathcal{F}_n}\left(\sum_{k\geq n} |\varphi_k|^2\right)$. Then $\{U_n, \mathcal{F}_n, n \geq 1\}$ is a nonnegative supermartingale. Hence by the Riesz decomposition (cf. II.5.5) we have $U_n = V_n + W_n$ uniquely where $\{V_n, \mathcal{F}_n, n \geq 1\}$ is a nonnegative martingale and $\{W_n, \mathcal{F}_n\}_{n\geq 1}$ is a potential so that $W_n \to 0$ a.e. as well as in $L^1(P)$, and $V_n \to V$ a.e. But $\{E^{\mathcal{F}_n}(g^2), n \geq 1\}$ is also a uniformly integrable martingale such that $E^{\mathcal{F}_n}(g^2) \to g^2$ a.e. and in $L^1(P)$. It dominates the supermartingale $\{U_n, \mathcal{F}_n, n \geq 1\}$, i.e. $U_n \leq E^{\mathcal{F}_n}(g^2), n \geq 1$. Hence $\lim_n U_n = V \leq g^2$ a.e. But by the mere domination, $g^2 = V$ may not be concluded. If $U^* = \sup_n U_n$ is integrable, then we see that $g^2 \leq U^*$ in (28). Thus we need the smallest g or a lower envelope of such g's satisfying (28) to define the norm. Such a norm for $\mathcal{K}^\infty = BMO$ was originally introduced by John and Nirenberg [1] in an equivalent form, and the generalization to $\mathcal{K}^q, q \geq 2$, was given by Garsia [2].

To prove the representation theorem, we start with an extension of (4) in the form of a general Hölder inequality.

6. Theorem. *Let $\mathcal{H}^p$ and $\mathcal{K}^q$ be as above, $1 \leq p \leq 2, p^{-1} + q^{-1} = 1$. Then for any $X = \{X_n, \mathcal{F}_n, n \geq 1\} \in \mathcal{H}^p, Y = \{Y_n, \mathcal{F}_n, n \geq 1\} \in \mathcal{K}^q$, we have*

$$|E(X_nY_n)| \leq \sqrt{\frac{2}{p}}\|X\|_p\|Y\|_q^*. \tag{29}$$

Proof. For $p = 1$, this is (4). So one may assume that $p > 1$. Since $X_0 = 0 = Y_0$, and $\varphi_n^2 = (X_n - X_{n-1})^2 \leq s_n^2(X) = s_n^2, (Y_n - Y_{n-1})^2 \leq E^{\mathcal{F}_n}(g^2)$, we see that $X_n \in L^p, Y_n \in L^q$ (in fact, $\mathcal{K}^q \subset L^q(\ell^2)$), and hence $X_nY_n \in L^1(P)$. Then (5) holds and we may express (6), with $h_n = Y_n - Y_{n-1}$, as:

$$\begin{aligned} |E(X_nY_n)| &\leq \sum_{i=1}^n E\left(\frac{|\varphi_i|}{s_i^{1-(p/2)}} \cdot s_i^{1-(p/2)}|h_i|\right) \\ &\leq \left[\sum_{i=1}^n E\left(\frac{\varphi_i^2}{s_i^{2-p}}\right)\right]^{1/2} \left[\sum_{i=1}^n E(s_i^{2-p}h_i^2)\right]^{1/2} \end{aligned} \tag{30}$$

Using the fact that $s_0 = 0, 0 \leq 2 - p \leq 1$, one simplifies the right side in the same way as (7).

Consider the first term: $\frac{\varphi_n^2}{s_n^{2-p}} = \frac{s_n^2 - s_{n-1}^2}{s_n^{2-p}} = s_n^p(1-a^2)$ where $a = \frac{s_{n-1}}{s_n} \le 1$, so the left side is less than s_n^p. We would like to bound this by a multiple of $(s_n^p - s_{n-1}^p)$. So it suffices to find a constant $\alpha_0 > 0$ such that $s_n^p(1-a^2) \le \alpha_0(s_n^p - s_{n-1}^p)$ or $(1-a^2) \le \alpha_0(1-a^p)$, since we can consider $s_n > 0$ only. Taking $a \to 0$ we see that $\alpha_0 \ge 1$, and letting $a \to 1$ one has $\alpha_0 \le \frac{2}{p}$. Since $(1-a^2)/(1-a^p)$ decreases as $p \to 2$, it follows that $\alpha_0 = \frac{2}{p}$ will be the maximum of this ratio. Hence $(\varphi_n^2/s_n^{2-p}) \le \frac{2}{p}(s_n^p - s_{n-1}^p)$ and the first term on the right of (30) is at most $\left(\frac{2}{p}\sum_{i=1}^n [E(s_i^p - s_{i-1}^p)]\right)^{1/2} = \left(\frac{2}{p}E(s_n^p)\right)^{1/2}$. For the second term, one has:

$$\begin{aligned}
\sum_{i=1}^n E(s_i^{2-p}h_i^2) &= \sum_{i=1}^n E\left(s_i^{2-p}\left(\sum_{j=i}^n h_j^2 - \sum_{k=i+1}^n h_k^2\right)\right) \\
&= \sum_{i=1}^n E[(s_i^{2-p} - s_{i-1}^{2-p})\sum_{j=i}^n h_j^2], \text{ since } s_0 = 0, \\
&= \sum_{i=1}^n E[(s_i^{2-p} - s_{i-1}^{2-p})E^{\mathcal{F}_i}(\sum_{j=i}^n h_j)^2], \\
&\qquad \text{since } h_j\text{'s are orthogonal,} \\
&= \sum_{i=1}^n E[s_i^{2-p} - s_{i-1}^{2-p})E^{\mathcal{F}_i}(Y_n - Y_{i-1})^2], \\
&\le \sum_{i=1}^n E[(s_i^{2-p} - s_{i-1}^{2-p})E^{\mathcal{F}_i}(Y_\infty - Y_{i-1})^2] \\
&\le \sum_{i=1}^n E[(s_i^{2-p} - s_{i-1}^{2-p})E^{\mathcal{F}_i}(g^2)], \\
&\qquad \text{since } Y \in \mathcal{K}^q \text{ and (28) holds,} \\
&= \sum_{i=1}^n E[(s_i^{2-p} - s_{i-1}^{2-p})g^2] = E(s_n^{2-p}g^2) \\
&\le [E(s_n^{(2-p)\alpha})]^{1/\alpha}[E(g^{2\beta})]^{1/\beta}, \alpha > 1, \alpha^{-1} + \beta^{-1} = 1.
\end{aligned} \tag{31}$$

Let $(2-p)\alpha = p > 1$, or $\alpha = \frac{p}{2-p} > 1$, and $\beta = \frac{\alpha}{\alpha-1} = \frac{1}{2}\frac{p}{p-1} = \frac{q}{2} > 1$. Substituting this and the previous bound from (30) in (31), one finds

$$|E(X_nY_n)| \le \sqrt{\frac{2}{p}} \cdot [\|X\|_p^{p/2} \cdot \|X\|_p^{p(2-p)/2p} \cdot \|g\|_q].$$

Since $g \in G_Y$, this implies (29) and the result follows. □

The desired representation of $(\mathcal{H}^p)^*$ is given by

7. Theorem. *For each functional $x^* \in (\mathcal{H}^p)^*, 1 \le p \le 2$, there exists uniquely an element $Y = \{Y_n, \mathcal{F}_n, n \ge 1\} \in \mathcal{K}^q, p^{-1} + q^{-1} = 1$, such that*

$$x^*(f) = E\left(\sum_{n=1}^{\infty} \varphi_n h_n\right) = \sum_{n=1}^{\infty} E(\varphi_n h_n), \quad f \in \mathcal{H}^p, \tag{32}$$

$$\sqrt{\frac{2}{p}}\|x^*\| \le \|Y\|_q^* \le \frac{2q}{q-1}\|x^*\|, \tag{33}$$

where $\varphi_n = f_n - f_{n-1}, h_n = Y_n - Y_{n-1}$ and $f_0 = 0 = Y_0 (\frac{2q}{q-1} = 2$ if $q = +\infty), f = \{f_n, \mathcal{F}_n, n \ge 1\} \in \mathcal{H}^p$. Thus $(\mathcal{H}^p)^*$ is topologically equivalent to $\mathcal{K}^q$, and $x^*(f) = E(\langle f, Y\rangle)$, using a duality pairing.

Proof. If $Y = \{Y_n, \mathcal{F}_n, n \ge 1\} \in \mathcal{K}^q$, then for each $X = \{X_n, \mathcal{F}_n, n \ge 1\} \in \mathcal{H}^p$, we have

$$|E(X_n Y_n)| \le \sqrt{\frac{2}{p}}\|X\|_p \|Y\|_q^*, \tag{34}$$

by the preceding theorem. On the other hand let us consider the "stopped" martingale $\{X_n, \mathcal{F}_n, n \ge 1\}$ where $X_{n_0} = X_{n_0}$ for $n \ge n_0; = X_n$ for $n < n_0$, so that $\|X_\infty - X_{n_0}\|_p$ can be made arbitrarily small by choosing n_0 large enough because all the elements of $\mathcal{H}^p$ are L^p-convergent martingales. So the stopped martingales are norm dense in $\mathcal{H}^P$. Consequently x^* can be defined on this subset by the equation:

$$x^*(\tilde{X}) = E\left(\sum_{n=1}^{\infty} g_n h_n\right), \quad X_n = \sum_{k=1}^{n} g_k, \quad Y_n = \sum_{k=1}^{n} h_k, \tag{35}$$

for each $Y = \{Y_n, \mathcal{F}_n, n \ge 1\} \in \mathcal{K}^q$ and $\tilde{X} = \{X_n, \mathcal{F}_n, n \ge 1\}$ in the above (dense) set. Here we used the fact that $E(X_n Y_n) = E\left(\sum_{k=1}^{n} g_k h_k\right)$, noted in (5). Since $X_n \to X_\infty$ in L^p, and $Y_n \to Y_\infty$ in L^q, by definition, so that (by the classical Hölder inequality) $X_n Y_n \to X_\infty Y_\infty$ in L^1, one can let $n \to \infty$ in this representation and (34) is well defined. We note however that, since $g_n = 0$ if $n > n_0$ for stopped martingales, the sum in (35) is actually finite although n_0 varies with

$\tilde{X} \in \mathcal{H}^p$. In any case x^* does not depend on n_0, and by (34) $\|x^*\| < \infty$. The linearity being evident, we deduce that $x^* \in (\mathcal{H}^p)^*$. The nontrivial part is the converse which will now be established.

Recall that $\mathcal{H}^p$ is a subspace of $L^p(\ell^2)$. If $x^* \in (\mathcal{H}^p)^*$, then the continuous linear functional x^* can be extended from $\mathcal{H}^p$ to $L^p(\ell^2)$ preserving the norm, by the Hahn-Banach theorem. We denote the extension by the same symbol. By the classical results in linear analysis as noted before, one has $(L^p(\ell^2))^* = L^q(\ell^2)$, and hence there exists a unique $\tilde{Y} \in L^q(\ell^2)$ such that x^* corresponds to $\tilde{Y}, \|x^*\| = \|\tilde{Y}\|_q, \tilde{Y} = \{\tilde{h}_n, n \geq 1\}$ (the last being the norm of $L^q(\ell^2)$),

$$x^*(X) = E(\langle X, \tilde{Y}\rangle) = \sum_{n=1}^{\infty} E(g_n \tilde{h}_n), \quad X = \{g_n = X_n - X_{n-1}, n \geq 1\}. \tag{36}$$

We shall now show that the $\tilde{h}_n$ can be replaced by martingale differences and that (32) and (33) follow from (36).

For each $k \geq 1$, define $h_k = (E^{\mathcal{F}_k} - E^{\mathcal{F}_{k-1}})(\tilde{h}_k)$. Thus $E^{\mathcal{F}_\ell}(h_k) = 0$ for $0 \leq \ell \leq k-1; = h_k$ for $\ell \geq k$ so that if $Y_n = \sum_{k=1}^{n} h_k$, then $\{Y_n, \mathcal{F}_n, n \geq 1\}$ is a martingale (let $Y_0 = 0$). Since $q \geq 2, Y_n \in L^2(P)$ and since $Q_k = E^{\mathcal{F}_k} - E^{\mathcal{F}_{k-1}}$ is also a projection in $L^2(P)$, we see that

$$\begin{aligned}\|Y_n\|_2^2 &\leq \int_\Omega \left(\sum_{k=1}^{\infty} h_k^2\right) dP = \sum_{k=1}^{\infty} \int_\Omega (Q_k \tilde{h}_k)^2 dP \\ &\leq \sum_{k=1}^{\infty} \int_\Omega \tilde{h}_k^2 dP \leq \|\tilde{Y}\|_q < \infty, \quad n \geq 1,\end{aligned}$$

because $L^2(\ell^2) \supset L^q(\ell^2)$. So $Y_n \to Y_\infty$ a.e. and in L^2. In order to show that (32) is true, it suffices to establish that $Y = \{Y_n, \mathcal{F}_n, n \geq 1\} \in \mathcal{K}^q$, or equivalently that there is a $g \in G_Y$ such that (cf. (28))

$$\sum_{k \geq n} E^{\mathcal{F}_n}(h_k^2) = E^{\mathcal{F}_n}(|Y_\infty - Y_{n-1}|^2) \leq E^{\mathcal{F}_n}(g^2), \text{ a.e.}, \quad n \geq 1. \tag{37}$$

We shall now exhibit a g satisfying (37). The idea of finding such a function g (or of a dominating martingale $E^{\mathcal{F}_n}(g^2)$) is to consider $|\tilde{Y}| = \|\tilde{Y}\|_{\ell^2} \in L^q(P)$ and take (a multiple of) the maximal function of the L^q (and hence L^2)-martingale $E^{\mathcal{F}_n}(|\tilde{Y}|)$. Here are the details.

By definition of the ℓ^2-norm,

$$|\tilde{Y}|^2 = \sum_{n=1}^{\infty} \tilde{h}_n^2, \quad \text{and let } |\tilde{Y}|^* = \sup_n E^{\mathcal{F}_n}(|\tilde{Y}|). \tag{38}$$

Then by (24), $|\tilde{Y}|^* \in L^2(P)$ (and $\in L^q(P)$ in fact). Also, if $k > n$,

$$\begin{aligned}
E^{\mathcal{F}_n}(h_k^2) &= E^{\mathcal{F}_n}[(E^{\mathcal{F}_k}(\tilde{h}_k))^2 + (E^{\mathcal{F}_{k-1}}(\tilde{h}_k))^2] \\
&\quad - 2E^{\mathcal{F}_n}[E^{\mathcal{F}_k}(\tilde{h}_k E^{\mathcal{F}_{k-1}}(\tilde{h}_k))], \\
&= E^{\mathcal{F}_n}[()^2 + ()^2] - 2E^{\mathcal{F}_n}[E^{\mathcal{F}_{k-1}}(\tilde{h}_k E^{\mathcal{F}_{k-1}}(\tilde{h}_k))], \\
&\quad \text{since } k-1 \geq n, \\
&= E^{\mathcal{F}_n}([E^{\mathcal{F}_k}(\tilde{h}_k)]^2) - E^{\mathcal{F}_n}([E^{\mathcal{F}_{k-1}}(\tilde{h}_k)]^2) \\
&\leq E^{\mathcal{F}_n}([E^{\mathcal{F}_k}(\tilde{h}_k)]^2) \leq E^{\mathcal{F}_n}(\tilde{h}_k^2), \text{ by Theorem II.1.8.}
\end{aligned}$$

Hence (37) becomes, with the above estimate and (38),

$$\begin{aligned}
\sum_{k\geq n} E^{\mathcal{F}_n}(h_k^2) &\leq h_n^2 + \sum_{k>n} E^{\mathcal{F}_n}(h_k^2) \\
&\leq (E^{\mathcal{F}_n}(\tilde{h}_n))^2 + (E^{\mathcal{F}_{n-1}}(\tilde{h}_n))^2 \\
&\qquad + 2E^{\mathcal{F}_n}(|\tilde{h}_n|)E^{\mathcal{F}_{n-1}}(|\tilde{h}_n|) + \sum_{k>n} E^{\mathcal{F}_n}(\tilde{h}_k^2) \\
&\leq E^{\mathcal{F}_n}(\sum_{k\geq n} \tilde{h}_k^2) + E^{\mathcal{F}_{n-1}}(\tilde{h}_n^2) + 2(|\tilde{Y}|^*)^2 \\
&\leq 4(|\tilde{Y}|^*)^2.
\end{aligned} \tag{39}$$

Now taking $g = 2|\tilde{Y}|^* \in L^2(P)$, and applying $E^{\mathcal{F}_n}$ on both sides of (39), we get (37). Hence, if $p > 1$ so that $q < \infty$, we have $|\tilde{Y}|^* \in L^q(P)$ and so $\|g\|_q < \infty$. If $p = 1$, then by the result of the representation (36) $|\tilde{Y}| = |\tilde{Y}|_{\ell^2} \in L^\infty(P)$ and so $|\tilde{Y}|^*$ is also bounded. Hence $\lim_{q\to\infty} \|g\|_q = \|g\|_\infty$ is finite. Thus in all cases, (37) or (2) is valid. Hence $Y \in \mathcal{K}^q$.

Finally we have

$$\begin{aligned}
\|Y\|_q^* = \inf\{\|g\|_q : g \in G_Y\} &\leq \|g\|_q \leq 2\||\tilde{Y}|^*\|_q \\
&\leq 2 \cdot \frac{q}{q-1} \||\tilde{Y}|\|_q \text{ if } q < \infty, \\
&\leq 2\||\tilde{Y}|\|_\infty, \text{ if } q = \infty,
\end{aligned} \tag{40}$$

where the last inequality follows from Theorem II.4.8. Moreover,

$$\begin{aligned}
E(g_n h_n) &= E(g_n E^{\mathcal{F}_n}(\tilde{h}_n)) - E(g_n E^{\mathcal{F}_{n-1}}(\tilde{h}_n)) \\
&= E(g_n \tilde{h}_n) - E(E^{\mathcal{F}_{n-1}}(g_n)\tilde{h}_n) = E(g_n \tilde{h}_n),
\end{aligned}$$

since $E^{\mathcal{F}}$ is self-adjoint and $E^{\mathcal{F}_{n-1}}(g_n) = 0$. This and (36) imply:

$$x^*(X) = E(\langle X, \tilde{Y}\rangle) = \sum_{n=1}^{\infty} E(g_n h_n) = E(\langle X, Y\rangle), \quad X \in \mathcal{H}^p, \tag{41}$$

with $Y \in \mathcal{K}^q$. By the density of stopped martingales in $\mathcal{K}^q$, Y is unique in that subspace (of $L^q(\ell^2)$). With (38) and (36), we deduce also that $\|Y\|_q^* \leq \frac{2q}{q-1}\|x^*\|$. This proves (32) and also (33) since the left side inequality is immediate from (40), (41), (29) and the definition of $\|x^*\|$. □

In the course of the above proof, between (38) and (40), we have actually established that each element $\tilde{Y} \in L^q(\ell^2)$ associates (in a many-to-one manner in general) an element of $\mathcal{K}^q$. This is analogous to the fact that (by the Hahn-Banach theorem) each element of $(\mathcal{H}^p)^*$ associates (in a one-to-many correspondence in general) an element of $(L^p(\ell^2))^*$. For reference we state this precisely as follows:

8. Proposition. *If $\tilde{Y} \in L^q(\ell^2), \tilde{Y} = \{\tilde{h}_n, n \geq 1\}$ and $q \geq 2, \|\tilde{Y}\|_q = \alpha$, then there exists (a martingale) an element $Y = \{Y_n, \mathcal{F}_n, n \geq 1\} \in \mathcal{K}^q$ defined as:*

$$Y_n = \sum_{k=1}^{n} h_k, \quad h_k = E^{\mathcal{F}_k}(\tilde{h}_k) - E^{\mathcal{F}_{k-1}}(\tilde{h}_k), \tag{42}$$

(so $Y_n \to Y_\infty$ a.e. and in $L^2(P)$) such that $\|Y\|_q^* \leq \frac{2q\alpha}{q-1}$. In particular, if $p = 1$, (or $q = \infty$) so that $\mathcal{K}^q = BMO$, we have $Y \in BMO$ and $\|Y\|_B \leq 2\alpha$.

Remark. Since $L^p(\ell^2), 1 < p \leq 2$, is reflexive, $\mathcal{H}^p \subset L^p(\ell^2)$ and $(\mathcal{H}^p)^* \cong \mathcal{K}^q \cong \mathcal{H}^q, 1 < p \leq 2$, by known classical results. Garsia [2] proved this equivalence with a straight computation $\left(\|X\|_q^* \leq \|X\|_p \leq \sqrt{\frac{q}{2}}\|X\|_q^*\right)$. Thus we have the duality for all $\mathcal{H}^p, 1 \leq p < \infty$. It will be interesting to extend this to the general $\mathcal{H}^\varphi$-spaces (or the Hardy-Orlicz spaces) for which (27) will be helpful. Indeed an extension of $\mathcal{H}^p$-spaces to semimartingales (the latter to be analyzed in Chapters V and VI) has been considered by Emery.

Let us briefly discuss another result for continuous parameter martingales in which the increments play a key role. If $I = [a, b]$, and $\{X(t), t \in I\}$ is a stochastic process on (Ω, Σ, P), and $D_j : a = t_{j1} <$

$\cdots < t_{jm_j} = b$ is a subdivision, then we recall that the *quadratic variation* Q_j relative to D_j (*not* to be confused with the φ-bounded variation of II.7.2 if $\varphi(x) = x^2$) is given by:

$$Q_j^2 = \sum_{i=1}^{m_j} [X(t_{ji}) - X(t_{ji-1})]^2, \quad X(t_{j0}) = 0, \text{ a.e.} \tag{43}$$

In the case that the process is a right continuous martingale, then for each j, the sequence $\{X(t_{ji}), \mathcal{F}(t_{ij}), 1 \le i \le m_j\}$ is also a martingale. Since Q_j is closely related to $s_n(X)$ of the functions in (3), we may establish:

9. Proposition. *Let $\{X(t), \mathcal{F}_t, t \in I\}$ be a right continuous martingale on (Ω, Σ, P), and Q_j be as in (43) for each subdivision. Then $Q_j \to Q$ in $L^1(P)$ implies $X^* = \sup_{t\in I} |X(t)| \in L^1(P)$. [$X^*$ is measurable by the right continuity of $X(t)$'s and the localizability of P, as discussed in Sections III.1 and III.5.]*

Remark. If $X^* \in L^p(P)$ for some $p > 1$, then the sequence Q_j can be shown to converge in $L^1(P)$ also. This implication is more involved and will not be proved here. (Results of the next chapter will be needed. Cf., Doléans [1].) In the above result, $\{Q_j\}_{j\ge1}$ converges in $L^p(P)$ means that the $\{D_j\}_{j\ge1}$ is preordered by inclusion (or refinement) and the Moore-Smith convergence is used.

Proof. Let $Q_j \to Q$ in $L^1(P)$ so that $\sup_j \|Q_j\|_1 < \infty$. If $f_{jn} = X(t_{jn})$, then, right continuity implying separability, we have

$$X^* = \sup_j \sup_n |f_{jn}| = \sup_j f_j^* \text{ (say)}, \tag{44}$$

and for a subsequence $f_{j'}^* \to X^*$. Since $\{f_{jn}, \mathcal{F}_{jn}, n \ge 1\}$ is a martingale, we have by (15)

$$\|f_{j'}^*\|_1 \le C\|s(f_j)\|_1 = C\|Q_{j'}\|_1. \qquad (C^2 = 10.)$$

Hence

$$\|X^*\|_1 \le \underline{\lim}_{j'} \|f_{j'}^*\|_1 \le \underline{\lim}_{j'} C\|Q_{j'}\|_1 \le C \sup_j \|Q_j\|_1 < \infty. \tag{45}$$

This implies the result. □

To analyze the continuous parameter case further, it is necessary to proceed with the decompositions initiated in Chapter II. That work leads at once to the stochastic integration theory. We shall therefore turn to this aspect in the following chapters. The Complements section contains other related results mostly for the discrete index case.

4.6 Complements and exercises

1. Let (Ω, Σ) be a measurable space, I, J be directed sets, and $\{\mathcal{F}_i, i \in I\}$ be a net of σ-subalgebras of Σ. Let T_j be a stopping time of the net for each $j \in J$, and let $\mathcal{G}(T) = \{A \subset \Omega : A \cap [T \leq i] \in \mathcal{F}_i, i \in I\}$. (a) Show that $\mathcal{G}(T_j)$ is a σ-algebra and that $\mathcal{G}(T_j) \subset \Sigma$, if, in addition, I has a countable cofinal subset. (b) In the general case, let $\mathcal{B}_j = \mathcal{G}(T_j) \cap \Sigma$. If $\{X_i, \mathcal{F}_i, i \in I\}$ is a bounded submartingale on the probability space (Ω, Σ, P) (I is even finite), show by an example that the adapted process $\{Y_j, \mathcal{B}_j, j \in J\}$ is not necessarily a submartingale where $Y_j = X \circ T_j$.

2. Let $\{X_n, \mathcal{F}_n, n \geq 1\}$ be a submartingale on a probability space, and T any stopping time of $\{\mathcal{F}_n, n \geq 1\}$. Let $A = \left[\limsup_n X_n < \infty\right]$. If $E(X^+ \circ T) < \infty$, then $X_n \to X_\infty$ a.e. and X_∞ is finite a.e. on A. [This is an analog of Theorem 4.1. Note that $A \neq \Omega$ in general, e.g., let $X_n = n$ a.e.]

3. Let $\{X_n, \mathcal{F}_n, n \geq 1\}$ be an adapted stochastic process such that $E|X \circ T| < \infty$ for each stopping time T of $\{\mathcal{F}_n, n \geq 1\}$. Let $\mathcal{T}$ be the collection of all finite stopping times of $\{\mathcal{F}_n, n \geq 1\}$.

(a) If $X^* = \limsup_n X_n, X_* = \liminf_n X_n$, and $Y_\tau = X \circ T_\tau, T_\tau \in \mathcal{T}$, then show that

$$E(X^*) \leq \limsup_\tau E(Y_\tau), \qquad E(X_*) \geq \liminf_\tau E(Y_\tau).$$

(b) If $E(X^*) < \infty$ and $E(X_*) > -\infty$ in the above, show that there is equality in the inequalities of (a). [Thus Fatou's inequalities become equalities in this context. Moreover, the above hypothesis holds if $|X_n| \leq Z$ a.e. for some integrable Z, so that the Dominated Convergence theorem is true under stopping time transformations.]

(c) Show by an example that the Vitali convergence fails under the stopping time transformations; i.e. if $\{X_n, \mathcal{F}_n, n \geq 1\}$ and $\mathcal{T}$ are as in (a) $X_n \to X$ a.e. and $\{X_n\}_1^\infty$ is uniformly integrable, so that $E(|X_n - X|) \to 0$, then it is not true that $E(|Y_\tau - X|) \to 0$ for every directed family $\{T_\tau, \tau \in J\} \subset \mathcal{T}$. [Hints: Let $0 < c < \infty, A_{n,c} = [X_n \leq cZ]$ for some $0 < Z \in L^1(\mathcal{F}_1, P)$. Let $X_n^c = X_n \chi_{A_{n,c}}, X_*^c, X_c^*$ be defined as 'inf' and 'sup' (cf. (a)). We may assume that X_c^* is integrable. If $\mathcal{G}_n = \sigma(X_1, \dots, X_n) \subset \mathcal{F}_n, V_n = E^{\mathcal{G}_n}(X_c^*), Z_\tau = V \circ T_\tau$, then

$$\int_\Omega Z_\tau dP = \sum_{n=1}^\infty \int_{[T_\tau = n]} E^{\mathcal{G}_n}(X_c^*) dP = \int_\Omega X_c^* dP. \tag{+}$$

Let $B_n^\epsilon = [V_n < X_n^c + \epsilon Z] \in \mathcal{F}_n, T_0 = \sum_{n=1}^\infty n \chi_{B_n^c}$ and $T_1 = T_0$, if $T_0 \geq T_{\tau_0}, = \infty$ if not. T_1 is a stopping time of $\{\mathcal{F}_n, n \geq 1\}, T_{\tau_0} \in \mathcal{T}$. Hence (+) becomes (since $T_{\tau_0} \leq T_1 < \infty$ a.e., because $V_n \to X_c^*$ a.e. and in $L^1(P)$)

$$\begin{aligned} \infty > \int_\Omega X^c \circ T_1 dP &= \sum_{n=1}^\infty \int_{[T_1 = n]} X_n^c dP \geq \sum_{n=1}^\infty \left[\int_{[T_1 = n]} (V_n - \epsilon Z) dP \right] \\ &= \int_\Omega X_c^* dP - \epsilon \int_\Omega Z dP. \end{aligned}$$

If $\epsilon \searrow 0$ and then $c \nearrow \infty$, we get the first inequality of (a) and the second inequality is similarly proved. If $W_k = \sup_{n \geq k} X_n$, then $(X \circ T)\chi_{[T \geq n]} \leq W_n$ for $T \in \mathcal{T}$. Hence

$$\limsup_\tau \int_\Omega X \circ T_\tau dP \leq \lim_n \int_\Omega W_n dP = \int_\Omega X^* dP.$$

This gives (a) and (b). Note that the measure P can be allowed to be infinite if each $\mathcal{F}_n$ is assumed to be P-rich. In the case of probability measures, one can take $Z = 1$ a.e.

For (c), an example can be constructed starting with a sequence of independent uniform random variables on the Lebesgue unit interval to satisfy the negative conclusion. Regarding this construction and other aspects of the problem, see Sudderth [1].]

4. Let $\{X_n, \mathcal{F}_n, n \geq 1\}$ be an adapted stochastic process; $\mathcal{F}_n \subset \mathcal{F}_{n+1}$ are σ-subalgebras in (Ω, Σ, P). Let $\{T_n, n \geq 1\}$ be a stopping time process of $\{\mathcal{F}_n, n \geq 1\}$.

(a) If X_n's are independent identically distributed and $Y_n = X \circ T_n$, show that $\{Y_n, n \geq 1\}$ are also independent and identically distributed. [*Hint*: If $B_i, 1 \leq i \leq k$ are Borel sets of $\mathbb{R}$, observing that $X_{i_{k+1}}$ is independent of $X_{i_1}, \ldots, X_{i_k}, i_j < i_{j+1}$, and of $[T_{n_k} = i_k]$, we get

$$P[Y_{n_1} \in B_1, \ldots, Y_{n_k} \in B_k] = \sum_{1 \leq i_1 < \cdots < i_k < \infty} P[T_{n_1} = i_1, X_{i_1} \in B_1, \ldots, T_{n_k} = i_k, X_{i_k} \in B_k]$$

with the right side having the factor $P[X_{i_1} \in B_1] = P[X_1 \in B_1]$. Thus the sum is $\prod_{i=1}^{k} P[X_i \in B_i]$.]

(b) If X_n above are not independent, suppose that $X_n \in L^1(P)$ and $E^{\mathcal{F}_n}(X_{n+1}) \geq 0$ a.e. Let $A_{n,k} = [T_k \geq n]$ and $E^{\mathcal{F}_n}(|X_{n+1}|) \leq \alpha < \infty$, on $A_{n,k}$. Then $E(|Y_n|) < \infty, E^{\mathcal{B}_n}(Y_{n+1}) \geq 0$, where $Y_n = X \circ T_n$, and $\mathcal{B}_n = \mathcal{B}(T_n)$, the σ-algebra of events prior to T_n. Show $\left\{\sum_{k=1}^{n} Y_k, \mathcal{B}_n, n \geq 1\right\}$ is a submartingale transformed from the submartingale $\left\{\sum_{k=1}^{n} X_k, \mathcal{F}_n, n \geq 1\right\}$. [*Hint:* Show that $E\left(\left|\sum_{k=1}^{n} Y_k\right|\right) < \infty$, and apply Proposition 1.11.]

5. Let $\{\mathcal{F}_t, t \geq 0\}$ be a net of right continuous σ-algebras in (Ω, Σ, P), and let $M(\Omega, \Sigma)$ be the space of all measurable scalar functions. Show that the class of all stopping times of $\{\mathcal{F}_t, t \geq 0\}$ does not form a (positive) cone in the above space.

6. Let $\{X_n, n \geq 1\}$ be independent random variables on a probability space (Ω, Σ, P), each with zero mean. Suppose that the partial sums $Y_n = \sum_{i=1}^{n} X_i$ lie in a ball of $L^{p_0}(P)$ for some $p_0 \geq 1$. Show that $Y_n \to Y_\infty$ a.e. and in $L^{p_0}(P)$, and that for each $p \geq 1$,

$$\left\| \sup_n |Y_n| \right\|_p \leq 5 \|Y_\infty\|_p, \tag{*}$$

where the right side may be infinite for $p > p_0$. [*Hints*: The martingale convergence implies that of the series $Y_n \to Y_\infty$ a.e., and in $L^{p_0}(P)$.

First suppose that each X_n is symmetrically distributed, i.e. $P[X_n \le a] = P[-X_n \le a], a \in \mathbb{R}$. If $\mathcal{F}_n = \sigma(X_1, \dots, X_n)$ and T is any stopping time of $\{\mathcal{F}_n, n \ge 1\}$, then $E^{\mathcal{B}(T)}(Y_\infty) = Y \circ T = Y_T$ (say). By the symmetry hypothesis, $P[|Y_T| \ge |Y_\infty|] = P[|Y_T| \le |Y_\infty|] \ge 1/2$. Let $A_n = [Y_n > \lambda, Y_i \le \lambda, 1 \le i \le n-1], A_1 = [Y_1 > \lambda]$, and let $T = T_1^\lambda = \sum_{n=1}^{\infty} n\chi_{A_n}$. Then $\bigcup_{n\ge1} A_n = \left[\sup_n Y_n > \lambda\right]$, and $|Y_{T_0}|(\omega) > 0$ iff $\omega \in \bigcup_{n\ge1} A_n$. If $B_\lambda = [|Y_{T_0}| \ge |Y_\infty|] \subset [|Y_\infty| > \lambda]$, so that $P(B_\lambda^c) \le 1/2 \le P(B_\lambda)$, we have

$$2P[|Y_\infty| > \lambda] \ge 2P(B_\lambda) \ge 1 \ge P[\sup_n |Y_n| > \lambda], \quad \lambda \ge 0.$$

Replacing λ by $\lambda^{1/p}$, one gets

$$E\left(\sup_n |Y_n|^p\right) = \int_0^\infty P[\sup_n |Y_n|^p > \lambda] d\lambda$$

$$\le 2\int_0^\infty P[|Y_\infty|^p > \lambda] dP = 2\|Y_\infty\|_p^p. \qquad (+)$$

This proves (*) in this case with a constant $2^{1/p}$. We now deduce the general case from (+). Let $\{\tilde{X}_n, n \ge 1\}$ be a process with the same distributions as the X_n's but independent of them. Using an adjunction procedure, if necessary, we may assume the existence of these families on the same space and let $Z_n = X_n - \tilde{X}_n$. Then $\{Z_n, n \ge 1\}$ are symmetrically distributed to which (+) applies since $\sum_{k=1}^{n} Z_k \to Z_\infty = Y_\infty - \tilde{Y}_\infty$ a.e. and in $L^{p_0}(P)$. But $|Y_n|^p \le 2^{p-1} \sup_m |Y_m - \tilde{Y}_m|^p + 2^{p-1}|\tilde{Y}_n|^p$. We get (*) with the constant $(2^{2p} + 2^{p-1})^{1/p} \le 5$. This result is due to Doob ([1], p. 337) if $E(X_1) = \alpha, p_0 = 1$, and with 8 for 5 in (*). The present estimate is due to Hunt [1], and the argument illustrates the improvements obtainable for II.4.8, in special cases.]

7. Let $\{X_t, t \in \bar{\mathbb{R}}^+\}$ be a stochastic process, $\mathcal{F}_t = \sigma(X_s, s \le t)$ and $\mathcal{F}_\infty = \sigma(X_t, t \in \bar{\mathbb{R}}^+)$.

(a) Show that $\mathcal{F}_t = \{A \in \mathcal{F}_\infty$: for each $\omega \in A, X_s(\omega) = X_s(\omega')$ for $0 \le s \le t$ implies $\omega' \in A\}$. [*Hints*: Clearly, $\mathcal{F}_t \subset \{\}$ and the latter is a σ-algebra. For the opposite inclusion, if $\alpha_t : \Omega \to \Omega$ is $(\mathcal{F}_t, \mathcal{F}_\infty)$-measurable, where $X_s \circ \alpha_t = X_{s\wedge t}$, then $X_{s\wedge t}(\omega') = X_s(\alpha_t(\omega)), \omega \in A \in \{\}, \omega' = \alpha_t(\omega), s \le t$, implies $\alpha_t^{-1}(A) = A$, so $A \in \mathcal{F}_t$.]

(b) If $T : \Omega \to \mathbb{R}^+$ is $\mathcal{F}_\infty$-measurable, then T is a stopping time of $\{\mathcal{F}_t, t \in \mathbb{R}^+\}$ iff for any ω, ω' in Ω, $T(\omega) \le t$ and $X_s(\omega) = X_s(\omega'), 0 \le s \le t$ imply $T(\omega') \le t$.

(c) Let $T : \Omega \to \bar{\mathbb{R}}^+$ be a stopping time of $\{\mathcal{F}_t, t \ge 0\}$ and $\mathcal{B}(T)$ be the σ-algebra of events prior to T. Then $\mathcal{B}(T) = \{A \in \mathcal{F}_\infty : \omega \in A, X_s(\omega) = X_s(\omega'), 0 \le s \le T(\omega)$, implies $\omega' \in A\}$. [Regarding this problem, see Courrège and Priouret [1]. Note that in the above $X_t : \Omega \to \Lambda$ with $X_t^{-1}(\mathcal{L}) \subset \Sigma$ is possible, where $(\Lambda, \mathcal{L})$ is an abstract measurable space, and (Ω, Σ) is the given measurable space and no measure intervenes.]

8. Let $\{X_t^n, \mathcal{F}_t, t \ge 0\}, t \ge 1$, be a sequence of right continuous supermartingales on a complete space (Ω, Σ, P), such that $X_t^n \le X_t^{n+1}$ a.e., $t \ge 0, n \ge 1$. If $X_t = \lim_n X_t^n$ a.e., then $\{X_t, \mathcal{F}_t, t \ge 0\}$ is a right continuous supermartingale. A similar result holds if "left continuous" replaces "right continuous" and $X_t^n \ge X_t^{n+1}$ a.e.

9. We sketch an example, announced in the remark following Proposition 3.4, regarding the existence of non-class (D) uniformly integrable continuous parameter supermartingales, with continuous sample paths. Let $\{X_t, \mathcal{F}_t, t \ge 0\}$ be a Brownian motion process on (Ω, Σ, P), with values in $\mathbb{R}^3$ and $P[X_0 = p_0] = 1$, where $p_0 = (1, 0, 0)$. If $r(p)$ is the distance between $p \in \mathbb{R}^3$ and the origin, then $u(p) = r(p)^{-1}$ is a concave increasing function (as $p \to 0$) and $\{u(X_t), \mathcal{F}_t, t \ge 0\}$ is a nonnegative supermartingale with continuous sample paths, and is uniformly integrable if $u(X_\infty) = 0$ a.e., but is not in class (D). [*Hints:* $E(u(X_t)) < \infty$ follows from a classical integral formula:

$$E(u(X_t)) = \int_\Omega u(x) dF_t(x) = \frac{4\pi}{(2\pi t)^{3/2}} \int_0^1 r^2 e^{-r^2/2t} dt + t\, e^{-(1/2)t}, \quad t > 0.$$

To see that $\{Z_t = u(X_t), t \ge 0\} \notin$ class (D), let $T_\alpha = \inf\{t > 0 : Z_t \ge \frac{1}{\alpha}\}$, with $\inf(\emptyset) = +\infty, 0 < \alpha < 1$. If $A_\alpha = \{(t, \omega) : Z_t(\omega) \ge \frac{1}{\alpha}\}$, then $T_\alpha = D_{A_\alpha}$, the debut of A_α, and is a stopping time of $\{\mathcal{F}_t, t \ge 0\}$. We assert that $\{Z \circ T_\alpha, 0 < \alpha < 1\}$ is not uniformly integrable (and hence is not in (D)). For, we can interpret $\int_{[Z \circ T_\alpha > 1/\alpha]} Z \circ T_\alpha dP$ as the "expected value of the X_t-process hitting the ball $B(0, \alpha) = \{p : r(p) \le \alpha\}$ for the first time," so that it is equal to

$$\frac{1}{\alpha}P\left\{\omega : X_t(\omega) \in B(0,\alpha), \text{ for } t \geq \frac{1}{\alpha}, t - \varepsilon < \frac{1}{\alpha} \text{ for every } \varepsilon > 0\right\}.$$

Using the theory of harmonic measures, this probability can be shown to be $\frac{1}{\alpha}$ so that for any $s < \frac{1}{\alpha}$, we have $\int_{[Z\circ T_\alpha > s]} Z \circ T_\alpha dP = 1$. Thus letting $\alpha \to 0$ and then $s \to \infty$, since the above value is unaltered, the desired conclusion follows. This example is due to Johnson and Helms [1].]

10. We sketch a proof of measurability of the debut function in Theorem 2.5. First recall the Souslin operation. If $\mathcal{P}_0(\Omega) = 2^\Omega$ is the power set, and $\mathcal{A} \subset \mathcal{P}_0(\Omega)$ is any nonempty collection, it is a *determining system* if for each finite sequence $(n_1, \ldots, n_k), n_i \geq 1$, of integers we can associate a set $A_{n_1,\ldots,n_k} \in \mathcal{A}$. Let $s = (n_1, \ldots, n_k)$, and S denote the class of all such finite sequences. Let $\sigma = (n_1, n_2, \ldots), n_i \geq 1$, be an infinite sequence, of integers and $\mathfrak{S}$ denote the set of all such σ. Partially order S as: for s_1, s_2 in S, say $s_1 \prec s_2$ iff s_1 is the initial segment of s_2 and similarly $s \prec \sigma$ iff s is the initial segment of $\sigma \in \mathfrak{S}$. Then a determining system is a mapping $\Delta : S \to \mathcal{A}$, where $\Delta(s)(= A_s^\Delta) \in \mathcal{A}$, such that $s_1 \prec s_2 \Rightarrow \Delta(s_1) \supset \Delta(s_2)$ (i.e. Δ orders subsets of $\mathcal{A}$). For $\sigma \in \mathfrak{S}$, let $A_\sigma^\Delta = \bigcap_{s \prec \sigma} A_s^\Delta$. The *nucleus* of $\mathcal{A}$ is $N^\Delta = \bigcup_{\sigma \in \mathfrak{S}} A_\sigma^\Delta$. The operation leading from Δ to N^Δ is called *operation* (A) (or *analytic*, or *Souslin operation*). The collection $\mathcal{U}(\mathcal{A}) = \{N^\Delta | \Delta : S \to \mathcal{A}, \text{ for all such } \Delta\text{'s}\}$ is the class of $\mathcal{A}$-analytic sets. We present, following Choquet, an alternative form of $\mathcal{U}(\mathcal{A})$. Thus let $\mathfrak{S}_s$ be a section of $\mathfrak{S}, \mathfrak{S}_s = \{\sigma \in \mathfrak{S} : \sigma \succ s\}$ so that $\mathfrak{S}_{s_1} \supset \mathfrak{S}_{s_2}$ if $s_1 \prec s_2$. If k is the length of s (i.e. the number of n_i's), set $|s| = k$ and if $B_s^\Delta = \mathfrak{S}_s \times A_s^\Delta \subset \mathfrak{S} \times \Omega$ for each $s \in S$ and Δ, define $C_k^\Delta = \bigcup_{|s|=k} B_s^\Delta$ (so $C_k^\Delta \supset C_{k+1}^\Delta$), $C^\Delta = \bigcap_{k \geq 1} C_k^\Delta$. If $\pi : \mathfrak{S} \times \Omega \to \Omega$ is the coordinate projection, then we have:

Lemma. *For each* $\Delta, \pi(C^\Delta) = N^\Delta$.

Proof. This follows from

$$\pi(C^{\Delta}) = \pi\left(\bigcap_{k\geq 1}\bigcup_{|s|=k} B_s^{\Delta}\right) = \bigcap_{k=1}^{\infty}\bigcup_{|s|=k}\pi(\mathfrak{S}_s \times A_s^{\Delta}) = \bigcap_{k=1}^{\infty}\bigcup_{|s|=k} A_s^{\Delta}$$
$$= \bigcup_{\sigma\in\mathfrak{S}}\bigcap_{s\prec\sigma} A_s^{\Delta} = N^{\Delta}, \tag{1}$$

provided we justify the commutativity of π and $\cap$ in the third term of the first line of (1), the rest of the operations being immediate from definitions. Since $\pi\left(\bigcap_k C_k^{\Delta}\right) \subset \bigcap_k \pi(C_k^{\Delta})$ always, we prove the opposite inclusion by using the special structure here.

Let $\omega \in \bigcap_k \pi(C_k^{\Delta})$. then for each k, there is $\tilde{\omega}_k \in C_k^{\Delta}$ with $\omega = \pi(\tilde{\omega}_k)$, and $\tilde{\omega}_k \in \pi^{-1}(\{\omega\}) \cap C_k^{\Delta}$, and similarly $\pi^{-1}(\{\omega\}) \cap \bigcap_{k=1}^{n} C_k^{\Delta} \neq \emptyset$ for each n. Since $\mathcal{A} \neq \emptyset, C^{\Delta} \neq \emptyset$ and $\pi^{-1}(\{\omega\}) = \mathfrak{S} \times \{\omega\}$. Hence

$$\emptyset \neq \pi^{-1}(\{\omega\}) \cap C_k^{\Delta} = (\mathfrak{S} \times \{\omega\}) \cap \bigcup_{|s|=k} (\mathfrak{S}_s \times A_s^{\Delta})$$
$$= \bigcup_{|s|=k} (\mathfrak{S}_s \times \{\omega\}), \text{ discarding the } A_s^{\Delta}\text{'s not containing the } \omega.$$
$$= \left(\bigcup_{|s|=k} \mathfrak{S}_s\right) \times \{\omega\}. \tag{2}$$

But $\mathfrak{S}_{s_1} \supset \mathfrak{S}_{s_2}$ for $s_1 \prec s_2$, so that we have from (2)

$$\pi^{-1}(\{\omega\}) \cap C^{\Delta} = \bigcap_{k=1}^{\infty}\left[\left(\bigcup_{|s|=k} \mathfrak{S}_s\right) \times \{\omega\}\right] \supset \{\sigma\} \times \{\omega\}, \tag{3}$$

where $\sigma(\succ s)$ is an element of the nonempty intersection of $\bigcap_{s\prec\sigma} \mathfrak{S}_s$ for $\sigma \in \mathfrak{S}$. So there is at least one $\tilde{\omega} \in \pi^{-1}(\{\omega\}) \cap C^{\Delta}$ satisfying $\pi(\tilde{\omega}) = \omega$ and hence $\pi(C^{\Delta}) \supset \bigcap_{k=1}^{\infty} \pi(C_k^{\Delta})$, giving the result. □

The point here is that there exists a set $\mathfrak{S}$, and a collection $\{\mathfrak{S}_s, s \in S\}$ of its subsets such that every countable subcollection with the finite intersection property has a nonempty intersection, implies $\{\pi^{-1}(\{\omega\}) \cap (\mathfrak{S}_s \times A_s^{\Delta}) : s \in S\}$ has the same property. Let us call this *"semi-compactness"* of the collection (temporarily). Then the *alternative definition* is: Any set $A \subset \Omega$ is said to be $\mathcal{A}$-*analytic* for a nonempty $\mathcal{A} \subset$

$\mathcal{P}_0(\Omega)$ (or $A \in \mathcal{U}(\mathcal{A})$), iff there exists an auxiliary pair $(E, \mathcal{E})$ such that $\mathcal{E}$ is (nonempty and) semi-compact and a set $C \subset E \times \Omega, C \in (\mathcal{E} \times \mathcal{A})_{\sigma\delta}$ with the property that $\pi(C) = A$ where $\pi : E \times \Omega \to \Omega$ is the coordinate projection. With the above lemma, it is seen that both definitions are the same. We use the following classical result of N. Lusin and W. Sierpiński (1918) (see Saks [1], p. 50 for a proof) and also Rao [11], p.46).

Proposition A. *Let μ be an outer measure on Ω and let $\mathcal{M}_\mu$ be the class of all Carathéodory measurable sets for μ (so $\mu|\mathcal{M}_\mu$ is σ-additive and μ has no further σ-additive extensions). If $\mathcal{A} \subset \mathcal{M}_\mu$ is nonempty, then $\mathcal{U}(\mathcal{A}) \subset \mathcal{M}_\mu$. [Thus $\mathcal{U}^n(\mathcal{A}) = \mathcal{U}(\mathcal{U}^{n-1}(\mathcal{A})) \subset \mathcal{M}_\mu$ for all $n \geq 2$.]*

The following properties are needed for a proof of Theorem 2.5, and they will be established after the theorem itself.

Proposition B. *(a) For any (nonempty) $\mathcal{A} \subset \mathcal{P}_0(\Omega)$, we have $\mathcal{A}_\sigma \subset [\mathcal{U}(\mathcal{A})]_\sigma = \mathcal{U}(\mathcal{A})$, and $\mathcal{A}_\delta \subset [\mathcal{U}(\mathcal{A})]_\delta = \mathcal{U}(\mathcal{A})$. Thus $\mathcal{U}(\mathcal{A})$ is a σ-lattice in $\mathcal{P}_0(\Omega)$. (b) (i) If $\mathcal{A} \subset \mathcal{P}_0(\Omega), \mathcal{B} \subset \mathcal{P}_0(\tilde{\Omega})$, then $\mathcal{U}(\mathcal{A}) \times \mathcal{U}(\mathcal{B}) \subset \mathcal{U}(\mathcal{A} \times \mathcal{B}) \subset \mathcal{P}_0(\Omega \times \tilde{\Omega})$. (ii) If $M \in \mathcal{U}(\mathcal{A} \times \mathcal{B})$ and $\pi : \Omega \times \tilde{\Omega} \to \tilde{\Omega}$ is the coordinate projection, and $\mathcal{A}$ is semi-compact, then $\pi(M) \in \mathcal{U}(\mathcal{B})$. (c) $\mathcal{U} : \mathcal{A} \to \mathcal{U}(\mathcal{A})$ is idempotent, i.e., $\mathcal{U}^2(\mathcal{A}) = \mathcal{U}(\mathcal{A})$. (d) $\sigma(\mathcal{A}) \subset \mathcal{U}(\mathcal{A})$ iff for $A \in \mathcal{A}$ we have $\{A, A^c\} \subset \mathcal{U}(\mathcal{A})$.*

Proof of Theorem 2.5. Let A be a progressively measurable set of $I \times \Omega$ with $\{\mathcal{F}_t, t \in I\}, I \subset \mathbb{R}$, a right continuous complete net of σ-algebras from (Ω, Σ, P) and D_A as the debut of A. Then $[D_A \leq t] \in \mathcal{F}_t$ is equivalent to $[D_A < t] \in \mathcal{F}_t (= \mathcal{F}_{t+}), t \in I$. Let $I_t = \{s \in I : s \leq t\}$, and $\mathcal{B}_t$ be the Borel σ-algebra of I_t. By definition $[D_A < t] = \pi[A \cap (I_t \times \Omega)]$. [Note $\omega \in \bigcup_{n=1}^{\infty} [D_A \leq t - \frac{1}{n}]$ iff $(t, \omega) \in A \cap (T_t \times \Omega)$.] Let $\mathcal{C}$ be the collection of all compact subsets of $\bar{I}_t$, the closure of I_t. Now let $\mathcal{A} = \mathcal{C} \times \mathcal{F}_t$, and $K \in \mathcal{A} (\subset \mathcal{U}(\mathcal{A}))$. If $K = C \times E, C \in \mathcal{C}, E \in \mathcal{F}_t$, then $K^c = (C^c \times \Omega) \cup (C \times E^c)$ and $C \times E^c \in \mathcal{A}$. But C^c is open, so $C^c = \bigcup_{n=1}^{\infty} C_n \cap \bar{I}_t$ (σ-compactness), $C_n \cap \bar{I}_t \in \mathcal{C}$. Hence $C^c \times \Omega = \bigcup_{n=1}^{\infty} [(C_n \cap I_t) \times \Omega] \in \mathcal{U}(\mathcal{A})$, by Proposition B(a). Thus $K^c \in \mathcal{U}(\mathcal{A})$. Proposition B(d) then implies $\sigma(\mathcal{A}) \subset \mathcal{U}(\mathcal{A})$. But $\mathcal{B}_t \otimes \mathcal{F}_t \subset \sigma(\mathcal{A})$. Since clearly $\mathcal{C}$ is a compact (hence semi-compact) collection, by Proposition B(b(ii)) we can conclude that

$[D_A < t] = \pi(A \cap I_t \times \Omega)) \in \mathcal{U}(\mathcal{F}_t) \subset \mathcal{F}_t$. Since $\mathcal{F}_t = \mathcal{F}_{t+}$, this yields Theorem 2.5. □

Proof of Proposition B. (a) Let $\{A_n\}_1^\infty \subset \mathcal{U}(\mathcal{A}), A = \bigcap_n A_n$, and $B = \bigcup_n A_n$. By definition of $\mathcal{A}$-analyticity, there exist $(E_n, \mathcal{E}_n), \mathcal{E}_n$ semi-compact over $E_n, C_n \subset E_n \times \Omega, C_n \in (\mathcal{E}_n \times \mathcal{A})_{\sigma\delta}$ such that $p_n(C_n) = A_n, p_n : E_n \times \Omega \to \Omega$ being the projection . Let $E = \times_n E_n, \mathcal{E} = \times_n \mathcal{E}_n$ be the Cartesian products, and $\pi = (p_1, p_2, \dots) : E \times \Omega \to \Omega$ be the combined projection. Thus $\mathcal{E}$ is semi-compact. If $D_n = \times_{\substack{i\geq 1\\ i\neq n}} E_i$, then $D_n \times C_n \subset E \times \Omega$, and is a cylinder with base C_n. So $\pi(D_n \times C_n) = p_n(C_n) = A_n$. We may assume $A \neq \emptyset$. Also for A_1, A_2 in $\mathcal{U}(\mathcal{A})$, letting $p_{12} : (E_1 \times E_2) \times \Omega \to \Omega$, we observe that

$$A_1 \cap A_2 = p_1(C_2) \cap p_2(C_2) = p_{12}(C_1 \times E_2) \cap p_{12}(E_1 \times C_2) = p_{12}(C_1 \times C_2). \tag{4}$$

Here p_{12} stands for the combined projection (p_1, p_2) in the earlier notation. Thus Equation (4) follows from definition of the various projections involved. Now by the semi-compactness of $\mathcal{E}, \{(D_n \times C_n) \cap \pi^{-1}(\{a\})\}_1^\infty$ is also semi-compact and

$$A = \bigcap_{n\geq 1} \pi(D_n \times C_n) = \pi\left(\bigcap_n (D_n \times C_n)\right). \tag{5}$$

But $C_n \in (\mathcal{E}_n \times \mathcal{A})_{\sigma\delta}$ and $D_n \times C_n \in (\mathcal{E} \times \mathcal{A})_{\sigma\delta} = \left(\times_{\substack{i\neq n\\ i\geq 1}} \mathcal{E}_i\right) \times (\mathcal{E}_n \times \mathcal{A})_{\sigma\delta}$. This and (5) imply $A \in \mathcal{U}(\mathcal{A})$, and $\mathcal{A}_\delta \subset (\mathcal{U}(\mathcal{A}))_\delta \subset \mathcal{U}(\mathcal{A})$.

Next let $\bar{E} = \bigoplus_{n\geq 1} E_n, \bar{\mathcal{E}} = \bigoplus_n \mathcal{E}_n$, the direct sums (i.e., $F = \bigoplus_n F_n \in \bar{\mathcal{E}}$ iff $F_n = \emptyset$ for all but finitely many $n, F_n \in \mathcal{E}_n$). We assert that $\bar{\mathcal{E}}$ is semicompact. For, let $\{H_n\}_1^\infty \subset \mathcal{E}$ with the finite subset property. Then $H_n = \bigoplus_{m=1}^\infty G_{nm}, G_{nm} \in \mathcal{E}_m, G_{nm} = \emptyset$ for all but finitely many m, each $n \geq 1$, and so $\bigcap_{n\geq 1} H_n = \bigoplus_m \bigcap_n G_{nm}$. Since $\bigcap_{n=1}^k H_n \neq \emptyset, 1 \leq k < \infty$, $\bigcap_{n=1}^k G_{nm} \neq \emptyset$ for at least one $m = m(k)$, for each $k \geq 1$, by the semi-compactness of $\mathcal{E}_m$. So by definition of direct sums, we deduce that $\bigcap_n H_n \neq \emptyset$. Thus $\mathcal{E}$ is semi-compact. Let $C = \bigoplus_n C_n \subset \bigoplus_n (E_n \times \Omega), \pi : \bar{E} \times \Omega \to \Omega$, and $\bar{C}_n \subset \bar{E} \times \Omega$ be identified with a set whose nth element $= E_n$, and $= \emptyset$ otherwise. Then $C = \bigcup_{n\geq 1} \bar{C}_n \subset E \times \Omega$, and $\pi(C) = \bigcup_n (\pi(\bar{C}_n)) = \bigcup_n p_n(C_n) = \bigcup_n A_n = B$. But $C_n \in (\mathcal{E}_n \times \mathcal{A})_{\sigma\delta}$ so $C_n = \bigcap_{m=1}^{\infty} K_{nm}, K_{nm} \in (\mathcal{E}_n \times \mathcal{A})_{\sigma}, m \geq 1$. Thus writting $\tilde{K}_{nm} = K_{nm}$ at the nth place, $= \emptyset$ elsewhere, we get $\tilde{K}_{nm} \in (\mathcal{E} \times \mathcal{A})_\sigma$ and hence $\bigcup_{n\geq 1} \tilde{K}_{nm} \in (\mathcal{E} \times \mathcal{A})_\sigma$ so that $C = \bigcup_n \tilde{C}_n = \bigcup_n \bigcap_m \tilde{K}_{nm} = \bigcap_m \bigcup_n K_{nm} \in (\mathcal{E} \times \mathcal{A})_{\sigma\delta}$. This implies $B \in \mathcal{U}(\mathcal{A})$, and $\mathcal{A}_\sigma \subset [\mathcal{U}(\mathcal{A})]_\sigma \subset \mathcal{U}(\mathcal{A})$. (b) (i) If $A \in \mathcal{U}(\mathcal{A}), B \in \mathcal{U}(\mathcal{B})$, so that by using the first definition, $A = \bigcup_\sigma \bigcap_{s \prec \sigma} A_s = \bigcup_\sigma H_\sigma, B = \bigcup_{\sigma'} \bigcap_{s' \prec \sigma'} B_{s'} = \bigcup_{\sigma'} G_{\sigma'}$. Then $H_\sigma \times G_{\sigma'} \in \mathcal{A}_\delta \times \mathcal{B}_\delta \subset (\mathcal{A} \times \mathcal{B})_\delta$, and so $A \times B \in (\mathcal{A} \times \mathcal{B})_{\sigma\delta} \subset \mathcal{U}(\mathcal{A} \times \mathcal{B})$ by (a). (ii) Under the present hypothesis, with the second definition, there exists $(Y, \mathcal{Y})$, $\mathcal{Y}$ semi-compact, and $G \subset Y \times (\Omega \times \tilde{\Omega}), G \in (\mathcal{Y} \times (\mathcal{A} \times \mathcal{B}))_{\sigma\delta}$ and $M = pr(G)$, where $pr : Y \times (\Omega \times \tilde{\Omega}) \to \Omega \times \tilde{\Omega}$, is a projection. But $\mathcal{Y} \times \mathcal{A}$ is also semi-compact, and since $\mathcal{Y} \times (\mathcal{A} \times \mathcal{B}) = (\mathcal{Y} \times \mathcal{A}) \times \mathcal{B}$ (isomorphically), we have $\tilde{pr}(G) \in \mathcal{U}(\mathcal{B})$ where $\tilde{pr} : (Y \times \Omega) \times \tilde{\Omega} \to \tilde{\Omega}$. Since $\tilde{pr}(G) = \pi(pr(G)) = \pi(M)$, the result follows. (c) Now $\mathcal{U}$ is monotone increasing. Let $A \in \mathcal{U}^2(\mathcal{A})$. Then there is $(E, \mathcal{E}), \mathcal{E}$ semi-compact, $C \subset E \times \Omega, \pi(C) = A (\pi : E \times \Omega \to \Omega)$, and $C \in (\mathcal{E} \times \mathcal{U}(\mathcal{A})_{\sigma\delta}$. But $\mathcal{E} \subset \mathcal{U}(\mathcal{E})$, and by (b), $\mathcal{E} \times \mathcal{U}(\mathcal{A}) \subset \mathcal{U}(\mathcal{E}) \times \mathcal{U}(\mathcal{A}) \subset \mathcal{U}(\mathcal{E} \times \mathcal{A})$. Thus by (a), $C \in \mathcal{U}(\mathcal{E} \times \mathcal{A})$. Since $\mathcal{E}$ is semi-compact, by (b) (ii), $A = \pi(C) \in \mathcal{U}(\mathcal{A})$. Hence $\mathcal{U}^2(\mathcal{A}) \subset \mathcal{U}(\mathcal{A})$, and (c) follows. Finally (d) is a consequence of (a) since $\mathcal{B} = \{A \subset \Omega : (A, A^c) \subset \mathcal{U}(\mathcal{A})\}$ is a σ-algebra. □

Remark. The above proof is adapted from Choquet [1] and Meyer [1]. Even though we deduced (c) from (a) and (b), an independent proof of (c) can be given (cf., Sierpiński [1]).

11. With the terminology introduced for Theorem 4.7, let $\{\mathcal{F}_n, n \geq 1\}$ be an increasing sequence of σ-algebras in (Ω, Σ, P). Let $\mathcal{F}_\infty = \lim_n \mathcal{F}_n$, and $\Omega^{(r)}$ be the regular part relative to this family.

(a) If $B \subset \Omega - \Omega^{(r)}$ and $B \in \mathcal{F}_\infty$, then there exists a martingale $\{X_n, \mathcal{F}_n, n \geq 1\}$ such that $X_n(\omega) \to X_\infty(\omega)$ exists and is finite for a.a. (ω) on B^c and $X_n(\omega) \to \infty$, a.a. (ω) on B. [*Hints*: Let $\delta_n > 0, \sum_{n=1}^{\infty} \delta_n < \infty$. Choose a sequence $\{n_k, B_k\}_0^\infty, B_k \in \mathcal{F}_{n_k}$ and $P(B \Delta B_k) < \delta_k$. Let $S_k(\delta_n)$ be a δ_n-splitting set of $\mathcal{F}_k$ relative to $\mathcal{F}_{k+1}$. If $T_0 = 1$ and for $m \geq 1, T_m(\omega) = \inf\{k > T_{m-1}(\omega) \vee n_{m-1} : \omega \in S_k(\delta_m)\}$, then $\{T_m, m \geq 0\}$ is a stopping time process of $\{\mathcal{F}_n, n \geq 0\}$ where $\mathcal{F}_0 = \{\emptyset, \Omega\}$. As noted just prior to Theorem 4.7, with each δ_n-splitting there corresponds a sequence $\{A_k(\delta_n), k \geq 1\}, A_k(\delta_n) \in \mathcal{F}_{k+1}$, and $A_k(\delta_n)$ goes with $S_k(\delta_n)$. Define $A_{T_n} = \bigcup_{k\geq 1} A_k(\delta_n) \cap [T_n = k]$ for $n \geq 1$. Let $\{X_n, n \geq 1\}$ be a sequence defined inductively as: $X_1 = 0, X \circ T_1 = 0$, and let

$$X \circ (T_k + 1) = \cdots = X \circ (T_{k+1}) = \begin{cases} (X \circ T_k) + 1 & \text{on } A_{T_k}^c \cap B_k \\ g_k & \text{on } A_{T_k} \cap B_k \quad (*) \\ X \circ T_k & \text{on } B_k^c \end{cases}$$

with $g_k = [X \circ T_k \cdot P^{\mathcal{B}(T_k)}(A_{T_k}) - P^{\mathcal{B}(T_k)}(A_{T_k})]/P^{\mathcal{B}(T_k)}(A_{T_k})$, on $A_{T_k} \cap B_k$. Then $\{X_n, \mathcal{F}_n, n \geq 1\}$ is a martingale with the stated properties. If moreover, in $(*)$ $(X \circ T_k) + 1$ is replaced by $(X \circ T_k) + (-1)^k$, then also $\{X_n, \mathcal{F}_n, n \geq 1\}$ is a martingale which will oscillate boundedly on B and converge on B^c.]

(b) Let $\{\mathcal{F}_n, n \geq 1\}$, in the above, be a d-sequence as defined in the remark following Theorem 4.7. Let f be an extended real valued $\mathcal{F}_\infty$-measurable function. Then there exists a martingale $\{X_n, \mathcal{F}_n, n \geq 1\}$ on (Ω, Σ, P) such that $\lim_n X_n = f$ a.e. iff $|f| < \infty$ a.e. on $\Omega^{(r)}$. [This uses Theorem 4.7 and (a) and some further computations of a similar nature. These representations are due to Lamb [1].

12. A stochastic base $\{\mathcal{F}_n, n \geq 1\}$ in (Ω, Σ, P) is *regular* if for each $A \in \mathcal{F}_n, n \geq 1$ there is a $B_n (= B_n(A))$ in $\mathcal{F}_{n-1}, B_n \supset A$, such that $P(B_n) \leq cP(A)$ for some fixed $c > 0$. Let $\{X_n, \mathcal{F}_n, n \geq 1\}$ be a martingale or a submartingale with $\{\mathcal{F}_n, n \geq 1\}$ forming a regular base, $X_n \in L^1(P)$. If $\varphi_n = X_n - X_{n-1}$, $A = \left\{\omega : \sum_{n=1}^{\infty} |\varphi_n(\omega)|^2 < \infty\right\}$ and $B = \left\{\omega : \underline{\lim}_n X_n(\omega) = \overline{\lim}_n X_n(\omega)\right\}$, then $P(A\Delta B) = 0$ for martingales and $B \supset \left\{\omega : \sup_n X_n(\omega) < \infty\right\}$ for submartingales. Moreover, if $C = \left\{\omega : \sum_{n=1}^{\infty} E^{\mathcal{F}_n}(\varphi_{n+1}^2)(\omega) < \infty\right\}, X_n \in L^2(\Sigma)$ and the process is a martingale then $P(C\Delta A) = 0$. The same result holds for infinite measures if $\mathcal{F}_n$'s are P-rich σ-algebras. [*Hints:* For each $\lambda > 0$, let $T^\lambda = \inf\{n \geq 1 : X_n \geq \lambda\}, A_n = [T^\lambda = n + 1]$. By regularity, there is a $B_n \in \mathcal{F}_n, B_n \subset A$ of the same measure, and let $\tau_m(\omega) = \inf\{n \geq$

$m : \omega \in B_n\}$, with $\inf(\emptyset) = +\infty, T^* = T^\lambda \wedge \tau_m$. Then T^λ, T^*, τ_m are stopping times of $\{\mathcal{F}_n, n \geq 1\}$ and if $Z_n = \sum_{k=1}^{n} \varphi_k \chi_{[T^* \geq k]}$, verify that (i) $\{Z_n, \mathcal{F}_n, n \geq 1\}$ is a (sub)martingale, and (ii) $\sup_n E(Z_n^+) < \infty$. The method of proof is the same as that of Theorem 4.9, and when (i) and (ii) are established, the conclusion can be obtained from the latter theorem. The result is essentially due to Doob [2] and Chow [3].]

13. Let $\{Z_n, \mathcal{F}_n, n \geq 1\}$ be any adapted process in $L^1(\Omega, \Sigma, P)$. Then there exists a submartingale $\{X_n, \mathcal{F}_n, n \geq 1\}$ and a predictable process $\{V_n, \mathcal{F}_{n-1}, n \geq 1\}$ such that (non-uniquely) $Z_n = (V \cdot X)_n = \sum_{k=1}^{n} V_k \varphi_k$, where $\varphi_k = X_k - X_{k-1}$. [*Hints:* Let $V_1 = \operatorname{sgn} E(Z_1)$ and for $k > 1, V_k = 1$ on $[E^{\mathcal{F}_{k-1}}(Z_k) > Z_{k-1}]$, and $= -1$ on $[E^{\mathcal{F}_{k-1}}(Z_k) \leq Z_{k-1}]$. If $X_n = (V \cdot Z)_n$, so that $E^{\mathcal{F}_{n-1}}(X_n) \geq X_{n-1}$ a.e., we have $(V \cdot X)_n = \sum_{k=1}^{n} V_k(V_k \cdot \varphi_k) = \sum_{k=1}^{n} \varphi_k = Z_n$ since $Z_0 = 0$, by definition. The predictable process constructed here is optimal among bounded predictables, as defined after Corollary 4.12. Cf., Millar [1].]

Bibliographical remarks

As indicated at the beginning of this chapter, stopping time transformations are used to get more refined results than those of Chapter II. Hence Sections 1 and 2 contain very general considerations. Most of the results here are due to Chow [1], but the initial treatment in subsections (a) and (b) of Section 1 is influenced by the work of Hunt [1]. The general technique in these results is well illustrated in Theorem 1.13 which is due to Chow [1], and the basic ideas go back to Bochner [3]. The calculus of real valued stopping times is treated in detail in Chung and Doob [1] (and also in Courrège and Priouret [1]) from whom many of these results are taken. Since most of the sources are given in the text, we shall only indicate the highlights below.

The results on class (D) martingales and the related treatment is influenced by the work of Meyer [1], and Johnson-Helms [1]. The regularity of martingales was introduced by Snell [1], but it was analyzed and clarified by Chow [1] to whom Theorem 3.12 is due. Theorem 3.14 follows from this and some considerations of Chapter I. The relation between the two classes of results was noticed apparently only in the present work. Theorem 4.1 is due to Chow [1], and its use provided a generalization of a decomposition due to Gundy [1] in Theorem 4.6. The results of Theorems 4.7 and 4.8 are due to Lamb [1] and Stein [1] respectively. The whole field of study of martingale difference processes was opened up by the simple but very useful observation given in Extension 4.10, due to Austin [1]. This is similar to the classical "Lusin s-function." (See Peterson [1] for a recent account of this and the $\mathcal{H}^p$, BMO theory from an analytic viewpoint.) This "Austin function" is basic to the probabilistic treatment of $\mathcal{H}^p$-spaces and BMO. Though the $\mathcal{H}^p$-spaces are certain subsets of L^p-spaces, they are not measurable subspaces, i.e., not of the form $L^p(\mathcal{B})$ for some $\mathcal{B} \subset \Sigma$ (cf. Chapter II). So the adjoint spaces $(\mathcal{H}^p)^*$ are not conveniently identified and the nonreflexive case $p = 1$ was an open problem for some time. The $(\mathcal{H}^1)^*$ was identified as BMO, a space introduced by John and Nirenberg [1], by Fefferman [1]. An exposition of these results was given by Garsia [2], and identifying the probabilistic $\mathcal{H}^p$ as subspaces of $L^p(\ell^2)$, we essentially tailored the presentation from the latter author and Herz [1].

We have included a number of complements to the work in the text as exercises with details to show the far reaching influence of stopping time tranformations. This is an effective tool for stochastic calculus as distinct from the classical works. We shall specialize the general theory for stochastic integration and derivations in the following chapters in which continuous (time) index will be prominent. It should also be remarked that the above treatment is presented with a view to applications in mathematical analysis, and hence it is extendable to infinite measures. This is noted by frequent reference to P-rich σ-algebras in the statements and proofs but could be ignored by the readers interested only in probability theory.

Chapter V

Martingale decompositions and integration

Continuing the work of the preceding chapter on continuous parameter (sub)martingales, we present a solution of the Doob decomposition problem, raised in Section 2.5, which is due to Meyer. We give an elementary (but longer) demonstration and also sketch a shorter (but more sophisticated) argument based on Doléans-Dade signed measure representation of quasimartingales. This decomposition leads to stochastic integration with square integrable martingales as integrators generalizing the classical Itô integration. The material is presented in this chapter in considerable detail, since it forms a basis for semimartingale integrals with numerous applications to be abstracted and treated in the following chapter. Orthogonal decompositions of square integrable martingales (of continuous time parameter), its time change transformation leading to a related Brownian motion process and the Lévy characterization of the latter from continuous parameter martingales, are covered. Stopping (or optional) times play a key role in all this work, and some classifications of these are given. The treatment also includes the Stratonovich integrals as well as an identification of the square integrable martingale integrators with spectral measures of certain normal operators in Hilbert space. Finally some related results appear as exercises in the Complements section.

5.1 Preliminaries on continuous parameter processes

We present here a few results, in the form of preliminaries, on continuous parameter processes related to martingales to be used in the fundamental decomposition theory of the next section. Let *class*(R) denote the set of supermartingales admitting the Riesz decomposition, given in Section 2.5. It was already remarked at the begining of Section 2.5 that the Riesz decomposition extends to the continuous parameter case without difficulty or additional work. As before, we call a positive right continuous supermartingale $X = \{X_t, \mathcal{F}_t, t \geq 0\}$ a *potential* if $X_t \to 0$ in $L^1(P)$ as $t \to \infty$, i.e., $E(X_t) \to 0$. Thus we now state the analog of Theorem II.5.5 as:

1. Theorem. *Let $\{X_t, \mathcal{F}_t, t > 0\}$ be a right continuous supermartingale on (Ω, Σ, P) with $\{\mathcal{F}_t, t > 0\}$ as a right continuous filtration. Then one has an* a.e. *unique decomposition:*

$$X_t = X_t^{(1)} + X_t^{(2)}, \tag{1}$$

with $\{X_t^{(1)}, \mathcal{F}_t, t > 0\}$ a right continuous martingale and $\{X_t^{(2)}, \mathcal{F}_t, t > 0\}$ a potential, iff it dominates a (right continuous) martingale $\{Y_t, \mathcal{F}_t, t > 0\}$ so that $Y_t \leq X_t$, a.e., *$t > 0$.*

The same argument as in the discrete parameter case applies, as noted, and it will be omitted. We now introduce another class, isolated by Itô and S. Watanabe [1], to be used in the general decomposition below and in other places in martingale analysis.

2. Definition. Let (Ω, Σ, P) be a complete probability space and $\{\mathcal{F}_t, t \geq 0\}$ be a right continuous filtration from Σ as before, so that each $\mathcal{F}_t$ is a complete σ-subalgebra and $\mathcal{F}_t = \mathcal{F}_{t+} \subset \mathcal{F}_{t'}$ for $t < t'$, termed the *standard filtration.* Then a right continuous adapted process $\{Y_t, \mathcal{F}_t, t \geq 0\}$ is a *local martingale* if there exists a sequence $\{T_n, n \geq 1\}$ of stopping times of the filtration satisfying: (i)$T_n \leq T_{n+1}$, (ii) $P[T_n \leq n] = 1$, (iii) $P[\lim_n T_n = \infty] = 1$, and (iv) if $\tau_t^n = T_n \wedge t, Z_t^n = Y \circ \tau_t^n$, then, for each $n, \{Z_t^n, \mathcal{F}_t, t \geq 0\}$ is a uniformly integrable martingale.

This concept, which is unmotivated, plays a key role in later analysis. It is slightly weaker than the martingale notion. In fact, a positive local martingale $\{Y_t, \mathcal{F}_t, t \geq 0\}$ is a supermartingale. To see this, let $A \in \mathcal{F}_s, 0 \leq s < t$. Now $\{Y \circ \tau_t^n, t \geq 0\}$ is uniformly integrable. Also for all $\alpha \geq 0$, the set $\{\omega : \lim_n T_n(\omega) = \infty\} = \bigcup_{n \geq 1} \{\omega : T_n(\omega) \geq \alpha\} = \bigcup_{n \geq 1} B_{n,\alpha}$ (say) differs from Ω by a P-null set. So $B_{n,\alpha} \in \mathcal{F}_\alpha$ for all $n \geq 1$, and

$B_{n,\alpha} \subset B_{n+1,\alpha} \uparrow \Omega$ a.e. Hence,

$$\int_A Y_t\,dP = \lim_n \int_{A\cap B_{n,t}} Y_t\,dP = \lim_n \int_{A\cap B_{n,t}} Y\circ\tau_t^n\,dP,$$

since $Y\circ\tau_t^n = Y_t$ on $B_{n,t}$,

$$\le \lim_n \int_{A\cap B_{n,s}} Y\circ\tau_t^n\,dP, \quad (B_{n,s}\supset B_{n,t} \text{ for } s<t \text{ and } Y_t\ge 0)$$

$$= \lim_n \int_{A\cap B_{n,s}} Y\circ\tau_s^n\,dP,$$

by the martingale property of $\{Y\circ\tau_s^n, \mathcal{F}_s, s\ge 0\}$,

$$= \lim_n \int_{A\cap B_{n,s}} Y_s\,dP = \int_A Y_s\,dP.$$

Thus $E^{\mathcal{F}_s}(Y_t) \le Y_s$, a.e., as asserted.

It is clear that the sum of two local martingales, with the same stochastic base, is a local martingale. Since each martingale is a local martingale (take $T_n = n$ in the definition), the sum of a martingale and a local martingale is a local martingale. We use these relations in the solution of the general problem.

To motivate the concepts in the continuous parameter case, recall that a supermartingale $\{X_n, \mathcal{F}_n, n\ge 1\}$ admits a unique decomposition:

$$X_n = X_n' - \sum_{i=1}^{n-1}(X_i - E^{\mathcal{F}_i}(X_{i+1})) = X_n' - A_n, \quad \text{(say)}$$

where $A_n \ge 0, A_1 = 0$ a.e., and A_n is $\mathcal{F}_{n-1}$-adapted. In the continuous parameter case the sum should be replaced by an integral and that A_n be $\mathcal{F}_{n-1}$-adapted should find a corresponding new concept. It was called the "predictable" property in the discussion following Corollary IV.4.12. Let us explain these ideas further.

Since $0\le A_n \le A_{n+1}$, let $A_\infty (= \lim_n A_n)$ be integrable. If $\{Y_n, \mathcal{F}_n, n\ge 1\}$ is any $L^\infty(P)$-martingale, so that $\sup_n E(|Y_n A_\infty|) < \infty$, and

$\lim_n Y_n = Y_\infty$, a.e. as well as in $L^p(P)$-mean, $p \geq 1$, consider

$$\begin{aligned} E\Big[\sum_{n=1}^{\infty}(Y_n - Y_\infty)(A_{n+1} - A_n)\Big] &= \sum_{n=1}^{\infty} E[E^{\mathcal{F}_n}\{(Y_n - Y_\infty)(A_{n+1} - A_n)\}] \\ &= \sum_{n=1}^{\infty} E((A_{n+1} - A_n)(Y_n - E^{\mathcal{F}_n}(Y_\infty))) \\ &= 0, \end{aligned} \tag{2}$$

since A_n is $\mathcal{F}_{n-1}$-adapted and the computation is clearly valid. The left side of (2) can be written as:

$$E\Big[\sum_{n=1}^{\infty} Y_n(A_{n+1} - A_n)\Big] = E(Y_\infty A_\infty). \tag{3}$$

Conversely, if (3) holds for every bounded martingale $\{Y_n, \mathcal{F}_n, n \geq 1\}$, then (2) is true implying that A_n is $\mathcal{F}_{n-1}$-adapted. Now (3) generalizes to the continuous parameter case as follows.

Let $\{A_t, \mathcal{F}_t, t \geq 0\}$ be a right continuous adapted process such that $A_0 = 0$ a.e., $A_{\cdot}(\omega)$ is increasing for a.a. $\omega \in \Omega$, and $\sup_t E(A_t) < \infty$, where the filtration $\{\mathcal{F}_t, t \geq 0\}$ is right continuous and completed in (Ω, Σ, P). If $\mathcal{R}$ is the semi-ring of half open intervals of $\mathbb{R}^+$ and $\mathcal{B} = \sigma(\mathcal{R})$ is the Borel σ-algebra, then let $\overline{\mu}_\omega((a,b]) = A_b(\omega) - A_a(\omega)$ for each $\omega \in \Omega$ so that $\overline{\mu}_\omega : \mathcal{R} \to \overline{\mathbb{R}}^+$ defines a Lebesgue-Stieltjes measure μ_ω on $\mathcal{B}$. Let $\mathcal{U} = \mathcal{B} \otimes \Sigma$ be the product σ-algebra of $U = \mathbb{R}^+ \times \Omega$. Since the A_t-process is measurable for $\mathcal{U}$, it follows that the mapping $\omega \mapsto \mu_\omega(B)$ is measurable (Σ) for each $B \in \mathcal{B}$ and $\{\mu_\omega, \omega \in \Omega\}$ is a family of σ-finite measures on $\mathcal{B}$. Hence by the Fubini theorem, we deduce the existence of a unique σ-finite $\nu : \mathcal{U} \to \overline{\mathbb{R}}^+$ such that

$$\nu(C \times B) = \int_C \mu_\omega(B)\, dP(\omega),\ B \in \mathcal{B},\ C \in \Sigma, \tag{4}$$

(cf. Prop. II.3.17). Further for any measurable $f : U \to \mathbb{R}$ which is integrable for ν, the mapping $\omega \mapsto \int_{\mathbb{R}^+} f(t,\omega)\mu(dt)$ is measurable (Σ) and

$$\int_U f(t,\omega)\, d\nu = \int_\Omega dP(\omega) \int_{\mathbb{R}^+} f(t,\omega)\, \mu_\omega(dt). \tag{5}$$

Since, by definition, we may write $\mu_{\cdot}(dt) = dA(t)$, (5) can be expressed for any measurable and ν-integrable process $\{X_t, \mathcal{F}_t, t \geq 0\}$ as:

$$\int_U X(t,\omega) d\nu = \int_\Omega \int_{\mathbb{R}^+} X(t,\omega)\, d_t A(t,\omega) dP = E\Big(\int_{\mathbb{R}^+} X(t,\cdot)\, dA(t)\Big). \tag{6}$$

By Theorem III.3.8 the right continuous $\{X_t, \mathcal{F}_t, t \geq 0\}$ is progressively measurable, and the same is true of $\chi_B X_t$ for each $B \in \mathcal{B}([0,t])$, the restriction Borel σ-algebra to [0,t]. In particular, one has

$$Y_t = \int_0^t X_s \, dA(s) (= \int_{\mathbb{R}^+} \chi_{[0,t]} X_s \, dA(s)), \; t \geq 0, \tag{7}$$

to be $\mathcal{F}_t$-adapted and may be taken to be right continuous (will be continuous iff $A.(\omega)$ is continuous for a.a. (ω)). Thus $\{Y_t, \mathcal{F}_t, t \geq 0\}$ is also progressively measurable. If $\{T_j, j \in J\}, J \subset \mathbb{R}^+$, is an increasing family of optionals of $\{\mathcal{F}_t, t \geq 0\}$ such that $\lim_{j \uparrow j_0} T_j = T_{j_0}, j_0 \in J$ and $Z_j = Y \circ T_j$ then we deduce from Prop. IV.2.3 that $\{Z_j, \mathcal{G}_j, j \in J\}$ is progressively measurable where $\mathcal{G}_j = \mathcal{B}(T_j)$. Here Z_j may be written (with $[\![0, T]\!] = \{(t, \omega) : 0 \leq t \leq T(\omega)\}$) as:

$$Z_j = Y \circ T_j = \int_0^{T_j} X_s \, dA(s) = \int_{\mathbb{R}^+} \chi_{[\![0,T_j]\!]} X_s \, dA(s), \; j \in J. \tag{8}$$

It is this pointwise "stochastic" integral that replaces (3). We can now state the analog of (3) precisely as follows.

3. Definition. An increasing right continuous adapted process $\{A_t, \mathcal{F}_t, t \geq 0\}$, with right continuous filtering, such that $A_0 = 0$ a.e., and $\sup_t E(A_t) < \infty$, is called *predictable* iff for every right continuous bounded positive martingale $\{X_t, \mathcal{F}_t, t \geq 0\}$ it is true that

$$E(\int_{\mathbb{R}^+} X_{t-} dA_t) = E(A_\infty X_\infty), \tag{9}$$

where $A_\infty = \lim_t A(t), X_\infty = \lim_t X_t$, and $X_{t\pm} = \lim_n X_{t \pm (1/n)}$ exist a.e. (see Theorem III.5.2).

5.2 The Doob-Meyer decomposition theory

With the above concepts, we can give the decomposition of a super-(sub-)martingale. The solution is mainly based on the works of Meyer[3] and Itô and S. Watanabe [1]. An alternative derivation will be included later, using a measure representation of a process. But the present proof is conceptually simpler and the details contain more information.

1. Theorem. *Let $\{X_t, \mathcal{F}_t, t \geq 0\}$ be a right continuous supermartingale of class (R) on a complete probability space (Ω, Σ, P) with $\mathcal{F}_t \subset \Sigma$ and the filtration is right continuous and complete. Then X_t decomposes into: (i) a right continuous local martingale $\{Y_t, \mathcal{F}_t, t \geq 0\}$, (ii) a predictable (integrable) increasing right continuous process $\{A_t, \mathcal{F}_t, t \geq 0\}$, and (iii) a right continuous martingale $\{Z_t, \mathcal{F}_t, t \geq 0\}$ such that*

$$X_t = Y_t + Z_t - A_t, \quad t \geq 0, a.e. \tag{1}$$

Moreover, the Y_t-process is a martingale iff the X_t- process is in class (DL). The decomposition is unique if $\{Y_t - A_t, t \geq 0\}$ is a potential.

A large part of this section will be devoted to a proof of this result. The argument is presented in a series of propositions, some of which are of independent interest. Here is the basic strategy of proof.

Since $\{X_t, \mathcal{F}_t, t \geq 0\}$ is in class(R), one may apply the Riesz decomposition (1.1) so that $X_t = M_t + \tilde{Y}_t$ where $\{M_t, \mathcal{F}_t, t \geq 0\}$ is a right continuous martingale and $\{\tilde{Y}_t, \mathcal{F}_t, t \geq 0\}$ is a potential. Then one shows that $\tilde{Y}_t = Z_t - A_t$ for a predictable increasing process $\{A_t, t \geq 0\}$ and a local martingale $\{Z_t, \mathcal{F}_t, t \geq 0\}$. The latter is obtained in stages when $\{\tilde{Y}_t, t \geq 0\}$ is in calss (D), then in class (DL), and finally the general case. Each of these subresults is nontrivial and has some independent interest. We prove them starting with the crucial class (D) supermartingales, and then extend. Hereafter we write X_t and $X(t)$, and similarly others, interchangeably. The next result is due to Meyer [3] with a different proof.

2. Theorem. *Let $\{Z_t, \mathcal{F}_t, t \geq 0\}$ be a right continuous potential. Then there exists an a.e. unique predictable increasing process $\{A_t, \mathcal{F}_t, t \geq 0\}$ such that*

$$Z_t = E^{\mathcal{F}_t}(A_\infty) - A_t, \qquad t \geq 0, \tag{2}$$

where $A_\infty = \lim_t A_t$ a.e., iff the process $\{Z_t, \mathcal{F}_t, t \geq 0\}$ belongs to class (D).

Proof. The plan is to "discretize the process", apply Theorem II.5.1, and then obtain the continuous parameter version by a careful limiting argument.

Let $n \geq 1$ be a fixed integer and let $t = m/2^n, m \geq 0$, so that t is a dyadic rational. Set $Y_m = Y_m^n = Z_t, \mathcal{F}_m = \mathcal{F}_m^n = \mathcal{F}_t$ for such t. Then for each fixed n, it is clear that $\{Y_m, \mathcal{F}_m, m \geq 0\}$ is a (discrete) potential. Hence there is a martingle $\{Y'_m, \mathcal{F}_m, m \geq 0\}$ and an increasing process $A_m(= A^n_m)$ such that

$$Y_m = Y'_m - A_m, \quad m \geq 0, \lim_m E(A_m) < \infty. \tag{3}$$

Since $Y_m \to 0$ in $L^1(P)$, and $A_m \uparrow A_\infty(= A^n_\infty)$ a.e. and in $L^1(P)$, it follows that both the Y_m- and A_m-processes are uniformly integrable so that the Y'_m-process is also. Hence $Y'_\infty = \lim_m Y'_m = 0 + A^n_\infty$ a.e. and $Y'_m = E^{\mathcal{F}_m}(Y'_\infty) = E^{\mathcal{F}_m}(A^n_\infty)$ a.e. Thus (3) becomes

$$0 \leq Z^n_m = Z(m/2^n) = Y_m = E^{\mathcal{F}_m}(A_\infty) - A^n_m, \text{a.e.} \tag{4}$$

This is (2) for dyadic t whether or not the Z-process is in class(D). The additional hypothesis is needed for the general case when we let both $m, n \to \infty$.

We establish a technical result, on the limit behavior of the A^n_m-process appearing in (4), in the following form.

3. Lemma. *The set $\{A^n_\infty, n \geq 1\} \subset L^1(P)$ is uniformly integrable iff the potential $\{Z_t, \mathcal{F}_t, t \geq 0\}$ is of class (D).*

Proof. In the direct implication, let $\{Z_t, t \geq 0\}$ be in class (D). For any $\lambda > 0$, and integer $n \geq 1$, define

$$T^\lambda_n = \inf\{\frac{i}{2^n} : A^n(\frac{i+1}{2^n}) > \lambda\}, \tag{5}$$

and set $A^n_{i+1} = A^n(\frac{i+1}{2^n})$ as in (4). Since A^n_{i+1} is $\mathcal{F}^n_i$-adapted, Theorem IV.2.5 implies that T^λ_n is an $\{\mathcal{F}^n_k, k \geq 0\}$ optional. Also $[T^\lambda_n < \infty] = [A^n_\infty > \lambda]$, from definition, and so the following computations hold.

Let $\mathcal{B}(T)$ denote the σ-algebra of events "prior to T" for an optional T of $\{\mathcal{F}_t, t \geq 0\}$. Then the positive (super) martingale $\{E^{\mathcal{F}_m}(A^n_\infty), m \geq 0\}$ satisfies the hypothesis of Corollary IV.1.12, so that taking $Z = A^n_\infty, T_1 = T^\lambda_n, T_2 = \infty$, and $X_m = Y_m + A^n_m$ of (4) in that result, we get $X \circ T_1 = E^{\mathcal{B}(T_1)}(Z)$, a.e. Hence in the present case

$$Z \circ T^\lambda_n = E^{\mathcal{B}(T^\lambda_n)}(A^n_\infty) - A^n \circ T^\lambda_n, \text{a.e.} \tag{6}$$

But $E(A^n_\infty) < \infty$. Thus for each n,

$$P[T^\lambda_n < \infty] \leq \frac{1}{\lambda} E(A^n_\infty) \to 0, \qquad \text{as } \lambda \to \infty. \tag{7}$$

Consequently, the uniform integrability of $\{A^n_\infty, n \geq 0\}$ follows if we show

$$\lim_{\lambda\to\infty} \int_{[A^n_\infty > \lambda]} A^n_\infty \, dP = 0, \qquad \text{uniformly in } n. \tag{8}$$

Since $[A^n_\infty > \lambda] = [T^\lambda_n < \infty] \in \mathcal{B}(T^\lambda_n)$, (8) simplifies, with (5)-(7), to

$$\begin{aligned}\int_{[A^n_\infty>\lambda]} A^n_\infty dP &= \int_{[A^n_\infty>\lambda]} E^{\mathcal{B}(T^\lambda_n)}(A^n_\infty) dP \\ &= \int_{[T^\lambda_n<\infty]} Z \circ T^\lambda_n dP + \int_{\bigcup_{i\geq 1}[T^\lambda_n = i2^{-n}]} A^n_i dP \\ &\leq \int_{[T^\lambda_n<\infty]} Z \circ T^\lambda_n dP + \lambda \int_{[T^\lambda_n<\infty]} dP.\end{aligned} \tag{9}$$

Hence we have

$$\begin{aligned}\lambda P[A^n_\infty > 2\lambda] &\leq \int_{[A^n_\infty>2\lambda]} (A^n_\infty - \lambda) dP \\ &\leq \int_{[T^\lambda_n<\infty]} Z \circ T^\lambda_n dP.\end{aligned} \tag{10}$$

Changing λ to 2λ in (9), and replacing the last term there with the estimate (10), one obtains (uniformly in n)

$$\begin{aligned}0 \leq \int_{[A^n_\infty>2\lambda]} A^n_\infty dP &\leq \int_{[T^{2\lambda}_n<\infty]} Z \circ T^{2\lambda}_n dP \\ &\quad + 2\int_{[T^\lambda_n<\infty]} Z \circ T^\lambda_n dP \to 0, \text{ as } \lambda \to \infty,\end{aligned} \tag{11}$$

since the Z-process is in class (D). This establishes (8).

Conversely, let $\{A^n_\infty, n \geq 1\}$ be uniformly integrable so that it is in a ball of $L^1(P)$. Since in a finite measure space the above sequence has the stated property iff it is relatively weakly compact, by a classical result of Dunford's, one has the conclusion: there is a subsequence $\{A^{n_k}_\infty, k \geq 1\}$, and an $A_\infty \in L^1(P)$ such that for each continuous linear functional x^* on $L^1(P)$, $x^*(A^{n_k}) \to x^*(A_\infty)$. But the conditional expectation $E^{\mathcal{B}}$ is contractive on this space and so is its adjoint. Thus if we let $y^* = x^* \circ E^{\mathcal{B}}$, then y^* is also a continuous linear functional on $L^1(P)$ and

$$x^*(E^{\mathcal{B}}(A_\infty^{n_k})) \to y^*(A_\infty) = x^*(E^{\mathcal{B}}(A_\infty)), \tag{12}$$

for any σ-subalgebra $\mathcal{B}$ of Σ. Let $Y_t = E^{\mathcal{F}_t}(A_\infty)$. Then by the right continuity of the filtration we can assume, after a modification if necessary (cf. Theorem III.5.2), that the martingale $\{Y_t, \mathcal{F}_t, t \geq 0\}$ is right continuous. We now show that $Y_t - Z_t = A_t$ defines an increasing integrable process and that the Z_t-process belongs to class (D).

Let $r \leq s$ be a pair of dyadic rationals in $\mathbb{R}^+$. Then for large n, A_r^n and A_s^n are defined, $A_r^n \leq A_s^n$, since (4) implies

$$A_r^n = E^{\mathcal{F}_r}(A_\infty^n) - Z_r^n \leq E^{\mathcal{F}_s}(A_\infty^n) - Z_s^n = A_s^n, \text{a.e.} \tag{13}$$

Taking limits in the weak topology when x^* ranges over the positive cone of the real $(L^1(P))^*$ we see by (12) that (13) holds when n is omitted. Thus, since $Y_r = E^{\mathcal{F}_r}(A_\infty)$, we get

$$A_r = Y_r - Z_r \leq Y_s - Z_s = A_s, \text{a.e.} \tag{14}$$

Since the Y_t- and Z_t-processes are right continuous, the increasing process A_t of (14) is also. But the dyadic rationals are dense in $\mathbb{R}^+$, so one concludes that $\{A_t, t \geq 0\}$ is defined on $\mathbb{R}^+$ and is a separable process. Hence $A_\infty = \lim_t A_t$ a.e. Thus

$$0 \leq Z_t = E^{\mathcal{F}_t}(A_\infty) - A_t \leq E^{\mathcal{F}_t}(A_\infty), \quad \text{a.e.} \tag{15}$$

The potential $\{Z_t, t \geq 0\}$ is nonnegative. So by Corollary IV.1.12 again

$$0 \leq Z \circ T \leq E^{\mathcal{B}(T)}(A_\infty), \text{ a.e.,} \tag{16}$$

for all optionals T of the filtration. Since $\{E^{\mathcal{B}}(A_\infty), \mathcal{B} \subset \Sigma\}$ is uniformly integrable, we deduce that the set $\{Z \circ T : \text{all such } T\}$ is uniformly integrable. Thus the Z_t- process is of class (D), as asserted. □

Completion of proof of Theorem 2. Using the preceding notations, we have

$$A_t = Y_t - Z_t, \qquad t \geq 0, \text{a.e.,} \tag{17}$$

where $\{A_t, t \geq 0\}$ is an increasing integrable right continuous process and $\{Y_t, \mathcal{F}_t, t \geq 0\}$ is a right continuous martingale ($Y_t = E^{\mathcal{F}_t}(A_\infty)$) iff $\{Z_t, \mathcal{F}_t, t \geq 0\}$is in class (D), by Lemma 3. It thus remains only to show that the A_t-process is predictable and the decomposition is unique, for the direct part.

To prove that the A_t-process is predictable, let $\{M(t), \mathcal{F}_t, t \geq 0\}$ be a bounded right continuous (can be taken càdlàg) process. One has to verify that

$$E(\int_0^\infty M(t-0)dA_t) = E(A_\infty M_\infty), \tag{18}$$

where $M_\infty = \lim_t M(t)$ a.e. We establish this by reducing it again to the discrete case and using Lemma 3. Thus the integral on the left of (18) may be approximated as:

$$\begin{aligned}
&E\left(\int_0^\infty M(t-0)dA_t\right) \\
&= \lim_{n\to\infty}\sum_{i=0}^\infty E\left(M\left(\frac{i}{2^n}\right)[A^n(\frac{i+1}{2^n}) - A^n(\frac{i}{2^n})]\right), \\
&\quad \text{by the Dominated Convergence,} \\
&= \lim_{n\to\infty}\sum_{i=0}^\infty E[M(\frac{i}{2^n})E^{\mathcal{F}_i^n}(Z(\frac{i}{2^n}) - Z(\frac{i+1}{2^n}))], \\
&\quad \text{by(4)}, \ (\mathcal{F}_i^n = \mathcal{F}_{i/2^n}^n), \\
&= \lim_{n\to\infty}\sum_{i=0}^\infty [E\{M(\frac{i+1}{2^n})A^n(\frac{i+1}{2^n})\} - E\{M(\frac{i}{2^n})A^n(\frac{i}{2^n})\}], \\
&\quad \text{since } E\{M(\frac{i}{2^n})A^n(\frac{i+1}{2^n})\} = E(E^{\mathcal{F}_i^n}\{M(\frac{i+1}{2^n})A^n(\frac{i+1}{2^n})\}), \\
&= \lim_{n\to\infty} E(A_\infty^n M_\infty) = E(A_\infty M_\infty),
\end{aligned} \tag{19}$$

since $M(\frac{i}{2^n}) \to M_\infty$ a.e. (independent of the sequence used), and $A^n(\frac{i}{2^n}) \to A_\infty^n$ a.e., as $i \to \infty (A^n(0) = 0)$, and by Lemma 3, $A_\infty^{n_k} \to A_\infty$ weakly ($M_\infty \in L^\infty(P)$). Thus (18) is true and $\{A_t, t \geq 0\}$ is predictable. Note that the computations leading to (19) actually show that the full sequence $\{A_\infty^n, n \geq 1\}$ converges weakly to A_∞.

To prove uniqueness, let $\{B_t, t \geq 0\}$ be another predictable process satisfying (17) so that $B_t = \tilde{Y}_t - Z_t$. Thus A_t and B_t are dominated by the same potential $\{Z_t, \mathcal{F}_t, t \geq 0\}$, and then $\{A_t - B_t, t \geq 0\}$ is a martingale. To show that $A_t = B_t$ a.e., it suffices to verify that $E(Y'A_t) = E(Y'B_t)$ for any bounded $\mathcal{F}_t$-measurable function $Y' \geq 0$.

Let $Y'(s) = E^{\mathcal{F}_s}(Y')$ be a right continuous modification of the martingale shown so that $Y'_\infty = Y'(t) = Y'$ a.e., and $Y'(s) = Y'(t)$ for $s \geq t$. Then

$$\begin{aligned} E(Y'A_\infty) &= E\left(\int_0^\infty Y'(s-0)dA_s\right) \\ &= E\left(\int_0^t Y'(s-0)\,dA_s\right) + E(Y'A_\infty) - E(Y'A_t), \end{aligned} \tag{20}$$

so that

$$E\left(\int_0^t Y'(s-0)\,dA_s\right) = E(Y'A_t). \tag{21}$$

Similarly

$$E\left(\int_0^t Y'(s-0)\,dB_s\right) = E(Y'B_t). \tag{22}$$

By our assumption, $B_t - A_t = \tilde{Y}(t) - Y(t)$ is a right continuous martingale for the filtration $\{\mathcal{F}_t, t \geq 0\}$, so that $E(B_t) = E(A_t)$, for all $t \geq 0$. With this we show that the integrals on the left side of (21) and (22) are equal by discretizing the Y'-process. Thus let $Y_0'^n = Y'(0)$, and $Y_s'^n = Y'(\frac{kt}{n}), k \geq 0, n \geq 1$ whenever $s \in (\frac{kt}{n}, \frac{k+1}{n}t]$, so that $Y_s'^n$ is $\mathcal{F}_{kt/n}$-measurable. Hence writing $\tau_k = \frac{kt}{n}$, $A_t = A(t)$ and similarly others, one has

$$\begin{aligned} E\left(\int_0^t Y_s'^n\,dA_s\right) &= \sum_{k=0}^{n-1} E[Y'(\frac{kt}{n})\{A(\frac{k+1}{n}t) - A(\frac{kt}{n})\}] \\ &= \sum_{k=0}^{n-1} E[Y'(\tau_k)E^{\mathcal{F}_{\tau_k}}(A(\tau_{k+1}) - A(\tau_k))] \\ &= \sum_{k=0}^{n-1} E[Y'(\tau_k)E^{\mathcal{F}_{\tau_k}}(B(\tau_{k+1}) - B(\tau_k))], \end{aligned}$$

since $B(t) - A(t)$ is a martingale with zero mean,

$$= E\left(\int_0^t Y_s'^n \, dB_s\right). \tag{23}$$

Since $Y_s'^n \to Y_{s-}'$ as $n \to \infty$, so the interval degenerates to s, and since Y' is bounded, we may take the limit inside the integral by the Lebesgue bounded convergence theorem. This shows the equaivalence of (21) and (22) and hence $A_t = B_t$ a.e. Thus the decomposition is unique.

Conversely, if the right continuous potential $\{Z_t, \mathcal{F}_t, t \geq 0\}$ admits a decomposition as (17) with $\{A_t, t \geq 0\}$ increasing, right continuous, and integrable satisfying $E(A_\infty) < \infty$, it follows that the A_t-process is uniformly integrable. Also $Y_t \to Y_\infty$, a.e. and in $L^1(P)$, and $Y_\infty = A_\infty$ a.e., so that $\{Y_t, \mathcal{F}_t, t \geq 0\}$ is a uniformly integrable right continuous martingle. These two facts together with Proposition IV.3.6 imply that both the Y_t and A_t-processes and hence the Z_t-process belong to class (D). $\square$

We should note here that the concept of "predictability" given in Definition 1.3 above is also called "naturalness" of the increasing process A_t considered in the decomposition, leaving the predictability concept for functions that are measurable relative to predictable σ-algebras, to be extensively employed in integration theory later. However, it turns out (cf. Prop. 17 below) that both notions coincide and no conflict is involved. Also it may be observed that we are using the same notation for a process and its modification, to limit the proliferation of symbols.

Resuming our theme, we note that the above theorem implies the following: when both $\{Z_t, \mathcal{F}_t, t \geq 0\}$ and $\{A_t, t \geq 0\}$ are in class (D), they determine each other uniquely. Then one says that the increasing process $\{A_t, t \geq 0\}$ *generates* the potential $\{Z_t, \mathcal{F}_t, t \geq 0\}$.

4. Remark. The above proof and Lemma 3, together with the uniqueness of decomposition, imply that the full sequence $\{A_n, n \geq 1\}$ is weakly convergent to A_∞, as indicated already. In particular, $E(A_\infty^n) \to E(A_\infty)$, since $(L^1(P))^* = L^\infty(P)$. Hence the sequence also converges in mean (cf., Dunford-Schwartz [1], IV.8.12) iff $A_\infty^n \to A_\infty$ in probability, which we do not have. Alternatively, if for some $p > 1, \|A_\infty^n\|_p \to$

$\|A_\infty\|_p$ and $A_\infty^n \to A_\infty$ weakly in $L^p(P)$, then the mean or norm convergence holds. Remark 10 below gives an instance of the latter property.

Let us extend Theorem 2 for a larger class of potentials, allowing a correspondingly weaker conclusion. This is due to Itô and S. Watanabe [1]. Another generalization of Theorem 2 is in Cornea and Licea[1].

5. Theorem. *Let $\{Z_t, \mathcal{F}_t, t \geq 0\}$ be a right continuous potential. Then there exist a right continuous local martingale $\{Y_t, \mathcal{F}_t, t \geq 0\}$ and a predictable increasing process $\{A_t, \mathcal{F}_t, t \geq 0\}$ such that one has*

$$Z_t = Y_t - A_t, \qquad a.e., t \geq 0. \tag{24}$$

The decomposition is a.e. unique.

Proof. We reduce this result to that of Theorem 2 by transforming the Z_t-process with a suitable sequence of optionals $\{T_n, n \geq 1\}$. Thus let

$$T_n(\omega) = \inf\{t : Z_t(\omega) \geq n\} \wedge n. \tag{25}$$

Since $P[\sup_t Z_t < \infty] = 1$ by the (super)martingale maximal inequality (cf., Theorem II.4.3) and the fact that $E(Z_t) \to 0$ as $t \to \infty$, each $T_n(\omega)$ is finite for a.a. (ω). By Theorem IV.2.5, each T_n is an optional of the filtration $\{\mathcal{F}_t, t \geq 0\}$ and $T_n \uparrow \infty$, as required by Definition 1.2. Since $0 \leq E(Z_t) \leq E(Z_0) < \infty$, let $X_n(t) = Z_0(T_n \wedge t)$, and $\tau_t^n = T_n \wedge t$. Then $\{X_n(t), \mathcal{B}(\tau_t^n), t \geq 0\}$ is also a supermartingale for each n, by Theorem IV.1.7 which applies here since rationals are dense in $\mathbb{R}^+$. We assert that the $X_n(t)$-process is of class (D).

Since the $X_n(t)$-process is a nonnegative right continuous supermartingale, it is of class (DL) and will be of class (D) if it is uniformly integrable, by Proposition IV.3.6. To see that it holds here, we have

$$0 \leq X_n(t) = Z \circ \tau_t^n \leq (Z \circ T_n) \vee n, \text{a.e.} \tag{26}$$

This is easily verified pointwise using the definition of the T_n in (25). But the last term in (26) does not involve t and is integrable. Hence $\{X_n(t), t \geq 0\}$ is uniformly integrable for each n. So it is of class (D), and Theorem 2 is applicable after a Riesz decomposition.

By the unique decomposition (2) and that of Riesz, we get:

$$X_n(t) = Y_n(t) - A_n(t), \qquad t \geq 0, \tag{27}$$

for a right continuous martingale $\{Y_n(t), \mathcal{B}(\tau_t^n), t \geq 0\}$ and a predictable integrable increasing process $\{A_n(t), \mathcal{B}(\tau_t^n), t \geq 0\}$. Note that

$$\begin{aligned}\sup_{n,t} E(A_n(t)) \leq \sup_{n,t} E(Y_n(t)) &= \sup_{n,t} E(Y_n(0)) \\ &= \sup_{n} E(X_n(0)) = E(Z_0) < \infty. \qquad (28)\end{aligned}$$

Since $X_{n+1}(t) = Z \circ (\tau_t^{n+1})$ and $T_n \leq T_{n+1}$, one has $X_{n+1} \circ (\tau_t^n) = Z \circ (T_{n+1} \wedge \tau_t^n) = X_n(t)$. This and the membership of class (D) imply (by uniqueness) that $Y_{n+1}(\tau_t^n) = Y_n(t)$ a.e., and similarly $A_{n+1}(t) = A_n(t)$, a.e., for all $t \geq 0$, since for any increasing predictable process such as $A(t)$ one has $A(t \wedge T) = B_t$, defining a similar process for any optional T of the filtration. This is seen as follows. Let $\{V_t, \mathcal{F}_t, t \geq 0\}$ be any bounded right continuous martingale. Then

$$\begin{aligned}E\Big(\int_0^\infty V_{s-}\, dB_s\Big) &= E\Big(\int_0^T V_{s-}\, dB_s + \int_T^\infty V_{s-}\, dB_s\Big), \text{ by (1.8)}, \\ &= E\Big(\int_0^T V_{s-}\, dA_s\Big) + 0, \\ &= E(V_T B_T), \text{by (22) with a simple calculation}, \\ &= E(V_\infty B_\infty), \text{ by definition of } B_t \\ &\qquad \text{and the integrability of } B_\infty. \qquad (29)\end{aligned}$$

Thus B_t-process is predictable.

Finally, define $Y(t) = Y_n(\tau_t^n)$ and $A(t) = A_n(\tau_t^n), n \geq 1, t \geq 0$. Then $\{Y(\tau_t^n), \mathcal{F}_t, t \geq 0\}$ is a uniformly integrable right continuous martingale for each n, and $\{A_t, t \geq 0\}$ is an integrable increasing predictable process with $A_0 = 0$ a.e. The first assertion is proved above and hence it is a local martingale by Definition 1.2. The truth about the A_t-process is verified as follows.

$$\begin{aligned}\int_\Omega A_t\, dP &= \lim_n \int_{[T_n \geq t]} A_t\, dP = \lim_n \int_{[T_n \geq t]} A_n(t)\, dP \\ &\leq \lim_n E(A_n(t)) \leq \lim_n E[Y_n(t) + X_n(0)] \\ &\leq 2E(X(0)) < \infty, \quad \text{since} \quad E(Y_n(0)) = E(X_n(0)). \qquad (30)\end{aligned}$$

This shows that $E(A_\infty) < \infty$, by the Monotone Convergence theorem. Next let $\{U_t, \mathcal{F}_t, t \geq 0\}$ be a right continuous bounded martingale. Then

$$
\begin{aligned}
E\Big(\int_0^t U_s\, dA_s\Big) &= \lim_n \int_\Omega \chi_{[T_n \geq t]} \Big(\int_0^t U_s\, dA_s\Big)\, dP, \qquad t > 0,\\
&= \lim_n E(\chi_{[T_n \geq t]} \int_0^t U_{s-}\, dA_n(s)),\\
&\qquad \text{since } A_n(s) \text{ is predictable,}\\
&= \lim_n E\Big(\int_0^t U_{s-}\, dA_s . \chi_{[T_n \geq t]}\Big),\\
&= E\Big(\int_0^t U_{s-}\, dA_s\Big). \qquad (31)
\end{aligned}
$$

Since t is arbitrary, we may let $t \to \infty$ in (31) by the Dominated Convergence and conclude that $\{A_t, t \geq 0\}$ is predictable.

For the uniqueness assertions, we have proved everything needed to apply the computations (22)-(23), and hence the decomposition is unique in the same sense. □

We are now in a position to quickly complete the:

Proof of Theorem 1. Let $\{X_t, \mathcal{F}_t, t \in \mathbb{R}\}$ be in class (R). Then it dominates a martingale so that it can be expressed uniquely as:

$$X_t = M_t + Z_t, \qquad t \geq 0 \text{ a.e.} \qquad (32)$$

where $\{M_t, \mathcal{F}_t, t \geq 0\}$ is a right continuous martingale and $\{Z_t, \mathcal{F}_t, t \geq 0\}$ is a right continuous potential. Consequently by Theorem 5 we have

$$Z_t = Y_t - A_t, \qquad t \geq 0 \text{ a.e.}, \qquad (33)$$

where $\{Y_t, \mathcal{F}_t, t \geq 0\}$ is a right continuous local martingale and $\{A_t, t \geq 0\}$ is an integrable increasing predictable process. Thus (32) and (33) imply the decomposition (1).

For uniqueness, suppose $X_t = M_t' + Y_t' - B_t$ is another such decomposition. Then, if $Z_t' = Y_t' - B_t$, it is a right continuous potential. The uniqueness of Riesz' decomposition implies that $M_t = M'_t$ a.e., and $Z_t = Z'_t$ a.e. Hence by the uniqueness of (24) $Y'_t - B_t = Y_t - A_t$ so that $Y_t = Y'_t$ a.e. and $A_t = B_t$ a.e. (This will not hold if A_t were not predictable as seen from Example 16 below.)

Finally the process $Y_t^* = M_t + Y_t$ for the filtration $\{\mathcal{F}_t, t \geq 0\}$ is a local martingale since a martingale is always a local martingale and the sum of two of the latter is of the same kind. We now show that $\{Y_t^*, \mathcal{F}_t, t \geq 0\}$ is a martingale (when the A_t-process is predictable) iff $\{X_t, \mathcal{F}_t, t \geq 0\}$ is of class (DL), which will then finish the proof of the theorem. This follows from the next result which is slightly more general. □

All class (DL) super martingales (without reference to class (R)) can be characterized using Theorem 2. This result is due to Meyer [3].

6. Theorem. *A right continuous supermartingale $\{X_t, \mathcal{F}_t, t \geq 0\}$ belongs to class (DL) iff it admits a decomposition:*

$$X_t = Y_t - A_t, a.e., \qquad t \geq 0, \tag{34}$$

where $\{Y_t, \mathcal{F}_t, t \geq 0\}$ is a right continuous martingale and $\{A_t, t \geq 0\}$ is an increasing process. If the latter is predictable, then the decomposition is unique.

Proof. Since $\{Y_t, \mathcal{F}_t, t \geq 0\}$ is a right continuous martingale, it is of class (DL) by Proposition IV.3.6, and since $\{A_t, \mathcal{F}_t, t \geq 0\}$ is a right continuous submartingale (because it is increasing) the same proposition implies that it is also of class (DL). Hence (34) shows that $\{X_t, \mathcal{F}_t, t \geq 0\}$ must then be of class (DL), and thus only the converse is nontrivial.

So let $\{X_t, \mathcal{F}_t, t \geq 0\}$ be in class (DL), to obtain the decomposition (34). Let $a > 0$ and consider $h : [0, a) \to \mathbb{R}^+$, a strictly increasing continuous onto mapping such as ,e.g., $h(x) = \frac{a}{a-x} - 1$. Let $g = h^{-1} : \mathbb{R}^+ \to [0, a)$. If $\mathcal{C}_t = \mathcal{F}_{g(t)}$, then $\{\mathcal{C}_t, t \geq 0\}$ is an increasing right continuous family of σ-subalgebras of Σ, and $Y_t = X_{g(t)}$ is $\mathcal{C}_t$-adapted. Indeed, $\{Y_t, \mathcal{C}_t, t \geq 0\}$ is a right continuous supermartingale of class (D). To see this, let $\mathcal{T}$ and $\mathcal{U}_a$ be the classes of all stopping times of the filtrations $\{\mathcal{C}_t, t \geq 0\}$ and $\{\mathcal{F}_t, 0 \leq t < a\}$. Then for any $T \in \mathcal{T}$, we have $U' = h \circ T \in \mathcal{U}_a$ so that it is a stopping time of $\{\mathcal{C}_{h(t)} =$

$\mathcal{F}_t, 0 \le t < a\}$. Hence $\{Y \circ T : T \in \mathcal{T}\} \subset \{X \circ U' : U' \in \mathcal{U}_a\}$. But the latter set is uniformly integrable by the membership of class (DL) so that $\{Y_t, \mathcal{C}_t, t \ge 0\}$ is in class (D). By the Riesz decomposition, which applies since the Y_t-process is a uniformly integrable supermartingale, we get $Y_t = M_t' + Z_t'$ uniquely where $\{M_t', \mathcal{C}_t, t \ge 0\}$ is a right continuous martingale and $\{Z_t, \mathcal{C}_t, t \ge 0\}$ is a right continuous class (D) potential. Hence by Theorem 2,

$$Y_t = M_t' + Z_t' = (M_t' + M_t'') - A_t = M_t - A_t, \text{a.e.}, \qquad t \ge 0, \tag{35}$$

where the M_t''- and hence M_t-process is a right continuous martingale and A_t is a predictable increasing integrable process.

Returning to the original process, let $A_t^a = A_{g(t)}, 0 \le t < a$. Then $\{A_t^a, 0 \le a\}$ is an increasing process and by the uniqueness of decomposition (2), if $a < a'$ one has $A_t^a = A_t^{a'}$ on $[0, a)$. Thus for any $a > 0, A_t = A_t^a$ is unambigously defined, and

$$X_t = Y_{g(t)} = M_{g(t)} - A_t, \qquad 0 \le t < a, \tag{36}$$

and $Z_t = X_t + A_t, 0 \le t < a$ is a right continuous martingale. Since $a > 0$ is arbitrary we deduce that the decomposition (36) holds. Regarding uniqueness, if $\tilde{A}_t = A_t^a$ on $[0, a)$ and $\tilde{A}_t = A_a^a$ for $t \ge a$, then $\tilde{A}_t$ is a predictable increasing integrable process. Moreover, for any bounded right continuous martingle $\{V_t, \mathcal{F}_t, t \ge 0\}$, we get for $a < \infty$,

$$\begin{aligned} E(A_a V_a) = E(\tilde{A}_\infty V_\infty) &= E\Big(\int_0^\infty V_{s-}\, d\tilde{A}_s\Big) \\ &= E\Big(\int_0^a V_{s-}\, dA_s^a\Big) = E\Big(\int_0^a V_{s-}\, dA_s\Big) \end{aligned} \tag{37}$$

and thus the computation for (23) applies to yield the uniqueness of the A_t-process. The case that $a = +\infty$ follows as in (22) - (23). □

As a consequence of Theorem 1 and Corollaries II.5.6 and II.5.8, we have the following result.

7. Corollary. *Let $\{X_t, \mathcal{F}_t, t \ge 0\}$ be a right continuous supermartingale. If either $X_t \ge 0$ a.e., or more generally $\sup_{t \ge 0} E(|X_t|) < \infty$, then the X_t-process admits the decomposition(1).*

Using the terminology introduced preceding Theorem 5, we may ask whether there are further useful relations between a right continuous potential and its generator A_t, both of which are now defined for class (DL) potentials by Theorem 6. In particular one may ask as to when is an increasing process A_t a continuous generator of a potential? Our main objective here is to show that this is the case for a class of potentials which are a subset of regular supermartingales given by Definition IV.3.7 and which we proceed to isolate. First we include some properties of generators to use in its proof.

8. Proposition. *Let $\{A_t, \mathcal{F}_t, t \geq 0\}$ be a predictable integrable increasing right continuous process. Then for every uniformly integrable (not necessarily bounded) positive right continuous martingale $\{Y_t, \mathcal{F}_t, t \geq 0\}$ we have*

$$E(Y_t A_t) = E\Big(\int_0^t Y_{s-}\, dA_s\Big) = E\Big(\int_0^t Y_s\, dA_s\Big), \; 0 \leq t \leq \infty. \tag{38}$$

Proof. The first half of (38) was already proved in (21) for bounded Y_t and the first and last terms are always equal (use approximating sums again). We prove (38) for unbounded Y_t by reducing it to the bounded case.

The uniform integrability of $\{Y_t, t \geq 0\}$ implies $Y_\infty = \lim_t Y_t$ a.e. and $Y_t = E^{\mathcal{F}_t}(Y_\infty)$ a.e. For each $n \geq 1$, let $Y_\infty^n = Y_\infty \wedge n$, and $Y_t^n = E^{\mathcal{F}_t}(Y_\infty^n) \geq 0$, a.e. By the right continuity of the filtration, one may assume that the Y_t^n-process is right continuous (or can take such a modification). Moreover, $\lim_n Y_t^n = Y_t$ a.e., $t \geq 0$. The equality of the first and the last terms of (38)follows from this.

Regarding the equality of the first two terms of (38), note that, using the left modifications, $0 \leq Y_{t-}^n \leq Y_{t-}^{n+1}$ a.e. These exist by Theorem III.5.1. Hence we may define $\tilde{Y}_t = \lim_{n\to\infty} Y_{t-}^n$, a.e. But $\{Y_t, \mathcal{F}_t, t \geq 0\}$ is a positive right continuous martingale (cf. Theorem III.5.2(iii)). Clearly $\tilde{Y}_t \leq Y_{t-}$ a.e. We now show that there is equality here to complete the proof. Thus

$$\begin{aligned} 0 \le Y_{t-} - \tilde{Y}_t &= \lim_n (Y_{t-} - Y^n_{t-}) = \lim_n E^{\mathcal{F}_{t-}}(Y_\infty - Y^n_\infty) \\ &= E^{\mathcal{F}_{t-}}\Big(\lim_n [Y_\infty - Y^n_\infty]\Big), \text{ by the conditional} \\ &\quad \text{Dominated Convergence, (cf. Prop. II.1.3(c)),} \\ &= 0. \end{aligned}$$

Hence $Y_{t-} = \tilde{Y}_t$ a.e., and by the predictability of $\{A_t, \mathcal{F}_t, t \ge 0\}$ one has

$$E(Y_t A_t) = E(E^{\mathcal{F}_{t-}}(Y_t A_t)) = E(Y_{t-} A_t) = E(\tilde{Y}_t A_t). \tag{39}$$

This establishes (38) in full. □

Let us recall a form of integration by parts formula for the Lebesgue-Stieltjes integrals to use it in the next proposition. Thus if f, g are nonnegative nondecreasing right continuous (deterministic) functions on $\mathbb{R}^+$ such that $f(0) = g(0), \lim_{t\to\infty} f(t) = f(\infty), \lim_{t\to\infty} g(t) = g(\infty)$ all exist and be finite, then we have

$$\begin{aligned} f(\infty)g(\infty) &= \int_0^\infty \int_0^\infty df(x)\, dg(y) \\ &= \int_0^\infty df(x) \int_x^\infty dg(y) + \int_0^\infty dg(y) \int_{y-}^\infty df(y), \text{ by Fubini's theorem,} \\ &= 2f(\infty)g(\infty) - \int_0^\infty g(x)\, df(x) - \int_0^\infty f(y-)\, dg(y), \end{aligned}$$

so that

$$f(\infty)g(\infty) = \int_0^\infty g(x)\, df(x) + \int_0^\infty f(y-)\, dg(y). \tag{40}$$

[Here the upper limit ∞ can clearly be replaced by any $a > 0$.]

9. Proposition. *Let $\{X_t, \mathcal{F}_t, t \ge 0\}$ be a right continuous potential of class (D) with its unique generator $\{A_t, \mathcal{F}_t, t \ge 0\}$. Then we have*

$$E(A^2_\infty) = E\Big(\int_0^\infty (X_t + X_{t-})\, dA_t\Big). \tag{41}$$

in the sense that both sides are finite and equal or both are infinite. In particular, if $|X_t| \leq c < \infty$ a.e., then the right side (hence the left side) is finite and (41) becomes

$$E(A_\infty^2) \leq 2cE(A_\infty) \leq 2cE(X_0) \leq 2c^2. \tag{42}$$

Proof. To establish (41) we may assume that at least one of the sides is finite. Suppose first that $E(A_\infty^2) < \infty$. Then by Theorem 2, $X_t = Y_t - A_t = E^{\mathcal{F}_t}(A_\infty) - A_t$ a.e., where $\{Y_t, \mathcal{F}_t, t \geq 0\}$ is a right continuous martingale with $A_\infty = \lim_t A_t = \lim_t Y_t$ a.e., and in $L^1(P)$. Hence

$$E(A_\infty^2) = E(Y_\infty A_\infty) = E\Big(\int_0^\infty Y_{t-}\, dA_t\Big) = E\Big(\int_0^\infty Y_t\, dA_t\Big), \tag{43}$$

by the predictability of A_t and (38). Consequently,

$$\begin{aligned} 0 &\leq E\Big(\int_0^\infty (X_{t-} + X_t)\, dA_t\Big) \\ &= E\Big(\int_0^\infty (Y_{t-} + Y_t)\, dA_t\Big) - E\Big(\int_0^\infty (A_{t-} + A_t)\, dA_t\Big) \\ &= 2E(A_\infty^2) - E\Big(\int_0^\infty (A_{t-} + A_t) dA_t\Big). \end{aligned} \tag{44}$$

Since $E(A_\infty^2) < \infty$, the second term on the right is finite. Thus $\{A_t, t \geq 0\}$ is integrable and the Lebesgue-Stieltjes integral of (44) can be simplified, by taking $f = g = A(\cdot, \omega)$ in (40) and substituting it in (44) to get (41).

Next suppose $E(\int_0^\infty [X_{t-} + X_t]\, dA_t) < \infty$. It suffices to show that $E(A_\infty^2) < \infty$ so that the preceding proof implies the truth of (41) again. As before consider the discretization procedure in the computation. Thus $K_0 < \infty$ where

$$K_0 = E\Big(\int_0^\infty X_{t-}\, dA_t\Big) = \lim_{n\to\infty} E\Big(\sum_{i=0}^\infty X(\tfrac{i}{2^n})[A^n(\tfrac{i+1}{2^n}) - A^n(\tfrac{i}{2^n})]\Big),$$

so that for any $n \geq 1$, if $B_i = A^n(\frac{i}{2^n})(\leq B_{i+1})$, we have

$$E(\sum_{i=0}^\infty X(\tfrac{i}{2^n})[B_{i+1} - B_i]) \leq K_0 < \infty. \tag{45}$$

Since $B_i \uparrow B_\infty (= A_\infty^n)$, in the notation of Lemma 3, $\{B_i, i \geq 1\}$ generates the potential $Z_n = \{X(\frac{i}{2^n}), i \geq 0\}$. If $\mathcal{F}_i = \mathcal{F}_{i/2^n}$ for fixed n, then

$$Z_m = E^{\mathcal{F}_m}(B_\infty) - B_m, \text{a.e.} \tag{46}$$

Since these functions may not be in $L^2(P)$, let $B_i^N = B_i \wedge N$. Then (a.e.) $B_i^N \uparrow B_\infty^N$ as $i \to \infty, B_i^N \leq B_{i+1}^N$ and that $B_\infty^N \uparrow B_\infty$ as $N \to \infty$. If $\{Z_i^N, i \geq 0\}$ is the potential generated by $\{B_i^N, i \geq 0\}$ (so $Z_i^N = E^{\mathcal{F}_i}(B_\infty^N) - B_i^N$ a.e.), then by Theorem 2, and conditional Monotone Convergence we get

$$\lim_{N\to\infty} Z_i^N = E^{\mathcal{F}_i}(\lim_N B_\infty^N) - \lim_N B_i^N = E^{\mathcal{F}_i}(B_\infty) - B_i = Z_i \text{ a.e.} \tag{47}$$

Since $\{B_i^N, i \geq 0\} \subset L^2(P), E((B_\infty^N)^2) < \infty$ so that we may use the discrete version of (41) which is true by the first case, one obtains:

$$E((B_\infty^N)^2) = E(\sum_{i=0}^{\infty}(Z_i^N + Z_{i+1}^N)(B_{i+1}^N - B_i^N))$$

$$\begin{aligned} E(B_\infty^2) &= \lim_{N\to\infty} E((B_\infty^N)^2) \leq E(\sum_{i=0}^{\infty}(Z_i + Z_{i+1})(B_{i+1} - B_i)) \\ &\quad \text{with } B_\infty^N \uparrow B_\infty, \text{ using (47) and Fatou's lemma,} \\ &= E(\sum_{i=0}^{\infty}[X(\tfrac{i}{2^n}) + X(\tfrac{i+1}{2^n})]\,[A^n(\tfrac{i+1}{2^n}) - A^n(\tfrac{i}{2^n})]) \\ &\leq K_0. \end{aligned} \tag{48}$$

But $B_\infty = A_\infty^n$. Hence (48) shows that $E((A_\infty^n)^2) \leq K_0 < \infty$, for all $n \geq 1$ by (45). This means (a) $A_\infty^n \to A_\infty$ weakly by Lemma 3, and (b) $\{A_\infty^n, n \geq 1\}$ is a bounded subset of $L^2(P)$ so that it has a weakly convergent subsequence with limit $\tilde{A}_\infty$. Since $(L^1(P))^* = L^\infty(P) \subset (L^2(P))^* = L^2(P)$ this convergence is also in the weak topology of $L^1(P)$, and hence $\tilde{A}_\infty = A_\infty$, a.e. But $\tilde{A}_\infty \in L^2(P)$ so that $E(A_\infty^2) < \infty$. By the earlier comment (41) holds. We also see that every convergent subsequence has the same weak limit A_∞ so that the sequence itself converges weakly. □

10. Remark. When $E(A_\infty^2) < \infty$, the sequence $\{A_\infty^n, n \geq 1\}$ not only converges to A_∞ weakly, but it converges in $L^2(P)$-norm also. To see this one only needs to check, by a classical result, that $\|A_\infty^n\|_2 \to \|A_\infty\|_2$ as $n \to \infty$. Since the following integral is finite we have:

$$E\Big(\int_0^\infty (A_t + A_{t-})\, dA_t\Big) = \lim_{n\to\infty} E\Big(\sum_{i=0}^\infty [A^n\Big(\frac{i+1}{2^n}\Big) + A^n\Big(\frac{i}{2^n}\Big)] \times$$
$$[A^n\Big(\frac{i+1}{2^n}\Big) - A^n\Big(\frac{i}{2^n}\Big)]\Big)$$
$$= \lim_{n\to\infty} E[(A_\infty^n)^2] = \|A_\infty^n\|_2^2. \tag{49}$$

By (40) with $f(\cdot) = A(\cdot,\omega) = g(\cdot)$ we get the left side of (49) to be $E(A_\infty^2)$. Hence the desired limit is verified and the $L^2(P)$-norm convergence holds.

It is possible to approximate certain potentials by bounded ones as follows.

11. Proposition. *Let $\{X_t, \mathcal{F}_t, t \geq 0\}$ be a right continuous potential of class (DL). Then there exists a sequence of bounded right continuous potentials $\{X_t^n, \mathcal{F}_t, t \geq 0, n \geq 1\}$ such that $X_t^n \preceq X_t^{n+1}$ (i.e.,$X_t^{n+1} - X_t^n = Y_t^n$ defines a potential), and $X_t^n \to X_t$, a.e.*

Proof. As in the proof of Theorem 6, we may assume that the X_t-process is of class (D) for this proof. Let $\{A_t, \mathcal{F}_t, t \geq 0\}$ be the generator of the X_t, and if $A_t^N = A_t \wedge N$, then $\{A_t^N, \mathcal{F}_t, t \geq 0\}$ is a predictable increasing integrable process. Let $A_\infty^N = \lim_{t\to\infty} A_t^N$ and define

$$Z_t^N = E^{\mathcal{F}_t}(A_\infty^N) - A_t^N, \text{a.e.}, \tag{50}$$

with $\{E^{\mathcal{F}_t}(A_\infty^N), t \geq 0\}$ as a right continuous modification, (same notation being used as before). Then $\{Z_t^N, \mathcal{F}_t, t \geq 0\}$ is a right continuous bounded potential.

Since $A_t^N \leq A_t^{N+1}$ so that $A_\infty^N \leq A_\infty^{N+1}$ a.e., and $E(A_\infty) < \infty$, we get, by the conditional Monotone Convergence, $Z_t^N \to X_t$ a.e. Indeed this is a monotone limit and $\{A_t^N, t \geq 0\}$ uniquely determines the Z_t^N-process, and therefore $A_t^{N+1} = A_t^N$ on the set $\{\omega : A_t(\omega) \leq N\}, t \geq 0$. Hence $Z_T^N = Z_T^{N+1}$ a.e. on this set. Since $Z_t \geq 0$, we get $Z_t^N \leq Z_t^{N+1}$ a.e. It is clear that each Z_t^N-process is of class (D), and the result on $Y_t^N = Z_t^{N+1} - Z_t^N$ is easy. □

12. Remark. The preceding may be stated differently: Every right continuous potential of class (DL) can be represented as a sum of a sequence of bounded right continuous potentials. Indeed, if $Y_t^N = Z_t^{N+1} - Z_t^N$ in the above notation,

$$X_t = \sum_{N=0}^{\infty} Y_t^N, \qquad t \geq 0, \text{ a.e.,} \tag{51}$$

and $\{Y_t^N, \mathcal{F}_t, t \geq 0, N \geq 1\}$ is the desired sequence of potentials.

We are now ready to characterize the class of potentials with a.e. continuous generators, after introducing a relevant concept for the purpose.

13. Definition. Let $\{X_t, \mathcal{F}_t, t \geq 0\}$ be a right continuous potential. It is then called *s-regular(=strictly regular)* if it is of class (DL) and for each increasing sequence $\{T_n, n \geq 1\}$ of stopping times of the filtration converging to a bounded stopping time T, we have

$$\lim_n E(X \circ T_n) = E(X \circ T). \tag{52}$$

14. Discussion. Since $\{X_t, \mathcal{F}_t, t \geq 0\}$ is a potential, $X_t \to X_\infty = 0$, a.e. and in $L^1(P)$. Hence $\{X_t, \mathcal{F}_t, t \in \mathbb{R}^+\}$ is a uniformly integrable supermartingale. If $\{T_j, j \in J\}$ is any stopping time process of the filtration and $T_0 = \infty$, then $T_j \leq T_0, j \in J$ and by Theorem IV.1.7, $\{X \circ T_j, \mathcal{B}(T_j), j \in J\}$ is also a supermartingale. As seen in Example IV.3.13, regularity is not enough of a restriction for supermartingales (and none for potentials). Thus each s-regular potential is regular. Since the process is now of class (DL), and is uniformly integrable, by Proposition IV.3.6(iv), it is of class (D) itself. Because of this, in (52), the limit is also valid for all $T_n \uparrow T \leq \infty$, even though we demanded it only for bounded T. Indeed, let $T_n \uparrow T$, and for each $N > 0$, let $T_n^N = T_n \wedge N \uparrow T \wedge N = T^N$, as $n \to \infty$. Then by (52), $E(X \circ T^N) \to E(X \circ T)$. Consider

$$\begin{aligned} E(X \circ T) &= \int_{[T \leq N]} X \circ T \, dP + \int_{[T > N]} X \circ T \, dP \\ &= E(X \circ T^N) + \int_{[T > N]} E^{\mathcal{F}_N}(X \circ T) \, dP. \end{aligned}$$

But by Theorem IV.1.7 applied to the optionals $\{N, T\}$, we get $E^{\mathcal{F}_N}(X \circ T) \leq X_N$ a.e., since the transformed process is a supermartingale. Now $X_N \to 0$, a.e. and in $L^1(P)$. Hence $E(X \circ T) = \lim_N E(X \circ T^N)$. Replacing T by T_n we get the conclusion that $T_n \uparrow T$ implies $E(X \circ T_n) \to E(X \circ T)$.

The following result gives the desired characterization and is due to Meyer [3] with a different argument.

15. Theorem. *Let $\{X_t, \mathcal{F}_t, t \geq 0\}$ be a right continuous potential of class (DL), or of class (D), with $\{A_t, t \geq 0\}$ as its predictable integrable increasing process, i.e., its generator. Then $\{A_t, t \geq 0\}$ has a.e. continuous sample paths iff the potential is s-regular.*

Proof. We give the details in steps for convenience.

I. Since $\{X_t, \mathcal{F}_t, t \geq 0\}$ is a potential of class (DL), or of class (D), by the preceding discussion $X_t = Y_t - A_t \geq 0$ a.e., and $E(A_t) \leq E(Y_t) \leq E(X_0)$, or $\sup_t E(A_t) < \infty$, so that A_t is automatically an integrable generator. Suppose that A_t has a.e. continuous sample paths. Then for any sequence $T_n \uparrow T$ of bounded stopping times of the filtration we have $A \circ T_n \uparrow A \circ T$ a.e., by the continuity of the generator. Hence by the Monotone Convergence theorem, $E(A \circ T_n) \uparrow E(A \circ T)$.

Since $\{A_t, t \geq 0\}$ is a generator of the X_t-process, one has

$$X_t = E^{\mathcal{F}_t}(A_\infty) - A_t, \qquad t \geq 0, \text{a.e.} \tag{53}$$

But $\{X_t, t \geq 0\}$ is also uniformly integrable, and $X_\infty = 0$ a.e. Thus by Theorem IV.1.7, $\{X \circ T_n, \mathcal{B}(T_n), n \geq 1\}$ is a positive (hence) convergent supermartingale. Since the A_t-process is also uniformly integrable, (53) becomes

$$X \circ T_n = E^{\mathcal{B}(T_n)}(A_\infty) - A \circ T_n = E^{\mathcal{B}(T_n)}[A_\infty - A \circ T_n], \text{a.e.} \tag{54}$$

But $(A_\infty - A \circ T_n) \downarrow (A_\infty - A \circ T) = X \circ T$ a.e., and hence (54) yields

$$\lim_n E(X \circ T_n) = \lim_n E[A_\infty - A \circ T_n] = E(X \circ T),$$

by the Dominated Convergence theorem. This is (52) and so the X_t-process is s-regular.

II. Conversely, let $\{X_t, \mathcal{F}_t, t \geq 0\}$ be s-regular. The right continuity of X_t implies the same of $f(t) = E(X_t)$, and f is monotone decreasing to zero. Choosing constant times $T_n = t_n \uparrow t = T$, in (52), we get $f(t_n) = E(X \circ T_n) \to E(X \circ T) = f(t)$ for each $t \in \mathbb{R}^+$ so that f is also left continuous at each t, implying that f is continuous. But $X_t = E^{\mathcal{F}_t}(A_\infty) - A_t$, since X_t-process is of class (DL). Hence if $g(t) = E(A_t)$, we get

$$f(t) = E(A_\infty) - g(t), \qquad t \in \mathbb{R}^+. \tag{55}$$

Since $E(A_\infty) < \infty$, and does not depend on t, (55) implies that $g : \bar{\mathbb{R}}^+ \to \mathbb{R}^+$ is continuous. Thus $0 \leq E(A_t - A_{t-}) = g(t) - g(t) = 0$, and $A_t = A_{t-}$ outside of $N_t \subset \Omega, P(N_t) = 0, N_t$ *depending on* t. However, we need to show that outside of a *fixed* set $N \subset \Omega, P(N) = 0, A_.(\omega), \omega \in N^c$, is continuous. For this stronger conclusion one has to show more; namely,

$$E\left(\int_0^\infty (A_t - A_{t-})\, dA_t\right) = 0. \tag{56}$$

III. Note that the continuity of g implies only that for each dyadic rational r of $\mathbb{R}^+$, one has $P[A_r = A_{r-}] = 1$. We first suppose that $|X_t| \leq c < \infty$ a.e., prove (56), and then deduce the general case from this. The boundedness of X_t implies by Proposition 9, that $E(A_\infty^2) \leq 2c^2$ and since $0 \leq A_t \leq A_\infty$ a.e., Proposition IV.1.10 which holds in the present case implies that $E((A \circ T)^2) = E(A^2 \circ T) \leq 6c^2 < \infty$. Hence the set $\{A \circ T : T$ an optional of the filtration$\}$ is uniformly integrable. [This also follows from the fact that now X_t-process is of class (D).] Using the definition of the integral for (1.6) or (1.7), we have

$$\begin{aligned} E\left(\int_0^\infty A_t\, dA_t\right) \\ &= \lim_{n\to\infty} \sum_{i=0}^\infty E\left[A^n\left(\frac{i+1}{2^n}\right)\left(A^n\left(\frac{i+1}{2^n}\right) - A^n\left(\frac{i}{2^n}\right)\right)\right] \\ &= \lim_{n\to\infty} \sum_{i=0}^\infty E\left(\int_{I_{n,i}} A_n(t)\, dA_t\right), \end{aligned} \tag{57}$$

where $I_{n,i} = \{t : \frac{i}{2^n} \leq t < \frac{i+1}{2^n}\}$ and $A_n(t) = A^n\left(\frac{i+1}{2^n}\right) = E^{\mathcal{F}_t}\left(A^n\left(\frac{i+1}{2^n}\right)\right)$, $t \in I_{n,i}, i \geq 0$ with the predictability of $\{A_t, t \geq 0\}$. Thus on $I_{n,i}$, for fixed $(n,i), \{A_n(t), t \in I_{n,i}\}$ is a (trivial) martingale and by Proposition 8, the integrand $A_n(t)$ can be replaced by $A_n(t-), t \in I_{n,i}$. This is true if only $P[A_r = A_{r-}] = 1$ even without predictability (cf.(55)). Hence (57) becomes after this replacement:

$$E\Big(\int_0^\infty A_t\, dA_t\Big) = \lim_{n\to\infty} E\Big(\int_0^\infty A_n(t-)\, dA_t\Big). \tag{58}$$

But if $\{T_n, 1 \le n < \infty\}$ is a stopping time process of the filtration, then on noting that $A_n(t) \ge A_t$ a.e., and if we choose (as we may) for fixed $t, I_{n,i} \supset I_{n+1,k}$ so that $A_n^i \ge A_{n+1}^i$, and $A_n(t) \downarrow A_t$ for $t \in \mathbb{R}^+$, we have

$$\begin{aligned}
E\Big(\int_0^\infty (A_t - A_{t-})\, dA_t\Big) &= \lim_{n\to\infty} E\Big(\int_0^\infty (A_n(t-) - A_{t-})\, dA_t\Big) \\
&\le \Big[E\Big(\int_0^{T_n} (A_n(t-) - A_{t-}) dA_t + \int_{T_n}^\infty A_n(t-)\, dA_t\Big)\Big], \\
&\quad \text{since } A_{t-} \ge 0 \text{ a.e. (cf.(1.8))}, \\
&\le \lim_n \Big\{E\Big(\int_0^{T_n} (A_n(t-) - A_{t-}) dA_t\Big) + E(A_\infty(A_\infty - A_{T_n}))\Big\}, \\
&\quad \text{since } A_n(t-) \le A_\infty \text{ a.e.}
\end{aligned} \tag{59}$$

IV. Let us choose $\{T_n, n \ge 1\}$ suitably and show that the right side of (59) is zero because of (52). So let $\varepsilon > 0$ be given and define $T_n = T_{n,\varepsilon}$ as:

$$T_{n,\varepsilon} = \inf\{t : A_n(t) - A_t \ge \varepsilon\}$$

where we take, as usual, that $\inf \emptyset = \infty$. The monotonicity of $A_n(t)'s$ implies that of $T_{n,\varepsilon}$ and each is a stopping time of $\{\mathcal{F}_t, t \ge 0\}$. If $T_\varepsilon = \lim_n T_{n,\varepsilon}$, then T_ε is also a stopping time (both by Theorem IV.2.5) and (59) becomes for this choice of $T_{n,\varepsilon}$,

$$\begin{aligned}
E\Big(\int_0^\infty (A_t - A_{t-}) dA_t\Big) &\le \lim_n \{\varepsilon E(A_{T_{n,\varepsilon}}) + E(A_\infty(A_\infty - A_{T_{n,\varepsilon}}))\}, \\
&\le \varepsilon E(A_\infty) + \lim_n E(A_\infty(A_\infty - A_{T_{n,\varepsilon}})),
\end{aligned} \tag{60}$$

since $A_t \uparrow A_\infty$ and $A_{T_{n,\varepsilon}} \le A_\infty$. This will be zero if we can show that

$\lim_n A_{T_{n,\varepsilon}} = A_\infty$, a.e., because $E(A_\infty^2) < \infty$. But for any $\varepsilon > 0$,

$$\varepsilon P[T_\varepsilon < \infty] = \lim_n \varepsilon P[T_{n,\varepsilon} < \infty] = \lim_n \varepsilon \int_{[T_{n,\varepsilon}<\infty]} dP,$$

$$\leq \lim_n \int_\Omega (A_n \circ T_{n,\varepsilon} - A \circ T_{n,\varepsilon})\, dP, \quad \text{by monotonicity,}$$

$$\leq \lim_n \int_\Omega [E^{\mathcal{B}(T_{n,\varepsilon})}(A \circ g_n(T_{n,\varepsilon})) - A \circ T_{n,\varepsilon}]\, dP,$$

$$\text{where } g_n(t) = \sum_{i=0}^{\infty} \frac{i+1}{2^n} \chi_{I_{n,i}} \text{ and } g_n(T) \geq T$$

(use Theorem IV.1.7 and (57)),

$$= \lim_n \int_\Omega (A \circ g_n(T_{n,\varepsilon}) - A \circ T_{n,\varepsilon})\, dP$$

$$\leq \lim_n \int_\Omega (A \circ g_n(T_\varepsilon) - A \circ T_{n,\varepsilon})\, dP$$

$$= \lim_n \int_\Omega (A \circ g_n(T_\varepsilon) - A \circ T_\varepsilon)\, dP + \lim_n \int_\Omega (A \circ T_\varepsilon - A \circ T_{n,\varepsilon})\, dP$$

$$= \lim \int_\Omega (A \circ T_\varepsilon - A \circ T_{n,\varepsilon}) dP, \tag{61}$$

since $g_n(t) \to t$ as $n \to \infty$, and $I_{n,k} \supset I_{n+1,k}$. We now use the s-regularity of the process for the first time. Thus by (52) and Discussion 14, $E(X \circ T_{n,\varepsilon}) \to E(X \circ T_\varepsilon)$ as $n \to \infty$. But $X_t = E^{\mathcal{B}_t}(A_\infty) - A_t$ so that $E(X \circ T_{n,\varepsilon}) = E(A_\infty) - E(A \circ T_\varepsilon) \to E(X \circ T_\varepsilon) = E(A_\infty) - E(A \circ T_\varepsilon)$. Hence $E(A \circ T_{n,\varepsilon}) \to E(A \circ T_\varepsilon)$. Thus (61) is zero, and since $\varepsilon > 0$ is arbitrary $P[T_\varepsilon < \infty] = 0$. Since $A_{T_{n,\varepsilon}} \to \tilde{A}$ weakly (for a subsequence) in $L^2(P)$ and then weakly in $L^1(P)$, we get in (60) that $E(A_\infty^2) = \lim_n E(A_\infty A_{T_{n,\varepsilon}})$. But $A_{T_{n,\varepsilon}} \to A_{T_\varepsilon} = A_\infty$ in $L^1(P)$ since $A_t \geq 0$. So $\tilde{A} = A_\infty$ a.e. It follows that (60) reduces to

$$E\left(\int_0^\infty (A_t - A_{t-})\, dA_t\right) \leq \varepsilon E(A_\infty) < \infty. \tag{62}$$

Since $\varepsilon > 0$ is arbitrary, $A_t = A_{t-}$ a.e., and $\{A_t, t \geq 0\}$ is sample continuous for almost all ω.

V. Suppose finally that X_t is not necessarily bounded. Then by Proposition 11, there exists a sequence $\{Z_t^N, \mathcal{F}_t, t \geq 0\}$ of bounded right continuous potentials $Z_t^N \leq Z_t^{N+1} \to X_t$ a.e., such that $\{Z_t^{N+1} - Z_t^N, t \geq 0\}$ is a potential. Each of the Z_t-processes is in class (D) and is also s-regular. In fact, by the construction used in Proposition 11, we choose $X_t = Y_t^N + Z_t^N$ where each component is also a potential. Then $Z_t^N \to 0$ a.e. and in $L^1(P)$. Hence by Theorem IV.1.7, $E(Z^N \circ T_n) \geq E(Z^N \circ T)$ and $E(Y^N \circ T_n) \geq E(Y^N \circ T)$ for any sequence of optionals $T_n \uparrow T$ as in the last step. Since $E(X \circ T_n) \to E(X \circ T) = E(Y^N \circ T) + E(Z^N \circ T)$ and the former quantities tend to the right side ones. Thus each $\{Z_t^N, \mathcal{F}_t, t \geq 0\}$ is s-regular. Hence by Step IV, each of the generators $\{A_t^N, t \geq 0\}$ is continuous. But $\lim_n E(A_t^N) = E(A_t) < \infty$, and A_t^N is increasing in N. Since $Z_t^{N+1} - Z_t^N$ is a potential, $A_t^{N+1} - A_t^N$ is also an increasing process. Hence $(A_t - A_t^N)(\omega) \to 0$ for all $\omega \in \Omega - D_0, P(D_0) = 0$, for each t, and by continuity of $A_\cdot^N(\omega)$, this convergence is uniform on compact intervals. It follows that $A_\cdot(\omega)$ for $\omega \in \Omega - D_0$ is continuous on $\mathbb{R}^+$. Since A_t is the generator of $\{X_t, \mathcal{F}_t, t \geq 0\}$, we are done. □

The uniqueness of the decompositions was used in the proofs above, and this in turn depended on the predictability of the increasing integrable process (or generator). The following example shows that the uniqueness assertion will not hold without such an additional condition.

16. Example. Let $\Omega = \{1, 2\}, \Sigma$=power set, $P(\{1\}) = P(\{2\}) = 1/2$. Let $\mathcal{F}_t = \{\emptyset, \Omega\}$ for $0 \leq t < 1, \mathcal{F}_t = \Sigma, t \geq 1$. Let $A_t = 0$ for $0 \leq t < 1, = 1$ for $t \geq 1$ and define another increasing integrable process$\{B_t, t \geq 0\}$ as: $B_t(\{1\}) = 0, 0 \leq t < \infty, B_t(\{2\}) = 0$ for $0 \leq t < 1, = 2$ for $t \geq 1$. Then it may be verified that both $\{A_t, \mathcal{F}_t, t \geq 0\}$ and $\{B_t, \mathcal{F}_t, t \geq 0\}$ are increasing processes of the potential $\{X_t, \mathcal{F}_t, t \geq 0\}$, where $X_t = 1$ for $0 \leq t \leq 1, = 0$ for $t \geq 1$. It may also be seen that $\{A_t, t \geq 0\}$ is predictable while $\{B_t, t \geq 0\}$ is not. Note that $0 \leq B_t \leq 2A_t$, and $(2A_t - B_t)$-process is increasing. Thus an increasing (nonpredictable) process may be dominated by a predictable one.

The filtration used in the above example has the following feature. If $T = 1$, then it is a stopping time of $\{\mathcal{F}_t, t \geq 0\}$, and $\mathcal{B}(T) = \Sigma$. But if $T_n = 1 - \frac{1}{n}$, it is also a stopping time of the same filtration satisfying $\mathcal{B}(T_n) = \{\emptyset, \Omega\}$. Hence $\sigma(\cup_{n\geq 1}\mathcal{B}(T_n)) = \{\emptyset, \Omega\} \subsetneqq \mathcal{B}(T)$ so that $\lim_{n\to\infty} \mathcal{B}(T_n) \subsetneqq \mathcal{B}(\lim_{n\to\infty} T_n)$, and the filtration is said to have a "time of discontinuity". If there is equality for *all sequences* of optionals

$T_n \uparrow T$, then the filtration has *no times of discontinuity*. A right continuous martingale $X = \{X_t, \mathcal{F}_t, t \geq 0\}$ is *quasi-left-continuous* if for any strictly increasing optionals $T_n \uparrow T$ of the filtration $\{\mathcal{F}_t, t \geq 0\}$, we have $X \circ T_n \to X \circ T$ a.e. on the set $\{\omega : T(\omega) < \infty\}$. Note that if further X is $L^2(P)$ bounded, it is uniformly integrable and hence is of class (D), by Proposition IV.3.6. Now by Definition 2.13, a quasi-left-continuous $L^2(P)$-bounded right continuous martingale is *s*-regular, although these two concepts are not the same. It may be shown that if each $L^2(P)$-bounded right continuous martingale is quasi-left-continuous then the filtration cannot have times of discontinuity.

Motivated by this discussion, we introduce a classification of optionals, and then name the corresponding filtrations to utilize in martingale integration starting in the next section. Thus let $\{\mathcal{F}_t, t \geq 0\}$ be a right continuous family satisfying the standard conditions so that it is a (usual) filtration of (Ω, Σ, P). A stopping time T of the filtration is called *predictable* (or *previsible*) if it is announced by a sequence of finite stopping times $T_n \uparrow T$ in the sense that one has:(i)$T_n \leq T_{n+1}$, (ii)$\lim_n T_n = T$, and (iii) on $[T > 0], T_n < T$, a.e. [Thus if $\mathbb{R}^+$ is replaced by the natural numbers $\mathbb{N}$, then T is predictable if $[T \leq n] \in \mathcal{F}_{n-1}$, for all $n \geq 1$.] Additionally, T is said to be an *accessible time* if there exists a sequence of predictable stopping times $\{S_n, n \geq 1\}$ communicating with T in the sense that $P\{\omega : S_n(\omega) = T(\omega) < \infty, \text{for some}\, n\} = 1$; and T is *totally inaccessible* if there is no predictable time S reaching T so that $P\{\omega : S(\omega) = T(\omega) < \infty\} = 0$, and T is merely *inaccessible* if it is equivalent to a totally inaccessible time τ on a set of positive probability.

To analyze further we introduce the concept of a *stochastic interval*: Thus if $T_1 \leq T_2$ are stopping times of a (standard) filtration, then the set $[\![T_1, T_2]\!]$ is a "closed" stochastic interval, defined as (cf. also the earlier Eq.(1.8))

$$[\![T_1, T_2]\!] = \{(t, \omega) : T_1(\omega) \leq t \leq T_2(\omega)\} \subset U = \mathbb{R}^+ \times \Omega.$$

The corresponding intervals $[\![T_1, T_2)\!)$, $(\!(T_1, T_2]\!]$, $(\!(T_1, T_2)\!)$ are similarly defined. Thus we have

$$[\![T_1, T_2)\!) = \bigcup_{\substack{t \in \mathbb{R} \\ t-rational}} \{[T_1 \leq t] \cap [T_2 \leq t]^c\} \in \mathcal{B}(\mathbb{R}^+) \otimes \mathcal{F}_\infty = \mathcal{U}' \subset \mathcal{U}$$

where as usual $\mathcal{F}_\infty = \sigma(\cup_t \mathcal{F}_t)$ and $\mathcal{U} = \mathcal{B}(\mathbb{R}^+) \otimes \Sigma$ the σ-algebra of U. Note that by definition $[\![T_1, T_2]\!] = [\![T_1, T_2)$ if $T_2(\omega) = \infty$ for all $\omega \in \Omega$, and $=\emptyset$ if $T_1(\omega) = T_2(\omega) = \infty$ for all $\omega \in \Omega$. The degenerate interval $[\![T, T]\!]$ is called the *graph* of T. All these stochastic intervals are clearly in $\mathcal{U}'$. Let $\mathcal{S}$ be the collection of all subsets of U which are stochastic intervals of the form $[\![T_1, T_2)$ for all pairs of stopping times of our filtration $\{\mathcal{F}_t, t \geq 0\}$. Then $\mathcal{S}$ is a semi-algebra, since $[\![0, \infty) = U \in \mathcal{S}$ and $[\![T_1, T_2) \in \mathcal{S} \Rightarrow [\![T_1, T_2)^c = [\![0, T_1) \cup [\![T_2, \infty)$ this being a (finite) disjoint union of stochastic intervals, and if $[\![T_i, S_i) \in \mathcal{S}$,i=1,2, we have

$$[\![T_1, S_1) \cap [\![T_2, S_2) = [\![T_1 \vee T_2, (T_1 \vee T_2) \vee (S_1 \wedge S_2)) \in \mathcal{S}.$$

Here one uses the fact that the minimum and maximum of a pair of stopping times are stopping times. Let $\mathcal{O} = \sigma(\mathcal{S})$, the σ-algebra generated by the semi-algebra of stopping times, called the *optional* (or *well-measurable*) σ-algebra. Here if we restrict the times to predictale or accessible classes, then the corresponding σ-algebras are called respectively the *predictable* or *accessible* ones denoted by $\mathcal{P}$ and $\mathcal{A}$. It is seen that $\mathcal{P} \subset \mathcal{A} \subset \mathcal{O}$; and one can verify that $\mathcal{P} = \mathcal{A}$ if the filtration has no times of discontinuity. The predictable σ-algebra is not only the smallest but is also the most important in the following work. We should observe here that (for the standard filtration) every element of $\mathcal{S}$ (and hence of $\mathcal{O}$) is a progressively measurable set. Indeed, let $A = [\![T_1, T_2)$ and set $X_t(\omega) = \chi_A(\omega)$. Then $\{X_t, t \geq 0\}$ is a right continuous process and if we define simple functions $X^n : U \to \mathbb{R}^+$ by setting $X_0(\omega) = X_0^n(\omega)$ and for $s > 0, n \geq 0$,

$$X_s^n(\omega) = X_{[k/2^n]t}(\omega), \quad \frac{k}{2^n}t \leq s < \frac{k+1}{2^n}t, k = 0, 1, \ldots, 2^n - 1, \quad (63)$$

then $X^n : [0,t] \times \Omega \to \mathbb{R}^+$ is $\mathcal{B}([0,t]) \otimes \mathcal{F}_\infty$-measurable for each n. Since $X^n \to X$ as $n \to \infty$, pointwise, it follows that X is progressively measurable. In particular, processes measurable relative to the predictable (accessible) σ-algebras are progressively measurable. In view of this one may ask whether an integrable increasing right continuous process measurable relative to a predictable process is predictable in the sense of Definition 1.3. As noted before, a positive answer is given by the following:

17. Proposition. *Let $\{A_t, \mathcal{F}_t, t \geq 0\}$ be an integrable increasing right continuous process. Then it is predictable in the sense of Definition 1.3 iff it is measurable relative to the predictable σ-algebra $\mathcal{P}$ of the standard filtration $\{\mathcal{F}_t, t \geq 0\}$, so that it is a predictable process in either sense.*

This result was originally established by C. Doléans-Dade [3], and a detailed discussion is in C. Dellacherie and P.-A. Meyer [1],Part B. We omit the proof here referring the reader to the latter, but will use either of the concepts according to convenience.

In view of the already noted importance of the predictable σ-algebras, we give an alternative formulation of the same.

18. Proposition. *Let $\{\mathcal{F}_t, t \geq 0\}$ be a standard filtration of (Ω, Σ, P), and $\mathcal{P}$ be its predictable σ-algebra. Then $\mathcal{P}$ is also generated by the sets of the form $(s,t] \times A \subset U = \mathbb{R}^+ \times \Omega, A \in \mathcal{F}_s, 0 \leq s \leq t$, or also equivalently by adapted a.e. continuous processes $\{X_t, \mathcal{F}_t, t \geq 0\}$. In fact every left continuous adapted process is predictable.*

Proof. First observe that if $\mathcal{P}'$ is the σ-algebra generated by the given collection, then $\mathcal{P}' \subset \mathcal{P}$ since every ordinary time is (trivially) a stopping time, and $(s,t] \times A = ((S,T]$ where $S = s\chi_A + t\chi_{A^c}, T = t$. On the other hand if $S \leq T$ are a pair of simple stopping times, then we can represent them as $S = \sum_{k=1}^n a_k \chi_{A_k}, A_k \in \mathcal{F}_{a_k}$ where the $a_k \uparrow$, $a_k \in \mathbb{R}^+$, and A_k's form a partition of Ω, and similarly T. Then the set $(a_k, t] \times A_k$, a finite union of generators of $\mathcal{P}'$, is in it for any $t > S$. Similarly $((T, t']$ is in $\mathcal{P}'$ and hence $((S,T] = ((S,t] - ((T,t], t \geq T$, is also. But we can approximate any stopping time by a decreasing sequence of simple times as above (e.g., if T is given let $T_n = \sum_{k \geq 1} k2^{-n} \chi_{[(k-1)2^{-n} \leq T < k2^{-n}]}$ with $T_n = \infty$ on the set $[T = \infty]$ so that $T_n \downarrow T$). Thus $\mathcal{P}' \supset \mathcal{P}$ hence there is equality.

On the other hand let $A \in \mathcal{F}_0$ and $T(\omega) = 0, \omega \in A; = \infty, \omega \in A^c$, so that $\{X_t = \chi_{[T,\infty)}(t), t \geq 0\}$ is a continuous predictable process, and for any predictable stopping time $T' > 0$ the process $X_t = t - T' \wedge t$ is an adapted predictable process. Thus $\mathcal{P}$ is contained in the σ-algebra generated by the continuous adapted processes. For the opposite inclusion we prove the more gerneral last statement. So let $\{X_t, \mathcal{F}_t, t \geq 0\}$ be a left continuous process. Define for each n the elementary function

$$X^n = X_0 \chi_{[0,0)} + \sum_{k \geq 1} X_{k/n} \chi_{((k/n,(k+1)/n]}$$

so that $\{X_t^n, \mathcal{F}_t, t \geq 0\}$ is predictable and $X^n \to X$ pointwise, implying that X is predictable. □

We now proceed to present an alternative derivation of the Doob-Meyer decomposition theorem by representing the process in terms of a signed measure, and analyze the consequences. This is due originally to C. Doléans-Dade. We give the representation for a more inclusive class of processes called semi(and quasi)-martingales. These are motivated by considering differences of martingales which contain the differences of increasing processes in their earlier decompositions, and which thus have (locally) bounded variation. Let us first introduce the relevant concepts.

19. Definition. Let $X = \{X_t, \mathcal{F}_t, t \in I\}, I \subseteqq \mathbb{R}^+$ be an adapted right continuous stochastic process on (Ω, Σ, P) with $\{\mathcal{F}_t, t \in I\}$ as a standard filtration. Then:

(a) X is called a *semimartingale* if $X_t = Y_t + Z_t$ where $\{Y_t, \mathcal{F}_t, t \in I\}$ is a (separable) martingale and $Z = \{Z_t, \mathcal{F}_t, t \in I\}$ is a (separable) adapted process whose sample functions are not only a.e. of bounded variation on each compact subset of I, but is the difference of two predictable increasing processes based on the filtration $\{\mathcal{F}_t, t \in I\}$. If in the above representation Y is only a local martingale, then X is called a *local semimartingale.*

(b) The process X given above is called a *quasimartingale* if for any $0 \leq a < b \leq \infty$, any partition $a \leq t_1 < t_2 < \ldots < t_{n+1} \leq b$, the following is true:

$$E\Big[\sum_{i=1}^{n} |E^{\mathcal{F}_{t_i}}(X_{t_{i+1}} - X_{t_i})|\Big] = \sum_{i=1}^{n} E[|X_{t_i} - E^{\mathcal{F}_{t_i}}(X_{t_{i+1}})|] \leq K_x^b < \infty, \tag{64}$$

where K_x^b is independent of the partition.

The concept and the name "quasimartingale" were introduced for a slightly smaller class of processes by Fisk [1], and the general "semimartingales" are thereafter defined by Meyer. Quasimartingales for right continuous processes were termed "F-processes" by Orey [2], and were called by the present name by K.M.Rao [2]. In the first edition of this book, they were termed "(*)-processes".

Note that every $L^1(P)$-bounded sub- and supermartingale is a quasimartingale. We find that essentially every quasimartingale is a (local) semimartingale. Although one can extend the proofs of the decompositions established above in the section, to quasimartingales, here we associate an additive set function with a quasimartingale and prove the related results, on the predictable σ-algebra of the filtration.

If U_a stands for $U = \mathbb{R}^+ \times \Omega$ when $\mathbb{R}^+$ is replaced by its subinterval $[0, a]$, let $\mathcal{S}_a, \mathcal{P}_a$ be the corresponding classes so that the predictable σ-algebras satisfy the inclusion relationships $\mathcal{P}_a \subset \mathcal{P}'_a, a < a'$, and similarly for others. Thus on $\mathcal{S}_a$ we define μ_a^x by the equation:

$$\mu_a^x((t, t'] \times A) = \begin{cases} \int_A (X_t - X_{t'})\, dP, & A \in \mathcal{F}_t, t' < a \\ \int_A X_a\, dP, & \text{for } t = a. \end{cases} \tag{65}$$

It will be verified that $\mu_a^x : \mathcal{S}_a \to \mathbb{R}$ is finitely additive, whenever $X = \{X_t, \mathcal{F}_t, t \geq 0\}$ is in $L^1(P)$. We take $X_\infty = 0$ if $a = \infty$.

The following result is the desired measure representation of a quasimartingale.

20. Theorem. *Let $\{\mathcal{F}_t, t \geq 0\}$ be a standard filtration from (Ω, Σ, P) and let $X = \{X_t, \mathcal{F}_t, t \geq 0\}$ be a right continuous process in $L^1(P)$. Let X be a quasimartingale and μ_a^x be its associated set function given by (65) for $a \geq 0$. Then there exists a unique signed measure $\mu^x : \mathcal{P} \to \mathbb{R}$ such that $\mu^x | \mathcal{S}_a = \mu_a^x$ iff X is of class (DL).*

Proof. We begin by noting that the μ^x given by (65) may also be expressed as follows. For any finite set $\{S_i, T_i, i \geq 1\}$, $S_i < T_i < S_{i+1}, S_i, T_i$ being finite stopping times of the filtration, let $B = \cup_{i=1}^n ((S_i, T_i] \cup (\{0\}) \times A)$ for $A \in \mathcal{F}_0$. First one observes that the predictable σ-algebra of the filtration is generated by sets of the form $[\![0_A), ((S, T], A \in \mathcal{F}_0$ where 0_A is the stopping time which is zero on A and infinity on A^c, and S, T are any stopping times. But the algebra generated by sets of the form B above determines the predictable σ-algebra. The representation of B here can be taken as: S_1 is the debut of B, and T_1 is the debut of $((S_1, \infty] \cap B^c$, and S_2 that of $((T_1, \infty] \cap B$ and so on. This makes the representation of the elements of the predictable algebra also unique, and we use this as a "canonical form" below. Then (65) can be expressed as (since $X_\infty = 0$):

$$\mu^x(B) = \sum_{i=1}^{n} E(X_{S_i} - X_{T_i}). \tag{66}$$

In fact, if $n = 1$ and S_i, T_i are simple, then (66) reduces to (65) and the general case follows by approximation. From either of the forms (65) or (66), it is verified (with a standard computation) that μ^x is (finitely) additive on the algebra $\mathcal{A}$ generated by $\mathcal{S}$. If for any $a > 0, \mathcal{P}_a, \mathcal{A}_a, \mu_a^x$ denote the corresponding symbols when $\mathbb{R}^+$ is replaced by the interval $[0, a)$, then μ_a^x is additive on $\mathcal{A}_a$ and $\mathcal{P}_a = \sigma(\mathcal{A}_a) = \sigma(\mathcal{S}_a) \subset \mathcal{P}_{a'}, \mu_a^x = \mu_{a'}^x|\mathcal{A}_a, a < a'$. Also it is clear that μ^x on the algebra $\mathcal{A} = \cup_{a>0}\mathcal{A}_a$ is given by $\mu^x(B) = \mu_a^x(B)$ for $B \in \mathcal{A}$ so that $B \in \mathcal{A}_a$ for some $a > 0$ and that $\mu_a^x(B)$ is defined. In view of the classical Carathéodory-Hahn extension theorem, it suffices to show that μ^x is σ-additive on $\mathcal{A}$. Note that $|\mu^x((t,t'] \times A)| \leq E(|X_t - E^{\mathcal{F}_t}(X_{t'})|) \leq K_x < \infty$ by (64), implying $|\mu^x|(U) \leq K_x$, whence μ^x is bounded. We now establish the σ-additivity with a simple modification of an argument of Doléans-Dade [3], (cf. Dellacherie [1]) giving the details in steps.

I. For an element $B_1 \in \mathcal{A}$,(which is always representable as a union used in (66)) let $\bar{B}_1$ stand for the given union where the left end point of the stochastic interval is also included. [Then of course $\bar{B}_1$ need not be in $\mathcal{A}$.] Now with the right continuity of the process, we assert that for each $\varepsilon > 0$, and $B \in \mathcal{A}$ there is a $C(= C_\varepsilon) \in \mathcal{A}$, $\bar{C} \subset B$ and $|\mu^x|(B - C) < \varepsilon$. (This technical point is needed in establishing the σ-additivity of μ^x below.) Note that $\bar{C} = \cup\{\bar{C}(\omega) : \omega \in C \subset B.\}$

Indeed, we may consider one interval $((S, T]$ and assume that $S(\omega) < \infty, \omega \in \Omega$. For $n \geq 1$, define $S_n = S + \frac{1}{n}$ if $S + \frac{1}{n} < T$; and $= \infty$ otherwise, so that $S_n \uparrow S$, i.e., the S_n announce S. If $C_n = ((S, T]$, then clearly $C_n \uparrow B$, and $\bar{C}_n \subset B, n \geq 1$. Observe now from (65) that

$$\lim_{\delta \downarrow 0} \mu^x((t, t+\delta] \times A) = -\lim_{\delta \downarrow 0} \int_A (X_{t+\delta} - X_t)\, dP, \qquad A \in \mathcal{F}_t,$$

and the right side is zero if we can interchange the limit and integral. [There is no measurability problem here since the X_t-process has a measurable modification.] But this follows from the fact that $\{X_t, t \in I\}$ is uniformly integrable on each compact interval $I \subset \mathbb{R}^+$, as in the case of a right continuous submartingale (cf. also Exercise 5.7(d)).Hence by the hypothesis that the process is of class (DL), we deduce that $X_{S_n} \to X_S$ in $L^1(P)$, and so $\mu^x(C_n) \to \mu^x(B)$ as $n \to \infty$. Thus there is an $n_0 = n_0(\varepsilon)$ such that $|\mu^x(B) - \mu^x(C_{n_0})| < \varepsilon/2$. It follows that for every set $A \subset B - C_{n_0}, |\mu^x(A)| < \varepsilon/2$. But a standard result in real analysis implies that $|\mu^x|(B - C_{n_0}) \leq 2|\mu^x(B) - \mu^x(C_{n_0})| < 2.\varepsilon/2 = \varepsilon$. Taking $C_\varepsilon = C_{n_0}$ our assertion follows. This is used to show that $|\mu^x|(\cdot)$, hence $\mu^x(\cdot)$ is σ-additive on $\mathcal{A}$ (and therefore on $\mathcal{P}$). Actually, in view of the hypothesis of class (DL), and by the classical measure theory as well as the boundedness of μ^x, we can and do replace $\mathbb{R}^+$ by any compact interval which we take it to be $[0, a]$ for $a > 0$ in the rest of the proof.

II. If $\{B_n \downarrow \emptyset\}$ is a sequence of elements from $\mathcal{A}_a, a > 0$, then $|\mu^x|(B_n) \to 0$.

For, by the preceding step, given $\varepsilon > 0$, there exist $C_n \in \mathcal{A}, \bar{C}_n \subset B_n$ such that $|\mu^x|(B_n - C_n) < \varepsilon/2^n$. Let $F_n = \cap_{k=1}^n C_k (\in \mathcal{A})$. Then $\bar{F}_n \subset B_n$, and since $F_n = C_n - \cup_{i=1}^{n-1}(B_i - C_i)$, one has

$$\begin{aligned} |\mu^x|(F_n) &\geq |\mu^x|(C_n) - \sum_{i=1}^{n} |\mu^x|(B_i - C_i) \\ &\geq |\mu^x|(B_n) - \frac{\varepsilon}{2^n} - \sum_{i=1}^{n-1} \frac{\varepsilon}{2^i} \geq |\mu^x|(B_n) - \varepsilon. \qquad (67) \end{aligned}$$

Thus it suffices to show that $|\mu^x|(F_n) \to 0$, and use (67). Since $\bar{F}_n$ is in $\mathcal{B}([0, a]) \otimes \Sigma$, if D_n is the debut of $\bar{F}_n$, then by Theorem IV.2.5, D_n is a stopping time of the filtration where $\inf(\emptyset) = a$. Moreover $\{B_n, n \geq 1\} \subset \mathcal{A} = \cup_{t>0}\mathcal{A}_t$ so that $B_n \in \mathcal{A}_{a_n}$, for some $0 < a_n < a$, whence $B_n \in \mathcal{A}_{a_n}$ for some a_n. Since $B_n \supset B_{n+1}$, and each B_n of $\mathcal{A}_{a_n}$ is of the form $(\{0\} \times A) \cup \bigcup_{i=1}^m ((S_i^n, T_i^n]$ by the fact that the algebra $\mathcal{A}_{a_n}$ is generated by $\mathcal{S}_{a_n}$, we see that all these stopping times take values in $[0, a]$. Further $\bar{F}_n(\omega)$ is a closed subset of $[0, a]$ for each n, and $\cap_n \bar{F}_n \subset \cap_n B_n = \emptyset$. Hence $D_n \uparrow a$ and the construction implies that $S_i^n \leq D_n \leq T_i^n, i = 1, \ldots, m$ for each n. [This inequality, which is needed here, could not be obtained if we take D_n as the debut of B_n.] Now the process is (or can be taken to be) separable and has no fixed discontinuities. So the right and left limits exist for $X(t, \omega)$ in t for almost all $\omega \in \Omega$,(cf. also Exercise 5.7(c)). Thus $X_{D_n} \to X_a$ a.e. as

$n \to \infty$, and also in $L^1(P)$, since the X_t-process is of class (D). It now follows that $|\mu^x|(F_n) \le E(|X_{D_n} - X_a|) \to 0$. By the earlier reduction this proves the σ-additivity of μ^x.

III. For the necessity, let μ^x be σ-additive on $\mathcal{P} \to \mathbb{R}$. Then the signed measure μ^x is bounded, i.e., $|\mu^x|(U) < \infty$ where U is the product space given for (65). We need to verify the bound (64) and then to show that the process is of class (DL). Regarding the bound, consider $A = \{\omega : X_s(\omega) > E^{\mathcal{F}_s}(X_t)(\omega)\}$ for $0 \le s < t \le \infty$ with $X_\infty = 0$ a.e. Then $A \in \mathcal{F}_s$ and by the additivity of μ^x,

$$\begin{aligned} |\mu^x|((s,t] \times \Omega) &= |\mu^x((s,t] \times A^c)| + |\mu^x((s,t] \times A)| \\ &\ge E(|X_s - E^{\mathcal{F}_s}(X_t)|). \end{aligned} \tag{68}$$

Since $|\mu^x|(U) < \infty$, (68) implies (64) with K_x no larger than the total variation of μ^x. But the reverse inequality follows immediately from (65) and the definition of $|\mu^x|$ so that $K_x = |\mu^x|(U)$. We need to show that for each $0 < a < \infty$, $\{X_t, t \in [0,a]\}$ is in class (D).

Let $\mu^x = (\mu^x)^+ - (\mu^x)^-$ be the (Jordan) decomposition of the signed measure μ^x. Then $(\mu^x)^\pm$ are bounded measures on $\mathcal{P}$. Since $\mu^x_t(A) = \mu^x((t,\infty) \times A)$ defines a signed measure on $\mathcal{F}_t, t \ge 0$, which is P-continuous, the same holds for $(\mu^x)^\pm$. Let X_t', X_t'' be the Radon-Nikodým derivatives on $\mathcal{F}_t$. The right continuity of X implies that we may take the right continuous modifications of X_t' and X_t'', denoted X_t^+, X_t^+. Then we get $X_t = X_t^+ - X_t^-$, and $\{X_t^\pm, \mathcal{F}_t, t \ge 0\}$ are right continuous nonnegative processes. For each $A \in \mathcal{F}_t$, we have $(\mu^x)^\pm((s,t] \times A) = \int_A (X_s^\pm - X_t^\pm)\, dP$; it follows that $\{X_t^\pm, \mathcal{F}_t, t \ge 0\}$ are supermartingales, and so the X-process can be expressed as a difference of two positive supermartingales. So far, the class (D) condition is not verified. Thus it suffices to show that $\{X_t^\pm, t \ge 0\}$ processes are of class (DL), and by symmetry we need only consider one of them, say $\{X_t^+, t \in [0,a]\}$ and show that it is in class (D)for any $a > 0$.

IV. Let $\tau_n = \inf\{t : X_t^+ > n\}$. Then τ_n is a stopping time of the filtration. If $s < a$, then $\tau_n \wedge s \uparrow s$ as $n \to \infty$ by the supermartingale maximal inequality. Hence

$$(\mu^x)^+(((\tau_n \wedge s, s)) = E(X_s^+ - X_{\tau_n \wedge s}^+) \to 0, \tag{69}$$

as $n \to \infty$, since the left side decreases to zero by the σ-additivity of μ^x. Let T be any stopping time of $\{\mathcal{F}_t, 0 \le t \le a\}$. Define $\tau_n' = T$ on $[X_T^+ > n]$, and $= s$ on $[X_T^+ \le n]$. Then examining cases, we find that $\tau_n \wedge s \le \tau_n'$, and with the supermartingale property (cf. Theorem IV.1.7(ii)),

$$E(X_{\tau_n \wedge s}^+) \ge E(X_{\tau_n'}^+) \ge \int_{[X_T > n]} X_T^+ \, dP + \int_{[X_T \le n]} X_s^+ \, dP.$$

Next transferring the last term to the left and simplifying:

$$\int_{[\tau \le s]} X_{\tau_n}^+ \, dP + \int_{[\tau_n > s]} X_s^+ \, dP - \int_{[X_T \le n]} X_s^+ \, dP \ge \int_{[X_T > n]} X_T^+ \, dP. \tag{70}$$

However, $[\tau_n > s] \subset [X_T^+ \le n]$ so that the middle integrals of (70) give a negative value. Hence, using the facts that $P[\tau_n \le s] \to 0$ as $n \to \infty$, and $\{X_{\tau_n \wedge s}^+, n \ge 1\}$ is uniformly integrable by (69), $X_{\tau_{n_j} \wedge s}^+ \to X_s$ a.e. (a subsequence also being a supermartingale), we get:

$$\int_{[X_T > n]} X_T^+ \, dP \le \int_{[\tau_{n_j} \le s]} X_{\tau_{n_j} \wedge s}^+ \, dP = E(X_{\tau_{n_j} \wedge s}^+ \chi_{[\tau_{n_j} \le s]}) \to 0, \tag{71}$$

as $n \to \infty$, uniformly in T. Hence $\{X_t^+, 0 \le t \le a\}$ is in class (D). Similarly the X_t^--process is in class (D). □

Remarks. 1. The σ-additivity of μ^x given by (65) can also be established for some nonclass (DL) processes X, if the σ-algebras $\mathcal{F}_t$ are considerably restricted. If they are a "standard Borel family" such a result was given by H. Föllmer [1] (cf. also C. Stricker [1]), and later he extended it for two parameter processes in [3]. A detailed analysis without such a restriction for a standard filtration has also been given by K. M. Rao [2] for a related result.

2. In the above proof, we actually have shown the following: The set of right continuous processes satisfying (64) and of class (DL), and the set of signed measures μ on $\mathcal{P}$ satisfying $\lim_{\delta \downarrow 0} \mu((t, t+\delta] \times A) = 0$ for each $A \in \mathcal{F}_t$ whose marginals μ_t on $\mathcal{F}_t$ are P-continuous, are in a one-to-one correspondence.

We note the following consequence for reference.

21. Corollary. *Let $X = \{X_t, \mathcal{F}_t, t \ge 0\}$ be a class (DL) right continuous quasimartingale and let μ^x be the associated signed measure* *on $\mathcal{P}$, the predictable σ-algebra of the (standard) filtration. Then X is: (i) a martingale iff μ^x is supported by $\{\infty\} \times \Omega$, (ii) a supermartingale iff $\mu^x((t,\infty] \times A)$ is nonnegative for all $t > 0$ and $A \in \mathcal{F}_t$, and (iii) a potential iff μ^x is positive and continuous at '∞', i.e., $\lim_{t\to\infty} \mu^x((t,\infty] \times A) = 0$ for all $A \in \cup_{t>0}\mathcal{F}_t$.*

Semimartingales and quasimartingales are related as follows. As usual the filtration is assumed to be standard without comment.

22. Theorem. *(a) Every quasimartingale is the difference of two nonnegative supermartingales with the same stochastic base as the given process. It can and will be called a* **generalized Jordan decomposition**.

(b) A quasimartingale is a semimartingale in the sense of Definition 17 iff it is of class (DL).

Proof. Part (a) has already been established above. (An independent proof of it is given by K. M. Rao.) For (b), let $X_t = X_t^+ - X_t^-$ given by (a) and let X be in class (DL). Then the equation $|\mu^x| = (\mu^x)^+ + (\mu^x)^-$ implies that $X^\pm = \{X_t^\pm, \mathcal{F}_t, t \geq 0\}$ satisfies (64). The σ-additivity of $(\mu^x)^\pm$ implies, by Theorem 20, that $X^\pm$ is in class (DL). Then by Theorem 6, there exist right continuous martingales $\{Y_t^i, \mathcal{F}_t, t \geq 0\}$ and predictable increasing processes $\{A_t^i, t \geq 0\}$, $i = 1, 2$, such that

$$X_t^+ = Y_t^1 - A_t^1, \qquad X_t^- = Y_t^2 - A_t^2, \qquad t \geq 0,$$

uniquely. Hence $X_t = (Y_t^1 - Y_t^2) + (A_t^1 - A_t^2), t \geq 0$, a.e. and since $B_t = A_t^2 - A_t^1, t \geq 0$ is a process with sample paths of bounded variation a.e. on compact sets, and $\{Y_t^1 - Y_t^2, \mathcal{F}_t, t \geq 0\}$ is a martingale, we deduce that X is a semimartingale. The converse assertion is an immediate consequence of the converse part of Theorem 6. □

We may state the second part of the above result differently, using Theorem 1. Since every positive supermartingale is of class (R) by Corollary II.5.6, we have the following result as a consequence of the first part.

23. Corollary. *A quasimartingale can be written as a sum of a local martingale and a difference process of two class (D)-potentials* **or a**

semipotential *(i.e. a supermartingale which tends to zero in $L^1(P)$). Thus a quasimartingale is a local semimartingale.*

From the preceding results we can present a characterization of $L^1(P)$-bounded semimartingales, which in case of a.e. continuous sample paths, is due to Fisk [1].

24. Theorem. *Let $X = \{X_t, \mathcal{F}_t, t \geq 0\}$ be a right continuous $L^1(P)$-bounded process. Then the following statements are equivalent:*

(i) X is a semimartingale [with continuous sample paths]

(ii) X is (a) a quasimartingale, and (b) in class (DL), [and (c) $\lim_{n\to\infty} nP[\sup_{t\in I}|X_t| > n] = 0$, for any compact interval $I \subset \mathbb{R}^+$, and X has continuous sample paths].

Proof. If (i) holds then $X_t = Y_t + Z_t$ where $Y = \{Y_t, \mathcal{F}_t, t \geq 0\}$ is an $L^1(P)$-bounded martingale and $Z = \{Z_t, \mathcal{F}_t, t \geq 0\}$ has paths of a.e. bounded variation. Consequently, Z is the difference of two increasing positive right continuous processes $Z^i = \{Z_t^i, \mathcal{F}_t, t \geq 0\}, i = 1,2$. By the Jordan decomposition of martingales (cf. Theorem II.5.2), $Y_t = Y_t^1 - Y_t^2$ where $\{Y_t^i, \mathcal{F}_t, t \geq 0\}, i = 1,2$ are positive martingales with the same properties and it is unique. If $X_t^i = Y_t^i + Z_t^i, i = 1,2$ so that $X_t = X_t^1 - X_t^2$, we get $\{X_t^i, \mathcal{F}_t, t \geq 0\}, i = 1,2$ to be positive supermartingales. Hence X is a quasimartingale of class (DL) by Theorem 22. If moreover X has continuous sample paths for its components Y and Z, then with the properties of various decompositions above, we deduce that the X^i also have a.a. continuous sample paths, so that each process is uniformly integrable on compact intervals of $\mathbb{R}^+$. This implies that the X^i are of class (D) on each such interval. Hence by Proposition IV.3.3, condition (c) holds for each X^i so that the same is true of X. This proves (ii).

Conversely, let (ii) hold. Then (a) and (b) imply (i) by Theorem 22. If (c) also holds, and X has continuous sample paths, let $I \subset \mathbb{R}^+$ be a compact interval. If μ^x is the associated signed measure given by Theorem 20, we note that $\mu^x([s - \delta_1, t + \delta_2] \times A) \to 0$ as $\delta_i \downarrow 0, i = 1,2$ and $t \to s$ for each $A \in \mathcal{F}_s (s < t)$ since X has continuous paths. The same is true (by the uniform integrability of X on compacts) of $(\mu^x)^{\pm}$. From this we may conclude that the positive supermartingales X^i have continuous paths on compact intervals and, since this is a local property, on $\mathbb{R}^+$ itself. The condition (c) further implies that on I (with separability),

$$nP[\sup_{t\in I} X_t^i > n] \leq nP[\sup_{t\in I} |X_t| > n] \to 0, \qquad i = 1, 2.$$

So $\{X_t^i, t \in I\}$ is of class (D) by Proposition IV.3.3 (whence X^i is of class (DL)). But then $X_t^i = Y_t^i - A_t^i, i = 1, 2$ and both the right side processes are continuous a.e. Hence Y, A have a.a. continuous sample paths, which is (i). □

Discussion. If $X = Y + Z$ is a right continuous semimartingale which is $L^1(P)$-bounded, suppose that $X = \tilde{Y} + \tilde{Z}$ is another such representation. Then $Y - \tilde{Y} = \tilde{Z} - Z$, so that the martingale $Y - \tilde{Y}$ has a.a. of its sample functions of bounded variation. Let $V = Y - \tilde{Y} = A^1 - A^2$ where each A^i is integrable, increasing, and zero at 0. Then $E(A_t^1) = EA_t^2) + c$ where $c = E(V_t)$ is a constant. Since $A_0^1 = 0 = A_0^2$ a.e., we must have $c = 0$. In case X has a.a. continuous sample paths, then by the above result V will also have the same property (and the same holds for $A^i, i = 1, 2$). If X and hence V were in $L^2(P)$, then $V_t = V_0$ a.e., and using a stopping time argument the same conclusion can be seen to hold even for the $L^1(P)$-case. Thus if $X_0 = 0$, a.e., then $V_0 = 0$ a.e., and we get uniqueness in the continuous case. Using the arguments of Theorem 6 to the general case the same uniqueness conclusion can be deduced since in $X^i = Y^i - B^i, X = X^1 - X^2, Y = Y^1 - Y^2$, and $Z = B^1 - B^2$, the B^i are predictable, increasing, and right continuous. Thus the earlier work of this section extends completely to semimartingales. We therefore can and will use this general class in stochastic integration later after analyzing the martingale case in the next section.

Before concluding this general treatment of optionals, we include a "section theorem", which will also be used in integration, as a final item. Thus let $\pi : U = \mathbb{R}^+ \times \Omega \to \Omega$ be the coordinate projection and let P^* be the outer measure generated by (Σ, P). Define $\tilde{P}$ on the subsets of U by the equation $\tilde{P}(A) = P^*(\pi(A))$. [Then $\tilde{P}$ is a capacity function on subsets of U. For a definition of capacity and related results cf. e.g., Rao[11], Proposition 7.2.2.] If X is a (real) random variable on Ω its graph is $G(X) = \{(y, \omega) : y = X(\omega) \in \mathbb{R}\}$ so that $G(X) \in \Sigma$ if X is measurable for Σ. Note that for an optional T of a standard filtration its *graph* is given by $[\![T]\!]$ which is simply $G(T)$ in the above notation. Also if $A \in \mathcal{U} = \mathcal{B}(\mathbb{R}^+) \otimes \Sigma$, then $\pi(A) \in \Sigma$ and the debut function $D_A(\cdot) = \inf\{t \geq 0 : (t, \cdot) \in A\}$ is measurable for Σ. Using some results of Capacity and Measure, (cf. e.g.,Rao [11],Chapter 7, Exercises 6 and 7), one has: for each $A \in \mathcal{U}$, and each $\varepsilon > 0$, there exists a measurable $f : \Omega \to \bar{\mathbb{R}}^+$ such that $G(f) \subset A$ and $\tilde{P}(G(f)) \geq \tilde{P}(A) - \varepsilon$, where by definition $\tilde{P}(G(f)) = P[f < \infty]$, and in fact $\pi(A) = [f < \infty]$, a.e. With this background we have the following *section theorem* due to Meyer [1], whose proof below follows Dellacherie [1].

25. Theorem. *Let $A \in \mathcal{P}$ [or$A \in \mathcal{O}$] (i.e., A is either a predictable or an optional set). Then for each $\varepsilon > 0$ there exists a predictable [or optional] stopping time T of the standard filtration $\{\mathcal{F}_t, t \geq 0\}$ on (Ω, Σ, P) such that $G(T) \subset A$ and $P(\pi(A) - [T < \infty]) < \varepsilon$. (A similar result is also true in the accessible case.)*

Proof. The argument is given for all the cases simultaneously until the end. If $\mathcal{C}$ is any one of the σ-algebras, let $A \in \mathcal{C}$. By the preceding (capacity-measure theoretical) result, for each $\varepsilon > 0$, there is a measurable function $f : U \to \mathbb{R}^+$ such that $G(f) \subset A$ and $\tilde{P}(G(f)) \geq \tilde{P}(A) - \varepsilon$. Define a new measure μ on $\mathcal{C}$ by the following equation for any bounded measurable h on U:

$$\int_U h \, d\mu = \int_{[f<\infty]} h(f(\omega), \omega) \, dP. \tag{72}$$

In particular, taking $h = \chi_B, B \in \mathcal{C}$, we see that μ vanishes outside of A, since $G(f) \subset A$. Thus $\mu(A) = P([f < \infty]) = P(\pi(A)), \mu(A \cup B) \geq \mu(B)$ for $B \in \mathcal{C}$ by the subadditivity of P. Note that $\mu = \tilde{P}$ on $\mathcal{C}$.

Since $\mathcal{C}$ is determined by the stochastic intervals $[\![S, T)$ for all optionals $\{S, T\}$ of the filtration, and since they form a semi-algebra $\mathcal{S}, \sigma(\mathcal{S}) = \mathcal{C}$, we deduce from the classical Carathéodory theory (cf.,e.g., Rao [11], page 41) that for each $\varepsilon > 0$ there exists a $B \in \mathcal{S}_\delta$ such that $B \subset A$ and $\mu(B) \geq \mu(A) - \varepsilon$, where $\mathcal{S}_\delta = \{\cap_{i=1}^n A_i : A_i \in \mathcal{S}, n \geq 1\}$. Then

$$P(\pi(A)) = P([f < \infty]) = \mu(A) \leq \mu(B) + \varepsilon = P(\pi(B)) + \varepsilon. \tag{73}$$

Let $T_0 = D_B$, the debut of B. We assert that T_0 is an optional of the filtration, and satisfies the requirements of the theorem.

It is immediate from Theorem IV.2.5 that D_B is measurable for $\mathcal{U}$; but we need to show a stronger result that it has the additional properties asserted above. Since by definition and construction $G(D_B) = \{(t,\omega) : t = D_B(\omega) < \infty\} \subset B$, we only need to show that D_B is a (predictable etc.) stopping time of the filtration.

Let $\mathcal{T}$ be the class of all stopping times of $\{\mathcal{F}_t, t \geq 0\}$ which are dominated by D_B. Then $\mathcal{T}$ is clearly nonempty and if $T_1, T_2 \in \mathcal{T}$ then $T_1 \vee T_2 \leq D_B$ and being a stopping time of the same filtration is in $\mathcal{T}$. If $T_n \leq T_{n+1}$ are in $\mathcal{T}$ then $\lim_n T_n \leq D_B$ so that it is also in $\mathcal{T}$. Let $T = \sup\{T : t \in \mathcal{T}\}$. Then $T \leq D_B$, and we see that for some $\tau_n \in \mathcal{T}, \tau_n \uparrow T$ a.e. So $T \in \mathcal{T}$ also. It therefore suffices to show that $T = D_B$ and T is a limit of a sequence of stopping times of the desired kind. But $B \in \mathcal{S}_\delta$ so that there is a sequence $B_n \in \mathcal{S}$, satisfying $B = \cap_{n=1}^\infty B_n$. Since $\mathcal{S}$ is a semi-ring, we may assume $B_n \supset B_{n+1}$. Define S_n as the debut of B_n so that $S_n \leq S_{n+1}$, and $[S_n \leq t] \in \mathcal{F}_t$ (recall that $B_n = [\![V_n, \tau_n)$ for some stopping times of the filtration). If $A_0 = \{\omega : T(\omega) = \infty\}$, then we have $B_0 = \{\infty\} \times A_0 \in \mathcal{S}$ so that $B_0^c \in \mathcal{S}$ and let $\tau_n = S_n \wedge D_{B_0^c}$. Clearly τ_n is a stopping time of the filtration and $D_B \geq T \geq \tau_n$, a.e. It follows that $G(\tau_n) \subset B_n$, and $\tau_n \in \mathcal{T}$. Since T is the supremum of $\mathcal{T}$, we conclude that $\tau_n \uparrow T$ a.e. Moreover,

$$B = \cap_{n=1}^\infty B_n \supseteq \cap_{n=1}^\infty G(\tau_n) = \lim_n G(\tau_n) = G(T). \tag{74}$$

Since $G(\tau_n) \supset G(D_B), n \geq 1$, we deduce that $G(T) = G(D_B)$ and so $T = D_B$ a.e.

Finally choosing $\mathcal{C}$ to be $\mathcal{P}$ or $\mathcal{O}$ etc., we obtain from this that $T = D_B$ is a predictable or general etc., optional. □

Remarks. (a) If $\{X_t, \mathcal{F}_t, t \geq 0\}$ is a uniformly integrable right continuous martingale, so that $X_t \to X_\infty$ a.e. and in $L^1(P)$, implying $X_t = E^{\mathcal{F}_t}(X_\infty)$ a.e., let T be a stopping time of the filtration and consider $X \circ T (= X_T)$. Letting $T_\infty = \infty$ a.e. and $\mathcal{F}_\infty = \sigma(\cup_{t\geq 0}\mathcal{F}_t)$, we see from Theorem IV.1.7(ii) that $E^{\mathcal{B}(T)}(X_\infty) = X_T$, a.e. If moreover T is predictable, then we even have $E^{\mathcal{B}(T-)}(X_\infty) = X_{T-}$ a.e. In fact, by definition there is a strictly increasing (on $0 < T < \infty$) sequence $T_n \uparrow T$ of optionals of the filtration (by the predictability of T), such that $\{X_{T_n}, \mathcal{B}(T_n), n \geq 1\}$ is a martingale (cf. Corollary IV.1.12), and the X_t-process is of class (D) (as noted in Proposition IV.3.6). Hence $X_{T_n} = E^{\mathcal{B}(T_n)}(X_\infty)$ a.e. But $\lim_n \mathcal{B}(T_n) = \mathcal{B}(T-)$ and since $X_{T_n} \to \tilde{Y}$ a.e. and in $L^1(P)$, we have $E^{\mathcal{B}(T-)}(X_\infty) = \tilde{Y}$ a.e. So $\tilde{Y} = X_{T-}$ a.e. by the preceding result. From this and the fact that $\mathcal{B}(T-) \subset \mathcal{B}(T)$ we have for each predictable $T, E^{\mathcal{B}(T-)}(X_T) = E^{\mathcal{B}(T-)}(E^{\mathcal{B}(T)}(X_\infty)) = E^{\mathcal{B}(T-)}(X_\infty) = X_{T-}$ a.e., or $E^{\mathcal{B}(T-)}(X_T - X_{T-}) = 0$ a.e.

(b) Recall that for any A from Σ and a stopping time T of a filtration $\{\mathcal{F}_t, t \geq 0\}$ the restriction $T_A (= T\chi_A + \infty\chi_{A^c})$ is also a stopping time of the filtration iff $A \in \mathcal{B}(T)$. Indeed $[T \leq t] = A \cap [T \leq t] \in \mathcal{F}_t, t \geq 0$, iff A is an event prior to T. Then we may also describe $\mathcal{P}$ as the σ-algebra generated by$\{[\![0_A]\!], A \in \mathcal{F}_0\}$ and the intervals $(\!(S, T]\!]$ and the graphs $G(T)$. The proof is set theoretical, and is left to the reader, (cf. also Dellacherie [1]).

5.3 Square integrable martingales and stochastic integration

The preceding decomposition theory will be specialized to obtain new insight for square integrable martingales to be used in the ensuing integration theory. Thus if $\{X_t, \mathcal{F}_t, t \geq 0\}$ is an $L^2(P)$-bounded martingale, then $\{-X_t^2, \mathcal{F}_t, t \geq 0\}$ is a supermartingale and we can apply the results of the last section to this particular class. The work enables us to find its relations with Brownian motion. The main results here are essentially based on the work of Kunita and S. Watanabe [1]. The theory of this section is a foundation for many extensions and applications of martingale processes. Indeed they play an analogous role as the class of simple functions in Lebesgue integration. Thereafter, we use a localization technique with stopping times for generalizing the theory to larger families.

(a) *General structure.* Let $\{X_t, \mathcal{F}_t, t \geq 0\}$ be a right continuous $L^2(P)$-bounded martingale on (Ω, Σ, P), where we use a standard filtration from Σ. Then $\{X_t^2, \mathcal{F}_t, t \geq 0\}$ is a uniformly integrable submartingale, $X_t^2 \to X_\infty^2$ a.e. and in $L^1(P)$. If $Y_t = E^{\mathcal{F}_t}(X_\infty^2)(\geq X_t^2 a.e.)$ is the right continuous martingale (here as before we use the same symbols for modifications) and $Z_t = Y_t - X_t^2$, then $Z_t \to 0$ a.e. and in $L^1(P)$. Clearly $\{Z_t, \mathcal{F}_t, t \geq 0\}$ is a potential. With its uniform integrability, it is of class (D) by Proposition IV.3.6, so that by Theorem 2.5 there is a unique increasing, integrable, predictable process $\{A_t, t \geq 0\}$ of the filtration such that

$$Z_t = E^{\mathcal{F}_t}(A_\infty) - A_t, \qquad t \geq 0, \text{a.e.} \tag{1}$$

Hence one has

$$X_t^2 - A_t = Y_t - E^{\mathcal{F}_t}(A_\infty) = \tilde{Y}_t(\text{say}), \, t \geq 0, \tag{2}$$

where $\{\tilde{Y}_t, \mathcal{F}_t, t \geq 0\}$ is a right continuous uniformly integrable martingale. Consequently, for any optional process $\{T_j, j \in J\}$ of the filtration the family $\{\tilde{Y} \circ T_j, \mathcal{B}(T_j), j \in J\}$ is a martingale by Corollary IV.1.12. In particular, if $J = \{1, 2\}$, then writing $\mathcal{B}_j$ for $\mathcal{B}(T_j)$, we see that

$$\begin{aligned} E^{\mathcal{B}_1}(A \circ T_2 - A \circ T_1) &= E^{\mathcal{B}_1}[X^2 \circ T_2 - X^2 \circ T_1], \\ &\quad \text{since}\{\tilde{Y} \circ T_i, \mathcal{B}_i, i = 1,2\}\text{is a martingale}, \\ &= E^{\mathcal{B}_1}[(X \circ T_2 - X \circ T_1)^2], \text{a.e.}, \\ &\quad \text{since}\{X \circ T_i, \mathcal{B}_i, i = 1,2\}\text{is a martingale}. \end{aligned} \tag{3}$$

Thus defining $\langle X, X\rangle_t = A_t$ (or $\langle X\rangle_t = A_t$), (3) implies,

$$E^{\mathcal{F}_{t_1}}(X_{t_2} - X_{t_1})^2 = E^{\mathcal{F}_{t_1}}(\langle X\rangle_{t_2}) - \langle X\rangle_{t_1}, \text{a.e.} \, (t_1 \leq t_2). \tag{4}$$

Now Equation (4) reminds the polarization identity in a Hilbert space, if we interpret $E^{\mathcal{F}}$ as an "integral". Thus if $\{X_t, \mathcal{F}_t, t \geq 0\}$ and $\{Y_t, \mathcal{F}_t, t \geq 0\}$ are two right continuous $L^2(P)$-martingales, their sum and difference are again of the same kind, so that they have unique predictable increasing integrable processes $\{(\langle X \pm Y\rangle_t, \langle X\rangle_t, \langle Y\rangle_t, \mathcal{F}_t, t \geq 0\}$. Now if one defines $\langle X, Y\rangle_t$ by the equation:

$$\langle X, Y\rangle_t = \frac{1}{4}\{\langle X+Y\rangle_t - \langle X-Y\rangle_t\}, \qquad t \geq 0, \tag{5}$$

then (4) can be expressed as follows:

$$\begin{aligned} E^{\mathcal{F}_{t_1}}(X_{t_2}Y_{t_2} - X_{t_1}Y_{t_1}) &= E^{\mathcal{F}_{t_1}}[(X_{t_2} - X_{t_1})(Y_{t_2} - Y_{t_1})] \\ &= E^{\mathcal{F}_{t_1}}[\langle X, Y\rangle_{t_2}] - \langle X, Y\rangle_{t_1}, \text{a.e.} \end{aligned} \tag{6}$$

In fact, this follows from (4) and (5), if we set $a = X_{t_2} - X_{t_1}, b = Y_{t_2} - Y_{t_1}$ and use the identity $4ab = [(a+b)^2 - (a-b)^2]$. Also with the parallelogram law and the martingale property of the $X \pm Y$-processes one gets:

$$\begin{aligned} & E^{\mathcal{F}_{t_1}}[(\langle X+Y, X+Y\rangle_{t_2} - \langle X+Y, X+Y\rangle_{t_1}) \\ & \quad + (\langle X-Y, X-Y\rangle_{t_2} - \langle X-Y, X-Y\rangle_{t_1})] \\ & \quad = 2E^{\mathcal{F}_{t_1}}[(\langle X, X\rangle_{t_2} - \langle X, X\rangle_{t_1}) + (\langle Y, Y\rangle_{t_2} - \langle Y, Y\rangle_{t_1})]. \end{aligned} \tag{7}$$

If we define $\|X\|^2_{t_2,t_1} = E^{\mathcal{F}_{t_1}}(\langle X, X\rangle_{t_2} - \langle X, X\rangle_{t_1}) \geq 0$ a.e., then for each $0 \leq t_1 \leq t_2$, it follows that $\|aX\| = |a|\|X\|_{t_2,t_1}$, and from (4),(6), and (7), that

$$\|X\|_{t_2,t_1} + \|Y\|_{t_2,t_1} \geq \|X+Y\|_{t_2,t_1}, \text{a.e.} \tag{8}$$

Thus $\|\cdot\|_{t_2,t_1}$ is a semi-norm and (7) can be written as:

$$\|X+Y\|^2_{t_2,t_1} + \|X-Y\|^2_{t_2,t_1} = 2(\|X\|^2_{t_2,t_1} + \|Y\|^2_{t_2,t_1}) \text{ a.e.} \tag{9}$$

This identity (cf. (5)) implies that $\langle X, X\rangle_{t,s} = \|X\|^2_{t,s}$ for $t \geq s \geq 0$ and hence that

$$2\langle X, Y\rangle^{\tilde{}}_t = \langle X+Y, X+Y\rangle_t - \langle X, X\rangle_t - \langle Y, Y\rangle_t, \text{a.e.} \tag{10}$$

Both (5) and (10) determine the same semi-definite sesquilinear functional $\langle X, Y\rangle_{t,s}, t \geq s \geq 0$, for all $L^2(P)$-martingales X, Y satisfying (4). Further the classical Hilbert space theory implies that $\langle X, Y\rangle$ is uniquely determined by X and Y. Also (6) yields that

$$E^{\mathcal{F}_{t_1}}(\langle \varphi X, Y\rangle_{t_2} - \langle \varphi X, Y\rangle_{t_1}) = \varphi(E^{\mathcal{F}_{t_1}}(\langle X, Y\rangle_{t_2} - \langle X, Y\rangle_{t_1})), \text{a.e.,} \tag{11}$$

for $t_2 \geq t_1 \geq 0$ and bounded $\mathcal{F}_{t_1}$-measurable φ.

Let $\mathcal{M}$ denote the class of all right continuous $L^2(P)$-martingales (not necessarily norm bounded)relative to a fixed standard filtration $\{\mathcal{F}_t, t \geq 0\}$. Let $\mathcal{M}_{loc} = \{\{X_t, \mathcal{F}_t, t \geq 0\}$: for some sequence$T_n \uparrow \infty$ of stopping times of the filtration, and$\{X^{(n)}_t = X(T_n \wedge t), \mathcal{F}_t, t \geq 0\} \in \mathcal{M}$, for each $n \geq 1\}$. Let $\mathcal{M}^c$ and $\mathcal{M}^c_{loc}$ be the sets of (a.e.) continuous elements of $\mathcal{M}$ and $\mathcal{M}_{loc}$ respectively. Also let $\mathcal{A}^+ = \{\{A_t, \mathcal{F}_t, t \geq$

$0\}$: $A_t$is a predictable increasing process with$E(A_t) < \infty$for each$t\}$. Similarly as in the martingale case let $\mathcal{A}^+_{loc}$ be defined, and set $\mathcal{A} = \mathcal{A}^+ - \mathcal{A}^+$, the vector difference, so that it is the set of differences of elements of $\mathcal{A}^+$. In an analogous manner define the sets of continuous elements of these classes and denote them as $\mathcal{A}^c, \mathcal{A}^c_{loc}, (\mathcal{A}^+)^c, (\mathcal{A}^+_{loc})^c$. Note that for $A \in \mathcal{A}, A_0 = 0$ a.e. from the definition of predictability, and also $\mathcal{F}_0$ is often taken as the trivial σ-algebra for simplicity. When this is not needed it will be assumed that $\mathcal{F}_{0-} = \mathcal{F}_0$ and that the latter contains all P-null sets.

Remark. The elements of $\mathcal{M}_{loc}$ are not local martingales in the sense of Definition 1.2 since we are not assuming that the processes $\{X_t^{(n)}, \mathcal{F}_t, t \geq 0\}$ are also uniformly integrable for each n. Note however that $\{X_t, t \in [0,a]\}, a \in \mathbb{R}^+$, is uniformly integrable and hence $\{X_t, t \in \mathbb{R}^+\}$ is in class (DL). So each member of $\mathcal{M}_{loc}$ may be said to be "locally in class (DL)". If predictability of the increasing process $\langle X, X\rangle_t$ is not demanded, then another (more general) process, called the quadratic (co-)variation process can be defined in lieu of the above procedure. It and its relation with $\langle X, X\rangle$ will be explained at the end of this subsection.

We begin the analysis with the observation:

1. Proposition. *For each pair X, Y from $\mathcal{M}$ there is an a.e. unique $\langle X, Y\rangle \in \mathcal{A}$ such that (6) holds for each $t \geq s \geq 0$. More generally, if X, Y are in $\mathcal{M}_{loc}$, then there exists an a.e. unique $\langle X, Y\rangle \in \mathcal{A}_{loc}$ such that the following equations hold a.e.:*

$$E^{\mathcal{F}_s}[X_t^{(n)}Y_t^{(n)} - X_s^{(n)}Y_s^{(n)}] = E^{\mathcal{F}_s}[\langle X, Y\rangle(T_n \wedge t) - \langle X, Y\rangle(T_n \wedge s)]$$
$$E^{\mathcal{F}_s}(\langle \chi_A, Y\rangle_t) - \langle \chi_A X, Y\rangle_s = \chi_A[E^{\mathcal{F}_s}(\langle X, Y\rangle_t) - \langle X, Y\rangle_s], A \in \mathcal{F}_s. \tag{12}$$

Proof. By the above remark, $\{X_t^2, \mathcal{F}_t, t \geq 0\}$ is in class (DL). Hence by Theorem 2.8, the computations (4)–(10) above prove the first part as well as the last equation.

Regarding the more general case, let X, Y be in $\mathcal{M}_{loc}$. Then there is a sequence $T_n \uparrow \infty$ of stopping times of the filtration such that $X_t^{(n)} = X(T_n \wedge t)$ and $Y_t^{(n)} = Y(T_n \wedge t)$ define elements of $\mathcal{M}$ for each $n \geq 1$ where we again write $X_t = X(t)$ etc. Hence by the preceding case, there exist $\langle X^{(n)}, Y^{(n)}\rangle \in \mathcal{A}, n \geq 1$, a.e. unique, such that for $n \geq m$,

$$\langle X^{(m)}, Y^{(m)}\rangle_t = \langle X^{(n)}, Y^{(n)}\rangle(T_m \wedge t), \qquad t \geq 0, \text{ a.e.}, \qquad (13)$$

since $X_t^{(m)} = X(T_m \wedge t) = X(T_n \wedge T_m \wedge t) = X^{(n)}(T_m \wedge t), t \geq 0, (T_m \leq T_n)$. So if the process $\{A_t, \mathcal{F}_t, t \geq 0\}$ is obtained by setting $A(T_n \wedge t) = \langle X^{(n)}, Y^{(n)}\rangle_t$, then (13) implies that A is unambiguously defined and is unique. Since the right side $\langle X^{(n)}, Y^{(n)}\rangle \in \mathcal{A}$ so that it is predictable, it follows as in the proof of Theorem 2.8 that $A \in \mathcal{A}_{loc}$. □

Regarding the process $\langle X, Y\rangle$, we record the following.

2. Proposition. *Let X, Y be in $\mathcal{M}$. Then $\langle X, Y\rangle \in \mathcal{A}^c$ in each of the following cases: (a)$\{X, Y\} \subset \mathcal{M}^c$, (b) the family $\{\mathcal{F}_t, t \geq 0\}$ has no times of discontinuities, (c) X, Y are quasi-left-continuous. In fact, $\langle X, X\rangle \in (\mathcal{A}^+)^c$ iff X is quasi-left-continuous.*

Proof. Since continuity is a stronger condition than quasi-left-continuity, (a) implies (c). Regarding (b), for each optional sequence $T_n \uparrow T$ of the filtration, by definition, $\mathcal{F}(T_n \wedge t) \uparrow \mathcal{F}(T \wedge t)$ for each $t \geq 0, t \in \mathbb{R}^+$; so the set $\{X(T \wedge t) : T \text{ any optional}\}$ is uniformly integrable since X is in class (DL). Thus $\{X(T_n \wedge t), \mathcal{F}(T_n \wedge t), n \geq 1\}$ is a uniformly integrable martingale, and even $\{X_s, 0 \leq s \leq t\}$ is uniformly integrable. Then $E^{\mathcal{F}(T_n \wedge t)}(X_t) = X(T_n \wedge t)$, and $X(T_n \wedge t) \to E^{\mathcal{F}(T \wedge t)}(X_t) = X(T \wedge t)$, a.e., (cf. IV.1.8 and II.6.2). Since $t \in \mathbb{R}^+$ is arbitrary, we conclude that $X(T_n) \to X(T)$ a.e. on $[T < \infty]$, and this shows that X is quasi-left-continuous. Thus (c) is implied, and it remains to establish (c). By (10) this will follow, if we prove the result with $X = Y$, and this is done now.

Since the process $\{X_t^2, \mathcal{F}_t, t \geq 0\}$ is of class (DL), we can and do use the same trick as in the proof of Theorem 2.6 and reduce the result to class (D) processes. Thus $X = X' + A$ where the martingale X' is also of class (D). But for each class (D) martingale Y, we have $Y \circ T_n \to Y \circ T$ a.e., and in $L^1(P)$ for any optional sequence $T_n \uparrow T$ of the filtration. Thus by the Riesz decomposition $X_t^2 = Y_t + Z_t$ with Y as a class (D) martingale and Z as a class (D) potential. Consequently Theorem 2.15 implies that the A_t-process $(=\langle X, X\rangle)$ which generates the Z_t-process, is continuous iff for each optional sequence $T_n \uparrow T$ one has

$E(X^2 \circ T) \to E(X^2 \circ T)$. Now X^2 is in class (D) implies (by the CBS inequality) that X is in class (D), so $X \circ T_n \to X \circ T$ a.e. and in $L^2(P)$. But $E(X \circ T_n - X \circ T)^2 = E(X^2 \circ T_n) - E(X^2 \circ T)$. So $X \circ T_n \to X \circ T$ in $L^2(P)$ iff $X^2 \circ T_n \to X^2 \circ T$ in $L^1(P)$, or iff X is quasi-left-continuous. □

Remark. We always have $\mathcal{M} \cap \mathcal{A} = \{0\}$. For, let X be as in this intersection. Then there exist $A^i \in \mathcal{A}^+$ such that $X_t = A_t^1 - A_t^2, t \geq 0$,and the X_t-process is a martingale. Hence $E(X_t) = E(A_t^1) - E(A_t^2)$ is a constant. Since $A_0^1 = 0 = A_0^2$ a.e., we see that $E(X_t) = 0$. However the predictability of the A_t^i implies $E(YA_t^1) = E(YA_t^2), t \geq 0$ for all measurable $Y \geq 0$ (cf. part (ii) of the proof of Theorem 2.2). Hence $A_t^1 = A_t^2$ a.e. and so $X_t = 0$ a.e., $t \geq 0$, or $X \equiv 0$. A similar reasoning shows that $\mathcal{M}_{loc} \cap \mathcal{A}_{loc} = \{0\}$ also.

The special structure of $\langle X, Y \rangle \in \mathcal{A}$ for $X, Y \in \mathcal{M}$ (especially the form (5) or (10) based on the square integrability of these processes) will be useful in introducing a general stochastic integral towards which we now proceed.

Let $A \in \mathcal{A}$ so that $A = A^1 - A^2, A^i \in \mathcal{A}, i = 1, 2$. Denote by $|A|_t = A_t^1 + A_t^2$, so that (as a pointwise Stieltjes integral)

$$|A|_t = \int_0^t dA_s^1 + \int_0^t dA_s^2 = \int_0^t d|A_s|. \tag{14}$$

Recall that for any optional T of the filtration $\{\mathcal{F}_t, t \geq 0\}$, $\mathcal{F}_T$ is the σ-algebra of events "prior to T"; thus it is determined by $A \cap [T \leq t] \in \mathcal{F}_t, t \geq 0$, and $A \in \mathcal{F}_\infty = \sigma(\cup_t \mathcal{F}_t)$. Similarly $\mathcal{F}_{T-}$ is the σ-algebra of events "strictly prior to T," i.e.,

$$\mathcal{F}_{T-} = \sigma\{\mathcal{F}_0, A \cap [T > t] : A \in \mathcal{F}_t, t \geq 0\}. \tag{15}$$

This notion is due to Chung and Doob [1]. It is seen that $\mathcal{F}_{T-} \subset \mathcal{F}_T$ and since $[T > t] \in \mathcal{F}_{T-}, T$ is $\mathcal{F}_T$-measurable. Similarly if $T_n \uparrow T$ is a family of optionals of the filtration, then $\mathcal{F}_{T-} = \sigma(\cup_n \mathcal{F}_{T_n-})$, since clearly $\mathcal{F}_{T_n-} \subset \mathcal{F}_{T_{n+1}-}$. Let $\mathcal{C}$ be the class of measurable processes $\{X_t, \mathcal{F}_t, t \geq 0\}$ for $\mathcal{B}(\mathbb{R}^+) \otimes \Sigma$ such that for each optional $T < \infty$ of the filtration, $X \circ T$ is $\mathcal{F}_{T-}$-adapted. The class $\mathcal{C}$ plays a key role below. Note that by the work in Section III.3 every right (left) continuous process is measurable and has a [separable and] progressively measurable modification. If $\{X_t, \mathcal{F}_t, t \geq 0\}$ is such a process, then $X \circ T$ is $\mathcal{F}_{T-}$-adapted for each finite optional T. Since $\mathcal{F}_{T-} \subset \mathcal{F}_T$, with possibly a strict inclusion, not all progressively measurable processes belong to $\mathcal{C}$. However, we shall find that every left continuous process X as above has a version which belongs to $\mathcal{C}$.

For any $A \in \mathcal{A}$, define $\mathcal{L}^1(A) = \{X \in \mathcal{C} : E(\int_0^t |X_s|\, d|A_s|) < \infty, t \geq 0\}$. If $X \in \mathcal{C}$ and Y is given by $Y_t = (X \cdot A)_t = \int_0^t X_s\, dA_s$ such that $E(\int_0^t |X_s|\, d|A_s|) < \infty, t \geq 0$, then it is seen that $Y \in \mathcal{L}^1(A)$. Moreover, from $X_t = X_t^+ - X_t^-, A_t = A_t^1 - A_t^2, \{A_t^i, t \geq 0\} \in \mathcal{A}^+$ (and $X^\pm \in \mathcal{C}$), we deduce that (using (1.7)) each of the integrals below is $\mathcal{F}_{T-}$-adapted for an optional T of the filtration. Thus

$$Y_t = \int_0^t X_s^+\, dA_s^1 + \int_0^t X_s^-\, dA_s - \int_0^t X_s^+\, dA_s^2 - \int_0^t X_s^-\, dA_s^1, \quad (16)$$

has each of its terms to be integrable increasing predictable processes, so that $Y \in \mathcal{A}$. This implies that $\mathcal{L}^1(A) \subset \mathcal{A}$. We may topologize $\mathcal{L}^p(A) = \{X \in \mathcal{C} : |X|^p = \{|X_t|^p, t \geq 0\} \in \mathcal{L}^1(A)\}, p \geq 1$, by the family of semi-norms:

$$\|X\|_{p,A(t)} = \left[E\left(\int_0^t |X_s|^p\, d|A_s|\right)\right]^{1/p}, \qquad t \geq 0, \quad (17)$$

and simply write $\|X\|_{2,A(t)}$ as $\|X\|_{A(t)}$. We develop a calculus starting with the simpler integral given for Definition 1.3, and then obtain the generalized Itô integral from this.

It is seen from (16) that $Y \in \mathcal{A}$. Since it is an indefinite (Lebesgue type) integral of A, we should be able to prove an analog of the Radon-Nikodým theorem for these "measures". We show that this is possible and then use it in the study of stochastic integrals.

If A, B are two increasing processes adapted to $\{\mathcal{F}_t, t \geq 0\}$, let $\mu_\omega, \tilde{\mu}_\omega$ be the associated measures, and $\nu, \tilde{\nu}$ the corresponding induced measures on $\mathcal{U}$, the σ-algebra of $\mathbb{R}^+ \times \Omega = U$, as in the preceding section. Thus for $C \times F \in \mathcal{U}$,

$$\nu(C \times F) = \int_F \mu_\omega(C)\, dP(\omega), \quad \tilde{\nu}(C \times F) = \int_F \tilde{\mu}_\omega(C)\, dP(\omega), \quad (18)$$

where $\mu_\omega([a, b)) = A_b(\omega) - A_a(\omega)$ and similarly $\tilde{\mu}_\omega$ is defined with the B_t-process, $0 \leq a < b < \infty$. Then B is said to be *absolutely continuous relative to* A iff $\tilde{\nu} \ll \nu$. The measures $\nu, \tilde{\nu}$ are σ-finite on U and finite iff $\sup_t E(A_t) < \infty, \sup_t E(B_t) < \infty$. We can now state:

3. Proposition. *Let* $\{A, B\} \subset \mathcal{A}^+$, *and suppose* B *is absolutely continuous for* A *as above. Then there is a positive process* $X \in \mathcal{L}^1(A)$ *such that, for* $t \geq 0, B_t = \int_0^t X_s dA_s$, *a.e., or equivalently,*

$$E\Big(\int_0^t Y_s\, dB_s\Big) = E\Big(\int_0^t Y_s X_s\, dA_s\Big), \qquad t \geq 0, \tag{19}$$

for any bounded progressively measurable $Y = \{Y_t, \mathcal{F}_t, t \geq 0\}$.

Proof. By definition of absolute continuity of $\tilde{\nu}$ and ν (induced by B and A) on $\mathcal{U}$, there exists a ν-unique $\mathcal{U}$-measurable function $X : U \to \bar{\mathbb{R}}^+$ such that (writing $X_t(\omega)$ for $X(t, \omega)$)

$$\tilde{\nu}(C \times F) = \int_{C \times F} X_t(\omega)\, d\nu(t, \omega), \qquad C \in \mathcal{B}(\mathbb{R}^+), F \in \Sigma. \tag{20}$$

Then $t \mapsto X_t$ is a measurable process, and has a progressively measurable modification $\hat{X}$. Indeed, if $\mathcal{U}_1 \subset \mathcal{U}$ is the σ-algebra relative to which each left continuous process is measurable, then the predictability of A, B and the right continuity of the filtration imply that $\nu, \tilde{\nu}$ are σ-finite on $\mathcal{U}_1$. Hence X is measurable relative to $\mathcal{U}_1$. Taking the right continuous modification $\hat{X}$, we see that $\hat{X} \in \mathcal{C}$. Thus

$$\begin{aligned}\int_F \tilde{\mu}_\omega(C)\, dP(\omega) = \tilde{\nu}(C \times F) &= \int_{C \times F} \hat{X}_t(\omega)\, d\nu(t, \omega) \\ &= \int_{C \times F} \hat{X}_t(\omega) \mu_\omega(C)\, dP(\omega).\end{aligned}$$

Since the set $C = [a, b]$ is a generator of $\mathcal{B}(\mathbb{R}^+)$, using Fubini's theorem we get:

$$\begin{aligned}\int_F \int_0^b dB_s(\omega)\, dP(\omega) &= \int_F \hat{X}_t(\omega) \int_0^b dA_s(\omega)\, dP(\omega) \\ &= \int_F \Big(\int_0^b \hat{X}_s(\omega)\, dA_s(\omega)\Big) dP(\omega).\end{aligned} \tag{21}$$

But the integrand is measurable for (Σ) and $F \in \Sigma$ is arbitrary. So (21) implies that:

$$B_t = \int_0^t dB_s = \int_0^t \hat{X}_s \, dA_s, \text{ a.e.}$$

Now for each relatively compact set $C, \tilde{\nu}(C \times F) < \infty$, and also $\{\hat{X}_s, s \geq 0\} \in \mathcal{L}^1(A)$. From this (19) follows. □

If $\{A, B\} \subset \mathcal{A}$, then $A = A^1 - A^2, B = B^1 - B^2$ for some $\{A^i, B^i\} \subset \mathcal{A}^+, i = 1, 2$. Hence we can deduce the general case from the positive one with the argument of (16) to conclude:

4. Corollary. *If $\{A, B\} \subset \mathcal{A}$ and B is absolutely continuous relative to A (i.e., for $|A|$), then there is a process $X \in \mathcal{L}^1(A)$ such that*

$$B_t = \int_0^t X_s \, dA_s, \text{ a.e.}, \qquad t \geq 0. \tag{22}$$

Using the above result, we can prove a CBS-type inequality:

5. Proposition. *Let $\{X, Y\} \subset \mathcal{M}$, and $A = \langle X \rangle, B = \langle Y \rangle$ be the corresponding elements of $\mathcal{A}^+$. Then for any $f \in \mathcal{L}^2(\langle X \rangle), g \in \mathcal{L}^2(\langle Y \rangle)$, the integral $\int_0^t f_s g_s d\langle X, Y \rangle_s$ defines an element of $\mathcal{A}$ and moreover*

$$E\left(\left| \int_0^t (f_s g_s) d\langle X, Y \rangle_s \right|\right) \leq \|f\|_{\langle X \rangle_t} \|g\|_{\langle Y \rangle_t}, \qquad t \geq 0. \tag{23}$$

Proof. Recall that $\langle X \rangle = \langle X, X \rangle$ so that it is in $\mathcal{A}^+$. For any real a, b, consider $\langle aX + bY \rangle$, which is an increasing process. Then

$$\langle aX + bY \rangle_t = a^2 \langle X, X \rangle_t + 2ab \langle X, Y \rangle_t + b^2 \langle Y, Y \rangle_t, \ t \geq 0, \text{a.e.} \tag{24}$$

But from Hilbert space theory we have $|\langle X, Y \rangle| \leq \langle X \rangle^{1/2} \langle Y \rangle^{1/2}$, a.e., (cf. (5)) and $\{\langle X \rangle, \langle Y \rangle\} \subset \mathcal{A}^+$, so that the process $\langle X, Y \rangle$ is absolutely continuous relative to both $\langle X \rangle$ and $\langle Y \rangle$. Hence the same holds relative to their sum, say, $\langle h\rangle$. By Corollary 4, there exist processes $Z^{(i)} \in \mathcal{L}^1(\langle h\rangle)$ such that $(i = 1,2,3, t \geq 0)$

$$\langle X\rangle_t = \int_0^t Z_s^{(1)} d\langle h\rangle_s, \langle Y\rangle_t = \int_0^t Z_s^{(3)} d\langle h\rangle_s,$$

$$\langle X, Y\rangle_t = \int_0^t Z_s^{(2)} d\langle h\rangle_s. \tag{25}$$

Hence (24) and (25) yield

$$\int_0^t (a^2 Z_s^{(1)} + 2ab Z_s^{(2)} + b^2 Z_s^{(3)})\, d\langle h\rangle_s \geq 0, \qquad t \geq 0, \text{ a.e.}$$

From this we conclude that the integrand is nonnegative a.e. $(d\mu^h \otimes dP)$, where $\mu^h([a_1, b_1)) = \langle h\rangle_{b_1} - \langle h\rangle_{a_1}$. Let $N_{rs} \subset U = \mathbb{R}^+ \times \Omega$ be the $d\mu^h \otimes dP$-null set outside of which the above result holds. If $N = \cup\{N_{rs} : r, s \text{ rational in } \mathbb{R}\}, (s, \omega) \in N^c = U - N$ and $a = uf(\omega), b = vg(\omega), \{u, v\} \subset \mathbb{R}$, where we take $v = k(\omega_0), \omega_0 \in \Omega$, k being a random variable, then

$$u^2 f_s^2(\omega) Z_s^{(1)}(\omega) + 2uv f_s(\omega) g_s(\omega) Z_s^{(2)}(\omega) + v^2 g_s^2(\omega) Z_s^{(3)}(\omega) \geq 0.$$

Integrating this relative to the measure $d\langle h\rangle$ on $[0, t)$ we get on using (25):

$$u^2 \int_0^t f_s^2 d\langle X\rangle_s + 2uk \int_0^t f_s g_s d\langle X, Y\rangle_s + k^2 \int_0^t g_s^2 d\langle Y\rangle_s \geq 0 \text{ a.e.} \tag{26}$$

Now let $k = \text{sgn}(\int_0^t f_s g_s d\langle X, Y\rangle)$ so $k^2 = 1$. The resulting expression in (26) holds a.e. iff the quadratic form in u has no real roots. This implies (23) at once. □

Remark. The difference between (23) and the classical CBS iequality is that we have a change of measures problem here. So the additional argument through a use of Proposition 3 is needed. This technique is, however, familiar in classical interpolation theory with change of measures (cf. e.g., the author [8]). The inequality was proved and used by Kunita and S. Watanabe [1] for this purpose.

It was noted at the end of Section IV.5, especially in Proposition IV.5.9, that for a right continuous martingale $X = \{X_t, \mathcal{F}_t, t \in I \subset \mathbb{R}^+\}$, if $\pi_j : t_{j1} < \ldots < t_{jm_j}$ are partitions of I ordered by refinement, and $Q_j^2 = \sum_{i=1}^{m_j}(X_{t_{ji}} - X_{t_{j(i-1)}})^2$, $(X_{t_{j0}} = 0$ a.e.), then the *quadratic variation* $Q = \lim_j Q_j$, of the process, if it exists in $L^1(P)$ as the π_j are refined, contains useful information on X. Moreover we saw that its discrete version is the "Luzin s-function", given as Extension IV.4.10 (due to Austin [1]). The resulting (discrete) analysis was the subject of Section IV.5. In the continuous parameter case, if the martingale is square integrable, then not only can we assert that the $L^1(P)$-limit of Q_j exists, but Q is closely related to $\langle X \rangle$, and in fact equals a.e. with the latter if the process has continuous sample paths. This also implies that we can define the *covariation* between two such processes X, Y just as for $\langle X, Y \rangle$, denoting Q as $[X, X]$ and then using the polarization identity to get $[X, Y]$ with $I = \mathbb{R}^+$. We now set down the precise formulation and then use it later in applications.

6. Theorem. *Let $X = \{X_t, \mathcal{F}_t, t \geq 0\}$ be a right continuous square integrable martingale on (Ω, Σ, P). Then for each $t > 0$, the* **quadratic variation** *of X on $[0, t]$, denoted $[X, X]_t$, defined by*

$$[X, X]_t = \lim_{n \to \infty} \sum_{k \geq 0} (X_{\frac{k+1}{2^n} \wedge t} - X_{\frac{k}{2^n} \wedge t})^2, \tag{27}$$

exists in $L^1(P)$, and $\{X_t^2 - X_0^2 - [X, X]_t, \mathcal{F}_t, t \geq 0\}$ is a right continuous martingale. Moreover, if a.a. the sample functions of X are continuous, then $[X, X]_t$ and $\langle X, X \rangle_t$ agree for all $t \geq 0$.

Proof. For any $t > 0$, fixed, consider a partition $\Pi_n : \{\frac{k}{2^n}, k = 0, \ldots, n\}$ of $[0, t]$, and let $I_{nj} = (\frac{j}{2^n} \wedge t, \frac{j+1}{2^n} \wedge t]$. Then

$$\begin{aligned} X_t^2 - X_0^2 &= \sum_{j \in \Pi_n} (X^2_{\frac{j+1}{2^n} \wedge t} - X^2_{\frac{j}{2^n} \wedge t}) \\ &= \sum_{j \in \Pi_n} [(X_{\frac{j+1}{2^n} \wedge t} - X_{\frac{j}{2^n} \wedge t})^2 + 2X_{\frac{j}{2^n} \wedge t}(X_{\frac{j+1}{2^n} \wedge t} - X_{\frac{j}{2^n} \wedge t})] \\ &= \sum_{j \in \Pi_n} [(Z(I_{nj}))^2 + 2X_{\frac{j}{2^n} \wedge t} Z(I_{nj})], \end{aligned}$$

where $Z(I) = X_b - X_a$ for $I = (a, b]$. Letting $f_n = \sum_{j \in \Pi_n} X_{\frac{j}{2^n} \wedge t} \chi_{I_{nj}}$, the simple function measurable relative to $\mathcal{F}_{\frac{j}{2^n} \wedge t} \otimes \mathcal{B}([0, t])$, we can write the above expression as

$$\sum_{j \in \Pi_n} (Z(I_{nj}))^2 = X_t^2 - X_0^2 - 2\int_0^t f_n(s)\, dZ(s). \tag{28}$$

The limit of the last expression, if it exists in $L^1(P)$, is precisely $[X,X]_t$, and this is equivalent to showing that the elementary integral term on the right of (28) has such a limit. We show that the latter exists and the other conclusions of the theorem then follow.

First consider $g_n = \sum_{i=0}^n a_i \chi_{(t_i,t_{i+1}]\times A_i}$, where $A_i \in \mathcal{F}_{t_i}, 0 \le t_1 < \ldots < t_{n+1} \le t, a_i$ is $\mathcal{F}_{t_i}$-adapted and bounded, and $Z(I) = X_b - X_a$, which defines an additive function on the semi-ring of such intervals $I = (a,b]$, as before with martingale differences. Then $E(Z(I)) = 0$ and $E(Z(I)^2) = E(X_b^2) + E(X_a^2) - 2E(X_a E^{\mathcal{F}_a}(X_b)) = E(X_b - X_a)^2$, using the martingale property. Hence if $J(g_n) = \int_0^t g_n\, dZ$ which is a finite sum and which is well-defined (does not depend on the partition points), we have

$$\begin{aligned} E(J(g_n)^2) &= \sum_{j=1}^n E\Big(a_j^2 \chi_{A_j}(X_{t_{j+1}}^2 - X_{t_j}^2)\Big) + 0, \\ &= \int_{[0,t]\times\Omega} |g_n|^2\, d\mu^x. \end{aligned}$$

Here the product term vanishes by the martingale property, and μ^x is the Doléan-Dade measure associated with the positive right continuous class (DL) submartingale $\{X_t^2, \mathcal{F}_t, t \ge 0\}$, (cf. Theorem 2.20). Also μ^x is finite on $[0,t]\times\Omega$ for each $t > 0$. Hence replacing g_n by f_n^m, which has the same expression as f_n but with X_t truncated at m so that it is a bounded $\mathcal{F}_t$-measurable function for which the above result is applicable, we see that the following limit exists:

$$\lim_{n\to\infty} E(J(f_n^k)^2) = \int_{[0,t]\times\Omega} |f^k(s-)|^2\, d\mu^x.$$

The same result shows that

$$E(J(f_n^k - f_m^k)^2) = \int_{[0,t]\times\Omega} |f_n^k - f_m^k|^2(s,\omega)\, d\mu^x \to 0,$$

as $n, m \to \infty$. Hence for each $k > 0$, $J(f_n^k(s)) \to J(f^k(s-))$ in $L^1(P)$. But by the submartingale inequality (cf. the proof of Theorem III.5.1), we get $P[\sup_{0\leq s\leq t} X_s^2 > k] \leq \frac{1}{k}E(X_t^2) \to 0$ as $k \to \infty$. Consequently we can let $k \to \infty$ in the above and conclude that $J(f^k(s-)) \to J(f(s-))$ a.e. and in $L^1(P)$. It follows from this that the limit in (28) exists in $L^1(P)$ and hence (27) is valid. We also note that $J(f_n^k)$ can be interpreted as a predictable transform of the martingale X (cf. the discussion following IV.4.12), and therefore it is a martingale. Thus the limit $J(f)$ of $J(f_n^k)$ is also a martingale. It follows from (27) that $\{X_t^2 - X_0^2 - [X,X]_t, \mathcal{F}_t, t \geq 0\}$ is a right continuous martingale.

Finally, if a.a. the sample paths of the X-process are continuous, then on each compact interval $[0,t]$, the convergence considered above is uniform so that $[X,X]_t$- process also has continuous sample paths, whence it is predictable. Then we have the Doob-Meyer decomposition of the submartingale as:

$$X_t^2 - X_0^2 = Y_t + \langle X, X\rangle_t = \tilde{Y}_t + [X,X]_t,\ t \geq 0.$$

Since both $\langle X,X\rangle_t, [X,X]_t$ are predictable, by the uniqueness of the decomposition $Y_t = \tilde{Y}_t$, and $\langle X,X\rangle_t = [X,X]_t$ outside of a P-null set for all $t \geq 0$, i.e., except for an evanescent set these processes agree. □

7. Remarks. (i) In the above proof, we could have replaced in f_n^k a stopped martingale $X \circ T_m$ instead of a truncated one, and obtained the same conclusion. This type of "localization" argument will be used in future calculations. Also it is necessary to employ a detailed analysis in discussing the boundedness of the integral of the type appearing in (28) for the ensuing work. A generalization of this idea will be considered in the next chapter. [See Proposition VI.2.12.]

(ii) In addition to Example 2.16, here is another instance of two decompositions of the submartingale X^2 having predictable and not necessarily having that property. But $\{[X,X]_t - \langle X,X\rangle_t, \mathcal{F}_t, t \geq 0\}$ is a right continuous martingale, which is the difference of two integrable increasing processes of which the subtracted one is predictable. The latter is sometimes called the *compensator* of the former to obtain a martingale. This difference often consists of certain jumps. Then the $[X,X]_t$-process is also called a (nonpredictable or) *raw* increasing process. It will be seen later that the square integrable martingales of the type considered admit decompositions into continuous and discrete parts. For this we need to develop the relevant integration theory to which we turn next.

(iii) Note also that in all cases, by Corollary 2.21, both $[X,X]_t$, $\langle X,X\rangle_t$-processes determine the same Doléans-Dade measure μ^x on $\mathcal{U}$ of $U = \mathbb{R}^+ \times \Omega$, since their difference determines a martingale whose measure is supported by $\{\infty\} \times \Omega$, and the process vanishes on this set.

We shall discuss later a close relation between a stochastic integral of one of the two types and the classical Bochner integral, after presenting a general boundedness principle for stochastic integrals (cf. Section VI.2). This will be appreciated only after considering both the types of stochastic integrals to be treated below.

(b) Stochastic integration. There are two types of stochastic integrals, often used in applications. The first one may be termed a Wiener type and the second one the Itô type, but both will be considered in a generalized form. The former, cronologically also the earliest one, deals with integration of nonstochastic integrands and stochastic integrators. The second one allows both integrands and integrators to be stochastic. We discuss both types here and then abstract the methods for a unified treatment in the next chapter. However, the second one does not completely contain the first type.

As seen in the proof of Theorem 6 above, a set function Z induced by a process X_t, by the formula $Z(I) = X_b - X_a$ for $I = (a,b]$, is (only) finitely additive on the semi-ring of such intervals of $\mathbb{R}^+$, but has a σ-additive extension to the Borel σ-algebra $\mathcal{B}(\mathbb{R}^+)$, if the process is a right continuous martingale. This is true for a Brownian motion as it satisfies the stated condition. More generally, if $Z : \mathcal{B}(\mathbb{R}^+) \to L^2(P)$ is found to be σ-additive (and there exist processes which need not be martingales but which satisfy this weaker hypothesis), we can use the Dunford-Schwartz integration of scalar functions relative to such Z, as discussed in Section II.3 for conditional measures. In the present general context, we again recall it for convenience as follows.Thus a σ-additive function Z is called a vector measure, if $Z(\cup_{i=1}^{\infty} A_i) = \sum_{i=1}^{\infty} Z(A_i)$, for disjoint $A_i \in \mathcal{B}(\mathbb{R}^+)$, the series converging in norm. Such a Z always has a finite semivariation $\|Z\|(\cdot)$, the latter being σ-subadditive and $\|Z\|(\mathbb{R}^+) < \infty$. A measurable $f : \mathbb{R}^+ \to \mathbb{R}(\mathbb{C})$ is D-S integrable relative to Z, if :(i) there exists a sequence of simple functions $f_n = \sum_{i=1}^{k_n} a_{ni}\chi_{A_{ni}}, A_{ni} \in \mathcal{B}(\mathbb{R}^+)$,disjoint, $f_n \to f$ except for a set $A_0, \|Z\|(A_0) = 0$, and (ii)$\{\int_{\mathbb{R}^+} f_n\, dZ = \sum_{i=1}^{k_n} a_{ni}Z(A_{ni}), n \geq 1\}$ forms a Cauchy sequence, whose limit is denoted by $\int_{\mathbb{R}^+} f\, dZ (= \lim_{n\to\infty} \int_{\mathbb{R}^+} f_n\, dZ)$. It is then verified that the D-S integral is well-defined, $B_0 = \{\omega : f(\omega) \neq g(\omega)\}$, satisfying $\|Z\|(B_0) = 0 \implies \int_{\mathbb{R}^+} f\, dZ = \int_{\mathbb{R}^+} g\, dZ$, and $f \mapsto \int_{\mathbb{R}^+} f\, dZ$ is a continuous linear mapping from the Banach space of bounded Borel functions, $B(\mathbb{R}^+)$, to $L^\infty(P) \subset L^2(P)$, and for which the dominated convergence theorem is valid. (For details, see Dunford and Schwartz [1], and for a streamlined treatment one should also refer to Gould [1].) When Z is determined by a stochastic process, as in the present work, we call the resulting entity the *classical stochastic integral* (or one of the *first kind*) in what follows. This was defined by Cramér and Kolmogorov using an extension of the Riemann-Stieltjes definition in the early 1940's, and was further studied by Karhunen [1]. These definitions are included in the D-S general construction above, and it also applies to the Wiener integral. As one may surmise, this integral is weaker than the classical Lebegue's definition.

8. Proposition. *Let $Z : \mathcal{B}(\mathbb{R}^+) \to L^p(P), p \geq 1$, be a vector measure, and $B(\mathbb{R}^+)$ denote the Banach space of real bounded Borel functions on $\mathbb{R}^+$. Then the mapping $\tau : f \mapsto \int_{\mathbb{R}^+} f\, dZ, f \in B(\mathbb{R}^+)$ is well-defined, satisfies $\|\tau(f)\|_2 \leq \|f\|_\infty \|Z\|(\mathbb{R}^+)$, and is a bounded linear operator (i.e.,$\tau \in B(B(\mathbb{R}^+), L^p(P))$). Moreover, the integral can be treated as a Lebesgue-Stieltjes integral iff $Z(\cdot)(\omega)$ has finite variation for a.a. (ω) on each compact interval $I \subset \mathbb{R}^+$, in the sense that if $\Pi_n : t_1 < \cdots < t_n, t_i \in I$, is a partition, then $|Z|(I)\omega) < \infty$ for a.a. (ω), where*

$$|Z|(I)(\omega) = \sup\{ \sum_{t_i \in \Pi_n} |Z(t_i, t_{i+1}]|(\omega) : \Pi_n, \textit{ is a partition of} I, n \geq 1\}. \tag{29}$$

Proof. The first part is a restatement of the standard properties of the D-S integral discussed above. For the last part, if $Z(\cdot)(\omega)$ has finite variation for a.a. (ω), (P-measure) then the integral can be defined by the classical method, (see e.g. Rao [11], Theorem 4.1.6). Thus we only need to establish the converse, which was remarked by Meyer ([7],p. 107).

Suppose then the integral can be treated in the sense of Stieltjes for a.a. (ω). If Π_n is a partition of $I = (0,t]$, then for any bounded Borel function $f : I \to \mathbb{R}$, we have on letting $f_{\Pi_n} = \sum_{t_i \in \Pi_n} f(t_i)\chi_{(t_i,t_{i+1}]}$ the following:

$$S_n(f) = \int_I f_{\Pi_n}\, dZ = \sum_{t_i \in \Pi_n} f(t_i)(Z(t_{i+1}) - Z(t_i)) \in L^2(P),$$

and $\|S_n(f)\|_2 \le \|f\|_\infty \|Z\|(I)$, and $S_n(f) \to \tau(f) = \int_I f\, dZ$ as $n \to \infty$. Hence $S_n : B(I) \to L^2(P)$ is a bounded linear operator for each n, and $\{S_n(f), n \ge 1\}$, being a Cauchy sequence, is bounded for each f. Since $(B(I), \|\cdot\|_\infty)$ is a Banach space, it follows by the uniform boundedness principle that $\sup_n \|S_n\| = \alpha_0 < \infty$. In particular, if we let, for each $\omega \in \Omega$,

$$h_n^\omega = \sum_{i=0}^{n} sgn\Big(Z(t_{i+1}) - Z(t_i)\Big)(\omega)\chi_{(t_i,t_{i+1}]},$$

then $h_n^\omega \in B(I)$ and $\|h_n^\omega\|_\infty \le 1, n \ge 1$. Also

$$\begin{aligned} S_n(h_n^\omega) &= \sum_{t_i \in \Pi_n} |Z(t_{i+1}) - Z(t_i)|(\omega) \\ &\le \|S_n\| \|h_n^\omega\|_\infty \le \alpha_0 < \infty. \end{aligned}$$

Taking the supremum over all partitions, we deduce that $|Z|(I)(\omega) < \infty$ for a.a.(ω), (P-measure). □

Remark. Since a Brownian motion process is a square integrable martingale, it satisfies the hypothesis of Theorem 6 as well as Proposition 8, for the D-S integral. But it does not have finite variation on any non-degenerate interval of $\mathbb{R}^+$. It follows that the corresponding stochastic integral cannot be treated in the Lebesgue-Stieltjes sense. It will be seen later that there is an alternative way of dealing with the above type of integrals (motivated by the "integration by parts" formula for the Stieltjes integral) which is related to the Bochner integral.

We now turn to the second type of integral where both the integrand and the integrator are stochastic functions. This is the general (Itô) integral. For this we need to analyze further the class $\mathcal{C}$ or $\mathcal{L}^p(A)$ introduced in the preceding subsection.

Let $\mathcal{W}$ be the set of all processes $\{X_t, \mathcal{F}_t, t \geq 0\}$, regarded as mappings on $\mathbb{R}^+ \times \Omega \to \mathbb{R}$, which are $(\mathcal{O}, \mathcal{B}(\mathbb{R}))$-measurable. They are termed *optional (or well-measurable) processes.* One has the inclusion $\mathcal{C} \subset \mathcal{W}$, and the latter may be described explicitly as follows. By the classical structure theorem of measurable functions, if $X \in \mathcal{W}$ there exists a sequence of $\mathcal{O}$-simple functions $f_n = \sum_{i=1}^n a_i^n \chi_{A_i}, A_i = [\![T_1^i, T_2^i]\!]$, such that $f_n \to X$, pointwise, where the sets A_i are generators of $\mathcal{O}$. Since each $f_n(\cdot, \omega)$ of the above form is a bounded right continuous function with left limits, $\mathcal{W}$ can alternatively be described as the family of processes which are pointwise limits of bounbed right continuous processes with left limits. As noted before, $\mathcal{W}$ is a subclass of progressively measurable processes. Thus let

$$L^p(A) = \Big\{X \in \mathcal{W} : E\Big(\int_0^t |X_s|^p \, d|A|_s\Big) < \infty, t \geq 0\Big\}, \, p \geq 1. \tag{30}$$

Then $\mathcal{L}^p(A) \subset L^p(A)$. Since $\mathcal{W}$ is closed under pointwise limits and $A(\cdot, \omega)$ is increasing, it may be verified that $\mathcal{L}^p(A)$ is a complete locally convex space for the semi-norms $\{\|\cdot\|_{p,A(t)}, t \geq 0\}$, (cf., e.g., Proposition 28 below). The above work with $\mathcal{L}^p(A)$ also holds for $L^p(A)$.

We can now present the first basic result for the set $L^p(A)$:

9. Theorem. *Let $X \in \mathcal{M}$ and $Y \in L^2(\langle X \rangle)$. Then there exists a P-unique element $I_y^x = I_y = I(Y) \in \mathcal{M}$ such that for all $Z \in \mathcal{M}$*

$$\langle I_y, Z \rangle_t = \int_0^t Y_s \, d\langle X, Z \rangle_s, \ a.e. \tag{31}$$

Based on this result we can introduce the desired concept as:

10. Definition. For each $X \in \mathcal{M}$ and $Y \in L^2(\langle X \rangle)$ the P-unique martingale $I_y \in \mathcal{M}$, guaranteed by (31), is denoted

$$(Y \cdot X)_t = (I_y^x)_t = \int_0^t Y_s \, dX_s, \tag{32}$$

and is called the *stochastic integral* of the optional process Y relative to the square integrable martingale X (and the result is a square integrable martingale). If X is a Brownian motion process, then, as will be seen later, I_y^x becomes the Itô integral.

Proof of Theorem 9. First observe that, since $\langle X, Z\rangle \in \mathcal{A}$, the integral on the right of (31) is well-defined. Next, if I_y^x exists then the mappings $Y \mapsto I_y^x = I(Y)$ and $Z \mapsto \langle I_y^x, Z\rangle$ are linear. This implies uniqueness since, if $I'_y \in \mathcal{M}$ is another element then $\langle I_y^x, Z\rangle = \langle I'_y, Z\rangle$ for all $Z \in \mathcal{M}$. Taking $Z = I_y^x - I'_y$ we get $\langle I_y^x - I'_y\rangle = 0$, so that for all $t \geq 0$, $(I_y^x)_t = (I'_y)_t$, a.e. if we identify two elements of $\mathcal{M}$ when they agree in t for a.a. (ω) (termed *indistinguishable*). Thus we need to prove the existence of I_y^x.

Since Y is an optional process, by its structure (cf. the discussion preceding (30)), for each $\varepsilon > 0$ there is a simple $h \in L^2(\langle X\rangle)$ such that $\|Y - h\|_{\langle X\rangle_t} < \varepsilon$, for all $t > 0$, where h is càdlàg so that it has no oscillatory discontinuities. To define the integral, we express h in a convenient form. Let $T_0 = 0$ and if $T_n = T_n^\varepsilon$ is defined, set

$$T_{n+1}(\omega) = \inf\{t > 0 : t > T_n(\omega), |h(t,\omega) - h \circ T_n(\omega)| > \varepsilon\},$$

where $\inf(\emptyset) = +\infty$. Since $\{h(t,\cdot), t \geq 0\}$ is right continuous, we see that $0 \leq T_n \leq T_{n+1}$ are stopping times of the filtration, and that $T_n \uparrow \infty$, a.e., as $n \to \infty$. Also if $T_n(\omega) < \infty$ for a.a.(ω), then $|h \circ T_{n+1} - h \circ T_n| \geq \varepsilon$. Thus if we define $\tilde{h}$ by

$$\tilde{h}(t,\omega) = \sum_{n=0}^{\infty} h \circ T_n(\omega)\chi_{A_n}(t,\omega), \qquad A_n = [\![T_n, T_{n+1}), \ t \geq 0, \quad (33)$$

then $|h - \tilde{h}|(t,\omega) \leq \varepsilon$ for all (t,ω). Hence we have

$$E\Big(\int_0^t |h_t - \tilde{h}_s|^2 \, d\langle X\rangle_s\Big)^{1/2} \leq (\varepsilon^2 E(\langle X\rangle_t))^{1/2}, \quad t \geq 0. \quad (34)$$

This implies $\tilde{h} \in L^2(\langle X\rangle)$ and $\|Y - \tilde{h}\|_{\langle X\rangle_t} < 2\varepsilon, t \geq 0$. So it suffices to define the integral I_t and prove (31) with $\tilde{h}$. Let

$$I_t = \sum_{n=1}^{\infty} \tilde{h}(T_{n-1} \wedge t)(X \circ (T_n \wedge t) - X \circ (T_{n-1} \wedge t)). \quad (35)$$

Since $\tilde{h}(T_n \wedge t)$ is $\mathcal{B}(T_n \wedge t)$-adapted, it follows that I_t is $\mathcal{F}_t$-adapted, and is right continuous. Note that in (35) the sum is finite since $T_n \wedge t = t$ on $[T_n \geq t]$, and that the process $\{\tilde{h}_t, t \geq 0\}$ is in a ball of $L^2(P)$. The random variable I_t is a predictable transform of the square integrable martingale $\{X \circ (T_n \wedge t), \mathcal{B}(T_n \wedge t), n \geq 1\}$ in the terminology (at the end) of Section 4.4, which follows from Corollary IV.1.12 and the uniform integrability of $\{X_s, 0 \leq s \leq t\}$. But h is simple, hence bounded, so that $\tilde{h}$, of (33), is also a bounded random variable. Consequently

$$E(I_t^2) = \sum_{n=1}^{\infty} E(\tilde{h}^2(T_{n-1} \wedge t) E^{\mathcal{B}(T_{n-1} \wedge t)}(X(T_n \wedge t) - X(T_{n-1} \wedge t))^2) < \infty,$$

since $\{X(T_n \wedge t), \mathcal{B}(T_n \wedge t), n \geq 1\}$ is an $L^2(P)$-martingale, and hence its increments are orthogonal. Moreover, $I = \{I_t, \mathcal{F}_t, t \geq 0\}$ is a martingale since it is a predictable transform of a martingale. Thus $I \in \mathcal{M}$.

It follows from (31) and Proposition 1, applied to I_t with $T = t$, that for each $Z \in \mathcal{M}, t \geq 0$,

$$\begin{aligned} E(\langle I, Z \rangle_t) &= E\Big(\sum_{n=1}^{\infty} \tilde{h}(T_{n-1} \wedge t) \langle X \circ (T_n \wedge t) - X \circ (T_{n-1} \wedge t), Z \rangle_t\Big) \\ &= E\Big(\int_0^t \tilde{h}_s \, d\langle X, Z \rangle_s\Big). \end{aligned} \tag{36}$$

Now let $Y \in L^2(\langle X \rangle)$. Then there exists $h^n \in L^2(\langle X \rangle)$ of the form (33) such that $\|Y - h^n\|_{\langle X \rangle_t} \to 0$ as $n \to \infty$, for each $t \geq 0$. If I^n is defined by (35) so that $\langle I^n, Z \rangle_t = \int_0^t h_s^n \, d\langle X, Z \rangle_s$, we assert that $\{I_t^n, n \geq 1\}$ is Cauchy in $\| \cdot \|_{\langle X \rangle_t}$. In fact,

$$\begin{aligned} E\Big(\Big| \int_0^t (Y_s - h_s^n) \, d\langle X, Z \rangle_s \Big|\Big) &\leq \|Y - h^n\|_{\langle X \rangle_t} \|1\|_{\langle Z \rangle_t}, (cf. (23)), \\ &= \|Y - h^n\| E(\langle Z \rangle_t^{1/2}) \to 0, \end{aligned}$$

as $n \to \infty$. Writing $I_t^n = \int_0^t h_s^n \, dX_s$ which is really a sum (cf. (35) if h_n is put in for $\tilde{h}$), one has

$$\begin{aligned} E(|I_t^n - I_t^m|^2) &= E(\langle I^n - I^m \rangle_t), \text{ by Proposition 1 with } s = 0, \\ &= \|h^n - h^m\|_{\langle X \rangle_t}^2 \to 0, \end{aligned} \tag{37}$$

since $\{h^n, n \geq 1\}$ is Cauchy. A direct computation, retaining the summation symbol, is also easy. Hence $I_t^n \to I_t$(say), in $L^2(P)$, as $n \to \infty$, for each $t \geq 0$. Then $I_t \in L^2(P)$, and has a right continuous version. It is a martingale since each I^n is. Thus $I^n \in \mathcal{M}$. Also by the CBS-inequality we have,

$$\begin{aligned} E(|\langle I - I^n, Z\rangle_t|) &\leq \|1\|_{\langle I-I^n\rangle_t} \|1\|_{\langle Z\rangle_t} \\ &= E(\langle I - I^n\rangle_t^{1/2}) E(\langle Z\rangle_t^{1/2}) \to 0, \end{aligned}$$

by (37). Then by the $L^1(P)$-convergence, we deduce that, for a subsequence, $\langle I, Z\rangle_t = \lim_{n_i\to\infty}\langle I^{n_i}, Z\rangle_t = \lim_{n_i\to\infty}\int_0^t h_s^{n_i}\, d\langle X, Z\rangle_s = \int_0^t Y_s\, d\langle X, Z\rangle_s$, a.e. This implies (36). □

The following consequences of the above computations are of interest.

11. Corollary. *Simple functions of the form (33) are dense in the space $L^p(\langle X\rangle)$[or in $\mathcal{L}^p(\langle X\rangle)$], $X \in \mathcal{M}$ for the topology defined by the family of semi-norms $\{\|\cdot\|_{\langle X\rangle_t}, t \geq 0\}, 1 \leq p < \infty$.*

Another one is:

12. Corollary. *If $X = \{X_t, \mathcal{F}_t, t \geq 0\}$ is a càdlàg process, then it can be approximated pointwise with processes of the form (35) so that $X \in \mathcal{W}$, i.e., X is optional.*

The next one is even more important.

13. Corollary. *If $X \in \mathcal{M}^c, Y \in \mathcal{L}^2(\langle X\rangle)$ then $I_y^x \in \mathcal{M}^c$. Further $\langle I_y^x, I_y^x\rangle_t = \int_0^t Y_s^2\, d\langle X\rangle_s$; and more generally, if X, Y are in $\mathcal{M}$ and $f \in \mathcal{L}^2(\langle X\rangle), g \in \mathcal{L}^2(\langle Y\rangle)$, then*

$$\langle I_f^x, I_g^y\rangle_t = \int_0^t (fg)_s\, d\langle X, Y\rangle_s, \tag{38}$$

and

$$E[(I_f^x)_t (I_g^y)_t] = E(\langle I_f^x, I_g^y\rangle_t), \quad t \geq 0. \tag{39}$$

Proof. If $X \in \mathcal{M}^c$ then, by Proposition 2, $\langle X\rangle \in (\mathcal{A}^+)^c$. But by Corollary 11, for $Y \in \mathcal{L}^2(\langle X\rangle)$, there exist simple $h^n \in \mathcal{L}^2(\langle X\rangle)$ such that $\|Y - h^n\|_{\langle X\rangle_t} \to 0, t \geq 0$, as $n \to \infty$. Thus if $I_t^n = \int_0^t h_s^n \, dX_t$, then $I_t^n \to I_t$ in $L^1(P)$ for each $t \geq 0$. Since X has continuous sample paths, and h_s^n is simple, it is clear that I_t^n also has a.a. continuous paths, so $I_t^n \in \mathcal{M}^c$. Consider the submartingale $\{|I_t - I_t^n|, \mathcal{F}_t, t \geq 0\}$. Using Theorem II.4.3 (separability may be assumed), we have for each $\varepsilon > 0, a > 0$,

$$P[\sup_{0 \leq s \leq a} |I_s - I_s^n|^2 > \varepsilon^2] \leq \frac{1}{\varepsilon^2} E(I_a - I_a^n)^2 = \frac{1}{\varepsilon^2} E(\langle I - I^n \rangle_a) \to 0,$$

as $n \to \infty$ by (37). This implies for a subsequence $I_t^{n_k} \to I_t$ a.e. uniformly in $0 \leq t \leq a$, and hence I_t is continuous on $[0, a]$ for a.a. (ω). Since $a > 0$ is arbitrary, we see that $\{I_t, \mathcal{F}_t, t \geq 0\} \in \mathcal{M}^c$. The above estimate also proves

$$E(\langle I_y^x \rangle_t^2) = \|Y\|_{\langle X\rangle_t}^2 = E\Big(\int_0^t Y_s^2 \, d\langle X\rangle_s\Big). \tag{40}$$

For the formula (40), the fact that $X \in \mathcal{M}$ is sufficient, as seen from the proof of the theorem. Now, if X, Y are in $\mathcal{M}$, then by the theorem I_f^x, I_g^y are in the same space for any $f \in \mathcal{L}^2(\langle X\rangle), g \in \mathcal{L}^2(\langle Y\rangle)$. Hence by (23), $fg \in \mathcal{L}^1(\langle X, Y\rangle)$. If $X = Y$ then by (40) and polarization, we deduce (39). Moreover, Proposition 1 yields (36) from (39). A direct argument can also be given. □

Remark. It is clear from (38) that the mapping $Y \mapsto I_y^x$ is linear on $\mathcal{L}^2(\langle X\rangle)$. The integral I_y^x can be introduced directly if X is a Brownian motion, without using the Doob-Meyer decomposition theory. We shall indicate this in the Complements section later which is Itô's original method.

We prove a generalization of the important "integration by parts" formula due to Itô. It helps to identify the functional I_y^x as a vector integral, and it is one of the fundamental results of the theory. This again shows the distinction between the current integral and the classical Stieltjes case very strikingly.

14. Theorem. *Let $f : \mathbb{R} \to \mathbb{C}$ be a twice continuously differentiable function and $X \in \mathcal{M}_{loc}^c, A \in \mathcal{A}_{loc}^c$. If $Z = X + A$ (a semimartingale),* *then for* $t \geq 0$,

$$f(Z_t) - f(Z_0) = \int_0^t \frac{df}{dx}(Z_s)dX_s + \frac{1}{2}\int_0^t \frac{d^2f}{dx^2}(Z_s)d\langle X\rangle_s + \int_0^t \frac{df}{dx}(Z_s)dA_s. \quad (41)$$

Proof. We first observe that (41) is precisely Itô's formula if we set $A = 0$ and take X_t as Brownian motion. Note that if X and A (hence Z) are also bounded, then the continuity of $\frac{df}{dx}, \frac{d^2f}{dx^2}$ on $[0,t]$ imply that $\frac{df}{dx}(Z) \in \mathcal{L}^2(\langle X,\rangle)$, $\frac{d^2f}{dx^2}(Z) \in \mathcal{L}^1(\langle X\rangle)$ and $\frac{df}{dx}(Z) \in \mathcal{L}^1(A)$. Now it suffices to prove (41) when X and A are bounded. For, let $N > 0$, and define the stopping time:

$$T_N = \inf\left\{t > 0 : |X_t| > \frac{N}{2}, \text{or} |A_t| > \frac{N}{2}\right\}$$

where as usual $\inf(\emptyset) = \infty$. If $X_t^N = X(t \wedge T_N), A_t^N = A(t \wedge T_N)$, then $\{X_t^N, t \geq 0\}$ and $\{A_t^N, t \geq 0\}$ are bounded elements of $\mathcal{M}^c$ and $\mathcal{A}^c$ so that $\{Z_t^N, t \geq 0\}$ is also bounded. Further, as $N \to \infty$, they tend pointwise and in $\|\cdot\|_{\langle X\rangle_t}$-norm to X_t, A_t, Z_t respectively. If (41) holds for X^N, A^N, and Z^N, then we claim that (41) holds in general. This is the localization procedure of importance in the current analysis.

In fact, $f(Z_t^N) \to f(Z_t), t \geq 0$, by continuity, and since $\langle X^N\rangle_s \uparrow \langle X\rangle_s$, and $|A_s^N| \uparrow |A_s|, s \geq 0$, as $N \to \infty$, the limit relations hold for the last two terms by the Monotone Convergence criterion. We now verify that the same holds for the first (i.e. stochastic) integral also. If $\in \mathcal{M}$, then for $X \in \mathcal{M}^c, A \in \mathcal{A}$, (the stochastic integral of $\frac{df}{dx}(Z_s)$ relative to X exists by the same token),

$$\left|\left\langle \int_0^t \frac{df}{dx}(Z_s^N)\, dX_s^N - \int_0^t \frac{df}{dx}(Z_s)\, dX_s, Y\right\rangle\right|$$

$$\leq \left|\int_0^t \left(\frac{df}{dx}(Z_s^N) - \frac{df}{dx}(Z_s)\right) d\langle X^N, Y\rangle_s\right| + \left|\int_0^t \frac{df}{dx}(Z_s)\, d\langle X^N - X, Y\rangle_s\right|$$

$$\leq \left\|\frac{df}{dx}(Z^N) - \frac{df}{dx}(Z)\right\|_{\langle X^N\rangle_t} \cdot [E(\langle Y\rangle_t)]^{1/2} + \left\|\frac{df}{dx}(Z)\right\|_{\langle Y\rangle_t} \cdot [E(\langle X^N - X\rangle_t)]^{1/2} \to 0,$$

as $N \to \infty$, since $E(\langle X^N - X\rangle_t) \to 0, \|\cdot\|_{\langle X^N\rangle_t} \leq \|\cdot\|_{\langle X\rangle_t}$ and hence $\left\|\frac{df}{dx}(Z^N) - \frac{df}{dx}(Z)\right\|_{\langle X\rangle_t} \to 0$ as $N \to \infty$. If $X \in \mathcal{M}^c_{loc}$ and $Y \in \mathcal{M}^c_{loc}$, then the same conclusions hold by Proposition 1, as is easily seen. Thus we may and do assume that $|X_t| \leq \frac{N_0}{2}$, and $|A_t| \leq \frac{N_0}{2}$, a.e. for some $N_0 > 0$ for this proof. We also can assume that $A \in (\mathcal{A}^+)^c$ by a Jordan decomposition.

One considers a "discretization" of the integrals to establish the result by refining the "stochastic partitions" somewhat in the classical manner for the Riemann-Stieltjes approximations since both X, A are continuous. To define the former, let $T_0 = 0$ and if $T_n \uparrow \infty$ is a sequence of optionals of a (given standard) filtration, then for each $\varepsilon > 0, \{T_n, n \geq 1\}$ is a *stochastic partition* of length ε (or an $\varepsilon -$ *partition*) for the X_t-process iff $\sup\{|X_t - X_s| : T_n \leq t, s < T_{n+1}\} < \varepsilon$ for a.a. (ω) and all $n \geq 1$. Since X_t is a separable process, this supremum is measurable for $\mathcal{F}_\infty$. We first assert that such partitions exist for all $X \in \mathcal{M}_{loc}$.

To see this, for each $\varepsilon > 0$, define $\{T_n^\varepsilon, n \geq 1\}$ as we did just above (33) with X_t in place of h_t there. Then X, being continuous, satisfies all the conditions so that $T_n^\varepsilon \uparrow \infty$, as $n \to \infty$. With $\varepsilon = 2^{-k}, k \geq 1$, one takes $\mathcal{C}_k = \{T_n^k, n \geq 1\}, (T_n^\varepsilon = T_n^k)$. Then $\mathcal{C}_k$ is a chain, and if $\mathcal{C}_k, \mathcal{C}_{k+1}$ are two such chains, there is a superposition of these two, denoted $\tilde{\mathcal{C}}_{k+1}$, which refines both (cf. Proposition IV.2.6). Hence starting with $\{\mathcal{C}_k, k \geq 1\}$ we may produce refinements $\{\tilde{\mathcal{C}}_{k+1}, k \geq 1\}$ and continue the process so that one has $\bar{\mathcal{C}}_k \prec \bar{\mathcal{C}}_{k+1}$ and $\sup_n |\bar{T}^k_{n+1} - \bar{T}^k_n|(\omega) \to 0$ for a.a.(ω), as $n \to \infty$ where $\bar{\mathcal{C}}_k = \{\bar{T}^k_n, n \geq 0\}, \prec$ being the refinement order noted above. Thus for each $k \geq 1$, we can produce a $\frac{1}{2^k}$-partition simultaneously for the following processes:$X \in \mathcal{M}^c_{loc}, \langle X\rangle, h, |A| \in (\mathcal{A}^+_{loc})^c$ and a function h_1 with $h_1(t) = t$, using a superposition, a finite number of (five) times, if necessary. Let $\bar{\bar{\mathcal{C}}}_k = \{T^k_n, k \geq 1\}$ be such a final $\frac{1}{2^k}$-partition to be used for (41). This is our last simplified form.

Set $\sigma_n = \sigma^k_n = T^k_n \wedge t$, and consider the left side of (41):

$$f(Z_t) - f(Z_0) = \sum_{n=1}^{\infty}[f(Z_{\sigma_n}) - f(Z_{\sigma_{n-1}})], \text{ the sum is finite since } \sigma_n \uparrow t,$$

$$= \sum_{n=1}^{\infty} f'(Z_{\sigma_{n-1}})(Z_{\sigma_n} - Z_{\sigma_{n-1}}) + \frac{1}{2}\sum_{n=1}^{\infty} f''(\theta_{n-1})(Z_{\sigma_n} - Z_{\sigma_{n-1}})^2,$$

by the Taylor expansion around $Z_{\sigma_{n-1}}(\omega)$,

and θ_{n-1} lies between $Z_{\sigma_{n-1}}$ and Z_{σ_n},

f', f'' being the derivatives of f,

$$= I_1^k + \frac{1}{2} I_2^k (\text{say}). \tag{42}$$

We shall now show that, as $k \to \infty, I_i^k, i = 1, 2$, tend to the right side of (41).

First consider I_1^k. Since f', f'' are continuous on $\mathbb{R}$, and $Z_{\sigma_n} - Z_{\sigma_{n-1}} = (X_{\sigma_n} - X_{\sigma_{n-1}}) + (A_{\sigma_n} - A_{\sigma_{n-1}})$, from the boundedness of X and A, we have $f'(Z) \in \mathcal{L}^2(\langle X \rangle) \cap \mathcal{L}^2(|A|)$. Hence

$$I_1^k = \sum_{n=1}^{\infty} f'(Z_{\sigma_{n-1}})(X_{\sigma_n} - X_{\sigma_{n-1}}) + \sum_{n=1}^{\infty} f'(Z_{\sigma_{n-1}})(A_{\sigma_n} - A_{\sigma_{n-1}}).$$

Since $[\![T_n^k, T_{n+1}^k)$ is converging pointwise to $[\![0]\!]$ as $k \to \infty$, we see that $E(|I_1^k - I_1^\ell|_t^2) \to 0$ as $k, \ell \to \infty$ (cf. (35) and (37)). Thus, in the L^2-sense, for each $t \geq 0$,

$$(I_1^k)_t \to \int_0^t f'(Z_s)\, dX_s + \int_0^t f'(A_s)\, dA_s, \text{ as } k \to \infty. \tag{43}$$

Next consider I_2^k of (42). This simplification demands more work. Consider

$$\begin{aligned} I_2^k &= \sum_{n=1}^{\infty} f''(Z_{\sigma_{n-1}})(Z_{\sigma_n} - Z_{\sigma_{n-1}})^2 \\ &\quad + \sum_{n=1}^{\infty} [f''(Z_{\sigma_{n-1}}) - f''(\theta_{n-1})](Z_{\sigma_n} - Z_{\sigma_{n-1}})^2 \\ &= \tilde{I}_2^k + \bar{I}_2^k \text{ (say)}. \end{aligned}$$

Since $|Z| \leq N_0$, f'' is continuous (hence uniformly so) on $[-N_0, N_0]$, then for $|a_1 - a_2| < \varepsilon = 2^{-k}, |f''(a_1) - f''(a_2)| < \delta(\varepsilon)$ where $\delta(\varepsilon) \to 0$ as $\varepsilon \to 0$ (independent of a_i). Thus:

$$\begin{aligned} E(|\bar{I}_2^k|) &\leq \sum_{n=1}^{\infty} E[|f''(Z_{\sigma_{n-1}}) - f''(\theta_{n-1})|(Z_{\sigma_n} - Z_{\sigma_{n-1}})^2] \\ &\leq \delta(\frac{2}{2^k}) \sum_{n=1}^{\infty} E(|Z_{\sigma_n} - Z_{\sigma_{n-1}}|^2), \text{ since} \end{aligned}$$

$$|Z_{\sigma_n} - Z_{\sigma_{n-1}}| \leq |X_{\sigma_n} - X_{\sigma_{n-1}}| + |A_{\sigma_n} - A_{\sigma_{n-1}}| < \frac{2}{2^k},$$
$$\to 0 \text{ as } k \to \infty.$$

We now simplify $\tilde{I}_2^k$. This can be expressed as:

$$\begin{aligned}
\tilde{I}_2^k &= \sum_{n=1}^{\infty} f''(Z_{\sigma_{n-1}})(X_{\sigma_n} - X_{\sigma_{n-1}})^2 \\
&+ \sum_{n=1}^{\infty} f''(Z_{\sigma_{n-1}})(A_{\sigma_n} - A_{\sigma_{n-1}})^2 \\
&+ \sum_{n=1}^{\infty} f''(Z_{\sigma_{n-1}})(X_{\sigma_n} - X_{\sigma_{n-1}})(A_{\sigma_n} - A_{\sigma_{n-1}}) \\
&= J_1^k + J_2^k + J_3^k \text{ (say).}
\end{aligned} \tag{44}$$

We assert that $E(|J_i^k|) \to 0, i = 2, 3$, as $k \to \infty$, and $J_1^k \to$ the desired integral. Consider J_2^k. Since $|f''| \leq \alpha$ for some α on $[-N_0, N_0]$, and if $\alpha_0 = \alpha \wedge N_0$, one has

$$\begin{aligned}
E(|J_2^k|) &\leq \sum_{n=1}^{\infty} E[|f''(Z_{\sigma_{n-1}})|(A_{\sigma_n} - A_{\sigma_{n-1}})^2] \\
&\leq \alpha_0 \frac{1}{2^k} E\Big(\sum_{n=1}^{\infty}(A_{\sigma_n} - A_{\sigma_{n-1}})\Big) \\
&\leq \alpha_0 \frac{1}{2^k} E(A_t) \to 0, \text{ as } k \to \infty.
\end{aligned}$$

Similarly,

$$\begin{aligned}
E(|J_3^k|) &\leq \alpha_0 \sum_{n=1}^{\infty} E((|X_{\sigma_n} - X_{\sigma_{n-1}}|)(A_{\sigma_n} - A_{\sigma_{n-1}}) \\
&\leq \alpha_0 \Big\{ E\Big[\sum_{n=1}^{\infty}(X_{\sigma_n} - X_{\sigma_{n-1}})^2\Big]^{1/2} \times \\
&\qquad E\Big[\sum_{n=1}^{\infty}(A_{\sigma_n} - A_{\sigma_{n-1}})^2\Big]^{1/2}\Big\},
\end{aligned}$$

by the CBS inequality,

$$\begin{aligned}
&\leq \alpha_0 [E(X_t - X_0)^2]^{1/2} \frac{1}{2^k} [E(A_t)]^{1/2}, \\
&\to 0, \text{ as } k \to \infty.
\end{aligned}$$

Finally consider J_1^k. Since $X \in \mathcal{M}$, using (4) one has

$$E\Big(|J_1^k - \sum_{n=1}^{\infty} f''(Z_{\sigma_{n-1}})(\langle X\rangle_{\sigma_n} - \langle X\rangle_{\sigma_{n-1}})|^2\Big)$$

$$= E\Big\{|\sum_{n=1}^{\infty} f''(Z_{\sigma_{n-1}})[(X_{\sigma_n} - X_{\sigma_{n-1}})^2 - (\langle X\rangle_{\sigma_n} - \langle X\rangle_{\sigma_{n-1}})]|^2\Big\}$$

$$\leq \sum_{n=1}^{\infty} E\{|f''(Z_{\sigma_{n-1}})|^2 \times [\quad\quad]^2\}, \text{ by expanding ,using the CBS}$$

inequality, and the orthogonality of martingale differences,

$$\leq 2E\Big(\sum_{n=1}^{\infty} |f''(Z_{\sigma_{n-1}})|^2[(X_{\sigma_n} - X_{\sigma_{n-1}})^4 + (\langle X\rangle_{\sigma_n} - \langle X\rangle_{\sigma_{n-1}})^2]\Big),$$

using$(a-b)^2 \leq 2(a^2+b^2)$,

$$\leq 2\alpha_0^2\Big[\frac{1}{2^{2k}}E\Big(\sum_{n=1}^{\infty}(X_{\sigma_n} - X_{\sigma_{n-1}})^2\Big) + \frac{1}{2^k}E\Big(\sum_{n=1}^{\infty}(\langle X\rangle_{\sigma_n} - \langle X\rangle_{\sigma_{n-1}})\Big)\Big]$$

$$= 2\alpha_0^2\Big[\frac{1}{2^{2k}}E\Big(\sum_{n=1}^{\infty}(\langle X\rangle_{\sigma_n} - \langle X\rangle_{\sigma_{n-1}})\Big) + \frac{1}{2^k}E(\quad)\Big], \text{ by (4)},$$

$$= 2\alpha_0^2\Big(\frac{1}{2^{2k}} + \frac{1}{2^k}\Big)E(\langle X\rangle_t - \langle X\rangle_0) \to 0, \text{ as } k \to \infty. \tag{45}$$

However, the second term on the left is a "simple" function. Hence as in (37) that term tends to $\int_0^t f''(Z_s)\,d\langle X\rangle_s$ in $L^2(P)$. So (45) implies that $\lim_k J_1^k = \int_0^t f''(Z_s)\,d\langle X\rangle_s$ in $L^1(P)$ and the latter integral belongs to $L^2(P)$. Thus, (43) and (42) prove (41) in the bounded case. By the initial reduction, (41) is true in general. □

It is of interest to note that there is some additional information in the simplification of J_1^k. We have shown that (taking a subsequence of ε_k's if necessary) for a.a (ω),

$$\lim_{k\to\infty} J_1^k = \int_0^t f''(Z_s)\,d\langle X\rangle_s = \lim_{k\to\infty}\sum_{n=1}^{\infty} f''(Z_{\sigma_{n-1}})(X_{\sigma_n} - X_{\sigma_{n-1}})^2. \tag{46}$$

Now suppose that $A = 0$ a.e., and $f(x) = \frac{1}{2}x^2$; so $X = Z \in \mathcal{M}_{loc}^c$, $f'' = 1$, and $\int_0^t f''(Z_s)\,d\langle X\rangle_s = \langle X\rangle_t - \langle X\rangle_0$. Hence we have the following:

15. Proposition. *For each $X \in \mathcal{M}_{loc}^c$, there exists a chain $C^k = \{T_n^k, n \geq 1\}$ of $\frac{1}{2^k}$-partitions, $C^k \prec C^{k+1}$, and with probability 1 we* *have:*

$$\lim_{k\to\infty}\sum_{n=1}^{\infty}[X(T_n^k \wedge t) - X(T_{n-1}^k \wedge t)]^2 = \langle X\rangle_t - \langle X\rangle_0. \tag{47}$$

The importance of this formula is that if $\langle X\rangle_t$ is not a constant for all t, then $Y_n^k = X \circ T_n^k$ obtained from $X(\in \mathcal{M}_{loc}^c)$ has its sample functions of unbounded variation because $\langle X\rangle_t < \infty$ a.e. for each $t \geq 0$, and increasing a.e. It is shown below that $\langle X\rangle_t = t$ a.e. iff X is Brownian motion, and in that case, taking $T_n^k \wedge t = \frac{nt}{2^k}$, the hypothesis of Proposition 15 is satisfied, and one gets from (47):

$$\lim_{k\to\infty}\sum_{n=1}^{2^k}\left[X\left(\frac{nt}{2^k}\right) - X\left(\frac{n-1}{2^k}t\right)\right]^2 = \langle X\rangle_t = t, \text{ a.e.} \tag{48}$$

Thus the Brownian motion has a.a. of its sample paths of unbounded variation. Another proof of (48) will be outlined in Complements and exercises.

The following result is an immediate extension of the formula (41) to n-dimensions. The proof, being the same except for notational adjustments, will be omitted.

16. Theorem. *Let $f : \mathbb{R}^n \to \mathbb{R}$ (or $\mathbb{C}$) be a twice contimuously differentiable function and $X \in \mathcal{M}_{loc}^c, A^i \in \mathcal{A}_{loc}^c, Z_t^i = X_t^i + A_t^i, i = 1, \ldots, n$. If $Z_t = (Z_t^1, \ldots, Z_t^n), t \geq 0$, then*

$$f(Z_t) - f(Z_0) = \sum_{i=1}^{n}\int_0^t \frac{\partial f}{\partial x_i}(Z_s)\,dX_s^i$$
$$+\frac{1}{2}\sum_{i,j=1}^{n}\int_0^t \frac{\partial^2 f}{\partial x_i \partial x_j}(Z_s)\,d\langle X^i, Y^j\rangle_s + \sum_{i=1}^{n}\int_0^t \frac{\partial f}{\partial x_i}(Z_s)\,dA_s^i. \tag{49}$$

This is symbolically expressed as:

$$df(Z_t) = \sum_{i=1}^{n}\frac{\partial f}{\partial x_i}(Z_t)\,dX_t^i +$$
$$\frac{1}{2}\sum_{i,j=1}^{n}\frac{\partial^2 f}{\partial x_i \partial x_j}(Z_t)\,d\langle X^i, X^j\rangle_t + \sum_{i=1}^{n}\frac{\partial f}{\partial x_i}(Z_t)\,dA_t^i. \tag{49'}$$

Remark. If $\mathbb{R}^n$ is replaced by a differentiable manifold, then there is a corresponding generalization of formula (49). Similar and more abstract extensions are also of current interest in both research and applications. A brief account of the latter is given in the last chapter.

We state Itô's original result for independent Brownian motions (so that $\langle X^i, X^j\rangle_t = 0$ for $i \neq j; = t$ for $i = j$ and take $A = 0$) from (49) for reference.

17. Corollary. *Let X^i be independent Brownian motions with variance parameter unity, $i = 1, \dots, n$; hence $X_0^i = 0$, and $X^i \in \mathcal{M}^c$ also. Then for any twice continuously differentiable $f : \mathbb{R}^n \to \mathbb{C}$ we have*

$$f(X_t) = f(0) + \sum_{i=1}^{n} \int_0^t \frac{\partial f}{\partial x_i}(X_s)\, dX_s^i + \frac{1}{2}\sum_{i=1}^{n} \int_0^t \frac{\partial^2 f}{\partial x_i^2}(X_s) ds, \tag{50}$$

where $X_t = (X_t^1, \dots, X_t^n), t \geq 0$.

To see the special nature of this integral, let $f(x) = x^2$ in (50) when $n = 1$. Then $\langle X\rangle_0 = 0, X_0 = 0$, a.e., and

$$X_t^2 = 2\int_0^t X_s\, dX_s + t. \tag{51}$$

This never reduces to the Lebesgue-Stieltjes integral for the Brownian motion. Moreover (51) shows an extra term for the "integration by parts" formula, which is different from the classical case, and is a peculiarity of the Itô integral.

(c) Relations with Brownian motion. In the preceding subsection, we have seen some special properties of the integral related to Brownian motion. The results are better understood with the following characterization theorem, due to P. Lévy, and related results.

18. Theorem. *Let $X \in \mathcal{M}_{loc}^c$ and $\langle X\rangle_t = t, t \geq 0$. Then X is a Brownian motion process. This means: (a) for $0 \leq s < t, X_t - X_s$ is independent of $\mathcal{F}_s$, and (b) $X_t - X_s$ is a Gaussian random variable with mean zero and variance $t - s$.*

Proof. Let $f_y : \mathbb{R} \to \mathbb{C}$ of (41) be given by $f_y(t) = e^{ity}, y \in \mathbb{R}$. Then taking $A = 0$ in (41) we have for $s < t$,

$$f_y(X_t) - f_y(X_s) = \int_s^t \frac{df_y}{dx}(X_u)\, dX_u + \frac{1}{2}\int_s^t \frac{d^2 f_y}{dx^2}(X_u)\, d\langle X\rangle_u. \quad (52)$$

Since $\langle X\rangle_t = t$ and $\frac{df_y}{dx}(X_u) = iye^{iyX_u} \in \mathcal{L}^2(\langle X\rangle)$, it follows that the first integral of (52) ($=I_{f,s,t}$, say) is a martingale increment so that $E^{\mathcal{F}_s}(I_{f,s,t}) = 0$ a.e. Hence for any $B \in \mathcal{F}_s$, integrating both sides of (52) on B we get,

$$\int_B e^{iyX_t}\, dP - \int_B e^{iyX_s}\, dP = \frac{1}{2}\int_B \int_s^t (-y^2 e^{iyX_u}) du\, dP.$$

Dividing by the $\mathcal{F}_s$-measurable non-zero random variable e^{iyX_u}, remembering the averaging property of the conditional expectation, and using Fubini's theorem on the right, the above becomes

$$\int_B e^{iy(X_t - X_s)}\, dP = P(B) - \frac{1}{2}y^2 \int_s^t \int_B e^{iy(X_u - X_s)} dP\, du. \quad (53)$$

Let $p_y(B : t, s) = \int_B e^{iy(X_t - X_s)} dP$. Then (53) gives the functional equation

$$p_y(B : t, s) = P(B) - \frac{1}{2}y^2 \int_s^t p_y(B : u, s)\, du.$$

For fixed B and s, this has a unique solution subject to the obvious boundary conditions: $p_y(B : s, s) = 0, p_0(B : t, s) = P(B)$, and $|p_y(B : t, s)| \leq P(B)$. Namely, $p_y(B : t, s) = P(B)\exp[-\frac{1}{2}y^2(t - s)]$. If $B = \Omega$, then $p_y(\Omega : t, s) = \exp[-\frac{1}{2}y^2(t - s)]$, which implies (by the uniqueness theorem for characteristic functions, cf. I.5.1) that $X_t - X_s$ is Gaussian with the properties in (b). But $E((X_t - X_u)(X_u - X_s)) = E((X_u - X_s)E^{\mathcal{F}_u}(X_t - X_u)) = 0$ for $s < u < t$. So the increments are uncorrelated and hence independent since they are Gaussian distributed. Thus (a) is also true. □

From this we can deduce a classical Brownian motion characterization, also due to P. Lévy.

19. Corollary. *If $X \in \mathcal{M}_{loc}^c, X = 0$ a.e., and $\{X_t^2 - t, \mathcal{F}_t, t \geq 0\}$ is a martingale then X is a Brownian motion process and conversely.*

Proof. If $f(x) = x^2$ in (41), using (52) with $s = 0$, we obtain $X_t^2 = 2\int_0^t X_u \, dX_u + \langle X \rangle_t$. By hypothesis $\{X_t^2 - t, \mathcal{F}_t, t \geq 0\}$ is a martingale and hence also is $2\int_0^t X_u \, dX_u$. Thus by subtraction $\{\langle X \rangle_t, \mathcal{F}_t, t \geq 0\}$ is a martingale so that its expectation is a constant, which is zero when $t = 0$. Hence $E(\langle X \rangle_t) = t$. But $\{\langle X \rangle_t, \mathcal{F}_t, t \geq 0\}$ is a continuous submartingale of class (DL). Hence we have two representations of this process as:

$$\langle X \rangle_t = (\langle X \rangle_t - t) + t = 0 + \langle X \rangle_t. \tag{54}$$

Since $\langle X \rangle \in (\mathcal{A}^+)^c$, so that $\{\langle X \rangle_t, t \geq 0\}$ admits two Doob-Meyer decompositions; but by the uniqueness part of the representation in Theorem 2.6, applied to $\{-\langle X \rangle_t, \mathcal{F}_t, t \geq 0\}$, we must have $\langle X \rangle_t = t$ a.e. Then by the above theorem, X is a Brownian motion process. The converse is immediate. □

Just as in Theorem 16, we have the following n-dimensional version of Theorem 18 which will be similarly stated.

20. Theorem. *Let $X^i \in \mathcal{M}_{loc}^c, i = 1, \ldots, n$, and $\langle X^i, X^j \rangle_t = t$, if $i = j$, and $= 0$ otherwise. Then the vector process $X = (X^1, \ldots, X^n)$ is an n-dimensional Brownian motion, i.e., for each $u \in \mathbb{R}^n$, if $Y_t = (u, X_t) = \sum_{i=1}^n u_i X_t^i$, then $Y = \{Y_t, \mathcal{F}_t, t \geq 0\}$ satisfies conditions (a) and (b) of the above theorem.*

The preceding result does not give all the information regarding the intimate relations between Brownian motions and square integrable martingales. It is an important and a relatively deep result stating that each nonconstant martingale of class $\mathcal{M}^c$ can be transformed into a Brownian motion under a suitable stopping time transformation. This was independently proved by Dambis [1], and Dubins and Schwarz [1]. Because of its interest in applications, we shall present the proof here essentially following Kunita and S.Watanabe [1].

21. Definition. Let $\{\mathcal{F}_t, t \in I \subset \mathbb{R}^+\}$ be a right continuous net of σ-subalgebras of a probability space (Ω, Σ, P). Let $\mathcal{T} = \{T_j, j \in J\}, J \subset \mathbb{R}^+$, be a stopping time process of $\{\mathcal{F}_t, t \in I\}$. Then $\mathcal{T}$ is called a *time change*, for a process $\{X_t, \mathcal{F}_t, t \in I\}$, if $T : J \times \Omega \to I$ is right continuous in j for a.a. (ω). The function T is an (*continuous* or *strict*) integrable time change if moreover it (is continuous or strictly increasing for a.a. (ω) and) satisfies $E(T_j) < \infty, j \in J$.

Note that any $A \in \mathcal{A}^+$ qualifies to be a time change transformation. If $\{T_j, j \in J\}$ is any time change for a process $X = \{X_t, \mathcal{F}_t, t \in I\}$ and if $Y_j = X \circ T_j, \mathcal{B}_j = \mathcal{B}(T_j)$, then $\{Y_j, \mathcal{B}_j, j \in J\}$ is the transformed process which is a martingale (by Corollary IV.1.12) when the X-process is a uniformly integrable martingale. To get some insight into the structure of the transformed processes, we begin with an elementary example of a Gaussian process which is taken into a Brownian motion.

22. Lemma. *Let $\{X_t, t \in [0,1]\}$ be a Gaussian process with $E(X_t) = m_t$ and covariance $r(s,t) = u(s \wedge t)v(s \vee t), 0 \leq s,t \leq a$, where u, v are continuous nondecreasing functions on $[0,a]$ such that $u(0) = 0$ and $\frac{u}{v}$ is real and strictly increasing (so r is a covariance function). Then, if $b^{-1}(t) = \frac{u(t)}{v(t)}$, and $Y_t = X_{b(t)}, Z_t = \frac{Y_t - m_{b(t)}}{v(b(t))}$, we have $\{Z_t, t \in [0, b^{-1}(a)]\}$ to be a Brownian motion [and $(b(t), t \in [0,a])$ is the strict time change] transformed from the $L^2(P)$-martingale $\{X_t - m_t, \mathcal{F}_t, t \in [0,a]\}$ where $a > 0$.*

Proof. The hypothesis implies that the mapping $t \mapsto b(t)$ is one-to-one, order-preserving, and continuous from $[0, b^{-1}(a)]$ onto $[0,a]$. Let $0 < t_1 < \ldots < t_n < b^{-1}(a)$ and $s_k = b(t_k)$. Then $Y_{t_i} = X_{s_i}$ and

$$P\{\omega : Y_{t_i}(\omega) < a_i, i = 1, \ldots, n\} = P\{\omega : X_{s_i} < a_i, i = 1, \ldots, n\}$$

so that the Y-process as well as the Z-process are Gaussian, and $E(Z_t) = 0$. Also for t_i, s_i as above we have

$$E(Z_{t_1} Z_{t_2}) = \frac{r(s_1, s_2)}{v(s_1)v(s_2)} = t_1 \wedge t_2.$$

Thus the Z-process is a Brownian motion.

If $\bar{Y}_t = X_t - m_t$, then it is a Gaussian process with independent increments. To see this, let $u_i = u(s_i), v_i = v(t_i)$ with $0 < s_1 < \ldots < s_n < a, s_i = b(t_i)$, and the n-dimensional Gaussian density of $\bar{Y}_s$'s is explicitly computed to be

$$f_{s_1,\ldots,s_n}(x_1, \ldots, x_n) = (2\pi)^{-n/2}(\det R_n)^{-1/2} \times \exp\Big\{-\frac{1}{2}\sum_{k=1}^{n} \frac{(v_{k-1}x_k - v_k x_{k-1})^2}{v_{k-1}v_k(u_k v_{k-1} - u_{k-1}v_k)}\Big\}, \qquad (55)$$

where $R_n = (r(s_i, s_j), 1 \le i, j \le n)$ and

$$\det R_n = \prod_{i=1}^{n+1} (u_i v_{i-1} - u_{i-1} v_i), \; u_0 = 0 = v_{n+1}, \; v_0 = 1 = u_{n+1}.$$

Then $\{\bar{Y}_t, \mathcal{F}_t, t \ge 0\}$ is an $L^2(P)$-martingale for the base $\mathcal{F}_t = \sigma\{\bar{Y}_s, s \le t\} = \sigma\{X_s, s \le t\}$. $\square$

Remark. If $u(t) = t, v(t) = 1, t \in [0, a]$, then one gets the density of the standard Brownian motion from (55).

We now present the general result on the characterization problem announced earlier.

23. Theorem. *Let $\{\mathcal{F}_t, t \ge 0\}$ be a standard filtration from a complete probability space (Ω, Σ, P). If $X_t = (X_t^1, \dots, X_t^n)$, then let $\{X_t, \mathcal{F}_t, t \ge 0\}$ be a (separable) process almost all of whose sample paths are continuous and $X_0 = a \in \mathbb{R}^n$ a.e. Assume the following conditions: (a)$X^i - X_0^i \in \mathcal{M}_{loc}^c, 1 \le i \le n$, (b) for $i \ne j$, $X^i X^j - X_0^i X_0^j \in \mathcal{M}_{loc}^c, 1 \le i, j \le n$, (c) for all $1 \le i, j \le n, [(X^i)^2 - (X^j)^2] - [(X_0^i)^2 - (X_0^j)^2] \in \mathcal{M}_{loc}^c$, and (d) $P[\sup_{s,t \in I} |X_s - X_t| > 0] = 1$ for each open interval $I \subset \mathbb{R}^+$. Then there exists a Brownian motion $\{Y_t, \mathcal{B}_t, t \ge 0\}$ such that the X-process is obtained from the Y-process by a continuous strict time change of $\{\mathcal{B}_t, t \ge 0\}$. Here (Ω, Σ, P) may be assumed rich enough (by an adjunction procedure if necessary) to support both these processes so that there exists $\mathcal{T} = \{T_t, t \ge 0\}$, a strict time change of $\{\mathcal{B}_t, t \ge 0\}$ and $X_t = Y \circ T_t$. This may be expressed symbolically as $X = Y \circ \mathcal{T}$ or $Y = X \circ \mathcal{T}^{-1}$, with $\mathcal{T}^{-1} = \{T_t^{-1}, t \ge 0\}$ being a stopping time process of $\{\mathcal{F}_t, t \ge 0\}$. [The symbol $|\cdot|$ in (d) is the norm of $\mathbb{R}^n$.]*

Proof. Note that conditions (b) and (c) are vacuous if $n = 1$, and the general formulation uses Theorem 16 to deduce the desired time change transformation by considering the inverses of $\langle X^i, X^j \rangle$, and then employing Theorem 20. We present the details in steps for convenience.

I. Let $f : \mathbb{R}^n \to \mathbb{R}$ be given by $f(x) = x_i x_j$. Then $\frac{\partial f}{\partial x_i} = x_j, \frac{\partial^2 f}{\partial x_i \partial x_j} = 1$ and hence (49) becomes (with $A = 0, i \ne j$ there):

$$\begin{aligned} f(X_t) - f(X_0) &= X_t^i X_t^j - X_0^i X_0^j \\ &= \int_0^t X_s^j \, dX_s^i + \int_0^t X_s^i \, dX_s^j + \frac{1}{2} \int_0^t d\langle X^i, X^j \rangle_s. \end{aligned} \tag{56}$$

By (b) and (49) the left side and the first two terms on the right define elements of $\mathcal{M}^c_{loc}$. Hence the last term is also in this space by linearity of the latter. But $\langle X^i, X^j\rangle \in \mathcal{A}^c_{loc}$ (by definition) and by the remark after the proof of Proposition 2, $\mathcal{M}_{loc} \cap \mathcal{A}_{loc} = \{0\}$, so that $\langle X^i, X^j\rangle = 0, i \neq j$. If $f(x) = x_i^2 - x_j^2$, then (49) yields

$$f(X_t) - f(X_0) = 2\Big(\int_0^t X^i_s\, dX^i_s - \int_0^t X^j_s\, dX^j_s\Big) + \langle X^i, X^i\rangle_t - \langle X^j, X^j\rangle_t. \tag{57}$$

By (c) the left side and the two integrals on the right define elements of $\mathcal{M}^c_{loc}$, and by linearity $\langle X^i, X^i\rangle - \langle X^j, X^j\rangle \in \mathcal{M}^c_{loc} \cap \mathcal{A}^c_{loc}$, so that as before this implies $\langle X^i, X^i\rangle = \langle X^j, X^j\rangle$. Hence this does not depend on i or j, and we let $A_t = \langle X^i, X^i\rangle_t = \langle X^j, X^j\rangle_t, t \geq 0$, and $A \in (\mathcal{A}^+_{loc})^c$. [We observe that if (b) is also assumed to hold for $i = j$, the result becomes X^i_t=constant a.e., and would contradict (d). Thus (b) and (c) are two different conditions.] It is to be shown that this A satisfies all the requirements.

II. We claim that $\{A_t, t \geq 0\}$ is strictly increasing for a.a. (ω). In fact, assuming this to be false we derive a contradiction. First let us make a simplification. Since $A \in (\mathcal{A}^+)^c_{loc}$, there exists (by definition) a stopping time process $\{T_n, n \geq 1\}$ of the filtration, $T_n \uparrow \infty$, such that if $A^n_t = A_{t\wedge T_n}$ then $A^n \in (\mathcal{A}^+)^c$ for $n \geq 1$. If $A_{t_2} - A_{t_1} (t_1 < t_2)$ is constant on a set of positive P-measure, and if $\tilde{T}_n = T_n \vee t_2, \tilde{A}^n_t = A_{t\wedge \tilde{T}_N}$, then $\tilde{A}^n \in (\mathcal{A}^+)^c, E(\tilde{A}^n_t) < \infty$ and $\tilde{A}^n$ is a constant on $[t_1, t_2]$ for an ω-set of positive P-measure. Hence for the present argument we may and do assume that $E(A_t) < \infty, t \geq 0$ and $A_{t_2} - A_{t_1}$ is constant on a set of positive P-measure. Thus there is a $\delta > 0$ and a rational $r \geq 0$ such that $P\{\omega : A_{r+\delta}(\omega) = A_r(\omega)\} > 0$.

Let T_r be the inverse of $A_t, t > r$, so that

$$T_r(\omega) = \inf\{t > r : A_t(\omega) > A_r(\omega)\},\ \inf(\emptyset) = \infty. \tag{58}$$

Then the continuity of A implies that $\{T_t, t \geq r\}$ is a stopping time process of $\{\mathcal{F}_t, t \geq r\}$ and $T_r \geq r + \delta$ on a set of positive measure, i.e., $P[T_r > r] > 0$. Note also that $A_{T_r} = A \circ T_r = r$ a.e. Given $\varepsilon > 0$, define

$$S^\varepsilon_r(\omega) = \inf\{t > r : |X_t - X_r|(\omega) \geq \varepsilon\}.$$

Then as before S_r^ε is an optional of the same filtration, and for any $r_1 > r$, the following inclusions hold a.e.

$$\begin{aligned}
[T_r > S_r^\varepsilon] \cap [S_r^\varepsilon \geq r_1, \forall \varepsilon > 0] \\
&\subset \Big[\sup_{r \leq s \leq r_1} |X_s - X_r| \leq \varepsilon, S_r^\varepsilon < T_r, \forall \varepsilon > 0\Big] \\
&\subset \Big[\sup_{r \leq s \leq r_1} |X_s - X_r| = 0\Big].
\end{aligned}$$

By (d) the last set on the right is P-null. Hence as $\varepsilon \downarrow 0, S_r^\varepsilon(\omega) \downarrow r$ a.e. This and the earlier fact that $P[T_r > r] > 0$ imply the existence of an $\eta > 0$ and some $r', r < r' < r + \delta$ such that if $B(r', \eta) = [T_r \wedge r' > S_r^\eta]$, then $B(r', \eta) \in \mathcal{B}(S_r^\eta)$ and $P(B(r', \eta)) > 0$. But we also have by (a) and Proposition 1 (with $A_t = A(t), X_t = X(t)$),

$$\begin{aligned}
E^{\mathcal{B}(S_r^\eta)}[A(T_r \wedge r' \wedge S_r^\eta) - A(r)] &= E^{\mathcal{B}(S_r^\eta)}[(X^i(T_r \wedge r' \wedge S_r^\eta) - X^i(r))^2] \\
&= \frac{1}{n} E^{\mathcal{B}(S_r^\eta)}[|X(T_r \wedge r' \wedge S_r^\eta) - X(r)|^2]
\end{aligned} \tag{59}$$

and the left side is zero a.e. on $B(r', \eta) \subset [A_{r+\delta} = A_r]$ since $r' \leq r + \delta$. Hence multiplying (59) by $\chi_{B(r', \eta)}$ and integrating we get

$$0 = \frac{1}{n} \int_{B(r', \eta)} |X(T_r \wedge r' \wedge S_r^\eta) - X(r)|^2 \, dP > \frac{\eta^2}{n} P(B(r', \eta)) > 0.$$

This contradiction proves our original assertion that A_t is strictly increasing for a.a.(ω).

III. To complete the construction, let $\{T_u, u > 0\}$ be the inverse of $\{A_t, t \geq 0\}$ as in (58), which exists by Step II. Thus for $u > 0$,

$$T_u = \inf\{t > 0 : A_t > u\}, \ (\inf(\emptyset) = \infty), \tag{60}$$

and $\{T_u, u > 0\}$ is a strict time change for $\{\mathcal{F}_t, t > 0\}$. The same is true of $\{T_u \wedge s, u \geq 0\}$ for each $s > 0, T_u \wedge s \uparrow s$ as $u \uparrow \infty$ since $T_u \uparrow \infty$. Then for $u_1 \leq u_2$, and $s_1 \leq s_2$ we have $T_{u_1} \wedge s_1 \leq T_{u_2} \wedge s_2$ and X^i transforms into a square integrable martingale under this (bounded) strict time change by Corollary IV.1.8. Thus

$$E^{\mathcal{B}(T_{u_1} \wedge s_1)}(X^i(T_{u_2} \wedge s_2)) = X^i(T_{u_1} \wedge s_1), \text{ a.e.,} \quad i = 1, \ldots, n. \tag{61}$$

Since $X^i(T_u \wedge s)$ is square integrable (X^i_s having that property), we have by Proposition 1,(δ_{ij} is the Kronecker delta)

$$\begin{aligned} E^{\mathcal{B}(T_{u_1}\wedge s_1)}[(X^i(T_{u_2}\wedge s_2) - X^i(T_{u_1}\wedge s_1))\times \\ (X^j(T_{u_2}\wedge s_2) - X^j(T_{u_1}\wedge s_1))] \\ = E^{\mathcal{B}(T_{u_1}\wedge s_1)}[\delta_{ij}(A(T_{u_2}\wedge s_2) - A(T_{u_1}\wedge s_1))], \text{ a.e.} \end{aligned} \tag{62}$$

However $A(T_u \wedge s) \uparrow A(T_u)$ a.e. as $s \uparrow \infty$, for each u, and by (60) $A \circ T_u = u \wedge A_\infty$ a.e. Thus in (62) if one sets $i = j, s_1 = 0$ then since $u_2 > A(0)$, it follows that ($X_0 = a \in \mathbb{R}^n$ is now used)

$$\begin{aligned} E[(X^i(T_{u_2}\wedge s_2) - a^i]^2 &= E(A(T_{u_2}\wedge s_2)) - E(A_0) \\ &\leq E(u_2 \wedge A_\infty) < \infty. \end{aligned} \tag{63}$$

Since $s_2 > 0$ is arbitrary, we get the transformed martingale to be L^2- bounded and hence uniformly integrable. So $\lim_{s\to\infty} X^i(T_u \wedge s) = \tilde{X}^i_u$ exists a.e. and in $L^2(P)$. If one sets $\tilde{\mathcal{B}}_u = \lim_{s\to\infty} \mathcal{B}(T_u \wedge s) = \sigma\big(\cup_{s>0}\mathcal{B}(T_u\wedge s)\big)$, then, on noting that the projections $E^{\mathcal{B}(T_u\wedge s)} \to E^{\tilde{\mathcal{B}}_u}$ strongly (in $L^p, p < \infty$), we have, with probability one, that

$$E^{\tilde{\mathcal{B}}_{u_1}}(\tilde{X}^i_{u_2}) = \tilde{X}^i_{u_1}, \text{ a.e., } u_1 \leq u_2. \tag{64}$$

The same argument applied to (62) yields that $\{\tilde{X}^i_u, \tilde{\mathcal{B}}_u, u > 0\}$ is a square integrable martingale and that for $1 \leq i, j \leq n$,

$$E^{\tilde{\mathcal{B}}_{u-1}}[(\tilde{X}^i_{u_2} - \tilde{X}^i_{u_1})(\tilde{X}^j_{u_2} - \tilde{X}^j_{u_1})] = \delta_{ij}E^{\tilde{\mathcal{B}}_{u_1}}(A_\infty \wedge u_2 - A_\infty \wedge u_1), \text{ a.e.} \tag{65}$$

By Proposition 1, it now follows that $\langle \tilde{X}^i, \tilde{X}^j\rangle = \delta_{ij}(A_\infty \wedge u)$ a.e. If $A_\infty = \infty$ a.e. then this and Theorem 20 would show that $\{\tilde{X}_t - \tilde{X}_0, \tilde{\mathcal{B}}_t, t \geq 0\}$ is a Brownian motion process where $\tilde{X}_t = (\tilde{X}^1_t, \ldots, \tilde{X}^n_t)$. It therefore remains to consider the case that $A_\infty < \infty$ on a set of positive P-measure.

IV. Let $\{Z_t, \mathcal{G}_t, t \geq 0\}$ be a standard n-dimensional Brownian motion on (Ω, Σ, P) which is independent of the given process. If the original probability space is not rich enough to support such a process as well as the original one, we enlarge it with an adjunction procedure (cf. II.3.6(ii)). Supposing this has already been done, we define a Y-process on Ω:

$$Y^i_t = Z^i_t - Z^i_{t\wedge A_\infty} + \tilde{X}^i_t, \; t \geq 0, \; 1 \leq i \leq n. \tag{66}$$

Since the Z_t and $\tilde{X}_t$ processes are independent and are in $L^2(P)$, with continuous sample paths, the same is true of $\{Y_t, t \geq 0\}$ as well. Also $\langle X^i, Z^j \rangle_u = 0, \langle Z^i, Z^j \rangle_u = \delta_{ij} u, 1 \leq i,j \leq n$,a.e., and $\{A_t, \mathcal{F}_t^*, t \geq 0\}$ is a stopping time process where $\mathcal{F}_t^* = \tilde{\mathcal{B}}_t \otimes \mathcal{G}_t$. Indeed, $[A_t > u] = [T_u < t] = \cap_{s>0} [T_u \wedge s < t] \in \sigma\left(\cup_{s>0} \mathcal{B}(T_u \wedge s)\right) = \tilde{\mathcal{B}}_u$ which may be identified (cylindrically) as a subalgebra of $\mathcal{F}_u^*$. Thus,

$$\begin{aligned} \langle Y^i, Y^j \rangle_u &= \langle Z^i - Z^i_{(\cdot)\wedge A_\infty}, Z^j - Z^j_{(\cdot)\wedge A_\infty} \rangle_u + \langle \tilde{X}^i, \tilde{X}^j \rangle_u \\ &= \delta_{ij} [u - u \wedge A_\infty + u \wedge A_\infty] = \delta_{ij} u \text{ a.e.} \end{aligned} \tag{67}$$

Here we have used the fact that $Z_t^i - Z_{t\wedge A_\infty}^i$ is a Gaussian process starting at $Z_{t\wedge A_\infty}^i$ so that $\langle Z^i - Z^i_{(\cdot)\wedge A_\infty} \rangle_u = u - u \wedge A_\infty$ a.e. (since A_∞ and the Z-process are independent). Thus by Theorem 20, $\{Y_t - Y_0, \mathcal{F}_t^*, t \geq 0\}$ is a Brownian motion. [With the (additional) hypothesis that $A_0 = a$ a.e., we have $Y_0 = a$ a.e.] Now

$$\begin{aligned} Y \circ A_t &= Z \circ A_t - Z \circ (A_t \wedge A_\infty) + \tilde{X} \circ A_t \\ &= X(T \circ A_t) = X_t \text{ a.e.} \end{aligned} \tag{68}$$

Hence the X-process is obtained from the Brownian motion Y by a (strict) time change function $\{A_t, t \geq 0\}$ of $\{\mathcal{F}_t^*, t \geq 0\}$. $\square$

The following consequences of the general result will be recorded for ready reference.

24. Corollary. *Let $\{X_t, \mathcal{F}_t, t \geq 0\}$ be a process with continuous sample paths defined on (Ω, Σ, P) and starting at $X_0 = a \in \mathbb{R}^n$. Suppose that for each $h : \mathbb{R}^n \to \mathbb{R}$ such that $\Delta^2 h = 0$, where $\Delta^2 = \frac{\partial^2}{\partial x_1^2} + \ldots + \frac{\partial^2}{\partial x_n^2}$ is the Laplacian (i.e. h is a harmonic function) $\{h(X_t) - h(X_0), \mathcal{F}_t, t \geq 0\} \in \mathcal{M}_{loc}^c$. If condition (d) of the theorem is assumed then the X_t-process is obtained from an n-dimensional Brownian motion starting at a, under a time change transformation.*

Since $h(x) = x_i x_j$, or $= x_i^2 - x_j^2$ is harmonic, the result is a restatement of the theorem. Note that if $n = 1, h(x) = ax + b$ then one can verify Lemma 22 is subsumed immediately. An extension of the above theorem is given by Knight [1].

25. Corollary. *Let $\{X_t, \mathcal{F}_t, t \geq 0\}$ be a continuous square integrable sub(or super)-martingale where the filtration is a standard one. Then almost every sample function $X(\cdot, \omega) : \mathbb{R}^+ \to \mathbb{R}$ is either of unbounded* *variation or monotone on every nondegenerate interval. If the process is a martingale, and is of bounded variation on* $\mathbb{R}^+$, *then* $X_t = X_0$ *a.e.*

In fact, by Theorem 6, $X_t = Y_t + A_t$ (or $= Y_t - A_t$) where $Y = \{Y_t, \mathcal{F}_t, t \geq 0\} \in \mathcal{M}^c$, and $\{A_t, t \geq 0\} \in (\mathcal{A}^+)^c$. If Y is nonconstant, then by the above theorem $Y_t = Z \circ A_t$, where Z is a Brownian motion, and $A_t = \langle Y, Y \rangle_t$. From this the result follows, and it is also due to Dambis [1].

The following special case has some interesting information.

26. Corollary. *Let* $X - X_0 \in \mathcal{M}^c$ *and suppose that its increasing process* $\langle X \rangle$ *has absolutely continuous paths so that* $\langle X \rangle_t = \langle X \rangle_0 + \int_0^t \varphi(s)\, ds$ *a.e. If* $\varphi(s) > 0$ *for a.a.(s) and a.a.* (ω), *then there is a Brownian motion* $\{X_t, \mathcal{F}_t, t \geq 0\}$ *such that*

$$X_t - X_0 = \int_0^t [\varphi(s)]^{1/2}\, dY_s, \text{ a.e.} \tag{69}$$

Proof. Since $A_t = \langle X \rangle_t = \int_0^t dA_s + \langle X \rangle_0$, taking $X_0 = 0$ for simplicity, we see that condition (d) of the theorem is satisfied by the fact that $\varphi(s) > 0$ (and Theorem II.4.3). Hence there is a $\{Y_t, t \geq 0\}$ as above with $X_t = Y \circ A_t$. However, if $B_t = A_t^{-1} = \int_0^t \varphi(s)^{-1}\, ds$, and $\mathcal{T} = \{A_t, t \geq 0\}, \mathcal{T}^{-1} = \{B_t, t \geq 0\}$, then $X_t = \mathcal{T}\mathcal{T}^{-1}X_t = \mathcal{T}Y_t = Y \circ A_t$ and $Y_t = \mathcal{T}^{-1}X_t$. Since A is predictable, it follows that $\varphi^{-1/2} \in \mathcal{L}(\langle X \rangle)$. In fact,

$$E\Big(\int_0^t \varphi^{-1/2}(s)\, dA_s\Big) \leq \Big[E\int_0^t \varphi^{-1}(s)\, d\langle X \rangle_s\Big]^{1/2} \Big[E\int_0^t 1 \cdot d\langle X \rangle_s\Big]^{1/2}$$
$$= t^{1/2} E(\langle X \rangle_t) < \infty.$$

So $Y_{(\cdot)} = \int_0^{(\cdot)} \varphi^{-1/2}(s)\, dX_s (\in \mathcal{M}^c)$ by Corollary 12. But $\langle Y, Y \rangle_t = t$; whence the result follows from the theorem. □

(d) Orthogonal decomposition. The preceding theory uses elementary parts of Hilbert space geometry. This will be further illuminated by presenting an orthogonal decomposition of $\mathcal{M}$ relative to its family of semi-norms $\{\|\cdot\|_t, t \geq 0\}$, where $\|X\|_t^2 = E(|X_t|^2)$.

27. Definition. If X, Y are in $\mathcal{M}$, we say that they are *orthogonal* martingales and denote $X \perp Y$, iff $\langle X, Y\rangle_t = 0$ a.e. for all $t \geq 0$, or equivalently, iff $\{X_t Y_t, \mathcal{F}_t, t \geq 0\}$ is a martingale. [The equivalence is a consequence of Theorem 16.]

Using the sesquilinear nature of $\langle\cdot,\cdot\rangle$, one has that $X \perp Y$ iff

$$\langle X+Y, X+Y\rangle_t = \langle X, X\rangle_t + \langle Y, Y\rangle_t, \text{ a.e.}, \quad t \geq 0. \tag{70}$$

In the martingale language we can state (70) as: if $X, Y \in \mathcal{F}$, then $\langle X\rangle + \langle Y\rangle$ is the increasing integrable process of $X+Y$ iff $X \perp Y$, or iff $\{(X+Y)_t^2 - \langle X\rangle_t - \langle Y\rangle_t, \mathcal{F}_t, t \geq 0\}$ is a martingale.

We already remarked that the semi-norms above can be used to introduce a topology in $\mathcal{M}$. The result of interest is the following:

28. Proposition. *The set $\{\mathcal{M}, \|\cdot\|_t, t \geq 0\}$ is a complete locally convex vector (in fact a Fréchet) space and $\mathcal{M}^c(\subset \mathcal{M})$ is a closed subspace. Moreover, for each pair $X, Y \in \mathcal{M}$, if $\mathcal{L}(X) = \{V : V = \int_0^t Z_s\, dX_s, Z \in \mathcal{L}^2(\langle X\rangle), t \geq 0\}$, then there exist $Y', Y'' \in \mathcal{M}$ such that $Y = Y' + Y''$ uniquely, where $Y' \in \mathcal{L}(X), Y'' \perp V$, for all $V \in \mathcal{L}(X)$.*

Proof. Let $\{X^n, n \geq 1\} \subset \mathcal{M}$ be a Cauchy sequence for the semi-norms so that $\|X^n - X^m\|_t \to 0$ for each $t \geq 0$ as $m, n \to \infty$. Since $\{X_t^2, \mathcal{F}_t, t \geq 0\}$ is a submartingale, we have by the maximal inequality, for $\lambda \in \mathbb{R} - \{0\}$

$$P\Big[\sup_{0 \leq s \leq t} |X_t^n - X_s^m|^2 > \lambda^2\Big] \leq \frac{1}{\lambda^2}\|X^n - X^m\|_t^2 \to 0, \tag{71}$$

as $n, m \to \infty$. Hence $X_t^n \to X_t$ a.e. uniformly in compact neighborhoods of t, and $X_t \in L^2(\mathcal{F}_t, P)$. If $0 \leq s < t$, then we conclude from this (for a subsequence, denoted by the same symbols) the following:

$$E^{\mathcal{F}_s}(X_t) = \lim_n E^{\mathcal{F}_s}(X_t^n) = \lim_n X_s^n = X_s, \text{ a.e.}$$

Thus $\{X_t, \mathcal{F}_t, t \geq 0\} \in \mathcal{M}$. Since the topology, determined by these semi-norms, is evidently locally convex and Hausdorff, it is clear that X is the limit of $\{X^n, n \geq 1\}$ so that $\mathcal{M}$ is a complete locally convex vector space. If $\{X^n, n \geq 1\} \subset \mathcal{M}^c$, then the same uniformity implies that $X \in \mathcal{M}^c$ so that it is a closed subspace. Note that for $t_1 < t_2$, $\|X\|_{t_1} \leq \|X\|_{t_2}$ since $\{X_t^2, \mathcal{F}_t, t \geq 0\}$ is a submartingale. Choosing a countable dense subset $\{r_i\}_1^\infty \subset \mathbb{R}^+$, and letting $\|\cdot\| = \sum_{i=1}^{\infty} \frac{1}{2^{r_i}} \frac{\|\cdot\|_{r_i}}{1+\|\cdot\|_{r_i}}$, we may also conclude that $\mathcal{M}$ is a complete metric space, or a Fréchet space.

Next since $\mathcal{L}(X)$ is clearly linear, we verify that it is a closed subspace of $\mathcal{M}$. Let $\{Y^n, n \geq 1\}$ be a Cauchy sequence of the space so that $Y_t^n = \int_0^t Z_s^n \, dX_s$ for some $Z^n \in \mathcal{L}^2(\langle X \rangle)$. Hence one has

$$\|Y^n - Y^m\|_t = \|Z^n - Z^m\|_{\langle X \rangle_t} = E\Big(\int_0^t (Z_s^n - Z_s^m)^2 \, d\langle X \rangle_s\Big)^{1/2}, \tag{72}$$

by (17). Hence $\{Z^n, n \geq 1\}$ is Cauchy in $\mathcal{L}^2(\langle X \rangle)$. But the latter space is complete (as seen with (71) above) for the semi-norms. So there is a $Z \in \mathcal{L}^2(\langle X \rangle)$ such that $Z^n \to Z$ in the topology there and thus $Y^n \to Y = \int_0^{(\cdot)} Z_s \, dX_s$ which establishes that the space is closed.

Finally, let X, Y be in $\mathcal{M}$ and $\langle X, Y \rangle \in \mathcal{A}$. If ν and μ are the associated measures for $\langle X, Y \rangle$ and $\langle X \rangle$, then ν is a signed and μ is a σ-finite measure dominating ν on $\Sigma \otimes \mathcal{B}(\mathbb{R}^+)$, since $E(|\int_0^t f_s g_s \, d\langle X, Y \rangle_s|) \leq \|f\|_{\langle X \rangle_t} \|g\|_{\langle Y \rangle_t}, t \geq 0$, by (23). Hence by Corollary 4, there exists a $Z \in \mathcal{L}^2(\langle X \rangle)$ such that

$$\langle X, Y \rangle_t = \int_0^t Z_s \, d\langle X \rangle_s, \text{ a.e.} \tag{73}$$

So if $Y_t' = \int_0^t Z_s \, dX_s, t \geq 0$, then $Y' = \{Y_t', \mathcal{F}_t, t \geq 0\} \in \mathcal{M}$, and by the preceding work $Y' \in \mathcal{L}(X)$. Let $Y'' = Y - Y'$. Then $Y'' \in \mathcal{M}$,and

$$\begin{aligned} \langle X, Y'' \rangle_t &= \langle X, Y \rangle_t - \langle X, Y' \rangle_t \\ &= \langle X, Y \rangle_t - \int_0^t Z_s \, d\langle X, X \rangle_s, \text{ by definition of } Y_t', \\ &= \langle X, Y \rangle_t - \langle X, Y \rangle_t = 0, \text{ a.e. } t \geq 0, \text{ by (73).} \end{aligned} \tag{74}$$

Thus $Y'' \perp \mathcal{L}(X)$. If $Y = \tilde{Y}' - \tilde{Y}'' = Y - Y''$ are two decompositions, then $\tilde{Y}' - Y' = \tilde{Y}'' - Y'' \in \mathcal{L}(X)$ and $(\tilde{Y}'' - Y'') \perp \mathcal{L}(X)$ also. Hence $\langle \tilde{Y}' - Y' \rangle = 0$ so that $Y' = \tilde{Y}'$ and $Y'' = \tilde{Y}''$, in that they are indistinguishable. This is the desired result. □

Remark. If (Ω, Σ, P) is a separable space, then $\mathcal{M}$ is a separable Fréchet space as one can verify. The computations between (72) and (73) further imply the following result: If $X \in \mathcal{M}$ and $\mathcal{L}(X)$ is as in the theorem, then it is also the smallest closed subspace $\mathcal{N}$ of $\mathcal{M}$ containing X and having the property,

$$Y \in \mathcal{N},\, Z \in \mathcal{L}^2(\langle Y \rangle) \Rightarrow \int_0^{(\cdot)} Z_s\, dY_s \in \mathcal{N}. \tag{75}$$

Indeed it is clear that $\mathcal{L}(X) \subset \mathcal{N}$ since by definition, (75) is true for it with $X = Y$. If $Y \in \mathcal{N} - \mathcal{L}(X)$, then $Y \neq aX$ for a constant a since otherwise $Y \in \mathcal{L}(X)$ by linearity. Thus $\mathcal{N} \supset \mathcal{L}(X) \cup \mathcal{L}(Y) \supsetneqq \mathcal{L}(X)$ since $Y \neq 0$. But $\mathcal{L}(X)$ has the property (75) and contains X so that it is smaller than $\mathcal{N}$, contradicting the minimal character of $\mathcal{N}$. Thus $\mathcal{N} = \mathcal{L}(X)$, and this space can be infinite dimensional!

Using the unique decomposition of the above proposition we may define a mapping $P_{\mathcal{N}} : Y \mapsto Y'$ on $\mathcal{M}$ onto $\mathcal{N} = \mathcal{L}(X)$. Then $P_{\mathcal{N}}$ is a linear idempotent (i.e. a projection) operator. Also for each $t \geq 0$,

$$\|P_{\mathcal{N}} Y\|^2_{\langle X \rangle_t} = \|Y'\|^2_{\langle X \rangle_t} \leq \|Y'\|^2_{\langle X \rangle_t} + \|Y''\|^2_{\langle X \rangle_t} = \|Y\|^2_{\langle X \rangle_t},$$

so that $P_{\mathcal{N}}$ is a continuous operator on $\mathcal{M}$. We can prove a general result on orthogonal decompositions as follows.

29. Theorem. *Let $\mathcal{N} \subset \mathcal{M}$ be a closed subspace with property (75). Then, (a) any $Y \in \mathcal{M}$ admits a unique decomposition $Y = Y' + Y''$ where $Y' \in \mathcal{N}, Y'' \perp \mathcal{N}$; (b) there exists a set $\{Y^i, i \in I\} \subset \mathcal{N}$ such that $Y^i \perp Y^j$ for $i \neq j$, and the smallest closed subspace containing $\{Y^i, i \in I\}$ and having property (75) is $\mathcal{N}$; and (c) there is a countable set $\{i_n, n \geq 1\} \subset I$ such that Y' of (a) is given as $Y' = \lim_n \sum_{m=1}^n P_{\mathcal{L}(Y^{i_m})} Y$, the limit being taken in the topology of $\mathcal{M}$.*

Proof. We first establish (b) and then use it to prove (a) and (c). To apply a classical Hilbert space argument, it may be assumed that $\mathcal{N} \neq \{0\}$. If X_1, X_2 are in $\mathcal{N}$, then let $Y_1 = X_1$, so that $Y_2 = X_2 - P_{\mathcal{L}(Y_1)} X_2$ satisfies $Y_1 \perp Y_2, Y_1 \in \mathcal{L}(Y_1)$. If $\mathcal{L}(X_1, X_2)$ is the smallest subspace containing $\{X_1, X_2\}$ and obeying (75), then $\mathcal{L}(X_1, X_2) \supset \mathcal{L}(Y_1) \cup \mathcal{L}(Y_2)$, and so $\mathcal{L}(Y_1, Y_2) \subset \mathcal{L}(X_1, X_2)$. But since $\{Y_1, Y_2\} \subset \mathcal{L}(Y_1, Y_2)$ we deduce that $\{X_1, X_2\} \subset \mathcal{L}(Y_1, Y_2)$ and hence $\mathcal{L}(X_1, X_2) \subset \mathcal{L}(Y_1, Y_2)$

so that there is equality between these two spaces. Let $\mathcal{K}$ be the set of all possible orthogonal systems in $\mathcal{N}$. Clearly $\mathcal{K}$ is nonempty. If $\{Y_1^i\}_i, \{Y_2^j\}_j$ are two elements of $\mathcal{K}$, let $\{Y_1^i\} \prec \{Y_2^j\}$ iff each Y_1^i is some Y_2^j. Then $\{\mathcal{K}, \prec\}$ is partially ordered and if $\mathcal{K}_1 \subset \mathcal{K}$ is any chain then its upper bound is the union which clearly belongs to $\mathcal{K}$. Hence by Zorn's lemma there is a maximal orthogonal system $\{Y^i, i \in I\} \in \mathcal{K}$; the smallest linear closure of this system satisfying (75) is $\mathcal{N}$. [If $\mathcal{N}$ is separable, then we could continue with the Gram-Schmidt type procedure employed, at the begining of this paragraph, avoiding Zorn's lemma.] This yields (b).

For (a) let $Y \in \mathcal{M}$ and set $Z^i = P_{\mathcal{L}(Y^i)}Y \in \mathcal{L}(Y^i), i \in I$. If $\tilde{Z}^n = \sum_{j=1}^n Z^{i_j} \in \mathcal{L}(Y^{i_1}, \ldots, Y^{i_n})$, then $\tilde{Y} = Y - \tilde{Z}^n \perp Y^{i_j}, j = 1, \ldots, n$, and since $\langle \cdot, \cdot \rangle$ is a sesquilinear form continuous for the semi-norms of $\mathcal{M}$ we have, from the decomposition $Y = \tilde{Y} + \tilde{Z}^n$ and orthogonality,

$$\langle Y \rangle = \langle Y, Y \rangle = \langle \tilde{Y}, \tilde{Y} \rangle + \sum_{j=1}^n \langle Z^{i_j}, Z^{i_j} \rangle \geq \sum_{j=1}^n \langle Z^{i_j}, Z^{i_j} \rangle, \tag{76}$$

for all n. Letting $n \to \infty$, it follows from (76) that atmost a countable set of $\{\langle Z^i \rangle\}_{i \in I}$ are non-vanishing. Let this set be $(i_1, i_2, \ldots)$. Since $\sum_{j=1}^n \langle Z^{i_j}, Z^{i_j} \rangle_t$ is increasing and the series converges by (76), we deduce that these partial sums form a Cauchy sequence in $L^2(P)$. Hence $Y' = \sum_{j=1}^\infty Z^{i_j}$ exists (since $t \geq 0$ is arbitrary in the above) and $Y' \in \mathcal{N}$. If $Y'' = Y - Y'$, from the fact that $Z^i = 0$ for i not in the above set, we conclude that $Y - Y' = Y'' \perp Y^i, i \in I$, implying that $Y'' \perp \mathcal{N}$. So $Y = Y' + Y''$ is a decomposition, and the uniqueness is (as in Proposition 28) simple. This establishes both (a) and (c). □

The result of (a) in the theorem shows that for $Y \in \mathcal{M}, Y = Y' + Y''$ uniquely with $Y' \in \mathcal{N}, Y'' \perp \mathcal{N}$ and so $P_{\mathcal{N}} : Y \mapsto Y'$ defines a projection operator on $\mathcal{M}$ with range $\mathcal{N}$. Let $\mathcal{N}^\perp = \{Y \in \mathcal{M} : Y \perp \mathcal{N}\}$. Then $\mathcal{N}^\perp$ is easily seen to be a closed linear subspace of $\mathcal{M}$ so that $\mathcal{N}^\perp = \mathcal{L}(\mathcal{N}^\perp)$ and satisfies (75). We then have the following:

30. Corollary. *If $\mathcal{N}$ is a closed subspace of the Fréchet space $\mathcal{M}$, satisfying (75), then $(\mathcal{N}^\perp)^\perp = \mathcal{N}$.*

Proof. If $Y \in \mathcal{N}, Z \in \mathcal{N}^\perp$ then $\langle Y, Z \rangle = 0 = \langle Z, Y \rangle$. Since Z is arbitrary, $Y \perp \mathcal{N}^\perp$ so $Y \in (\mathcal{N}^\perp)^\perp$, or $\mathcal{N} \subset (\mathcal{N}^\perp)^\perp$. Conversely, if $Y \in (\mathcal{N}^\perp)^\perp \subset \mathcal{M}$, then $Y' = P_{\mathcal{N}}Y \in \mathcal{N}$. So $Y'' = Y - Y' \perp \mathcal{N}$ by the theorem, and $Y'' \in \mathcal{N}^\perp$. Since $Y \in \mathcal{N} \subset (\mathcal{N}^\perp)^\perp$, and $Y \in (\mathcal{N}^\perp)^\perp$ we deduce that $Y'' \in (\mathcal{N}^\perp)^\perp$, on using $\mathcal{L}(\mathcal{N}^\perp) = \mathcal{N}^\perp$. Thus $Y'' \in \mathcal{N}^\perp \cap (\mathcal{N}^\perp)^\perp = \{0\}$ and $Y'' = 0$. Hence $Y = Y' \in \mathcal{N}$ and $\mathcal{N} = (\mathcal{N}^\perp)^\perp$. □

Since $X \in \mathcal{M}^c, Y \in \mathcal{L}^2(\langle X \rangle)$ implies $\{\int_0^t Y_s \, dX_s, \, t \geq 0\} \in \mathcal{M}^c$, it is evident that $\mathcal{M}^c$ satisfies (75). Hence we have:

31. Corollary. *Let $\mathcal{M}$ and $\mathcal{M}^c$ be as above. Set $\mathcal{M}^d = (\mathcal{M}^c)^\perp$. Then $\mathcal{M} = \mathcal{M}^c \oplus \mathcal{M}^d$.*

If $X \in \mathcal{M}$, then $X = Y + Z, Y \in \mathcal{M}^c, Z \in \mathcal{M}^d$, and this result states that Y is the continuous part of X, while Z is orthogonal to every martingale $V \in \mathcal{M}$ without common discontinuities. In fact, $V = V_1 + V_2$ with $V_1 \in \mathcal{M}^c$ so that $V_1 \perp Z$ and if V has no common discontinuities with Z, the same must be true between V_2 and Z. But V_2 and Z have a.e. non-continuous sample paths so that $\{(V_2 Z)_t \geq 0\}$ has the same property. However, $\{(V_2 Z)_t, t \geq 0\}$ must be continuous since otherwise each of its discontinuity points will be a common discontinuity of V_2 and Z. But $(V_2 Z)_0 = 0$ so that $(V_2 Z)_t = 0$, a.e., $t \geq 0$. Thus $V_2 \perp Z$ and $V \perp Z$.

Recall that all the discontinuities of a martingale X are jumps except for at most a countable set of t-points which are fixed discontinuities.

However the latter set in the present case is empty. We thus have the following:

32. Corollary. *Let $\mathcal{M} = \mathcal{M}_1 \oplus \mathcal{M}_2$ where $\mathcal{M}_1 (\subset \mathcal{M})$ satisfies (75), and the decomposition is given by the theorem. Then for any strict time change $\mathcal{T} = \{T_t, t \geq 0\}$ of the filtration $\{\mathcal{F}_t, t \geq 0\}$ such that $X \in \mathcal{M}_1$ implies $\mathcal{T}X \in \mathcal{M}$ (which is denoted $\mathcal{T}(\mathcal{M}_1) \subset \mathcal{M}$), $T_0 = 0$, we also have $\mathcal{T}(\mathcal{M}_2) \subset \mathcal{M}_2$. In particular, let $\mathcal{M}_1 = \mathcal{M}^c$ and $\mathcal{T}$ be continuous, in that $T_{(\cdot)}(\omega) : \mathbb{R}^+ \to \mathbb{R}^+$ is continuous for a.a. (ω). Then $\mathcal{T}(\mathcal{M}^c) \subset \mathcal{M}^c$ and $\mathcal{T}(\mathcal{M}^d) \subset \mathcal{M}^d$.*

It may be noted by applying the Doob-Meyer decomposition to X^2 if $X \in \mathcal{M}$ that $\langle \mathcal{T}X, \mathcal{T}X \rangle = \mathcal{T}\langle X, X \rangle$ and hence $\langle \mathcal{T}X, \mathcal{T}Y \rangle = \mathcal{T}\langle X, Y \rangle$ for $\{X, Y\} \subset \mathcal{M}$. Since $\mathcal{T}$ has a unique inverse, $\mathcal{T}(\mathcal{M}_1) = \mathcal{M}_1$ and $\mathcal{T}(\mathcal{M}_2) = \mathcal{M}_2$ can be deduced. The details will be omitted.

These results show that one can analyze the members of $\mathcal{M}^d$ further, depending on whether a discontinuity is predictable, optional or other. Then one may study pairs of processes from $\mathcal{M}$ using their common (or lack of) discontinuities and their relation to quasi-left-continuous, or s-regular L^2-martingales and so on. (See Dambis [1], Kazamaki [1], and Meyer [4].) We leave this line of analysis at this point, and proceed to consider some related stochastic integrals and later discuss their connections with vector integration.

(e) *Stratonovich and related integrals.* There exist some modifications of the Itô or martingale integrals that are found to be well-adapted in applications. Of these, Stratonovich's became the most popular definition. It is closely related to the classical Riemann integral and is introduced as follows.

Let $Y = \{Y_t, \mathcal{F}_t, t \geq 0\}$ be a bounded predictable process and $X = \{X_t, \mathcal{F}_t, t \geq 0\} \in \mathcal{M}^c$. We have seen that the general Itô integral of Y relative to X can be obtained as the following $L^2(P)$-limit: if $0 = t_0 < t_1 < \ldots < t_n \leq a$ and $\pi_n = \max_i |t_i - t_{i-1}|$ then consider the limits,

$$\lim_{\pi_n \to 0} \sum_{i=0}^{n} Y_{t_i}(X_{t_{i+1}} - X_{t_i}), \ \lim_{\pi_n \to 0} \sum_{i=0}^{n} (Y_{t_i} - Y_{t_i})(X_{t_{i+1}} - X_{t_i}). \quad (77)$$

It can be easily verified that the limits exist as stated and hence in probability and they are respectively $\int_0^a Y_t \, dX_t$ and $\int_0^a d\langle Y, X\rangle_t$. (A more revealing procedure is outlined in Exercise 11, if X is Brownian motion. The above result follows by the same method as in Proposition 15 above. In fact, it is valid if $Y \in \mathcal{M}^c$ after using the polarization identity.)

A noteworthy point about the expressions in (77) is that the Y_t are evaluated at the left boundary of $[t_i, t_{i+1})$ multiplied by $(X_{t_{i+1}} - X_{t_i})$. This is not quite the same as the classical Riemann approximation of the integral which typically involves using $Y_{t'}, t' \in [t_i, t_{i+1})$. Using the midpoint evaluation of Y_t of the interval, R.L. Stratonovich (and independently D. L. Fisk) suggested the approximating sums in lieu of (77) as follows:

$$\lim_{|\pi_n| \to 0} \sum_{t_i \in \pi_n} \frac{1}{2}(Y_{t_i} + Y_{t_{i+1}})(X_{t_{i+1}} - X_{t_i}) = \lim_n I_n', \text{ (say)},$$

if the limit exists in probability. But this can be expressed as:

$$I_n' = \sum_{i=1}^{n} Y_{t_{i+1}}(X_{t_i} - X_{t_{i-1}}) + \frac{1}{2}\sum_{i=1}^{n} (Y_{t_i} - Y_{t_{i-1}})(X_{t_i} - X_{t_{i-1}}), \quad (78)$$

and by (77) both expressions on the right have limits in probability and therefore one has (cf. also Theorem 6 on the quadratic covariation expression):

$$\lim_{n\to\infty} I'_n = \int_0^t Y_s\, dX_s + \frac{1}{2}\int_0^t d\langle Y, X\rangle_s,$$

where the first one is the Itô integral and the second with quadratic covariation (hence of locally bounded variation as a member of $\mathcal{A}$) is a pathwise Stieltjes integral. We denote the left side by the symbol $\int_0^t Y_s \circ dX_s$, and call it the *symmetric* or the *Stratonovich* integral. Thus we have

$$\int_0^t Y_s \circ dX_s = \int_0^t Y_s\, dX_s + \frac{1}{2}\langle Y, X\rangle_t. \tag{79}$$

The usefulness of this apparently simple modification is revealed by the following result that contrasts with (41) which is an unusual "integration by parts" formula.

33. Proposition. *Let $X = \{X_t, \mathcal{F}_t, t \geq 0\} \in \mathcal{M}^c_{loc}, A \in \mathcal{A}^c_{loc}$ and $f : \mathbb{R} \to \mathbb{R}$ be a thrice continuously differentiable function. If $Z = X + A$, then we have for any $t > 0, Y_t = f(Z_t)$:*

$$Y_t - Y_0 = \int_0^t f'(Z_s) \circ dX_s. \tag{80}$$

Proof. Fist observe that Z is a semimartingale and hence so is $f'(Z)$, by Proposition 15, since f' is twice continuously differentiable so that it is a sum of a local martingale and a process of locally bounded variation.

Then we have by the definition of the symmetric integral (cf. (79)),

$$\int_0^t f'(Z_s) \circ dX_s = \int_0^t f'(Z_s)\, dX_s + \frac{1}{2}\langle f'(Z), X\rangle_t$$

$$= \int_0^t f'(Z_s)\, dX_s + \frac{1}{2}\{\langle f'(X_0 + A_0), X\rangle_t +$$

$$\langle f''(X+Z), (X+Z)(X+Z)\rangle_t + 0\}$$

using the Taylor expansion of $f,'$

$$= \int_0^t f'(Z_s)\, dX_s + \frac{1}{2}\{0+0\},$$

since the angle brackets of the terms indicated vanish,

$$= \int_0^t f'(Z_s)\, dX_s = Y_t - Y_0,$$

which is the expression for the ordinary integration by parts. □

This attractive property of the Stratonovich integral is one of the reasons for its use in applications as well as extensions to integation in smooth manifolds. However, the martingale and other (especially Markovian when X is Brownian motion) properties are lost in the latter extension.

Expressing (80) in the symbolic differential notation, as done in (49'), we have

$$dY_t = f'(Z_t) \circ dX_t. \tag{80'}$$

The relation between the Itô and Stratonovich formulations comes out quickly using the symbolic differential calculus which we now present following Itô [3] (cf. also Itô and S. Watanabe [1]). Since by Theorem 16, for any twice continuously differentiable f, $f(Z)$ is a (local) semimartingale whenever Z is, it is evident from (80) that the integrand in Stratonovich integral is restricted to semimartingales in general. Observing that the quadratic variation of elements of $\mathcal{A}$ vanishes, since on $[0,t], 0 \le t_1 < \ldots < t_n \le t$, one has

$$\sum_{i=1}^{n}(A(t_{i+1}) - A(t_i))^2 \le \max|A(t_i) - A(t_{i+1})| \sum_{i=1}^{n}|A(t_i) - A(t_{i+1})|$$

$$\to 0,$$

as $n \to \infty$ (A(t) being right continuous, and having finite local variation), we now present the "multiplication table" for the various differentials.

Let $\mathcal{A}^c_{loc}, \mathcal{M}^c_{loc}$, and $\mathcal{S}^c_{loc}$ be the continuous elements from the classes $\mathcal{A}, \mathcal{M}$, and local semimartingales $\mathcal{S}_{loc}$, adapted to the standard filtration $\{\mathcal{F}_t, t \geq 0\}$. If $\mathcal{B}_{loc}$ denotes the set of all locally bounded predictable processes (for the same filtration), then as a consequence of Theorem 9, and Definition 10, one has that for each $Y \in \mathcal{B}_{loc}$, the Itô integral $(Y \cdot X)_t = \int_0^t Y_s \, dX_s$ is defined and the process $Y \cdot X \in \mathcal{M}^c_{loc}$. Moreover the mapping $Y \mapsto (Y \cdot X)$ is linear. On the other hand from (77) (or (78)) we can directly deduce, for $X, Y \in \mathcal{M}^c$, that $\lim_{\pi_n} \sum_{i=1}^n (X_{t_{i+1}} - X_{t_i})(Y_{t_{i+1}} - Y_{t_i}) = \langle X, Y \rangle_t$ exists, π_n being a partition of $[0, t]$ and the limit is taken as $n \to \infty$. Hence one can introduce the operations of addition, multiplication, and symmetric (or Stratonovich) multiplication with (79), for the elements of $\mathcal{S}^c_{loc}$ using the symbolic differential notation:

(i)Addition: for $Z_1, Z_2 \in \mathcal{S}^c_{loc}$, $d(Z_1 + Z_2) = dZ_1 + dZ_2$;

(ii)Multiplication: $dZ_1 \cdot dZ_2 = d\langle X_1, X_2 \rangle, Z_i = X_i + A_i \in \mathcal{S}^c_{loc}, i = 1, 2$;

(iii)$\mathcal{B}_{loc}$- multiplication: for $Y \in \mathcal{B}_{loc}, Z = X + A \in \mathcal{S}^c_{loc}$, $Y \cdot Z = Y \cdot X + Y \cdot A$;

(iv)Symmetric multiplication: for $Z_1, Z_2 \in \mathcal{S}^c_{loc}$, $Z_1 \circ dZ_2 = Z_1 \cdot Z_2 + \frac{1}{2} dZ_1 \cdot dZ_2$.

With these operations we have the *multiplication table* which serves as a mnemonic device in applications. The relations are restatements of the above definitions and of Proposition 33.

34. Proposition. *If $d\mathcal{A}', d\mathcal{M}'$, and $d\mathcal{S}'$ denote the classes of differential elements introduced above, then one has:*

(a) $d\mathcal{S}' \cdot d\mathcal{S}' \subset d\mathcal{A}'$, (b) $d\mathcal{S}' \cdot d\mathcal{A}' = 0$, and (hence)(c) $d\mathcal{S}' \cdot d\mathcal{S}' \cdot d\mathcal{S}' = 0$.

In fact, $d\mathcal{S}'$ is a commutative algebra over $\mathcal{B}_{loc}(\mathcal{S}^c_{loc})$ for the operations (i)-(iii),((i),(ii), and (iv)). In detail, this may be expressed as:

for $Y_1, Y_2 \in \mathcal{B}_{loc}$ *and* $Z_1, Z_2, Z_3 \in \mathcal{S}^c_{loc}$ *one has*

$$(I) Y_1 \cdot (dZ_1 + dZ_2) = Y_1 \cdot dZ_1 + Y_1 \cdot dZ_2,$$
$$(Y_1 + Y_2) \cdot dZ_1 = Y_1 \cdot dZ_1 + Y_2 \cdot dZ_2,$$
$$Y_1 \cdot (dZ_1 \cdot dZ_2) = (Y_1 \cdot dZ_1) \cdot dZ_2, \; (Y_1 Y_2) \cdot dZ_1 = Y_1 \cdot (Y_2 \cdot dZ_1).$$
$$(II) Z_1 \circ (dZ_2 + dZ_3) = Z_1 \circ dZ_2 + Z_1 \circ dZ_3,$$
$$(Z_1 + Z_2) \circ dZ_3 = Z_1 \circ dZ_3 + Z_2 \circ dZ_3,$$
$$Z_1 \circ (dZ_2 \cdot dZ_3) = (Z_1 \circ dZ_2) \cdot dZ_3 = Z_1 \cdot (dZ_2 \cdot dZ_3),$$
$$(Z_1 Z_2) \circ dZ_3 = Z_1 \circ (Z_2 \circ dZ_3).$$

It is possible to consider a common extension of both the Itô and Stratonovich definitions by using the Riemann sums in the definition with an evaluation at an intermediate point of the intervals $[t_i, t_{i+1})$ in lieu of the extreme left or the middle points. The following is such an extension discussed by Yor [2]. We present the result for comparison and completeness.

The expression for I'_n of (78) leading to the Stratonovich integral can be written for a partion π_n of $[0, t]$ as:

$$I'_n = \sum_{i=1}^{n} ((1-s)Y_{t_i} + sY_{t_{i+1}})\Delta X_{t_i} = \sum_{i=1}^{n} (Y_{t_i} + s\Delta Y_{t_i})\Delta X_{t_i}, \qquad (81)$$

where $\Delta X_{t_i} = X_{t_{i+1}} - X_{t_i}$ and similarly ΔY_{t_i}, $0 \leq s \leq 1$ ($s = \frac{1}{2}$ in (78)). This may be abstracted as follows. If $f : \mathbb{R} \to \mathbb{R}$ is continuously differentiable, then we could use the Taylor expansion for f together with a convex combination of the points of the interval for evaluation to get: (μ is some probability measure on the unit interval)

$$\begin{aligned} I'_n(f) &= \sum_{i=1}^{n} \Big(\int_0^1 f(Y_{t_i} + s\Delta Y_{t_i})\, d\mu(s) \Big) \Delta X_{t_i} \\ &= \sum_{i=1}^{n} \int_0^1 [f(Y_{t_i}) + s\Delta Y_{t_i} f'(Y_{t_i}) + o(s\Delta Y_{t_i})] d\mu(s)\, \Delta X_{t_i} \\ &= \sum_{i=1}^{n} f(Y_{t_i})\Delta X_{t_i} + \int_0^1 s\, d\mu(s) \cdot \sum_{i=1}^{n} f'(Y_{t_i})\Delta Y_{t_i} \Delta X_{t_i} \\ &\quad + o(s\Delta Y_{t_i} \Delta X_{t_i}). \end{aligned} \qquad (82)$$

Since under our conditions $o(\Delta Y_{t_i} \Delta X_{t_i}) \to 0$ in probability as $n \to \infty$, and the other two terms tend respectively to stochastic integrals(cf. also Proposition 34), one gets

$$\lim_{n\to\infty} I_n'(f,\mu)_t = \int_0^t f(Y_s)\, dX_s + (\int_0^1 s\, d\mu(s)) \cdot \int_0^t f'(Y_s)\, d\langle Y, X\rangle_s. \quad (83)$$

We thus have the following result due to Yor [2], who gave a more elaborate argument.

35. Proposition. *Let X, Y be elements of S_{loc}^c relative to a standard filtration $\{\mathcal{F}_t, t \geq 0\}$ of (Ω, Σ, P) and $f : \mathbb{R} \to \mathbb{R}$ be a continuously differentiable function. Let α be a probability distribution on the unit interval with mean α_1. Define for any partition π_n of $[0,t]$:*

$$I_n'(f,\alpha)_t = \sum_{\pi_n} \int_0^1 f(X_{t_i} + s\Delta X_{t_i})\, d\alpha(s) \cdot \Delta Y_{t_i}. \quad (84)$$

*Then $\lim_{n\to\infty} I_n'(f,\alpha)_t$ exists in probability, and is denoted by $\int_0^t f(X_s) \circ_\alpha dY_s$ and termed the α-***stochastic integral***. It is related to the Itô integral as :*

$$\int_0^t f(X_s) \circ_\alpha dY_s = \int_0^t f(X_s)\, dY_s + \alpha_1 \int_0^t f'(X_s)\, d\langle X, Y\rangle_s. \quad (85)$$

Further, this α-integral is again a continuous local semimartingale.

Remark. If α concentrates at the point $s = 0$, then the α-integral becomes the Itô integral, if it concentrates at $s = \frac{1}{2}$, then it reduces to the (symmetric or) Stratonovich integral, and if $\alpha(s = 1) = 1$ it becomes what is called the *backward integral.* We shall see later in the next chapter how all these integrals satisfy a common generalized domination principle.

5.4 Stochastic integrators as Hilbertian spectral measures

The stochastic integration theory of the preceding section not only used many ideas of Hilbert spaces, but in fact it can be obtained from the classical spectral representation related to normal operators on such spaces. The details of this approach have been worked out by Cuculescu [1], and we present a central result to exemplify this point since it also leads to an $L^p(P)$-stochastic calculus. In the following chapter more general processes as integrators will be considered, but the ideas of the present treatment form a basis for that study.

Let $\mathcal{A}(\subset B(\mathbb{R}^+ \times \Omega)$, the space of all bounded complex functions on $\mathbb{R}^+ \times \Omega$ in the following applications) be an algebra of complex functions containing the identity e, on a space $\mathfrak{S}$, which is closed under conjugation and complete under the uniform norm $\|\cdot\|_u$. A homomorphism $T \mapsto T_f$ (i.e., a linear multiplicative mapping) of $\mathcal{A}$ into a subalgebra $\mathcal{L}$ of bounded linear operators on a Hilbert space $\mathcal{H}$, closed under adjoints, is called a *-representation if $(T_f)^* = T_{f^*}$ where $(T_f)^*$ is the adjoint of T_f, and f^* is the complex conjugate of f ($T_e = id.$). It can be verified that $\|T_f\| \leq \|f\|_u$ so that T is contractive. Then the function F_f defined by $F_f(x,y) = (T_f x, y)$ for $x, y \in \mathcal{H}$ is a continuous sesquilinear form, $(,)$ being the inner product of $\mathcal{H}$, and $F_{(\cdot)}(x,y) : \mathcal{A} \to \mathbb{C}$ is in the adjoint space $\mathcal{A}^*$ so that by the Riesz representation theorem there is a unique bounded additive set function $\mu_{x,y}$ on the σ-algebra $\mathcal{S}$ of $\mathfrak{S}$ relative to which every element of $\mathcal{A}$ is measurable, such that

$$F_f(x,y) = (T_f x, y) = \int_{\mathfrak{S}} f(s)\, d\mu_{x,y}(s), \qquad (1)$$

and $\mu_{x,y}$ is linear in x (antilinear in y), $|\mu_{x,y}|(\mathfrak{S}) = \|F_{(\cdot)}(x,y)\| \leq \|x\|\|y\|$. The set function $\mu_{x,y}$ is not in general σ-addtive. In case it is, and $\mu_{x,x} \geq 0$, it is called the *spectral measure* of T. Conversely if such a set function is given, then for each $f \in \mathcal{A}$, the functional $(\cdot,\cdot)_f$ defined by the right side of (1) is continuous (by the CBS inequality) and hence there exists a unique $T_f \in \mathcal{L}$ such that the left side of (1) holds. Since $f^{**} = f$ and from definition of a *-representation we have

$$(T_{f^* f} x, y) = (T_f^* T_f x, y) = (T_f x, T_f y) = \int_{\mathfrak{S}} |f|^2\, d\mu_{x,y}, \qquad (2)$$

it is clear that $\|T_f x\|^2 = \int_{\mathfrak{S}} |f|^2\, d\mu_{x,x}$ and the mapping $f \mapsto T_f$ is an isometric isomorphism of $\mathcal{A}$ into $L^2(\mathfrak{S}, \mathcal{S}, \mu_{x,x}), x \in \mathcal{H}$. The proof of existence of such representations is not obvious, and may be found, for instance, in Loomis ([1],p.93). It may also be noted that (1) and (2) imply the *normality* of the operator T_f in that T_f and $(T_f)^* = T_{f^*}$ commute; and $\mu_{x,y}$ determines the spectral measure of this operator in classical Hilbert space theory. This connection is essentially exploited in the following work.

The idea now is to consider $\mathcal{H} = L^2(\Omega, \Sigma, P)$ and treat the $L^2(P)$-bounded martingales X, i.e. $X \in \mathcal{M}^c_{loc}$, by "identifying" $\mu_{x,x}$ with $E(\langle X, X\rangle_{(\cdot)})$ and T_f with I_f of Theorem 3.9. However, the key point here is to show that $I_{fg} = I_f I_g$ for *bounded* f, g using the special kind of σ-algebras (e.g. the predictable ones), and this is nontrivial. We recall some facts for this purpose on stopping times of the net. If T_1, T_2 are any two stopping times of the standard filtration $\{\mathcal{F}_t, t \geq 0\}$, let $\mathcal{B}_i, i = 1, 2$, be the σ-algebras prior to $T_i, i = 1, 2$ (cf. (18) of Section IV.1). They may be alternatively given as:

$$\mathcal{B}_1 = \{A_1 \cup A_2 : A_i \in \mathcal{B}_i; A_1 \subset [T_1 \leq T_2], A_2 \subset [T_2 < T_1]\}, \quad (3)$$

and

$$\mathcal{B}_2 = \{A_1 \cup A_2 : A_i \in \mathcal{B}_i; A_1 \subset [T_2 < T_1], A_2 \subset [T_1 \leq T_2]\}. \quad (4)$$

Using these we have for any $f \in L^1(P)$, the following *key* identities of conditional expectations:

$$E^{\mathcal{B}(T_1)}(E^{\mathcal{B}(T_2)}(f)) = E^{\mathcal{B}(T_2)}(E^{\mathcal{B}(T_1)}(f)) = E^{\mathcal{B}_1}(f), \text{ a.e.}, \quad (5)$$

and

$$E^{\mathcal{B}(T_1)}(f) + E^{\mathcal{B}(T_2)}(f) = E^{\mathcal{B}_1}(f) + E^{\mathcal{B}_2}(f), \text{ a.e..} \quad (6)$$

This result will not be true for arbitrary σ-subalgebras, and the fact that they are determined by the stopping times T_i is essential. (See Section IV.2 and also Chung and Doob [1] on their calculus.)

For a proof of the identities, first consider (5). Let $A_1 \cup A_2 \in \mathcal{B}_1$ be a generator (cf. (3)) so that $A_1 \cap A_2 = \emptyset$. Using the observation that $\mathcal{B}_1 = \mathcal{B}(T_1) \cap \mathcal{B}(T_2)$ we deduce the following set of equations for conditional expectations:

$$\begin{aligned}\int_{A_1 \cup A_2} E^{\mathcal{B}_1}(f)\, dP &= \int_{A_1 \cup A_2} f\, dP = \int_{A_1} E^{\mathcal{B}(T_1)}(f)\, dP + \int_{A_2} E^{\mathcal{B}(T_1)}(f)\, dP,\\ &= \int_{A_1 \cup A_2} E^{\mathcal{B}(T_2)}(E^{\mathcal{B}(T_1)}(f))\, dP, \ A_1 \cup A_2 \in \mathcal{B}(T_i),\\ &= \int_{A_1 \cup A_2} E^{\mathcal{B}(T_1)}(E^{\mathcal{B}(T_2)}(f))\, dP, \text{ by symmetry.} \quad (7)\end{aligned}$$

Since the extreme integrands are $\mathcal{B}_1$-measurable and the integrals are equal for all $A_1 \cup A_2 \in \mathcal{B}(T_i)$, we deduce (5) from (7). To prove (6), note that if $Q_i = E^{\mathcal{B}(T_i)}, i = 1, 2$, then by (5) $Q_1Q_2 = Q_2Q_1$ and hence $Q = Q_1 + Q_2 - Q_1Q_2$ is also an orthogonal (= contractive) projection operator on $\mathcal{H}$ with range given by:

$$\mathcal{R}_Q = \overline{\text{sp}}(\mathcal{R}_{Q_1} \cup \mathcal{R}_{Q_2}) = \overline{\text{sp}}((L^2(\mathcal{B}(T_1)) \cup L^2(\mathcal{B}(T_2))) = L^2(\mathcal{B}_2). \quad (8)$$

But then $Q = E^{\mathcal{B}_2}$ by Theorem II.2.2. Since $E^{\mathcal{B}_1} = Q_1Q_2$ by (5), we deduce that (7) is simply $Q + Q_1Q_2 = Q_1 + Q_2$ which is thus true.

We consider a subclass of $\mathcal{M}^c$, namely the uniformly integrable elements, so that $X \in \mathcal{M}^c$ can be uniquely identified with an element $f \in L^2(\mathcal{F}_\infty)$ such that $X_t = E^{\mathcal{F}_t}(f), t \geq 0, X_0 = 0$ a.e. Thus $\mathcal{H} = L^2(\mathcal{F}_\infty) = \{f \in L^2(\mathcal{F}_\infty) : E^{\mathcal{F}_0}(f) = 0\}$ which is then isomorphic to the above subspace of $\mathcal{M}$. We now present a general result in the following form, which may be extended to one of Theorem 3.9.

1. Theorem. *Let $\mathcal{A} = B(U, \mathcal{P})$, the algebra of bounded complex $\mathcal{P}$-measurable functions on $U = \mathbb{R}^+ \times \Omega$ where $\mathcal{P}$ is the predictable σ-algebra (so that $\mathcal{A}$ is complete under the uniform norm and closed under conjugation). Then a *-representation of $\mathcal{A}$ into $B(\mathcal{H})$ exists and is determined by the equation ($\mathcal{H} = L_0^2(\mathcal{F}_\infty)$):*

$$I : f \mapsto I_f \in B(\mathcal{H}),\ I_g X = E^{\mathcal{B}(T)}(X),\ X \in \mathcal{H}, \quad (9)$$

with $g = \chi_A$, $A = [\![0, T)$ being a generator of $\mathcal{P}$. More explicitly,

$$(I_f X, Y) = \int_U f(t,\omega)\, d\mu_{x,y}(t,\omega) = \int_\Omega \int_{\mathbb{R}^+} f(s,\omega)\, d\langle X, Y\rangle_s\, dP(\omega), \quad (10)$$

for all $f \in \mathcal{A}$ with $\langle X, Y\rangle$ as the quadratic covariation of X, Y where X, Y are in $\mathcal{H}$, and $\mu_{x,y}$ is a spectral measure.

Remark. The element $I_f X \in \mathcal{H}$, for each $X \in \mathcal{H}$; it is the stochastic integral of f relative to X and is denoted in the old notation as:

$$E^{\mathcal{F}_t}(I_f X) = (I_f X)_t = \int_0^t f(s,\cdot)\, dX_s,\ t \geq 0,\ X_s = E^{\mathcal{F}_s}(X), \text{ a.e.} \quad (11)$$

This will become clear from the proof, when $\{X_t, \mathcal{F}_t, t \geq 0\}$ and $X \in \mathcal{H}$ are identified.

Proof. Let $X \in \mathcal{H}$ and T be a predictable stopping time of the filtration. Since $A = [\![0, T)$ is a generator of $\mathcal{P}$, we define $\mu_{x,x}(\emptyset) = 0$ and for A,

$$\mu_{x,x}(A) = (E^{\mathcal{B}(T)}(X), X) = (E^{\mathcal{B}(T)}(X), E^{\mathcal{B}(T)}(X)),$$
$$\text{since } E^{\mathcal{B}(T)} \text{ is a self-adjoint projection on } L^2(P),$$
$$= (X_T, X_T) = \int_\Omega |X_T|^2 \, dP. \tag{12}$$

Since $[\![0, T_1) \cap [\![0, T_2) = [\![0, T_1 \wedge T_2)$ and $[\![0, T_1) \cup [\![0, T_2) = [\![0, T_1 \vee T_2)$ for any stopping times $T_i, i = 1, 2$ of the filtration, and a similar one is true for countable collections $\{T_n, n \geq 0\}$, the set $\mathcal{S}_0$ of all such intervals is a σ-lattice, so that $\mathcal{S}_0$ is closed under countable unions and intersections containing $\emptyset$ and U. It is also seen that $\sigma(\mathcal{S}_0) = \mathcal{P}$. By (12), $\mu_{x,x}(\cdot)$ is well defined and taking $T = \infty$, it follows that $\mu_{x,x}(U) < \infty$. We may and do assume for this proof, by a normalization, that $\mu_{x,x}(U) = 1$, and assert that $(i) \mu_{x,x}(A) \leq \mu_{x,x}(B)$ for $A \subset B$; $(ii) \mu_{x,x}(\emptyset) = 0$, $\mu_{x,x}$ is strongly additive in the sense that $\mu_{x,x}(C \cup D) + \mu_{x,x,}(C \cap D) = \mu_{x,x}(C) + \mu_{x,x}(D)$; and $(iii) A_n \uparrow A, A_n \in \mathcal{S}_0$, then $\mu_{x,x}(A_n) \uparrow \mu_{x,x}(A)$. Now by a well-known argument in Measure Theory, $\mu_{x,x}$ can be uniquely extended to $\mathcal{P}$ to be a spectral measure (cf. e.g., Rao [11],p.469 ff). This gives (9), and we present the details in steps for convenience.

I. It follows from the fact that $\{X_T^2, \mathcal{B}(T_i)\}_{i=1}^2$ is a submartingale for $T_1 \leq T_2$, that $\mu_{x,x}(A_1) \leq \mu_{x,x}(A_2)$ for $A_i = [\![0, T_i), i = 1, 2$. So (i) is true. In the same way using the uniform integrability of $\{X_{T_i}^2, \mathcal{B}(T_i)\}_{i=1}^\infty$, if $T_i \uparrow T$ (cf. the remark at the end of Section 2 above), then $\mu_{x,x}(A_i) \uparrow \mu_{x,x}(A)$ where $A_i = [\![0, T_i)$. Thus (iii) is also true. (Here we used Proposition IV.3.1 and Corollary IV.1.12.) Since $\mu_{x,x}(\emptyset) = 0$ by definition, we need to prove the strong additivity of $\mu_{x,x}$. Thus let T_1, T_2 be a pair of stopping times of the filtration. Then by (6) (and Proposition IV.2.2) we have $(*) X_{T_1} + X_{T_2} = X_{T_1 \vee T_2} + X_{T_1 \wedge T_2}$, and the martingale property of the transformed process implies that $X_{T_i} = (X_{T_i} - X_{T_1 \wedge T_2}) + X_{T_1 \wedge T_2}, i = 1, 2$ is an orthogonal decomposition. So by (5) one has $:(+) E^{\mathcal{B}(T_1 \wedge T_2)}(X_{T_i}) = X_{T_1 \wedge T_2}$, and using a property of martingale differences, it follows that

$$E((X_{T_2} - X_{T_1 \wedge T_2})(X_{T_1} - X_{T_1 \wedge T_2})) = E([E^{\mathcal{B}(T_1 \wedge T_2)}(X_{T_2} - X_{T_1 \wedge T_2})] \times (X_{T_1} - X_{T_1 \wedge T_2})) = 0. \tag{13}$$

Hence one has, with $A_i = [\![0, T_i), i = 1, 2$ and (12):

$$\begin{aligned}
\mu_{x,x}(A_1 \cup A_2) - \mu_{x,x}(A_1 \cap A_2) &= \int_\Omega |X_{T_1 \vee T_2}|^2 \, dP - \int_\Omega |X_{T_1 \wedge T_2}|^2 \, dP \\
&= \int_\Omega [X_{T_1 \vee T_2} - X_{T_1 \wedge T_2}]^2 \, dP, \\
&\quad \text{by the martingale property,} \\
&= \int_\Omega [(X_{T_1} - X_{T_1 \wedge T_2}) + (X_{T_2} - X_{T_1 \wedge T_2})]^2 \, dP, \\
&\quad \text{by } (*), \\
&= \int_\Omega [X_{T_1} - X_{T_1 \wedge T_2}]^2 \, dP + \\
&\quad \int_\Omega [X_{T_2} - X_{T_1 \wedge T_2}]^2 \, dP, \text{ by}(13), \\
&= \int_\Omega [X_{T_1}^2 - 2X_{T_1 \wedge T_2}^2 + X_{T_2}^2] \, dP, \text{ by}(+), \\
&= \mu_{x,x}(A_1) + \mu_{x,x}(A_2) - 2\mu_{x,x}(A_1 \cap A_2). \qquad (14)
\end{aligned}$$

This shows that (ii) is true on (the generators and hence on all of) $\mathcal{S}_0$.

II. Now for each set $F \subset U$ define a function $\nu_{x,x}$ as:

$$\nu_{x,x}(F) = \inf\{\mu_{x,x}(A) : A \supset F, A \in \mathcal{S}_0\}. \qquad (15)$$

Then, it is seen that $\nu_{x,x}|\mathcal{S}_0 = \mu_{x,x}$ and $0 \leq \nu_{x,x}(F) \leq \mu_{x,x}(U) = 1$. We claim that $\nu_{x,x}$ is strongly subadditive. For, if $F_i \subset U$, and $\varepsilon > 0$, select $A_i \in \mathcal{S}_0$ such that $\nu_{x,x}(F_i) \geq \mu_{x,x}(A_i) - \varepsilon/2, i = 1, 2$. Hence

$$\begin{aligned}
\nu_{x,x}(F_1) + \nu_{x,x}(F_2) &\geq \mu_{x,x}(A_1) + \mu_{x,x}(A_2) - \varepsilon \\
&= \mu_{x,x}(A_1 \cup A_2) + \mu_{x,x}(A_1 \cap A_2) - \varepsilon, \text{ by}(14), \\
&\geq \nu_{x,x}(F_1 \cup F_2) - \varepsilon, \text{ by}(15). \qquad (16)
\end{aligned}$$

Thus the strong subadditivity follows since $\varepsilon > 0$ is arbitrary. Taking $F_2 = F_1^c$, and noting that $\nu_{x,x}(\emptyset) = 0, \nu_{x,x}(U) = \mu_{x,x}(U) = 1$, one has

$$\nu_{x,x}(F_1) + \nu_{x,x}(F_1^c) \geq 1. \qquad (17)$$

It is clear from (15) that $\nu_{x,x}(\cdot)$ is increasing. Let $\mathcal{D} = \{F \subset U : \nu_{x,x}(F) + \nu_{x,x}(F^c) = 1\}$. Then $\mathcal{D}$ is closed under complements and contains $U, \emptyset$, and we proceed with a familiar and well-known argument to show that it is an algebra. Indeed, if $F_i \in \mathcal{D}, i = 1, 2$, then by (17)

$$\nu_{x,x}(F_1 \cup F_2) + \nu_{x,x}(F_1^c \cap F_2^c) \geq 1, \tag{18}$$
$$\nu_{x,x}(F_1 \cap F_2) + \nu_{x,x}(F_1^c \cup F_2^c) \geq 1. \tag{19}$$

From these two relations and (16) we get

$$2 \leq \nu_{x,x}(F_1 \cup F_2) + \nu_{x,x}(F_1 \cap F_2) + \nu_{x,x}(F_1^c \cup F_2^c) + \nu_{x,x}(F_1^c \cap F_2^c) \leq 2. \tag{20}$$

Hence there is equality in (16)–(19) so that $F_1 \cup F_2, F_1 \cap F_2 \in \mathcal{D}$. Thus $\mathcal{D}$ is an algebra and $\nu_{x,x}$ is additive on it. To show that $\mathcal{D}$ is a σ-algebra, it suffices to verify that it is closed under increasing limits. For this we need to show that $\nu_{x,x}$ inherits property (iii) of $\mu_{x,x}$. Note that by the strong additivity of $\nu_{x,x}$ on $\mathcal{D}$ and the fact that $\nu_{x,x}|\mathcal{S}_0 = \mu_{x,x}$, we have $\mathcal{S}_0 \subset \mathcal{D}$.

Let $F_n \subset U, F_n \uparrow F$. If $\varepsilon > 0$, then by (15) for each $n \geq 1$, there exists $A_n \in \mathcal{S}_0$ such that $\nu_{x,x}(F_n) > \mu_{x,x}(A_n) - \frac{\varepsilon}{2^n}$. Since A_n need not be increasing, let $B_n = \cup_{i=1}^n A_i \in \mathcal{S}_0$. So $F_n \subset B_n, B_n \uparrow A$, and $F_n = F_n \cap F_{n+1} \subset B_n \cap A_{n+1}$. We assert that for all $n \geq 1$ (dropping the suffixes for μ and ν)

$$\nu(F_n) > \mu(B_n) - \varepsilon(1 - \frac{1}{2^n}). \tag{21}$$

This is clear for $n = 1$, and assuming it for n, we use induction:

$$\begin{aligned}
\mu(B_{n+1}) &= \mu(B_n \cup A_n) \\
&= \mu(B_n) + \mu(A_{n+1}) - \mu(B_n \cap A_n), \text{ by (14)}, \\
&\leq \nu(F_n) + \varepsilon(1 - \frac{1}{2^n}) + \nu(F_{n+1}) + \frac{\varepsilon}{2^{n+1}} - \mu(B_n \cap A_n), \\
&\quad \text{by the induction hypothesis,} \\
&\leq \nu(F_{n+1}) + \varepsilon(1 - \frac{1}{2^{n+1}}), \text{ since} A_n \cap B_n \supset F_n, \\
&\quad \text{and } \mu(B_n \cup A_n) \geq \nu(F_n).
\end{aligned}$$

Letting $n \to \infty$, we see that by property (iii) of μ, (21) yields $\nu(F) \leq \mu(A) \leq \lim_n \nu(F_n) + \varepsilon$. So $\nu(F) \leq \lim_n \nu(F_n) \leq \nu(F)$, since $\nu(F_n) \leq$

$\nu(F)$. This establishes the continuity property of ν, and thus its σ-additivity.

III. The class $\mathcal{D}$ is a σ-algebra and $\nu_{x,x}$ is σ-additive on it. For, let $F_n \uparrow F, F_n \in \mathcal{D}$. Then $F_n^c \downarrow F^c, \nu_{x,x}(F_n^c) \geq \nu_{x,x}(F^c)$ and by (18)-(19), we have $\nu_{x,x}(F_n)+\nu_{x,x}(F_n^c) = 1 \geq \nu_{x,x}(F_n)+\nu_{x,x}(F^c)$. Letting $n \to \infty$ we get $\lim_n \nu_{x,x}(F_n^c) = \nu_{x,x}(F^c)$. Hence $F = \cup_n F_n \in \mathcal{D}$, and $\mathcal{D}$ is a σ-algebra. Since $\nu_{x,x}$ is finitely additive and, by the preceding step, σ-subadditive, it is σ-additive on $\mathcal{D}$.

IV. Since $\nu_{x,x}|\mathcal{S}_0 = \mu_{x,x}$, if $\mathcal{P} = \sigma(\mathcal{S}_0) \subset \mathcal{D}$ and $\tilde{\mu}_{x,x} = \nu_{x,x}|\mathcal{P}$, then $\tilde{\mu}_{x,x}$ is a σ-additive extension of $\mu_{x,x}$, and to see that this is unique, let $\mu'_{x,x}$ be another extension of it to $\mathcal{P}$. Then by (15), if $B \in \mathcal{P}$,

$$\begin{aligned}\mu_{x,x}(B) &= \inf\{\mu_{x,x}(A) : A \in \mathcal{S}_0, A \supset B\}\\ &= \inf\{\mu'_{x,x}(A) : A \in \mathcal{S}_0, A \supset B\}\\ &\geq \mu'_{x,x}(B), \quad \text{since}\, \mu'_{x,x}(\cdot)\, \text{is monotone.}\end{aligned} \tag{22}$$

Hence also $1 = \tilde{\mu}_{x,x}(B)+\tilde{\mu}_{x,x}(B^c) \geq \mu'_{x,x}(B)+\mu'_{x,x}(B^c) = \mu_{x,x}(U) = 1$, using (22). So there is equality through out and $\tilde{\mu}_{x,x} = \mu'_{x,x}$, and we denote this extension by the same symbol $\mu_{x,x}$. Now (5) implies, if $g_i = \chi_{A_i}, A_i = [\![0, T_i)$, and $I_{g_i}X = E^{\mathcal{B}(T_i)}(X), X \in \mathcal{H}$, then $I_{g_i} \in B(\mathcal{H})$, and $I_{g_1 g_2}(X) = I_{g_1}I_{g_2}X$ so that by linearity, $f \mapsto I_f$ is defined for all $f \in \mathcal{A}$ and is a *-representation. The polarization then implies that we can define $\mu_{x,y}$ for $X, Y \in \mathcal{H}$ as:

$$\mu_{x,y} = \frac{1}{4}(\mu_{x+y,x+y} - \mu_{x-y,x-y} + i\mu_{x+iy,x+iy} - i\mu_{x-iy,x-iy}), \tag{23}$$

and $\mu_{x,y}$ is sesquilinear (in x, y). This gives (9) and the first half of (10).

V. The representation (10), the second half, is obtained through an application of a Radon-Nikodým theorem. We sketch the standard argument for completeness. Thus for $t \in \mathbb{R}^+$, define $\zeta_t : \mathcal{F}_\infty \to \mathbb{R}^+$ as: for each $F \in \mathcal{F}_\infty$, let $\zeta_t(F) = \mu([\![0, t) \times F)$ where μ stands for $\mu_{x,x}$. Then ζ_t is a measure and $0 \leq \zeta_t(F) \leq \mu(U) < \infty$, for all $t \geq 0, \zeta_0 \equiv 0$. It follows that ζ_t is P-continuous since the sets $[0, t] \times F$ with $P(F) = 0$ can be identified with $[\![0, T_F)$ for $T_F = 0$, a.e. Hence restricting to $\mathcal{F}_t$, we see by the Radon-Nikodým theorem that there exists an $\mathcal{F}_t$-measurable $P_{\mathcal{F}_t}$-unique function $A_t = \frac{d\zeta_t}{dP}$, a.e., such that $(i)A_t \geq 0$

a.e., $A_0 = 0$ a.e., and $(ii)\zeta_t(F) \leq \zeta_{t'}(F) \Rightarrow A_t \leq A_{t'}, t \leq t'$ a.e. Since moreover $t_n \downarrow t \geq 0 \Rightarrow A_{t_n} \to A_t$ in $L^1(P)$ and a.e., we may define a process $\{A_t, t \geq 0\}$ which is increasing and (by the standard reasoning) also $\mathcal{P}(\subset \mathcal{B}(\mathbb{R}^+) \otimes \mathcal{F}_\infty)$-measurable. One can express this as follows:

$$\mu_{x,x}(B) = \int_U \chi_B \, dA_s \, dP(\omega), \quad B \in \mathcal{P}. \tag{24}$$

It remains to show that $\{A_t, t \geq 0\}$ is predictable. Since $\sup_t E(A_t) = \mu_{x,x}(U) < \infty$, it follows that for each stopping time T of the filtration, A_∞ is defined (since $\lim_t A_t = A_\infty$ a.e. and in $L^1(P)$) and is $\mathcal{B}(T)$-adapted. But the spectral measure $\mu_{x,x}$ is defined on $\mathcal{P}$ (cf. Steps III and IV) and by the above procedure $\{A_t, t \geq 0\}$ is uniquely defined and the function $A = \{A_t(\omega) : t \in \mathbb{R}^+, \omega \in \Omega\}$ is $\mathcal{P}$-measurable. Since the A_t-process is adapted, this easily implies that A is predictable. Then writing $A_t = \langle X, X \rangle_t$, which is uniquely determined by $\mu_{x,x}$ and hence by X, we deduce the last half of (10) from (9) at once. Thus (10) is also established. □

2. Remark. It is possible to prove the predictability of A independently by first showing that it is accessible (i.e., for each totally inaccessible time T of the filtration $\{\mathcal{F}_t, t \geq 0\}, A_T - A_{T-} = 0$ a.e.) and then for each predictable time τ, A_τ is $\mathcal{B}(\tau-)$-measurable. Next one shows that these two properties imply Definition 1.3, but the details are nontrivial and the section theorem (Theorem 2.25) is employed in these computations. One can find the details in Cuculescu [1] and Meyer [4].

It is worth separaring a technical result contained in Steps II and III of the above proof. We state this as a proposition for a convenient reference.

3. Proposition. *Let $\mathcal{S}$ be a lattice of a set S such that $\{\emptyset, S\} \subset \mathcal{S}$. If $\mu : \mathcal{S} \to \mathbb{R}^+$ is a strongly additive function with $\mu(\mathcal{S}) < \infty, \mu(\emptyset) = 0$, and μ is increasing, then μ can be uniquely extended to be an additive (bounded) function on the algebra generated by $\mathcal{S}$. If, moreover, $\mathcal{S}$ is a σ-lattice, μ is right continuous (i.e., $A_n \in \mathcal{S}, A_n \uparrow A \in \mathcal{S}$ implies $\mu(A_n) \uparrow \mu(A)$), then μ has a unique σ-additive extension to $\sigma(\mathcal{S})$, the σ-algebra generated by $\mathcal{S}$. The same conclusion holds even if μ is not bounded but if there is a sequence $\{S_n, n \geq 1\} \subset \mathcal{S}, S_n \uparrow S$ and $\mu(S_n) < \infty$ for each n.*

This result implies that if two measures μ_1, μ_2 on a σ-algebra $\mathcal{A}$ agree on a lattice $\mathcal{S} \subset \mathcal{A}$, and the μ_i are σ-finite on $\mathcal{S}$, then they agree on $\sigma(\mathcal{S})$. The proposition was given by Cuculescu [1] with a different proof, and our argument is patterned after one of Neveu [1] in a related context.

The above theorem shows a possible generalization of stochastic integrals from the Hilbert space to some Banach spaces. The key sesquilinear form $\mu_{x,y}$ used here can be replaced by a vector integral and then it is possible to admit certain processes more general than martingales. We shall briefly consider these points in the next chapter.

5.5 Complements and exercises

1. Let $I \subset \mathbb{R}^+$ be an infinite set and $\{\mathcal{F}_t, t \in I\}$ be a standard filtration from (Ω, Σ, P). Let $\{X_t, \mathcal{F}_t, t \in I\}$ be a right continuous supermartingale. We then have the following assertions:

(a) The statements (i) and (ii) are equivalent, where: (i) there exists a martingale $\{Y_t, \mathcal{F}_t, t \in I\}$ such that $X_t \geq Y_t$ a.e. for $t \in I$, and (ii) there exist two processes $\{Z_t^{(i)}, \mathcal{F}_t, t \in I\}, i = 1, 2$ such that the $Z_t^{(1)}$-process is a martingale and the $Z_t^{(2)}$-process is a potential (i.e., if $\beta = \sup I, t_i \in I$ with $t_i \to \beta$ then $0 \leq Z_t^{(2)} \to 0$ in $L^1(P)$), and $X_t = Z_t^{(1)} + Z_t^{(2)}$ a.e., for $t \in I$. When this holds, the decomposition is unique.

(b) Suppose $\inf I = \alpha \in I$ and the X_t-process is in class (DL). Then there exist a martingale $\{Y_t, \mathcal{F}_t, t \in I\}$ and an increasing predictable, integrable process $\{A_t, \mathcal{F}_t, t \in I\}$ such that $X_t = Y_t - A_t$ a.e., $t \in I$, and this decomposition is unique.

[*Remark.* (a) and (b) are the "generalized" versions of the Riesz and the Doob-Meyer decompositions, and the proofs are essentially the same as in the text. The point here is that β is a boundary point of I, and I need not be all of $\mathbb{R}^+$. For a classical treatment of these questions, see Radó [1].]

2. Let $\{A_t, B_t, \mathcal{F}_t, t \geq 0\}$ be two increasing right continuous processes on (Ω, Σ, P) with $E(A_t) < \infty, E(B_t) < \infty, t \geq 0$. Let $C_t = A_t - B_t$, and suppose that $\{C_t, \mathcal{F}_t, t \geq 0\}$ is a martingale. If $\{Y_t, \mathcal{F}_t, t \geq 0\}$ is any nonnegative process having left limits a.e., then $E(\int_0^t Y_s \, dA_s) =$

$E(\int_0^t Y_s\, dB_s), t \geq 0$. In particular, if A_s, B_s are also predictable, then $C_t = 0$ a.e. for all t.

3. Let $\{\mathcal{F}_t, t \geq 0\}$ be a standard filtration of (Ω, Σ, P) and $\mathcal{T} = \{T_t, t \geq 0\}$ be a stopping time process for it. Define $\mathcal{S} = \{\tau_t, t \geq 0\}$ where $\tau_t = \inf\{u \geq 0 : T_u > t\}$, with $\inf\{\emptyset\} = \infty$. Show that $\mathcal{S}$ is a stopping time process, and that $T_t = \inf\{u > 0 : \tau_u > t\}$. Deduce that there is a one-to-one correspondence between $\mathcal{T}$ and $\mathcal{S}$, and if $T_t \uparrow$, then $\tau_t \uparrow$ also, and if one increases strictly, so does the other. (Observe that $[\tau_t \leq u] = \cap_n [T_{u+(1/n)} > t] \in \mathcal{F}_u, u \geq 0$.) In particular, if $\{A_t, t \geq 0\}$ is the generator of a potential $\{X_t, \mathcal{F}_t, t \geq 0\}$ (cf. the observation preceding Remark 2.4), it defines a stopping time process of the filtration, and its inverse is also such, where the latter is defined as the family $\mathcal{S}$ from $\mathcal{T}$ above.

4. (a) Let $X = \{X_t, \mathcal{F}_t, t \geq 0\}$ be a local martingale in the sense of Definition 1.2, with $X_0 = 0$ a.e. Show that there exists a continuous time change $\mathcal{T} = \{T_t, t \geq 0\}$ of the filtration such that $\mathcal{T}(X) = \{X \circ T_t, \mathcal{B}(T_t), t \geq 0\}$ is a martingale. Moreover, if $X \in \mathcal{M}_{loc}^c$ then $\mathcal{T}(X) \in \mathcal{M}_{loc}^c$. [*Hints:* By Definition 1.2 there is a stopping time process $\{T_n, n \geq 0\}$, with $T_0 = 0, T_n \uparrow \infty, T_n \leq n$ a.e., and $\{X(T_n \wedge t), t \geq 0\}$ is uniformly integrable, for each n. Let $\tau_t^n = T_n \vee (T_{n+1} \wedge t)$. If $f_n : [n, n+1) \to \mathbb{R}^+$ is a one-to-one increasing map (e.g., $f_n(x) = (x-n)(n+1-x)^{-1}$), then $S_t = \tau_{f_n(t)}^n, t \in [n, n+1)$, defines a bounded continuous time change. Then apply Theorem IV.1.7.]

(b) Let $T_n \uparrow \infty$ a.e. be a stopping time sequence as in (a). An adapted right continuous process $X = \{X_t, \mathcal{F}_t, t \geq 0\}$ is a *'weak' martingale* if there exist uniformly integrable martingales $\{Y_t^n, \mathcal{F}_t, t \geq 0\}, n \geq 1$, such that $X_t = Y_t^n, 0 \leq t < T_n$ for some $T_n \uparrow \infty$. Thus on the sets $[T_n = t]$, this condition need not hold. Show a weak martingale X transforms into a weak martingale $\mathcal{T}(X) = \{X \circ \tau_t, \mathcal{B}(\tau_t), t \geq 0\}$ under a time change function $\mathcal{T} = \{\tau_t, t \geq 0\}$. [*Hints*: Let $\zeta_n = \inf\{t > 0 : \tau_t \geq T_n\}$ so that $t < \zeta_n$ implies $t < T_n$. If $Z_t = X \circ \tau_t$, then $Z_t^n = X(\tau_t \wedge T_n \wedge t)\chi_{[\zeta_n > t]} = X(\tau_t \wedge t)\chi_{[\zeta_n > t]}, t \geq 0$, defines a uniformly integrable martingale for each n, relative to $\{\mathcal{B}(\tau_t), t \geq 0\}$. Now apply Theorem IV.1.7 as above. This is a weaker concept than the local one, and even class (D) weak martingales need not be martingales. For further properties of these processes, see Kazamaki [1].]

5. Let $X = \{X_t, \mathcal{F}_t, t \geq 0\}$ be a right continuous process relative to

a standard filtration of (Ω, Σ, P).

(a) Then X is a quasimartingale, in the sense of Definition 2.19, iff it can be uniquely decomposed as: $X_t = Y_t + Z_t, t \geq 0$, where $Y = \{Y_t, \mathcal{F}_t, t \geq 0\}$ is a right continuous martingale with $\sup_t E(|Y_t|) < \infty$, and $Z = \{Z_t, \mathcal{F}_t, t \geq 0\}$ is a right continuous "semipotential" in the sense that it is a quasimartingale such that $E(|Z_t|) \to 0$ as $t \to \infty$.

(b) If Z is a semipotential as in (a), then it can be expressed as a difference of two potentials, i.e., $Z_t = Z_t^1 - Z_t^2, t \geq 0$, a.e., and Z^i are potentials in the sense of Section 1.

(c) Deduce Theorem 2.22 as a consequence of (a) and (b) above. [*Hints*: If X satisfies the conditions of Definition 2.19, and $0 < t_1 < t_2 < \cdots \to \infty$,given $\varepsilon > 0$, choose $i_0 = i_0(\varepsilon)$ such that $\sum\limits_{i \geq i_0} E(|X_{t_i} - E^{\mathcal{F}_{t_i}}(X_{t_{i+1}})| < \varepsilon$. Let $Y_t^n = E^{\mathcal{F}_{t_i}}(X_{t_n})$. Verify that $Y_t^n \to Y_t$ in $L^1(P)$, and let $Z_t = X_t - Y_t$. This gives (a) after some computations. For (b), consider the dyadic rationals $t_n^k = k2^{-n} \geq 0$ and we have $\sum\limits_{n,k} E(|Z_{t_n^k} - E^{\mathcal{F}_{t_n^k}}(Z_{t_{n+1}^k})|) \leq K_z < \infty$. If $[i]$ stands for the integral part of i, define: $Z_n^1(t) = \sum\limits_{k \geq [2^n t]+1} E^{\mathcal{F}_t}(Z_{t_n^k} - E^{\mathcal{F}_{t_n^k}}(Z_{t_{n+1}^k})^+$ and similarly $Z_n^2(t)$ with $()^-$. Verify that, after an analysis, the $\{Z_n^i(t), t \geq 0\}$ are potentials and that $\lim\limits_{n \to \infty} Z_n^i(t) = Z_t^i$. The method is similar to that of Section 2, and is detailed in K.M.Rao [2].]

6. If $\{V_t, \mathcal{F}_t, t \in [a,b]\}$ is a process with continuous sample paths, the filtration being standard, and $\{X_t, \mathcal{F}_t, t \in [a,b]\}$ is a continuous square integrable martingale, consider $S_i = \sum_{i=1}^{m_i} V(t_{j,j-1})[X(t_{j,i}) - X(t_{j,i-1})]^2$ where $a = t_{j,1} < \cdots < t_{j,m_j} = b$ and $V(t) = V_t, X(t) = X_t$. If $V = 1$, then $\lim_j S_j$ is the quadratic variation. Generalizing this, let the X-process be a semimartingale, $X_t = Y_t + Z_t$ being its decomposition (cf. Definition 2.19) with Y- and Z-processes both sample continuous. Suppose moreover that the Y-process is a square integrable martingale and the Doob-Meyer decomposition of the Y^2-process be denoted by $Y_t^2 = M_t + A_t, t \geq 0$. Then verify that the S_i defined above with this X converges in probability to the random variable $\int_a^t V(s)\, dA_s$, the pathwise Riemann-Stieltjes integral. The convergence actually takes place in $L^1(P)$ if the V-process is bounded. [*Hints*: First assume that all are bounded processes, and use the conditions of Definition 2.19, and then extend the result by using a stopping time

argument. This result is due to Fisk [2] who also discussed some other properties.]

7. (a) Let $\{X_t, \mathcal{F}_i, 1 \leq i \leq n\}$ be an adapted integrable process and $\lambda \geq 0$. Then establish

$$\lambda P\Big[\max_k X_k \geq \lambda\Big] \leq E\Big[\sum_{i=1}^{n-1}(X_i - E^{\mathcal{F}_i}(X_{i+1}))^+\Big] + \int_{[\max_k X_k > \lambda]} X_n \, dP.$$

(b) Let $X = \{X_t, \mathcal{F}_t, t \in I\}, I \subset \mathbb{R}^+$, be a separable quasimartingale of class (DL). Show that for $\lambda \geq 0$, and K_x of Definition 2.19,

$$\lambda P\Big[\sup_{t\in I} X_t \geq \lambda\Big] \leq \lim_{t\uparrow\alpha} \int_\Omega X_t^+ \, dP + K_x = K_x^0 < \infty, \ \alpha = \sup I,$$

and

$$\lambda P\Big[\inf_{t\in I} X_t \leq -\lambda\Big] \leq \lim_{t\uparrow\alpha} \int_\Omega X_t^- \, dP + K_x = K_x^0 < \infty.$$

(c) If X is a process as in (b), show that a.a. sample paths have finite left and right limits at each $0 < t < \infty$. [*Hints*: Both (a) and (b) are proved as in Proposition IV.1.4, using the same type decompositions. For (c),use the upcrossings inequality (cf. II.7.14); if β_n is the number of upcrossings of $[a, b]$ by $\{X_{t_1}, \ldots, X_{t_n}\}$, verify that $E(\beta_n) \leq [E(X_{t_n} - a)^+ + K_x]/(b-a)$.]

(d) For the process X of (b) with $I = \mathbb{R}^+$ for any decreasing t_n, verify that $\{X_{t_n}, n \geq 1\}$ is uniformly integrable. Hence deduce that if $X_t = X_{t+0}$ then $X_{t+(1/n)} \to X_t$ in $L^1(P)$. [*Hint:* Use (b).]

(e) Let $X = \{X_t, \mathcal{F}_t, t \geq 0\} \subset L^1(P)$ be right continuous in $L^1(P)$. Show that then X is a quasimartingale iff $\frac{1}{h}\int_0^t E(|X_t - E^{\mathcal{F}_t}(X_{t+h})|)dt$ is bounded as $h \to 0^+$, and it is a martingale iff the bound is zero. [This depends on a careful estimate of the growths of integrals. The details may be found in Orey [2], and (e) is in Stricker [2].]

8. Let $\{X_t, \mathcal{F}_t, t \geq 0\}$ be a right continuous quasimartingale of class (DL). Then in the decomposition $X_t = Y_t + Z_t$, the Z-process can be constructed as follows. Let $\tilde{Z}_t^k = \frac{1}{k}\int_0^t E^{\mathcal{F}_s}(X_{s+k} - X_s)ds (\in L^1(P))$. Then $\tilde{Z}_t^k \to Z_t$ in the weak topology of $L^1(P)$, as $k \to \infty$, for each $t \geq 0$. Show that this is true and it satisfies the requirements of the Z-process in that representation. (The details are similar to those in Meyer [3].)

9. We now sketch the original construction of the stochastic integral for a Brownian motion, due to Itô, leaving some computations to the reader to complete.

(a) Let $\{X_t, \mathcal{F}_t, t \in [0,a)\}$ be a Brownian motion, and suppose $\mathcal{M}_0$ is the set of adapted measurable processes $Y = \{Y_t, \mathcal{F}_t, t \in [0,a)\}$ such that $\int_0^t E(|Y_t|^2)\,dt < \infty$, for each $a > 0$. If $0 = a_0 < a_1 < \cdots < a_n = a$, and $f_i \in L^2(\mathcal{F}_{a_i}, P)$, let $g_n = \sum_{i=0}^{n-1} f_i \chi_{[a_i, a_{i+1})}$. Then $g_n \in \mathcal{M}_0$ and if $Y \in \mathcal{M}_0$, there exists a sequence $\{g_n, n \geq 1\} \subset \mathcal{M}_0$ such that $g_n \to Y$ in $L^2([0,a) \times \Omega, dt \otimes dP)$ so that such simple g_n are dense in $\mathcal{M}_0$ in the L^2-topology.

(b) For each $g_n \in \mathcal{M}_0, g_n = \sum_{i=0}^{n-1} f_i^n \chi_{[a_i^n, a_{i+1}^n)}$, define an integral relative to the X-process as:

$$Z_n(a) = \int_0^a g_n(t)\,dX_t = \sum_{i=0}^{n-1} f_i^n (X_{a_{i+1}^n} - X_{a_i^n}). \qquad (*)$$

Verify that $Z_n(a)$ is well-defined, $Z_n(a) \in L^2(P), E(Z_n(a)) = 0$. If $g, h \in \mathcal{M}_0$ are any two simple functions as above, and Z_f, Z_g are the corresponding functions defined by $(*)$, then (even for a complex case)

$$\|Z_g - Z_h\|_{2,P}^2 = E[(\int_0^a (g(t) - h(t))dX_t)(\int_0^a (g(t) - h(t))dX_t)^*]$$

$$= \int_0^a E(|g(t) - h(t)|^2)dt = \|g - h\|_{2,dt\otimes dP}^2,$$

by Fubini's theorem, where $\|\cdot\|_{2,P}$ is the norm in $L^2(P)$ and the last one is that of $L^2(dt \otimes dP)$, and hence of $\mathcal{M}_0$. The mapping $g \mapsto Z_g$ is an isometry on simple functions. Since $\mathcal{M}_0$ is a closed subspace of $L^2(dt \otimes dP)$ by the first part, if $\{g_n, n \geq 1\} \subset \mathcal{M}_0$ is Cauchy with limit Y, then $\{Z_{g_n}, n \geq 1\} \subset L^2(P)$ is Cauchy with limit Z_Y (say), and denote it as:

$$Z_Y(a) = \int_0^a Y(t)\,dX_t. \qquad (+)$$

Now show that $Z_Y(a)$ is independent of the g_n-sequence used, and is determined by Y and X (and a). Show also that the mapping $Y \mapsto Z_Y$ is linear and if Y is replaced by $Y\chi_{[0,t)}$ and $Z_Y(t)$ is the corresponding value of (+), then $\{Z_Y(t), \mathcal{F}_t, t \geq 0\}$ is a martingale which has a modification with continuous sample paths. [For the last statement use the first Borel-Cantelli lemma.]

(c) If X is an $L^2(P)$-martingale, then $\{X_t^2, \mathcal{F}_t, t \in [0,a), a > 0\}$ is a submartingale so that $F(t) = E(X_t^2)$ defines a nondecreasing function, and $E((X_t - X_s)^2) = F(t) - F(s)$ for $s < t$. Let $\mathcal{M}_1$ be the subclass of $L^2(P)$-martingales for which $E^{\mathcal{F}_s}((X_t - X_s)^2) = F(t) - F(s)$ a.e. This class includes the Brownian motion with $F(t) = ct$, for some constant $c > 0$. Show that, replacing $L^2(dt \otimes dP)$ by $L^2(dF \otimes dP)$ in (b), the resulting process $\{Z_Y(t), \mathcal{F}_t, t \in [0,a)\}$ has a separable martingale version for each $Y \in \mathcal{M}_0 \subset L^2(dF \otimes dP)$. The sample functions are a.a. continuous if F is continuous, and in general the fixed discontinuities of the Z-process correspond to the discontinuities of F. [This generalization of (b) is due to Doob ([1], Chapter IX). The hypothesis on the Y-process can be weakened, and for a coprehensive account of Itô's calculus, one may consult McKean [1].]

10. (a) We give a direct proof of Proposition 3.15 when X is a Brownian motion. Thus if $X = \{X_t, \mathcal{F}_t, t \in [0,a)\}, X_0 = 0$, is the (standard) Brownian motion, then show that

$$\lim_{n\to\infty} \sum_{k=1}^{2^n} (X(\frac{na}{2^k}) - X(\frac{n-1}{2^k}a)) = a, \text{ a.e.}$$

[*Sketch*: Let $f_{kn} = X(\frac{na}{2^k}) - X(\frac{n-1}{2^k}a)$. Then $f_{kn}, k = 1, \ldots, 2^n$ are independent and Gaussian each with mean 0, variance $\frac{a}{2^k}$. Hence f_{kn}^2 are also independent and $\sum_{k=1}^{2^n} E(f_{kn}^2) = a$ and variance of $\sum_{k=1}^{2^n} f_{kn}^2$ is $a^2/2^{n-1}$. If we let $A_{n,\varepsilon} = \{\omega : |\sum_{k=1}^{2^n} f_{kn}^2(\omega) - a| > \varepsilon > 0\}$, then $P[A_{n,\varepsilon}] \leq \frac{a^2}{\varepsilon^2} 2^{-n+1}$ so that $\sum_{k=1}^{\infty} P(A_{n,\varepsilon}) < \infty$. Hence $P[\limsup_n A_{n,\varepsilon}] = 0$, implying the result.]

(b) Let X be as above, and $\mathcal{T} = \{T_t, t \geq 0\}$ be a continuous time change process of the filtration. Let $Y_t = X \circ T_t$ $(T_0 = 0)$ and $\mathcal{B}_t = \mathcal{B}(T_t)$. Then $\{Y_t, \mathcal{B}_t, t \geq 0\}$ is a continuous square integrable martingale iff $E(T_t) < \infty, t \geq 0$. [*Sketch*: Since X^2 is a class (DL) submartingale, $X_t^2 = M_t + A_t, t \geq 0$ uniquely, where $A_t = \langle X, X\rangle_t = t$ and M_t is a continuous martingale. Let $T_t^n = T_t \wedge n, M_t^n = M \circ T_t^n$. Then $\{M_t^n, \mathcal{B}_t, t \geq 0\}$ is a martingale (cf. Theorem IV.1.7). If $Y_t^n = X \circ T_t^n$, then for the "if" part $E((Y_t^n)^2) = E(M_t^n) + E(T_t^n) = E(T_t^n) < \infty$. Since $Y_t^2 = \lim_n (Y_t^n)^2$ a.e., we get $E(Y_t^2) \leq E(T_t) < \infty$ by Fatou's lemma. For the martingale property of the Y-process it suffices to show, by Theorem IV.1.13, that $\overline{\lim}_s \int_{[T_t>s]} |X_s|\, dP < \infty$, and the latter is bounded by $\|X_s\|_2 \|\chi_{[T_t>s]}\|_2 = [\int_{T_t>s]} s\, dP]^{1/2} \le [E(T_t)]^{1/2} < \infty$. For the converse, let $E(Y_t^2) < \infty$, the Y-process being in class (DL), we have $Y_t^2 = T_t + M \circ T_t = (X \circ T_t)^2$. Hence $E(T_t) \le E(Y_t^2) + E(|M \circ T_t|) < \infty$.]

11. Let $\{a^i(t), \mathcal{F}_t, t \ge 0\}, i = 1, 2$, be adapted measurable processes and $X \in \mathcal{M}^c$ (as in Section 3). For each t, suppose that $\int_0^t |a^i(s)|^2 ds < \infty, i = 1, 2$, a.e. A process Y is said to satisfy a *stochastic differential equation*: $(*)\, dY_t = a^1(t)\, dt + a^2(t)\, dX_t, t \ge 0$, iff one has

$$Y_t = \int_0^t a^1(s)\, ds + \int_0^t a^2(s)\, dX_s, \quad t \ge 0,$$

where the first one is a pointwise Lebesgue integral and the second one is the stochastic integral (cf. Exercise 9 above or Theorem 3.9). The definition extends to vector processes. Let $X^i \in \mathcal{M}^c, i = 1, \dots, n, \langle X^i, X^j \rangle_t = \delta_{ij} t$ (this means in the Brownian case independence of the components). If $Z = (Z^{ij}, 1 \le i \le n, 1 \le j \le m)$ is a matrix of random processes, $Z^{ij} \in L^2(\langle X^j \rangle)$, and $Y = (Y^1, \dots, Y^m)$ suppose that $dY_t = a_t\, dt + \sum_{i=1}^n Z_t^i\, dX_t^i$ for $a = (a^1, \dots, a^n), Z^i = (Z^{ij}, j = 1, \dots, m)$ holds. If $F : [0, a] \times \mathbb{R}^m \to \mathbb{R}$ is twice continuously differentiable, then show that $F(t, Y_t) = G_t$ defines a process adapted to $\{\mathcal{F}_t, t \ge 0\}$ and has a differential given by:

$$dG_t = \Big[\frac{\partial F}{\partial t}(t, Y_t) + \sum_i \frac{\partial F}{\partial x^i}(t, Y_t) a_t^i + \frac{1}{2} \sum_{i,j,k} \frac{\partial^2 F}{\partial x^i \partial x^j}(t, Y_t)\, Z_t^{ik} Z_t^{jk}\Big] dt + \sum_{i,k} \frac{\partial F}{\partial x^i}(t, Y_t) Z_t^{ik}\, dX_t^k.$$

[*Sketch*: In the formula of Theorem 3.16 take $\sum_i \int_0^t Z_s^{ik}\, dX_s^i$ for X_t^k and $\int_0^t a_s^k ds$ for A_t^k there. Since $E(\langle \int_0^{(\cdot)} Z_s^{ik} dX_s^i, \int_0^{(\cdot)} Z_s^{jk} dX_s^j \rangle_t) = \delta_{ij} \int_0^t E(Z_s^{ik} Z_s^{jk}) ds$, substitution and simplification in the cited theorem yields the result. This formula was directly (and differently) proved by Itô [1].]

12. Using the notation of Section 3, we consider some analogs of the Fundamental Theorem of Calculus for stochastic integrals.

(a) Let $X \in \mathcal{M}^c$ and $Y \in \mathcal{L}^2(\langle X \rangle)$. If $Z_t = \int_0^t Y_s\, dX_s$, so that $Z = \{Z_t, \mathcal{F}_t, t \geq 0\} \in \mathcal{M}^c$, suppose that for each $t \in \mathbb{R}^+, h > 0$, we have $P[(X_{t+h} - X_t)^2 \leq ah] = o(a)$; whence the distribution of $\frac{1}{h}(X_{t+h} - X_t)^2$ is nonatomic near the origin. Then $\frac{\Delta Z_t}{\Delta X_t} \to Y_t$ as $h \to 0$ in probability for almost all t (Leb.), where $\Delta Z_t = Z_{t+h} - Z_t$ and similarly ΔX_t. [*Hints*: $\frac{\Delta Z_t}{\Delta X_t} - Y_t = \frac{1}{\Delta X_t}\int_t^{t+h}(Y_s - Y_t)dX_s = \frac{1}{\Delta X_t} I_h^t$ (say), and using the condition on Y we deduce that $\{I_h^t, \mathcal{F}_t, t \geq 0\} \in \mathcal{M}^c$. Hence by the maximal inequality,(with separability) for each $\delta > 0, c > 0$ we have

$$P[\sup_{0 \leq h \leq \delta} |I_h^t|^2 > \delta c^2] \leq \frac{1}{\delta c^2} E(I_\delta^t)^2$$

$$= \frac{1}{c^2} E[\frac{1}{\delta} \int_t^{t+h} (Y_s - Y_t)^2\, d\langle X \rangle_s]. \qquad (*)$$

If $d\tilde{P} = d\langle X \rangle \otimes dP$ is the product measure on the space $[t, t+\delta] \times \Omega$, then it is absolutely continuous relative to $dt \otimes dP$, and by the classical Lebesgue differentiation theorem we conclude that the right side of $(*)$ tends to 0, as $\delta \to 0$ for almost all t (Leb.). But for $k > 0, 0 < h < \delta$ we have

$$P[\frac{h^{1/2}}{|\Delta X_t|} \cdot h^{-1/2}|I_h^t| > \varepsilon] \leq P[\frac{1}{h}(I_h^t)^2 > \frac{\varepsilon^2}{k^2}] + P[\frac{h}{(\Delta X_t)^2} > k^2].$$

By hypothesis on the distribution of ΔX_t, the second term is $o(\frac{1}{k^2})$. Choosing k large and then using $(*)$ the first term is $o(\delta)$. This implies the result. If X is Brownian motion, then $\frac{\Delta X_t}{\sqrt{h}}$ is Gaussian distributed with mean zero and unit variance, and the condition of nonatomicity is automatically satisfied.]

(b) Suppose now that $X \in \mathcal{M}^c$ is a Brownian motion. Show that there exists uniquely a martingale $Y \in \mathcal{L}(\langle X \rangle)$ such that $Y_t = 1 + \int_0^t Y_{s-}\, dX_s$,a.e. [*Hints*: If there is a solution, then by (a), $\{Y_t - 1, t \geq 0\}$ must be a continuous martingale, and $\frac{\Delta Y_t}{\Delta X_t} \to Y_t$ in probability. Such a process must be unique, since if $\tilde{Y}$ is another solution, then $Z_t = Y_t - \tilde{Y}_t = \int_0^t Z_s\, dX_s$ and $\varphi(t) = E(Z_t^2) = E(\int_0^t Z_s^2 d\langle X \rangle_s) = \int_0^t \varphi(s)ds$. Since $\varphi(\cdot) \uparrow, \varphi(0) = 0$ we get on iteration that $0 < \varphi(t) \leq \frac{C}{n!}$ where $C = \max_{0 \leq s \leq t} \varphi(s)$. Hence $\varphi(t) \equiv 0$ and $Y = \tilde{Y}$. By substituting $Y_t = \exp(X_t - c)$ and noting that $E(Y_t - 1) = 0$, we see that $E(e^{X_t}) = e^c$ so that $c = t/2$. This gives the unique solution of the equation.]

(The result of (a) was discussed , for the Brownian motion, by Issacson [1] using different methods, and (b) was noted by Doléans [2] where further extensions to semimartingales were discussed. This type of functional (or directional) differentiation is found to play a part in extending the stochastic integration to more general (not necessarily predictable) integrands in recent studies, cf. e.g., Nualart and Pardoux [1]. See also Maisonneuve [1].)

13. Let $dZ_t = a_t\, dX_t + b_t\, dt$ where $\{X_t, \mathcal{F}_t, t \geq 0\}$ is a Brownian motion, and $a \in \mathcal{L}^2(\langle X \rangle), b \in \mathcal{L}^1(\langle X \rangle)$. Suppose $\alpha_t = 1 + \int_0^t \alpha_s \varphi_s\, dX_s$ with $\varphi \in \mathcal{L}^2(\langle X \rangle)$. If P is the original measure for the X-process, and if $\tilde{P}$ is determined such that $\tilde{P}(A) = \int_A \alpha_s\, dP, a \in \mathcal{F}_s, \tilde{P}(\Omega) = 1$, then verify that $\{\tilde{X}_t, \mathcal{F}_t, t \geq 0\}$ on $(\Omega, \mathcal{F}_\infty, \tilde{P})$ is again a Brownian motion where $\tilde{X}_t = X_t - \int_0^t \varphi_s ds$. Moreover, conclude that P and $\tilde{P}$ are equivalent measures and the Z-process can be expressed as: $dZ_t = a_t d\tilde{X}_t + (a_t \varphi_t + b_t)dt$. [*Hints*: By Corollary 3.19, it suffices to show that $\{\tilde{X}_t^2 - t, \mathcal{F}_t, t \geq 0\}$ is a martingale on $(\Omega, \mathcal{F}_\infty, \tilde{P})$. Since by definition $\tilde{P}$ is P-continuous, and if $\mathcal{B}_1 \subset \mathcal{B}_2 \subset \mathcal{F}_\infty$, and $f_i = \frac{d\tilde{P}_{\mathcal{B}_i}}{dP_{\mathcal{B}_i}}, i = 1, 2$, we have for any $A \in \mathcal{B}_1$ and $\tilde{P}$-integrable g:

$$\int_A E^{\mathcal{B}_1}(g) d\tilde{P}_{\mathcal{B}_1} = \int_A g f_2 dP_{\mathcal{B}_2} = \int_A E^{\mathcal{B}_1}\left(\frac{g f_2}{f_1}\right) d\tilde{P}_{\mathcal{B}_1}.$$

Hence we can identify the extreme integrands $\tilde{P}$ a.e. and make calculations with the latter. The equivalence of measures can be carried out by a stopping time argument assuming the contrary. The result is due to Girsanov [1], and the equivalence problem was discussed by Kailath and Zakai [1]. See also Orey [1]. Note that by 12(b) above, $\alpha_t = \exp[\int_0^t \varphi_s\, dX_s - \frac{1}{2}\int_0^t \varphi_s^2 ds]$ and then $\frac{d\tilde{P}}{dP}|_{\mathcal{F}_t} = \alpha_t$ a.e. $[P]$.]

14. Theorem 2.1 can be extended as follows. Consider the standard filtration of (Ω, Σ, P), and let $\{X_t, \mathcal{F}_t, t \geq 0\}$ be a supermartingale such that $(i) \sup_t E(|X_t|) < \infty$, and (ii) the sample paths of the X_t-process are upper semicontinuous to the right. Then the following assertions hold.

(a) There exists a right continuous supermartingale Y and a left continuous integrable increasing process B, both adapted to the same standard filtration $\{\mathcal{F}_t, t \geq 0\}$, such that $X_t = Y_t - B_t, t \geq 0$. In fact $B_t = \sum_{s<t}(X_s - X_{s-})$ a.e. (and the sum is atmost countable). In particular, X_t is a pointwise limit of a decreasing sequence of right con-

tinuous supermartingales. [*Remarks.* This is a technical result needing a detailed analysis, and can be found in J.-F. Mertens [1]. As shown there, conditions (i) and (ii) are always satisfied if the X-process is optional (=well-measurable) and $X \circ T \in L^1(P)$ for all stopping times T of the filtration. See also Cornea and Licea [1] in this connection.]

(b) If also the X_t-process dominates a right continuous martingale adapted to the same filtration, then there exist a right continuous martingale Z, a local martingale M, and a predictable increasing integrable process A such that $\{M_t - A_t, t \geq 0\}$ is a potential so that the representation is a.e. unique:

$$X_t = M_t + Z_t - A_t, \qquad t \geq 0.$$

Here $\{A_t - B_t, t \geq 0\}$ is increasing and predictable; the M_t-process is a martingale iff the X_t-process is of class (DL). [*Remarks.* By (a), $Y_t = X_t + B_t$, and $Y_t \geq \tilde{Y}_t + B_t \geq \tilde{Y}_t + B_0$ where $\{\tilde{Y}_t, \mathcal{F}_t, t \geq 0\}$ is the dominated martingale by X_t of the hypothesis. Theorem 2.1 implies $Y_t = Z_t + M_t - A'_t, t \geq 0$. If $A_t = A'_t + B_t$ then (by the left continuity of the B-process) the A-process qualifies to be the desired predictable process, and the decomposition follows. The uniqueness is proved as in Theorem 2.1 and the last assertion is then obtained as in Theorem 2.6.]

15. Let $X = \{X_t, \mathcal{F}_t, t \geq 0\}$ be a right continuous martingale in $L^p(P), p \geq 1$. For each $k \geq 1$, let $\mathcal{C}_k = \{T_n^k, n \geq 1\}$ be a stochastic partition of length 2^{-k} relative to this martingale, as defined in the proof of Theorem 3.14. Let $S^2(X, \mathcal{C}^k) = \sum_{n \geq 1}[X \circ T_n^k - X \circ T_{n-1}^k]^2$ with $X \circ T_0^k = 0$. Thus $S(X, \mathcal{C}^k)$ is the (generalized) quadratic variation of X relative to $\mathcal{C}^k$, as defined in Proposition IV.5.9, the optionals in the latter being constants. Suppose that $\sup_{t \geq 0} |X_t| = X^* \in L^p(P)$. Then, for each $1 \leq r < p, S(X, \mathcal{C}^k) \to S(X)$ in $L^r(P)$ as $k \to \infty$, and the result holds for $p = 1$ if the convergence is "in probability". In any case, we have

$$c_{1p}\|S(X)\|_p \leq \|X^*\|_p \leq c_{2p}\|S(X)\|_p, \; 1 \leq p < \infty,$$

for some absolute constants $c_{ip} > 0, i = 1, 2$, depending only on p. [*Hints:* As in Extension IV.4.10, we may express $S(X, \mathcal{C}^k)$ as:(since $X_t \to X_\infty$ a.e. and in $L^1(P)$)

$$S^2(X, \mathcal{C}^k) = X_\infty^2 - 2\sum_{n \geq 0} X \circ T_n^k (X \circ T_{n+1}^k - X \circ T_n^k). \qquad (*)$$

Now using a truncation argument, first assume that the X-process is bounded a.e. and proceed as in the proof of Theorem 3.14 to show that the last term in (*) converges in norm to the $L^2(P)$-stochastic integral $\int_{\mathbb{R}^+} X_s\, dX_s$. Next let the assumed bound of the X-process increase to infinity and deduce the general case by an appropriate extension. From this, Theorem IV.5.4, and the Fatou property of the norm, deduce the inequality. Thus one gets for each $t \geq 0$

$$S^2(X)_t = X_t^2 - 2\int_0^t X_s\, dX_s.$$

This leads to martingale integrals in $L^p(P), p \neq 2$. In this connection see Millar [2], and Doléans [1].]

16. We indicate here a generalization of quasimartingales, showing the lattice and convergence properties clearly. Let $\mathcal{T}$ be the class of all bounded optionals of a standard filtration of (Ω, Σ, P) and X be a process in $L^1(P)$. Then it is called an *asymptotic martingale, or amart,* if for any sequence $\{\tau_n, n \geq 1\} \subset \mathcal{T}, \tau_n$ simple, and $\tau_n \leq \tau_{n+1}$, the net $\{E(X \circ \tau_n), n \geq 1\} \subset \mathbb{R}$ is convergent. [A reversed sequence is similarly obtained if $\tau_n \geq \tau_{n+1}$.] Verify the following assertions:

(a) If X, Y are amarts, then so are $X \vee Y, X \wedge Y$.

(b) Every quasimartingale is an amart. [*Hints*: If $\tau_1 < \tau_2 < \tau_3$ are simple, show that for each $A \in \mathcal{B}(\tau_1)$ we have,

$$\int_A |E^{\mathcal{B}(\tau_1)}(X \circ \tau_3) - X \circ \tau_1| dP \leq \int_A |E^{\mathcal{B}(\tau_2)}(X \circ \tau_3) - X \circ \tau_2| dP + \int_A |E^{\mathcal{B}(\tau_1)}(X \circ \tau_2 - X \circ \tau_1| dP.$$

Using this verify by induction that for any sequence $\tau_1 < \cdots < \tau_n$ whose (finite) set of values are $t_1, \ldots, t_k$, one has:

$$\sum_{j=1}^{n-1} E(|E^{\mathcal{B}(\tau_j)}(X \circ \tau_{j+1}) - X \circ \tau_j|) \leq \sum_{i=1}^{k-1} E(E^{\mathcal{F}_{t_i}}(X_{t_{i+1}} - X_{t_i}|).$$

Deduce that $\{E(X \circ \tau_n), n \geq 1\}$ is convergent.]

(c) If $X = \{X_n, \mathcal{F}_n, n \geq 1\}$ is an adapted sequence and $E(\sup_n |X_n|) < \infty$, then X is an amart iff it converges a.e. [Thus it is nontrivial to decide when a sequence is an amart.]

(d) If X is an adapted sequence as in (c), and $X^* = \limsup_n X_n, X_* = \liminf_n X_n$, then

$$\sup_{\{\sigma,\tau\}\subset\mathcal{T}} E(X\circ\tau - X\circ\sigma) \geq E(X^* - X_*) \geq 0.$$

Moreover, X^*, X_* are integrable if $\{X\circ\tau, \tau\in\mathcal{T}\}$ is $L^1(P)$-bounded. [Regarding this result and extensions, see Chacón [1] and Edgar and Sucheston [1]. Many other aspects and applications are given in the latter authors' recent monograph [2].]

Bibliographical remarks.

The fundamental continuous parameter decomposition of submartingales, as a generalization of the elementary discrete Doob decomposition of Chapter II, was an open problem for a decade and it was solved by Meyer [3]. This with its further extension for a larger class, given by Itô and S.Watanabe [1], is the content of Theorem 2.1. Doob [3] calls it the Meyer decomposition; and Meyer calls it the Doob decomposition in his writings. We termed it the Doob-Meyer decomposition in the first edition of this book, and retained the same terminology here. The present (first) proof, based on the discrete parameter case of Chapter II, is essentially taken from K.M.Rao [1] since it is more elementary than the original one. The key formula of Proposition 3.1 was noted by Meyer ([3],[1]) but was not utilized immediately. Its importance was noted for, and the result formed a basis of, the fundamental paper by Kunita and S.Watanabe [1]. Almost all the results on square integrable martingales are taken from this article, and only further details are added in the text. Theorem 3.14 is one of the high lights of the theory of stochastic integration, and it is originally due to Itô for the Brownian motion, whose proof for this case may be found in McKean [1]. The result in Theorem 3.23, proved differently, was first obtained independently by Dambis [1] and Dubins-Schwarz [1]. To illustrate the power and usefulness of the notion of time change, we included several results in the Complements section.

The decomposition of quasimartingales, using measure theoretic ideas, given as Theorem 2.20, is due to Doléns-Dade [3]. See also Föllmer [1]. Theorem 2.22 was noted by K.M.Rao [2], and Theorem 2.24 is due to Fisk [1]. We shall discuss further the stochastic integration relative to semimartingales, and their extensions in the next chapter. It should be noted that some of the results related to (and extensions of) the Doob-Meyer decomposition for the vector (or Banach space) valued case have been considered by Dinculeanu [3] in a series of papers and some in collaboration with Brooks [1]. The treatment there is in part motivated by and is an extension of that of Métivier and Pellaumail [2] and of Métivier [1]. For space reasons we do not consider the Banach space valued case in any detail in this volume, although the ideas pervade in the development.

Stratonovich integrals, as symmetrizations of the Itô integrals, had been considered by Stratonovich [1] and Fisk [1] independently, and the former author exploited the resulting convenience fully. The symbolic calculus for differentials was given by Itô [3] and further explained by Itô and S.Watanabe [2]. The common extension of both these concepts, as given in Proposition 3.35, was discussed by Yor [2].

The key role played by the Hilbert space geometry in the Kunita and S.Watanabe treatment is implicit. Its essential identity with certain spectral integrals was clarified and made explicit by Cuculescu [1], and the work of Section 4 follows his paper where some further results may be found. As seen in Exercise 5.9, the original method of Itô's is very close to the Dunford-Schwartz work, discussed in Chapter II, using the theory of vector measures. However, Itô's work predates the vector analysis just referred to, and is more general then the latter in the stochastic context.

Another approach to stochastic integration, without martingale decomposition theory, has been developed by McShane (cf. [3] and [4]), from fundamentals with modifications of the Riemann-Stieltjes method. In the next chapter we shall comment more on this point. Special features of the stochastic integral (distinguishing it from the classical Lebesgue point of view) are illuminated by Theorem 3.14, which shows that it cannot be treated in a simple way. See also Theorem 3.8, and Meyer [4] for a nice exposition. Many of the other authors' contributions referred to in the text should also be consulted.

Chapter VI

Stochastic integrals and differential systems

We abstract and extend the stochastic integration with martingale integrators to more general processes for which the dominated convergence theorem is still valid. The motivation here is to obtain a unified treatment of several different stochastic integrals, available in the literature, by means of a generalized boundedness principle based on a fundamental idea formulated by S. Bochner. After presenting the semimartingale integrals in the next section, to serve as a key example, the desired boundedness principle is treated in detail in Section 2. It is also shown there, and in Section 3, that the earlier integrals fit in this frame work; and several applications are worked out to exhibit the universality of the principle, including some vector and multiparameter cases. The rest of the chapter is devoted to the existence (and unicity) of solutions of both linear and nonlinear higher order stochastic differential equations and its progression to stochastic flows for the $L^{2,2}$-bounded case. This work takes up Sections 4 and 5 below, and most of Section 4 appears in book form for the first time. Several other results are included in the Complements section.

6.1 Semimartingale integrals

Recall that a (local) semimartingale is a process $X = \{X_t, \mathcal{F}_t, t \in T \subset \mathbb{R}^+\}$, adapted to a standard filtration, that is representable as: $X = Y + Z$ where $Y = \{Y_t, \mathcal{F}_t, t \in T\}$ is a (local) martingale and $Z = \{Z_t, \mathcal{F}_t, t \in T\}$ is a process with a.a. of its sample functions of bounded variation (on each compact subset of T), (cf. Definition V.2.19). Here we can take all processes to have separable versions. As a prelude to the work of the next section and with some extensions of the material of Section V.3, we present integration relative to the X-process, where

Z is not always predictable. When Z is not necessarily predictable, we express it as $Z \in \tilde{\mathcal{A}}_{\mathrm{loc}}$ (instead of $\mathcal{A}_{\mathrm{loc}}$). A high point in all the extensions below is to get analogs of Theorem V.3.14. Let Y be a local martingale. Then there is a sequence $\{T_n \uparrow \infty\}$ of finite stopping times such that $\{Y(T_n \wedge t), \mathcal{F}_t, t \geq 0\}, n \geq 1$, is a uniformly integrable martingale. Let $\tilde{Y}_t^n = E^{\mathcal{F}_t}(Y \circ T_n)$ be a right continuous version and define S_n (a stopping time or optional of $\mathcal{F}_t, t \geq 0\}$) as:

$$S_n = (\inf\{t > 0 : \tilde{Y}_t^n \geq a_n\}) \wedge T_n \tag{1}$$

where $a_n > 0$ are chosen such that $P\left[S_n < T_n - \frac{1}{n}\right] \leq \frac{1}{2^n}$. This is possible since $\tilde{Y}_t^n \to Y \circ T_n$ a.e. and in $L^1(P)$ as $t \to \infty$, and S_n may be taken to be close to T_n. Thus $S_n \uparrow \infty$, and $\{Y(S_n \wedge t), \mathcal{F}_t, t \geq 0\}$ is a bounded martingale on $[\![0, S_n)$. If we set $\tau_n = \inf\{S_k : k \geq n\}$, then $\{\tau_n \uparrow \infty\}$ is a sequence of optionals of $\{\mathcal{F}_t, t \geq 0\}$ so that $\{Y \circ \tau_n, n \geq 1\}$ is bounded on $[\![0, \tau_n)$ for each n. Thus in the definition of a local martingale, one may choose the sequence $\{T_n, n \geq 1\}$ to have this additional (boundedness) property. Such a sequence is said to *(strongly) reduce* Y.

To deduce another property of Y consider the sequence $\{\tau_n, n \geq 1\}$ of the above paragraph. Then $\{Y(\tau_n \wedge t), \mathcal{F}_t, t \geq 0\}$ is a bounded martingale and hence is in $\mathcal{M}$ on $[\![0, \tau_n)$. Thus by Corollary V.3.31, there exist uniquely on $[\![0, \tau_n)$ a $Y^1 \in \mathcal{M}^c$ and a $Y^2 \in \mathcal{M}^d$ such that $Y \circ \tau_n = Y_n^1 + Y_n^2$. Since $\tau_n \leq \tau_{n+1}$, by the uniqueness of this representation, $Y_n^i (= Y^i(\tau_n))$ and $Y_{n+1}^i, i = 1, 2$, agree on $[\![0, \tau_n)$. Hence there exists a Y^c and a Y^d, respectively continuous and discontinuous local martingales, such that $Y = Y^c + Y^d$ uniquely. (For the last assertion, if $\bar{Y}^c + \bar{Y}^d = Y^c + Y^d$ are two representations then $\bar{Y}^c - Y^c = Y^d - \bar{Y}^d$ and are orthogonal to themselves for each τ_n, and $(\bar{Y}^c - Y^c)^2 = (\bar{Y}^c - Y^c)(Y^d - \bar{Y}^d)$ is a local martingale vanishing at $t = 0$. So $\bar{Y}^c = Y^c$ and $\bar{Y}^d = Y^d$.) Since $Y^d \circ \tau_n$ is bounded on $[\![0, \tau_n)$ and τ_n is finite, we deduce that on writing $\Delta Y_s = Y_s - Y_{s-}$, the jump at $s, Y^d \circ \tau_n = \sum_{s \leq \tau_n} \Delta Y_s$ (when s is considered as a constant stopping time), or that $\sum_{s \leq t} (\Delta Y_s)^2$ converges a.e. for each $t \in [0, \infty)$. This decomposition is needed below. It is useful to note that the "times" at which the jumps occur are not predictable, and so the process Y^d is not predictable. If there are only finitely many jumps so that $\sum_{s \leq t} \Delta Y_s$, is a.e. convergent, then $Y^d \in \mathcal{M}^d \cap \tilde{\mathcal{A}}$ and still not zero. (Compare this with the remark following the proof of Proposition V.3.2.)

We recall that, if Y is a continuous local martingale, then $\langle Y \circ \tau_n, Y \circ \tau_n \rangle$ defines a unique predictable increasing integrable process of $Y \circ \tau_n \in \mathcal{M}$ where $\{\tau_n \uparrow \infty\}$ is as in the preceding paragraphs. By the uniqueness assertion and the fact that $\tau_n \leq \tau_{n+1}$, we may deduce the existence of an element $\langle Y, Y \rangle \in \tilde{\mathcal{A}}_{\text{loc}}$ such that $\langle Y, Y \rangle \circ \tau_n = \langle Y \circ \tau_n \rangle$. This is a consequence of the calculus of stopping time substitutions discussed in Chapter IV. However, for right continuous Y one cannot use this procedure. So we use the above decomposition and discuss a new process, $[Y, Y]$, called the *raw* increasing process (cf. Remark V.2.7). Consider the unique decomposition $(*)$ $Y = Y^c + Y^d$. Then we have:

$$[Y]_t = [Y, Y]_t = \langle Y^c, Y^c \rangle_t + \sum_{s \leq t} (\Delta Y_s)^2, \quad t \geq 0. \tag{2}$$

If Y_1, Y_2 are any two local martingales define, as in Section V.2,

$$[Y_1, Y_2] = \frac{1}{2}([Y_1 + Y_2, Y_1 + Y_2] - [Y_1, Y_1] - [Y_2, Y_2]), \tag{3}$$

or equivalently

$$[Y_1, Y_2] = \frac{1}{4}([Y_1 + Y_2, Y_1 + Y_2] - [Y_1 - Y_2, Y_1 - Y_2]). \tag{3'}$$

This functional has analogous properties of $\langle Y, Y \rangle$ (or $\langle Y, Y \rangle$) and $[Y_1, Y_2] \in \tilde{\mathcal{A}}_{\text{loc}}$. It will follow from the general result below (cf. Corollary 3) that $Y_1 Y_2 - [Y_1, Y_2]$ is a local martingale and then the theory of integration relative to such elements proceeds as in the case of the work with $\mathcal{M}$.

Let $U = \{U_t, t \geq 0\}$ be a locally bounded predictable process relative to $\{\mathcal{F}_t, t \geq 0\}$, i.e., there exist $\{S_n \uparrow \infty\}_1^\infty$, finite optionals of $\{\mathcal{F}_t, t \geq 0\}$, such that for each $n \geq 1$, $\{U \circ (S_n \wedge t) \chi_{[S_n > 0]}, t \geq 0\}$ is bounded. If $Y \in \tilde{\mathcal{M}}_{\text{loc}}$, and $\{\tau_n \uparrow \infty\}$ is a (strongly) reducing sequence of Y then it is seen that the $\{S_n\}_1^\infty$ may be taken to be the same as the $\{\tau_n\}_1^\infty$. Hence $U \circ \tau_n$ is predictable, and integrable relative to $Y \circ \tau_n$, by the theory of Section V.3 and we get a unique element $(U \circ \tau_n, Y \circ \tau_n)$, also denoted $\int_0^{(\cdot)} (U \circ \tau_n)(s-) d(Y \circ \tau_n)$. (Decompose Y as $Y^c + Y^d$ and define this.) Since $\tau_{n-1} \leq \tau_n$, one can define a unique element $(U \circ Y)$

such that $(U \circ Y) \circ \tau_n = (U \circ \tau_n, Y \circ \tau_n)$. This is denoted also as:

$$(U \circ Y)_t = \int_0^t U_{s-} dY_s, \quad t \geq 0. \tag{4}$$

To see that this formalism is in fact exact, we record the following:

1. Proposition. *For each locally bounded predictable U and each local martingale Y, the process $U \circ Y$ defined by (4) is also a local martingale. Moreover, this is the unique element which satisfies for all local martingales V, $[U \circ Y, V] = U \circ [Y, V]$ where the right side element is an ordinary pointwise Stieltjes integral.*

The unicity is obvious. The assertion about $U \circ Y$ follows from a stopping time argument used several times and the fact that the stochastic integrals of Section V.3 are again martingales. If Y, V are local martingales then we may clearly assume that there is a common sequence $\{\tau_n \uparrow \infty\}$ which reduces both Y, V and U by an easy extension of the earlier result. The decomposition into continuous and discrete parts noted earlier allows us to conclude the desired result by an application of Corollary V.3.13 and a simple computation. This implies all the statements of the proposition.

We now present a useful extension of the Itô formula in this context. For a semimartingale $X_t = X_0 + Y_t + Z_t$ (Y local martingale, $Z \in \mathcal{A}_{\mathrm{loc}}, Y = Y^c + Y^d$, so that $\langle Y^c, Y^c \rangle \in \mathcal{A}^c_{\mathrm{loc}}$), we set $\dot{X}^c = Y^c$ and $\langle \dot{X}^c, \dot{X}^c \rangle = \langle Y^c, Y^c \rangle$. With this notation, the following is the desired extension of Theorem V.3.16 to semimartingales. We include the details for completeness and because of its importance in the work that follows:

2. Theorem. *Let $X = (X^1, \dots, X^n), X^i$ being a (real) semimartingale, $i = 1, \dots, n$. Let $f : \mathbb{R}^n \to \mathbb{C}$ be a twice continuously differentiable function. Then for each $t \geq 0$, one has a.e.,*

$$f(X_t) - f(X_0) = \sum_{i=1}^n \int_0^t \frac{\partial f}{\partial x_i}(X_{s-}) dX_s^i + \frac{1}{2} \sum_{i,j=1}^n \int_0^t \frac{\partial^2 f}{\partial x_i \partial x_j}(X_{s-}) \\ d\langle \dot{X}^{ic}, \dot{X}^{jc} \rangle_s + \sum_{s \leq t} [f(X_s) - f(X_{s-}) - \\ \sum_{i=1}^n \frac{\partial f}{\partial x_i}(X_{s-})(X_0^i - X_{s-}^i)], \tag{5}$$

where the sum in the last term of (5) converges a.e., *for each t. Moreover the process $f(X)$ is a semimartingale.*

Proof. As in the earlier proof, we consider the case $n = 1$ for simplicity. Replacing X_0 by $X_0\chi_{A_k}, A_k = [|X_0| \leq k] \in \mathcal{F}_0$, if necessary, we may and do assume that X_0 is bounded. The argument will be divided into steps for convenience. Note that the elements of $\tilde{\mathcal{A}}$ need *not* be predictable, unless they are continuous.

I. Observe that in $X = X_0 + Y + Z$, if $Y \in \mathcal{M}^c_{\mathrm{loc}}, Z \in \tilde{\mathcal{A}}^c_{\mathrm{loc}}$, then the last term in (5) vanishes a.e. From this, with a stopping time device one sees that the result reduces to that of Theorem V.3.14 and hence the theorem is true in that case. Also by a standard approximation argument (replacing T by $T \wedge t$ etc.) it is clear that (5) holds if 0 and t are replaced by a pair of stopping times S, T of $\{\mathcal{F}_t, t \geq 0\}$ where $S \leq T$. With this, we extend the result successively to the general Y and Z in the following.

II. Let Y be a local martingale and $Z(\in \tilde{\mathcal{A}}_{\mathrm{loc}})$ have finitely many jumps in $[0, t]$. Let $T_0 = 0, T_{n+1} = t$ and for $1 \leq i \leq n$ set $T_i = \inf\{t > T_{i-1} : Z_t \neq Z_{t-1}\}$, where $\inf(\emptyset) = t$. Then $T_0 \leq T_1 \leq \cdots \leq T_n \leq T_{n+1}$, and T_i is the instance of the ith jump of the Z-process in $[0, t]$. If $Z = Z^c + Z^d, Z^c \in \tilde{\mathcal{A}}^c_{\mathrm{loc}}, Z^d \in \tilde{\mathcal{A}}^d_{\mathrm{loc}}$ (which is purely discontinuous, and it exists since Z is of bounded variation a.e. on each compact interval) one considers the continuous semimartingale $X_0 + Y + Z^c$ on $[\![T_i, T_{i+1})$ and uses the preceding case. In fact $Z^d_t = \sum\limits_{s \leq t}(Z_s - Z_{s-})$ and this sum is a.e. absolutely convergent since $Z \in \tilde{\mathcal{A}}_{\mathrm{loc}}$, and $Z^c = Z - Z^d$. Summing the resulting formulas over $0 \leq i \leq n$, and adding the contributions over the jumps at T_i, we get:

$$f(X_t) - f(X_0) = \int_0^t \frac{\partial f}{\partial x}(X_{s-})d(Y + Z^c)_s + \frac{1}{2}\int_0^t \frac{\partial^2 f}{\partial x^2}(X_{s-})d\langle Y, Y\rangle_s + \sum_{i=1}^n [f(X_{T_i-})]. \tag{6}$$

However, using an elementary property of Stieltjes integrals we may add and substract $\int_0^t \left(\frac{\partial f}{\partial x}\right)(X_{s-})dZ^e_s = \sum\limits_{i=1}^n \frac{\partial f}{\partial x_i}(X_{T_i-})\Delta Z_{T_i}$ to the right side. The resulting expression is precisely (5).

III. Let us reduce the general case to that of the above step. Consider the given semimartingale $X = X_0 + Y + Z$ where X_0 is bounded, by the reduction in the first paragraph. Let T be an optional of $\{\mathcal{F}_t, t \geq 0\}$ which strongly reduces Y. By the earlier discussion, we may assume that this T also has the property that $Z \circ T$ is absolutely integrable, i.e., if $\tilde{Z}_t = Z_t \chi_{[t<T]} + Z(T-) \cdot \chi_{[t \geq T]}, t \geq 0$, then $\tilde{Z} \in \tilde{\mathcal{A}}$. (In fact $E^{\mathcal{F}_t}(Y \circ T)$ and $\int_0^t |dZ_s|$ are both bounded for $0 \leq t < T$.) Similarly let $\tilde{Y}_t = Y_t \chi_{[t<T]} + Y(T-) \cdot \chi_{[t \geq T]}$ so that the process $\tilde{Y}$ is in $\mathcal{M}$. Hence $\tilde{X} = X_0 + \tilde{Y} + \tilde{Z}$ is a semimartingale and the two stopped processes X and $\tilde{X}$ agree on $[\![0, T)$, while $\tilde{X}$ is continuous at T. Since $X_s = \tilde{X}_s$ on $[s < T]$, we have for each $t \geq 0$

$$f \circ \tilde{X}_t = f \circ X_t + (f(\tilde{X} \circ T) - f(X \circ T))\chi_{[t \geq T]}. \tag{7}$$

Consequently, if (5) is true for $\tilde{X}_t$ then by substitution (7) implies that it also holds for X_t since for $[t \geq T]$ the last term is a fixed $\mathcal{F}_t$-adapted random variable. We thus need only prove (5) for the bounded semimartingale $\tilde{X}$, and f of the theorem vanishing off a compact set. This will be the standing assumption from now on.

IV. Since $Y \in \mathcal{M}$, by Corollary V.3.31, $Y = Y^c + Y^d$ and using an orthogonal decomposition (relative to a basis in the Hilbert space $\mathcal{M}^d$) we have

$$Y = Y^c + \sum_{i=1}^{\infty} V^i, \tag{8}$$

where V^i are pairwise orthogonal martingales and each V^i has only a single jump. [Recall that the elements of $\mathcal{M}^d$ are totally discontinuous, i.e., each has a zero continuous part if we apply the above decomposition.] Also $E\left[\sum_{i=1}^{\infty}(V^i_\infty)^2\right] = E((Y^d_\infty)^2) < \infty$. Hence one may choose a subsequence $\left\{\sum_{j=1}^{n_i} V^j = U^i, i \geq 1\right\}, V^j = \{V^j_t, t \geq 0\}$, with $Y^i = Y^c + U^i$ and $\sum_{i=1}^{\infty} E(|Y^i_\infty - Y_\infty|^2) < \infty$. Consequently by Theorem II.4.8,

$$E\left[\sup_{t \geq 0} |Y^i_t - Y^{i+n}_t|^2\right] \leq 4E[|Y^i_\infty - Y^{i+n}_\infty|^2] \to 0 \tag{9}$$

as $i \to \infty$ for each $n \geq 1$. Thus $Y^i_t(\omega) \to Y_t(\omega)$ for a.a. (ω) uniformly in t. Next one can apply the same procedure for $Z = Z^c + Z^d$ where

$Z_t^d = \sum_{s \leq t} \Delta Z_s$ and $\sum_{0 \leq s \leq \infty} |\Delta Z_s| < \infty$ a.e. Then choosing a sequence $W_t^n = \sum_{s \leq t} \Delta Z_s$ having n-jumps such that, if $Z^n = Z^c + W^n$, we have $E(\int_0^\infty |d(Z_s - Z_s^n)|) \to 0$ as $n \to \infty$ (so $Z_t^n(\omega) \to Z_t(\omega)$ uniformly in t for a.a. (ω)). Let $X^n = X_0 + Y^n + Z^n = X_0 + Y^c + (U^n + Z^n)$, so that $U^n + Z^n \in \tilde{A}$. (It is not hard to show that $U^n \in \mathcal{M} \cap \tilde{A}$ for each n, since $V^i \in \mathcal{M} \cap \tilde{A}, V^i$ being totally discontinuous.) Then X^n is a semimartingale such that $X_t^n(\omega) \to X_t(\omega)$ for a.a. (ω) uniformly in t, and the X^n satisfies the hypothesis of Step II for each $n \geq 1$. Thus using the fact that $(\dot{X}^n)^c = \dot{X}^c$ (by construction), we get from (6)

$$\begin{aligned} f(X_t^n) - f(X_0) = &\int_0^t \frac{\partial f}{\partial x}(X_{s-}^n) dX_{s-}^n + \frac{1}{2}\int_0^t \frac{\partial^2 f}{\partial x^2}(X_{s-}^n) d\langle \dot{X}^c, \dot{X}^c \rangle_s \\ &+ \sum_{s \leq t} \Big[f(X_s^n) - f(X_{s-}^n) - \\ &\frac{\partial f}{\partial x}(X_{s-}^n)(X_s^n - X_{s-}^n) \Big]. \end{aligned} \qquad (10)$$

To deduce (5) from (10), it suffices to take the limits as $n \to \infty$ and use $X_t^n \to X_t$, a.e. $[P]$ uniformly in t. We only need to justify this interchange of limits to complete the demonstration.

V. Since f is continuous, it is clear that the left side of (10) tends to that of (5). Also $\langle \dot{X}^c, \dot{X}^c \rangle$ defines a bounded measure and $\frac{\partial^2 f}{\partial x^2}(X_{s-}^n) \to \frac{\partial^2 f}{\partial x^2}(X_{s-})$ a.e., and boundedly. So by the Bounded Convergence the second integral of (10) tends to the corresponding term of (5). Since $\frac{\partial^2 f}{\partial x^2}$ is uniformly bounded, the last term of (10) is bounded by $C(X_s^n - X_{s-}^n)^2$ by the mean-value theorem where C is a constant, and the latter converges boundedly a.e., to $C(X_s - X_{s-})^2$ (uniformly on $[0, t]$). But $\sum_{s \leq t}(X_s - X_{s-})^2$ is a.e. convergent, and it is dominated by the convergent series $\sum_{s \leq t}(\Delta Y_s)^2$ and $\sum_{s \leq t}|\Delta Z_s|$. Hence the last term of (10) converges a.e. to the last term of (5) as $n \to \infty$. Only the first integral of (10) remains to be justified in this interchange.

Consider $\int_0^t \frac{\partial f}{\partial x}(X_{s-}^n) dX_s^n = \int_0^t \frac{\partial f}{\partial x}(X_{s-}^n) dY_s^n + \int_0^t \left(\frac{\partial f}{\partial x}\right)(X_{s-}^n) dZ_{s-}^n$.

Now,

$$
\begin{aligned}
&E\left(\left|\int_0^t \left(\frac{\partial f}{\partial x}\right)(X_{s-}^n)dY_s^n - \int_0^t \left(\frac{\partial f}{\partial x}\right)(X_{s-})dY_s\right|\right) \\
&\leq E\left(\left|\int_0^t \left(\frac{\partial f}{\partial x}(X_{s-}^n) - \frac{\partial f}{\partial x}(X_{s-})\right) dY_s^n\right|\right) \\
&+ E\left(\left|\int_0^t \frac{\partial f}{\partial x}(X_{s-})d(Y_s - Y_s^n)\right|\right), \\
&\leq \left[E\left(\int_0^t \left(\frac{\partial f}{\partial x}(X_{s-}^n) - \frac{\partial f}{\partial x}(X_{s-})\right)^2 d\langle Y^n, Y^n\rangle_s\right)\right]^{1/2} \\
&+ \left[E\left(\int_0^t \left|\frac{\partial f}{\partial x}(X_{s-})\right|^2 d\langle Y - Y^n, Y - Y^n\rangle_s\right)\right]^{1/2}, \qquad (11)
\end{aligned}
$$

by the CBS inequality and Theorem V.3.9. But $\langle Y^n, Y^n\rangle_s \leq \langle Y, Y\rangle_s$ and $\frac{\partial f}{\partial x}$ is bounded, and $\frac{\partial f}{\partial x}(X_{s-}^n) - \frac{\partial f}{\partial x}(X_{s-}) \to 0$ a.e. boundedly, as $n \to \infty$. Hence the right side of (11) tends to zero as $n \to \infty$. A similar conclusion holds about $E\left(\left|\int_0^t \frac{\partial f}{\partial x}(X_{s-}^n)dZ_s^n - \int_0^t \frac{\partial f}{\partial x}(X_{s-})dZ_s\right|\right)$ since Z_s defines a (bounded) signed measure and $Z_s^n \to Z_s$ a.e. Thus the first integral of (10) tends to the corresponding term of (5) in $L^1(P)$ and hence a.e. for a subsequence. This proves (5), and the last statement is now immediate. □

Taking $f(x,y) = xy$ and $X = (X^1, X^2)$ in the above theorem where the X^i are local martingales, and noting that $\int_0^{(\cdot)} X_{s-}^i dX_s^j$ is a local martingale, we get the following result, noted after (3′).

3. Corollary. *If $X^i, i = 1, 2$ are two local martingales (so $X_0^i = 0$, a.e.), then*

$$
(X^1X^2)_t - [X^1, X^2]_t = \int_0^t X_{s-}^1 dX_s^2 + \int_0^t X_{s-}^2 dX_s^1, \qquad (12)
$$

and the process $X^1X^2 - [X^1, X^2]$ is a local martingale.

We consider some applications of formula (5), and indicate some connections with the McShane integral which also extends the classical Itô integral.

4. Definition. Let $\{\mathcal{F}_t, 0 \le t \le a\}$ be an increasing net of σ-subalgebras in (Ω, Σ, P) and $X = \{X_t, \mathcal{F}_t, 0 \le t \le a\}$ an adapted process in $L^2(P)$. Then:

(a) X satisfies a *$K \cdot \delta t$ condition* if there exists an absolute constant K such that for each $0 \le u \le v \le a$ we have

$$|E^{\mathcal{F}_u}(X_v - X_u)| \le K(v-u), \quad E^{\mathcal{F}_u}(X_v - X_u)^2 \le K(v-u), \text{ a.e.} \tag{13}$$

(b) X satisfies *nearly a $K \cdot \delta t$ condition* if for each $\varepsilon > 0$ there exists an adapted process X^ε satisfying a $K \cdot \delta t$ condition such that

$$P\{\omega : X_t^\varepsilon(\omega) = X_t(\omega), 0 \le t \le a\} > 1 - \varepsilon. \tag{14}$$

We show how these processes form a subclass of those considered above. Replacing $\mathcal{F}_t$ by $\mathcal{F}_{t+}$ if necessary, one may conveniently assume that the net $\{\mathcal{F}_t, t \ge 0\}$ is right continuous. Also taking expectations in (13), it follows that X is $L^2(P)$–mean continuous and hence may be assumed right continuous (on replacing by a modification if necessary). In particular such a process is separable and (14) is meaningful. With these adjustments, it follows at once that (by (13)) the $K \cdot \delta t$ condition implies the quasimartingale property of Definition V.2.19. Hence by Corollary V.2.23, the process X of (13) is a *local semimartingale*. Now to assert that it is a semimartingale, we need to show that X is of class (DL) by Theorem V.2.22(b). In general this need not be true. However the $K \cdot \delta t$ condition implies this stronger conclusion. By (13),

$$K(u-v) \le E^{\mathcal{F}_u}(X_v) - X_u \le K(v-u), \text{ a.e.} \tag{15}$$

Thus, if $V_t = X_t + Kt$ then $\{V_t, \mathcal{F}_t, t \ge 0\}$ is a right continuous submartingale (by taking a right continuous version). Hence $V_t^+ = X_t^+ + Kt, t \ge 0$, defines a right continuous nonnegative submartingale (cf. simple Property (iii) of Section II.4). So by Proposition IV.3.6, V_t^+-process is of class (DL). Since $-X$ is also a quasimartingale, the above result applied to this implies that (or $V_t^- = X_t - Kt$ is a supermartingale) $X_t^- - Kt$ defines a process which is of class (DL) and hence (by addition) X is of class (DL). *Thus every right continuous adapted process, satisfying a $K \cdot \delta t$ condition, is a semimartingale.* (This was already noted by McShane in [4].) Next by Theorem 2 we deduce that the product of a finite number of $K \cdot \delta t$ processes is again a (local) semimartingale. Also in (13) $K(v-u)$ may be replaced by $F(v)-F(u)$ where $F(\cdot)$ is an increasing real right continuous nonnegative function, bounded on compact intervals.

Generalizing the Itô procedure and that of the real variable theory McShane has defined a "belated" stochastic integral which we call a *McShane integral.* (See his monograph [3] and a later paper [4] for a thorough discussion of this subject.) The definition proceeds initially with Cauchy partitions and in the approximating sums for the integral, $\sum_{j=1}^{n} f(t_j)\tilde{\Delta}X_{t_{j+1}}$, where $\tilde{\Delta}X_{t_{j+1}} = X_{t_{j+1}} - X_{t_j}, f(t)$ is $\mathcal{F}_t$-adapted and the evaluation points t_j precede the intervals $[t_j, t_{j+1})$. But this corresponds to the predictable concept used in our earlier definition of the stochastic integral. With this understanding we may present the existence theorem of the McShane integral as a consequence of Proposition 1 and Theorem 2. (The above sum then converges in probability to the integral.)

5. Theorem. *Let $f = \{f_t, \mathcal{F}_t, 0 \leq t < a\}$ be a predictable process with $E(|f_t|^2)$ Lebesgue integrable on $[0,a]$, and $X = \{X_t, \mathcal{F}_t, 0 \leq t < a\}$ be a right continuous process satisfying the $K \cdot \delta t$ condition. Then the stochastic integral $f \circ X$, defined by $(f \circ X)_t = \int_0^t f(s-)dX_s$ exists and is a semimartingale belonging to $L^2(P)$, so that it is locally square integrable if f is such.*

Thus far we have not used the second inequality of (13). With this McShane obtained an estimate, which generalizes Corollary V.3.13, as follows:

$$E(f \circ X)_t^2 \leq C \int_0^t E(f(s-)^2)ds, \quad 0 \leq t \leq a, \tag{16}$$

where C depends only on K. The processes of Definition 4(b) in which $K = K_\varepsilon$, correspond to local quasimartigales and the thus generalized theory proceeds along the work for local semimartingales (with some weakening in Definition V.2.19). Since processes with $K \cdot \delta t$ (and nearly $K \cdot \delta t$) conditions are a subclass of semimartingales (and local semimartingales), it will be interesting to see the extent to which the stochastic integral given by McShane (in [3] and [4]) admits a generalization to local semimartingales.

The following result is deduced from (5). We discuss the special character of these processes below.

6. Theorem. *Let* $\mathbf{X} = (X^1, \ldots, X^n)$ *be a McShane process on* $[0,1]$ *(i.e., each* X^i *satisfies Definition 4(a)) with continuous sample paths. Let* $f^i, g^i_\alpha, h^i_{\alpha\beta}(i = 1, \ldots, m; \alpha, \beta = 1, \ldots, n)$ *be predictable processes into* $L^2[0,a]$, *for* a.a. *sample paths. Define* $V = (V^1, \ldots, V^m)$ *by*

$$V^i_t = \int_0^t f^i(s-)ds + \sum_{\alpha=1}^m \int_0^t g^i_\alpha(s-)dX^\alpha_s + \sum_{\alpha,\beta=1}^m \int_0^t h^i_{\alpha\beta}(s-)d\langle X^\alpha, X^\beta\rangle_s, \tag{17}$$

$i = 1, \ldots, m$. *If* $F : [0,a] \times \mathbb{R}^m \to \mathbb{C}$ *is twice continuously differentiable let* $F_0 = \frac{\partial F}{\partial t}$, $F_i = \frac{\partial F}{\partial x_i}$, *and* $F_{ij} = \frac{\partial^2 F}{\partial x_i \partial x_j}, i, j = 1, \ldots, m$. *Then for each* $0 \le t \le a$ *we have*

$$F(t, X_t) = \int_0^t F_0(s, X_{s-})ds + \sum_{i=1}^m \int_0^t F_i(s, X_{s-})dV^i_s + \frac{1}{2}\sum_{i,j=1}^m \int_0^t F_{ij}(s, X_{s-})d\langle V^i, V^j\rangle_s. \tag{18}$$

Moreover, the process is a semimartingale with continuous sample paths.

Let us note that the first and last integrals of (17) define elements of $\mathcal{A}$, and the middle term is a semimartingale. Hence V^i is a semimartingale and has continuous sample paths. Also if $V^i = V_0 + Y^i + Z^i$ is the decomposition (cf., Definition V.2.19), then $\langle V^i, V^i\rangle = \langle Y^i, Y^i\rangle$ where $Y^i \in \mathcal{M}^c$. Hence all the terms are well defined, and (18) follows from (5). We must note, however, that an explicit expression for $\langle X^\alpha, X^\beta\rangle$ and $\langle V^i, V^j\rangle$ will be essential to apply this formula in most problems of interest. Using the belated integrals, McShane has established such a result by first defining the second order integrals $\int_0^{(\cdot)} h^i_{\alpha\beta}(s-)dX^\alpha_s dX^\beta_s$, and then obtaining (18) in the following form (cf. McShane [4], p. 147):

$$F(t, X_t) = \int_0^t F_0(s, X_{s-})ds + \sum_{i=1}^m \int_0^t F_i(s, X_{s-})f^i(s-)ds + \sum_{i=1}^m \sum_{\alpha=1}^n \int_0^t F_i(s, X_{s-})g^i_\alpha(s-)dX^\alpha_s$$

$$+\sum_{i=1}^{m}\sum_{\alpha,\beta=1}^{n}\int_0^t \{F_i(s,X_{s-})h^i_{\alpha,\beta}(s-)$$
$$+\frac{1}{2}\sum_{j=1}^{m}F_{ij}(s,X_{s-})g^i_\alpha(s-)g^j_\beta(s-)\}dX^\alpha_s dX^\beta_s. \tag{19}$$

7. Discussion of McShane's approach. The virtue of McShane's method is that one need not invoke the Doob-Meyer decomposition theory, as it does not depend on the general analysis of processes. For the subclass of quasimartingales treated, he has shown (by direct evaluation) that the higher order integrals beyond the second (cf. (19)) vanish, and we can now see that the second order ones correspond to quadratic (co) variation of the processes involved. He has noted that the same covariation of a process of finite quadratic variation and one of a finite variation is zero. The $K.\delta t$ (and near $K.\delta t$) hypothesis also implies the absolute continuity of the quadratic (co) variation. Thus the table given in Proposition V.3.24 is seen to be true for some processes more general then Brownian motion. The estimates of certain moments of the process of the integral are useful for solving stochastic differential equations later. The major drawback of this method is that one does not have a clear idea of the class of processes [the subclass of quasimartingales] thus admitted, since most of the results given in his book [3] and paper [4] can be read off from the general theory. The latter article, however, gives an overview of the McShane approach and is a good source for a general study. The subject can moreover, be learned in a relatively short time.

It will follow from the work of the next section that the McShane integrals possess a boundedness property, as are the other stochastic integrals studied so far. This will be analyzed in detail and extended, since it gives a unified approach to the subject.

6.2 Bochner's boundedness principle and its extensions

After analyzing Wiener's rigorous development of the stochastic integral, Bochner formulated in early 1950's a boundedness principle which applies to integrators more general then the Brownian motion. This was before the martingale integrals (treated in the preceding chapter) appeared on the scene. He verified that both the Brownian motion and the Lévy (or stable)-processes satisfy these principles which were termed $L^{\rho,p}$-bounded, with $\rho > 0, p \geq 1$. He expressed confidence that this principle has importance for future developments, but did not come back for a complete study. We shall now present an extended version of this concept so that it applies and unifies all the stochastic integrals we have considered so far and indeed seems to take all the (known) integrals into its fold. This together with several applications will occupy the following pages.

(a) *The boundedness principles.* We start with the simplest of these principles and present its analysis as a motivation for its extension.

1. Definition. Let $X = \{X_t, t \in I \subset \mathbb{R}\}$ be a process, $X_t \in L^2(P)$. Then X is $L^{2,2}$-*bounded* if there is a constant $C > 0$ such that for each simple function $f : I \to \mathbb{R}$, measurable relative to the Borel σ-algebra $\mathcal{B}_I$ of I, one has

$$E\left(\left|\int_I f(t)dX_t\right|^2\right) \leq C\int_I |f(t)|^2 dt, \tag{1}$$

where $\int_I f(t)dX_t = \sum_{i=0}^{n-1} a_i(X_{t_{i+1}} - X_{t_i})$ for $f = \sum_{i=0}^{n-1} a_i\chi_{[t_i,t_{i+1})}$, $a_i \in \mathbb{R}$. Replacing 2 and 2 in the exponents of (1) by $p(\geq 1)$ and $\rho(> 0)$, one gets the $L^{\rho,p}$-*boundedness of* X for $X_t \in L^p(P)$.

It is a standard exercise to show that $\int_I f(t)dX_t$ is unambigously defined and that it is linear in f. Then by (1), $\tau : f \mapsto \int_I f(t)dX_t$, becomes a bounded linear mapping on the simple functions of $L^2(I, dt)$ into $L^2(P)$ and has a unique bound preserving extension to all of $L^2(I, dt)$ (by the density of simple functions in $L^2(I, dt)$ and uniform continuity of τ), since the range is a complete space. [A similar statement holds for $\tau : L^\rho(I, dt) \to L^p(P)$ since $L^\rho(I, dt)$ is also a complete metric space.] The extended mapping is again denoted by the same symbol, called the *stochastic integral of the first type*, $\tau(f) = \int_I f dX_t, f \in L^2(I, dt)$, and if $Z : \mathcal{B}_I \to L^2(P)$ is defined by $Z(A) = \tau(\chi_A)$, then we verify that Z is a σ-additive, $L^2(P)$-valued set function called a *stochastic measure.* It is the condition, given by (1), that guarantees the vector measure property for $Z(\cdot)$ which is induced by the process X. In fact, while one can associate a finitely additive

$Z : \mathcal{S} \to L^2(P)$ on the semi-ring $\mathcal{S}$ of intervals of the type $[a, b)$ (or $(a, b]$) by the obvious definition $Z([a, b)) = X_b - X_a$, its countable additivity does not obtain for X in the absence of some condition such as (1). Let us record an interesting consequence for comparison: [we use the properties of vector integrals recalled just prior to Prop. V.3.8.]

2. Proposition. *Let $X = \{X_t, t \in I\}$ be $L^{2,2}$-bounded where I is a subinterval of $\mathbb{R}$. Then the process X induces a vector measure $Z : \mathcal{B}_I \to L^2(P)$ such that $Z([a, b)) = X_b - X_a$, and moreover there exists a process $Y : I \to L^2(P)$ such that*

$$\tau(f) = \int_I f(t)dZ(t) = \int_I f(t)Y(t)dt, f \in L^2(I, dt), \tag{2}$$

where the second integral on the right is the ordinary vector (also called the Bochner) integral, while the first one is the stochastic integral of the first type.

Proof. Since X is $L^{2,2}$-bounded, by (1), the mapping $\tau : L^2(I, dt) \to L^2(P)$ is bounded an linear, $\tau(f) = \int_I f(t)dX_t$, and hence τ maps bounded sets into bounded (= relatively weakly compact by the reflexivity of $L^2(P)$) sets. Then by a classical (Dunford's extension of) Riesz representation theorem, there exists a vector measure $Z : \mathcal{B}_I \to L^2(P)$ such that

$$\tau(f) = \int_I f dZ, \quad f \in L^2(I, dt). \tag{3}$$

But we can also consider $I = [a, b) \subset \mathbb{R}$ as a locally compact abelian group (with the group operation addition modulo $(b-a)$), and let $\tilde{I}$ be its dual group. Then by Plancherel's theorem on $L^2(I, dt)$ we have

$$E(|\tau(f)|^2) \leq C \int_I |f|^2 dt = C \int_{\hat{I}} |\hat{f}|^2 d\mu. \tag{4}$$

where $\hat{f} = F(f)$, the (L^2-) Fourier transform of f; and μ is the (normalized) translation invariant measure (= a multiple of the Lebesgue measure). Let $T = \tau \circ F^{-1}$, so that $T(\hat{f}) = \tau(f)$. Since F is unitary, T is continuous, and $T : L^2(\hat{I}, d\mu) \to L^2(P)$ is also representable (by the above recalled Riesz theorem) as:

$$T(\hat{f}) = \int_{\hat{I}} \hat{f}(t)\tilde{Z}(dt), \quad \hat{f} \in F(L^2(I, dt)) = L^2(\hat{I}, d\mu), \tag{5}$$

where $\tilde{Z} : \mathcal{B}_I \to L^2(P)$ is again a stochastic measure. Hence (3) and (5) give the following in which $e(x,t) = e^{ixt}$:

$$\begin{aligned}\int_I f(t)dX_t &= \tau(f) = T(\hat{f}) \\ &= \int_{\hat{I}} (\int_I e(x,t)f(t)dt)\tilde{Z}(dx) \\ &= \int_I f(t)\left[\int_{\hat{I}} e(x,t)\tilde{Z}(dx)\right] dt \\ &= \int_I f(t)Y(t)dt, \quad f \in L^2(I,dt),\end{aligned} \tag{6}$$

where $Y(t) = \int_{\hat{I}} e(x,t)Z(dx)$ is the Fourier transform of the stochastic measure $\tilde{Z}$, and is called a *weakly harmonizable process.* The integral in (6) is then in the standard vector (or Bochner) sense. Since X can be Brownian motion and hence is nondifferentiable on any interval, it is not possible to conclude from (6) that $dX_t = Y(t)dt$, but (6) is precisely (2). □

Remark. It can be observed that (2) is analogous to the alternative Paley-Wiener-Zygmund [1] definition of the classical Wiener integral with a formal integration by parts, and then using it as a definition for a rigorous treatment. However, our method above is different, and the procedure is rigorous from the beginning. (But see Exercise 6.3.)

A further extension of Bochner's principle is motivated by problems of the following type. Let $X = \{X_t, t \in \mathbb{R}\}$ be a stationary process (in the sense of Khintchine) so that $E(X_t) =$ constant (taken $= 0$ for simplicity) and $r(s,t) = cov(X_s, X_t) = \rho(s-t)$ where $\rho(\cdot)$ is a continuous function. Then the classical results due to Cramér, Kolmogorov and others imply that X admits a representation:

$$X_t = \int_{\mathbb{R}} e^{it\lambda} Z(d\lambda). \tag{7}$$

where $Z : \mathcal{B}_{\mathbb{R}} \to L^2(P)$ is a stochastic measure with the property that $E(Z(A)) = 0, E(Z(A)\bar{Z}(B)) = \mu(A \cap B), A, B$ being Borel sets, and μ is a positive bounded measure on $\mathcal{B}_{\mathbb{R}}$. The integrator Z satisfies the following analog of (1):

$$E(|\int_I f dZ|^2) = \int_I |f(t)|^2 d\mu(t). \tag{8}$$

Thus Z or the process determined by Z, namely $\tilde{X}_t = Z([a,t))$ for $t \geq a$, and $= -Z((t,a])$ for $t < a$, is not $L^{2,2}$-bounded unless μ is the Lebesgue measure, although it has similar properties. In order to include this and other (stochastic) integrators considered in the preceding chapter, and to obtain a unified theory we have to generalize the boundedness principle.

First let us recall some concepts from Orlicz space theory as it proves to be *essential* for the desired unification of various stochastic integrals. Thus let $\varphi : \mathbb{R} \to \bar{\mathbb{R}}^+$ be a monotone nondecreasing function such that $\varphi(x) = 0$ iff $x = 0$, *called a generalized Young function.* Let $L^\varphi(P)$ be the set of all scalar measurable functions f on (Ω, Σ, P) such that $\|f\|_\varphi < \infty$, where

$$\|f\|_\varphi = \|f\|_{\varphi,P} = \inf\{k > 0 : E(\varphi(|f|/k) \leq k\}. \tag{9}$$

Then $\{L^\varphi(P), \|\cdot\|_\varphi\}$ can be seen to be a linear complete metric space of equivalence classes of such f, (i.e., f, g are equivalent iff $\|f-g\|_\varphi = 0$ and then they are identified as usual). If moreover φ is convex, then the $L^\varphi(P)$-space becomes a Banach space, called an *Orlicz space.* In this case one has an equivalent metric $\|\cdot\|'_\varphi$ (now a norm) given by

$$\|f\|'_\varphi = \|f\|'_{\varphi,P} = \inf\{k > 0 : E(\varphi(|f|/k) \leq 1\}. \tag{10}$$

Let M^φ be the closed subspace of $L^\varphi(P)$ determined by the class of all simple functions, relative to the metric $\|\cdot\|_\varphi$ (or $\|\cdot\|'_\varphi$). If φ is moderating (i.e., satisfies the so called Δ_2-condition: $\varphi(2x) \leq k_0\varphi(x)$ for $x \geq x_0$ and some $k_0 > 0$) then $M^\varphi = L^\varphi(P)$. For a detailed analysis of these spaces and their properties, one can refer to the book by Rao and Ren [1].

Let $I \subset \mathbb{R}$ and $\mathcal{B}_I$ be its Borel σ-algebra. If (Ω, Σ, P) is the given probability space, let $\mathcal{O}$ be a σ-subalgebra of the product σ-algebra $\mathcal{B}_I \otimes \Sigma$. An $\mathcal{O}$-simple function $f : I \times \Omega \to \mathbb{R}$ is an $\mathcal{O}$-measurable function of the form $f = \sum\limits_{i=1}^{n} f_i\chi_{A_i}, A_i \in \mathcal{B}_I$, disjoint, so that $f_i\chi_{A_i}$ is $\mathcal{O}$-measurable and f_i is a bounded random variable. If $X : I \times \Omega = \Omega' \to \mathbb{R}$ is a measurable function, so $X = \{X_t, t \in I\}$ is a process, and $X_t \in L^{\varphi_1}(P), t \in I$, then the mapping, $(\tau f) : \Omega \to \mathbb{R}$ given by,

$$\tau : f \mapsto \int_I f dX_t = \sum_{i=1}^{n} f_i(X_{\sup(A_i)} - X_{\inf(A_i)}) \tag{11}$$

is unambiguously defined as an element of $L^{\varphi_1}(P)$ and is (finitely) additive.

The desired extension of Definition 1 on "boundedness" is as follows:

3. Definition. Let φ_1, φ_2 be generalized Young functions on $\mathbb{R}$, and $X = \{X_t, t \in I\}$ be a measurable process relative to $\mathcal{B}_I \otimes \Sigma$. If $\mathcal{O} \subset \mathcal{B}_I \otimes \Sigma$ is a σ-subalgebra, then X is said to be *L^{φ_1,φ_2}-bounded relative to $\mathcal{O}$, and φ_2* if there exists a σ-finite measure $\alpha : \mathcal{O} \to \bar{\mathbb{R}}^+$ and a constant $K(= K^{\alpha}_{\varphi_1,\varphi_2} > 0)$ such that for each $\mathcal{O}$-simple f the function τf of (11) satisfies

$$E(\varphi_2(|\tau f|)) \leq K \int_{\Omega'} \varphi_1(|f|) d\alpha. \tag{12}$$

If $\varphi_1(x) = |x|^{\rho}, \varphi_2(x) = |x|^p, p \geq 1, \rho > 0, \mathcal{O} = \mathcal{B}_I \otimes \{\emptyset, \Omega\}$ and $\alpha = \lambda \otimes P, \lambda =$ Lebesgue measure on $\mathcal{B}_I$, it is called *$L^{\rho,p}$-boundedness*, $\rho = p = 2$ giving (1).

4. Remarks. This definition is flexible and general enough that it includes all the stochastic integrals in use at this time. Indeed, it will be seen in the next theorem that (12) is essentially an optimal condition for these integrals to admit the dominated convergence assertions. In the classical $L^{2,2}$-boundedness case (as well as $L^{\rho,p}$) the mention of $\alpha(= \lambda \otimes P)$ is dropped since it is derived from the 'familiar' Lebesgue measure. Also the condition (12) is sometimes replaced by its essentially equivalent form based on the norm functionals (9) and (10) conveniently as:

$$\|\tau f\|_{\varphi_2,P} \leq K \|f\|_{\varphi_1,\alpha}. \tag{13}$$

This is a consequence of (12) if $f(\neq 0)$ is replaced by $f/K_0 (K_0 = \|f\|_{\varphi_1,\alpha})$ so that it becomes

$$E\left(\varphi_2\left(\frac{|\tau f|}{K_0}\right)\right) \leq K \int_{\Omega'} \varphi_1\left(\frac{|f|}{K_0}\right) d\alpha \leq K K_0 = K \|f\|_{\varphi_1,\alpha}. \tag{13'}$$

Note also that the crucial point of (12) or (13) is the existence of the dominating measure α on $\mathcal{O} \subset \mathcal{B} \otimes \Sigma$, and there is no requirement of the predictability of f relative to some filtration. In fact, we shall see in applications below that (12) applies both to the familiar nonanticipative and less familiar anticipative integrands. Only the additivity of τ on simple functions suffices. The (generalized) boundedness principle is an abstraction of all these cases.

We now indicate how a filtration associated with X and the σ-algebra $\mathcal{O}$ in the above conditions (1) and (12) are related. Let $\{\mathcal{F}_t, t \in I\}$ be a filtration of Σ and $X = \{X_t, \mathcal{G}_t, t \in I\}$ be an adapted process. Suppose that $\mathcal{F}_t \subset \mathcal{G}_t, t \in I$ and both are right continuous completed (for P) filtrations. Let $\mathcal{P}$ be the σ-algebra generated by sets $\{(s,t] \times A, A \in \mathcal{F}_s, s \in I\}$, together with $\{\{a_0\} \times B, B \in \mathcal{F}_{a_0}\}$ if $a_0 \in I$ is the first element, and similarly $\mathcal{P}'$ is defined with the $\mathcal{G}$-family. Then $\mathcal{P}$ and $\mathcal{P}'$ are the predictable σ-algebras determined by the $\mathcal{F}$ and $\mathcal{G}$ filtrations. For the Bochner $L^{2,2}$-boundedness one takes $\mathcal{O} = \mathcal{P}$ with $\mathcal{F}_t = \{\emptyset, \Omega\}, t \in I$ and $\mathcal{G}_t = \sigma(X_s, s \leq t, s, t \in I)$ and completed, where X is a *càdlàg* process with $X_t \in L^2(P)$ and $\alpha = \lambda \otimes P$. One can essentially take $\mathcal{O} = \mathcal{B}_I \otimes \{\emptyset, \Omega\} = \mathcal{P}$ in this case. In the more general situation of Section V.3, we take $\mathcal{P}$ as described above with $\mathcal{F}_t = \mathcal{G}_t$ where X is a square integrable martingale ($I = \mathbb{R}^+$), and other cases where $\mathcal{F}_t$ is a proper nontrivial σ-subalgebra of $\mathcal{G}_t$ will be discussed later.

We now introduce precisely the concept of a stochastic integrator.

5. Definition. Let $X = \{X_t, \mathcal{G}_t, t \in I\}$ be an adapted càdlàg process with $X_t \in L^{\varphi}(P)$ where φ is a generalized Young function. If $\mathcal{O} \subset \mathcal{B}_I \otimes \Sigma$ is a σ-subalgebra and $\mathcal{S}(\Omega', \mathcal{O})$ is the set of $\mathcal{O}$-measurable simple functions $f : \Omega' = I \times \Omega \to \mathbb{R}$, then X is termed a *stochastic integrator* on $\mathcal{S}(\Omega', \mathcal{O})$ if the following two conditions are met:

(i) The set $\{\tau(f) : f \in \mathcal{S}(\Omega', \mathcal{O}); \|f\|_\infty \leq 1\}$ is in a ball of $L^{\varphi}(P)$,

(ii) $f_n \in \mathcal{S}(\Omega', \mathcal{O}), |f_n| \downarrow 0$ a.e. $\Rightarrow \lim\limits_n \tau(f_n) = 0$, in probability.

The second condition above incorporates the dominated convergence property. A general characterization of stochastic integrals can be given in the following result, unifying all the cases considered.

6. Theorem. *Let φ_1, φ_2 be generalized Young functions, (Ω, Σ, P) be a probability space, $X = \{X_t, \mathcal{G}_t, t \in I\}$ be an adapted càdlàg process with $X_t \in L^{\varphi_2}(P), t \in I$, and $\mathcal{S}(\Omega', \mathcal{O})$ be the space of simple functions relative to a σ-subalgebra $\mathcal{O}$ of $\mathcal{B}_I \otimes \Sigma$. If X is L^{φ_1, φ_2}-bounded relative to a σ-finite measure $\alpha : \mathcal{O} \to \bar{\mathbb{R}}^+$ for $\mathcal{S}(\Omega', \mathcal{O})$, then the elementary stochastic integral τ of (11) extends from $\mathcal{S}(\Omega', \mathcal{O})$ to $M^{\varphi_1}(\alpha) = \bar{sp}\{\mathcal{S}(\Omega', \mathcal{O}), \|\cdot\|_{\varphi_1}\}$ into $L^{\varphi_2}(P)$ for which the dom-*

inated convergence theorem holds. Conversely, suppose that φ_2 is a moderating (generalized) Young function, (i.e., $\varphi_2(2x) \leq C_0\varphi_2(x), x \geq x_0$), (Ω, Σ, P) is separable, and X is a stochastic integrator on $L^{\varphi_2}(P)$ in the sense of Definition 5. Then there exists a (convex) Young function $\varphi_1, \frac{\varphi_1(x)}{x} \uparrow \infty$ as $x \uparrow \infty$, and a σ-finite measure $\alpha : \mathcal{O} \to \bar{\mathbb{R}}^+$ relative to which X is L^{φ_1,φ_2} bounded in the form (13) and then the integral $\tau(f)$ extends from $\mathcal{S}(\Omega', \mathcal{O})$ to $M^{\varphi_1}(\alpha)$ for which the dominated convergence criterion again holds.

The direct part is easy and the converse needs more work. Before presenting the proof we give some applications, and specializations to the previous cases. This will illuminate the structure of general integration vividly, and illustrate the techniques of finding the measure α on $\mathcal{O}$.

Let us consider the stochastic measures for the integrals of the first kind, discussed for Proposition V.3.8. The boundedness question is answered in the following result, as an example of the above theorem.

7. Proposition. *Let $Z : \mathcal{B}_I \to L^p(P), 1 \leq p \leq 2$ ($I \subset \mathbb{R}$ an interval) be a vector measure, i.e., Z is σ-additive in the norm topology of $L^p(P)$. If $X = \{X_t = Z((a_0, t]), t \in I\}$, for some $a_0 \in I$, then X is $L^{2,p}$-bounded relative to a measure $\mu : \mathcal{B}_I \to \mathbb{R}^+$. In particular, if $p = 2$ and Z is orthogonally scattered so that X has orthogonal increments, it is $L^{2,2}$-bounded relative to a finite measure $\tilde{\mu}$ on $\mathcal{B}_I$, or $\mu \otimes P$ or $\tilde{\mu} \otimes P$ on $\mathcal{O}$.*

Proof. The work depends on a result of Grothendieck and we sketch the esssential details. Thus let $\mathcal{X} = B(\Omega, \Sigma)$, the space of bounded measurable real functions on Ω with uniform norm. Using the $D - S$ integration, the mapping $\tau : f \mapsto \int_\Omega f dZ$ is defined, bounded and linear (cf. Proposition V.3.8) into $\mathcal{Y} = L^p(P)$. Classical results show that $\mathcal{X}$ is isometrically isomorphic to $\mathcal{X}_0 = C(S_0)$ the space of real continuous functions on a compact Hausdorff space S_0 where the implementing isomorphism J is also algebraic. Then $T = \tau \circ J^{-1} : \mathcal{X}_0 \to \mathcal{Y}$ is bounded and linear. But the space $\mathcal{Y}$ is an $L^p(P), 1 \leq p \leq 2$, and $\mathcal{X}$ is an "L^∞-space". So by a theorem of Grothendieck, extended by Pietsch and further perfected by Lindenstrauss and Pelczyński ([1], Cor. 2 to Theorem 4.3, and Prop. 3.1) there exists a regular (finite) measure μ on the Borel σ-algebra of S_0 such that $\|Tf\|_\mathcal{Y} \leq \|f\|_{2,\mu}, f \in \mathcal{X}_0$. We need to return to $\mathcal{X}$.

Now $f \in \mathcal{X}$ implies $\tilde{f} = J(f) \in \mathcal{X}_0$. Hence we have

$$\begin{aligned}
\|\tau f\|_{\mathcal{Y}}^2 &= \|T\tilde{f}\|_{\mathcal{Y}}^2 \leq \|\tilde{f}\|_{2,\mu}^2, f \in \mathcal{X}, \\
&= \langle \tilde{f}^2, \mu \rangle, \langle \cdot, \cdot \rangle \text{ is the duality pairing}, \\
&= \langle (J(f))^2, \mu \rangle, \text{ since } J \text{ is also algebraic}, \\
&= \langle f^2, J^*(\mu) \rangle, J^* : \mathcal{X}_0^* \to \mathcal{X}^*, \text{ the adjoint mapping}, \\
&= \int_\Omega |f|^2 d\mu_1, \qquad (14)
\end{aligned}$$

where $\mu_1 = J^*(\mu) \in \mathcal{X}^* = ba(\Omega, \Sigma)$ is the image set function, so that μ_1 is a bounded additive function and the integral in (14) is defined in the standard manner (cf., e.g., Dunford and Schwartz [1], p. 108 ff). The proposition follows when we show that the inequality (14) remains valid with μ_1 replaced by a σ-additive (finite) μ_2. This is done as follows.

Let μ_2^* be the outer measure generated by (μ_1, Σ) using the classical Carathéodory process, and let $\Sigma_{\mu_2^*}$ be the class of μ_2^*-measurable sets. It is a σ-algebra containing Σ and $\mu_2^* | \sum_{\mu_2^*} = \mu_2$ (say), is a measure, such that $\mu_2(A) \leq \mu_1(A), A \in \Sigma$, (cf., e.g., Rao [1], p. 41). Thus if $f = \sum_{i=1}^m a_i \chi_{A_i}$ is Σ-simple, (A_i-disjoint, $a_i \neq 0$), then for each $\varepsilon > 0$, there exist $A_{in}^\varepsilon \in \Sigma_{\mu_2^*}, A_i \subset \bigcup_{n=1}^\infty A_{in}^\varepsilon$, such that

$$\mu_2(A_i) + \varepsilon[m|a_i|^2]^{-1} > \sum_{n=1}^\infty \mu_1(A_{in}^\varepsilon). \qquad (15)$$

Let $f_N^\varepsilon = \sum_{i=1}^m a_i \chi_{\bigcup_{k=1}^N A_{ik}^\varepsilon}$. Then $f_N^\varepsilon \in \mathcal{X}$ and $f_N^\varepsilon \to f$ pointwise, and boundedly as $\varepsilon \downarrow 0$. Hence (14) becomes

$$\begin{aligned}
\|\tau f_N^\varepsilon\|_{\mathcal{Y}}^2 = \|\int_\Omega f_N^\varepsilon dZ\|_{\mathcal{Y}}^2 &\leq \int_\Omega |f_N^\varepsilon|^2 d\mu_1 \\
&\leq \sum_{i=1}^m |a_i|^2 \sum_{k=1}^N \mu_1(A_{ik}^\varepsilon), \text{ since } \mu_1 \text{ is subadditive},
\end{aligned}$$

Letting $N \to \infty$ on both sides and using the bounded convergence property of $D-S$ integrals (cf. Dunford-Schwartz [1], IV.10.10) and

(15) one gets

$$\begin{aligned}\|\tau f\|_{\mathcal{Y}}^2 &= \|\int_\Omega f dZ\|_{\mathcal{Y}}^2\\ &\le \sum_{i=1}^m |a_i|^2[\mu_2(A_i) + \varepsilon(m|a_i|^2)^{-1}]\\ &= \|f\|_{2,\mu_2}^2 + \varepsilon.\end{aligned}$$

Since $\varepsilon > 0$ is arbitrary, this shows that Z (and hence X) is $L^{2,p}$-bounded. In case $p = 2$ and $Z(\cdot)$ is orthogonally scattered, then it is evident that $(Z(A), Z(B)) = \mu(A \cap B)$ and the $L^{2,2}$-boundedness relative to μ follows. Here $\mathcal{O} = \mathcal{B}_I \otimes \{\phi, \Omega\}$ and so μ can be replaced by $\mu \otimes P$ to put it in the form of Definition 5. □

Remark. Comparing this result with Theorem 6, we see that $\varphi_2(x) = |x|^p, 1 \le p \le 2$, and $\varphi_1(x) = |x|^2$. Thus for each given φ_2, the existence of φ_1 and μ need a specialized (often nontrivial) analysis. The basic filtration here is the trivial one $\mathcal{F}_t = \{\emptyset, \Omega\}, t \in I$, and μ_2 is not the Lebesgue measure.

We now verify the principle for the semimartingale (hence the Itô) integrals of Section 1. Since by localization one can extend the work from that of bounded processes to the locally integrable case, we shall assume for simplicity that the processes below are bounded. To streamline the treatment, consider $X = \{X_t, \mathcal{G}_t, t \ge 0\}, (I = \mathbb{R}^+)$ as a square integrable càdlàg martingale and let $\{\mathcal{F}_t, t \ge 0\}$ be another standard filtration with $\mathcal{F}_t \subset \mathcal{G}_t$, and $\mathcal{P}, \mathcal{P}'$ be the corresponding predictable σ-algebras from $\mathcal{B}_{\mathbb{R}^+} \otimes \Sigma$. Let $f : \Omega' = \mathbb{R}^+ \times \Omega \to \mathbb{R}$ be a simple function so that it is of the form:

$$f = \sum_{i=0}^n a_i \chi_{(t_i, t_{i+1}] \times A_i}, \quad A_i \in \mathcal{F}_{t_i}, \quad 0 \le t_0 < \cdots < t_{n+1} \le t.$$

Then we have

$$
\begin{aligned}
E(|\int_{\mathbb{R}^+} f dX|^2) &= \sum_{i=0}^{n} a_i^2 E(\chi_{A_i} E^{\mathcal{G}_i}(X_{t_{i+1}} - X_{t_i})^2) + \\
&\quad 2 \sum_{0\le i<j\le n} a_i a_j E[\chi_{A_i\cap A_j}(X_{t_{i+1}} - X_{t_j}) E^{\mathcal{G}_{t_j}} \\
&\qquad (X_{t_{j+1}} - X_{t_j})] \\
&= \sum_{i=0}^{n} a_i^2 E(\chi_{A_i}(X_{t_{i+1}}^2 - X_{t_i}^2)) + 0 \\
&= \sum_{i=0}^{n} a_i^2 \mu((t_i, t_{i+1}] \times A_i), \text{ (say)}, \\
&= \int_{\Omega'} |f|^2 d\mu, \qquad (16)
\end{aligned}
$$

where μ is the Doléans-Dade measure determined by the L^1-bounded submartingale $\{X_t^2, \mathcal{G}_t, t \ge 0\}$ on $\tilde{\mathcal{P}}$. We use this expression below.

8. Proposition. *A càdlàg square integrable semimartingale* $X = \{X_t, \mathcal{G}_t, t \ge 0\}$ *is* $L^{2,2}$*-bounded relative to a* σ*-finite measure* $\beta : \mathcal{P}' \to \bar{\mathbb{R}}^+$ *for any filtration* $\{\mathcal{F}_t, t \ge 0\}$ *with* $\mathcal{F}_t \subset \mathcal{G}_t$*, where* $\mathcal{P}'$ *is the predictable* σ*-algebra determined by the other filtration* $\{\mathcal{G}_t, t \ge 0\}$ *as in Definition 5, so that* X *is a stochastic integrator.*

Proof. Let $f : \Omega' \to \mathbb{R}$ be a simple function, measurable relative to $\mathcal{P} \subset \mathcal{P}'$. Also $X = M + B$, where M and B are càdlàg bounded processes, $M = \{M_t, \mathcal{G}_t, t \ge 0\}$ being a martingale (square integrable) and $B = \{B_t, \mathcal{G}_t, t \ge 0\}$ a process of bounded variation (pointwise), so that $|B_t| \to |B_\infty|$, a.e. and in $L^1(P)$. Then by definition,

$$
\begin{aligned}
E(|\tau f|^2) &= E\left(\left|\int_{\mathbb{R}^+} f dM + \int_{\mathbb{R}^+} f dB\right|^2\right) \\
&\le 2E\left(\left|\int_{\mathbb{R}^+} f dM\right|^2 + \left(\int_{\mathbb{R}^+} |f| d|B|\right)^2\right) \\
&\le 2\left[\int_{\Omega'} |f|^2 d\mu + E(|B_\infty| \int_{\mathbb{R}^+} |f|^2(t) d|B_t|\right],
\end{aligned}
$$

by the Jensen inequality to the last term and μ being the Doléan-Dade measure of M^2 as in (16),

$$
= 2\int_{\Omega'} |f|^2 d\beta, \qquad (17)
$$

where $\beta(A) = \int_A \int [d\mu(t,\omega) + |B_\infty|(\omega)d_t|B|(t,\omega)dP(\omega)]$, is a σ-finite measure on $\mathcal{P}$ (so also on $\mathcal{P}'$). Hence X is $L^{2,2}$-bounded relative to β. □

We next verify that the α-stochastic (hence also the Stratonovich) integrals (cf. Proposition V.3.35) satisfy our boundedness criterion. Let X, Y be a pair of continuous semimartingales and α be a probability measure on the Borel sets of the unit interval $[0,1]$. If α_1, is the mean of α, then for any continuously differentiable $f : \mathbb{R} \to \mathbb{R}$, we have the α-stochastic integral of $f(X)$ relative to Y as:

$$\int_0^t f(X_s) \circ_\alpha dY_s = \int_0^t f(X_s)dY_s + \alpha_1 \int_0^t f'(X_s)d\langle X, Y\rangle_s, \qquad (18)$$

where the right side integrals are the semimartingale (as in Section 1) and pathwise Stieltjes integrals, as given by the above noted proposition. [Thus (18) again defines a continuous (locally integrable) semimartingale.] This is the Stratonovich's (symmetric) integral if α concentrates at $s = \frac{1}{2}$, since then it becomes (with $f(x) = x$),

$$\int_0^t X_s \circ dY_s = \int_0^t X_s dY_s + \frac{1}{2}\langle X, Y\rangle_t. \qquad (19)$$

For the class of integrals (18) we have:

9. Proposition. *For each continuously differentiable $f : \mathbb{R} \to \mathbb{R}$, the α-stochastic integral of $f(X)$ relative to Y, where X and Y are continuous semimartingales, is locally $L^{2,2}$-bounded for a σ-finite measure β so that β is defined on $\mathcal{B}_{[0,t]} \otimes \Sigma \to \bar{\mathbb{R}}^+$ for each $t > 0$. In fact, if $\tau_\alpha(f(X_t))$ denotes the left side of (18), we have with $\Omega'_t = [0,t] \times \Omega$,*

$$E(|\tau_\alpha(X_t))|^2) \le \int_{\Omega'_t} |||f(X_s)|||^2 d\beta(s,\omega). \qquad (20)$$

where $|||f(X_s)||| = [|f(X_s)|^2 + |f'(X_s)|^2]^{\frac{1}{2}}$ is a "Sobolev" type norm.

Proof. This follows from the preceding work. Indeed consider

$$E(|\tau_\alpha(f(X_t))|^2 \le 2E\left[\left|\int_0^t f(X_s)dY_s\right|^2 + \left(\int_0^t |f'(X_s)|d|\langle X,Y\rangle_s|\right)^2\right],$$

since $\langle X,Y\rangle$ has locally finite variation and $\alpha_1^2 \le 1$,

$$\le 2\left[\int_{\Omega_t'} |f(X_s|^2 d\beta_y(s,\omega) + E(|\langle X,Y\rangle_t| \int_0^t |f'(X_s)|^2 \right.$$

$$\left. d|\langle X,Y\rangle_s|)\right],$$

β_y being the measure for the $L^{2,2}$-bounded Y-process,

and use Jensen's inequality for the second term,

$$= 2\left[\int_{\Omega'} (|f(X_s)|^2 + |f'(X_s)|^2)d\beta_y(t,\omega)\right.$$

$$\left. + \int_{\Omega'} |\langle X,Y\rangle_t|(|f'(X_s)|^2 + |f(X_s)|^2 d|\langle X,Y\rangle_s|\right]$$

$$= 2\int_{\Omega'} |||f(X_s)|||^2 d\beta(s,\omega),$$

where $\beta(A) = \int_A [d\beta_Y(s,\omega) + |\langle X,Y\rangle_t| d|\langle X,Y\rangle_s|]$ for each predictable set $A \subset \Omega'$. Clearly β is a measure on the predictable σ-algebra on Ω' determined by the common filtration of X and Y. □

Remark. The above result shows that $E(|\tau(f)|^2) \le K \int_{\Omega'} \bar\sigma(f)^2 d\beta$ where $\bar\sigma(\cdot)$ is a norm functional of $\mathbb{R}^2$, and hence we need to generalize the boundedness concept for vector valued process. This will be done later. But the universality of the generalized principle is already illuminated.

Having seen the unification property of the theorem, it is time to present its demonstration. In preparation for its converse direction, we first dispose of a technical problem related to Orlicz spaces using the expression (10) (dropping the prime, i.e., $\|f\|_\varphi$ is the norm). For each Young function φ, there is a (Young) complementary function $\psi : \mathbb{R} \to \mathbb{R}^+$ given by $\psi(y) = \sup\{x|y| - \varphi(x) : x \ge 0\}$. It is again a convex function with similar properties. If $\nu : \Sigma \to \mathcal{X}$ (a Banach space) is σ-additive, the φ-semivariation, $\|\nu\|_\varphi(\cdot)$ of ν is defined as:

$$\|\nu\|_{\varphi,\mu}(A) = \sup\{\|\int_\Omega f(\omega)\nu(d\omega)\|_{\mathcal{X}} : \|f\|_{\psi,\mu} \le 1\},$$

for $A \in \Sigma, f \in L^{\psi}(\mu)$ on (Ω, Σ, μ) and the integral is in the $D-S$ sense. If $\varphi(x) = |x|^p$, one has the p-semivariation ($\psi(y) = k|y|^q$). The desired technical result is the following:

10. Proposition. *Let (Ω, Σ) be a measure space, $\mathcal{X}$ a Banach space, and $\nu : \Sigma \to \mathcal{X}$ a vector measure. Then there exists a measure $\mu : \Sigma \to \mathbb{R}^+$ and a continuous Young function $\varphi : \mathbb{R} \to \mathbb{R}^+$, $\frac{\varphi(x)}{x} \uparrow \infty$ as $x \uparrow \infty$, such that $\|\nu\|_{\varphi,\mu}(\Omega) < \infty$ relative to the pair (φ, μ).*

Proof. Since ν is σ-additive in $\mathcal{X}$, we have for each $A_n \in \Sigma$, disjoint,

$$\begin{aligned} 0 &= \lim_{n\to 0} \|\nu(\bigcup_{n=1}^{\infty} A_n) - \sum_{k=1}^{n} \nu(A_k)\|_{\mathcal{X}} \\ &= \limsup_{n\to\infty}\{|x^* \circ \nu(\bigcup_{n=1}^{\infty} A_n) - \sum_{k=1}^{n} x^* \circ \nu(A_k)| : \|x^*\| \le 1\}, \end{aligned} \tag{21}$$

where $x^* \in \mathcal{X}^*$, the adjoint space. Thus the family of scalar measures $\{x^* \circ \nu : \|x^*\| \le 1, x^* \in \mathcal{X}^*\}$ is uniformly σ-additive on Σ. Hence by a classical result (cf. Dunford-Schwartz [1], IV.10.5) there exists a finite positive measure μ on Σ such that $x^* \circ \nu$ is μ-continuous for all x^*. By the Radon-Nikodym theorem, $g_{x^*} = \frac{d(x^* \circ \nu)}{d\mu}$ exists and (21) implies

$$\lim_{\mu(A)\to 0} \left| \int_A g_{x^*}(\omega) d\mu(\omega) \right| = \lim_{\mu(A)\to 0} |x^* \circ \nu(A)| = 0$$

uniformly in $x^* \in \mathcal{X}^*, \|x^*\| \le 1$. Consequently the set $\{g_{x^*} : \|x^*\| \le 1\} \subset L^1(\mu)$ is bounded and uniformly integrable. Then by the classical de la Vallée Poussin theorem (cf., e.g., Rao and Ren [1], p. 3) there is a convex function, as given in the statement, such that $\int_\Omega \varphi(|g_{x^*}|) d\mu \le k_0 < \infty$, for all $x^* \in \mathcal{X}^*, \|x^*\| \le 1$. If ψ is the complementary Young function to φ, then

$$\begin{aligned} \|\nu\|_{\varphi,\mu}(\Omega) &= \sup\{\| \int_\Omega f d\nu \|_{\mathcal{X}} : \|f\|_{\psi,\mu} \le 1\} \\ &= \sup\{\sup\{ \left| \int_\Omega f (x^* \circ \nu)(d\omega) \right| : \|x^*\| \le 1\} : \|f\|_{\psi,\mu} \le 1\} \\ &= \sup\{\sup\{ \left| \int_\Omega f(\omega) g_{x^*}(\omega) d\mu \right| : \|x^*\| \le 1\} : \|f\|_{\psi,\mu} \le 1\} \\ &\le 2\sup\{\|g_{x^*}\|_{\varphi,\mu} : \|x^*\| \le 1\} \le 2k_0, \quad \text{by Hölder's inequality} \\ &\quad \text{in Orlicz spaces (cf. } ibid \text{ [1], pp. 58-59).} \end{aligned}$$

The bound on $\|g_{x^*}\|_{\psi,\mu}$ obtained above is also used in this step. □

With this preparation we can now present the

Proof of Theorem 6. As noted already, the sufficiency part is the most useful one and it is relatively easy. Indeed let X be L^{φ_1,φ_2} bounded as in Definition 3 so that (12) holds. Since $\tau : f \mapsto \int_I f dX$ defined by (11) is linear, condition (12) implies that τ is continuous on the set $\mathcal{S}(\Omega', \mathcal{O})$ of simple functions. Then by the principle of extension (by uniform continuity) in a metric space into a complete metric space, τ has a unique extension to the closure of this set, namely $M^{\varphi_1}(\alpha)$ of $L^{\varphi_1}(\alpha)$, and this again satisfies the inequality (12) since the range of τ is also a complete metric space. But $f_n \in \mathcal{S}(\Omega', \mathcal{O}), |f_n| \downarrow 0$ implies $\varphi_1(|f_n|) \downarrow 0$ in the same sense (i.e., in measure or pointwise) so that by the dominated convergence theorem the right side (and hence the left side) of (12) tends to zero. This shows that X satisfies the conditions of Definition 5, since $\varphi_1(x) = 0$ iff $x = 0$, and it is therefore a stochastic integrator.

We now consider the converse. Since X is a stochastic integrator, and τ is linear, by Definition 5(i), τ is bounded on $\mathcal{S}(\Omega', \mathcal{O})$. Let $G = B(\Omega', \mathcal{O})$ be the closure of $\mathcal{S}(\Omega', \mathcal{O})$ in the uniform norm and set $F = L^{\varphi_2}(P)$. Then by a known (generalization of Riesz's) representation theorem (cf., e.g., Dinculeanu [4], Theorem 1; see also Dunford-Schwartz [1], VI.7.2) there is a unique additive (operator) function $M : \mathcal{B}_I \to B(F, G)$ such that

$$\tau(f) = \int_I f(t) dM(t), \quad f \in G, \|\tau\| = \|M\|(I). \tag{22}$$

Now let $f_n \in \mathcal{S}(\Omega', \mathcal{O}), |f_n| \downarrow 0$ so that by Definition 5(ii), $\tau(f_n) \to 0$ in probability (hence also in norm since φ_2 is moderating). Taking $f_n = \chi_{A_n} f_0, A_n \in \mathcal{B}_I, A_n \downarrow \emptyset$, and f_0 is a bounded random variable, we get from (22) that $M(A_n) f_0 \to 0$ in $L^{\varphi_2}(P)$ as $n \to \infty$. Hence $M(\cdot) f_0$ is σ-additive on the algebra generated by all such A_n and has a σ-additive extension to $\mathcal{B}_I$, since φ_2 is moderating implies that $L^{\varphi_2}(P)$ does not contain an isomorphic copy of c_0 (cf. Rao and Ren [1], Prop. 10.1.3) and classical results apply to admit such extensions. Since $\mathcal{B}_I$ is countably generated and $L^{\varphi_2}(P)$ is separable by the conditions on $(\Omega, \Sigma, P), \mathcal{S}(\Omega', \mathcal{O})$ is also separable. So there is a countable dense set of elements f_n of the special form considered and using a diagonal procedure, we conclude that $M(\cdot)$ is σ-additive on $\mathcal{B}_I$, so that it is a vector measure. We can therefore invoke the techical result, Proposition 10, by which there exist a Young function φ_1 and a measure $\alpha_0 : \mathcal{B}_I \to \mathbb{R}^+$ relative to which $K_0 = \|M\|_{\psi_1,\alpha_0}(I) < \infty$, where ψ_i is the Young complementary function to $\varphi_i (i = 1, 2)$ and the constant is the ψ_1-semivariation of M_1 of that proposition relative to (ψ_1, α_0).

To complete the argument, we invoke the Hölder inequality for Orlicz spaces of vector valued functions, to get:

$$\begin{aligned}\|\tau(f)\|_{\varphi_2,P} &= \left\| \int_I f \, dM \right\|_{\varphi_2,P} \\ &\leq \big\| \|f(\cdot,\cdot)\|_{\psi_2,P} \big\|_{\varphi_1,\alpha_0} \|M\|_{\psi_1,\alpha_0}(I) \\ &= K_0 \big\| \|f(\cdot,\cdot)\|_{\psi_2,P} \big\|_{\varphi_1,\alpha_0}. \end{aligned} \tag{23}$$

This is a form of (13) with $\alpha = \alpha_0 \otimes P$ restricted to $\mathcal{O}$, and thus the pair (φ_1, α) satisfies the requirements. (Here φ_1 depends on φ_2 and M, hence α_0.) Finally, for the last statement, let $f_n \in \mathcal{S}(\Omega', \mathcal{O})$ and $|f_n| \uparrow |f|$ pointwise. Then by a property (called the Fatou property) of the norm, we get $\|f_n(t,\cdot)\|_{\psi_2,P} \uparrow \|f(t,\cdot)\|_{\psi_2,P}, t \in I$, and $\big\| \|f_n(\cdot,\cdot)\|_{\psi_2,P} \big\|_{\varphi_1,\alpha_0} \uparrow \big\| \|f(\cdot,\cdot)\|_{\psi_2,P} \big\|_{\varphi_1,\alpha_0} \leq K_1$ assumed finite for the dominated convergence statement. Then by (22), $\|\tau(f_n)\|_{\varphi_2,P} \leq K_0 K_1$, for all $n \geq 1$. It implies easily that $\tau(f_n) \to \tau(f)$ in $L^{\varphi_2}(P)$-norm. Thus X is L^{φ_1,φ_2}-bounded, and is extendable to all of $M^{\varphi_1}(\alpha)$ by (uniform) continuity. Replacing f by f/K_2 for a suitable $0 < K_2 < \infty$, ($K_2 = K_0 K_1$ will do) we find the right side of (23) to be bounded by 1. So

$$\sup\{E(\varphi_2(\frac{1}{K_2}|\tau(f)|)) : \big\| \|f(\cdot,\cdot)\|_{\psi_2,P} \big\|_{\varphi_1,\alpha_0} \leq 1\} < \infty. \tag{24}$$

This shows that X obeys the hypothesis of the first part so that $\tau(f)$ satisfies the dominated convergence statement as desired. Since $\mathbb{R}$ is a countable union of intervals I ($\bar{I}$ compact), we can replace α by a σ-finite measure in the general case, μ being $\alpha \otimes P$. □

11. Remark. The point of this theorem is that the stochastic integrals exist as long as one can find the dominating measure α and φ_1, φ_2 so that X is L^{φ_1,φ_2}-bounded relative to μ, and this is essentially the best condition. It includes both deterministic and stochastic anticipative as well as nonanticipative integrands. Since in any specific case, for a given φ_2, the actual construction of the pair (φ_1, μ) is generally nontrivial, one considers the predictable integrands, and some classes of anticipative ones so as to establish the existence of stochastic integrators and the consequent integrals. Propositions 7-9 vividly illustrate this phenomenon. The unification achieved (and structure revealed) is the impact of the generalized boundedness principle. Hereafter "$L^{2,2}$-boundedness" stands for the generalized Bochner concept as given in Definition 3.

(b) *Itô-type formula for $L^{2,2}$-bounded processes.* We present here an analog of Itô's differential formula for a class of integrators that are $L^{2,2}$-bounded, which uses the properties of the filtration. Thus let $X = \{X_t, \mathcal{G}_t, t \in I\}$ be an $L^{2,2}$-bounded process and $\{\mathcal{F}_t, t \in I\}$ be a second (standard) filtration and $\mathcal{P}$ be the predictable σ-algebra from $\mathcal{B}_I \otimes \Sigma$ determined by the latter. Then there is a σ-finite $\alpha : \mathcal{P} \to \mathbb{R}^+$ such that for each $\mathcal{P}$-simple f, one has

$$\| \int_I f dX \|_2^2 \leq K \int_{\Omega'} |f|^2 d\alpha$$

for some constant $K > 0$ and $\Omega' = I \times \Omega$. Thus, as seen before, X induces a vector measure $Z : \mathcal{B}_I \to L^2(P)$. If $\mathcal{X}$ is a Banach spae and $Z : \mathcal{B}_I \to \mathcal{X}$ is a vector measure, then its *quadratic variation* on I, denoted [Z] (or [X] if $X_t = Z((a,t])$ for a fixed $a \in I$), is defined as:

$$[Z](I) = \lim_{n\to\infty} \sum_{j=1}^{k_n} |Z(I_j^n)|^2, \tag{25}$$

if this limit exists strongly in $\mathcal{X}$, as the partitions I_j^n of $I(= \bigcup_{j=1}^{k_n} I_j^n)$ are refined so that $\max_{1\leq j\leq k_n} |I_j^n| \to 0$ as $n \to \infty$. If the variation of $Z(\cdot)$ exists then factoring out one of the product terms in (25), we see that $[Z](I) = 0$. If $\mathcal{X} = L^\varphi(P)$, then we can weaken the limit in (25) to limit in probability for this concept. In general this limit need not exist. We have the following result which is an analog of Theorem V.3.6 for square integrable martingales. But the argument is essentially the same. It is reproduced, in the present context, for completeness.

12. Proposition. *Let $X = \{X_t, \mathcal{G}_t, t \in [0,b)\}$ be an $L^{2,2}$-bounded càdlàg process. Then the quadratic variation $[X]$ exists on $I = [0,b)$.*

Proof. Let $\mathcal{P}$ be the predictable σ-algebra generated by the (standard) filtration $\{\mathcal{G}_t, t \in I\}$ in $\mathcal{B}_I \otimes \Sigma$. If $I_j^n = (\frac{j}{2^n} \wedge b, \frac{j+1}{2^n} \wedge b)$ is the dyadic partitions of I, then we have

$$\begin{aligned} X_b^2 - X_0^2 &= \sum_{i=0}^{n} [X_{\frac{j+1}{2^n}\wedge b} - X_{\frac{j}{2^n}\wedge b}]^2 + \\ &\quad 2\sum_{j=0}^{n} X_{\frac{j}{2^n}\wedge b}(X_{\frac{j+1}{2^n}\wedge b} - X_{\frac{j}{2^n}\wedge b}) \\ &= \sum_{j=0}^{n} \left|Z(I_j^n)\right|^2 + 2\sum_{j=0}^{n} X_{\frac{j}{2^n}\wedge b} Z(I_j^n) \end{aligned} \tag{26}$$

where $Z((c,d]) = X_d - X_c$, determines a stochastic measure by the fact that X is $L^{2,2}$-bounded. Let $f_n = \sum_{j=1}^{n} X_{\frac{j}{2^n}\wedge b} \chi_{I_j^n}$. This is a $\mathcal{P}$-measurable simple function and (26) becomes

$$\sum_{j=1}^{n} \left|Z(I_j^n)\right|^2 = X_b^2 - X_0^2 - 2\int_I f_n dZ. \tag{27}$$

Now by Theorem 6, the right side of (27) has a limit as $n \to \infty$ since Z is $L^{2,2}$-bounded for some μ on $\mathcal{P}$, and is càdlàg. So the left side tends to $[Z]_b = [X]_b$ as asserted. $\square$

In the same way, we can extend Theorem 1.2 to the present case. We shall include the result leaving the details of proof which are quite similar.

13. Theorem. *Let $X = \mathbb{R}^+ \to L^2(P), X = \{X_t, \mathcal{G}_t, t \geq 0\}$ be càdlàg and $L^{2,2}$-bounded relative to a σ-finite $\alpha : \mathcal{P} \to \bar{\mathbb{R}}^+$ where $\mathcal{P}$ is the predictable σ-algebra determined by the filtration $\{\mathcal{G}_t, t \geq 0\}$. If $f : \mathbb{R} \to \mathbb{R}$ is a twice continuously differentiable function, then one has for $t > 0$,*

$$\begin{aligned} f(X_t) - f(X_0) &= \int_{0+}^{t} f'(X_{s-})dX_s + \frac{1}{2}\int_{0+}^{t} f''(X_{s-})d[X]_s + \\ &\quad \sum_{0<s<t} [f(X_s) - f(X_{s-}) - f'(X_{s-})\Delta X_s] - \\ &\quad \frac{1}{2}\sum_{0<s\leq t} f''(X_s)(\Delta X_s)^2, \end{aligned} \tag{28}$$

where the sums are a.e. convergent and $\Delta X_s = X_s - X_{s-}$ *is the jump of* X *at* s.

The argument to establish this result extends even to multidimensions, if various products and derivatives are suitably interpreted. In fact, it is valid for infinite dimensional processes also. An aspect of the latter will be formulated in the next section.

The above result has an interesting "converse" implication. While $L^{2,2}$-boundedness, relative to a σ-finite $\alpha : \mathcal{O} \to \bar{\mathbb{R}}^+$ for a general $\mathcal{O} \subset \mathcal{B}_I \otimes \Sigma$, is widely applicable to various classes, its restriction to predictable σ-algebras $\mathcal{O}$ determined by standard filtrations (as in Theorem 13) gives us the following conclusion which is perhaps not surprising in view of Proposition 8. Thus considering the sample continuous case, for simplicity, for twice continuously differentiable $f : \mathbb{R} \to \mathbb{R}$, (28) implies that $\{f(X_t), \mathcal{G}_t, t \in I\}$ is $L^{2,2}$-bounded for each (bounded) interval $I \subset \mathbb{R}^+$ relative to some measure α, if X is $L^{2,2}$-bounded relative to $\alpha_0 : \mathcal{P} \to \mathbb{R}^+$ where α is determined by $\alpha_0, [X]$ and P, (cf. (17)). More generally, if $X = \{X, \mathcal{G}_t, t \geq 0\}$ is $L^{2,2}$-bounded relative to a σ-finite $\alpha : \mathcal{P} \to \bar{\mathbb{R}}^+$, define a set function ν for $s < t$ and $A \in \mathcal{G}_s$ as :

$$\begin{aligned} |\nu((s,t] \times A)| &= |E(\chi_A (X_t - X_s))| = |E(\tau(\chi_A))| \\ &\leq K \left[\int_{(s,t)\times\Omega} \chi_A^2 \, d\alpha \right]^{\frac{1}{2}} \end{aligned} \tag{29}$$

for some constant $K > 0$. Since ν is additive on the semialgebra of those sets from $\mathcal{P}$, (29) shows that $\nu(\cdot)$ is σ-additive and has an extension to the generated σ-algebra, namely $\mathcal{P}$. Fixing $s(= 0$ for covenience), and $t > s(= 0)$is arbitrary, then $\nu^t(\cdot) = \nu((0,t] \times (\cdot))$ is P-continuous and by the Radon-Nikodym theorem (since $\nu^t(\cdot)$ is a signed measure), $\frac{d\nu^t}{dP} = Q_t$ (say) is $\mathcal{G}_{t-}$ measurable, and we take a càdlàg version, say B_t of Q_t. Since ν^t has finite variation and $\mathcal{G}_t \subset \mathcal{G}_{t'}$, for $t \leq t'$, we conclude that $|Q_t| \uparrow$ as $t \uparrow$. Thus $\{B_t, t \geq 0\}$ has finite variations a.e. on each bounded interval. Letting $M_t = X_t - B_t$, one has for each $A \in \mathcal{G}_s (s < t)$

$$\begin{aligned} E(\chi_A (M_t - M_s)) &= E(\chi_A (X_t - X_s)) - E(\chi_A (B_t - B_s)) \\ &= E(\chi_A (X_t - X_s)) - E(\chi_A (Q_t - Q_s)) = 0, \end{aligned}$$

by definition of Q_t (and B_t being its modification). Thus $\{M_t, \mathcal{G}_t, t \geq 0\}$ is a martingale in $L^2(P)$ so that $X_t = M_t + B_t,\ t \geq 0$, is a semimartingale. This discussion clearly extends to (finite) vector X_t-processes. This result may be summarized as follows:

14. Theorem. *Let $X = \{X_t, \mathcal{G}_t, t \geq 0\}$ be an adapted càdlàg process which is $L^{2,2}$-bounded relative to a σ-finite $\alpha : \mathcal{P} \to \bar{\mathbb{R}}^+$ where $\mathcal{P}$ is the predictable σ-algebra determined by the filtration $\{\mathcal{G}_t, t \geq 0\}$. Then X is a semimartingale and the result also holds if X_t is a (finite) vector process. On the other hand a càdlàg square integrable semimartingale (of any dimension) is always $L^{2,2}$-bounded relative to a σ-finite α : $\mathcal{P} \to \bar{\mathbb{R}}^+$.*

This theorem, with Theorem 6, recovers the Dellacherie [2] and Bichteler [1] result on a characterization of stochastic integrators as semimartingales, when the filtrations are restricted, and it is more general if the latter are not so restricted. It will be interesting to consider extensions of the stochastic integration in all its specializations for L^{φ_1,φ_2}-bounded processes which are more general than semimartingales. Such a study will constitute a great digression, and will not be undertaken here. However, we include a brief account of multiparameter (and vector) analogs in the next section, and in Section 7.3 later.

6.3 Multidimensional and multiparameter analogs

The significance of the boundedness principle is understood better if its multivariate and multiparameter versions are also discussed. A brief account of these extensions will be included. First a direct analog of the multivariate processes (i.e., indexing is linear) is considered. Thus let $\Omega' = \mathbb{R}^+ \times \Omega$ as before, and let $\mathcal{X}, \mathcal{Y}, \mathcal{Z}$ be Banach spaces such that there is a continuous bilinear form of $\mathcal{X} \times \mathcal{Y} \to \mathcal{Z}$ with the product topology of $\mathcal{X} \times \mathcal{Y}$. For instance $\mathcal{Y}$ may be taken as a subspace of $B(\mathcal{X}, \mathcal{Z})$, the space of bounded linear operators from $\mathcal{X}$ to $\mathcal{Z}$.

We set up the frame-work with a standard filtration $\{\mathcal{F}_t, t \in I\}$ for an interval $I \subset \mathbb{R}$, from (Ω, Σ, P), and let $A_1, \ldots, A_n$ be disjoint elements of $\mathcal{B}_I$. If $E_t = B(\Omega, \mathcal{F}_t; \mathcal{Y})$ denotes the space of bounded (strongly) $\mathcal{F}_t$-measurable $\mathcal{Y}$-valued functions on Ω, and $\mathcal{E}_{\mathcal{Y}} = \{E_t, t \in I\}$, then $f \in \mathcal{E}_{\mathcal{Y}}$ iff for each $t \in I, f(t,\cdot) \in E_t$. The set $\mathcal{E}_{\mathcal{Y}}$ becomes a vector space (and an algebra) if for $f, g \in \mathcal{E}_{\mathcal{Y}}, a, b \in \mathbb{R}$, we define $af + bg = \{af(t,\cdot) + bg(t,\cdot), t \in I\} \in \mathcal{E}_{\mathcal{Y}}$ (and $\{(f \cdot h)(t,\cdot) = f(t,\cdot)h(t,\cdot), t \in I\} \in \mathcal{E}_{\mathcal{Z}}$). One can use the norm $|||f||| = \sup_t \|f(t,\cdot)\|_u$ where $\|\cdot\|_u$ is the uniform norm in E_t. Let $\mathcal{S}(\Omega', \mathcal{E}_{\mathcal{Y}})$ denote the set of simple functions (also called "vector fields") f of the form, for $A_i \in \mathcal{B}_I$, selected above:

$$f = \sum_{i=1}^{n} f_i \chi_{A_i}, \quad f_i \in E_{\inf(A_i)} = B(\Omega, \mathcal{F}_{\inf(A_i)}; \mathcal{Y}). \tag{1}$$

Then $\{\mathcal{S}(\Omega', \mathcal{E}_{\mathcal{Y}}), |||\cdot|||\}$ becomes a normed linear space. Note that when $\mathcal{X} = \mathcal{Y} = \mathcal{Z} =$ scalars, we get $\mathcal{S}(\Omega', \mathcal{E}_{\mathcal{Y}})$ to be $\mathcal{S}(\Omega', \mathcal{O})$ of the preceding section. In a straight forward manner one defines an Orlicz space of $\mathcal{X}$ valued functions $L^{\varphi}_{\mathcal{X}}(P)$ as the set of (strongly) measurable functions $f : \Omega \to \mathcal{X}$ such that $f \in L^{\varphi}_{\mathcal{X}}(P)$ iff $\|f\|_{\varphi} < \infty$, where

$$\|f\|_{\varphi} = \inf\{k > 0 : \int_{\Omega} \varphi\left(\frac{\|f\|_{\mathcal{X}}}{k}\right) dP \leq 1\}. \tag{2}$$

It is routine to verify that $\{L^{\varphi}_{\mathcal{X}}(P), \|\cdot\|_{\varphi}\}$ is a complete normed linear space. We can now present an analog of Definition 2.3.

1. Definition. If φ_1, φ_2 are a pair of Young functions and $X = \{X_t, \mathcal{G}_t, t \in I\}, X_t \in L^{\varphi_2}_{\mathcal{X}}(P)$ (càdlàg) let $\{\mathcal{F}_t, t \in I\}$ be another standard filtration. Then X is said to be L^{φ_1,φ_2}-*bounded* relative to a σ-finite measure $\alpha : \mathcal{P} \to \bar{\mathbb{R}}^+$ and φ, whenever

$$\tau : f \mapsto \tau(f) = \sum_{i=1}^{n} f_i(X_{\sup(I_i)} - X_{\inf(I_i)}) = \int_I f dX, \tag{3}$$

for $f \in \mathcal{S}(\Omega', \mathcal{E}_{\mathcal{Y}}), I_i \in \mathcal{B}_I$ disjoint, we have

$$E(\varphi_2(\|\tau(f)\|_{\mathcal{Z}})) = \int_{\Omega} \varphi_2(\|\tau(f)\|_{\mathcal{Z}}) dP \leq K \int_{\Omega'} \varphi_1(\|f\|_{\mathcal{Y}}) d\alpha, \tag{4}$$

for some fixed $K (= K^{\alpha}_{\varphi_1,\varphi_2} > 0)$. Here $\mathcal{P}$ is the predictable σ-algebra of the filtration. [In fact $\mathcal{P}$ can be replaced by any σ-algebra $\mathcal{O} \subset \mathcal{B}_I \otimes \Sigma$ here if $\mathcal{S}(\Omega', \mathcal{O}; \mathcal{Y})$ replaces $\mathcal{S}(\Omega', \mathcal{E}_{\mathcal{Y}})$ and τ is additive.]

It is to be observed that this includes the multivariate case studied by Cramér, Kolmogorov, Rozanov and others when we take $\varphi_1(x) = \varphi_2(x) = x^2, \mathcal{X} = \mathcal{Z} = \mathbb{R}^n, \mathcal{Y} = \mathbb{C}$ and $\mathcal{F}_t = \{\emptyset, \Omega\}$ giving a (vector) $L^{2,2}$-boundedness concept. The general case corresponding to Theorem 2.6 can be stated after we introduce a vector stochastic integrator.

2. Definition. A process $X = \{X_t, \mathcal{G}_t, t \in I \subset \mathbb{R}\}, X_t \in L^{\varphi}_{\mathcal{X}}(P)$, is a *vector stochastic integrator* on $\mathcal{S}(\Omega', \mathcal{E}_{\mathcal{Y}})$, using the preceding notations, if the following conditions hold:

(i) $\{\tau(f) : f \in \mathcal{S}(\Omega', \mathcal{E}_{\mathcal{Y}}), |||f||| \leq 1\}$ is a bounded set in $L^{\varphi}_{\mathcal{Y}}(P)$,

(ii) $f_n \in \mathcal{S}(\Omega', \mathcal{E}_{\mathcal{Y}}), \|f_n\|_{\mathcal{Y}} \downarrow 0$, a.e., implies $\|\tau(f_n)\|_{\mathcal{Z}} \to 0$ in probability as $n \to \infty$.

Using the argument of the proof of Theorem 2.6 with simple modifications we get the following result in which it is assumed that $\mathcal{Z}$ is reflexive (or at least is an adjoint space) so that $L^{\varphi}_{\mathcal{Z}}(P)$ does not contain a copy of c_0 whenever φ obeys a Δ_2-condition. We employ the notations introduced above without repetitions.

3. Theorems. *Let $\{(\Omega, \Sigma, \mathcal{F}_t, P), t \in I\}$ be a (standard) filtered space, $X = \{X_t, \mathcal{G}_t, t \in I\}, X_t \in L^{\varphi_2}_{\mathcal{Z}}(P)$ be a càdlàg process and be L^{φ_1,φ_2}-bounded relative to a σ-finite $\alpha : \mathcal{P} \to \bar{\mathbb{R}}^+$, a Young function φ_1 and separable Banach spaces $\mathcal{X}, \mathcal{Y}$ and $\mathcal{Z}$, the latter being an adjoint space in addition. Then X is a stochastic integrator on $M^{\varphi_1}_{\mathcal{Y}}(\alpha) = \bar{sp}\{\mathcal{S}(\Omega, \mathcal{E}_{\mathcal{Y}}) \cdot \|\cdot\|_{\varphi_1}\}$ so that the dominated convergence criterion holds for $\tau : M^{\varphi_1}_{\mathcal{Y}}(\alpha) \to L^{\varphi_2}_{\mathcal{X}}(P)$, defined by (3). Conversely, if φ_2 is a Young function satisfying a Δ_2-condition, (Ω, Σ, P) and $\mathcal{X}, \mathcal{Y}$ are also separable and X is a stochastic integrator on $\mathcal{S}(\Omega', \mathcal{E}_{\mathcal{Y}})$, then there are a Young function φ_1 with $\frac{\varphi_1(x)}{x} \uparrow \infty$ as $x \uparrow \infty$, a σ-finite $\alpha : \mathcal{P} \to \bar{R}^+$ relative to which X is L^{φ_1,φ_2}-bounded in the normed form, (i.e., the analog of Eq. (2.13) holds) and the dominated convergence criterion is valid for the mapping $\tau : M^{\varphi_1}_{\mathcal{Y}}(\alpha) \to L^{\varphi_2}_{\mathcal{X}}(P)$.*

To see the usefulness of this version, we consider an application and then compare it with several existing works on the subject. The vector version of a semimartingale is as follows:

4. Definition. A process $X = \{X_t, \mathcal{G}_t, t \in I\}, X_t \in L^2_{\mathcal{X}}(P)$ and càdlàg, is a (vector) *semimartingale* if it can be expressed as:

$$X = M + B, \tag{5}$$

where $M = \{M_t, \mathcal{G}_t, t \in I\}$ is a martingale such that $E(\|X_t\|^2) < \infty, t \in I$ (i.e., square integrable), and $B = \{B_t, \mathcal{G}_t, t \in I\}$ is a process of integrable variation so that

$$|B|(I) = \sup_{n,\pi_n}\{\sum_{t\in\pi_n} \|B_{t+1} - B_t\|_{\mathcal{X}} : \pi_n = (a_0 \leq t_0 < \cdots < t_n \leq b)\}, \tag{6}$$

when $I = (a,b)$, satisfies $E(|B|(I)) < \infty$.

One can show, for reflexive $\mathcal{X}$, that a quasimartingale admits such a decomposition as (5). If $Q((s,t)) = B_t - B_t$, then Q has finite variation for a.a. (ω), and using the right continuity of B, it defines a vector measure on $\mathcal{B}_I$ with $L^1_{\mathcal{X}}(P)$ of finite total variation. Hence if $\nu(\cdot)$ is the varition measure of A, then $\alpha = \nu \otimes P$ is a finite measure on $\mathcal{B}_I \otimes \Sigma$ and for each $f \in \mathcal{S}(\Omega', \mathcal{E}_{\mathcal{Y}})$ one has

$$E\left(\left\|\int_I f\ dQ\right\|_{\mathcal{Z}}\right) \leq \int_{\Omega'} \|f\|_{\mathcal{Y}}\ d\alpha, \tag{7}$$

using some standard elementary properties of such integrals (cf., e.g., Dinculeanu [1], p. 122). We now illustrate the vector boundedness principle, taking $\mathcal{X} = \mathcal{H}$, a Hilbert space, for simplicity. Even here we need to recall some properties of (tensor) products, so that the argument of Proposition 2.8 can be taken over.

If 'tr' denotes the *trace* functional and for each $x, y \in \mathcal{H}, x \otimes y$ is the tensor product so that $x \otimes y \in B(\mathcal{H}, \mathcal{H}) = B(\mathcal{H})$ and $tr(x \otimes y) = (x, y)$, the inner product of $\mathcal{H}$, then for each $A \in B(\mathcal{H})$, $tr(A) = \sum_{i=1}^{\infty}(Af_i, f_i)$ whenever the series converges absolutely for some (and then all) complete orthonormal sequences $\{f_i, i \geq 1\}$ of the separable $\mathcal{H}$. If A is positive (semi) definite, then $tr(A) \geq 0$, and $= 0$ only when $A = 0$. Moreover, $tr(x \otimes x) = \|x\|^2$, $tr(AB_0) = tr(B_0 A) \leq \|A\| tr(B_0), A \in B(\mathcal{H})$ and B_0 of positive finite trace, since $tr(\cdot)$ is a positive linear functional on $B(\mathcal{H})$. The details of these facts can be found, e.g., in Schatten ([1], Ch. V). We now establish

5. Proposition. *A square integrable càdlàg semimartingale X of (5) (with $X_t \in L^2_{\mathcal{H}}(P)$), $\sup_t E(\|X_t\|^2_{\mathcal{H}}) < \infty$, is $L^{2,2}$-bounded relative to some $\alpha : \mathcal{P} \to \mathbb{R}^+$ where $\mathcal{P}$ is the predictable σ-algebra generated by the standard filtration $\{\mathcal{G}_t, t \in I\}$.*

Proof. If $f \in \mathcal{S}(\Omega', \mathcal{E}_{\mathcal{Y}})$, then consider

$$\begin{aligned} E(\|\tau_1(f)\|^2_{\mathcal{Z}}) &= E\left(\left\|\int_I f\ dM\right\|^2_{\mathcal{Z}}\right) \\ &= E\left(\left\|\sum_{i=1}^{n} f_i(M_{t_{i+1}} - M_{t_i})\right\|^2_{\mathcal{Z}}\right) \end{aligned}$$

$$= E\left(tr\left[\sum_{0\le i,j\le n} f_i(M_{t_{i+1}} - M_{t_i}) \otimes f_j(M_{t_{j+1}} - M_{t_j})\right]\right)$$

$$= E\left(tr\left[\sum_{i=0}^{n}(f_i(M_{t_{i+1}} - M_{t_i}))^{\otimes 2}\right]\right) + 0,$$

writing $x \otimes x$ as $x^{\otimes 2}$, and using the martingale property of the M_t's,

$$= tr\left(\sum_{i=0}^{n} E[f_i(M_{t_{i+1}}^{\otimes 2} - M_{t_i}^{\otimes 2})f_i]\right)$$

by the martingale property again,

$$\le \sum_{i=0}^{n} E(\|f_i\|_{\mathcal{Y}}^2 \; tr(M_{t_{i+1}}^{\otimes 2} - M_{t_i}^{\otimes 2})),$$

since $f_i \in B(\mathcal{H})$ and 'tr' is linear,

$$\le \sum_{i=0}^{n} E(\|f_i\|_{\mathcal{Y}}^2(\|M_{t_{i+1}}\|_{\mathcal{X}}^2 - \|M_{t_i}\|_{\mathcal{X}}^2))$$

$$= \int_{\Omega'} \|f\|_{\mathcal{Y}}^2 d\alpha, \tag{8}$$

since $\{\|M_{t_i}\|_{\mathcal{X}}^2, \mathcal{G}_{t_i}, i \ge 0\}$ is a positive submartingale, and hence $\alpha_1 : \mathcal{P} \to \mathbb{R}^+$ is determined by this process as in Prop. 2.8. But $\tau(f) = \int_I f dX = \int_I f dM + \int_I f dB = \tau_1(f) + \int_I f dB$. Then (7) and (8) can be used to reduce the problem to the scalar case and apply Prop. 2.8 itself. $\square$

As in the earlier work, the L^{φ_1,φ_2}-boundedness implies that τ has a unique extension to $M_{\mathcal{Y}}^{\varphi_1}(\alpha)$ with values in $L_{\mathcal{Z}}^{\varphi_2}(P)$, and it is the (vector) stochastic integral that one can obtain. Let us present some specializations of Theorem 3 to compare with the classical results.

6. Comparison. (i) If $\mathcal{F}_t = \{\emptyset, \Omega\} \subsetneqq \mathcal{G}_t$, then this $L^{2,2}$-version of the stochastic integral includes the vector versions of Cramér and Kolmogorov integrals designed for representing stationary processes.

(ii) If $\Omega = \{\omega\}$, a singleton, so that $L_{\mathcal{Z}}^{\varphi_2}(P)$ may be identified with $\mathcal{Z}$, and $\mathcal{B}_I \otimes \Sigma$ can now be replaced by $\mathcal{B}_I$, then (4) reduces to the (*)-condition of Bartle's [1], countably additive theory of the bilinear vector integral, thereby implying that it is essentially optimal. Another form of this condition of Bartle's results is if the filtration $\{\mathcal{F}_t, t \in I\}$ is general but $\mathcal{G}_t = \{\emptyset, \Omega\}, t \in I$. Thus the freedom of two filtrations in our boundedness principle is useful for applications.

(iii) If $\Omega = \{\omega\}, \mathcal{X} = \mathbb{R}$ (or $\mathbb{C}$), $\mathcal{Y} = B(\mathcal{X}, \mathcal{Z}) = \mathcal{Z}$, then the $L^{2,2}$-boundedness reduces to X being a signed measure so that α is its (total) variation measure ($= |X|(\cdot)$), and (4) becomes

$$\| \int_I f dX \|_{\mathcal{Z}}^2 \leq K \int_I \|f\|_{\mathcal{Z}}^2 d|X|, \tag{9}$$

with $K = |X|(I)^2$. This is the classical Bochner integral, using the CBS-inequality for finite measures, the σ-finiteness extension being easy. Other reductions and comparisons can be made.

We now state the vector analog of the Itô-type formula. For simplicity, only the Hilbert space $\mathcal{H}$ will be considered. We take $f : \mathcal{H} \to \mathcal{H}$ as a mapping which is twice continuously differentiable in the norm topology, and if $\mathcal{Y} \subset B(\mathcal{H})$ is the set of linear operators $A(: \mathcal{H} \to \mathcal{H})$ such that $tr(AA^*) < \infty$, then for each $x \in \mathcal{H}$, one sees that $f(x) \in \mathcal{H}$, $f'(x) \in B(\mathcal{H})$ and $f''(x) \in B(\mathcal{Y}, \mathcal{H})$. Note that $A \in \mathcal{Y}$ iff $\|A\|^2 = (A, A) = \sum_{i=1}^{\infty} \|Ax_i\|^2 < \infty$ for some (and hence all) complete orthonormal sets $\{x_i, i \geq 1\} \subset \mathcal{H}$. Then $(\mathcal{Y}, (\cdot, \cdot))$ is a Hilbert space of Schmidt class of operators on $\mathcal{H}$ (cf., e.g., Schatten [1], p. 110). In this notation the quadratic variation of a process $X = \{X_t, \mathcal{G}_t, t \in I\}, X_t \in L^2_{\mathcal{H}}(P)$, denoted as $[\![X]\!]$, has range in $\mathcal{Y}$ and $[\![X]\!](J)$ is positive (semi) definite for each $J \in \mathcal{B}_I$. The multidimensional analog of Prop. 2.12 gives the existence of $[\![X]\!](J)$, and Theorem 2.13 takes the form:

7. Theorem. *Let $\mathcal{H}, \mathcal{Y}$ be separable Hilbert spaces, $\mathcal{Y} \subset B(\mathcal{H})$ being the schmidt class of operators on $\mathcal{H}$, and $f : \mathcal{H} \to \mathcal{H}$ be twice continuously differentiable with its derivatives uniformly continuous on the bounded sets of $\mathcal{H}$ and $\mathcal{Y}$. If $X : \mathbb{R} \to L^2_{\mathcal{H}}(P)$ is an $L^{2,2}$-bounded càdlàg process relative to its natural filtration $\{\mathcal{G}_t, t \geq 0\}, (I = \mathbb{R}^+)$ and a measure $\alpha : \mathcal{P}(= \bar{\mathcal{P}}) \to \bar{\mathbb{R}}^+$, then for any $t > 0$,*

$$\begin{aligned} f(X_t) - f(X_0) = \int_{0+}^t f'(X_{s-}) dX_s + \frac{1}{2} \int_{0+}^t f''(X_{s-}) d[\![X]\!]_s + \\ \sum_{0<s\leq t} [f(X_s) - f(X_{s-}) - f'(X_{s-})\Delta X_s] - \\ \frac{1}{2} \sum_{0<s\leq t} f''(X_s)(\Delta X_s)^{\otimes 2}, \end{aligned} \tag{10}$$

where the last two sums are strongly convergent a.e. and $\Delta X = X_s - X_{s-}$ is the jump of $\{X_s, \mathcal{G}_s, s \in \mathbb{R}^+\}$ at $s > 0$.

In a similar manner Theorem 2.14 can be extended to the vector case, but we shall not present a corresponding statement. Instead, we proceed to consider the multiparameter case which raises new questions since the index set (say $I \subset \mathbb{R}^k, k > 1$) will not have a (linear) ordering. Consequently the process X itself has to be taken as an additive set function. The following is a way of handling this problem.

Let $X = \{X(I), \mathcal{F}_I, I \in \mathcal{T}\}$ be a random family on (Ω, Σ, P), where $\mathcal{T}$ is a semialgebra of an index set T and for each $I \in \mathcal{T}$, there is a P-complete σ-algebra $\mathcal{F}_I \subset \Sigma$, satisfying $\mathcal{F}_I \subset \mathcal{F}_J$ for $I \supset J$ in $\mathcal{T}$. The σ-algebra $\mathcal{P}$ generated by the semiring $\mathcal{S} = \{I \times A : I \in \mathcal{T}, A \in \mathcal{F}_I\}$, will be called *predictable* and $f : T \times \Omega \to \mathbb{R}$ is a P-simple function if it is given for disjoint $A_i \in \mathcal{F}_{I_i}$,

$$f = \sum_{i=1}^{n} a_i \chi_{S_i}, \; S_i = I_i \times A_i, \; a_i \in \mathbb{R}. \tag{11}$$

Then

$$\tau(f) = \int_T f \, dX = \sum_{i=1}^{n} a_i \chi_{A_i} X(I_i), \tag{12}$$

is well-defined and is additive, $\mathcal{P} = \sigma(\mathcal{S})$. We can state the corresponding boundedness principle in the following:

8. Definition. Let φ_1, φ_2 be generalized Young functions and $X = \{X(I), \mathcal{G}_I, I \in \mathcal{T}\}$ be an additive (scalar valued) random field $X(I) \in L^{\varphi_2}(P)$. Then X is L^{φ_1,φ_2}*-bounded* relative to a filtration $\{\mathcal{F}_I, I \in \mathcal{T}\}$ of Σ and a σ-finite measure $\alpha : \mathcal{P} \to \bar{\mathbb{R}}^+$ (or $\alpha : \mathcal{O} \to \bar{\mathbb{R}}^+$ where $\mathcal{O} \subset \mathcal{B}_T \otimes \Sigma$, $\mathcal{B}_T = \sigma(\mathcal{T})$) if there exists a constant $K (= K^{\alpha}_{\varphi_1,\varphi_2} > 0)$ such that for all $\mathcal{S}$-simple (or $\mathcal{S}(\Omega', \mathcal{O})$-simple, $\Omega' = T \times \Omega$) functions $f, \tau(f)$ of (12) satisfies

$$E(\varphi_2(|\tau(f)|) \leq K \int_{\Omega'} \varphi_1(|f|) d\alpha. \tag{13}$$

This concept is general enough to include several multiple stochastic integrals available in the literature, but the verifications in various cases are themselves nontrivial. The analog of Theorem 2.6 will not be difficult. We shall illustrate (13) for some classical cases.

Let $T = \mathbb{R}^k, \mathcal{T} = \delta$-ring of bounded Borel sets and $\{Z_A, A \in \mathcal{T}\}$ be a real Gaussian random field with mean zero and covariance given by $E(Z_A Z_B) = \mu(A \cap B)$ where μ is the Lebesgue measure on $\mathcal{T}$. Then $Z_{(\cdot)}$ is called a 'white noise' and it is $L^{2,2}$-bounded relative to $\alpha = \mu \otimes P$ in (13). On the other hand a process $\{X_t, t \in T\}$ is a *Lévy-Brownian motion* if it is a real Gaussian process with mean zero, $X_0 = 0$ a.e., and for, $t, t' \in T, E((X_t - X_{t'})^2) = \|t - t'\|$ so that $C_X(t,t') = E(X_t X_{t'}) = \frac{1}{2}(\|t\| + \|t'\| - \|t - t'\|)$, where $\|\cdot\|$ is the norm of T. It is nontrivial to show that $C_X(\cdot,\cdot)$ is a covariance function, and hence the existence of the random field X_t. We show below that it is representable as the $D - S$ integral as:

$$X_t = \int_T f_t(u) dZ(u), t \in T. \tag{14}$$

Indeed the positive definiteness of C_X is discussed at considerable length by Gangolli [1] for a large class of homogeneous spaces T. To understand this interesting extension we give a direct proof of the existence of Lévy-Brownian motion (already established by Lévy [1]) when T is a Hilbert space, following Cartier [1]. In general such a process is not a martingale relative to its natural stochastic base.

9. Theorem. *Let T be a real Hilbert space. Then there exists a probability space (Ω, Σ, P) carrying a Lévy-Brownian motion $X = \{X_t, t \in T\}$, in the sense that it is Gaussian with mean zero and covariance C_X.*

Proof. We divide the proof into parts for convenience.
I. Any Hilbert space $\mathcal{H}_0$ is isomorphic to an $L^2(\Omega, \Sigma, P)$ where P is Gaussian.

For, let $\{\varphi_i, i \in I\}$ be an orthonormal basis of $\mathcal{H}_0$. Let $(\Omega_i, \Sigma_i, P_i)$ be a probability space with P_i determined by a standard normal random variable (i.e., we may take $\Omega_i = \mathbb{R}, \Sigma_i =$ Borel σ-algebra of $\mathbb{R}$, and $P_i(A) = \int_A \exp\left[-\frac{t^2}{2}\right] \frac{dt}{\sqrt{2\pi}}, A \in \Sigma_i$). Let $(\Omega, \Sigma, P) = \bigotimes_{i \in I} (\Omega_i, \Sigma_i, P_i)$, the product space which exists by the Fubini-Jensen theorem (or Theorem I.2.1). If $X_i(\omega) = \omega(i)$, the coordinate function, then $\{X_i, i \in I\}$ is a family of independent Gaussian random variables on (Ω, Σ, P) with means zero and variances one. Also $\{X_i, i \in I\}$ is an orthonormal basis of $L^2(\Omega, \Sigma, P)$ and the correspondence $\tau : \varphi_i \mapsto X_i$ defines an isomorphism of $\mathcal{H}_0$ into $L^2(\Omega, \Sigma, P)$. [Here we use the fact that a Σ-measurable function depends only on a countable collection of X_i's.]

Thus for the proof, it suffices to show that there exists a Hilbert space $\mathcal{H}$ and a mapping $f : T \to \mathcal{H}$ such that $f(0) = 0, \|f(t) - f(t')\|^2_{\mathcal{H}} = \|t - t'\|_T$, since then $\{(\tau \circ f)(t), t \in T\}$ will be a Lévy-Brownian motion. [$f(0) = a$ may be demanded here since $\tilde{f} = f - a$ will again give the desired process as $\{\tau(\tilde{f})(t), t \in T\}$.] We first need a technical computation.

II. Observe that $f : T \to \mathcal{H}$, as above, exists iff the function $g : (t, t') \mapsto \|t - t'\|_T$ is *conditionally negative definite*, i.e., for each set $\{a_i\}_1^n \subset \mathbb{R}, \sum_{i=1}^n a_i = 0$, and $\{t_i\}_1^n \subset T$ one has (writing $\|\cdot\|$ for $\|\cdot\|_T$)

$$g(t_i, t_i) = 0, \sum_{i,j=1}^n g(t_i, t_j)a_i a_j \leq 0 \text{ and } g(-t_i, -t_j) = g(t_j, t_i). \quad (15)$$

In fact, let $C(t, t') = \frac{1}{2}\left[\|t\| + \|t'\| - \|t - t'\|\right]$. Then $C(\cdot, \cdot)$ is positive definite if (15) is true, since

$$2 \sum_{i,j=2}^n C(t_i, t_j)a_i a_j = - \sum_{i,j=1}^n g(t_i, t_j)a_i a_j \geq 0, \quad (16)$$

where $t_1 = 0$ $\left(\text{and } \sum_{i=1}^n a_i = 0\right)$. So an inner product can be defined on T by using $C(\cdot, \cdot)$. Let $\mathcal{H}_2$ be the completion of T in this inner product. Then there is an $h_1 : T - \{0\} \to \mathcal{H}_2$ such that $(h_1(t), h_1(t')) = C(t, t')$ (so h_1 is continuous). Extend h_1 to $\mathcal{H}_2$ by setting $h_1(0) = 0$ (call it h). Then $g(t, t') = \|h(t) - h(t')\|^2_{\mathcal{H}_2}$. If $f = h$, then f satisfies the conditions of the preceding paragraph with $\mathcal{H} = \mathcal{H}_2$. Conversely, let there be an $f : T \to \mathcal{H}$ such that $\|f(t) - f(t')\|^2_{\mathcal{H}} = \|t - t'\|_T$. Then (15) must be true. Indeed, writing $(\cdot, \cdot)$ for the inner product of $\mathcal{H}$ we have

$$\begin{aligned} \sum_{i,j=1}^n g(t_i, t_j)a_i a_j &= \sum_{i,j=1}^n (f(t_i) - f(t_j), f(t_i) - f(t_j))a_i a_j \\ &= 2 \sum_{i,j=1}^n (f(t_i))a_i a_j - 2 \sum_{i,j=1}^n (f(t_i), f(t_j))a_i a_j \leq 0 \end{aligned}$$

since the first term vanishes $\left(\text{by } \sum_{i=1}^n a_i = 0\right)$ and the inner product of $\mathcal{H}$ is positive definite for the second term.

III. This is the key step. In view of the equivalence of (15) with the existence of the desired f, it is sufficient to exhibit such an f on $T_0 = \mathrm{sp}\{t_i, 1 \leq i \leq n\} \subset T$, an arbitrary finite (say, $d-$) dimensional subspace. Since $T_0(\cong \mathbb{R}^d)$ is a locally compact abelian group, let μ be the Haar measure on T_0 normalized to have unit mass for the unit cube. (Thus μ corresponds to the Lebesgue measure on $\mathbb{R}^d$.) Define $\mathcal{H}_1 = L^2_{\mathbb{C}}(T_0, \mu)$, where the real inner product is given by $\langle\langle g, h\rangle\rangle = \int_{T_0} Re(g(t)h(t)^*)d\mu(t)$ so that $\mathcal{H}_1$ is a real Hilbert space. Define $H_\alpha : T_0 - \{0\} \to \mathbb{C}$, by writing (α, t) for $\sum_{i=1}^{n} \alpha_i t_i$, and

$$h_\alpha(t) = \|t\|^{-(1/2)(d+1)}(1 - e^{i(\alpha,t)}), \quad t \in T_0 - \{0\}, \alpha \in T_0. \tag{17}$$

To see that $h_\alpha \in \mathcal{H}_1$, observe that

$$|h_\alpha(t)|^2 \leq \|\alpha\|^2 \|t\|^{1-d} \left(\frac{\sin \frac{1}{2}(\alpha, t)}{\frac{1}{2}(\alpha, t)}\right)^2, \tag{18}$$

where we used the fact that $|1 - e^{i(\alpha,t)}|^2 = 2(1 - \cos(\alpha, t))$. Now (18) is integrable (seen by a change of coordinates into spherical polars in $\mathbb{R}^d$). It is here we used crucially the Hilbertian nature of T_0. Thus $h_\alpha \in \mathcal{H}_1$. Let $V(\alpha) = \int_{T_0} |h_\alpha(t)|^2 d\mu(t)$. Then $V(0) = 0$ and for any $a \in \mathbb{R} - \{0\}, V(a\alpha) = |a|V(\alpha)$. Moreover, for any automorphism $U : T_0 \to T_0, V(U\alpha) = V(\alpha)$. Now from classical analysis, such a mapping V must be of the form $V(\alpha) = C_d\|\alpha\|$ for some fixed constant $C_d > 0$ where d is the dimension of T_0. But $|(h_\alpha - h_{\alpha'})(t)| = |h_{\alpha-\alpha'}(t)|$ from (17). Integrating,

$$\|h_\alpha - h_{\alpha'}\|^2_{\mathcal{H}_1} = \|h_{\alpha-\alpha'}\|^2_{\mathcal{H}_1} = V(\alpha - \alpha') = C_d\|\alpha - \alpha'\|_T. \tag{19}$$

Setting $f_\alpha = C_d^{-1/2} h_\alpha : T_0 \to \mathcal{H}_1$, one sees that $f : \alpha \mapsto f_\alpha$ satisfies all the requirements. In view of the earlier reduction, this gives the desired result. □

In the early part of this and last chapter, we considered the (Wiener)-Brownian motion, and the "white noise measure" $\{Z_A, A \in \mathcal{B}_0\}$ where $\mathcal{B}_0$ is the δ-ring of Borel sets of T, of finite μ-measure (the Haar measure of T discussed above), while T has dimension $d(< \infty)$. In terms of this measure we may define the stochastic integrals $f \mapsto \int_T f dZ_t$ for $f \in L^2(T, \mu)$ using the Dunford-Schwartz procedure as noted before. Moreover, $E(\int_T f(t)dZ_t)^2 = \int_T |f|^2(t)d\mu(t)$. Hence as a consequence of the above theorem, we may state the following:

10. Corollary. *If T is of finite dimension, say k, then the Lévy-Brownian motion $\{X_t, t \in T\}$ admits a representation, in terms of the white noise process $\{Z_A, A \in \mathcal{B}_0\}$ where $\mathcal{B}_0$ is the δ-ring of bounded Borel sets of T, as a Dunford-Schwartz integral:*

$$X_t = \int_T f_t(u) dZ_u, \quad t \in T, \tag{20}$$

where $f_t(u) = C_k^{-1/2} h_t(u)$, with $h_t(\cdot)$ and C_k given by (17) and (19).

Proof. If $\tilde{X}_t$ is the random variable given by the integral in (20), then it is clear that $\tilde{X} = \{\tilde{X}_t, t \in T\}$ is a Gaussian process with mean zero and

$$E(\tilde{X}_t \tilde{X}_{t'}^*) = \int_T f_t(u) f_{t'}^*(u) d\mu(u). \tag{21}$$

Taking real parts on both sides, we deduce that the covariance $C_{\tilde{X}}(t, t')$ is the same as that discussed above. Hence both the X and $\tilde{X}$ are Gaussian processes with the same means and covariance functions. By enlarging the probability spaces (or adjoining the two spaces as discussed in II.3.6(ii)), if necessary, we may assume that both processes are defined on the same space. Since a Gaussian process is uniquely determined by its mean and covariance functions (as noted in Chapter I), we conclude that $X_t = \tilde{X}_t, t \in T$, as asserted. □

11. Discussion. The above formula (20) is sometimes called a spectral representation of the process, and since X_t depends on the "whole history" of the Z-process, it is not a martingale if $k > 1$. Note that, if $k = 1$, we get from (20) the familiar formula for the Fourier transform of the classical Brownian motion where $X_t = Z_{[0,t)}$ or $= -Z_{[t,0)}$ according to whether $t \geq 0$ or ≤ 0. Thus

$$X_t = \frac{1}{\sqrt{2\pi}} \int_{\mathbb{R}} |u|^{-1}(1 - e^{iut}) dZ_u \tag{22}$$

is the classical representation. An excellent account of the theory of Brownian motion and the above extension (and related research through the middle 1960's) may be found in the monograph by P. Lévy [1]. The formula (20) seems to have been first used by Molchan [1] who also shows the intimate connections with, and applications of, Sobolev spaces in this study. Since the Lévy-Brownian motion is a locally homogeneous Gaussian random field [i.e., it has stationary increments since for any $\{t,t',s,s'\} \subset T$ we have $E((X_t - X_{t'})(X_s - X_{s'})) = \frac{1}{2}(\|t-s'\| + \|t'-s\| - \|t-s\| - \|t'-s'\|)$], one can also deduce (20) from the representation theory of Yaglom's [1]. Another aspect of multidimensional (Wiener-) Brownian motion is discussed in Section 7.3 below.

6.4 Stochastic differential equations

As noted in the first two sections above, if $X = \{X_t, t \in [a,b]\}$ is an $L^2(P)$-bounded semimartingale, or, more generally, an $L^{2,2}$-bounded process, with values in $\mathbb{R}^k$, relative to a σ-finite measure α on $\mathcal{B}_{[a,b]} \otimes \Sigma$, and $f : [a,b] \times \Omega \to \mathbb{R}^{m,n}$ (the space of m-by-n matrices), is a function in $L^2(\alpha)$, then for each $t \in [a,b]$ we have the well-defined stochastic integral.

$$Y_t - Y_a = \int_a^t f(s,\cdot)dX_s \tag{1}$$

which can be expressed using the differential symbolism as:

$$dY_t = f(t,\cdot)dX_t, \quad t \in [a,b]. \tag{2}$$

It is to be noted that (2) is *always* understood in its "integrated form" (1). The formal equation (2) is the stochastic differential equation, the subject of this section, where we specialize f and X to obtain more detailed information of the Y-process, called the solution of (2). Thus the (symbolic) equation

$$dY(t) = m(t,Y(t))dt + \sigma(t,Y(t))dX(t) \tag{3}$$

can be regarded as (2) with $f(t,\cdot) = (m(t,Y(t)), \sigma(t,Y(t)))$ and $\tilde{X}(t) = (X(t),t)^*$ ($*$-for transpose) so that f and $\tilde{X}$ take values in $\mathbb{R}^2$ and $\tilde{X}$ is $L^{2,2}$-bounded if X is. Thus (2) specializes to (3). However, the form (3) already represents a large class of (nonlinear) equations and one can present conditions on m and σ to guarantee existence and uniqueness of solutions Y if X is an (a subclass of) $L^{2,2}$-bounded process. We now consider the *linear* and *nonlinear* cases sepearately, treating the higher order equations at the same time. Thus in the above Y is a vector process containing the desired elements.

(a) *Linear differential systems.* We first discuss an elementary case to clarify the issues. Now the simplest higher order linear stochastic differential equation is one of the following form, describing movement of a "simple harmonic oscillator", symbolically expressed as:

$$\frac{dU}{dt} = -\beta U(t) + A(t) - \omega X(t) \tag{4}$$

where ω is the angular frequency, U is the velocity $(= \frac{dX}{dt}), X(\cdot)$ being the position, β is the friction coefficient, and $A(t)$ is the random disturbance (the "white noise"), all at time t. This is expressed on writing $dB(t) = A(t)dt$, as:

$$d\dot{X}(t) + \beta(t)\dot{X}(t)dt + \omega^2(t)X(t)dt = dB(t). \tag{5}$$

Here $\{B(t), t \geq 0\}$ is usually taken to be Brownian motion, so that $A(t) = \frac{dB(t)}{dt}$ is a (nonexistent) derivative. If we let

$$W(t) = \begin{bmatrix} B(t) \\ 0 \end{bmatrix}, \quad Z(t) = \begin{bmatrix} U(t) \\ X(t) \end{bmatrix}, \quad \alpha(t) = \begin{bmatrix} \beta(t) & \omega(t) \\ -1 & 0 \end{bmatrix},$$

where $U(t) = \dot{X}(t)$, then (5) becomes the vector equation:

$$dZ(t) + \alpha(t)Z(t)dt = dW(t). \tag{6}$$

This is a 2-dimensional "Langevin-type" equation. [The word "type" is omitted when $\beta(t) \equiv \beta$, $\omega(t) \equiv \omega$ are constants.]

To solve (6) consider the associated homogeneous (2-by-2 matrix) equation

$$dY(t) = Y(t)\alpha(t)dt, \quad t \geq 0, \quad \det(Y(t_0)) \neq 0. \tag{7}$$

Multiplying (6) by $Y(t)$ and rearranging we get

$$d(Y(t)Z(t)) = Y(t)dW(t). \tag{8}$$

It follows from the classical theory of ordinary differential equations ODE, (cf., e.g., Coddington and Levinson [1], p. 28) that (7) implies

$$\det(Y(t)) = \det(Y(t_0)) \exp\left(\int_{t_0}^{t} tr(\alpha(u)du)\right) \neq 0.$$

Hence (8) gives with the initial conditions $Z(t_0)$:

$$Z(t) = Y(t)^{-1} \int_{t_0}^{t} Y(u)dW(u) + Y(t)^{-1}(Y(t_0)Z(t_0)), \qquad (9)$$

which is a well-defined stochastic integral of the first type. Moreover, the process Z defined by (9) is the unique solution of (6). Indeed, if $\tilde{Z}$ is another solution, then $v(t) = (Z - \tilde{Z})(t)$ satisfies the homogeneous equation (7) with the initial value $v(t_0) = 0$. The classical ODE theory then implies that $\frac{dv}{dt} = \alpha(t)v(t)$ has only the trivial solution $v(t) \equiv 0$, so that $Z = \tilde{Z}$. Thus in (9) we need to find $Y(t)$ using (6). With the Picard method, this is given by (*-denoting again the transpose):

$$Y^*(t) = Y^*(t_0) + \int_{t_0}^{t} \alpha^*(t_1)Y^*(t_0)dt_1 + \cdots +$$
$$\int_{t_0}^{t} \alpha^*(t_1)dt_1 \int_{t_0}^{t_1} \alpha^*(t_2)dt_2 \cdots \int_{t_0}^{t_{n-1}} \alpha^*(t_n)Y^*(t_0)dt_n + R_n, \qquad (10)$$

where R_n is the nth term. By hypothesis, $B(\cdot), \omega(\cdot)$ and hence $\alpha(\cdot)$ are continuous on $[t_0, t]$ so that

$$\|\alpha^*(\tau)\| = [tr(\alpha^*(\tau)\alpha(\tau))]^{\frac{1}{2}} \le M < \infty, \quad t_0 \le \tau \le t,$$

and similarly $\|Y(\tau)\| \le N < \infty$. Thus

$$\|R_n(t)\| \le M^n N \frac{(t-a)^n}{n!} \to 0 \text{ as } n \to \infty,$$

for each $t \ge t_0$. So (10) defines Y as a function of α and $Y(t_0)$. If we set $a_1(t, t_0) = \int_{t_0}^{t} \alpha^*(\tau)d\tau, a_2(t, t_0) = \int_{t_0 \le t_2 \le t_1 \le t} \alpha^*(\tau_1)\alpha^*(\tau_2)d\tau_1 d\tau_2 \cdots,$ and $Y^*(t) = \sum_{n=1}^{\infty} a_n(t, t_0)Y^*(t_0)$, then (9) gives the unique solution Z of (6) with this Y. We can state this result as follows, where in no special properties of W are used:

1. Proposition. *Consider the linear stochastic differential equation (5) where the noise $B(\cdot)$ is an $L^{2,2}$-bounded process relative to some σ-finite measure $\mu : \mathcal{B}_{[t_0,t]} \otimes \Sigma \to \bar{\mathbb{R}}^+$. Then there is a unique 2-vector process $Z = \begin{bmatrix} \dot{X} \\ X \end{bmatrix}$ given by (9) with the initial condition $Z(t_0) = z_0 = \begin{bmatrix} x_1 \\ x_0 \end{bmatrix}$, or $X(t_0) = x_0, \dot{X}(t_0) = x_2$.*

The above result used only the existence of stochastic integrals in (9) with the techniques of the ODE. Since $W(t)$ is an element of $L^2(P)$, Z is an element of (in general) an infinite dimensional space $L^2(P;\mathbb{R}^2)$. This observation allows us to present an abstract version. It can be given as follows, and is due to Goldstein [3]. The mean square derivative is now termed the "strong" (or in the norm) sense. All the functions are considered in the same topology and the integrals will be taken as the appropriate (Bochner type) vector integrals. Thus we can replace $L^2(P)$ by a Banach space $\mathcal{X}$ and assert the following:

2. Proposition. *Let $J = [a,b], Z : J \to \mathcal{X}, \beta, \omega : J \to B(\mathcal{X})$ be mappings where $B(\mathcal{X})$ is the space of bounded linear operators on $\mathcal{X}$. Suppose Z is strongly continous at $t_0 \in J$, measurable and essentially bounded on J (relative to Lebesgue measure). If β, ω are (Bochner) integrable on J, then the equation:*

$$d\dot{X}(t) + (\beta(t)\dot{X}(t) + \omega(t)X(t))dt = dZ(t), \tag{11}$$

with the boundary (or initial) condition $X(t_0) = x_0, \dot{X}(t) = x_1$, has a unique solution on J so that $X : J \to \mathcal{X}$ verifies (11) and takes the given value at t_0.

Proof. Let $\mathcal{Y} = \mathcal{X} \times \mathcal{X}$ with norm $\|y\|' = \|[\begin{smallmatrix} x_1 \\ x_2 \end{smallmatrix}]\| = \|x_1\| + \|x_2\|, y \in \mathcal{Y}$, so that $\{\mathcal{Y}, \|\cdot\|'\}$ is a Banach space. Let $g(t) = \left[\begin{smallmatrix} 0 \\ f(t) \end{smallmatrix}\right] \in \mathcal{Y}, t \in J$, where $f(t) = -\beta(t)Z(t) - \omega(t)\int_a^t Z(s)ds$, and define:

$$\alpha(t) = \begin{bmatrix} 0 & -1 \\ \omega(t) & \beta(t) \end{bmatrix} \in B(\mathcal{Y}), \ t \in J,$$

so that for each $y = \left[\begin{smallmatrix} u \\ v \end{smallmatrix}\right] \in \mathcal{Y}$ we have:

$$\alpha(t)y = \begin{bmatrix} -v \\ \omega(t)u + \beta(t)v \end{bmatrix}.$$

If we set $U(t) = \begin{bmatrix} u(t) \\ v(t) \end{bmatrix}$, then (11) becomes

$$d\dot{U}(t) + \alpha(t)U(t)dt = g(t)dt = dG(t), \text{ say.} \tag{12}$$

This is of the form (6). Also denoting $\|\cdot\|$ in lieu of $\|\cdot\|'$ in $\mathcal{Y}$,

$$\begin{aligned} \|\alpha(t)\| &= \sup\{\|\alpha(t)y\| : \|y\| \le 1\} \\ &\le \sup\{\|v\| + \|\omega(t)\|\|u\| + \|\beta(t)\|\|v\| : \|u\| + \|v\| \le 1\} \\ &\le 1 + \|\omega(t)\| + \|\beta(t)\| < \infty, \ t \in J. \end{aligned}$$

Hence we can apply the same procedure as before to conclude that (12) has a unique solution with the given initial values. But (12) is equivalent to the system:

$$\begin{aligned}\dot{u}(t) &= v(t)\\ \dot{v}(t) + \omega(t)u(t) + \beta(t)v(t) &= f(t),\ t \in J.\end{aligned}$$

Define $x(t) = u(t) = \int_a^t Z(s)ds$. Then the above system becomes:

$$\begin{aligned}x(t_0) &= u(t_0) + \int_a^{t_0} Z(s)ds,\\ \dot{x}(t_0) &= \dot{u}(t_0) + Z(t_0) = v(t_0) + Z(t_0),\end{aligned}$$

and taking the initial value $\tilde{X}(t_0) = \left[x_0 - \int_a^{t_0} Z(s)ds\right]$, we get the X-process satisfying the requirements of the proposition. □

Restricting $\mathcal{X} = L^p(P), p > 1$, in the above, we can also obtain the sample continuity of the solution process:

3. Corollary. *Let $\mathcal{X} = L^p(P), 1 < p < \infty$ and $X = \{X(t), t \in J = [a,b]\}$ be a process satisfying the conditions of the preceding proposition. Then almost all sample functions of X are continuous.*

Proof. By the preceding result, the solution process is (strongly) differentiable in $\mathcal{X}$. Hence, in particular, for $t, t' \in J$, there is a $\delta(t) > 0$ such that $|t - t'| < \delta(t)$ implies

$$E\left(\left|\frac{X(t) - X(t')}{t - t'} - \dot{X}(t)\right|^p\right) \leq 1,$$

so that

$$E(|X(t) - X(t') - (t - t')\dot{X}(t)|^p) \leq |t - t'|^p. \tag{13}$$

Consequently, since $\varphi(\frac{t-t'}{2}) \leq \frac{1}{2}(\varphi(t) + \varphi(t'))$ for $\varphi(t) = |t|^p$,

$$\begin{aligned}E(|X(t) - X(t')|^p) &\leq 2^p\{E(|X(t) - X(t') - (t - t')\dot{X}(t)|^p) +\\ &\qquad |t - t'|^p E(|\dot{X}|^p)\}\\ &\leq 2^p(M_p^p + 1)|t - t'|^p,\end{aligned}$$

where $M_p = \sup\{\|\dot{X}(t)\|_p, t \in J\} < \infty$, since $\dot{X}(\cdot)$ is continuous in $\mathcal{X}$. But J is compact and $\{t' : |t - t'| < \delta(t)\}$ covers J, so that there is a finite subcover for $t = t_1, \ldots, t_n$. If we set $C = 2^p(M_p^p + 1)$ then one has

$$E(|X(t) - X(t')|^p) \leq C|t - t'|^p, 1 < p < \infty.$$

Hence by (Kolmogorov's) Theorem III.4.4, the function $t \mapsto X(t)$ is continuous with probability one. □

We specialize the last proposition further to gain some insight into the finer structure of the solution process and distinguish it from the classical ODE results. Thus consider the nth order equation:

$$P(D)Y_t = (D^n - a_1(t)D^{n-1} - \cdots - a_n(t))Y(t) = \dot{B}_t, \qquad (14)$$

where $P(x) = \sum_{i=0}^{n}(-1)^i a_i(t)x^{n-i}, (a_0 \equiv 1), D = \frac{d}{dt}$, and $\dot{B}_t$ is the (fictitous) white noise. Let $Z(t) = (Y(t), \ldots, Y^{(n-1)}(t))^*$ $dW(t) = (0, \ldots, 0, dB(t)), Z(0) = \mathbf{c} = (c_1, \ldots, c_n)^*, a_i(\cdot)$ being continuous, and consider the structure matrix

$$\alpha(t) = \begin{bmatrix} 0 & 1 & 0 & \cdots & 0 \\ 0 & 0 & 1 & \cdots & 0 \\ \vdots & & & & \\ a_n(t) & & a_{n-1}(t) & \cdots & a_1(t) \end{bmatrix}. \text{ (*-is transposition)}$$

In matrix notation, (14) can be expressed as

$$dZ(t) = \alpha(t)Z(t)dt + dW(t), \quad Z(0) = \mathbf{c}. \qquad (15)$$

Then (15) is of the form (12) and satisfies the same conditions. Hence it has a unique solution given by

$$Z(t) = Y^*(t)^{-1}\Big(\int_0^t Y^*(u)dW(u) + Y^*(0)Z(0)\Big), \qquad (16)$$

where $Y^*(\cdot)$ is an n-by-n matrix with $\det Y(0) \neq 0$ so that $Y(t)$ is nonsingular for all $t > 0$, and is given by an infinite series as in (10). Here again the $W(t)$-process is just an $L^{2,2}$-bounded vector process on $[0, a], a > 0$.

Suppose now that the coefficients $a_i(t) \equiv a_i$, constants, in (14) so that the structure matrix $\alpha(t) = A^*$ (say) is also a constant. Then Y^* of (16) becomes

$$Y^*(t) = \sum_{n=0}^{\infty} A^n \frac{(t - a_0)^n}{n!} = e^{A(t-a_0)}. \qquad (17)$$

Taking $a_0 = 0$ and $Y(0) =$ identity, for simplicity, we get (16) as:

$$Z(t) = \int_0^t e^{(t-u)A} dW(u) + e^{tA}\mathbf{c}, \tag{18}$$

where $Z(0) = \mathbf{c}$. The preceding two results reduce to:

4. Corollary. *Let the linear n^{th} order stochastic differential equation be defined by (14) with $W(\cdot)$ as an $L^{2,2}$-bounded process, relative to some σ-finite measure μ, and the coefficients $a_i(t)$ being continuous on $[0, a], a > 0$. Then the system has a unique solution given by (16) and moreover, almost all its sample paths are continuous. In particular, if $a_i(t) \equiv a_i$ are constants then the resulting solution is given by (18), for the initial value $Z(0) = \mathbf{c}$.*

An alternative method of (14) using Green's function representation is possible; it is not considered here (but see Exercise 7). We calculate the covariance function when $W(\cdot)$ is Brownian motion and use it for later analysis. For simplicity, let $Z(0) = \mathbf{c}$, a constant. In the general case, we can use these by replacing the expectation E with the conditional expectation $E^{\mathcal{B}}, \mathcal{B} = \sigma(Z(0))$. Now $\mathbf{m}(t) = E(Z_t) = e^{At}\mathbf{c}$, and if $\sigma(t) = \text{cov}(Z(t), Z(t))$, we find

$$\begin{aligned}\sigma(t) &= E((Z(t) - \mathbf{m}(t))(Z(t) - \mathbf{m}(t))^*) \\ &= \int_0^t e^{(t-u)A} e^{(t-u)A^*} du.\end{aligned}$$

Since $Z^i(t) = DZ^{(i-1)}(t)$, we get $\sigma_{ij}(t) = \text{cov}(Z^i(t), Z^i(t))$, the $(i,j)^{th}$ element of $\sigma(t)$, using the special form of (15), as:

$$Y^{(i)}(t) = \int_0^t (A\, e^{A(t-v)})_{in} dB(t) + (e^{At}\mathbf{c})_i, i = 1, \dots, n-1. \tag{19}$$

If $G^{(1)}(t) = (e^{At})_{in}, G^{(i)}(t) = (D^{(i-1}G)(t) = (e^{At})_{in}, i > 1$, then

$$\sigma_{ij}(t) = \int_0^t G^{(i-1)}(v) G^{(j-1)}(v) dv. \tag{20}$$

From this one concludes

5. Proposition. $\sigma_{ij}(t) = t^{(2n+1-i-j)}[(2n+1-i-j)(n-i)!(n-j!]^{-1} + o(t)$ *as* $t \to 0^+$. *Hence* $\det \sigma(t) \le ct^{n^2}$ *for some* $c > 0$

Proof. Since $\sigma_{ij}(\cdot)$ is evidently real analytic, consider

$$\sigma_{ij}(t) = \sum_{m=0}^{\infty} \sigma_{ij}^{(m)}(t)\frac{t^m}{m!}.$$

By (20) $D\sigma_{ij}(t) = \sigma'_{ij}(t) = G^{(i-1)}(t)G^{(j-1)}(t)$, so that on differentiating it we have:

$$D^{(m)}(\sigma'_{ij}(t)) = \sum_{k=0}^{m} \binom{m}{k} G^{((i-1)+k)}(0+)G^{((j-1)+m-k)}(0+)\frac{t^m}{m!}.$$

But $G^{(k)}(0+) = 0$ for $0 \le k \le m-2$, and the first nonzero term in the above is when $i-1+k = m-1$, and $j-1+m-k = n-1$ so that $k = n-i$ and $m = n+k-j = 2n-i-j$. Hence the first nonzero term is:

$$\begin{aligned} &= \frac{t^{2n-i-j+1}}{(2n-i-j+1)!}\binom{2n-i-j}{n-i}, \quad \text{since } G^{(n-1)}(0+) = 1, \\ &= \frac{t^{2n-i-j+1}}{(2n-i-j+1)(n-i)!(n-j)!}. \end{aligned}$$

The other terms are of smaller order of magnitude as $t \downarrow 0$.

Since $\sigma(\cdot)$ is nonnegative definite, as a covariance, its determinant is nonnegative and $\sigma_{ii}(t) > 0$. But $\det(\sigma(t))$ is equal to the volume of the parallelopiped in $\mathbb{R}^n$ with edges as the column vectors of $\sigma(t)$ and hence has largest volume when it is rectangular with vertices at distance $\sigma_{ii}(t)$ units on i^{th} axis, $1 \le i \le n$. Thus we have

$$0 \le \det(\sigma(t)) \le \sigma_{11}(t)\sigma_{22}(t)\ldots\sigma_{nn}(t). \tag{21}$$

However,

$$\prod_{i=1}^{n} \sigma_{ii}(t) \sim \frac{t^{2n-1)+(2n-2)+\cdots+1}}{\prod_{i=1}^{n}(2n-2i+1)((n-j)!)^2} \sim t^{n^2}. \tag{22}$$

Hence (21) and (22) imply, $\det(\sigma(t)) \le ct^{n^2}$ for some $c > 0$. □

These estimates are useful in studying the regularity properties of the sample paths of the solution process of (14). For this one needs to analyze other aspects (e.g. the Markovian character) of the process which we shall study later.

(b) *Nonlinear differential analysis.* We now generalize the preceding considerations to higher order systems that are nonlinear, in the form (3). The coefficients m and σ are now functions of t and (a vector) Y. We need to impose some conditions on m and σ to make the integrals meaningful. For simplicity again, we consider second order equations.

Let $X = \{X_t, t \in I\}$ be a process, $X_t \in L^p(P), I = [a, b)$, an interval, and Y_t be an $L^p(P)$-mean derivative of X_t at $t \in [a, b)$, (one sidedly at a and b). It is possible to select a version such that the derived process $Y = \{Y_t, t \in I\}$ is also separable and measurable (relative to $\mathcal{B}_I \otimes \Sigma$), following the general results of Section III.3; so $Y_t \in L^p(P)$ also. We consider such X which solves the equation (in what follows we *always* can and will take separable and measurable versions without comment):

$$d\dot{X}(t) = q(t, X(t), \dot{X}(t))dt + \sigma(t, X(t), \dot{X}(t))dZ(t), \tag{23}$$

where Z is an $L^{2,2}$-bounded process and q, σ satisfy certain conditions to be formulated below. This equation is similar to (3) but its dependence on the derived process Y is made explicit. Note that (23) is understood to be:

$$X(t) = A + Bt + \int_a^t \left[\int_a^s q(u, X(u), \dot{X}(u)du + \int_a^s \sigma(u, X(u), \dot{X}(u)dZ(u) \right] ds \tag{24}$$

where A, B are the initial values so that $X(a) = A$ and $\dot{X}(a) = B, a \leq s \leq t \leq b$. Here we impose restrictions on q, σ so that the integrals in (24) exist. The desired conditions (seemingly unmotivated, but shown later to be optional) are incorporated in the following [main] result.

6. Theorem. *Let $Z = \{Z(t), t \in I = [a, b) \subset \mathbb{R}\}$ be an $L^{2,2}$-bounded (N-vector) process relative to a σ-finite measure $\mu : \mathcal{B}_I \otimes \Sigma \to \bar{\mathbb{R}}^+$. Suppose mappings H and σ are given satisfying:*

(i) $(\Delta_s^t Hf) = (Hf)(t) - (Hf)(s), s \leq t$, *and* $f : \mathbb{R} \to \mathcal{L}^0(P)$ *is measurable for* $\mathcal{F}_s^t = \sigma(f(u), s \leq u \leq t)$, *i.e.,* H *is definable by the (observation of the) process* f *on* $[s,t]$ *and not on the s-past or t-future;*

(ii) *for each absolutely continuous process* $g, (Hg)$ *is right continuous without fixed points of discontinuity;*

(iii) *for all* $\alpha \leq t < \beta$, *processes* $f, g : \mathbb{R} \to \mathcal{L}^0(P)$, *one has the growth conditions to be satisfied:* ($\mathcal{L}^0(P)$ *is* $\mathcal{L}^p(P)$, *with* $p = 0$)

(a) $E(|(hf)(t) - (Hg)(t)|^2) \leq K_J(E(f(a) - g(a))^2 + \int_{[\alpha,t]\times\Omega} |f' - g'|^2(s,\omega)d\mu),$

(b) $E((Hf)^2(t)) \leq K_J(1 + E(|f(a)|^2) + \int_{[\alpha,t)\times\Omega} |f'|^2(s,\omega)d\mu)$
for each compact $J = [\alpha, \alpha'] \subset I, f', g'$ *being pointwise derivatives of* f, g *when they exist (thus* Hf *satisfies a local Lipschitz condition, and for simplicity the same measure* μ *as in* $L^{2,2}$ *boundedness is used, since otherwise their sum will suffice).*

(c) $\lim_{t \downarrow a} E(|Hf|^2(t)) = 0;$

(d) $\mu \ll B \otimes P$ *with a bounded density, where* β *is a Borel measure on* I,

(iv) $\sigma(= (\sigma_{ij}))$ *is an* N*-by-*N *matrix such that*

(a) $\sigma_{ij}(t, X, Y) - \sigma_{ij}(t, \tilde{X}, \tilde{Y})|^2 \leq \tilde{K}(\|X - \tilde{X}\|^2 + \|Y - \tilde{Y}\|^2)$

(b) $|\sigma_{ij}(t, X, Y)| \leq \tilde{K}(1 + \|X\|^2 + \|Y\|^2)$, *where* $\|\cdot\|$ *is the Euclidean norm of* $\mathbb{R}^N$;

(v) *there are random vectors* $A, B : \Omega \to \mathbb{R}^N$, *which are square integrable and adapted to the remote past of the* Z*-process,* i.e., *measurable relative to* $\bigcap_{t\in I} \sigma(Z(s, s \leq t))$.

Let $\mathcal{F}_t = \sigma(Z_s, s \leq t)$, *and be completed for* P. *Then there exists a unique process* $X = \{X_t, \mathcal{F}_t, t \in I\}$ *satisfying* (23) *or* (24), i.e.,

$$X(t) = A + \int_a^t (B - HX)(s)ds + \int_a^t \int_a^s \sigma(u, X(u), \dot{X}(u))dZ(u)ds, \quad (25)$$

with $X(a) = A, \dot{X}(a) = B$ *and* $(HX)(t)$ *replaces* $\int_a^t q(s, X(s), \dot{X}(s)ds$. *If, moreover,* Z, *has independent increments (i.e., each component* Z^i *of* Z *is independent of* $Z^j (i \neq j)$ *and* Z^i *has such increments) then writing* $V_{s,X} = \begin{bmatrix} X(s) \\ \dot{X}(s) \end{bmatrix}, V_{a,X} = \begin{bmatrix} A \\ B \end{bmatrix}$, *we have*

$$E(\sup_{s\leq t\leq u} \|V_{s,X}(t) - V_{s,Y}(t)\|^2 \leq C\|X - Y\|^2,$$

$C > 0$ *being a constant, where* $V_{s,X}(t)$ *is* $V_{t,X}$ *with the starting point* $s(\leq t)$ *and so* $V_{s,X}(s) = V_{s,X}$.

Remark. The (complicated looking) conditions on H are satisfed if:

$$(Hf)(t) = \int_a^t h(s, f(s))ds, \quad \text{or} \quad (Hf)(t) = F(f(t)\chi_{[a,c]}(t)) \tag{26}$$

for some $c > a$, where h and F are suitable Borel functions. These forms appear in many applications and, in particular, $\frac{d}{dt}(Hf)(t) = h(t; f)$ will be considered later. It will be found from the ensuing work that the conditions of this theorem, for H of (26) and Z as Brownian motion, are nearly the best.

Proof. The argument consists of showing the existence of a unique vector process $(X, \dot{X})^*$ satisfying (25) [so that X is absolutely continuous], and that the second component is the (mean-square) derivative of the the first. We sketch the details which are somewhat involved.

I. Since Z is $L^{2,2}$-bounded relative to a σ-finite μ and conditions (iii) and (iv) hold, the integrals in (25) exist for each compact interval $[a, t] \times [a, s], a \leq s \leq t < \infty$. Further the X-process is absolutely continuous. The hypothesis also implies that the integrand of the outer integral in (25) is Bochner integrable on $[a, t]$ and using a form of the Lebesgue differentiation theorem for such integrals, we can conclude that $\dot{X}(t)$ exists and equals a.e. with the integrand of $\int_a^t [\quad]ds$ evaluated at t. (See, e.g., Hille-Phillips [1], p. 87, Cor. 1 of Theorem 3.8.5.) A direct verification of the statement is also possible, and it was given by Borchers [1] in the case that Z is a square integrable martingale and his result extends to the $L^{2,2}$-bounded case. In fact for predictable integrands, as in this case, one can conclude by Theorem 2.14 that an $L^{2,2}$-bounded Z is just a semimartingale, and Borcher's method extends to this case. [Express Z as an $L^2(P)$-martingale and a process of b.v. on I etc.]

II. The argument now proceeds, by using the Picard method of successive approximations starting with a process $\{X_t^0, t \geq a\}$ satisfying (the right side of) (25), calling the result $X_t^{(1)}$ and iterating. Hence it suffices to show that (i) $X_t^{(n)} \to \bar{X}_t$, a.e. for each t, (ii) $\dot{X}_t^{(n)} \to \tilde{X}_t$ a.e., (iii) $\dot{\bar{X}}_t = \tilde{X}_t$ a.e., and finally (iv) that $\bar{X}_t$ and $\dot{\bar{X}}_t$ satisfy (25). The uniqueness is then established with relative ease.

Thus let $X_t^0 = A + Bt$ (so $\dot{X}_t^0 = B$), and define inductively

$$X_t^{(n+1)} = A + \int_a^t \left[B - (HX^{(n)})_s + \int_a^s \sigma(u, X_u^{(n)}, \dot{X}_u^{(n)}) dZ(u) \right] ds. \quad (27)$$

By the preceding paragraph, for each $n \geq 1, X^{(n)}$ is well-defined and $\dot{X}_t^{(n)}$ exists so that

$$\dot{X}_t^{(n+1)} = B - (HX^{(n)})_t + \int_a^t \sigma(u, X_u^{(n)}, \dot{X}_u^{(n)}) dZ(u). \quad (28)$$

To establish the desired convergence of $X^{(n)}$ and $\dot{X}^{(n)}$, let

$$\Delta^{(n+1)} X(t) = X_t^{(n+1)} - X_t^{(n)}, \Delta^{(n+1)} \dot{X}_t = \dot{X}_t^{(n+1)} - \dot{X}_t^{(n)},$$

and similarly $\Delta^{(n+1)}(HX)(t)$ (or $\Delta^{(n+1)}H(t)$ for short), and $\Delta^{(n+1)}\sigma(t)$. Applying the Δ operation to (28), using the Jensen inequality for convex functions to get $\left(\sum_{i=1}^{m} x_i\right)^2 \leq m \sum_{i=1}^{n} x_i^2$, and taking the ith component of these vector equations, we obtain:

$$E(|\Delta^{n+1} \dot{X}_i(t)|^2) \leq (N + 1\{E(|\Delta^{(n)} H_i(t)|^2) + \sum_{j=1}^{N} E\left(\left| \int_a^t \Delta^{(n)} \sigma_{ij}(s) dZ_j(s) \right|^2 \right) \}. \quad (29)$$

Using conditions (iii) and (iv) of the hypothesis to the right side terms of (29) and adding both sides over i, we get the key relation:

$$E(\|\Delta^{(n+1)} \dot{X}(t)\|^2) \leq \alpha_0 \int_{[a,t]\times\Omega} (\|\Delta^{(n)} \dot{X}(s)\|^2) d\mu(s,\omega), \quad (30)$$

where α_0 is a constant depending on $K_J, \tilde{K}, N$ and $b(a \leq t \leq b)$. But $\mu \ll \beta \otimes P$ with a bounded density $f \leq \alpha_1 < \infty$ on $[a, b] \times \Omega$. Substituting this in (30) and writing $\alpha_2 = \alpha_0 \alpha_1$ there, we obtain by Fubini's Theorem (since β is at most σ-finite):

$$\begin{aligned} E(\|\Delta^{(n+1)} \dot{X}(t)\|^2) &\leq \alpha_2 \int_a^t E(\|\Delta^{(n)} \dot{X}(s)\|^2) d\beta(s) \\ &\leq \alpha_2 \int_{\beta(a)}^{\beta(t)} E(\|\Delta^{(n)} \dot{X}(\beta^{-1}(s))\|^2 ds, \end{aligned} \quad (31)$$

where we used β both for the measure and the point function on $\mathbb{R}$ determined by it. Also β^{-1} is then its (generalized) inverse. The last formula of (31) is a standard one betweeen Lebesgue and Lebesgue-Stieltjes integrals, (cf., e.g., Riesz-Nagy [1], p. 124). By iteration of (31), for $n = 0, 1, \ldots$,, we get the useful bound:

$$E(\|\Delta^{(n+1)}\dot{X}(t)\|^2) \leq \alpha_2 \frac{(\beta(b) - \beta(a))^n}{n!}, \quad a \leq t \leq b. \tag{32}$$

With this it will be shown that $\{\dot{X}_t^{(n)}, n \geq 1\}$ forms a Cauchy sequence in $L^2(P)$ and hence has a limit whose indefinite integral is shown to be the $L^2(P)$-limit of $X_t^{(n)}$.

III. Since Z is a semimartingale and $\sigma(\cdot,\cdot,\cdot)$ is Lipschitz by (iv), it follows that $\dot{X}^{(n)}$ is also a semimartingale in $L^2(P)$. But by condition (iii) of the hypothesis and the fact that $X^{(n)}(a) = A$ for all $n \geq 0$, we get, for some absolute constant $\tilde{K}_J > 0$,

$$\begin{aligned} P[\sup_{a\leq t\leq b} |\Delta^{(n)} H_i(t)| \geq 2^{-n}] &\leq P\Big[\tilde{K}_J \| \int_a^b \|\Delta^{(n)}\dot{X}(s)\|^2 d\beta(s) \geq 4^{-n}\Big] \\ &\leq 4^n \tilde{K}_J \int_a^b E(\|\Delta^{(n)}\dot{X}(s)\|^2) d\beta(s), \\ &\qquad \text{by Čebyšev's inequality,} \\ &\leq \tilde{K}_J \alpha_2 4^n \frac{(\beta(b) - \beta(a))^n}{n!}, \quad \text{by (32).} \end{aligned} \tag{33}$$

Since the right-side is the general term of a convergent series, by the (first) Borel-Cantelli lemma, $\sup_{a\leq t\leq b} |\Delta^n H_i(t)| \leq 2^{-n}$, a.e. for large n. Regarding the second term of (29), we first note that the square integrable semimartingale $Z = M + V$, where M is a (local) martingale and V is a process of (locally) bounded variation. Here, since we are considering only compact intervals and Z is in $L^2(P)$, we can assume M to be a square integrable martingale and V to have an integrable variation, or that $|V|$ to be increasing and a.e. bounded on compact sets and is dominated by μ. Thus

$$\begin{aligned}
E(\left|\int_a^b \Delta^{(n)}\sigma_{ij}(s)dV(s)\right|^2 &\leq \int_{[a,b]\times\Omega} |\Delta^{(n)}\sigma_{ij}(s)|^2 d\mu(s,w) \\
&\leq K_J \int_a^b E(|\Delta^{(n)}\sigma_{ij}(s)|^2)\beta(s) \\
&\leq \bar{K} \int_a^b \int_a^t E(\|\Delta^{(n)}\dot{X}(s)\|^2) ds\ d\beta(t) \\
&+ \int_a^b E(\|\Delta^{(n)}\dot{X}(s)\|^2) d\beta(s), \\
&\text{since } X^{(n)}(t) = A + \int_a^t \dot{X}(s)ds, \\
&\leq K' \int_a^b E(\|\Delta^{(n)}\dot{X}(s)\|^2) d\beta(s), \qquad (34)
\end{aligned}$$

where K' is a constant depending only on $(b-a), K_J$ and not on n. Consequently, we get an analog of (33) here as:

$$\begin{aligned}
P[\sup_{a\leq t\leq b}\left|\int_a^t \sigma_{ij}(s)dV(s)\right| \geq 2^{-n}] &\leq 4^n E\left(\left|\int_a^b \sigma_{ij}(s)dV\right)^2 \\
&\leq 4^n K' \int_a^b E(\|\Delta^{(n)}\dot{X}(s)\|^2) d\beta(s), \\
&\text{by (34)}, \\
&\leq 4^n K' \alpha_2 \frac{(\beta(b)-\beta(a))^n}{n!}, \quad \text{by (32)}. \qquad (35)
\end{aligned}$$

On the other hand $\{\int_a^t \sigma_{ij}(s)dM(s), \mathcal{F}_t, t \geq a\}$ is a square integrable martingale. Hence by its maximal inequality (cf. Theorem III.5.1 and its proof), we get (as in (35)):

$$\begin{aligned}
P[\sup_{a\leq t\leq b}\left|\int_a^t \sigma_{ij}(s)dM(s)\right| \geq 2^{-n}] &\leq 4^n E\left(\left|\int_a^b \sigma_{ij}(s)dM(s)\right|^2\right) \\
&\leq 4^n K'' \int_a^b E(\|\Delta^{(n)}\dot{X}(s)\|^2) d\beta(s) \\
&\leq 4^n K'' \alpha_2 \frac{(\beta(b)-\beta(a))^n}{n!}, \qquad (36)
\end{aligned}$$

where K'' is a constant independent of n. Hence we deduce from (35) and (36) that $\sup_{a\leq t\leq b}\left|\int_a^t \sigma_{ij}(s)dZ(s)\right| < 2^{-n}$ a.e. for large n. It follows from this simplification and (28) that

$$\sup_{a\le t\le b} |\Delta^{(n+1)}\dot{X}(t)| \le (N+1)2^{-n}, a.e., \tag{37}$$

for large n. Hence for large m and n, $m < n$ we have a.e.

$$\sup_{a\le t\le b} |\dot{X}_i^{(m)} - X_i^{(n)}|(t) \le \sum_{k=m}^{n-1} \sup_{a\le t\le b} |\Delta^{k+1}\dot{X}_i(t)| \le (N+1) \sum_{k\ge m} 2^{-k}. \tag{38}$$

It follows from this that there is a P-null set Ω_0, depending on J, such that $\dot{X}_i^{(n)}(t) \to \bar{X}(t)$ uniformly in $t \in J$, with probability one, and also in $L^2(P), 1 \le i \le N$. Hence, returning to the vector form,

$$\int_a^t \dot{X}^{(n)}(s)ds \to \int_a^t \bar{X}(s)ds$$

uniformly for $t \in J$. Define the X-process now as:

$$X(t) = A + \int_a^t \bar{X}(s)ds.$$

Since

$$X^{(n)}(t) = A + \int_a^t \dot{X}^{(n)}ds \to A + \int_a^t \bar{X}(s)ds = \bar{X}(t),$$

the convergence being uniform in $t \in J$, with probability 1, the X-process has absolutely continuous paths, and its derived process can be identified with the $\bar{X}$-process and $\bar{X}$ has no fixed points of discontinuity, using the hypothesis. To see that $\bar{X}$ is also the mean square derivative, we can verify that $(n \ge m)$

$$\begin{aligned} E(\|\dot{X}^{(m)} - \dot{X}^{(n)}\|^2(t)) &= E\left(\left\|\sum_{k=m}^{n-1} 2^{-k}\Delta^{k+1}\dot{X}(t)\right\|^2\right) \\ &\le \sum_{k=m}^{n-1} 4^{-k} \sum_{k=m}^{n-1} E(\|\Delta^{(k+1)}\dot{X}(t)\|^2) \\ &\le 4^{-m} \sum_{k=m}^{n-1} \frac{(4k_0)^k}{k!} \le 4^{-n}e^{4k_0} \to 0. \end{aligned} \tag{39}$$

as $m, n \to \infty$. It follows from this that $\dot{X}^{(n)}(t) \to \bar{X}(t)$ in mean, uniformly in $t \in J$. We thus conclude that $\bar{X} = \dot{X}$ a.e. for the X-process defined above. We assert that this $(X, \dot{X})$-solves (25).

IV. By part (i) of the hypothesis and the computations for (33), one gets for the $(X, \dot{X})$,

$$\begin{aligned}\sup_{a\leq t\leq b} \|(HX^{(n)}) - (HX)\|^2(t) &\leq \tilde{K} \int_a^b \|\dot{X}^{(n)} - \dot{X}\|^2(s)d\beta(s) \\ &\leq (\beta(b) - \beta(a))\tilde{K} \sup_{a\leq t\leq b} \|\dot{X}^{(n)} - \dot{X}\|(t), \\ &\to 0 \text{ as } n \to \infty.\end{aligned}$$

Hence, uniformly in $t \in J$, we have

$$(HX^n)(t) \to (HX)(t), \int_a^t (HX^{(n)}(s)ds \to \int_a^t (HX)(s)ds,$$

with probability one. The earlier computations therefore show also that $\int_a^b E(\|\dot{X}(t)\|^2)d\beta(s) < \infty$, and the processes have separable (and measurable) modifications. Moreover, one has

$$\begin{aligned}P[\sup_{a\leq t\leq b} \left| \int_a^t (\sigma_{ij}(u, X(u), \dot{X}(u)) - \sigma_{ij}(u, X^{(n)}(u), \dot{X}^{(n)}(u)))dZ_j(u) \right| \\ \geq \frac{1}{n}] \\ \leq n^2 \int_a^b E((\sigma_{ij}(u, X(u), \dot{X}(u)) - \sigma_{ij}(u, X^{(n)}(u)\dot{X}^{(n)}(u)))^2)d\beta(u) \\ \leq n^2\tilde{K} \int_a^b E(\|X(s) - X^{(n)}(s)\|^2 + \|\dot{X}(s) - \dot{X}^{(n)}(s)\|^2)d\beta(s) \\ \leq n^2K^2(b-a) \int_a^b \int_0^s E(\|\dot{X}(s) - \dot{X}^{(n)}(s)\|^2)dud\beta(s) + \\ n^2K'^2 \int_a^b E(\|\dot{X}(u) - \dot{X}^{(n)}(s)\|^2)d\beta(s) \\ \leq n^2K'' \int_a^b E(\|\dot{X}(s) - \dot{X}^{(n)}(s)\|^2)d\beta(s) \\ \leq n^2K'''(\beta(b) - \beta(a))4^{-n}e^{4K_0},\end{aligned} \tag{40}$$

by (39) and the K's are constants depending only on a, b and β. It follows by the Borel-Cantelli lemma again, that

$$\sup_{a\leq t\leq b} \left| \int_a^b (\sigma_{ij}(u, X(u), \dot{X}(u)) - \sigma_{ij}(u, X^{(n)}(u), \dot{X}^{(n)}(u)))dZ_j(u) \right|^2 < \frac{1}{n}$$

for large n, with probability 1. Hence letting $n \to \infty$, we deduce that $(X, \dot{X})$ has all the asserted properties and solves (25).

V. Finally for the uniqueness of the X-process, suppose $\hat{X}$ is another process satisfying the given conditions and solving (25). Then letting $(\Delta H)(t)(HX - H\hat{X})(t), (\Delta\dot{X})(t) = \dot{X}(t) - \dot{\hat{X}}(t)$, and similarly $\Delta\sigma$, we get

$$\Delta\dot{X}(t) = -\Delta H(t) + \int_a^b (\Delta\sigma)(s)dZ(s). \tag{41}$$

Squaring both sides (in component form) and rearranging, we get:

$$E(\|\Delta\dot{X}(t)\|^2) \leq L \int_a^t E(\|\Delta\dot{X}(s)\|^2)d\beta(s), t \in J, \tag{42}$$

where $L \geq 0$ is an absolute constant depending only on J and β. But this is a classical Gronwall inequality which in this case states that for any integrable $\varphi : J = [a,b] \to \mathbb{R}$ satisfying $\varphi(t) \leq \alpha_0 + \beta_0 \int_a^t \varphi(s)ds, t \in J$, then $\varphi(t) \leq \alpha_0 e^{\beta_0 t}, t \in J$, so that with $\varphi(t) = E(\|\Delta\dot{X}(t)\|^2), \alpha_0 = 0, \beta_0 = L$, we get $\Delta\dot{X}(t) = 0$ a.e. for all $t \in J$. Hence $P[X(t) = \hat{X}(t), t \in J] = 1$, for any compact internal J, and this gives uniqueness. □

The following remarks are noteworthy in relation to the above result.

7. Discussion. 1. For first order equations, we can simplify the proof somewhat by considering X and never looking at the derived process, whose existence is not demanded. But in the higher order case the additional work and the proof that the second component is a derived process of the first cannot be omitted and thus the theorem is not a consequence of the standard results in the literature, (cf., e.g., Ikada-S. Watanable [1]).

2. The hypothesis that μ is dominated by the product measure is automatic if the Z-process is an $L^2(P)$-martingale with independent increments. In particular, this holds if Z is a Brownian motion. We used this domination condition for the computations of (31) - (42). However, the maximal inequalities involved in these calculations hold by employing a more detailed analysis with stopping times since such a result holds for semimartingales, as noted by Orey [2], which we outlined as Exercise V.6.7. For the existence theorem then one needs

to use a slightly different procedure. For this alternative approach with semimartingales and first order equations, see Métivier and Pellaumail [2].

We specialize the operator H of the above theorem and state the following form for reference purposes.

8. Corollary. *Let $Z = \{Z_t, t \in J = [a,b]\}$ be an $L^{2,2}$-bounded n-vector process relative to a σ-finite measure $\mu \ll \beta \otimes P$, and $q : J \times \mathbb{R}^n \times \mathbb{R}^n \to \mathbb{R}^n$ be a Borel function such that*

(i) $\|q(t,X,Y) - q(t,\tilde{X},\tilde{Y}\|^2 \leq K(\|X-\tilde{X}\|^2 + \|Y-\tilde{Y}\|^2)$,

(ii) $\|q(t,X,Y)\|^2 \leq K(1 + \|X\|^2 + \|Y\|^2), \quad X, Y \in \mathbb{R}^n$.

and $\sigma : I \times \mathbb{R}^n \times \mathbb{R}^n \to \mathbb{R}^{n^2}$ be as in theorem, where $K > 0$ is a constant. Then there is a unique absolutely continuous process $X = \{X_t, t \in J\}$ with a derived process $\dot{X} = \{\dot{X}_t, t \in J\}$ such that

$$\begin{aligned} d\dot{X}(t) &= q(t, X(t), \dot{X}(t))dt + \sigma(t, X(t), \dot{X}(t))dZ(t) \\ \dot{X}(a) &= A, \; X(a) = B \end{aligned} \tag{43}$$

for any given initial $\mathcal{F}_a(= \sigma(X_a))$-measureable and square integrable random variables, A and B.

This is the form of the stochastic differential equation which we consider later on. For more detailed information about the solution process when the Z-process is specialized (e.g. Brownian motion), it will be necessary to refer to some basic concepts and properties of Markovian elements. We now recall these and then proceed with the structural analysis of the solutions of (43), in the rest of this section.

(c) Markovian elements and differential equations. We first recall the general concept as:

9. Definition. A process $X = \{X_t, t \in T \subset \mathbb{R}\}$, with values in $\mathbb{R}^n$, on (Ω, Σ, P) is a *Markov process* if for any finite set of points $t_i \in T, t_1 < t_2 < \cdots < t_{n+1}$, and for any $A \in \sigma(X_{t_{n+1}})$ the following (set of) equations hold:

$$E(\chi_A | \sigma(X_{t_1}, \ldots, X_{t_n})) = E(\chi_A | \sigma(X_{t_n})), \text{ a.e.,} \tag{44}$$

or equivalently for any Borel set $B \subset \mathbb{R}^n$, if $A = [X_{t_{n+1}} \in B]$, then

$$P[X_{t_{n+1}} \in B | X_{t_1}, \ldots, X_{t_n}] = P[X_{t_{n+1}} \in B | X_{t_n}] \;\; a.e. \tag{45}$$

A symmetrical form of the above definition, of interest in applications, is given by the following:

10. Proposition. *Let $X = \{X_t, t \in J \subset \mathbb{R}\}$ be an $\mathbb{R}^n$-valued process on (Ω, Σ, P). Then it is Markovian iff for any integers $k, m \geq 1, t, s_i, t_j \in J$ such that $s_1 < s_2 < \cdots < s_k < t < t_1 < \cdots < t_m$, and Borel sets $B_i, C_j \subset \mathbb{R}^n$, one has*

$$P[X_{s_i} \in B_i, X_{t_j} \in C_j, 1 \leq i \leq k, 1 \leq j \leq m | X_t] = \\ P[X_{s_i} \in B_i, 1 \leq i \leq k | X_t] P[X_{t_j} \in C_j, 1 \leq j \leq m | X_t] \text{ a.e.} \quad (46)$$

Remark. Note that if $A = [X_{s_i} \in B_i, 1 \leq i \leq k], \tilde{A} = [X_{t_j} \in C_j, 1 \leq j \leq m], \mathcal{F} = \sigma(X_t)$, then (46) says that $P^{\mathcal{F}}(A \cap \tilde{A}) = P^{\mathcal{F}}(A) P^{\mathcal{F}}(\tilde{A})$ a.e., which, in words, means that, *when the conditional probability is regular*, the past (i.e. $A \in \sigma(X_{t_i}, 1 \leq i \leq k)$) and future (i.e., $\tilde{A} \in \sigma(X_{t_j}, 1 \leq j \leq m)$) are conditionally independent given the present, namely $\sigma(X_t)$. We also observe that if the range or *state space* $\mathbb{R}^n$ of the process X is replaced by a countable set, then the process is termed a (Markov) *chain*, and the regularity of $P^{\mathcal{F}}$ is automatic. It is of interest to record that (46) implies that the *Markovian character of X is still valid, if the ordering of J is reversed.*

Proof. Suppose the process X is Markovian in the sense of Definition 9 and let $A, \tilde{A}$ be as in the above remark. Using the (elementary) properties of conditional expectations (cf. Sec. II.1), we have

$$\begin{aligned} E^{\mathcal{F}}(\chi_A) E^{\mathcal{F}}(\chi_{\tilde{A}}) &= E^{\mathcal{F}}(\chi_A P(\tilde{A}|\mathcal{F})), \text{ by the averaging identity,} \\ &= E^{\mathcal{F}}(\chi_A P(\tilde{A}|\mathcal{F}, \sigma(X_{s_1}, \ldots, X_{s_k})))), \text{ by (45),} \\ &= E^{\mathcal{F}}(E(\chi_A \chi_{\tilde{A}} | \sigma(X_t, X_{s_1}, \ldots, X_{s_k}), \\ &\qquad \text{since } A \in \sigma(X_t, X_{s_1}, \ldots, X_{s_k}), \\ &= E(\chi_A \chi_{\tilde{A}} | \mathcal{F}), \text{ a.e.} \end{aligned}$$

This establishes (46).

For the converse, let (46) hold so that $E^{\mathcal{F}}(\chi_A \chi_{\tilde{A}}) = E^{\mathcal{F}}(\chi_A E^{\mathcal{F}}(\chi_A))$, a.e. If $C \in \mathcal{F}$ and A is as above, then $C \cap A$ is a generator of $\mathcal{F}_0 = \sigma(\mathcal{F}, X_{s_1}, \ldots, X_{s_k}) = \sigma(X_t, X_{s_1}, \ldots, X_{s_k}) \supset \mathcal{F}$. Then (46) gives

$$\begin{aligned} \int_{C \cap A} E^{\mathcal{F}_0}(\chi_{\tilde{A}}) dP &= \int_C \chi_{A \cap \tilde{A}} dP = \int_C E^{\mathcal{F}}(\chi_{A \cap \tilde{A}}) dP \\ &= \int_C E^{\mathcal{F}}(\chi_A E^{\mathcal{F}}(\chi_{\tilde{A}})) dP = \int_{C \cap A} E^{\mathcal{F}}(\chi_{\tilde{A}}) dP. \end{aligned}$$

Since $C \cap A$ is a generator of $\mathcal{F}_0$ and the extreme integrands are both $\mathcal{F}_0$-measurable, they can be identified a.e. But then this gives (44) so that X is Markovian. □

An important property of Markov processes is continued in:

11. Proposition. *Let $X = \{X_t, t \in J \subset \mathbb{R}\}$ be a Markov process and $u < t < v$ be points in J. Then the following relation, called the* **Chapman-Kolmogorov equation,** *holds for any Borel set $B \subset \mathbb{R}^n$,*

$$P[X_v \in B|X_u] = E(P[X_v \in B|X_t]|X_u), \text{ a.e.} \tag{47}$$

Moreover, if the conditional probabilities are regular, so that $p(\xi, u : B, v) = P[X_v \in B|X_u](\xi)$ where $\xi = X_u(\omega)$, is σ-additive in B and Borel measurable in ξ, then (47) can be expressed as:

$$p(\xi, u : B, v) = \int_{\mathbb{R}^n} p(\zeta, y; B, v) p(\xi, u; d\zeta, t). \tag{48}$$

for all $\xi \in \mathbb{R}^n$.

Proof. The relation (47) follows from (45) by writing t for u and applying $E^{\mathcal{F}}$ on both sides ($\mathcal{F} = \sigma(X_u)$), and using that $E^{\sigma(X_u)}E^{\sigma(X_s, s \leq t)} = E^{\sigma(X_u)}$ for $u < t < b$. Then (48) is a consequence of (47) by employing the integral representation for conditional measures (cf. Proposition II.3.4). The integrals in (48) will now be in Lebesgue's sense because of the regularity hypothesis, and the result holds, in general, if the Dunford-Schwartz integration is used. □

Remark. One should be warned that the above proposition only states that every Markov process (or chain) satisfies (47) (or (48)), but the converse is not implied. Indeed, there exist nonMarkov processes for which (47) holds. A simple instructive example to this effect can be found in Feller [2].

We observe that the existence of Markov processes, in quite general form, follows from Theorem III.6.1 and the reader is referred to that result for a detailed discussion. Using the regularity as a fundamental assumption, many refined assertions can be considered. We indicate a few, showing the far reaching nature of the concept, and use them in our work below.

12. Definition. A mapping $p : \mathbb{R}^n \times J \times \mathcal{R} \times J \to \mathbb{R}^+$ is a (*Markov*) *transition probability function* if for $u \leq v$ one has:

(i) $p(.,u;B,v)$ is Borel (or $\mathcal{R}$)-measurable, (ii) $p(\xi,u;.,v)$ is a probability, and (iii) p satisfies the Chapman-Kolmogorov equation (48) for $u < s < v$, together with the initial condition $p(\xi,u;B,u) = 1$ for $\xi \in B$ and $= 0$ for $\xi \notin B$ or $u > v$. Moreover p is termed *stationary* if $J \subseteq \mathbb{R}^+$ is a semigroup and $p(\xi,u;B,v) = \tilde{p}(\xi,v-u;B)$ for all $u \leq v$ in J. A Markov process whose finite dimensional distributions determine these transition functions, is itself stationary if the transition probabilities are stationary and each X_t has the same distribution for $t \in J$. [Thus the definition of transitions incorporates the regularity of conditioning by fiat!]

Using this concept, one can transform (48) with an operator identity on the space $B(\mathbb{R}^n)$ of bounded Borel functions, and this will relate to an important part of abstract analysis which enhances the structure theory of these processes. Here then is the desired connection:

13. Proposition. *Let $\{p(\cdot,u;\cdot v), u, \in J, u < v\}$ be a family of transition probability functions and $B(\mathbb{R}^n)$ be the Banach space of real bounded Borel functions under the uniform norm. Then for each $s < t$, the operators $U(s,t) : B(\mathbb{R}^n) \to B(\mathbb{R}^n)$ given by*

$$(U(s,t)g)(x) = \int_{\mathbb{R}^n} g(y)p(x,s;dy,t), \ \ g \in B(\mathbb{R}^n), \tag{49}$$

form a family of positive identity preserving contractive linear **evolutions** *so that for $s < r < t$ one has: $U(s,t)g \geq 0$ for $g \geq 0$, and that*

$$U(s,t)1 = 1, (1 = \chi_{\mathbb{R}^n}), U(s,r)U(r,t) = U(s,t), \ \ and \ \ \|U(s,t)g\| \leq \|g\|.$$

Conversely, every such class of evolution operators uniquely determines a family of transition probability functions (hence a Markov process).

Proof. Since $p(x,s;\mathbb{R}^n,t) = 1$ and $p \geq 0$, it is clear that $U(s,t)$ of (49) is a positive linear contraction with $U(s,t)1 = 1$. Also by (48)

$$(U(s,t)g)(x) = \int_{\mathbb{R}^n} g(u)\Big[\int_{\mathbb{R}^n} p(y,r;du,t)p(x,s;dy,r)\Big], g \in B(\mathbb{R}^n),$$

$$= \int_{\mathbb{R}^n} [\int_{\mathbb{R}^n} g(u)p(y,r;du,t)]p(x,s;dy,r),$$
by Fubini's theorem,
$$= \int_{\mathbb{R}^n} [U(r,t)g](y) \cdot p(x,s;dy,r), \text{ by (49)},$$
$$= U(s,r)[U(r,t)g](x), \quad x \in \mathbb{R}^n. \tag{50}$$

This sows that $U(s,t) = U(s,r)U(r,t)$ since $g \in B(\mathbb{R}^n), x \in \mathbb{R}^n$, are arbitrary.

For the converse, let $U(s,t) : B(\mathbb{R}^n) \to B(\mathbb{R}^n)$ with the given properties. Then by the classical Riesz representation theorem there is a unique positive measure $p(x,s;\cdot,t)$ which is a Borel probability on $\mathbb{R}^n$. Here we use the fact that $\|U(s,t)g_n\| \leq \|g_n\| \to 0$ for $g_n \downarrow 0$ uniformly if g_m is chosen (by Uryshon's lemma) to be compactly based on $A_m \subset \mathbb{R}^n, A_m \downarrow \emptyset$. But the evolution identity implies, by reversing the computation in (50), that p satisfies (48). □

We observe that, for the stationary transitions (since $p(\xi, s : B, s) = 1$ if $\xi \in B$ and $= 0$ if $\xi \notin B$), if $\tilde{U}(t-s) = U(s,t)$ then $\tilde{U}(0) =$ identity, $\tilde{U}(q)\tilde{U}(r) = \tilde{U}(q+r)$ for all $q, r \in J$ with $q + r \in J$. Thus $\{\tilde{U}(r), r \in J\}$ forms a positive contractive semigroup of linear operators on $B(\mathbb{R}^n)$. It is now natural to consider subclasses of Markov processes corresponding to various subclasses of semigroups of the above. Thus a Markov process $X = \{X_t, t \in J\}, J(\subset \mathbb{R})$ a semigroup, is said to be a *Feller process* if its transition probabilities are stationary and the corresponding semigroup of operators on $B(\mathbb{R}^n)$ maps continuous functions into themselves. The latter semigroup is also termed a *Feller semigroup*, since Feller [3] analyzed this class in great detail. We shall see in the next section that the concept of evolution mappings, motivated by the above proposition, plays a key role in the study of stochastic flows.

Let us describe two more properties of Markov processes, possesed by solutions of a large class of stochastic differential equations. The first result is on its relation with martingales.

14. Proposition. *Let $X = \{X_t, \mathcal{F}_t, t \geq 0\}$ be a Markov process with stationary transition probabilities and $X_t = X_{t+0}$, a.e., $t \geq 0$. Then, for each continuous f in $B(\mathbb{R})$ if $Y_t^f = f \circ X_t$, the martingale $\{E^{\mathcal{F}_s}(Y_t^f), 0 \leq s \leq t\}$ is representable as $Y_s^f = (V_{t-s}f)(X_t)$, a.e.,* *where* $\{V_t, t \geq 0\}$ *is the semigroup of operators associated with* X*. Moreover, if* $(V_s f)(x) \to f(x)$ *uniformly as* $s \to 0^+$*, and* $\{V_t, t \geq 0\}$ *is Feller, then* $Y^f_{s+0} = Y^f_s$ *a.e.,* $s \geq 0$*, so that* $\{X_t, \mathcal{F}_{t+}, t \geq 0\}$ *is also Markovian where* $\mathcal{F}_{t+} = \bigcap_{u>t} \mathcal{F}_u$*.*

Proof. Since $\mathcal{F}_t \uparrow$ as $t \uparrow$, the Markovian property implies, for each $u > t$ and $f \in B(\mathbb{R})$, that

$$E^{\mathcal{F}}(f(X_u)) = E^{\sigma(X_t)}(f(X_u)), \text{ a.e.}$$

Fix such an f in $B(\mathbb{R})$, and consider the martingale $\{Y^f_s, \mathcal{F}_s, 0 \leq s \leq t\}, Y^f_t = f(X_t)$. Then

$$\begin{aligned} Y^f_s &= E^{\sigma(Y_s)}(Y^f_t) = (V_{t,s} f)(X_s) \\ &= (V_{t-s} f)(X_s), \text{a.e.}, \end{aligned}$$

since X has stationary transitions. This is the first assertion.

For the second, with the additional hypothesis for $0 \leq s, s' < t$,

$$\begin{aligned} |Y^f_s - Y^f_{s'}|(\omega) \leq\, & |(V_{t-s} f)(X_s) - (V_{t-s} f)(X_{s'})|(\omega) + \\ & |(V_{t-s} f)(X_s) - (V_{t-s'} f)(X_s)|(\omega). \end{aligned}$$

The last term tends to 0, uniformly as $s \to s'$, since V_t is (strongly) continuous at 0. If f is continuous, then $V_{t-s} f$ has the same property since $\{V_t\}_t$ is a Feller semigroup. By the right continuity of the X process, we then get $Y^f_s = Y^f_{s+0}$ a.e. Hence $\{Y^f_s, \mathcal{F}_s, 0 \leq s < t\}$ is a right continuous martingale, by Theorem III.5.2. The rest also follows from these observations. □

A stronger property of the processes given by Definition 12 is motivated by the preceding result and is stated as:

15. Definition. Let $X = \{X_t, \mathcal{F}_t, t \geq 0\}$ be a measurable process relative to $\mathcal{B}(\mathbb{R}^+) \otimes \Sigma$. The process X is *strongly Markovian* if for each finite optimal T of the filtration $\{\mathcal{F}_t, t \geq 0\}$ and $f \in B(\mathbb{R})$, one has

$$E^{\mathcal{B}(T)}(f(X_{t+T})) = E^{\sigma(X_T)}(f(X_{t+T})), \text{ a.e.}, t > 0, \qquad (51)$$

where $\mathcal{B}(T)$ is the σ-algebra of events prior to T(so X_T is $\mathcal{B}(T)$-adapted).

If T is constant ($= s$ say) so that $\mathcal{B}(T) = \mathcal{F}_s$, (51) implies that a strong Markov process is Markovian with stationary transitions. On the other hand, replacing T by $T+s$ in (51) and writing $\mathcal{G}_s = \mathcal{B}(T+s), Y_s = X_{T+s}$, then (51) becomes, for $s \geq 0, t > 0$,

$$E^{\mathcal{G}_s}(f(Y_{s+t})) = E^{\sigma(Y_s)}(f(Y_{t+s})) = (V_t f)(Y_s), \quad \text{a.e.} \tag{52}$$

Hence $\{Y_s, \mathcal{G}_s, s \geq 0\}$ is a Markov process and (52) states that a strong Markov process X stopped at T, starts afresh as a Markov process with initial positions at $Y_0 = X_T$ and has the same transition semigroup $\{V_t, t \geq 0\}$ as that of X. This key property is not present for all Markov processes (if the index set is uncountable), but it is available for Brownian motion, Poisson process and more generally, as we show, for solutions of most of our stochastic differential systems. This class has been isolated by Dynkin and Yushkevich, and Hunt independently. The filtrations of these processes may be taken to be right continuous, or standard. All right continuous Feller processes are stongly Markovian (cf. Dynkin [1], Thm. 5.10). Also one can show that if X is such a process and if $\{V_t, t \geq 0\}$ is its associated semigroup, then for any (finite) stopping times T_1, T of the filtration $T_1 \geq T$, we have $\omega \mapsto (V_{T_1(\omega)-T(\omega)} f)(X_{T(\omega)}(\omega))$ to be $\mathcal{B}(T)$-measurable (cf. Meyer [5]). Thus these processes also have nice measurability properties. Many other specialized studies can be found in the works of Dynkin [2], Blumenthal-Getoor [1], Dellacherie-Meyer [1], and Meyer [5].

We present an application to the solutions of linear and nonlinear (higher order) differential systems to round out our study of these classes begun early in this section.

16. Proposition. *Suppose $Z = \{Z_t, \mathcal{F}_t, t \in J = [a,b]\}$ is an $L^{2,2}$-bounded process relative to $\mu \leq k(\beta \otimes P)$ and with, moreover, independent increments. If either (i) $\alpha : J \to \mathbb{R}^{n^2}$ is a matrix such that $\{U(t), \mathcal{F}_t, t \in J\}$ solves the (linear) equation, ($U_t = (u, \dot{u})_t$)*

$$dU(t) = \alpha(t)U(t)dt + dZ(t), \quad U(0) = \mathbf{c}, \tag{53}$$

or (ii) the operator H and the mapping $\sigma : I \times \mathbb{R}^n \times \mathbb{R}^n \to \mathbb{R}^{n^2}$ satisfy the conditions of Theorem 6 so that $\{U(t), t \geq 0\}$ solves (25), with $U(a) = A, U(b) = B$, then the vector process $\{U(t), \mathcal{F}_t, t \in J\}$ is a Markov process, and if Z has continuous sample paths, so does $U(t)$.

Proof. (i) The linear case. As shown in Corollary 4, the solution vector $U(t)$ is given by (cf. (16))

$$U(t) = Y^*(t)^{-1}\Big(\int_a^t Y^*(s)dZ(s) + Y^*(a)U(a)\Big), \tag{54}$$

where Y is an n-by-n matrix with $\det(Y(a)) \neq 0$. The present hypothesis implies that $U(a)$ is independent of $(Z(s) - Z(a)), s > a$. Also the stochastic integral in (54) can be approximated by the elements of the form (by $L^{2,2}$-boundedness) for $a = t_0 < t_1 < \cdots < t_{n+1} = t$,

$$U_n(t) = \sum_{i=1}^{n} Y^*(\tau_i)(Z(t_{i+1}) - Z(t_i)) \tag{55}$$

where $\tau_i \in [t_i, t_{i+1})$. Since the Z-process has independent increments this sum of independent random vectors forms a Markov process as is easily verified. Since the right side of (54) can be approximated in $L^2(P)$-mean by such processes, it defines a Markov process, establishing the first assertion.

(ii) The nonlinear case. For $-\infty < a < t < b < \infty$ we have by Theorem 6

$$\begin{aligned} u(t) - u(a) &= \int_a^t \dot{u}(s)ds, \\ \dot{u}(t) - \dot{u}(a) &= \Delta_a^t(Hu) + \int_a^t \sigma(s, u(s), \dot{u}(s))dZ(s), \end{aligned} \tag{56}$$

where $\Delta_n^t(Hu) = (Hu)(t) - (Hu)(a)$. Then the random vector $U(t) = (u(t), \dot{u}(t)), t \geq a$, depends only on the Z-increments between a and t, and $\Delta_a^t(Hu)$. The hypothesis on H implies that $\Delta_s^t(Hu)$ is $\mathcal{B}_s^t$-measurable where $\mathcal{B}_s^t = \sigma(U_r, s \leq r \leq t)$. It follows that $u(s), \dot{u}(s)$ are $\mathcal{B}_a^t$-adapted and $U(a) = (A, B)^*$, is independent of $(Z(t) - (a))$ for $t > a$. Also $Z_t - Z_s$ is independent of $\mathcal{B}_a^s$ which is determined by $U(a)$ and $Z(s)$. So if Γ is a Borel subset of $\mathbb{R}^{2n}$, then for $a \leq \alpha < t$,

$$P[U_t \in \Gamma | U_s, s \leq \alpha] = P[U_t \in \Gamma | \mathcal{B}_a^\alpha] = P[U_t \in \Gamma | U_\alpha], \text{ a.e.}$$

Hence the U_t-process is Markovian.

Finally, the continuity of the Z-process implies that of U, because of their forms (54) and (56), as is easily seen. □

In the case that the Z-process is Brownian motion, then, in addition to the above properties, one can present further interesting analysis of the solution process. We again treat the linear and nonlinear cases separately where in the former the coefficients are *assumed to be constants* so that the U_t-process has stationary transitions. There is a corresponding condition for the nonlinear case at least when H is specialized.

Using the hypothesis that Z is Brownian motion, we have an explicit expression for $U(t)$, (cf. (18)) in this case:

$$U(t) = e^{-(t-a)A}Y^{-1}(a)\int_a^t e^{-(s-a)A}dZ(s) + e^{-(t-a)A}U(a),$$

where the matrix $A(=\alpha(t))$ is independent of t. Then

$$U(t+h) = e^{-hA}[U(t) + \int_t^{t+h} e^{-(s-a)A}dZ(s)],$$

since A is a constant matrix. Let $\hat{U}(h) = U(t+h), \hat{Z}(h) = Z(t+h)$ so that $\{\hat{Z}(h), h \geq 0\}$ is again a Brownian motion, and then

$$\hat{U}(h) = e^{-hA}[\hat{U}(0) + \int_0^h e^{-(s-a)A}d\hat{Z}(s)],$$

which is of the same form as the original equation. Hence by the preceding proposition $\{\hat{U}(h), h \geq 0\}$ is also a Markov process from the point t onwards. Since $\hat{U}$ depends only on 0 and h, we get its distribution, for each Borel set Γ, as:

$$p(\xi, t; \Gamma, t+h) = p(\xi, 0; \Gamma, h), \quad U(t) = \xi.$$

Hence the $U(t)$ process has stationary transitions, and we denote this as $p(\xi, t; \Gamma, t+h) = \tilde{p}(\xi, h; \Gamma)$. Consider

$$(V_h f)(\xi) = \int_{\mathbb{R}^n} f(y)\tilde{p}(\xi, h; dy), f \in B(\mathbb{R}^n), h \geq 0, \tag{57}$$

and deduce that $\{V_h, h \geq 0\}$ is a strongly continuous positive contractive semigroup of linear operators on $B(\mathbb{R}^n)$. Similarly the adjoint semigroup V_h^* given by

$$(V_h^* f)(\Gamma) = \int_{\mathbb{R}^n} f(x)\tilde{p}(x, h; \Gamma)dx, f \in B(\mathbb{R}^n), h \geq 0, \tag{58}$$

defines $V_h^* : B(\mathbb{R}^n) \to (B(\mathbb{R}^n))^*$, the space of regular additive set functions on the σ-algebra of Borel sets of $\mathbb{R}^n$. Regarding these operators one has the following properties:

17. Theorem. *If $g_f(t,x) = (V_t f)(x)$, then $g_f : (0,\infty) \times \mathbb{R}^n \to \mathbb{R}$ is a real analytic function, and in particular $V_t(B(\mathbb{R}^n)) \subset C_b(\mathbb{R}^n)$, the Banach space of real bounded continuous functions. Thus $\{U(t), t \geq 0\}$ is a continuous Feller process, and hence is strongly Markovian.*

Proof. In this work we use the form of $\tilde{p}$ decisively, namely,

$$\frac{\partial \tilde{p}}{\partial v}(\xi, t; dv) = f(\xi, v)$$
$$= (2\pi)^{-\frac{n}{2}}(\det(\sigma(t)))^{-\frac{1}{2}} \exp[-\frac{1}{2}(v - e^{At}\mathbf{c}; \sigma^{-1}(t)(v - e^{At}\mathbf{c}))]$$

where $\sigma(t) = E(U(t) - E(U(t)))(U_t - E(U_t))^*)$, and $\xi = \mathbf{c}, E(U(t)) = e^{At}\mathbf{c}$.

We first show that $g(t,\cdot)$ and $g(\cdot, a)$ are infinitely differentiable and that they have series expansions which allow analytic continuations to the complex plane from which the result follows. Here the details are adapted from Dym [1].

To simplify the notation, let $\mathbf{a} = (a_1, \dots, a_n), \mathbf{b} = (b_1, \dots, b_n), \mathbf{k} = (k_1, \dots, k_n), \mathbf{m} = (m_1, \dots, m_n), |\mathbf{k}| = k_1, + \cdots + k_n, k_i, m_i \geq 0$ integers, and $\mathbf{a}^{\mathbf{k}} = \prod_{i=1}^{n} a_i^{k_i}$. Since $\tilde{p}$ is Gaussian with density f, the latter can be expressed as:

$$f(\mathbf{a}, t; \mathbf{b}) = \sum A_{i,\mathbf{k},\mathbf{m}}(t - t_0)^i \mathbf{a}^{\mathbf{k}} \mathbf{b}^{\mathbf{m}}, t_0 > 0. \tag{59}$$

This series converges absolutely and uniformly on each compact neighborhood of $(t_0, 0, 0) \in (0,\infty) \times \mathbb{R}^n \times \mathbb{R}^n$. Consider for $N > 0$,

$$g_N(t, \mathbf{a}) = \int_{\{\mathbf{b}:\|\mathbf{b}\| \leq N\}} h(\mathbf{b}) f(\mathbf{a}, t; \mathbf{b}) d\mathbf{b}, \quad h \in B(\mathbb{R}^n),$$
$$= \sum A_{i,\mathbf{k},\mathbf{m}}(t - t_0)^i \mathbf{a}^{\mathbf{k}} \int_{\{\mathbf{b}:\|\mathbf{b}\|\mathbf{b} \leq N\}} h(\mathbf{b}) \mathbf{b}^{\mathbf{m}} d\mathbf{b}.$$

This converges on compact neighborhoods of $(t_0, 0) \in (0,\infty) \times \mathbb{R}^n$. So g_N is real analytic for each N on $(0,\infty) \times \mathbb{R}^n$. Consider

$$|g_N(t, c_1 + ic_2)| \leq \int_{\mathbb{R}^n} |h(\mathbf{v})| \, |f(c_1 + ic_2, t; \mathbf{v})| d\mathbf{v}.$$

Expanding the exponent of f we get (after a simplication):

$$LHS \leq \exp[\frac{1}{2}\langle \tilde{A}c_2, R^{-1}\tilde{A}c_2 \rangle] \int_{\mathbb{R}^n} |h(\mathbf{v})| f(c_1, t, \mathbf{v}) d\mathbf{v},$$

where $\tilde{A} = e^{At}$ and R is the covariance matrix of U_t. Thus $g_N(t, \cdot)$ is uniformly bounded on a complex neighborhood of $\mathbf{c}$ in $\mathbb{C}^n$ and is analytic there, hence is analytic on $\mathbb{C}^n$.

Next consider $g_N(r + is, \mathbf{a})$. It is similarly verified that g_N is bounded and analytic on a complex neighborhood of $\mathbb{C}$ for each $\mathbf{a}$. Since $\det(R) \neq 0$, and is bounded by t^{n^2} (cf., Proposition 5) it follows that $R^{-1}(t)$ is analytic for $t > 0$. Then one can expand the exponent as:

$$Re\langle \eta + i\xi, R^{-1}(r + is, \eta + i\xi)\rangle \geq \langle \eta, R^{-1}(r)\eta\rangle - \langle \xi, R^{-1}(r)\xi\rangle \\ - \sqrt{M}(\|\eta\|^2 + \|\xi\|^2)^{1/2},$$

for a constant $M > 0$, satisfying

$$|\langle \eta + i\xi, R^{-1}(r + is) - R^{-1}(r)(\eta + i\xi)\rangle|^2 \leq M(\|\eta\|^2 + \|\xi\|^2).$$

This implies on taking $\eta = b - e^{Ar}(\cos A\zeta)a, \xi = -e^{Ar}(\sin A\zeta)a$, that $(\det R(\cdot))^{\frac{1}{2}} g_N(\cdot, a)$ is uniformly bounded and is analytic on a complex neighborhood of $t > 0$ so that its pointwise limit as $N \to \infty$, namely $(\det R(\cdot))^{\frac{1}{2}} g(\cdot, a)$ has the same property. Then by the classical Hartog theorem of (several) complex variables (cf. Cartan [1], p. 134), we deduce that g is analytic jointly in both variables as desired. This also implies that $\{U_t, t \geq 0\}$ is a (right continous) Feller process and then it is strongly Markovian by an earlier noted result. $\square$

Using a similar computation one can show, for the adjoint semigroup V_t^*, that $(V_t^* h)(A) = \mu^h(t, A)$ is a real analytic measure kernel, i.e., if $A = (-\infty, \mathbf{x})$ and $g_h(t, \mathbf{x}) = \mu^h(t, t - 0, \mathbf{x}))$, then $g_h(t, \mathbf{x})$ has the stated property for each $h \in B(\mathbb{R}^n)$. Several additional results on the sample paths of the solution process $\{U_t, t \geq 0\}$ can be obtained by considering the generators of $\{V_t, t \geq 0\}$ and $\{V_t^*, t \geq 0\}$. We include a typical consequence to indicate the possibilities.

Let us calculate the generator G of the semigroup $\{V_t, t \geq 0\}$ which is defined for $f \in C_b^{(2)}(\mathbb{R}^n) \subset C_b(\mathbb{R}^n)$, since $V_t(C_b(\mathbb{R}^n))$ is a subspace of $C_b(\mathbb{R}^n)$, the space of twice continuously differentiable functions. The first line below is the definition of G:

$$(Gg)(x) = \lim_{t \to 0} \frac{V_t g - g}{t}(x), \quad \text{limit in the norm of } C_b(\mathbb{R}^n),$$

$$= \lim_{t\to 0} \frac{1}{t}\{ \int_{\mathbb{R}^n} g(y) f(x,t;y)dy - g(x)\}, \text{ by (57)},$$

$$= \lim_{t\to 0} \frac{1}{t}\{\sum_{i=1}^{n} \frac{\partial g}{\partial x_i}(x) \int_{\mathbb{R}^n} (y_i - x_i) f(x,t;y)dy_i +$$

$$\frac{1}{2} \sum_{i,j=1}^{n} \frac{\partial^2 g}{\partial x_i \partial x_j} \int_{\mathbb{R}^n} \int_{\mathbb{R}^n} (y_i - x_i)(y_j - x_j) f(x,t;y) dy_i dy_j$$

$$+ o(|y-x|^2)\}$$

$$= \sum_{i=1}^{m} \frac{\partial g}{\partial x_i}(x) \lim_{t\to 0} \frac{(e^{At}x - x)_i}{t} +$$

$$\frac{1}{2} \sum_{i,j=\partial}^{n} \frac{\partial^2 g}{\partial x_i \partial x_j}(x) \lim_{t\to 0} \frac{R_{i,j}(t)}{t},$$

$$= \sum_{i=1}^{n} \frac{\partial g}{\partial x_i}(x)(Ax)_i - \frac{1}{2} \sum_{i,j=1}^{n} \frac{\partial^2 g}{\partial x_i \partial x_j}(x)\delta_{ijn}, \text{ by Prop. 5},$$

where $(\ )_i$ denotes the ith element of the vector $(\ )$ and $\delta_{ijn} = 1$ if $i = j = n$; $= 0$, if not. But $(Ax)_i = x_{i+1}, i = 1, \ldots, n-1$, and $(Ax)_n = a_n x_1 + a_{n-1} x_2 + \cdots + a_1 x_n$, in our case. Hence one has for $g \in C_b^{(2)}(\mathbb{R}^n)$,

$$Gg = [\frac{1}{2} \frac{\partial^2}{\partial x_n^2} + \sum_{i=1}^{n-1} x_{i+1} \frac{\partial}{\partial x_i} + (x_1 a_n + \cdots + x_n a_1)]g. \qquad (60)$$

Thus the generator G is a degenerate elliptic operator and its domain $\mathcal{D}_G$ contains $C_b^{(2)}(\mathbb{R}^n)$. One can similarly calculate the generator $\tilde{G}$ of the adjoint semigroup $\{V_t^*, t \geq 0\}$ to be:

$$\tilde{G}h = [\frac{1}{2} \frac{\partial^2}{\partial x_n^2} - \sum_{i=1}^{n-1} x_{i+1} \frac{\partial}{\partial x_i} - (x_1 a_n + \cdots + x_n a_1) \frac{\partial}{\partial x_n} - a_1]h, \quad (61)$$

for $h \in C_b^{(2)}(\mathbb{R}^n) \subset \mathcal{D}_{\tilde{G}}$.

With the above form of the generators, we have:

18. Proposition. *If $\{U_t, t \geq 0\}$ is the solution process of eqution (14) with constant coefficients and Brownian disturbance $\{B_t, \mathcal{B}_t, t \geq 0\}$, then the process $\{Y_t^g, \mathcal{B}_t, t \geq 0\}, g \in \mathcal{D}_G$, defined by*

$$Y_t^g = g(U_t) - g(U_a) - \int_0^t (Gg)(B_s)ds, \qquad (62)$$

satisfies $E^{\mathcal{B}_s}(Y_t^g) = Y_s^g$ *a.e.* $0 < s < t$, *so that it is a martingale. In particular if* $Gg = 0$, *then* $\{g(U_t), \mathcal{B}_t, t \geq 0\}$ *is itself a martingale.*

Proof. In the present case, we have seen that

$$E^{\mathcal{B}_t}(g \circ U(t+h)) = (V_h f)(U(t)) \tag{63}$$

and for each $t_0 > 0$, and $g \in \mathcal{D}_G$,

$$\lim_{t \downarrow 0} \frac{1}{t}\left[V_t(V_{t_0}g) - V_{t_0}g\right] = \lim_{t \downarrow 0} V_{t_0}(V_t g - g) = V_{t_0}(Gg),$$

so that $V_t g \in \mathcal{D}_G$. It also follows that $G(V_{t_0}g) = V_{t_0}(Gg)$, and then $\frac{d}{dt}(V_t g) = GV_t g = V_t Gg$. Since $t \mapsto \int_0^t V_s(Gg)ds$ is (strongly) differentiable with derivative $V_t(Gg)$, we have

$$\begin{aligned}\int_0^t G(V_s g)ds &= \int_0^t V_s(Gg)ds \\ &= \int_0^t \frac{d}{ds}(V_s g)ds = V_t g - g, \text{ a.e. (Leb.)}\end{aligned} \tag{64}$$

Since $g \circ U_t$ is bounded, it is in $L^1(P)$. So consider,

$$\begin{aligned}Y_t^g &= g \circ U_t - g \circ U_s + g \circ U_s - g \circ U_0 + \int_0^s G(g \circ U_r)dr + \\ &\int_s^t G(g \circ U_r)dr \\ &= Y_s^g + (g \circ U_t - g \circ U_s) + \int_s^t G(g \circ U_r)dr.\end{aligned}$$

Since Y_t^g is $\mathcal{B}_t$-adapted, we have from the above,

$$E^{\mathcal{B}_s}(Y_t^g) = Y_s^f + E^{\mathcal{B}_s}([\quad]) \tag{65}$$

and this establishes the martingale property when the last term is shown to vanish. So consider,

$$\begin{aligned}E^{\mathcal{B}_s}([\quad])(x) &= \int_{\mathbb{R}^n} (g(y) - g(x))p(\alpha, s; t, dy) - \\ &\int_{\mathbb{R}^n} \int_s^t (Gg)(v)(x, s; u, dv)du \\ &= \int_{\mathbb{R}^n} (g(y) - g(x))p(x, 0; t-s, dy) - \\ &\int_{\mathbb{R}^n} \int_s^t (Gg)(v)p(x, a; u-s, dv)du \\ &= E^{\mathcal{B}_0}[g \circ U_{t-s} - g \circ U_0) - \int_0^{t-s} (Gg)(U_t)dv](x) \\ &= (V_{t-s}g - g)(x) - \int_0^{t-s} G(V_v g)(x)dv \\ &= 0, \text{ by (64)}, \quad x \in \mathbb{R}^n. \quad \square\end{aligned}$$

We next turn to the analogous problem in the *nonlinear case* (of higher order). For this also, the existence, uniqueness, and the Markovian character of the solutions have already been established. As is to be expected, the analysis now is more complicated than the linear problem. First we consider the infinitesimal behavior of the solution process, and then refine the computations when Z is taken to be Brownian motion.

19. Theorem. *Let the U_t-process be the solution of (24) and that the coefficients q, σ satisfy conditions (ii) - (v) of Theorem 6, but (i) is replaced by the following:*

$$(i')\begin{cases} \|q(t,X,Y) - q(t,\tilde{X},\tilde{Y})\|^2 \le K(\|X - \tilde{X}\|^2 + \|Y - \tilde{Y}\|^2) \\ \|q(t,X,Y)\|^2 \le K(1 + \|X\|^2 + \|Y\|^2), \; X, Y, \tilde{X}, \tilde{Y} \in \mathbb{R}^n, \end{cases}$$

for t in compact intervals and $K (= K_t > 0)$, a constant. Suppose also that the $L^{2,2}$ bounded Z process, relative to $\mu \le \beta \otimes P$, has independent increments and without fixed points of discontinuities.

Then we have the following estimates for the U_t-process of (25) *as $r \to 0$:*

$$1.(a) \quad E(\sup_{t \le s \le t+r} \|u(s) - u(t)\|^2) = o(r^2)(1 + E(\|u(t)\|^2 + \|\dot{u}(t)\|^2)),$$

(b) $E(\sup_{t\leq s\leq t+r} \|\dot{u}(s)-\dot{u}(t)\|^2) = o(r)(1+E(\|u(t)\|^2+\|\dot{u}(t)\|^2))$,

(c) $E^{\sigma(A,B)}(\sup_{t\leq s\leq t+r} \|u(s)-A\|^2) = o(r)(1+\|\begin{pmatrix}A\\B\end{pmatrix}\|^2)$, *a.e.*,

(d) $E^{\sigma(A,B)}(\sup_{t\leq s\leq t+r} \|\dot{u}(s)-B\|^2) = o(r)(1+\|\begin{pmatrix}A\\B\end{pmatrix}\|^2)$, *a.e.*,

2.(a) $E^{\sigma(A,B)}(u(t+r)-A) = rB + o(r^{\frac{3}{2}})(1+\|\begin{pmatrix}A\\B\end{pmatrix}\|^2)^{\frac{1}{2}}$, *a.e.*,

(b) $E^{\sigma(A,B)}(\dot{u}(t+r)-B) = \int_t^{t+r} q(s,A,B)ds +$

$o(r^{\frac{3}{2}})(1+\|\begin{pmatrix}A\\B\end{pmatrix}\|^2)^{\frac{1}{2}}$, *a.e.*,

(c) $E^{\sigma(A,B)}(\|u(t+r)-A\|^2) = o(r^2)(1+\|\begin{pmatrix}A\\B\end{pmatrix}\|^2)$, *a.e.*,

(d) $E^{\sigma(A,B)}(u(t+r)-A)(\dot{u}(t+r)-B)^*)$

$= o(r^{\frac{3}{2}})(1+\|\begin{pmatrix}A\\B\end{pmatrix}\|^2)$, *a.e.*,

(e) $E^{\sigma(A,B)}(\|\dot{u}(t+r)-B\|^2) = \int_t^{t+r} \sigma(s,A,B)^2\, d\mu +$

$o(r^{\frac{3}{2}})(1+\|\begin{pmatrix}A\\B\end{pmatrix}\|^2)^2$, *a.e.*,

for t ranging on compact subsets of $\mathbb{R}^+$, and the o-terms are uniform for $(t,A,B)\in[0,a]\times\mathbb{R}^n\times\mathbb{R}^n$ for any $a>0$.

Finally if the Z process is Brownian motion, then each o-term is uniform on $\mathbb{R}^+\times\mathbb{R}^n\times\mathbb{R}^n$ whenever the Lipschitz constants in (i') and (iv) are uniform.

Remark. These estimates are needed to calculate the generator of the associated evolution operator of the $(u(t),\dot{u}(t))$-process which is (vector) Markov. The calculations are tedius and we shall only indicate the necessary effort by spelling out the details of just one of them.

Proof. For ease of the notation we treat the case that $n=1$. Then let us consider $1(b)$ as a typical example. By hypothesis for $t\leq s\leq t+r$,

$$\Delta(u(s) = u(s)-u(t) = (s-t)\dot{u}(t) + \int_t^s [\dot{u}(v)-\dot{u}(t)]dv.$$

Hence

$$|\Delta u(s)|^2 \leq 2r^2\dot{u}(t)^2 + 2r\int_t^{t+r} |\dot{u}(v)-\dot{u}(t)|^2 dv.$$

Writing $\Delta q(s) = q(s, u(s), \dot{u}(s)) - q(s, u(t), \dot{u}(t))$ and similarly $\Delta\sigma(s)$, $\Delta u(s), \Delta\dot{u}(s)$, and if $\Delta\beta = \beta(t+r) - \beta(t)$ in $\mu \leq \beta \otimes P$, for $r > 0$ fixed, we get the above equation as:

$$\begin{aligned}|\Delta u(s)|^2 \leq 4r \int_t^{t+r} |\Delta q(v)|^2 dv + 4(\int_t^s \Delta\sigma(v) dZ(v))^2 + \\ 4r \int_t^{t+r} |q(v, u(v), \dot{u}(v))|^2 dv + 4(\int_t^{t+r} \sigma(v, u(v), \dot{u}(v)) dz(v))^2 \\ = 4r\alpha_1 + 4\alpha_2(s) + 4\alpha_3 + 4\alpha_4(s) \quad \text{(say)}. \end{aligned} \tag{66}$$

Using Part (i) of the hypothesis we can get bounds for (66). Thus with K_i as some constants,

$$\begin{aligned}E(\alpha_1) \leq 2K_1^2 r^3 E(\dot{u}(t)^2) + 2K_1^2 r^2 \int_t^{t+r} E(|\Delta\dot{u}(v)|^2) dv + \\ K_1^2 \int_t^{t+r} E(|\Delta\dot{u}(v)|^2 dv. \end{aligned} \tag{67}$$

Since $\{\alpha_2(s), \mathcal{B}_s, t \leq s \leq t+r\}$ is a submartingale because $Z(\cdot)$ has also independent increments, we get on taking a separable version,

$$\begin{aligned}E(\sup_{t\leq s\leq t+r} \alpha_2(s)) &\leq 4E(\alpha_2(t+r)) \\ &\leq 8K_2^2 r^2 \Delta\beta(t) E(\dot{u}(t)^2) + 8K_1^2 r \Delta\beta(s) \times \\ &\int_t^{t+r} E(|\Delta\dot{u}(v)^2) dv + K_1 K_2 \int_t^{t+r} E(|\Delta\dot{u}(v)|^2) d\beta(r). \end{aligned} \tag{68}$$

A similar reasoning is used for the submartingale $\{\alpha_4(s), \mathcal{B}_s, t \leq s \leq t+r\}$ to get

$$\begin{aligned}E(\sup_{t\leq s\leq t+r} \alpha_4(s)) &\leq 4E(\alpha_4(t+r)) \\ &\leq 4K_3^2 \Delta\beta(t)(1 + E[u(t)^2 + \dot{u}(t)^2]),\end{aligned}$$

and then

$$E(\alpha_3) \leq K_1^2 r(1 + E[U(t)^2 + \dot{u}(t)^2]).$$

Putting these estimates together in (66) we get

$$\begin{aligned}E(\sup_{t\leq s\leq t+r} |\Delta\dot{u}(s)|^2) \leq \Delta\beta(t)(4K_1^2 r + 16K_s^2) + \\ \Delta\beta(t)(4K_1^2 r + 16K_2^2) E(u(t)^2) + \\ \Delta\beta(t)(8K_1^2 r^3 + 32K_2^2 r^2 + 4K_1^2 r + 16K_2^2) \times \\ E(\dot{u}(t)^2) + (8K_1^2 r^3 + 4K_1^2 r + 32K_2^2 r \Delta\beta(t) + \\ 16K_2^2) \int_t^{t+r} E(|\Delta\dot{u}(v)|^2) d\beta(v).\end{aligned}$$

If C_1 is the largest of the first three terms and C_2 is that multiplying the last integral, one sees that the left side satisfies:

$$LHS \leq C_1 \Delta\beta(t)(1 + E[u(t)^2 + \dot{u}^2(t)]) + C_2 \int_t^{t+r} E|\dot{u}(v) - \dot{u}(t)|^2) d\beta(v)$$
$$\leq C\Delta\beta(t)(1 + E(u(t)^2 + \dot{u}(t)^2)),$$

after some routine simplification, where $C > 0$ is a constant. This gives the estimate since $\Delta\beta = o(r)$ for 1(b).

The others are similarly verified after nontrivial computations. □

The details of this theorem, and the higher order nonlinear case, have been abstracted from Borcher's unpublished work [1]. Our slight extension here is to $L^{2,2}$-bounded processes relative to a σ-finite measure $\mu \leq K\beta \otimes P$, whereas he considered the square integrable (independent increment) martingale case, but the computations follow his work.

We now specialize the results when the Z-process is in fact the standard Brownian motion. (Again, for simplicity, the case $n = 1$ is treated.) In this case, the solution $u(t)$ of (24) is called a (2nd order) *diffusion process* and the functions q, σ are termed the *drift* and *diffusion* coefficients. Thus $(u(t), \dot{u}(t))$ is a (vector) Markov process (although $u(t)$ itself is not), and has a finer structure which we now explain. The solution process $u(t)$ is also called (when Z is Brownian motion) an (second order) *Itô process.* Hereafter q, σ are assumed to satisfy a locally uniform (i.e., on compact t-sets) Lipschitz condition and are locally bounded:

$$\begin{aligned} &\{q(t,x,y)|, |\sigma(t,x,y)|\} \leq K_T, \\ &|q(t,x,y) - q(t,\tilde{x},\tilde{y})|^2 + |\sigma(t,x,y) - \sigma(t,\tilde{x},\tilde{y})| \\ &\qquad\qquad \leq K_T(\|x-\tilde{x}\|^2 + \|y-\tilde{y}\|^2, \end{aligned} \tag{69}$$

for some constant K_T depending only on compact sets $T \subset \mathbb{R}$.

To get estimates of moments of the process, analogous to those of Theorem 19, but using the Gaussian distribution crucially, the following lemma due to Maruyama [1] is employed.

20. Lemma. *Let* $u = \{u(t), t \geq 0\}$ *be a second order Itô process with* $u(0) = c_1, \dot{u}(0) = c_2$, *constants* a.e. *Then for any real locally bounded (*i.e., *on compacts of* $\mathbb{R}^+$*) Borel function* f *on* $\mathbb{R}^+ \times \mathbb{R} \times \mathbb{R}$, *we have* *(i)* $E\{\exp[\rho \int_0^t f(s,u(s),\dot{u}(s))dz(s) - \frac{1}{2}\rho^2 \int_0^t f(s,u(s),\dot{u}(s))^2 ds]\} = 1,$ *and more generally*

(ii) $E\{\exp[\rho \int_) ^t \int_0^s f(r,u(r),\dot{u}(r))dz(r)ds -$

$\frac{1}{2}\rho^2 \int_0^t (t-s)^2 f(s,u(s),\dot{u}(s))^2 ds]\} = 1$, *for a real* ρ *and* $t > 0$. *The same formulas hold if* E *is replaced by* $E^{\mathcal{F}_\tau}, \mathcal{F}_\tau = \sigma(u(\tau),\dot{u}(\tau)), \tau \le s \le t$, *when the initial values* $u(\tau), \dot{u}(\tau)$ *are not constants.*

Proof. Since (ii), reduces to (i) if f is replaced by $\tilde{f}$ where

$$\tilde{f}(s,x,y) = (t-s)f(s,x,y), 0 \le s \le t = 0 \text{ for } s > t, \ x,y \in \mathbb{R},$$

it suffices to verify (i). By approximation we need only consider (i) if f is a simple function. But then the result follows from the known formula for the moment generating function of a Gaussian density. The details are omitted. (See Exercise 6(b).) □

Expanding the exponential in (i) of the above lemma, and transposing the terms, we get a useful moment relation:

$$E([\int_0^t f(s,u(s),\dot{u}(s))]^4) = 6E([\int_0^t f(s,u(s),\dot{u}(s))dz(s)]^2 \times \int_0^t f(s,u(s),\dot{u}(s))^2 ds) - 3E([\int_0^t f(s,u(s),\dot{u}(s))^2 ds]^2). \quad (70)$$

We can get truncated moments, with the formula (70). Let $\varepsilon_1, \varepsilon_2 > 0$ be given and consider the rectangle $R(\varepsilon_1, \varepsilon_2, x)$ of $\mathbb{R}^2$ as:

$$R(\varepsilon_1,\varepsilon_2,x) = \{(y_1,y_2) \in \mathbb{R}^2 : |y_1 - x_1| < \varepsilon_1, |y_2 - x_2| < \varepsilon_2\}, x = (x_1,x_2).$$

Define the truncated moments $\alpha_{ij}(i,j = 0,1,2)$ as:

$$\alpha_{ij}(k) = E^{\sigma(u_t,\dot{u}_t)}[(u_{t+k} - x_1)^i(\dot{u}_{t+k} - x_2)^j \chi_{R(\varepsilon_1,\varepsilon_2,x)}(u_{t+k}, \dot{u}_{t+k})].$$

Then Theorem 19, in this case, takes the following form, given for $n = 1$ (the scalar process) for simplicity:

21. Theorem. *For the vector process $(u_t, \dot{u}_t)$, with initial value (x_1, x_2) and $\varepsilon_1, \varepsilon_2 > 0$, we have as $k \to 0$,*

(i) $\alpha_{10}(k) = x_2 k + o(k^{\frac{3}{2}})(1 + x_2^2)$,

(ii) $\alpha_{01}(k) = \int_t^{t+k} q(s, u(s), \dot{u}(s))ds + o(k^{\frac{5}{4}})(1 + x_2^2)$,

(iii) $\alpha_2(k) = o(k^2)(1 + x_2^2)^{\frac{3}{2}}$,

(iv) $\alpha_{11}(k) = o(k^{\frac{3}{2}})(1 + x_2^2)$,

(v) $\alpha_{02}(k) = \int_t^{t+k} \sigma(s, u(s), \dot{u}(s))^2 ds + o(k^{\frac{3}{2}})(1 + x_2^2)$.

The proof uses Maruyama's lemma and (70), and the details are analogous to those of Theorem 19 (involving similar long computations but not new ideas). They will not be reproduced.

The conditions in Theorems 19 and 20 are essentially optimal. This is made precise by the following "converse" assertion.

22. Proposition. *Let $\{V(t) = \binom{x(t)}{\bar{x}(t)}, \mathcal{F}_t, t \geq 0\}$ be an adapted vector process such that:*

(i) *almost all sample paths of $x(t)$ are absolutely continuous and those of $\bar{x}(t)$ are continuous,*

(ii) *$E^{\mathcal{F}_s}(\|x(t)\|^2) = \varphi_s$ a.e. $(s \leq t)$, where φ_s is $\mathcal{F}_s$-adapted and integrable,*

(iii) *there exist Borel functions q, σ on $\mathbb{R}^+ \times \mathbb{R} \times \mathbb{R}$ such that $\{|q(t, x, \bar{x})|, \sigma(t, x, \bar{x})\} \leq k(1 + \|\binom{x}{\bar{x}}\|^2)^{\frac{1}{2}}$,*

(iv) (a) $E^{\mathcal{F}_s}(|x_{(s+r)} - x(s) - r\bar{x}(s)|) \leq rg(r)$, *a.e.*,

(b) $|E^{\mathcal{F}_s}(\bar{x}(s + r) - \bar{x}(s) + \int_s^{s+r} q(v, x(v), \bar{x}(v))dv| \leq rg(r)$, *a.e.*,

(c) $|E^{\mathcal{F}_s}[\bar{x}(s + r) - \bar{x}(s)]^2 - \int_s^{s+r} \sigma(v, x(v), \bar{x}(v))dv| \leq rg(r)$, *a.e.*,

where $g(r) = f(r)(1 + \|V(s)\|^2), f(r) \uparrow$ and $\lim_{r \downarrow 0} f(r) = 0$. Then the process $\{V(s) - V(0), \mathcal{F}_s, \geq 0\}$ is a Markov process and $\dot{x}$ exists in mean, $P[\bar{x}(t) = \dot{x}(t)] = 1$. Moreover, if $\sigma > 0$, then there is a Brownian motion Z on the same (Ω, Σ, P) such that x is a solution of a second order stochastic differential equation of the form (24) with the coefficients q, σ as given. If $\sigma = 0$ on a set of positive (Leb $\otimes P$) *- measure, then the same statement holds on an enlarged probability space to which a Brownian motion can be adjoined.*

This result is an extension, to the second order equations, of a classical result due to Itô as abstracted by Doob ([1], Thm. VI.3.3). It is first verified by showing that the y-process defined by

$$y(t) = \bar{x}(t) + \int_0^t q(s, x(s), \bar{x}(s))ds,$$

is a martingale with continuous sample paths, and that the $\bar{x}$ process is the derived process of x under the given hypothesis. It then follows after an application of Corollary V.3.26, that $y(t) = y(0) + \int_0^t \sigma(s, x(s), \dot{x}(s))dZ(s)$ for a Brownian motion Z. The nontrivial computations (given in Borcher [1]) are omitted.

Our interest in presenting these results is to show that the (higher order) Itô processes are of Feller type and also have nice structural properties. We now specialize by taking q, σ to be uniformly (bounded) continuous functions of the state variables and not of time t, but still satisfy a uniform Lipschitz condition. The choice of the function space now is slightly different. It is the set of real Borel functions $B(= B(\mathbb{R}^2))$ such that $f \in B$ iff $\|f\| = \sup\{|f(x)|e^{-|x_2|} : x = (x_1, x_2) \in \mathbb{R}^2\} < \infty$. A standard computation shows that $\{B, \|\cdot\|\}$ is a Banach space. This unusual (instead of the uniform) norm is prompted by the fact that the associated generator turns out to be a differential operator (instead of a more complicated integro-differential one if the uniform norm is used). We show below that the associated semigroup of the (vector) Markov process $(u(t), \dot{u}(t))$, with stationary transitions, maps continuous functions C of B into C, although the semigroup is no longer strongly continuous at the origin.

Let $V_{s,X}(t) = \binom{u_t}{\dot{u}_t}$ be the soluton of (24) (Z as Brownian motion) on $[s, \infty)$ with $V_{s,X}(s) = X = \binom{A}{B}$ as the initial condition ($V_{0,X} = V_X$). Then $\{V_{s,X}(t), t \geq 0\}$ is a Markov process, and

$$p(t, x, \Gamma) = P[V_{s,X}(s+t) \in \Gamma] = P[V_X(t) \in \Gamma]$$

where Γ is a planar Borel set. It follows from the Chapman-Kolmogorov equation that if

$$\begin{aligned}(T_t f)(x) &= E^{\sigma(V_X(0))}(f(V_X(t))(x) \\ &= \int_{\mathbb{R}^2} f(y)p(t, x, dy), \quad x \in \mathbb{R}^2, f \in B, \end{aligned} \tag{71}$$

then $T_{s+t} = T_s T_t, t \geq 0, s \geq 0$. Also $T_t(C) \subset C$, because

$$\begin{aligned}|(T_t f)(x)| &= |E(e^{|\dot{u}(t)|}(f(V_X(t)))e^{-|\dot{u}(t)|})| \\ &\leq \|f\| E^{\sigma(V_X(0))}(e^{|\dot{u}(t)|})(x) \\ &\leq L\|f\|e^{|x_2|},\end{aligned}$$

since $E(e^{\rho|\dot{u}(t)|}) \leq Le^{\rho|x_2|}$, as a consequence of Maruyama's lemma with some computations where L is a constant depending only on t and the bounds of q, σ. Also $V_{0,x}(t) \to V_{0,x'}(t)$ in mean as $x \to x'$. Hence $T_f f \in B$ for each $f \in B$, and using the above continuous dependence on x, one shows that (by another consequence of Lemma 20)

$$\begin{aligned} E(f(V_{0,x}(t))^2) &= E(e^{2|\dot{u}(t)|}(f(V_{0,x}(t))e^{-|\dot{u}(t)|})^2) \\ &\leq \|f\|^2 E^{\sigma(V_x(0))}(e^{2|\dot{u}(t)|})(x) \\ &\leq \tilde{L}\|f\|^2 e^{2|x_2|}, \end{aligned}$$

for some constant $\tilde{L}$. Hence by the bounded convergence theorem for each $f \in C$ we get $f(V_x(t)) \to f(V_{x'}(t))$ as $x \to x'$, in mean so that $(T_t f)(x) \to (T_t f)(x')$ as $x \to x'$, and hence $T_t f \in C$.

With this information we can find the generator of $\{T_t, t \geq 0\}$. If $f_{ij}(x) = \frac{\partial^2 f}{\partial x_1^i \partial x_2^j}(x)$, $f_{00} = f$, let us define a set C_2 as:

$$C_2 = \{f \in C : \sup_x |f_{ij}(x)|e^{-c_f|x_2|} < \infty, 0 \leq i+j \leq 2,\ 0 \leq c_f = c < 1,$$

$$f \text{ has compact support } S_f, \text{ for } i+j=3, \sup_{x \in S_f^c} |f_{ij}(x)|e^{-c|x_i|} < \infty\}.$$

Evidently $C_2 \subset C \subset B$, and C_2 is also a linear set. Our main assertion on the semigroup is given by:

23. Theorem. *Let the second order Itô process $u(t)$ solving (24) with q, σ and Z as described above be given. Then $(u(t), \dot{u}(t))$ determines a semigroup $\{T_t, t \geq 0\}$ whose generator G with domain $\mathcal{D}_G \supset C_2$, is obtained as*

$$Gf = [\frac{1}{2}\sigma(x_1, x_2)^2 \frac{\partial^2}{\partial x_2^2} - q(x_1, x_2)\frac{\partial}{\partial x_2} + x_2 \frac{\partial}{\partial x_1}]f = \lim_{h \to 0} \frac{(T_h - I)f}{h}, \tag{72}$$

a degenerate elliptic operator. The vector $(u(t), \dot{u}(t))$ is a continuous Feller (hence strong Markov) process.

Proof. We outline the essential ideas, since the detail is in many ways similar to that of Theorem 17, although the computations are much more involved. Thus for each $\varepsilon_i > 0, i = 1, 2$, consider a rectangle $R(\varepsilon_1, \varepsilon_2)$ with center at the origin of $\mathbb{R}^2$, and for $f \in C_2$ with support

$S(=S_f)$ such that, for $y \in R(\varepsilon_1, \varepsilon_2)$, and $x \in S$, we have

$$\begin{aligned} f(y) &= f_{00}(x) + (y_1 - x_1)f_{10}(x) + (y_2 - x_2)f_{01}(x) + \\ &\frac{1}{2}\{(y_1 - x_1)^2 f_{20}(x) + 2(y_1 - x_1)(y_2 - x_2)f_{11}(x) + \\ &(y_2 - x_2)^2 f_{02}(x)\} + r(x,y)\|x - y\|^2, \end{aligned} \tag{73}$$

with $r(x,x) = 0$ and $\lim_{y \to x} r(x,y) = 0$ uniformly for $x \in S$. Next consider

$$[\frac{1}{h}(T_h - I)f](x) = \frac{1}{h}\int_{R(\varepsilon_1,\varepsilon_2)} [f(y) - f(x)]p(h,x,dy) + \beta(h,x), \tag{74}$$

where $\beta(h,x)$ is the value of the integral on $\mathbb{R}^2 - R(\varepsilon_1, \varepsilon_2)$. Now substituting for the integrand from (73) in (74) and then using the expression for the truncated moments from Theorem 21, we can simplify. Here the estimates employ the Hölder inequalities, the norm of c_2 and a careful analysis shows that $\lim_{h\to 0} \sup_x |\beta(h,x)|e^{-|x_2|} = 0$. The integral on $R(\varepsilon_1, \varepsilon_2)$ simplifies, after considerable computation, to give that $\lim_{h\to 0} \sup_x |\frac{(T_h - I)f}{h}(x) - (Gf)(x)|e^{-|x_2|} = 0$, where G is the differential operator given in (72). Since $T_t(C) \subset C$ the result is that $\{T_t, t \geq 0\}$ is a Feller semigroup, and then the other conclusions follow. □

Remark. It is a characteristic feature of the higher order stochastic differential equations (both linear and nonlinear) that the generator of the associated semigroup (under conditions as employed in our work) is always a *degenerate* elliptic operator. In the first order case one can find conditions so that the corresponding generator is (even) strictly elliptic. But for the higher order equations this is not the case, and it appears to be one of the basic differences showing up in this study.

Using the operator G of (73) we can get information on the path behavior of the u_t and $\dot{u}_t$ processes. The following result on this question has been established by Goldstein [2] which we include, omitting the long but accessible proof.

24. Theorem. *Suppose $u(t)$ is a second order Itô process satisfying the hypothesis of Theorem 23, G is given by (27), and $E(e^{|\dot{u}(0)|}) < \infty$. Suppose that $g \in \mathcal{D}_G$ and either $Gg = (\geq)\lambda$, or $Gg = (\geq)\lambda g$, for some real λ. Then the process $\{g(V_t) - \lambda t, \mathcal{F}_t, t \geq 0\}$, or $\{e^{-\lambda t} g(V(t)), \mathcal{F}_t, t \geq 0\}$ respectively is a (sub) martingale, where $V(t) = \binom{u_t}{\dot{u}_t}$. If* *moreover* $p(x_2) = \frac{q(x_1, X_2)}{\sigma^2(x_1, x_2)}$ *is a function of* x_2 *alone, is bounded, and* $\frac{dp}{dy}$ *exists, bounded and is continuous on the set* $\{x_2 : |x_2| \geq \beta_0\}$ *for some* $\beta_0 > 0$, *let* $g(x) = g(x_2) = \int_0^{x_2} \exp\{-2\int_0^t p(s)ds\}dt$. *Then* $g \in C_2$ *and* $Gg = 0$ *so that* $\{g(V_t), \mathcal{F}_t, t \geq 0\}$ *is a martingale. If* $\lim_{t\to\infty} g(\dot{u}(t))$ *exists a.e., then* $\dot{u}(t) \to \dot{u}(\infty)(\in \bar{\mathbb{R}})$ *a.e., and for each* $s \in \mathbb{R}$, $\lim_{t\to\infty}(u(t+s) - u(t)) = s\dot{u}(\infty)$ *exists a.e.*

A proof of this result together with the sample path and moment behavior of the $u(t)$ (and $\dot{u}(t)$) processes under various hypotheses on q and σ, can be found in Goldstein's paper referred to above. Further analysis depends on the study of the partial differential equations $Gg = \lambda g$ (or $= \lambda$) and these are nontrivial problems in the PDE theory.

This section will be concluded by discussing one more result which indicates a further potential avenue of fruitful investigation for the $L^{2,2}$ -bounded processes (relative to some measure μ), in lieu of a Brownian motion, in Hilbert space. This shows the far reaching nature of the (generalized) boundedness principle for both the theory and applications.

Motivated by Proposition 18, suppose that $X^i = \{X_t^i, \mathcal{G}_t, t \geq 0\}, i = 1, \ldots n$, are $L^{2,2}$ -bounded continuous processes relative to a σ-finite μ determined by the given filtration (cf. Thm. 2.13), and let $f : \mathbb{R}^n \to \mathbb{R}$ be a twice continuously differentiable function. Then for the vector $X = (X^1, \ldots, X^n)$, we have

$$f(X_t) - f(X_0) - \frac{1}{2}\sum_{i,j=1}^{m}\int_0^t \frac{\partial^2 f}{\partial x_i \partial x_j}(X_s)d[X^i, X^j]_s$$

$$= \sum_{i=1}^{n}\int_0^t \frac{\partial f}{\partial x_i}(X_s)dX_s^i, \quad (75)$$

where $[X^i, X^j]$ is the quadratic (co-)variation of (X_t^i, X_t^j)-processes. Thus the right side is really an $L^{2,2}$-bounded process. Now consider

$$dZ_t^i = a_t^i dX_t^i + b_t^i dt, \quad a^i, b^i \in \mathcal{L}^2([X^i]), \quad [X^i, X^j]_t = a^{ij}t, \qquad (76)$$

and define the operator

$$L = \frac{1}{2}\sum_{i,j=1}^{n} a^{ij}\frac{\partial^2}{\partial x_i \partial x_j} + \sum_{i=1}^{n} b^i \frac{\partial}{\partial x_i},$$

with the n-by-n matrix (a^{ij}) being positive (semi-)definite. Then for $Z_t = (Z_t^1, \ldots, Z_t^n)$, (75) can be stated as:

$$f(Z_t) - f(Z_0) - \int_0^t (Lf)(Z_t)dt \tag{77}$$

is an $L^{2,2}$-bounded process, relative to a σ-finite measure for each bounded twice continuously differentiable function f. But (77) gives a stochastic differential equation which has a unique solution under certain conditions, such as those of Theorem 6, and if the X-process has independent increments, the Z is a Markov process with continuous paths, and (77) is still a valid statement. Now let us consider the converse problem in the following sense. Let $\mathcal{H}$ be a separable Hilbert space, and $(i, \mathcal{H}, \mathcal{X})$ be an abstract Wiener triple in the sense of Section I.4. Let $\{X_t, \mathcal{G}_t, t \geq 0\}$ be an $L^{2,2}$ bounded $\mathcal{H}$-valued continuous process relative to a σ-finite μ. If $a, b : \mathcal{H} \to B(X)$ are bounded (strongly) measurable functions, $a(x)$ being positive definite and of trace class, let

$$(Lf)(x) = \frac{1}{2} tr(a(x)(D^2 f)(x)) + ((Df)(x), b(x))_{\mathcal{H}}$$

be an operator defined for all twice strongly (or Fréchet) differentiable $f : \mathcal{H} \to \mathbb{R}$, whose derivatives are bounded on bounded sets. Then a family $\{P_x, x \in \mathcal{H}\}$ is called a solution of the equation (76) relative to (x, a, b) if

$$M_t^r = f(z_t) - f(z_0) - \int_0^t (Lf)(z_0)ds \tag{78}$$

is a continuous $L^{2,2}$- bounded process on (Ω, Σ, P_x) and a σ-finite μ_x, for each such f and the given X, with $P_x[Z_0 = x] = 1$. The process $\{Z_t, \mathcal{G}_t, t \geq 0\}$ is a solution of the equation $Z_t = x + \int_0^t b(Z_s)ds + \int_0^t \sigma(Z_s)dX_s$, if $\|b\|, \|\sigma\| \in L^2([0, r], dt)$ and σ is Hilbert-Schmidt. The existence of solutons to (75) and assertion (78) are related. If the X-process is Brownian motion in $\mathcal{H}$ (i.e., $X_0 = 0, X_t$ has independent increments, and for each $x^* \in \mathcal{H}^*, x^*(X_t)$ is Graussian with mean zero and variance $\|x^*\|^2 t$), then we have:

25. Theorem. *With the above notation and assumptions the following equivalent statements (X being Brownian motion!) hold:*

(i) $\{M_t^r, \mathcal{G}_t, t \geq 0\}$ *is a continuous local martingale for each* f,

(ii) for each $h \in \mathcal{H}(= \mathcal{H}^*)$ $Y_t^h = (h, Z_t - Z_0 - \int_0^t b(Z_s)ds)_{\mathcal{H}}$ *defines an element of* $\mathcal{M}_{\text{loc}}^c$ *and its increasing process is in* $\mathcal{A}_{\text{loc}}^c$, *given by* $A_t^h = \int_0^t (h, a(Z_s)h)_{\mathcal{H}} ds$,

(iii) the process $\{\exp(Y_t^h - \frac{1}{2}A_t^h), \mathcal{G}_t, t \geq 0\}$ *is a local martingale for each* $h \in \mathcal{H}$, *using the notation of (ii).*

If $a = \sigma\sigma^*$ *in the above (so a is trace class), then the family* $\{P_x, x \in \mathcal{H}\}$ *associated with L gives the process (i) to be strongly Markovian when and only when the unique solution Z relative to* (x, b, a) *is such that* $x \mapsto \int_\Omega f(Z_t)dP_x$ *is measurable for each such function f.*

This result when $\mathcal{H} = \mathbb{R}^n$ was first established by Stroock and Varadhan ([1], [2]), and the present extension to infinite dimensions is due to Yor [1], and Krinik [1]. (We shall not include a proof here.) Some aspects of the above result when a is degenerate was considered by Bonami *et al* [1]. It was extended further (with $\mathcal{H} = \mathbb{R}$) by Anderson [1] when the X-process is a submartingle, instead of being a martingale which is the hypothesis of the earlier results. The natural generalization to the $L^{2,2}$-bounded processes, relative to some measure μ, will be interesting indeed.

6.5 Progression to stochastic flows

We include a brief discussion of an immediate follow up of the preceding work for stochastic flows. Consider a (vector) stochastic differential equation: ($U(t)$ is a $2n$-vector)

$$dU(t) = q(t, U(t))dt + \sigma(t, U(t))dZ(t), \quad U(0) = x \in \mathbb{R}^{2n}, \tag{1}$$

where q is a vector and σ is a $2n$-by-$2n$ matrix, the "drift" and "diffusion" coefficients. The solution (under conditions of Theorem 4.6) depends on both x and t. If we denote it $U(t, x)$, then for each x it is a process in t. However, we can also think of this as a random field with the parameter set $T_s = [s, \infty) \times \mathbb{R}^{2n}$, so $U : T \to L^2(P)$, under suitable moment conditions. If the initial time is s, (so $U(s) = x$) the solution process is represented as $U_{s,t}(x)$, and the study of stochastic flows is the study of the behavior of $x \mapsto U_{s,t}(x)$ as $0 \leq s \leq t < \infty$ varies. If the $L^{2,2}$-process $Z(t)$, is Brownian motion and q, σ are infinitely differentiable with bounded derivatives (in addition to the Lipschitz conditions), then one shows that, with (nontrivial) additional work, the mapping $x \mapsto U_{s,t}(x,\omega)$ is a diffeomorphism, for each $t \geq s$, of $\mathbb{R}^{2n} \to \mathbb{R}^{2n}$ for a.a.(ω). If the conditions on q, σ are relaxed, then one can study the weaker properties of the above mapping, in that it is a homeomorphism, or only one-to-one, and the like.

A related (but more important and deeper) problem is the converse. Namely, suppose we are given a family $\{U_{s,t}(x), s \leq t\}$ of stochastic processes such that (i) U is jointly measurable in (t,x,ω) (ii) $(s,t,x) \mapsto U_{s,t}(x)$ is continuous in probability, (iii) $U_{s,s} = id.$, a.e. for all $s \geq 0$, and (iv) for $s < r < t, U_{s,t}(x) = U_{s,r}(U_{r,t}(x))$ a.e. for each $x \in \mathbb{R}^{2n}$, i.e., $U_{s,t} = U_{s,r} o U_{r,t}$ a.e. Then we call the family $\{U_{s,t}, 0 \leq s < t\}$ a *(forward) stochastic flow* on $\mathbb{R}^{2n} \to L^2(P)$. [A *backward flow* $\{U_{s,t}, 0 \leq t \leq s\}$ is similarly defined.] If moreover for $0 \leq t_0 < t_1 < \cdots < t_{n+1} \leq a, x_i \in \mathbb{R}^n, 0 \leq i \leq n, U_{t_i,t_{i+1}}(x_i), i = 0, \ldots, n$ are independent random variables, then the flow is said to have *independent increments.* Note that $U_{s,t}$ generally involves products of $Z's$, i.e., a nonlinear functional of $Z(t)'s$. However, taking a constant initial value x, we can connect the previous study with the new developments.

To understand the last comment, consider the process $X_t(x)$ defined as: $dX_t(x) = q(t,x)dt + \sigma(t,x)dZ_t$, so that (1) becomes ($U_s = U_{0,s}$)

$$U_{s,t}(x) = x + \int_s^t dX_r(U_r), \quad (U_s = x \in \mathbb{R}^{2n}) \tag{2}$$

If the above integral is defined for more general (nonlinear) integrators $\{X_s(x), s \geq 0\}$, depending on a parameter, than the one given by (1) which is locally $L^{2,2}$-bounded, then the solution process $\{U_{s,t}(x), t \geq s \geq 0\}$ will be a (homeomorphic) stochastic flow when the process $x \mapsto U_{s,t}(x)$ defines a homeomorphism of $\mathbb{R}^{2n}$ into itself, such that $U_{s,s}(x) = x, s \geq 0$, and is a càdlàg process in $t \geq s$, taking values in $C(\mathbb{R}^{2n}, \mathbb{R})$. An extensive discussion of nonlinear stochastic integrators is given by Carmona and Nualart [1], including the conditions for the existence of solutions of (2) in a general setup. Motivated by Lévy's characterization of a Brownian motion (cf. V.3.19), one says that $\{U_{s,t}(x), t \geq s \geq 0\}$ of the type discussed above is a *Brownian flow* with values in $C(\mathbb{R}^{2n}, \mathbb{R})$, if it is also continuous (in t) and has independent values. (Note, however, that a Brownian flow need not be Gaussian.) For instance, if $\mathbf{x}^{(m)} = (x_1, \ldots, x_m), x_i \in \mathbb{R}^n$, and if $U_{s,t}(\mathbf{x}^{(m)}) = (U_{s,t}(x_1), \ldots, U_{s,t}(x_m))$, so that for each s and $U_{s,t}(\mathbf{x}^{(m)}) = \mathbf{x}^{(m)}$, suppose it is a Brownian flow. Let $\mathcal{F}_{s,t} = \cup_{\varepsilon > 0} \sigma(U_{p,q} : s - \varepsilon \leq p \leq q \leq t + \varepsilon)$, so that it is a (two parameter) filtration of Σ, generated by the flow. Then the following result is a consequence of the concept of Markov property as given in Propositions 4.11 and 4.13.

1. Proposition. *The Brownian flow* $\{U_{s,t}(\mathbf{x}^{(m)}), \mathcal{F}_{s,t}, t \geq s \geq 0\}$ *has the Markov property with transition probabilities given by*

$$P_{s,t}^{(m)}(\mathbf{x}^{(m)}, B) = P[U_{s,t}(\mathbf{x}^{(m)}) \in B], \tag{3}$$

where B *is a Borel subset of* $\mathbb{R}^{mn}$, *and the related family of evolution operators* $\{T^{(m)}, t \geq s \geq 0\}$ *on* $C_0(\mathbb{R}^{mn}, \mathbb{R})$ *is given by*

$$(T_{s,t}^{(m)} f)(\mathbf{x}^{(m)}) = \int_{\mathbb{R}^{mn}} f(\mathbf{y}^{(m)}) P^{(m)}(\mathbf{x}^{(m)}, d\mathbf{y}^{(m)}). \tag{4}$$

Thus the corresponding study of the Brownian flow can be accomplished from the associated operator analysis of the family.

A natural problem now is to find conditions on the existence (through the projective limit results of Sections 1.5, and 3.6; see also Darling [1]) of general flows which arise as solutions of suitable stochastic differential equations relative to $L^{2,2}$-bounded processes Z (and some measure μ). This area grew into a separate field, and it also leads to several new studies. Here $\mathbb{R}^n$ can be replaced by a differentiable manifold. We refer to the monographs on the subject by Kunita[1], and Ikeda-S. Watanane [1]. For space reasons, we shall not include any more of these results on this facinating branch.

6.6 Complements and exercises

1. Let $X = \{X_t, \mathcal{F}_t, t \geq 0\}$ be a (right continuous) local martingale, and T be any finite stopping time of $\{\mathcal{F}_t, t \geq 0\}$ strongly reducing the process. If $\tilde{X} = \{X(T \wedge t), \mathcal{F}_t, t \geq 0\}$ denotes the resulting martingale, then show that $\tilde{X} = Y + V$ where $Y = \{Y_t, \mathcal{F}_t, t \geq 0\}$ is an $L^p(P)$-bounded martingale for all $1 \leq p < \infty$, and $V \in \mathcal{M}^1 \cap \tilde{\mathcal{A}}$ is a martingale of integrable total variation. Moreover, $V_t = (X \circ T)\chi_{[T \leq t]} + A_t$ where $A = \{A_t, t \geq 0\} \in \mathcal{A} \cap L^p(P), 1 \leq p < \infty$.

2. If $\{\mathcal{F}_t, t \geq 0\}$ is a complete stochastic base from a probability space (Ω, Σ, P), verify the following:

(a) Consider $X = \{X_t, \mathcal{F}_t, t \geq 0\}$, a local semimartingale. Let $Y = \{Y_t, \mathcal{F}_t, t \geq 0\}$ satisfy the equation (*) $Y_t = 1 + \int_0^t Y_{s-} dX_s, t \geq 0$. Then the solution process of (*) is "explicitly" given by the following expression with $\Delta X_s = X_s - X_{s-}$

$$Y_t = (\exp[X_t - \frac{1}{2}\langle \dot{X}^c, \rangle \dot{X}_t^c]). \prod_{s \leq t} (1 + \Delta X_s) e^{-\Delta X_s},$$

where the product on the right converges a.e. for each $t \geq 0$ ($\dot{X}^c$ being as in Theorem 1.2).

(b) If in the above, X is a continuous local martingale, then so is the solution Y of (*). Moreover, we have $Y_t = \sum_{n\geq 0} X_t^{(n)}/n!$ where $(X_t^{(0)} = 1$ and) for $n \geq 1$,

$$X_t^{(n)} = n! \int_0^t dX_{s_1} \int_0^{s_1} dX_{s_2} \cdots \int_0^{s_{n-1}} dX_{s_n} = n \int_0^t X_s^{(n-1)} dX_s, \text{(say)}, \tag{++}$$

so that $\{X_t^{(n)}, \mathcal{F}_t, t \geq 0\}$ is a continuous local martingale.

[*Sketch:*(a) From the fact that $\prod_{n\geq 1}(1+a_n)$ converges if $\sum_{n\geq 1} |a_n| < \infty$, we see that $Z_t = \prod_{s\leq t}(1 + \Delta X_s)e^{-\Delta X_s}$ exists for a.a. (ω) when $\sum_{s\leq t} |(1 + \Delta X_s)e^{-\Delta X_s} - 1|(\omega) < \infty$. But this is dominated (Mean-Value theorem) by $C \sum_{s\leq t}(\Delta X_s)^2(\omega)$ which is finite since X is a semimartingale (cf. last part of Step V of the proof of Theorem 1.2), since $|\Delta X_s| \geq 1$ holds only for finitely many jumps. Thus $\sum_{s\leq t} |\Delta Z_s|(\omega) < \infty$ for a.a.(ω) so that $Z = \{Z_t, \mathcal{F}_t, t \geq 0\}$ is a local semimartingale. But $T_t = \exp(X_t - \langle \dot{X}^c, \dot{X}^c\rangle_t)$ clearly defines a similar process. Now in Theorem 1.2 take $f(x,y) = e^u v$ and consider $f(Y_t, X_t)$. This yields (+). For unicity, first suppose that $X \in \tilde{\mathcal{A}}$, so it is of bounded variation a.e. If $Y, \tilde{Y}$ are two solutions, then $V_s = Y_s - \tilde{Y}_s = \int_0^s V_{u-} dX_u$ and hence by iteration it satisfies (++), so that

$$|V_{s-}| \leq \int_0^{s-} |dX_{s_1}| \cdots \int_0^{s_{n-1}-} |V_{u-}||dX_u| \leq \sup_{u\leq s} |V_{u-}||X_s^{(n)}|/n!, a.e.$$

Since $\sup_{u\leq s} |V_{u-}|(\omega)$ is finite, we see that $V_s = 0$, a.e., whence $Y = \tilde{Y}$, a.e. For the general case, let $\{W_t, \mathcal{F}_t, t \geq 0\}$ be another semimartingale satisfying (+). Let $T_1 = \inf\{t : |\Delta X_t| \geq \frac{1}{2}\}$ and $T_n = \inf\{t \geq T_{n-1} : |\Delta X_t| \geq \frac{1}{2}\}$. Thus $T_n \uparrow \infty$, and if $A_t = \sum_{n=1}^{\infty}(\Delta X_{T_n})\chi_{[T_n\leq t]}$, then $A \in \tilde{\mathcal{A}}$. Now $X - A = U$ and A have no common discontinuities, and the jumps of U lie in $(-\frac{1}{2}, \frac{1}{2}), U^c = \dot{X}^c$ being the continuous part. Hence $K_t = \exp(U_t - \langle \dot{X}^c, \dot{X}^c\rangle_t) \cdot \prod_{s\leq t}(1 + \Delta U_s)e^{-\Delta U_s}$ is defined and is a submartingale. Further, $K_t = 1 + \int_0^t K_{s-} dU_s$. Since $K_s \neq 0$(a.e.), there exists a chain of optionals $\{S_n \uparrow \infty\}$ such that $|K_{t-}| \geq \frac{1}{2n}$ on $[S_n \geq t]$. So $|K_t| = |K_{t-}||1 + \Delta U_t| \geq \frac{1}{2n}$ on $[S_n \geq t]$. Let $f(x,y) = x/y$ for $|y| \geq \frac{1}{2n}$, and apply Theorem 1.2 for the processes $W(S_n \wedge t), K(S_n \wedge t)$ with this f. If $B_t = W_t K_t^{-1}$, then letting $n \to \infty$ in the above application, one gets $B_t = 1 + \int_0^t B_{s-} dA_s$. By the special case $B_t = e^{A_t} \prod_{s\leq t}(1 + \Delta A_s)e^{-\Delta A_s}$ is the unique solution of this equation. Since $\Delta A_t.\Delta U_t = 0$, a.e. and $A_t = X_t - U_t$, it follows by substitution in $W_t = K_t B_t$ that $W_t = Z_t$, a.e. for $t \geq 0$.

(b) The infinite product for Y_t in (a) is now unity, and $Y(\cdot,\omega)$ is also continuous for a.a. (ω). Let $Y_t^\lambda = \exp[\lambda(X_t - \langle X, X\rangle_t)] = 1 + \lambda\int_0^t Y_{s-}^\lambda dX_s$. Thus $Y_t^\lambda = 1+\lambda\int_0^t dX_s+\cdots+\lambda^n\int_0^t dX_{s_1}\cdots\int_0^{s_{n-1}} dX_{s_n}+\lambda^{n+1}\int_0^t dX_{s_1}\cdots\int_0^{s_n} Y_{s_{n+1}-}^\lambda dX_{s_{n+1}}$. But the last term tends to zero as $n\to\infty$, by the estimate in (a), so that (++) is true for any $\lambda\in\mathbb{R}$. The identity (+) was noted and effectively used by Stroock and Varadhan [1]. The uniqueness of the solution of (+) was discussed in detail by Maisonneuve [1], and the generalization in (a) is due to Doléans-Dade [2].]

3. By a formal integration by parts, show that if f is a real continuously differentiable function on $[a,b]$, $f(a)\neq f(b)$, then Prop. 2.2 can be expressed as $:\int_a^b f(t)dZ(t) = \int_a^b \tilde{Y}(t)f'(t)dt$, for some integrable $\tilde{Y}$. Here the right side is defined as a Bochner integral and the left side is equal to it (thus the representation of that proposition is not unique). Discuss the properties of this alternative definition for non-stochastic integrands. Can we extend this to all bounded Borel f as in that propositition?

4. Show that an arbitrary vector measure $Z:\mathcal{B}(\mathbb{R}^+)\to L^2(P)$ is $L^{2,2}$-bounded relative to some σ-finite $\mu:\mathcal{B}(\mathbb{R}^+)\otimes\Sigma\to\bar{\mathbb{R}}^+$. Deduce that an $L^{2,2}$-bounded process is not less general than a semimartingale integrator. (*Hint:*Use Prop. 2.7.)

5. Here we define the Skorokhod integral and show that it satisfies the generalized boundedness principle. The definition is based on Itô's version of the multiple Wiener integral (cf. Itô[4]), which we recall, as the concept involves an elaborate structure. Thus $L^2(P)$ admits a direct sum decomposition: $L^2(P) = \otimes_{n=0}^\infty H_n$, where H_n is a set each of whose elements h is an Itô multiple integral, i.e., $h = I_n(f_n), f_n\in L^2([0,1]^n,\mathbb{R}^{dn})$ and that $f_n(t_1,\cdots,t_n)^{j_1\cdots j_n}$ is symmetric in the n-tuples $((t_1,j_1),\ldots,(t_n,j_n))$. [Here $f_2(t_1,t_2)^{j_1j_2} = \frac{\partial^{j_1+j_2}f}{\partial x_1^{j_1}\partial x_2^{j_2}}$, and similarly others.] Let $I_0(f_1(t)^j) = f_1(t)^j$. Then define $D_t^j(I_n(f_n)) = nI_{n-1}(f_n(\cdot,t)^{\cdot j})$ and $D_t^j(f) = \sum_{i=1}^n \frac{\partial f}{\partial x_{ij}}(W(t_1),\cdots,W(t_n))\chi_{[0,t_i]}$ for $t\in[0,1], j=1,\ldots,d$. Thus a process $\mathbf{u} = (u_t, 0\le t\le 1)$ is square integrable and d-dimensional means $\mathbf{u}\in L^2([0,1]\times\Omega,\mathbb{R}^d)$, the underlying measure space being $([0,1],\mathcal{B},dt)\otimes(\Omega,\Sigma,P)$. Now u_t can be decomposed (cf. Itô's paper *loc.cit*) as:

$$u_t = \sum_{m=0}^\infty I_m(f_m(\cdot,t)), \qquad (*)$$

called the Wiener-chaos, with $f_m(s_1,\cdots,s_m,t)^{j_1\cdots j_m\cdot j}\in L^2([0,1]^{m+1},\mathbb{R}^{d(m+1)})$ as a symmetric function of m-couples $((s_1,j_1),\ldots,(s_m,j_m))$. The symmetrization $\tilde{f}$ of an f_m in $(m+1)$-couples $(s_i,j_i), 1\le i\le m$

is given by

$$\tilde{f}_m(s_1, \cdots, s_m, t)^{j_1 \cdots j_m \cdot j} = \frac{1}{m+1}[f_m(s_1, \cdots, s_m, t)^{j_1 \cdots j_m \cdot j} + \sum_{i=1}^{m} f_m(s_1, \cdots, s_{i-1}, t, s_{i+1}, \cdots, s_m, s_i)^{j_1 \cdots j_m \cdot j_i}].$$

The *Skorokhod integral* of the **u**-process relative to the multiple Wiener integral is defined as:

$$\begin{aligned} S(\mathbf{u}) &= \sum_{m=0}^{\infty} I_{m+1}(\tilde{f}_m) \\ &= \sum_{m=0}^{\infty} \sum_{j_1, \ldots, j_m, j=1}^{d} \int_{[0,1]^{m+1}} \tilde{f}_m(s_1, \cdots, s_m, t)^{j_1 \cdots j_m \cdot j} \times \\ &\qquad dW(s_1)^{j_1} \cdots dW(s_m)^{j_m} dW(t)^j, \end{aligned}$$

whenever the series converges in $L^2(P)$, where each of the above terms is an Itô integral. This is denoted as:

$$S(\mathbf{u}) = \int_0^1 u_t \cdot dW_t = \sum_{i=1}^{d} \int_0^1 u_t^i dW^i(t).$$

Let $\mathcal{L}^{2,1}$ be the set of $\mathbf{u} = \{u_t, 0 \leq t \leq 1\}$ from $L^2([0,1] \times \Omega, dt \otimes dP)$ such that it is given by (*), and the following holds:

$$E(\int_0^1 \int_0^1 |D_s u_t|^2 ds dt) < \infty$$

Using a key inequality of Nualart and Pardoux ([1], Prop. 3.5) one has

$$\| \int_0^1 u_t . dW(t) \|_p \leq c_p [\int_0^1 (|E(u_t)|^2 dt)^{\frac{1}{2}} + \| (\int_0^1 \int_0^1 |D_s u_t|^2 ds dt)^{\frac{1}{2}} \|_p],$$

where c_p is a constant. We verify that $S(\mathbf{u})$ is $L^{2,2}$-bounded relative to $\mu = Leb \otimes P$, with the following computation ($K_0 = 2c_p > 0$, a constant):

$$\begin{aligned} E(S(\mathbf{u})^2) &\leq K_0 [\int_0^1 |E(u_t)|^2 dt + E(\int_0^1 \int_0^1 |D_s u_t|^2 ds dt)] \\ &\leq K_0 [\int_0^1 \int_0^1 E(|u_t|^2) dt ds + \int_0^1 \int_0^1 E(|D_s u_t|^2) ds dt] \\ &\leq K_0 \int_0^1 \int_0^1 E[|u_t|^2 + |D_s u_t|^2] ds dt \\ &= \int_{[0,1]^2 \times \Omega} |||\mathbf{u}|||_2^2 d\mu, \end{aligned}$$

where $|||\cdot|||$ is the Sobolev norm, just as in the case of (generalized) Stratonovich integral (cf. Prop. 2.9).

6. (a) Complete the details of proof of Theorem 2.13.

(b) Prove Maruyama's lemma (4.20) using the following sketch: Let $f = \sum_{j=0}^{n} \varphi_j \chi_{[t_j,t_{j+1})}, 0 = t_0 < t_1 < \cdots < t_{n+1} = t$, with φ_j as $\mathcal{F}_{t_j}$-adapted and bounded, so

$$J_m = \int_0^t f dZ = \sum_{j=0}^{m} \varphi_j \Delta Z_j; L_m = \int_0^t f^2 ds = \sum_{j=0}^{m} \varphi_j^2 \Delta s_j,$$

where the $\Delta Z_j = Z(t_{j+1}) - Z(t_j)$ are independent $N(0, \Delta t_j), \Delta t_j = t_{j+1} - t_j$. If $Q_m = \rho J_m - \frac{1}{2}\rho^2 L_m$, $r_m = \rho\varphi_{m-1}\Delta Z(t_m) - \frac{1}{2}\rho^2\varphi_{m-1}^2 \Delta t_m$, then $E(\exp \alpha Q_m) = E[\exp(\alpha Q_{m-1}).E^{\mathcal{F}(t_{m-1})}(\exp \alpha r_m)], \alpha \geq 1$. Also

$$E^{\mathcal{F}(t_{m-1})}(\exp \alpha r_m) \leq \exp[\frac{1}{2}\alpha(\alpha-1)\rho^2 M_0^2 \Delta t_n]$$

with equality for $\alpha = 1$, where M_0 is the bound of f. Since $\sum_{m=0}^{n} \Delta t_m = t$, deduce that

$$1 \leq E(\exp \alpha Q_m) \leq \exp[\frac{1}{2}\alpha(\alpha-1)\rho^2 M_0^2 t], \quad \alpha \geq 1.$$

This gives the lemma for simple f, and shows that $\{\exp \alpha Q_m, m \geq 1\}$ is uniformly integrable. The general case follows by the usual approximation of f by simple functions f_n after replacing Q_m by Q_{mn} in the above uniform integrability condition.

7. Using the notation of Corollary 4.3, for the second order linear case with $n = 1$, let $(d_{ij}(t,u)) = Y(t)^{-1}Y(u)$. Show that, in this case,

$$X(t) = \int_0^t d_{21}(t,u)dB(u) + d_{21}(t,0)C_1 + d_{22}(t,0)C_2,$$

and that the covariance matrix $r(t,t)$ of the vector $(X(t), \dot{X}(t))$ is:

$$r(t,t) = \begin{pmatrix} \int_0^t d_{11}^2(t,u)du & \int_0^t d_{11}(t,u)d_{21}(t,u)du \\ \int_0^t d_{11}(t,u)d_{21}(t,u)du & \int_0^t d_{21}^2(t,u)du \end{pmatrix}.$$

Deduce, in particular, that if we have constant coefficients, and the second one is zero (= no friction) and the third one is $\omega > 0$ (= positive angular velocity), then the corresponding covariance r is given by $r(s,t) = \int_0^{s\wedge t} \sin \omega(t-u) \sin \omega(s-u)du$.

8. We give an adjunct to Theorem 5.25 of an infinite dimensional differential equation. Thus let $\{Z_t, \mathcal{B}_t, t \geq 0\}$ be a Brownian motion

on a separable Hilbert space $\mathcal{H}$ and let $(i, \mathcal{H}, \mathcal{X}^*)$ be the associated abstract Wiener space (cf. Sec. I.4). Let q, σ be the "drift" and "diffusion" coefficients on $\mathbb{R}^+ \times \mathcal{X}$ into $\mathcal{H}$ and $B(\mathcal{X}, \mathcal{X}^*)$ satisfying the following growth conditions:

(i) $(\sigma(t, x) - I)$ is continuous in $t \geq t_0 \geq 0$ for each $x \in \mathcal{X}$ in the Schmidt norm $\|\cdot\|_2$ in $B(\mathcal{X}, \mathcal{X}^*)$ of continuous linear operators (thus $\|A\|_2 =$ sum of the eigenvalues of AA^*, assumed finite);

(ii) $q(t, x)$ is continuous in $t \geq t_0 \geq 0$ for each $x \in \mathcal{X}$ in the norm topology of $\mathcal{H}$; and

(iii) for any $x, y \in \mathcal{X}$ one has

$$\|\sigma(t, x) - \sigma(t, y)\|_2 + \|q(t, x) - q(t, y)\| \leq K_0 \|x - y\|,$$
$$\|\sigma(t, x) - I\|_2^2 + \|q(t, x)\|^2 \leq K_0(1 + \|x\|^2).$$

Then the stochastic differential equation

$$dY_t = Q(t, Y_t)dt + \sigma(t, Y_t)dZ_t, \quad Y_{t_0} \in L^2(P),$$

has a unique solution $\{Y_t, \mathcal{B}_t, t \geq t_0\}$ which is a square integrable continuous Markov process. If also q, σ are independent of time t, then it is a strong Markov process and, in any case, it is a semi-martingale. [*Hints*: Let $\mathcal{N}_0$ be the class of all adapted square integrable $Y = \{Y_t, \mathcal{F}_t, t_0 \leq t \leq t_1\}$ for some fixed $t_1 > t_0$, such that $\|Y\|^2 = \sup_{t \leq t_1} E(\|Y_t\|_{\mathcal{H}}^2) < \infty$. Then $(\mathcal{N}_0, \|\cdot\|)$ is a Banach space, and if $T : \mathcal{N}_0 \to \mathcal{N}_0$ is defined by

$$(TY)_t = Y_{t_0} + \int_{t_0}^t q(s, Y_s)ds + \int_{t_0}^t \sigma(s, Y_s)dZ_s,$$

the norm of the second term being a submartingale, verify that

$$\|TY\| \leq 5[E(\|Y_{t_0}\|^2) + 2CT_1^{\frac{1}{2}} + C_2(1 + t_1 - t_0)(t_1 - t_0)(1 + \|Y\|^2)],$$

where C_2 is a constant and $E(|Z_t|_{\mathcal{H}}^2) = Ct$. Thus $TY \in \mathcal{N}_0$ and for $Y, Y' \in \mathcal{N}_0$,

$$E(\|(TY - TY')_t\|^2) \leq C_3 \int_{t_0}^t E(\|Y_s - Y_s'\|^2)ds,$$

after some computation. Replacing T by T^n in the above one gets

$$\|T^n Y - T^n Y'\| \leq [C_3(t_1 - t_0)^n n!]^{\frac{1}{2}} \|Y - Y'\|,$$

and hence T^n is a contraction mapping for large n on $\mathcal{N}_0$. Then by Banach's contraction mapping theorem (cf. eg.,Dunford-Schwartz

[1],Chap. V) there is a unique $Y_0 \in \mathcal{N}_0$ such that $TY_0 = Y_0$, and this is the solution on $[t_0, t_1]$. The remaining conclusions are proved with a similar argument as in the last part of Prop. 4.16. In connection with this result see Kuo [1]. (More details were included in the first edition of this book; see also the related works of Cabaña[1], and Dalecky [1].)]

9. This exercise gives an expression for the Radon-Nikodým density of the measures $P_x = P_{x,q}$ and $P_{x,0}$ (i.e., $q = 0$). Using the conditions that σ is positive definite and continuous, q is bounded measurable on $\mathcal{H} = \mathbb{R}^n$, we have for each $t > 0$ the density $\frac{dP^c_{x,q,t}}{dP_{x,0,t}}$ ($P^c_{x,q,t}$ being the absolutely continuous part relative to $P_{x,0,t}$, both restricted to $\mathcal{F}_t$) given by the expression:

$$\exp\Big[\int_0^t (q(u, Z_u), \sigma^{-1}(u, Z_u)dZ_u) - \frac{1}{2}\int_0^t (q(u, Z_u), \sigma^{-1}(u, Z_u))du\Big]$$

where we employed the notation of the preceding exercise, $(\cdot, \cdot)$ being the duality pairing. (This result is a culmination of a series of detailed computations, and the reader should consult Stroock-Varadhan [1]. An extension to a Hilbert space setting is in Krinik [1].)

10. Let $\{\mathcal{F}_t, t \geq 0\}$ from (Ω, Σ, P) be a quasi-left-continuous filtration (cf. Proposition V.3.2) and let $\{X_t, \mathcal{F}_t, t \geq 0\}, \{A_t, \mathcal{F}_t, t \geq 0\}$ be a locally square integrable and a continuous increasing process respectively. If f, g are continuously differentiable real functions on $\mathbb{R}$, with bounded derivatives, $X_0 = 0 = A_0$, a.e., then show that $(*) Y_t = x + \int_0^t f(Y_u)dX_u + \int_0^t g(Y_u)dA_u, x \in \mathbb{R}$, has a unique solution upto indistinguishability. [*Hints*: By V.3.2, $\langle X \rangle \in \mathcal{A}^c$ and $X^2 - \langle X \rangle$ is a local martingale. If $a_t = t + \langle X \rangle_t + A_t, b_t = \inf\{s : a_s > t\}$, then $\{a_s, \mathcal{F}_t, t \geq 0\}, \{b_t, \mathcal{F}_t, t \geq 0\}$ are strict time change transformations, and if $Z_t = Y \circ b_t, U_t = X \circ b_t$ and $B_t = A \circ b_t$, then B is continuous, $U \in \mathcal{M}$, and $(*)$ becomes $(+) Z_t = x + \int_0^t f(Z_s)dU_s + \int_0^t g(Z_s)dB_s$. Now use the Picard method; see also Kazamaki [2].]

11. Complete the proof of Proposition 5.1 after remembering the facts that $U_{s,t}, t \geq s \geq 0$, has the evolution property for a.a (ω), and that $U_{s,t}$ and $\mathcal{F}_{t,r}$ are independent for $s < t < r$.

Bibliographical remarks

The main thrust of this chapter has been an elaboration and generalization of Bochner's boundedness principle for stochastic integration and differential analysis together with some important applications. Theorem 2.2 for semimartingales is due to Doléans-Dade and Meyer [1], and it extends the corresponding martingale result due to Kunita-S.

Watanabe [1]. Although a McShane process is a semimartingale, its associated measure is absolutety continuous (relative to the Lebesgue measure), a property characteristic of Brownian motion. However here McShane([3], cf. also [4]) exploited this fact and developed the integration theory from definitions, without appeal to the general martingale analysis. The multiparameter version of his approach has not yet been well settled to match it with Brownian motion.

The $L^{2,2}$-boundedness principle in the classical case is due to Bochner [2] and [4]. It seems to predate McShane's $K\Delta t$ condition, which may be considered as a sharpening of the former. But these authors seem to have formulated their conditions independently of each other and there is no reference to earlier works. For a really unified approach of all the available integration processes, it is necessary to generalize the concept allowing Stieltjes (or Borel) measures. This has been done only recently by the author (cf. Rao [15]– [17]), and this chapter shows that all the currently known stochastic integrals come under this formulation with or without the general structure theory of the processes. The work of Section 2 follows the author(Rao [16]) and slightly extends it. (We now allow Sobolev norms.) Earlier Brennan [2] used a similar idea for a specialized class using elementary (instead of simple) functions and showed that it includes Meyer's [6] definition. See also Métivier-Pellaumail [2] and a more recent treatment in Protter [3]. It is interesting to note that the $L^{2,2}$-boundedness plays the same role in stochastic integration (admitting the dominated convergence criterion), as the simple functions in the classical Lebesgue-Carathéodory theory (cf.,e.g., Rao [11], Chapter 4). We emphasized its use in a variety of applications in Sections 3 and 4. The multiparameter aspect is not yet as well developed, but we shall include some nontrivial results in the next chapter. Here we gave a glimpse by an application to the Lévy-Brownian motion.

A substantial account of linear as well as nonlinear stochastic differential equations of higher order is the content of Section 4.There one finds some novelties, not apparent in the first order case, and not dealt with in most books on the subject. The first order case can be obtained by an analogous, but somewhat simpler, treatment and we leave that specialization to the reader. Our analysis is largely based, in the linear case, on the results of Dym [1] and Goldstein [3] (cf. also Sec. 4 of Rao [9]). In the nonlinear case, the work is considerably involved, and we presented a slightly extended (to the $L^{2,2}$-bounded processes) and simplified version from Borcher's [1] unpublished thesis. Here the main result is Theorem 4.19. The infinitesimal behavior of the process in 4.21–4.23 is also due to him. We have condensed the treatment to conserve space, and perhaps the full account should be given elsewhere in order to help readers who are pressed for time. Theorem 4.24 is due to Goldstein [2] and the proof is long. We have to refer the reader to his paper.

It is interesting to note that one has to study the evolution equations in this work when the "drift" and "diffusion" coefficients depend on time. Although a study of the abstract case has been considered (cf. Goldstein [1] and others) motivated by probability theory, its implications to the subject, which would generalize the classical work of Yosida-Kakutani [1], have yet to be completed. The nonlinear analogs of these (evolution) equations appear again in the analysis of flows of diffeomorphisms and their extensions to manifolds. The ideas briefly noted in Section 5, have been studied by Kunita [1] and Ikeda-S. Watanabe [1] and others, as part of a new branch of mathematics called stochastic differential geometry. The material in Sections 2–5 therefore is very useful and it leads to an active area of current research in stochastic analysis. We shall include a brief account of the latter on manifolds in the final chapter. Several results in the complements, especially the Skorokhod integral as being a constituent of the boundedness principle, should also be noted.

Chapter VII

Stochastic analysis on differential structures

In most of the preceding work on stochastic integral and differential analysis, the study was based on the real line as its range space as well as its parameter set. However, modern developments demand that we consider the analogous theory on differential structures, more general than the real line. In this final chapter we first discuss the conformal properties of martingales with values in the complex plane and then devote Section 2 for a formulation and an extension of the work on certain (differential or smooth) manifolds, leading to Malliavin's approach to the hypoellipticity problem. The study continues in Section 3 by discussing semimartingales with $\mathbb{R}^n, n \geq 2$, as the parameter set. There are substantial difficulties in this extension due to a lack of linear ordering. We observe that, even here, the $L^{2,2}$-boundedness principle relative to a (σ-finite) measure is at work and corresponding stochastic integrals can be defined. Finally we indicate in Section 4 how one considers stochastic partial differential equations in this context. Then some complements to the preceding work are given, as in earlier chapters, in the last section to conclude this monograph.

7.1 Conformal martingales

Recall that a complex Gaussian random process Z is defined as $Z_t = X_t + iY_t (i = \sqrt{-1})$ with X_t, Y_t as real Gaussian processes which additionally are assumed to satisfy $E(Z_s Z_t) = E(Z_s)E(Z_t)$ for $s, t \geq 0$. This condition is somewhat arbitrary, but it is imposed to reduce the number of parameters allowing its characterization as in the real case. This implies $cov(X_s, X_t) = cov(Y_s, Y_t)$ and $cov(X_s, Y_s) = 0$. Being Gaussian, these conditions show that X_t and Y_t are independent. The process Z_t is then called *strongly normal.* A complex Brownian motion is thus a complex Gaussian process whose real and imaginary parts are independent Brownian motions. Hence $E(Z_s \bar{Z}_s) = 2\sigma^2 s$ where

$E(X_s^2) = E(Y_s^2) = \sigma^2 s, s > 0$. ¿From the earlier theory the covariation of the processes X_t, Y_t vanishes a.e. We abstract this fact:

1. Definition. Let $X^j = \{X_t^j, \mathcal{F}_t, t \geq 0\}, j = 1, 2$, be a pair of continuous locally integrable martingales. Then $Z = Z_0 + X^1 + iX^2$ where Z_0 is $\mathcal{F}_0$-measurable, $\{\mathcal{F}_t, t \geq 0\}$ being a standard filtration, is called a *conformal martingale* if $\langle X^1 \rangle = \langle X^2 \rangle$ and $\langle X^1, X^2 \rangle = 0$, a.e. If $Z_0 = 0$, a.e., here, then X^2 is termed a *conjugate* of X^1.

Let $\mathcal{C}[\mathcal{C}_0]$ be the class of all continuous conformal martingales (and $Z_0 = 0$ a.e.). Then we get the following characteristic properties from definition:

2. Lemma. *(a)$Z \in \mathcal{C}$, (b)$\bar{Z} \in \mathcal{C}$, (c)$\langle Z, \bar{Z} \rangle = 0$ are equivalent statements.*

Here $\langle Z, \bar{Z} \rangle$ is defined by the sesquilinearity of the functional $\langle \cdot, \cdot \rangle$ on locally square integrable martingales. Also repeating the arguments of Section V.3 (see e.g., Theorem VI.1.2) one gets:

3. Theorem. *Let D be a domain in the complex space $\mathbb{C}$ and $Z^j \in \mathcal{C}, j = 1, \ldots, n$. Suppose that $P[Z_t^j \in D, t \geq 0, j = 1, \ldots, n] = 1$, and let $f : U \to \mathbb{C}$ be a holomorphic function where U is a neighborhood of D^n. If $Z_t = (Z_t^1, \cdots, Z_t^n), t \geq 0$, then*

$$f(Z_t) - f(Z_0) = \sum_{j=1}^{n} \int_0^t \frac{\partial f}{\partial z_j}(Z_t) dZ_s^j + \frac{1}{2} \sum_{j,k=1}^{n} \int_0^t \frac{\partial^2 f}{\partial z_j \partial z_k}(Z_s) d\langle Z^j, \bar{Z}^k \rangle_s. \tag{1}$$

This is a complex process, and reduces to a martingale if $\langle Z^j, \bar{Z}^k \rangle = 0$ for all j,k. In that case we see that $f(Z) \in \mathcal{C}$. Generally a semimartingale process (i.e., the one representable as $Z = Z_0 + X + A$ with values in a domain D as in the theorem), is a *conformal semimartingale* iff its martingale component X is conformal in the sense of Definition 1. Thus Z is a conformal semimartingale iff Z^2 (or equivalently X^2) is such, so that conformal (semi)martingales on a given filtration need not form a vector space.

We now define an L^p-norm for complex local martingales $Z = \{Z_t, \mathcal{F}_t, t \geq 0\}$ (i.e., the real and imaginary parts of Z are local martingales) as:

$$\|Z\|_p = \sup\{\|(Z \circ T)\|_p : T \text{ a finite optional of } \mathcal{F}_t, t \geq 0\}. \quad (2)$$

It is remarked that the set of T's can be replaced by $T_n \uparrow \infty$ and then the 'sup' in (2) can be replaced by 'lim'. We have the following special property of conformal martingales.

4. Proposition. *Let $Z \in \mathcal{C}, p > 0$ be given. Then $\{|Z_t|^p, \mathcal{F}_t, t \geq 0\}$ is a local submartingale, and*

$$E(\sup_{0 \leq s \leq t} |Z_t|^p) \leq 4E(|Z_t|^p), \quad t \geq 0. \quad (3)$$

Proof. Using a stopping time argument, we may assume for this proof that Z is bounded so that $|Z|^2 = |X|^2 + |Y|^2$ is a bounded submartingale. Since $\alpha + Z \in \mathcal{C}$ for any $\alpha > 0$, taking $f(u) = u^{\frac{p}{2}}$ in (1) we get $f(\alpha + Z) \in \mathcal{C}$ whence $|\alpha + Z| > 0$ a.e. Letting $\alpha \downarrow 0$ through a sequence such that $\alpha + Z \neq 0$ a.e., one gets the first statement.

Regarding the second, let $\{T_n, n \geq 1\}$ be a sequence for (2), such that $Z \circ T_n$ is bounded. But $\{|Z \circ T_n|^{\frac{p}{2}}, n \geq 1\}$ is a submartingale in $L^2(P)$ by the first part. Then the maximal martingale inequality gives (3) with $Z \circ T_n$ in place of Z. The boundedness in $L^2(P)$ implies uniform integrability so that

$$E(\sup_{0 \leq s \leq t} |Z(T_n \wedge s)|^p) \leq 4E(|Z_t|^p) \leq 4\|Z\|_p^p < \infty. \quad (4)$$

Since $t > 0$ and $n \geq 1$ are arbitrary, this shows $\sup_t |Z_t| \in L^2(P)$, and so Z is uniformly integrable. Letting $n \to \infty$ in (4), we get (3) in general. □

Conformal martingales were considered by Getoor and Sharpe [1] to extend the $(\mathcal{H}^1, BMO)$-duality for the continuous parameter. For the "square function" $s_n^2(X) = \sum_{j=1}^n \varphi_j^2, \varphi_j = X_j - X_{j-1}, X_0 = 0, a.e., n \geq 1$, one sees that $E(s_n^2(X)) = E(\langle X, X \rangle_n)$ if $X = \{X_n, n \geq 1\}$ is a square integrable martingale. With this motivation, the corresponding spaces for the continuous parameter martingales can be introduced, using the notations of Chapter IV ($X \in \mathcal{M}_{loc}, \langle X, X \rangle \in \mathcal{A}_{loc}$, etc.) as:

$$\|X\|_p^p = E(\langle X, X \rangle_\infty^{\frac{p}{2}}), 0 < p < \infty. \quad (5)$$

and if $p = \infty$, also assuming uniform integrability,

$$\|X\|_B = \sup_t \|E^{\mathcal{F}_t}(\langle X, X\rangle_\infty - \langle X, X\rangle_{t-})^{\frac{1}{2}}\|_\infty. \tag{6}$$

Let $\mathcal{H}^p = \{X \in \mathcal{M}^c{}_{loc} : \|X\|_p < \infty\}$, and $BMO = \{X \in \mathcal{M}^c : \|X\|_B < \infty\}$. As before, one can show routinely that $\{\mathcal{H}^p, \|\cdot\|_p\}$ and $\{BMO, \|\cdot\|_B\}$ are Banach spaces.

Note that if $Z \in \mathcal{C}_0$, then $\langle Z, Z\rangle = \langle X, X\rangle + \langle Y, Y\rangle = 2\langle X, X\rangle$ (cf. Definition 1). Hence (5) and (6) become

$$\|Z\|'_p = \sqrt{2}\|X\|_p, p \geq 1; \quad \|Z\|'_B = \sqrt{2}\|X\|_B. \tag{7}$$

It is also useful to observe that $\|\cdot\|_p, \|\cdot\|_B$ norms have the Fatou property in that for monotone increasing sequences the limits and norms can be interchanged. Although the present considerations extend (nontrivially) to right continuous martingales, we restrict ourselves, for simplicity, to processes with *continuous sample paths.*

The following analogs of Hölder and other inequalities of Section IV.5 admit extensions without much difficulty.

5. Theorem. *(a) If $X \in \mathcal{H}^1(\subset \mathcal{M}^c_{loc}), Y \in BMO(\subset \mathcal{M}^c)$, then*

$$E(|\langle X, Y\rangle_\infty|) \leq \sqrt{2}\|X\|_1\|Y\|_B. \tag{8}$$

Hence if $Z, W \in \mathcal{C}_0$, then

$$E(|\langle Z, W\rangle|) \leq 2\sqrt{2}\|Z\|'_1\|W\|'_B. \tag{8'}$$

(b) If $X \in \mathcal{M}^c_{loc}$, $\frac{1}{2} \leq p < \infty$, and $X^*_t = \sup_{0 \leq s \leq t} |X_s|$, then there exist constants $0 < c_{jp} < \infty$, j=1,2, such that

$$c_{1p}\|\langle X, X\rangle_t\|_p \leq \|X^*_t\|_p \leq c_{2p}\|\langle X, X\rangle_t\|_p, t \geq 0, \tag{9}$$

and a similar inequality is valid if X is replaced by $Z \in \mathcal{C}_0$.

These inequalities are true in the discrete parameter case, and by using dyadic rationals, they can be extended to the continuous time with the procedure of Section V.2. They can also be derived directly by other methods (cf., Getoor-Sharpe [1] for this procedure). Here we present the key duality result of $(\mathcal{H}^1)^*$ as BMO.

6. Theorem. *Let* $\{\mathcal{F}_t, t \geq 0\}$ *be a standard filtration from* (Ω, Σ, P) *and* $\mathcal{H}^1, BMO$ *be as defined above. Then for each* $x^* \in (\mathcal{H}^1)^*$*, there is a unique* $Y \in BMO$ *such that* $x^*(X) = E(\langle X, Y\rangle_\infty), X \in \mathcal{H}^1$*, and*

$$c_1\|x^*\| \leq \|Y\|_B \leq c_2\|x^*\|, \tag{10}$$

where the costants $c_j > 0$ *may be taken as* $c_1 = \frac{1}{\sqrt{2}}, c_2 = 1$*, and* $\|\cdot\|$ *is the adjoint norm to* $\|\cdot\|_1$ *which is thus equivalent to the BMO norm* $\|\cdot\|_B$.

The proof is again an extension of the discrete case, which we sketched in the first edition of this book, and which for space reasons will be omitted here.

We isolate a subclass of conformal martingales, called holomorphic processes, (considered by Föllmer [2]); a related concept appears in some aspects of multiparameter stochastic integration to be discussed later. The notation of Section V.4 will again be employed.

7. Definition. (a) Let $\mathcal{E} = \{X^n, n \geq 1\} \subset \mathcal{M}$ be a set of mutually orthogonal processes (i.e., $\langle X^n, X^m\rangle_t, t \geq 0, n \neq m$). It is an *orthonormal basis* of $\mathcal{M}$ if each $Y(\in \mathcal{M})$ can be expressed as:

$$Y_t = \sum_{n=1}^{\infty}(H^n \cdot X^n)_t = \sum_{n=1}^{\infty}\int_0^t H_s^n dX_s^n, \tag{11}$$

for some $\{H^n, n \geq 1\} \subset \mathcal{L}^2(\langle X^n\rangle)$, and if $\langle X^n, X^m\rangle = \delta_{mn}\langle X^n\rangle = \delta_{mn}A$, for a fixed $0 \neq A \in \mathcal{A}^+$. (The last is the normalization condition, δ_{mn} being Kronecker's symbol.)

(b) A set $\mathcal{Z} = \{Z^n = X^n + iY^n, n \geq 1\}$, is a *conformal basis* of $\mathcal{M} + i\mathcal{M}$ if $\{X^n, Y^n, n \geq 1\}$ form separately orthonormal bases of the real space $\mathcal{M}$.

(c) A complex process $G = \{G_t, \mathcal{F}_t, t \geq 0\}$ is *holomorphic* relative to a conformal basis $\mathcal{Z}$ if we have the (mean) convergent sum:

$$G_t = \sum_{n=1}^{\infty}\int_0^t H_t^n dZ_s^n, t \geq 0, \tag{12}$$

for a suitable $H^n = H^{1n} + iH^{2n}, H^{jn} \in \mathcal{L}^2(A), j = 1, 2$ where $A = \langle X^n, X^n\rangle = \langle Y^n, Y^n\rangle, Z^n = X^n + iY^n$.

We denote by $\mathcal{H}(\mathcal{Z})$ the *class of complex holomorphic processes* relative to some fixed conformal basis, and call H^n the *complex derivative* of G relative to Z^n, denoted also by G^n_z. Real derivatives are defined similarly. [Real holomorphic processes also exist as two-parameter processes, and their analysis is more involved. See Section 3 below and especially Theorem 3.13 for a comparison with (12) above; but only complex processes are discussed presently.]

Remark. From the theory of stochastic integration presented in Chapters V and VI, we find that each term in (12) is a martingale. However an analogous concept may also be formulated if X^n, Y^n are $L^{2,2}$-bounded processes, instead of martingales, since their quadratic (co)variation is well-defined and then the conformality and holomorphic notions can be given for these larger classes of processes. Thus each term of (12) defines a martingale (respectively an $L^{2,2}$-bounded process) so that $G(\in \mathcal{H}(\mathcal{Z}))$ is a martingale of class (DL) (or of an $L^{2,2}$-bounded process).

In the following μ^A is the (Doléans-Dade) measure associated with a semimartingale. Here we restrict only to martingale classes, for space reasons.

8. Proposition. *Let $G = U + iV$ be a martingale where $U, V \in \mathcal{M}$. Then G is a holomorphic process relative to a conformal basis $\mathcal{Z}$ iff U, V satisfy the "Cauchy-Riemann" equations a.e.[μ^A], in that*

$$U^n_x = V^n_y;\ U^n_y = -V^n_x,\ \text{a.e.}[\mu^A], n \geq 1, \tag{13}$$

where $U^n_x(V^n_y)$ is the real derivative of $U(V)$ relative to $X^n(Y^n)$ with $Z^n = X^n + iY^n$ of the conformal basis. In particular, each $G \in \mathcal{H}(\mathcal{Z})$ is a conformal martingale so that $\mathcal{H}(\mathcal{Z}) \subset \mathcal{C}_0$.

Proof. If $G \in \mathcal{H}(\mathcal{Z})$, then $G = U + iV$ can be represented by (12). Consequently the series expansions are given by (see (11)):

$$U_t = \sum_{n=1}^{\infty}\left(\int_0^t U^n_{xs}dX^n_s + \int_0^t U^n_{ys}dY^n_s\right) \tag{14}$$

and similarly for V. Considering analogous expansions for G (see (12)), we get on comparing the real and imaginary parts for $\langle G, X^n\rangle, \langle G, Y^n\rangle$:

$$\begin{aligned}\int_0^t H_s^{1n} dA_s &= \int_0^t U_{xs}^n dA_s; \quad -\int_0^t H_s^{2n} dA_s = \int_0^t U_{ys}^n dA_s, t \geq 0\\ \int_0^t H_s^{1n} dA_s &= \int_0^t V_{ys}^n dA_s; \quad \int_0^t H_s^{2n} dA_s = \int_0^t V_{ys}^n dA_s, t \geq 0,\end{aligned} \tag{15}$$

since $\langle X^n, Y^n\rangle_t = 0$. Thus (15) gives (13). Conversely, (13) implies (15) and then (12). Note that by (13), $\langle X, X\rangle = \langle Y, Y\rangle$, and $\langle X, Y\rangle = 0$. So $\mathcal{H}(\mathcal{Z}) \subset \mathcal{C}_0$ is also obtained immediately. □

The existence of a conjugate martingale of a given one is answered by the following result if $\mathcal{M}$ has a conformal basis.

9. Corollary. *For each $U \in \mathcal{M}$, the latter having a conformal basis $\mathcal{Z}$, there is a $V \in \mathcal{M}$ such that $U + iV \in \mathcal{H}(\mathcal{Z})$.*

Proof. By (12), $U_x^n, U_y^n, n \geq 1$, exist where $\{Z^n = X^n + iY^n, n \geq 1\} = \mathcal{Z}$. Let V be an element defined by (12) after setting $V_x^n = -U_y^n, V_y^n = U_x^n$. Then it is seen that $V \in \mathcal{M}$ and because of (13), $G = U + iV \in \mathcal{H}(\mathcal{Z})$, as asserted. □

We now use these ideas for a more general structure.

7.2 Martingales in manifolds

Some of the most recent important applications are for processes which take values in spaces that look like "smooth" subsets of $\mathbb{R}^n$, i.e., manifolds. Thus it is necessary to recall the concept of a (differentiable) manifold, and then introduce various martingales on it, since a manifold need not have a linear structure and addition or multiplication operations will be undefined in general. A typical example is a sphere (or a hypersurface) in $\mathbb{R}^n$. We can only demand that each neighborhood of a point is like $\mathbb{R}^n$(or $\mathbb{C}^n$) or a Banach space, for the most part. We consider both real and complex manifolds here.

More precisely, a Hausdorff space M is a *C^p-(or analytic) manifold*, $p \geq 0$, if there is an open covering $\{U_\alpha, \alpha \in I\}$ of M and a collection $\{\varphi_\alpha, \alpha \in I\}$ of homeomorphisms on the former such that :(i) $\varphi_\alpha : U_\alpha \to \varphi_\alpha(U_\alpha)$, open subsets of $\mathbb{R}^n$(or $\mathbb{C}^n$), and each φ_α is a bijection of U_α onto its image, (ii)for any $\alpha, \beta, \varphi_\alpha(U_\alpha \cap U_\beta)$ is open in

$\mathbb{R}^{d_\alpha}$(or $\mathbb{C}^{d_\alpha}$), and (iii) $\varphi_\beta \circ \varphi_\alpha^{-1} : \varphi_\alpha(U_\alpha \cap U_\beta) \to \varphi_\beta(U_\alpha \cap U_\beta)$ is a C^p-(or analytic) isomorphism for each α, β in I. The pair $(U_\alpha, \varphi_\alpha)$ is called a *chart*, and the collection $\{(U_\alpha, \varphi_\alpha), \alpha \in I\}$ is termed an *atlas*. Also one says that, for $x \in U_\alpha$, U_α is a *coordinate neighborhood*, and $\varphi_\alpha(x) = (x'_1, \dots, x'_{d_\alpha}) \in \mathbb{R}^{d_\alpha}$(or $\mathbb{C}^{d_\alpha}$) is a *local coordinate* of x. When all $\mathbb{R}^{d_\alpha}$(or $\mathbb{C}^{d_\alpha}$) are the same or isomorphic for all α, which holds by (ii) if M is connected, then M is said to be *modelled* after $\mathbb{R}^d$(or $\mathbb{C}^d$), a *d-dimensional manifold*, for short. In what follows we assume, for simplicity, that *M is separable and the φ_α are C^∞- (or holomorphic) isomorphims, so that M is a C^∞-(or holomorphic) manifold of finite fixed real(complex) dimension d. Moreover our manifolds are taken to be without boundaries and are connected.*

First we give some motivation for the manifold results that are needed. Let $X = \{X_t, \mathcal{F}_t, t \geq 0\}$ be a process on (Ω, Σ, P) with values in a manifold M. Let $F(M)$ be the collection of all real functions f on M such that for each $\{U_\alpha, \varphi_\alpha\}$, one has $f \circ \varphi_\alpha^{-1} : \varphi_\alpha(U_\alpha) \to \mathbb{R}$ is infinitely differentiable. Then X is said to be $L^{2,2}$-*bounded* if for each $f \in F_c(M) \subset F(M)$, of compactly supported elements, the real process $\{f \circ X_t, \mathcal{F}_t, t \geq 0\}$ is $L^{2,2}$-bounded relative to a σ-finite measure μ_f. Now to study the properties of such X it is necessary to attach some vector spaces at each point $x \in M$, and the natural objects here are the tangent and cotangent spaces which must be considered before continuing with the desired stochastic analysis.

(a) *Basic formulation.* If $x \in M$, the set of all *tangent vectors* to M at x, [denoted by $T_x(M)$] is the collection of (real) linear functionals ℓ^x on $F(M)$ that satisfy the additional condition (of "derivation"):

$$\ell^x(fg) = f(x)\ell^x(g) + \ell^x(f)g(x). \tag{1}$$

Taking $f = 1$, and $g = c$, a constant, this and linearity imply that $\ell^x(c) = 0$, $x \in M$. Also, for each pair $\ell_1^x, \ell_2^x \in T_x(M)$, and $a \in \mathbb{R}, (a\ell_1^x + \ell_2^x)(f) = a\ell_1^x(f) + \ell_2^x(f)$, and hence $a\ell_1^x + \ell_2^x \in T_x(M)$, so that $T_x(M)$ is a vector space, called the *tangent space* of M at x. If $(x_1, \dots, x_d)$ is a point of the coordinate neighborhood U of x and $f \in F(M)$, then $\ell_i^x : f \mapsto \frac{\partial f}{\partial x_i}(x)$ are linearly independent tangent vectors, or we may identify ℓ_i^x with $(0, \cdots, \frac{\partial}{\partial x_i}, \cdots, 0)_x$ and each is regarded as an element of $T_x(M)$. [In detail, if $V \subset \mathbb{R}^d$ is an open set, $\varphi : U \to V$, and $g = f \circ \varphi^{-1}$, then $\partial_i f = \frac{\partial f}{\partial x_i}(\cdot) = \frac{\partial g}{\partial x_i} \circ \varphi = \frac{\partial (f \circ \varphi^{-1})}{\partial x_i} \circ \varphi(\cdot)$.]

Thus the dimension of $T_x(M)$ is the same as dim(M)=d, for all x of M. Its adjoint space $T_x^*(M)$ is called the *cotangent space* at x of M which is also of dimension d. The set $\{\ell_1^x, \ldots, \ell_d^x\}$ forms a basis of $T_x(M)$. The collection $T(M) = \{T_x(M), x \in M\}$, called the *tangent bundle*, is the space of *vector fields* where $\ell \in T(M)$ iff $\ell(f)(x) = \ell^x(f), x \in M$, and ℓ is a C^∞-vector field iff $x \mapsto \ell^x(f)$ is a C^∞-function on M, i.e., the coefficients $a_i(\cdot)$, in the definition of ℓ^x using the basis elements, are C^∞-functions. The class of all C^∞-vector fields is also denoted by $\mathcal{C}(M)$. Now the d-coordinates of a neighborhood U_α of M and d-components of the tangent vector at that point determine an element of the tangent bundle, and then one can show that $T(M)$ is of dimension $2d$ for the d-dimensional manifold M. The thus obtained correspondence with $\mathbb{R}^{2d}$ will be injective, and one topologizes $T(M)$ so that this correspondence is a homeomorphism, and the bundle space becomes a $2d$-dimensional manifold. We may iterate this procedure for higher order structures when needed.

There is an interesting theorem of Whitney's which states that every manifold of dimension d is diffeomorphic (i.e. a smooth homeomorphism in both directions) to a submanifold of $\mathbb{R}^{2d+1}$, and hence we may identify our M with such a space and think of the U_α's as the corresponding neighborhoods of the numerical d-space $\mathbb{R}^d$ for simplicity, although this may not always be appropriate. Another key theorem in the subject is that every C^∞-manifold (of dimension d) can be given a Riemannian metric (recalled below) so that we may assume the existence of such a metric in our spaces as desired.

If M is a d-manifold and $T_x(M)$ is the tangent space of M at $x \in M$, and $\ell_1^x, \ell_2^x \in T_x(M)$, the latter being diffeomorphic to $\mathbb{R}^d$, let $\langle \ell_1^x, \ell_2^x \rangle$ be the inner product. If U is an open set (or neighborhood) of x, let $\{\ell_1, \ldots, \ell_d\}$ be vector fields which, using Gram-Schmidt procedure, can be taken orthonormal. From linear algebra, we can represent $\langle \ell_1, \ell_2 \rangle$ as follows. If $(x_1, \ldots, x_d)$ are the coordinates of $x \in U$, and since $\{dx_1, \ldots, dx_d\}$ is a basis of the cotangent space $T_x^*(M)$, [note the clear distinction in dx_d, the differential of the dth coordinated x_d], we have

$$\langle \ell_1, \ell_2 \rangle(x) = \sum_{i,j=1}^{d} g_{ij}(x) \ell_1(dx_i)(x) \ell_2(dx_j)(x). \tag{2}$$

If $x \mapsto \langle \ell_1, \ell_2 \rangle(x)$ is differentiable, then $\langle \ell_1, \ell_2 \rangle(x)$ is called a *Riemannian metric* on U, and the positive definite matrix $(g_{ij}(x), 1 \le i, j \le d) =$

$g(x)$ is also termed the *metric tensor*. Writing $\xi^t(x) = (\ell^x(dx_i), i = 1, \dots, d)$ for the vector (t for transpose), we see that $\langle \ell_1, \ell_2 \rangle(x) = \xi_1^t(x) g(x) \xi_2(x)$ or $g(\xi_1, \xi_2)(x)$ and hence $g(x) : T_x(M) \times T_x^*(M) \to \mathbb{R}$ is a (positive definite) bilinear form.

As an example, let $\gamma : I \to U$ be a differentiable parametrized curve on an interval I and suppose that $\dot{\gamma}(x) \neq 0, x \in U$. Then $\ell(t) = \dot{\gamma}(t), t \in U$, and $\langle \ell(t), \ell(t) \rangle \neq 0, t \in I$. We can reparametrize γ such that the tangent vectors all have length 1 relative to $\langle \cdot, \cdot \rangle$. In fact, if $I = [a, b]$, consider the arc length

$$s(t) = \int_a^t \sqrt{\langle \ell(u), \ell(u) \rangle_{\gamma(u)}} du, \tag{3}$$

then $s(\cdot)$ is strictly increasing and differentiable. Let $t(\cdot)$ be its inverse $(0 \leq s \leq s(b))$ so that $t(s(u)) = u$ and differentiating $t'(s(u))s'(u) = 1$ or $t'(s) = (s')^{-1}$, and if $\tilde{\gamma} \mapsto \gamma(t(s))$, then the tangent vector to $\tilde{\gamma}$ is

$$\dot{\gamma}(t(s))t'(s) = \ell(t)(\langle \ell(t), \ell(t) \rangle_{\gamma(t)})^{-1}.$$

Thus the tangent of $\tilde{\gamma}$ has length 1. Note that $s(\cdot)$ is the arc length function of γ relative to the metric tensor $\langle \ell_1, \ell_2 \rangle_x$. In the simplest case with $d = 2, M = \mathbb{R}^2, T_x(M)$ is the space generated by $\partial/\partial x_1, \partial/\partial x_2$ and $\alpha, \beta \in T_x(M)$ are given by $\alpha_x = a_1(x)(\partial/\partial x_1)_x + a_2(x)(\partial/\partial x_2)_x$, and similarly for β_x with $b_i(x)$ in place of $a_i(x)$ we then have $\langle \alpha, \beta \rangle_x = (a_1 b_1)(x) + (a_2 b_2)(x)$.

The following further concept is also needed to introduce martingales in a manifold. This preparation is necessary since a stochastic function need not be smooth (as illustrated by the Brownian motion) and new substitutes are desirable to extend the theory to these functions. They solve some problems which otherwise could not be, as seen from Malliavin's work (below) on hypoellipticity. Now to introduce the martingale concept in manifolds, we recall the assertion of Theorem VI.4.25, namely, if $\{X_t, \mathcal{F}_t, t \in I \subset \mathbb{R}\}$ is an adapted process with values in the manifold $M = \mathbb{R}^d$ and $f : M \to \mathbb{R}$ is a compactly supported smooth function, then that $f \circ X_t$ is a martingale is the same as saying (letting $I = [0, a]$)

$$f \circ X_t - f \circ X_0 - \int_0^t (Lf)(X_s) ds \in \mathcal{M}_{loc}^c, \tag{4}$$

i.e., a local martingale. Here L is an elliptic operator without a constant term. This formulation admits the desired extension if M is a smooth manifold carrying second order differentials, or an Hessian, or a connection operation. This will be outlined here in order to specialize the $L^{2,2}$- (or semimartingale) processes to (conformal) martingales or to define a Brownian motion.

(b) *Conformal (semi)martingales and Brownian motions.* Comparing (4) with the original Itô formula, it becomes clear that the quadratic variation of the process X should be defined in terms of bilinear forms in M. Since the manifolds under discussion carry them which are elements of tensor products of cotangent bundles, we need to define the quadratic variation of a process relative to these forms. Thus let $b \in T^*(M) \otimes T^*(M)$ be a bilinear form. As a consequence of another theorem due to H. Whitney, there exists a finite family of smooth functions $(f_i, 1 \leq i \leq d)$ on M such that $b(\cdot,\cdot) = \sum_{i,j=1}^d b_{ij} df_i \otimes df_j$, and the b_{ij} depend on b smoothly. Then for each semimartingale X in M, we have $b_{ij} \circ X(s)$ to be adapted and the square bracket $[f_i \circ X, f_j \circ X](\cdot)$ well-defined so that our earlier theory implies that

$$\int_0^t b(dX_s, dX_s) = \sum_{i,j=1}^d \int_0^t b_{ij}(X_s) d[f_i \circ X, f_j \circ X]_s, \tag{5}$$

is meaningful. If b has two representations then one needs to verify that their difference which is a zero form, gives a zero integral by the above procedure. This can be shown just as in the theory of any integration process, although it depends on some, not entirely trivial, computation. The thus obtained integral is called the *b-quadratic variation* of X. From (5) we also have for each smooth $f : M \to \mathbb{R}$, of compact support, the square bracket functional satisfies $[f \circ X, g \circ X]_t = \int_0^t df \otimes dg(dX, dX)_s$, and

$$\int_0^t (fb)(dX, dX)_s = \int_0^t f \circ X_s b(dX, dX)_s. \tag{6}$$

These symbols will be used for special bilinear forms, namely the Hessians, which are the d-by-d matrices given by $Hess(f) = (D_{ij} f, 1 \leq$

$i, j \leq d$) where $D_{ij} = \partial^2/dx_i dx_j$. Since we are dealing with smooth functions $D_{ij} = D_{ji}$ so that our *Hessian is a symmetric bilinear form*, and the operator Hess($\cdot$) is a linear mapping from $C^\infty(M)$ into symmetric bilinear forms on M. Such an operator is called a *(torsion free) connection* if moreover

$$Hess f^2 = 2f Hess f + 2df \otimes df,$$

for each $f \in C^\infty(M)$.

1. Definition. A semimartingale or more genarally an $L^{2,2}$-bounded process X into M with a connection, is called a *martingale* if for each $f \in C^\infty(M)$ (5) holds in the form

$$f \circ X_t - f \circ X_0 - \frac{1}{2}\int_0^t Hess f(dX, dX)_s \in \mathcal{M}^c_{loc}(M), \qquad (7)$$

so that it is a local (real) martingale.

Although a bilinear form on M need not be symmetric and hence a connection defined through a nonsymmetric Hessian will not be torsion-free, the martingale analysis itself depends only on the symmetric part of a bilinear functional, and hence our restriction to the latter entails no loss of generality. On a coordinate chart the above connection is defined by a symmetric bilinear mapping $\Gamma(x) = (\Gamma^k_{ij}(x))$ such that if $\{x_i, i = 1, \ldots, d\}$ is a local coordinate system (so $f(x) = x_i$) then

$$Hess\, x_k = -\sum_{i,j=1}^{d} \Gamma^k_{ij}(x) dx_i \otimes dx_j, k = 1, \ldots, d, (\Gamma^k_{ij} = \Gamma^k_{ji}). \qquad (8)$$

These $\Gamma^k_{ij}(\cdot)$ are called *Christoffel symbols* which are components of the connection. Writing f, for $f(x_1, \cdots, x_d)$ we get

$$(Hess f)_{ij} = (D_{ij} - \sum_{k=1}^{d} \Gamma^k_{ij} D_k) f. \qquad (9)$$

This concrete formula will be useful in computations later, particularly for diffusion processes. If M is considered as a Riemannian manifold with the metric tensor $g(\cdot)(= (g_{ij}(\cdot), 1 \leq i, j \leq d))$, we have the following result connecting a Hess operator and g. The Lie derivative along a vector field A is defined as: $\mathcal{L}_A f = \frac{d}{dt}(f(\gamma(t)))$ where

$t \mapsto \gamma(t)$ is a curve on $\mathbb{R}$ determining A, i.e., $\dot{\gamma}(t) = A(\gamma(t))$, and thus $\mathcal{L}_A f = Af$, and if A, B are vector fields, one has the formula,

$$\mathcal{L}_A(Bf) = (\mathcal{L}_A B)f + B(\mathcal{L}_A f). \tag{10}$$

In this way we get $Hess f = \frac{1}{2}\mathcal{L}_{grad\, f}\, g$, for the metric tensor g, since

$$\begin{aligned} Hess f^2 &= \frac{1}{2}\mathcal{L}_{2f\, grad\, f}\, g = \mathcal{L}_{f\, grad\, f}\, g \\ &= f\mathcal{L}_{grad\, f}\, g + df \otimes g\langle grad\, f, \cdot\rangle + g\langle \cdot, grad\, f\rangle \otimes df \\ &= 2f Hess f + df \otimes df, \end{aligned} \tag{11}$$

which agrees with the expression given prior to Definition 1, since $g\langle \cdot, grad\, f\rangle = df = g\langle grad\, f, \cdot\rangle$.

Using these relations, one can introduce the Brownian motion in such a manifold (M, g) through a concrete form of Hess in terms of the Laplace-Beltrami operator (see (12) below). If $g(\cdot) = (g_{ij}(\cdot))$, let $g^{-1} = (g^{ij}(\cdot))$, the inverse of the metric tensor g of M. Then in local coordinates (9) can be reduced to:

$$\Delta f = \sum_{i,j=1}^{d} g^{ij}(D_{ij} f - \sum_{k=1}^{d} \Gamma_{ij}^k D_k f), f \in C^\infty(M), \tag{12}$$

and this Δ is called the *Laplace-Beltrami operator*. If $M = \mathbb{R}^d$, then $\Gamma_{ij}^k = 0$ and Δ reduces to the classical Laplacian. We now have

2. Definition. Let $\{X_t, \mathcal{F}_t, t \geq 0\}$ be an M-valued semimartingale. Then it is called a *Brownian motion* on (M, g) if $t \mapsto X_t$ is continuous and for each $f \in C_c^\infty(M)$

$$f(X_t) - f(X_0) - \frac{1}{2}\int_0^t (\Delta f)(X_s) ds \in \mathcal{M}^c{}_{loc}. \tag{13}$$

To relate these concepts we present the following simple charactrization of Brownian motion on a manifold.

3. Proposition. *Let X be an M-valued semimartingale where M is a d-dimensional smooth manifold. Then it is a Brownian motion iff X is a martingale and for each $f \in C^\infty(M)$ the quadratic variation of $f \circ X$ satisfies:*

$$[f \circ X, f \circ X]_t = \int_0^t \|grad\, f\|^2 dt, \tag{14}$$

Proof. Suppose first that the semimartingale X satisfies (14). In order to conclude that X is a Brownian motion condition (13) should be verified. So let us calculate $\|grad\, f\|^2 = \langle grad\, f, grad\, f\rangle$ in $T_x(M)$. If $\{e_i, i = 1, \ldots, d\}$ is an orthonormal basis of $T_x(M)$, then we have

$$\begin{aligned}\|grad\, f\|^2 &= \sum_{i=1}^{d} \langle grad\, f, e_i\rangle^2 = \sum_{i=1}^{d} \langle df, e_i\rangle^2 \\ &= \sum_{i=1}^{d} (df \otimes df)\langle e_i, e_i\rangle = trace(df \otimes df). \end{aligned} \tag{15}$$

This shows that (13) holds for "simple" bilinear forms after polarization, whence for $b = df_1 \otimes df_2$. Then by approximation of the integral, discussed for (6), (14) holds for general b. Consequently, taking $b \circ f$ as $Hess f$, we have on using (7),(11), and (12)

$$\begin{aligned} f \circ X_t - f \circ X_0 - &\frac{1}{2}\int_0^t Hess f(dX, dX)_s \\ &= f \circ X_t - f \circ X_0 - \frac{1}{2}\int_0^t (\Delta f) \circ X_s ds \in \mathcal{M}^c{}_{loc}, \end{aligned}$$

as desired.

For the converse, if (13) holds, on using (10) and (11), so that

$$\Delta f^2 = 2f\Delta f + 2\|grad\, f\|^2$$

we get for (13),

$$f^2 \circ X_t - f^2 \circ X_0 - \int_0^t (f\Delta f)(X_s)ds - \int_0^t \|grad\, f\|^2(X_s)ds, \tag{16}$$

to be a local continuous martingale. Since $f^2 \circ X$ is a semimartingale, by Theorem VI.1.2 (the generalized Itô formula for the continuous $f \circ X$), we get

$$\begin{aligned} f^2 \circ X_t - f^2 \circ X_0 &= 2\int_0^t f \circ X_s d(f \circ X)_s + [f \circ X, f \circ X]_t \\ &= \int_0^t f \circ X_s[(\Delta f)(X_s)]dt \\ &\quad + [f \circ X, f \circ X]_t + a\, local\, martingale, \end{aligned} \tag{17}$$

by (13). Since $f \in C^\infty(M)$ is arbitrary, (16) and (17) imply (14). Moreover, we get $\int_0^t Hess f(dX, dX)_s = \int_0^t (\Delta f)(X_s) ds$, for any $f \in C^\infty(M)$, and hence by (7) X must be a martingale. □

It is possible to continue the semimartingale analysis without invoking second order differentials on M, if we use Stratonovich integrals. The interplay between various stochastic integrals and differential geometry of M will be of importance in the analysis of stochastic differential equations and stochastic flows. We shall not undertake this study here for space reasons. [See Emery (1989) on these points.]

We should note the special properties available when the complex structures on M are involved so that it is a d-dimensional complex analytic manifold as defined at the begining of this section. In this context it is appropriate to indicate the associated conformal martingales discussed in the last section. Thus if $T_z(M), T_z^*(M)$ are the tangent and cotangent spaces of the complex (smooth) manifold M at $z \in M$, let $\{z_1, \ldots, z_d\}$ be the local coordinates of (U, φ) at z. This means letting $z_j = u_j + iv_j \in \mathbb{C}, i = \sqrt{-1}$ hereafter, we find that $T_z(M)$ and $T_z^*(M)$ to be of $2d$-real dimension with bases $\{\partial/\partial u_j, \partial/\partial v_j, j = 1, \ldots, d\}_z$ and $\{du_j, dv_j, j = 1, \ldots, d\}_z$ respectively. Hence with the complex differential notations, used in Theorem 1.3, we have

$$(\frac{\partial}{\partial z_j})_z = \frac{1}{2}(\frac{\partial}{\partial u_j} - i\frac{\partial}{\partial v_j})_z, \quad (dz_j)_z = (du_j + idv_j),$$

and similarly replacing z by its conjugate with $-i$ in place of i in the above one gets the corresponding bases for the complexified $T_z^c(M)(= T_z(M) + iT_z(M))$ and $T_z^{*c}(M)$. Let $J_z : T_z(M) \to T_z(M)$ be defined by

$$J_z(\frac{\partial}{\partial u_j})_z = (\frac{\partial}{\partial v_j})_z, \quad J_z(\frac{\partial}{\partial v_j})_z = -(\frac{\partial}{\partial u_j})_z,$$

for $j = 1, \ldots, d$, so that $J \mapsto J_z$ is a mapping attached to M at $T_z(M)$. If g_z is a Riemannian metric of M at z, it is termed *Hermitian* if $g_z(J_z A, J_z B) = g_z(A, B)$ for $A, B \in T_z(M)$ so that J_z is an isometry under g_z at z. This is used to define conformal (sub)martingales in such a manifold M.

A point to remember for conformality in the context of analytic compact manifolds is that there are no nonconstant holomorphic functions to use in a direct extension of the concept of (semi)martingales in contrast to the real case. Hence only a local definition is meaningful to take into account of the nonlinearity of such manifolds. Thus a process $X = \{X_t, \mathcal{F}_t, t \geq 0\}$ into an analytic manifold M is termed a *conformal (semi)martingale* if for each open set $U \subset M$ and for any smooth (or at least C^2) holomorphic function $f : U \to \mathbb{C}$, $f \circ X$ is a conformal (semi)martingale in the sense of Definition 1.1 (and the remark following Theorem 1.3) for almost all $\omega \in \Omega$. It should be noted that if $f_j : U_j \to \mathbb{C}, j = 1, 2$ are as above, then $f_1 \circ X$ and $f_2 \circ X$ agreeing on $U_1 \cap U_2$, are possibly different (semi)martingales in general; but if $M = \mathbb{C}^n$, it reduces to the earlier concept. Several properties of (real) semimartingales and their integrals, considered in Chapter VI, can be extended to (analytic) manifolds with suitable modifications (and additional arguments or restrictions). Here we discuss some further ideas merely to indicate the directions, by restricting to Brownian motion in these manifolds.

(c) *Diffusions on special manifolds.* We defined above a Hermitian metric g_z on $T_z(M)$ at each point $z \in M$. This may be extended to $T_z^c(M)$, the complexified tangent space, in the usual way: $(X_j, Y_j \in T_z(M), j = 1, 2, i = \sqrt{-1})$

$$\begin{aligned} g_z(X_1 + iY_1, X_2 + iY_2) = g_z(X_1, Y_1) - g_z(Y_1, Y_2) \\ + i(g_z(X_1, Y_2) + g_z(Y_1, X_2)), \end{aligned}$$

and let its components $g_{jk}, g_{j'k}, g_{jk'}, g_{j'k'}$ at $z \in M$ for $1 \leq j, k \leq d$ be:

$$g_{jk}(z) = g_z((\frac{\partial}{\partial z_j})_z, (\frac{\partial}{\partial z_k})_z); \quad g_{jk'}(z) = g_z((\frac{\partial}{\partial z_j})_z, (\frac{\partial}{\partial \bar{z}_k})_z), \tag{18}$$

and similarly others (prime going with the conjugate derivative). The symmetry of g_z implies that $g_{jk} = g_{kj}, g_{j'k'} = g_{k'j'}, g_{jk'} = g_{k'j}, \overline{g}_{jk} = g_{j'k'}, \overline{g}_{jk'} = g_{j'k}$ and that $g_{jk} = g_{j'k'} = 0$ for the $g_z \circ J_z = g_z$ property, $1 \leq j, k \leq d$.

We now define a subclass of analytic manifolds, suitable for the diffusion processes, as follows. For $X, Y \in T_z(M)$, let $\omega_z(X, Y) = g_z(J_z X, Y)$, so that $\omega : z \mapsto \omega_z$ defines a real differential 2-form (attached to the Hermitian metric g). If ω is a closed form, i.e., $d\omega = 0$, then g is called a *Kählerian metric.* Thus the Hermitian condition signifies the isometry of J_z of M at $z \in M$, and the Kählerian condition stipulates that J_z in addition be invariant under parallel displacement. Briefly, the latter means if $X \in T_z(M)$ and $X = \sum_{j=1}^d a_j e_j$, e_j

are orthonormal in $T_z(M)$ under the metric, $e_j = \sum_{k=1}^{d} \omega_{jk} e_k, \omega_{jk}$ being one-forms on M, then $dX = \sum_j da_j e_j + \sum_j \sum_k a_j \omega_{jk} e_j = \sum_k (da_k + \sum_j a_j \omega_{jk}) e_k$, and *parallel displacement* means $dX = 0$ so that $da_k + \sum_j a_j \omega_{jk} = 0$ for all j. Hence if X, Y move by parallel displacement then $(X, Y) =$ constant, so that the differential $d(X, Y) = (dX, Y) + (X, dY) = 0$. Thus a curve $\gamma : I \to M$, expressed in arc length form, is a *geodesic* if its unit tangent $\dot{\gamma}(s)$ moves by parallel displacement (cf. (3)), and it is maximal if it is not a restriction of another geodesic. It can be shown that for a manifold with a connection for any $0 \neq X \in J_x(M)$, there is a unique maximal geodesic $\gamma(\cdot)$ in M such that $\gamma(0) = x$ and $\dot{\gamma}(0) = X$. A complex manifold M with a Kähler metric g is termed a *Kähler manifold.* We want to discuss diffusion processes on this class of spaces.

In terms of the above conditions the Laplace-Beltrami operator Δ (12) takes the following simpler form:

$$\Delta f = \sum_{j,k=1}^{d} g^{jk'}(z) \frac{\partial^2 f}{\partial z_j \partial \bar{z}_k} \quad (g^{jk'} = g^{k'j}), \tag{19}$$

where the $g^{jk'}(z)$ and $g^{k'j}(z)$ satisfy for $1 \leq j, k \leq d$,

$$\sum_{\ell'} g_{j\ell'}(z) g^{\ell' k}(z) = \delta_{jk} = \delta_{jk'} = \sum_{\ell} g^{j'\ell}(z) g_{\ell k'}. \tag{20}$$

Thus a *Brownian motion on a Kähler manifold* (M, g) is given by Definition 2 where the Δ there is replaced by that of (19). With these concepts we now introduce diffusions on a Kähler manifold as the solution of a stochastic differential equation with driving force as a Brownian motion.

It will be convenient to restate the (affine) connection for a Riemannian manifold (M, g). We denote it $\nabla : T(M) \times T(M) \to T(M)$, a bilinear mapping which additionally satisfies for $X, Y \in T_x(M)$:

$$(\nabla_X(fY))(h) = f(\nabla_X Y)(h) + (Xf)(Yh), f, h \in C^\infty(M). \tag{21}$$

One calls ∇_X the *covariant differentiation* relative to X. In terms of the components of the connection $\Gamma_{ij}^k(\cdot)$, using local coordinates, one has with $X = \sum_j X^j(x) D_j, Y = \sum_j Y^j(x) D_j$ where $D_j = \frac{\partial}{\partial x_j}$, the following:

$$\nabla_X Y = \sum_k [\sum_{i,j} (\Gamma_{ij}^k(x) X^i(x) Y^j(x) + \sum_i X^i(x) D_i Y^k(x)] D_k. \tag{22}$$

Hence $Y(t)$ is *parallel* along the curve $\gamma(t)$, if the following system of differential equations holds, where in local coordinates of the neighborhood $U, x_i(t) = x_i(\gamma(t)), X^i(t) = X^i(x(t)) = \dot{x}_i(t)$, and $Y^i(t) = Y^i(\gamma(t)), t \in I$:

$$\frac{dY^k}{dt} + \sum_{i,j} \Gamma_{ij}^k \frac{dx_i}{dt} Y^j = 0, \ t \in I. \tag{23}$$

For the complex case, the definition is extended as: if $Z_j = X_j + iY_j, X_j, Y_j \in T_x(M), j = 1, 2$, then ∇_Z is given by,

$$\nabla_{Z_1} Z_2 = \nabla_{X_1} X_2 - \nabla_{Y_1} Y_2 + i(\nabla_{X_1} Y_2 + \nabla_{Y_1} X_2). \tag{24}$$

For a Kähler manifold, the coefficients Γ_{jk}^{ℓ} of ∇ are determined by the equations:

$$\sum_{\ell} g_{\ell p} \Gamma_{jk}^{\ell} = \frac{\partial g_{jp}}{\partial z_k}, \ \sum_{\ell'} g_{\ell' p'} \Gamma_{j' k'}^{\ell'} = \frac{\partial g_{p' j'}}{\partial \bar{z}_k}. \tag{25}$$

To consider a stochastic differential or integral equation on a manifold, we need to attach at each point a linear structure so that these equations are well defined, since otherwise addition and multiplication may not be possible. This will be done through tangent spaces with appropriate mappings between them. Thus corresponding to the natural unit vectors of $\mathbb{R}^d$, we consider an orthonormal set $e(x) = \{e_1, \dots, e_d\}_x$ in $T_x(M)$ for each $x \in M$, and call $e(x)$ an *orthogonal frame* at x of M. The set of all pairs $\{(x, e(x)), x \in M\}$ is denoted by $GL(M)$. This space can be made a smooth manifold by assigning a system of charts based on those of M as follows. Thus if $\{(U_i, \varphi_i), i \in J\}$ is an atlas of M, then let $(\tilde{U}_i, \tilde{\varphi}_i)$ be defined as:

$$\tilde{U}_i = \{v_x \in GL(M) : x \in U_i\}, \tilde{\varphi}_i(v_x) = (\varphi_i(x), e_{jk}, 1 \leq j, k \leq d),$$

for $i = 1, \dots, d$ where $e_j(x) = \sum_{k=1}^d e_{jk}(x)(\frac{\partial}{\partial x_k})_x, j = 1, \dots, d$. It may be verified that $GL(M)$ becomes a smooth manifold, with the atlas $\{(\tilde{U}_i, \tilde{\varphi}_i)\}_i$, which can be mapped into $\mathbb{R}^{d+d^2}$ so that it is of dimension $d+d^2$, and is termed an *orthogonal frame bundle* of M. Then the general linear group $GL(d, \mathbb{R})$ of d-by-d matrices acts on it on the right, by the rule:

$$v_x . g = (x, e(x))g = (x, e(x)g). \tag{26}$$

The coordinate projection $\pi : GL(M) \to M$ is that given by $\pi(v_x) = x$. Here $GL(M)$ is acted on by $GL(d, \mathbb{R})$ and is often called the *principal bundle*. If M is a complex manifold, then the frames will be *unitary vectors* and if M is also Riemannian with metric tensor g_z, then the $e_i(z)$ will be taken as orthonormal relative to g_z, i.e., $g_z(e_i(z), e_j(z)) = \delta_{ij}, z \in M$, and we demand that each map $z \mapsto e_j(z)$ be holomorphic as well; and denoting $GL(M)$ by $UL(M)$, we call it a *unitary frame bundle*. Now if M is a manifold with a connection $\nabla = (\Gamma_{ij}^k(x), x \in M)$, and if $T_{v_x}(GL(M))$ is the tangent space, consider a subspace H_v, called the *horizontal space*, defined by

$$H_{v_x} = \{X = \sum_i [a_i(x)(\frac{\partial}{\partial x_i})_x - \sum_{j,k} \Gamma_{ij}^k(x) a_k(x) e_{ij}(x) \frac{\partial}{\partial e_{ij}}\}, \quad (27)$$

where $a_i(x) \in \mathbb{R}$. In the complex case, we take a_i's, e_{ij}'s to be holomorphic. The vectors X of H_{v_x} are termed *horizontal*. If X is a C^∞-vector field at $x \in M$, then there exists a unique C^∞-vector field $\tilde{X}$ on $GL(M)$ such that $\tilde{X} \in T_{v_x}(GL(M))$, called a *horizontal lift* of X satisfying $\pi(v_x) = x$ and $\tilde{\pi}_{v_x}(\tilde{X}) = X$, where $\tilde{\pi}_{v_x}$ is the projection of $T_{v_x}(GL(M))$ onto $T_x(M)$. This $\tilde{X}$ is given as an element of H_{v_x} in (27), which is why it is called a horizontal space. Similarly a curve $\gamma : I \to M$ is said to have a (horizontal)*lift* to $\tilde{\gamma} : I \to GL(M)$ if $\pi(\tilde{\gamma}(t)) = \gamma(t)$, and $\dot{\tilde{\gamma}}(t) \in H_{v_x}$, so it is horizontal for each $t \in I$. In local coordinates, these can be stated in terms of differential operators. If $v_x = (x, e(x))$ with $e(x) = (e_1(x), \ldots, e_d(x))$ as a frame, at $x \in M$, so that $e_i(t) \in T_{\gamma(t)}(M)$ is true for $e_i(t)$ by parallel displacement along $\gamma(t)$, then $\tilde{\gamma}$ is the horizontal lift of $\gamma(\cdot)$ at v_x. The C^∞-vector field L_j on $GL(M)$ is then given by

$$L_j = \sum_i e_{ij} \frac{\partial}{\partial x_i} - \sum_{j,r,p,k} \Gamma_{jr}^k e_{jr} e_{rp} \frac{\partial}{\partial e_{pk}}, \quad j = 1, \ldots, d, \quad (28)$$

and $(L_1, \ldots, L_d)$ is a system of horizontal vector fields.

Let the *bundle of orthogonal frames* be defined as:

$$O(M) = \{v_x = (x, e_x) \in GL(M) : e(x) \quad ortho.frame\, of\; T_x(M)\}.$$

Then v satisfies $\sum_{m=1}^d e_{im} e_{jm} = g^{ij}$, where (g^{ij}) is the inverse of the metric tensor (g_{ij}) of M.

If $Z(t) = (Z_1(t), \dots, Z_d(t))$ is an $L^{2,2}$-bounded process, as defined earlier, and if $X : [0,\infty] \to M$ is a process given before, then we say that the stochastic differential equation has $v_{X(x)}(\cdot)$ as a solution, of the formal (i.e., symbolic) equation, starting at x:

$$dv_{X(t)}(t) = \sum_{j=1}^{d} L_j(v_{X(t)}(t)) \circ dZ_j(t), \;\; v_{X(0)} = v_x. \tag{29}$$

The solution $v_{X(t)}(t,x)$ is a stochastic flow on $O(M)$ relative to the canonical horizontal vector fields $(L_1, \dots, L_d)$. In more detail, we have (29) expressed as (Stratonovich integrals)

$$\begin{aligned} dX^i(t) &= \sum_{j=1}^{n} e^i_j(t) \circ dZ_j(t), i = 1, \dots, d, \\ de^i_j(t) &= -\sum_{r,k} \Gamma^i_{rk}(X(t)) e_{jr}(t) \circ dX_r(t), i,j = 1, \dots, d, \end{aligned} \tag{30}$$

with $v_{X(t)}(x) = (x, e(t))$. The flow $v_{X(t)}(x) \in O(M)$ if $v_{X(0)}(x) \in O(M)$, at least when the Z-process is a semimartingale.

In the complex case we consider $\tilde{Z}_j = X_j + \sqrt{-1}Y_j$ where X_j, Y_j are $L^{2,2}$-bounded and the bundle $O(M)$ is replaced by the *bundle of unitary frames*, $U(t) = \{v_z = (z, e_z) : e_z \, unitary \, frame \, of \, T_z, z \in M\}$. For Kähler manifolds, with Z as Brownian motion, we usually normalize by taking X_j, Y_j as independent Brownian motions and dividing each by $\sqrt{2}$. The resulting equation is

$$dv_{X(t)}(t,z) = \sum_{j} L_j(v_{X(t)}(t)) \circ d\tilde{Z}_j(t), \tag{31}$$

where $L_j(v_z) \in T^c_{v_z}(U(M))$. Replacing $O(M)$ by $U(M)$ here the corresponding statements obtain. We can therefore present the following general statement:

4. Theorem. *Let (M,g) be a Riemannian manifold with an affine connection, which is compatible with g. If $L_j, j = 1, \dots, d$ are canonical horizantal vector fields and $(Z_j, j = 1, \dots, d)$, is a Brownian motion process, then the solution $v_{X(t)}(x)$ of (29) or (30) is a stochastic flow in $O(M)$, and its natural projection $X(t) = \pi(v_{X(t)})$ defines a diffusion process on M corresponding to the generator $L = \frac{1}{2}\Delta_M + b$ where Δ_M*

is the Laplace-Beltrami operator, (cf.(12)):

$$\Delta_M f = \sum_{i,j} [g^{ij}(\frac{\partial^2 f}{\partial x_i \partial x_j} - \sum_k \Gamma^k_{ij} \frac{\partial f}{\partial x_k})], \quad f \in C^\infty(M),$$

and b is the vector field corresponding to the drift on $O(M)$, given by:

$$b = \sum_j b_j(x) \frac{\partial}{\partial x_j},$$

in which $b_j(\cdot)$ are functions of the components of the connection.

The complete detail in the case that Z_t is Brownian motion is given in Ikeda-S. Watanabe(1989) for the Riemannian as well as the Kähler manifolds. [The former result seems to be due to Eells and Elworthy, and the latter is the (conformal) reformulation due to these authors.] The existence and uniqueness of solutions of (30) is based on the corresponding result for (semi)martingales. Since we also have an analogous version in Theorem VI.4.6 if the Z_t-process is $L^{2,2}$-bounded relative to a σ-finite measure, a similar extension seems entirely possible. Indeed higher order equations can be considered. The details have not yet been completed. The above version is included to indicate the type of results that one wants to formulate for the stochastic differential equations in manifolds using the generalized boundedness principles.

(d) *Malliavin's approach.* Consider the abstract Cauchy problem:

$$\frac{\partial u(x,t)}{\partial t} = (\frac{1}{2}\Delta + A_0)u(x,t); \;\; \lim_{t \downarrow 0} u(x,t) = f(x), x \in \mathbb{R}^d, \qquad (32)$$

where $\Delta = \sum_{i=1}^m A_i^2$ and $A_i = \sum_{j=1}^d a_{ij} \frac{\partial}{\partial x_j}, i = 0, 1, \ldots, m$ are smooth vector fields. Then the classical theory implies that (32) has a smooth solution which is representable as:

$$u(x,t) = \int_{\mathbb{R}^d} K(t,x,y) f(y) dy, \quad (t,x) \in (0,\infty) \times \mathbb{R}^d,$$

where $K(t,\cdot,\cdot) : \mathbb{R}^d \times \mathbb{R}^d \to \mathbb{R}$ is a smooth kernel for each $t > 0$, if the (positive definite) matrix $(a_{ij})(a_{ji})$ is invertable (i.e.,Δ is an elliptic operator). Then Hörmander(1967) showed that the same conclusion holds if the operators A_i satisfy the following far weaker condition; namely, the vector fields

$$\{A_i, [A_i, A_j], [[A_i, A_j], A_k], \ldots, 1 \le i \le n, 0 \le j, k, r, \ldots, \le n\}, \quad (33)$$

span the tangent space $T_x(\mathbb{R}^d)$ for each $x \in \mathbb{R}^d$, and the resulting operator Δ is called *hypoelliptic*. Here $[A_i, A_j] = A_iA_j - A_jA_i$, the Lie bracket, which is a derivation and is used as a vector field. As we know before, the Cauchy problem is closely related to Brownian motion, and hence the general case is also expected to be related to Brownian functionals, i.e., to solutions of Itô (or Stratonovich) stochastic differential equations with Brownian driving force, so that the Itô solution is a Markov process. Then the associated semigroup (or evolution family) of operators shall have the infinitesimal generator $\mathcal{G}$ which is essentially $(\frac{1}{2}\Delta + A_0)$ above. This was shown to be the case, with even a possibly weaker condition than (33), by Malliavin (1976) who in the course of the solution has created stochastic differential geometry and a calculus of variation. The preceding work thus forms an introduction to this new area and we can only include a brief sketch of the subject primarily for space reasons. Here then is Malliavin's formulation of the problem.

Let $(B, \mathcal{B}, P)$ be an abstract Wiener space in the sense of Section I.4, and $\{W_t, t \ge 0\}$ be an n-dimensional Wiener (or Brownian motion) process on it. If $\{X_t, t \ge 0\}$ is a solution of the equation (under conditions similar to those of Exercise VI.6.8)

$$X_t = x + \int_0^t A_0(X_s)ds + \sum_{i=1}^n \int_0^t A_i(X_s) \circ dW_i(s), \quad (34)$$

so that X_t is a Brownian functional, let $Q_t = P \circ X_t^{-1}$ be the image measure of P under X_t, where $\{A_0, \ldots, A_n\}$ are smooth vector fields. The problem is to show that Q_t has a smooth density (relative to the Lebesgue measure) for each t where B is replaced by its canonical representation $B = C([0,1], \mathbb{R}^d)$. One sees, as in Example 2 after Theorem I.4.5, that B is the closure in the uniform norm (which is a measurable norm in the sense defined there) of the Hilbert space of absolutely continuous functions f satisfying $\|f\|^2 = \int_0^1 \cdots \int_0^1 \|D_x^2 f\|^2 dx_1 \cdots dx_n < \infty$. Then one uses a result from Fourier analysis which states that for any $\mathbb{R}^d$-valued random variable X, with distribution F, such that for a multiindex $\alpha = (\alpha_1, \ldots, \alpha_n), \alpha_i \geq 0$ integers, and $|\alpha| = \alpha_1 + \cdots + \alpha_n$, there is a constant $C_n > 0$ satisfying for $|\alpha| \leq n - d - 1$,

$$E(|(D^\alpha f)(X)|) \leq C_n \|f\|_\infty, \quad f \in C_b^\infty(\mathbb{R}^d), \tag{35}$$

we have a C^{n-d-1}-smooth density. A proof of this statement is a consequence of the Lévy inversion formula when for the characteristic function $\widehat{F}$ of F the map $u \mapsto |u|^{d+1}\widehat{F}(u)$ is in $L^1(\mathbb{R}^d)$. Thus the verification of inequality (35) becomes critical in our problem and all the work and ideas of differential geometry go into finding the "best" conditions for the purpose. It may be shown that (33) is a good condition in this case. A better (or weaker) hypothesis is devised by Malliavin as follows. It comes up in the proof but otherwise has little motivation.

Using the notation of (33) and (34), let Z_t be a d-by-d matrix process defined as a solution of the (Stratonovich) stochastic differential equation with X_t being a solution of (34):

$$Z_t = I + \int_0^t Z_s(DA_0)(X_s)ds - \sum_{i=1}^n \int_0^t Z_s(DA_i)(X_s) \circ dW_i(s),$$

for $t \geq 0$, where I is the identity matrix. Define another matrix valued process C_t, with $A = (A_1, \ldots, A_n)$, as:

$$C_t = Z_t^{-1}\Big[\int_0^t Z_s(AA^*)(X_s)Z_s^*\Big](Z_t^{-1})^*. \tag{36}$$

Here Z^* denotes the transpose of the matrix Z. The $C_t, t \geq 0$, is called the *Malliavin covariance* of Z_t. We then have the following fundamental result:

5. Theorem. *Suppose $\{X_t, t \in [0,1]\}$ is a solution of (34) with $\{W_t, t \in [0,1]\}$ as a d-dimensional Brownian motion driving force. Then the distribution of the random variable X_t has a C^∞-smooth density relative to the Lebesgue measure if the C_t-process of (36) satisfies* $\det(C_t)^{-1} \in L^p(P)$ *for all* $1 \leq p \leq \infty$, *and the last condition holds automatically when (33) is valid.*

The statement does not give a clue about the stated restrictions, and the proof is based on a series of results which are of independent interest. This is given fully, albeit in compact form, by Malliavin (1976) himself, and numerous additions and extensions appeared later. We refer to Ikeda-S.Watanabe (1989) for a somewhat different argument, and Norris (1986) for another proof using only the classical notation, by untangling the differential geometric symbolism, using the standard probability estimates. The differential geometric ideas are, however, basic for the discovery, elaboration, and extension of this fundamental result which created a new area in Probability. It would appear that some of the work that goes into the abstract argument extends to McShane processes from Brownian motion, and perhaps eventually to the general $L^{2,2}$-bounded classes. The results of $\mathbb{R}^d$-valued processes can be localized and formulated for smooth manifolds. Indeed such an extension is given by Taniguchi (1983) and various generalizations are actively persued in the literature. Even an infinite dimensional extension using the methods of Belopol'skaya and Dalecky (1990) seem not only possible, but is an interesting topic for future research since the necessary work on manifolds modelled after Banach spaces is also available. We do not discuss the subject further for the reasons noted earlier.

7.3 Extensions to multiparameters

Except for some work in parts of Chapters V and VI, most of the preceding analysis was restricted to processes with index sets contained in $\mathbb{R}$. There arise new problems in extending the theory to multidimensions just as in the several-variable-function-theory in classical analysis. Because of its importance in many applications, we outline some aspects of this theory. Research in this area is also actively pursued.

It is natural to start with Brownian motion in several dimensions. We already discussed the Lévy-Brownian motion which is not a martingale. However, there is another extension which retains the latter property and is as follows. Let T be a finite dimensional Euclidean subspace (say $T \subset \mathbb{R}^d$), and μ be a Haar measure on it. Set $\mathcal{B}_0 = \{A \subset T : A \quad Borel, \mu(A) < \infty\}$. Then $\mathcal{B}_0$ is a δ-ring which is directed by inclusion. i.e., $A_1, A_2 \in \mathcal{B}_0, A_1 \prec A_2$ iff $A_1 \subset A_2$. Consider a Gaussian process $Z = \{Z_A : A \in \mathcal{B}_0\}$ with means zero and covariances $\{\mu(A \cap B) : A, B \in \mathcal{B}_0\}$. By Theorem I.2.1 such a process exists if we take $\Omega = \mathbb{R}^{\mathcal{B}_0}, Z_A(\omega) = \omega(A) \in \mathbb{R}$, and for each $n \geq 1, (A_1, \ldots, A_n)$ from $\mathcal{B}_0$, we assign to $(Z_{A_1}, \ldots, Z_{A_n})$ an n-dimensional Gaussian distribution with means zero and the positive definite matrix $(\mu_{ij} = \mu(A_i \cap A_j), 1 \leq i, j \leq n)$ as its covariance matrix, which then determines a probability measure P on Σ, the σ-algebra generated by the cylinder sets of Ω. Thus on (Ω, Σ, P) the desired process is defined. Also $A \cap B = 0 \Rightarrow Z_A + Z_B = Z_{A \cup B}$. Moreover, if $\{A_n, B\} \subset \mathcal{B}_0, A_n$, disjoint, $B = \cup_{n=1}^{\infty} A_n$, then $Z_B = \sum_{n=1}^{\infty} Z_{A_n}$, the series converging in $L^2(P)$, because the Z_{A_n} are Gaussian (cf., I.5.3).

If $T = \mathbb{R}$ and $\{X_t, t \in T\}$ is the standard Brownian motion, let $Z_{[0,t]} = X_t, t \geq 0$, and $= Z_{[t,0]}, t \leq 0$, making Z the "white noise". In case T is the positive orthant of $\mathbb{R}^k, A = [0, t_1) \times \cdots \times [0, t_k)$, then the process $X_{t_1, \ldots, t_k} = Z_A$ is called the n-dimensional *Wiener-Brownian motion.* With a separable version, it can be shown to have a.a. continuous sample paths. It is also a martingale with the above ordering, using the properties of the Z-process. Indeed, let $\mathcal{F}_S$ be the σ-algebra generated by $\{Z_A : A \in \mathcal{B}_0, A \prec S\}$. Then $\{\mathcal{F}_A, A \in \mathcal{B}_0\}$ is a stochastic base and for $A_1 \prec A_2$ (from $\mathcal{B}_0$), Z_{A_1} and $Z_{A_2} - Z_{A_1}$ are independent since $E(Z_{A_1}(Z_{A_2} - Z_{A_1})) = \mu(A_1 \cap A_2) - \mu(A_1) = 0$, the Z_{A_i} being Gaussian. So $E^{\mathcal{F}_{A_1}}(Z_{A_2}) = Z_{A_1} + E^{\mathcal{F}_{A_1}}(Z_{A_2} - Z_{A_1}) = Z_{A_1}$,a.e. The X-process is also a martingale under coordinate ordering (with μ as the Lebesgue measure). Under the latter ordering $\{\mathcal{F}_t, t \in T\}$ in (Ω, Σ, P) does *not* satisfy the Vitali condition "V_0" (cf., Section III.5).

To gain some insight, we present a maximal inequality and a pointwise convergence assertion for martingales with the componentwise ordering, complementing the earlier work in Section III.5.

1. Theorem. *Let (Ω, Σ, P) be a probability space and $\{\mathcal{F}_s, s \in \mathbb{N}^k\}$, with componentwise ordering of the positive integer lattice $\mathbb{N}^k$, be an increasing family of σ-subalgebras of Σ which are complete. Let $\{X_s, \mathcal{F}_s, s \in \mathbb{N}^k\}$ be a martingale. Then, for each $\lambda \geq 0$,*

$$\lambda P[\sup_s |X_s| \geq \lambda] \leq C_k \sup_s E(|X_s|(lg^+|X_s|)^{k-1}) + B_k, \tag{1}$$

and for $p > 1$,

$$E(\sup_s |X_s|^p) \leq C_{p,k} \sup_s E(|X_s|^p), \tag{2}$$

where $C_k, C_{p,k}, B_k$ are constants depending only on k, p. If also the right sides of (1) or (2) is finite, then $\lim_{s \to \infty} X_s$ exists a.e. where the limit is taken in the coordinatewise ordering.

Remark. Cairoli [1] gave an example of a uniformly integrable martingale $\{X_s, \mathcal{F}_s, s \in \mathbb{N}^k\}$ on a product Lebesgue space and a nonnegative convex function φ_0 which grows slower than $t(\lg^+ t)^{k-1}, t \geq 0$, such that $\sup_s E(\varphi_0(|X_s|) < \infty$, but $\lim_s X_s$ does not exist on a set of positive measure (and hence the filtration cannot satisfy the V_0-condition). Also if $(\Omega, \Sigma) = \otimes_{i=1}^k (\Omega_i, \Sigma_i)$ and $\mathcal{F}_s = \otimes_{i=1}^k \mathcal{F}_{s_i}, \{\mathcal{F}_{s_i} \uparrow\} \subset \Sigma_i$, then the hypothesis on $\{\mathcal{F}_s, s \in \mathbb{N}^k\}$ is automatic. In particular, (1) and (2) give with $p = 2, \lambda > 0$,

$$P[\sup_{s \leq t} |X_s| \geq \lambda] \leq C_{k,2}\lambda^{-2} E(|X_t|^2), \tag{1'}$$

and similarly for $p > 1$,

$$E(\sup_{s \leq t} |X_s|^p) \leq C_{p,k} E(|X_t|^p), \tag{2'}$$

in which one can take $C_{2,2} = 4, C_{p,2} = (\frac{p}{p-1})^{2p}$. The argument essentially follows Cairoli [1].

Proof. Both the inequalities are true for $k = 1$, by Section II.4. We show that they are also true for $k = 2$, and then use induction on k for the general case. For nontriviality the right sides of (1) and (2) are assumed finite.

Consider (1). By definition of the ordering, $\{|X_{s_1,s_2}|, \mathcal{F}_{s_1,s_2}, s_1 \geq 0, s_2 \geq 0\}$ is a submartingale if s_1 is varied and s_2 is arbitrarily fixed. Also for $1 \leq r_2 \leq m$, if $Y_{r_1,r_2} = \sup_{r_2} |X_{r_1,r_2}|$, then $\{Y_{r_1,m}, \mathcal{F}_{r_1,m}, r_1 \geq 0\}$ is also a submartingale, (cf.,Section II.4(b)). Hence letting $Y_r = \lim_{m\to\infty} Y_{r,m} = \sup_s |X_{r,s}|$, we deduce that $\{Y_r, \mathcal{F}_{r,\infty}, r \geq 0\}$ is again a submartingale. So by Corollary IV.1.5 and Theorem II.4.8, we get for any $\lambda > 0$

$$\begin{aligned} \lambda P\big[\sup_{r_1,r_2} |X_{r_1,r_2}| \geq \lambda\big] &\leq \sup_r E(Y_r) = \sup_r E(\sup_s |X_{r,s}|) \\ &\leq \sup_r \big[C_1 \sup_s E(|X_{r,s}(\lg^+ |X_{r,s}|)) + B\big], \end{aligned} \tag{3}$$

where $C_1 = B = \frac{e}{e-1}$ since $\{|X_{r,s}|, \mathcal{F}_{\infty,s}, s \geq 0\}$ is a positive submartingale for each $r \geq 1$. This shows that (1) is true for $k = 2$.

Now to use induction, suppose (1) is true for $k = m \geq 2$. Thus fixing $r_1, \ldots, r_m$, let $Y_{r_1,\ldots,r_m} = \sup_r |X_{r_1,\ldots,r_m,r}|$, so that

$\{Y_{r_1,\dots,r_m}, \mathcal{F}_{r_1,\dots,r_m,\infty}, r_i \geq 0, i = 1,\dots,m\}$ is a submartingale. Hence by the inductive hypothesis there exist two constants C_m, B_m such that for $\lambda > 0$

$$\lambda P[\sup_{r_1,\dots,r_m} Y_{r_1,\dots,r_m} \geq \lambda] \leq C_m \sup_{r_1,\dots,r_m} E(Y_{r_1,\dots,r_m}(\lg^+ Y_{r_1,\dots,r_m})^{m-1}) + B_m. \tag{4}$$

But $\varphi(t) = t(\lg^+ t)^{m-1}$ is convex increasing and $\varphi(0) = 0$. Hence $\varphi(\max_i Z_i) \geq \max_i \varphi(Z_i)$ for any $Z_i \geq 0$, so that (4) becomes

$$\lambda P[\sup_{r_1,\dots,r_{m+1}} |X_{r_1,\dots,r_{m+1}}| \geq \lambda] \leq C_m \sup_{r_1,\dots,r_m} E(\sup_{r_{m+1}} \varphi(|X_{r_1,\dots,r_{m+1}}|)) + B_m. \tag{5}$$

Since $\{\varphi(|X_{r_1,\dots,r_m,r_{m+1}}|), \mathcal{F}_{r_1,\dots,r_m,r_{m+1}}, r_{m+1} \geq 0\}$ is a submartingale for fixed $(r_1,\dots,r_m)$ by Section II.4(b), we may apply Theorem II.4.8 again to get

$$E(\sup_{r_{m+1}} \varphi(|X_{r_1,\dots,r_m,r_{m+1}}|)) \leq C'_m \sup_{r_{m+1}} E(\varphi(|X_{r'_1\dots,r_{m+1}}|)(\lg^+ \varphi(|X_{r_1,\dots,r_{m+1}}|))) + B'. \tag{6}$$

However if $u = \lg^+ t (\leq t \text{ for } t > 1)$, we have

$$tu^{k-1} \lg^+(tu^{k-1}) \leq ktu^k.$$

Thus (6) becomes

$$E(\sup_{r_{m+1}} \varphi(|X_{r_1,\dots,r_{m+1}}|)) \leq C'_m \sup_{r_{m+1}} E[|X_{r_1,\dots,r_{m+1}}|(\lg^+ |X_{r_1,\dots,r_{m_1}}|)^m] + B'. \tag{7}$$

Then (5) and (7) imply (1) for $k = m+1$, if $C_{m+1} = C_m C'_m, B_{m+1} = B_m + C_m B'$.

The proof of (2) is similar, and we omit the details of computation.

Let us turn to the convergence assertion. First consider the case $p > 1, \sup_{r_1,\dots,r_k} E(|X_{r_1,\dots,r_k}|^p)$ is finite. Then a bounded set in $L^p(P)$ is uniformly integrable, and Theorem III.5.5 implies that $X_s \to \tilde{X}$

in L^p-norm. Moreover $X_s = E^{\mathcal{F}_s}(\tilde{X})$ since the terminal uniformity is clearly implied. If $s^{n_0} = (s_1^{n_0}, \ldots, s_k^{n_0})$ is arbitrarily fixed, and $s \succ s^{n_0}$, then $\{X_s - X_{s^{n_0}}, \mathcal{F}_s, s \succ s_{n_0}\}$ is a martingale. Hence by (2) we get with $\lambda = \frac{1}{n_0}$, where s^{n_0} is chosen such that $E(|\tilde{X} - X_{s^{n_0}}|^p) \leq \frac{1}{n_0^2}$, that

$$P[\sup_{s \succ s^{n_0}} |X_s - X_{s^{n_0}}|] \leq C_{p,k}\frac{1}{n_0}. \tag{8}$$

Letting $s \to \infty$ (i.e., $s_i \to \infty, i = 1, \ldots, k$), and then $n_0 \to \infty$, (8) implies the conclusion in this case. We outline the other part.

Let the right side of (1) be finite, and $\varphi(t) = t(\lg^+ t)^{k-1}, t \geq 0$. Then φ is a Young function and the Orlicz space $L^\varphi(P)$ is contained in $L^1(P)$. Moreover, since $k > 1$ (for $k = 1$ the result was already proved in Theorem II.6.1), $\varphi'(t) \uparrow \infty$, so that the complementary Young function ψ is also continuous and by the Hölder inequality for these spaces ($\|\cdot\|_\varphi, \|\cdot\|_\psi$ are congugate norms, cf., e.g, Rao and Ren [1],p.58),

$$\lim_{P(A)\to 0} \int_A |X_s|dP \leq 2 \lim_{P(A)\to 0} \|X_s\|_\varphi \|\chi_A\|_\psi \leq K_0 \lim_{P(A)\to 0} \|\chi_A\|_\psi = 0, \tag{9}$$

where K_0 is a constant determined by the right side of (1). Hence by Theorem III.5.5, there exists an $X \in L^1(P)$ such that $E^{\mathcal{F}_s}(X) = X_s$ and $X_s \to X$ in $L^1(P)$. It is then also true that $X_s \to X$ in $L^\varphi(P)$. (This is an easy computation using Fatou and Jensen inequalities to conclude that $X \in L^\varphi(P)$ and then one shows that $(X_s - X) \to 0$ in $L^\varphi(P)$.) Hence we can apply the same argument as in the preceding paragraph with (1) to deduce the pointwise convergence. □

2. Note. It should be emphasized that while the $L^1(P)$ convergence of X_s to X is true if $\sup_s E(|X_s|) < \infty$, and terminal uniform integrablity holds, the pointwise convergence is false if (1) is not finite, and this fact distinguishes the multiparameter case from the linear parameter result.

To develop further, recall that an adapted integrable process $X = \{X_n, \mathcal{F}_n, n \geq 1\}$ is a submartingale iff for each $n \geq 1$, $E^{\mathcal{F}_n}(X_{n+1} - X_n) \geq 0$, a.e. With a view to get an analog of a Doob-Meyer decomposition, we now introduce some relevant concepts.

3. Definition. Let (Ω, Σ, P) be a probability space and $T \subset \mathbb{R}^k$ with a componentwise ordering. If $\{\mathcal{F}_t, t \in T\}$ is an increasing (in each component as before) family of σ-subalgebras of $\Sigma, X = \{X_t, \mathcal{F}_t, t \in T\}$ is an adapted integrable process, $h_i > 0$, and $\Delta_h X_t$ is the increment of X_t (so for instance $k = 2, \Delta_h X_t = (X_{t_1+h_1,t_2+h_2} - X_{t_1,t_2+h_2} - X_{t_1+h_1,t_2} + X_{t_1,t_2})$, if $t + h \in T$), then X is called a *subprocess* provided

$$E^{\mathcal{F}_t}(\Delta_h X_t) \geq 0, a.e., t \in T, h_i > 0, t + h \in T. \tag{10}$$

It is a *superprocess* if the inequality in (10) is reversed. If X is both a sub- and a superprocess then it is termed an M *– process*. [When the parameter is in a subset of $\mathbb{R}^k, k > 1$, we should call these classes, *random fields* instead of processes. We are following the popular usage, but may use both terms according to convenience.]

As examples of the above classes, we cite the following: (i) The Wiener-Brownian motion process introduced earlier in this section, whose increments $\Delta_h X_t$ are Gaussian distributed with mean zero, variance $\prod_{i=1}^k h_i$, and are independent of $\mathcal{F}_t$, is an M-process as well as a martingale; (ii) If $(\Omega, \Sigma, P) = \otimes_{i=1}^k(\Omega_i, \Sigma_i, P_i), \mathcal{F}_t^i \subset \Sigma_i$ is increasing with $\mathcal{F}_t = \otimes \mathcal{F}_t^i \subset \Sigma$, and $\{X_{t_i}^i, \mathcal{F}_t^i, t_i \geq 0\}$ is a submartingale for each i, then $\{X_t, \mathcal{F}_t, t \geq 0\}$ is a subprocess, where $X_t = X_{t_1}^1 X_{t_2}^2 \cdots X_{t_k}^k$ and $t = (t_1, \ldots, t_k)$; (iii) If $\{X_t, \mathcal{F}_t, t \in T\}$ is a square integrable martingale, then $\{X_t^2, \mathcal{F}_t, t \in T\}$ is a subprocess whenever (Ω, Σ, P) is a product space as in (ii) or the increments $\Delta_h X_t$ are independent of $\mathcal{F}_{t+h}$ for each $h > 0$ with $T = \mathbb{R}_+^k$, the positive orthant of $\mathbb{R}^k$. Note that the sub- and supermartingales need not be sub- and super processes (but a martingale is evidently an M-process).

The statement of (iii) may be less evident. To verify it, taking $k = 2$ for simplicity, note that

$$\begin{aligned} E^{\mathcal{F}_t}((\Delta_h X_t)^2) &= E^{\mathcal{F}_t}(X_{t_1+h_1,t_2+h_2}^2 - X_{t_1,t_2+h_2}^2 - X_{t_1+h_1,t_2}^2 + X_{t_1,t_2}^2) \\ &= E^{\mathcal{F}_t}(\Delta_h X_t^2), \end{aligned} \tag{11}$$

since $E^{\mathcal{F}_{t_1,t_2+h_2}}(X_{t_1+h_1,t_2}) = E^{\mathcal{F}_{t_1,t_2}}(X_{t_1+h_1,t_2}) = X_{t_1,t_2}$ by the fact that the measure space is a product of its components in the first case, and $X_{t_1+h_1,t_2} - X_{t_1,t_2}$ is independent of $\mathcal{F}_{t_1,t_2+h_2}$ in the second. We also remark that the latter occurs if $\{X_t, \mathcal{F}_t, t \in T\}$ is a Wiener-Brownian motion which is a martingale so that $\{X_t^2, \mathcal{F}_t, t \in T\}$ is a submartingale as well as a subprocess.

4. Definition. A process $\{A_t, \mathcal{F}_t, t \geq 0\}$ on (Ω, Σ, P) with right continuous $\mathcal{F}_t (= \cap_{s>t} \mathcal{F}_s)$ is *increasing* if (a) $A_{t_1,\ldots,t_{j-1},0,t_{j+1},\ldots,t_k} =$

0,a.e., $2 \leq j \leq (k+1)$, (b)$\Delta_h A_t \geq 0$ a.e. for each $h > 0$, and (c) $A_{t+0} = A_t$ (i.e., is right continuous). A process $\{X_t, t \geq 0\}$ is *predictable* if it is $\mathcal{P}$-measurable as a mapping of $\mathbb{R}_+^k \times \Omega \to \mathbb{R}$, where $\mathcal{P}$ is the σ-algebra generated by all the left-continuous $\mathcal{F}_t$-adapted real processes on $\mathbb{R}_+^k \times \Omega$, or equivalently by the class $\{(s,t] \times F, F \in \mathcal{F}_s\}$,for $s < t$ in $\mathbb{R}_+^k$, with $(s,t] = \otimes_{i=1}^k (s_i, t_i]$, a left-open right-closed interval. [The above A may also be equivalently called predictable if $E(\int_0^t X_{s-} dA_s) = E(X_s A_t), t = (t_1, \ldots, t_k), t \in \mathbb{R}_+^k$, for each bounded X with $X_{t-} = \lim_{h \to 0} X_{t-h}, h > 0$.]

In the discrete case the last statement above reduces to the condition that $A_{n_1+1,\ldots,n_k+1}$ is $\mathcal{F}_{n_1,\ldots,n_k}$-measurable, as one can verify. With these concepts we want to extend the decomposition results of Chapters II and V. We start with the case $k = 2$ by giving an analog of Theorem II.5.1, for the discrete parameter processes.

5. Theorem. *Let $\{X_n, \mathcal{F}_n, n \in \mathbb{N}^2\}$ be a subprocess on (Ω, Σ, P). Then it admits a unique decomposition as:*

$$X_n = Y_n + A_n, n \in \mathbb{N}^2, \tag{12}$$

where $\{Y_n, \mathcal{F}_n, n \in \mathbb{N}^2\}$ is an M-process and $\{A_n, n \in \mathbb{N}^2\}$ is a predictable increasing integrable process. If moreover (Ω, Σ, P) is a product space (or the filtration $\mathcal{F}_n$ satisfies $E^{\mathcal{F}_{n_1,n_2}} E^{\mathcal{F}_{m_1,m_2}} = E^{\mathcal{F}_{n_1 \wedge m_1, n_2 \wedge m_2}}$, the conditional independence) then, in (12) Y can be expressed as

$$Y_{n_1,n_2} = Y'_{n_1,n_2} + Y''_{n_1,n_2} + Y'''_{n_1,n_2} \tag{13}$$

where Y' is a martingale and the last two are one parameter martingales in the first (respectively the second) parameters when the other is fixed. This decomposition is also unique if we choose $Y''_{n_1,n_2}(Y'''_{n_1,n_2})$ to be $\mathcal{F}_{n_1,n_2-1}(\mathcal{F}_{n_1-1,n_2})$-adapted, where by definition $\mathcal{F}_{n_1,-1} = \mathcal{F}_{n_1,0}, \mathcal{F}_{-1,n_2} = \mathcal{F}_{0,n_2}$, and $\mathcal{F}_{-1,-1} = \mathcal{F}_{0,0}$. Finally, if the X-process is a submartingale (as well as a subprocess) lying in a ball of $L^1(P)$, then in (12) $E(A_{n,m}) \leq K_0 < \infty$ and

$$\sup_{n_1} \sum_{n_2=0}^{\infty} (|Y''_{n_1,n_2+1} - Y''_{n_1,n_2}|) < \infty; \sup_{n_2} \sum_{n_1=0}^{\infty} E(|Y'''_{n_1+1,n_2} - Y'''_{n_1,n_2}|) < \infty. \tag{14}$$

Sketch of Proof. Consider the first part. If X is a subprocess, define $A_{n_1,n_2} = \sum_{i=0}^{n_1} \sum_{j=0}^{n_2} a_{ij}$, where

$$a_{ij} = E^{\mathcal{F}_{i-1,j-1}}(X_{i,j} - X_{i-1,j} - X_{i,j-1} + X_{i-1,j-1}), i \geq 1, j \geq 1,$$

and $a_{i,0} = a_{0,i} = 0, i \geq 1$. Let $Y_{n_1,n_2} = X_{n_1,n_2} - A_{n_1,n_2}$. Then it is easily seen that the Y is an M-process, the A is an increasing integrable process and the decomposition (12) holds. That A is predictable in this case is clear, and we need to deduce the uniqueness of decomposition in (12), since the sum of an M-process and a predictable increasing integrable process defines a subprocess. It involves some computation. Details of this and the other parts can be referred to Cairoli [2]. □

We now discuss the continuous parameter version of the above result which naturally involves further assumptions and work. For simplicity of notation, we restrict to $k = 2$. In this extension we follow largely Brennan [1] (cf. also Walsh [2]).

Hereafter we use the notation: for $s, t \in \mathbb{R}_+^2$, $s \prec t$ means that each coordinate of s is strictly less than the corresponding one of t, and $s \preceq t$ satisfies the weaker inequality, in this ordering.

A natural (broad) class of multiparameter processes for which the stochastic calculus can be extended may be based on quasimartingales of Definition V.2.19. The condition there is motivated by the classical concept of a function of bounded variation $f : [a, b] \to \mathbb{R}$, and in multidimensions, even in $\mathbb{R}^2$, there are several distinct definitions of this notion which were discussed and interrelations explored by Clarkson and Adams [1]. From these we find the analogs of Vitali, Hardy, and Arzelà to be most appropriate here, although the Fréchet variation is more general and important for the spectral analysis of second order processes (called "harmonizable") it will not be effectively used in the present context. The above three variations have the following stochastic formulation.

For simplicity we take $T = [0, 1]^2$, the unit square, as the parameter set with its coordinate ordering as its partial order, and if $0 = t_1^i < t_2^i < \cdots < t_{n_i}^i = 1, i = 1, 2$, let $t_{ij} = (t_i^1, t_j^2) \in T$ and similarly $s_{ij} = (s_i^1, s_j^2) \in T$ be the vertices of the rectangle $(s_{ij}, t_{ij}]$. The set of all such vartices s_{ij} of T will be termed a *grid* $g = (g_{n_1,n_2})$. For a process $X = \{X_t, t \in T\}, (\Delta X)(s_{ij})$, is the two dimensional increment of X on the rectangle $(s_{ij}, s_{i+1,j+1}]$. Also let Γ_1, Γ_2 be the left and bottom boundaries of T

and Γ_3 be the remaining boundary so that $\partial T = \cup_{i=1}^{3}\Gamma_i$. With this notation, we introduce quasimartingales based on Vitali, Arzelà, and Hardy variations as follows:

6. Definition. Let $\{X_t, \mathcal{F}_t, t \in T\}$ be an adapted integrable process on (Ω, Σ, P). Then it is called:

(a) a *V-quasimartingale process* (*V-martingale, V-submartingale*) if $K_V^x < \infty$, where

$$K_V^x = \sup\{E(\sum_{t\in g} |E^{\mathcal{F}_t}(\Delta_g X)(t)|) : g\, grid\, on\, T\},$$

$(K_V^x = 0, E^{\mathcal{F}_s}(\Delta X(s,t)) \geq 0$, a.e. for all $s \preceq t)$;

(b) an *A-quasimartingale process (A-martingale, A-submartingale)* if $K_A^x < \infty$, where

$$K_A^x = \sup\{E(\sum_{i=0}^{n-1} |E^{\mathcal{F}_{t_i}}(X_{t_{i+1}} - X_{t_i}))| : 0 \preceq t_1 \prec \cdots \prec t_n \preceq (1,1)\},$$

$(K_A^x = 0, E^{\mathcal{F}_s}(X_t) \geq X_s$ a.e., for all $s \preceq t)$;

and (c) an *H-quasimartingale process* if it satisfies (a) and $\{X_t, \mathcal{F}_t, t \in \tilde{\Gamma}_i\}, i = 1, 2$, are one parameter quasimartingales (in the sense of Definition V.2.19) where $\tilde{\Gamma}_1, \tilde{\Gamma}_2$ are the upper and right boundaries of T.

In the literature, a V-martingale is also called a *weak martingale*, as introduced in Cairoli and Walsh [1]. Moreover, the V-submartingale is a subprocess and a martingale is an M-process, in terms of Definition 3 above. For brevity, we often omit the word "quasimartingale" in the above definition, and call them V-, A- (and H-) processes.

Let us record the following properties of V-processes, leaving their verifications to the reader.

(i) A process $\{X_t, \mathcal{F}_t, t \in T\}$ is simultaneously a V- and an A-process iff it is an H-process.

(ii) If X and Y are weak martingales for the same filtration $\{\mathcal{F}_t, t \in T\}$, and if they agree on the upper boundary $\tilde{\Gamma}_1$ then they are identical except for an evanescent set.

(iii) A weak martingale $X = \{X_t, \mathcal{F}_t, t \in T\}$ is a martingale iff the boundary pieces of $\{X_t, \mathcal{F}_t, t \in \Gamma_3\}$ are one parameter martingales.

(iv) If $X = \{X_t, \mathcal{F}_t, t \in T\}$ is an A-process it has a modification Y which is right continuous with left limits in $L^1(P)$-mean in $T^0 = T - \partial T$.

In fact writing limit for the mean limit, we have

$$Y_t = \begin{cases} \lim_{s \to t, s \succ t} X_s, t \in T - (\tilde{\Gamma}_1 \cup \tilde{\Gamma}_2), \\ X_{1,1},\ t = (1,1), \\ \lim_{h \to 0+} X_{t_1+h,0}, t \in \tilde{\Gamma}_1 - (1,0), \\ \lim_{h \to 0+} X_{1,t_2+h}, t \in \tilde{\Gamma}_2 - (1,1). \end{cases}$$

The last property needs a more detailed computation. The desired arguments for all these assertions can, however, be found in Brennan [1]. An analysis of these quasimartingales is facilitated by associating a (signed) measure, extending the one-parameter Doléans-Dade representation given in Theorem V.2.20, which we now formulate. For this purpose, it will be useful to enlarge the given multiparameter filtration as follows. If $t = (t_1, t_2) \in T$, let $\mathcal{F}_t^1 = \sigma(\cup_{t_2} \mathcal{F}_{t_1,t_2})$, and similarly define $\mathcal{F}_t^2$ with t_1 in place of t_2, and set $\mathcal{F}_t^* = \sigma(\mathcal{F}_t^1 \cup \mathcal{F}_t^2)$. We then say that an integrable process $X = \{X_t, \mathcal{F}_t, t \in T\}$ is a *strong martingale* if for any $s \prec t$ in T for the increment over $(s,t]$ one has $E^{\mathcal{F}_s^*}(\Delta X((s,t])) = 0$ a.e. Since $\mathcal{F}_t \subset \mathcal{F}_t^*$, it follows that a strong martingale is a martingale which, in turn, is a weak martingale. The process X is an *i-martingale*, $i = 1, 2$, if $\{X_t, \mathcal{F}_t^i, t \in T\}$ is adapted and $E^{\mathcal{F}_s^i}(X(s,t]) = 0, s \prec t$. Similarly a V-submartingale defined earlier will be termed a *V-strong submartingale* if the $\mathcal{F}_t$-filtration is replaced by the $\mathcal{F}_t^*$-filtration. Analogously, with the predictable σ-algebra $\mathcal{P}$ of the given filtration, one has the enlarged predictable σ-algebra $\mathcal{P}^*$ using the $\mathcal{F}_t^*$-filtration. A reason to consider the latter objects is that the strong (sub)martingales have several properties that hold in the one-parameter case. It can be verified that a martingale X, above, is an i-martingale for $i = 1, 2$ simultaneously whenever the filtration is conditionally independent (cf. Theorem 5 or (18) below).

If X is an integrable process, define μ_x on the generators of $\mathcal{P}$ (i.e., on the sets of the form $(s,t] \times A, A \in \mathcal{F}_s$) by:

$$\mu_x((s,t] \times A) = \int_A (\Delta X(s,t])dP. \tag{15}$$

Then μ_x is additive on the ring generated by such sets, since if $(s,t] \times A = \cup_{i=1}^n (s_i, t_i] \times A_i$ is a disjoint union, and $I(\omega) = \{i : \omega \in A_i\}$, so that $(s,t] \times \{\omega\} = \cup\{(s_i,t_i] \times \{\omega\} : i \in I(\omega)\}$, we have

$$\chi_A(\Delta X(s,t]) = \sum_{i=1}^{n} \chi_{A_i}(\Delta X(s_i,t_i]),$$

and integration relative to P yields the result.

Using a slight modification of a familiar argument (cf.,e.g.,Rao [11],p.217), one can show that a process X is a V-quasimartingale iff the variation $|\mu_x|(T \times \Omega) = K_V^x < \infty$, where K_V^x is given by Definition 6 above. Consider $Q_t(A) = \mu_x((t,(1,1)] \times A), A \in \mathcal{F}_t$. Since Q_t is a P-continuous (signed) measure so that $\tilde{X}_t = \frac{dQ_t}{dP}, t \in T$, defines a process, one sees that the set functions $\mu_{\tilde{x}} = \mu_x$ on $\mathcal{F}_t$. Also $\tilde{X}_t = 0, t \in \tilde{\Gamma}_1 \cup \tilde{\Gamma}_2$ so that $\tilde{X}$ is an H-process with the additional vanishing property on the right boundary, called *normalization.*

We can now present the following general decomposition of a V-process.

7. Theorem. *Let $X = \{X_t, \mathcal{F}_t, t \in T\}$ be a right continuous V-quasimartingale. Then X admits a decomposition: $X = M + Y - Z$ where M is a right continuous (in $L^1(P)$) weak martingale and Y, Z are similarly right continuous nonnegative normalized H-submartingales, all for the same filtration.*

Proof. Since X is a V-process, there is a normalized H-process $\tilde{X}$ such that $\mu_x = \mu_{\tilde{x}}$ as noted above. By the right continuity (in $L^1(P)$) of $X, \tilde{X}$ can also be chosen to have the same property. Let $M = X - \tilde{X}$. Then $\mu_M = 0$ and hence M is a right continuous weak martingale. We now decompose $\tilde{X}$ to obtain the desired Y, Z by employing an approximation procedure and the fact that X is separable (using the dyadic rationals of T as a universal separating set $S = \{s_{ij}^n = (\frac{i}{2^n}, \frac{j}{2^n}) : 0 \le i,j \le 2^n, n \ge 1\}$).

Consider $Q_s^n(A)$ for $A \in \mathcal{F}_s$, by setting:

$$Q_s^n(A) = \sup\{\sum_{i,j=0}^{2^n-1} |\int_{A_{ij}} \Delta X(s_{ij}^n) dP| : A_{ij} \in \mathcal{F}_{s_{ij}^n}(A), s \prec s_{ij}^n \in S\}.$$

Then $Q_s^n : \mathcal{F}_s \to \mathbb{R}^+$ is a measure which is P-continuous, and $Q_s^n(A) \le Q_s^{n+1}(A) \le |\mu_x|((s,(1,1)] \times A), A \in \mathcal{F}_s$. If $Q_s(A) = \lim_n Q_s^n(A)$, which exists, then by the classical Vitali-Hahn-Saks theorem (cf.,e.g., Rao [11],p.176), Q_s is σ-additive and P-continuous. Hence for any $\varepsilon >$

$0, s \in T, A \in \mathcal{F}_s$ we can choose a grid $g = \{r_{ij}\}$ of T and $A_{ij} \in \mathcal{F}_{r_{ij}}(A)$ for $s \prec r_{ij}$, such that

$$|\mu_x|((s,(1,1)] \times A) \leq \sum_{i,j} |\int_{A_{ij}} \Delta X(R_{ij}) dP| + \varepsilon. \tag{16}$$

If $r_{ij} \in (s^n_{k-1,p-1}, s^n_{k,p}]$, then take $r^n_{ij} = s^n_{kp}$. Hence by the right continuity of X, there is an n_0 such that for $n \geq n_0$

$$\sum_{i,j} |\int_{A_{ij}} \Delta X(r_{ij}) dP| \leq \sum_{i,j} |\int_{A_{ij}} \Delta X(r^n_{ij}) dP| + \varepsilon \leq Q^n_s(A) + \varepsilon. \tag{17}$$

From (16) and (17) we deduce that $|\mu_x|((s,(1,1)] \times A) \leq Q^n_s(A) + 2\varepsilon$ for $n \geq n_0$, and hence $|\mu_x|((s,(1,1)] \times A) = Q_s(A), A \in \mathcal{F}_s$, defines a measure.

If $\mu_x = \mu_x^+ - \mu_x^-$ is the (Jordan) decomposition (cf.,e.g., Rao [11],p.179) then $\mu_x^\pm$ are P-continuous on $\mathcal{F}_s$, and if Y, Z are the (Radon-Nikodým) derived processes, one verifies at once from $\mu_x((s,(1,1)] \times A) = \mu_{\tilde{x}}((s,(1,1)] \times A), A \in \mathcal{F}_s$ that $\tilde{X}_s = Y_s - Z_s$ and the latter are normalized positive H-submartingales. Thus $X = M + Y - Z$, as asserted. □

Note. Although the above decomposition is not unique, it is the most efficient one as in the classical Jordan result (cf., e.g., Rao [11],p.180). In this representation M is a weak martingale which, by known counterexamples, is not a stochastic integrator in the sense of Section VI.3, and also need not obey our boundedness principles. The V-class should even be restricted for the σ-additivity of the associated μ_x's. We therefore present conditions for μ_x to be σ-additive on $\mathcal{P}$, and then give strong versions of the corresponding decompositions of X which satisfies an $L^{2,2}$-boundedness condition relative to a σ-finite measure on $\mathcal{P}^*$.

An analog of class (D) in the one-parameter processes is needed for solving the above problem on the measure representation.

8. Definition. Let $X = \{X_t, \mathcal{F}_t, t \in T\}$ be a V-process and $S \subset T$ be a universal separating set of dyadic rationals. If $I_x = \{\sum_{i,j=0}^{2^n-1} |E^{\mathcal{F}_{s^n_{ij}}}(\Delta X_{s^n_{ij}})| : n \geq 1\}$ is uniformly integrable, then X is said to be of *class* (D').

If T is a linear set or interval, then one may verify that class (D') reduces to class (D) of Definition IV.2.3. With this concept we have the following result about μ_x.

9. Theorem. *Let $X = \{X_t, \mathcal{F}_t, t \in T\}$ be a right continuous (in $L^1(P)$), V-quasimartingale. Then X belongs to class (D') iff the associated measure μ_x is σ-additive (on $\mathcal{P}$), and this holds iff X admits a decomposition (*) $X = M + A$ where M is a weak martingale and A is a process of bounded variation vanishing on the lower boundary, both for the same filtration, so that A is the difference of two right continuous (in $L^1(P)$) increasing (i.e., $\Delta A((s,t]) \geq 0$) processes essentially bounded on (compact subsets of) T. Moreover, if A is chosen predictable (i.e., measurable for $\mathcal{P}$) then the decomposition of X is also unique.*

We omit a proof of this result, for space reasons, and refer to Brennan([1],p.477). In the above decomposition of X, it is desirable to have M to be a strong martingale so that its quadratic variation exists when it is ("locally") square integrable. For this we need to restrict the filtration to satisfy an additional condition, namely *conditional independence*, of $\mathcal{F}_s$ and $\mathcal{F}_t$ given $\mathcal{F}_{s\wedge t}$. This is expressed as, for each $s, t \in T$, and bounded measurable random variable Y

$$E^{\mathcal{F}_s}(Y) = E^{\mathcal{F}_{s\wedge t}}(Y), \tag{18}$$

so that $E^{\mathcal{F}_s}, E^{\mathcal{F}_t}$ commute. Then we have the following strengthening of the above result with $\mathcal{F}_t^*$ in place of $\mathcal{F}_t$ in the definition of strong martingale, (see also Dozzi[1],p.32, for this form).

10. Proposition. *Let $\{\mathcal{F}_t, t \in T\}$ be a right continuous (completed) filtration which also satisfies (18) and let $\{\mathcal{F}_t^*, t \in T\}$ be the associated family. If $X = \{X_t, \mathcal{F}_t^*, t \in T\}$ is a V-process as in the above theorem, then μ_x is σ-additive on $\mathcal{P}^*$ iff $X = M + A$ where M is a strong martingale and A is as in the theorem which can be chosen measurable relative to $\mathcal{P}^*$ to get uniqueness. When this decomposition holds, then X is $L^{2,2}$-bounded on $\mathcal{P}^*$ relative to a σ-finite measure on it whenever $X_t \in L^2(P), t \in T$.*

The fact that X in this case satisfies the $L^{2,2}$-boundedness follows exactly as in the one-paremeter case (cf., Proposition VI.2.8), on using the fact (established without much difficulty) that a square integrable strong martingale has this property. Thus when we consider processes that obey a boundedness condition as above, for integration purposes, this class automatically includes the V-quasimartingales of the type described in the proposition. We discuss, therefore, only the two-parameter integrals to extend the work of Chapter VI. Similar integration is evidently possible for the L^{φ_1,φ_2}-bounded processes.

Let $X : T \times \Omega \to \mathbb{R}$ be $L^{2,2}$-bounded relative to a σ-finite measure on $\mathcal{O} \subset \mathcal{B}(T) \otimes \Sigma$, where $\mathcal{B}(T)$ is the Borel σ-algebra of T. For instance $\mathcal{O} = \mathcal{P}^*$ and $X = \{X_t, \mathcal{F}_t^*, t \in T\}$, a (strong) V-quasimartingale with filtration satisfying (18), will do. If $f : T \times \Omega \to \mathbb{R}$ is a simple function, define as usual with $\Omega' = T \times \Omega$,

$$\tau(f) = \int_{\Omega'} f dX = \sum_{i=1}^{n} a_i \mu_x^*((s_i, t_i] \times A_i), \tag{19}$$

where f is an obvious $\mathcal{O}$-simple function generated by the sets shown and $\mu_x^*((s,t] \times A) = \chi_A X((s,t])$ which is an additive function in general but under the $L^{2,2}$-boundedness is a (vector) measure with $\alpha : \mathcal{O} \to \mathbb{R}^+$ as a σ-finite measure. Further we have

$$E(|\tau(f)|^2) \leq C \int_{\Omega'} |f(t,\omega)|^2 d\alpha(t,\omega), \tag{20}$$

for some $0 < C < \infty$. Thus τ is defined by linearity on $L^2(\mathcal{O}, \alpha) \to L^2(P)$, and is a stochastic integral having the dominated convergence property. Restricting f to $(0,t] \subset T$, we get for bounded $\mathcal{O}$-measurable $f (\in L^2(\mathcal{O}, \alpha))$

$$\tau(f)_t = \int_0^t f(s) dX_s, \; t \in T. \tag{21}$$

This is a double integral, and under only further conditions one can get a repeated integral equivalent to it, since in general $dX_t = dX(t_1, t_2) \neq dX_{t_1} \otimes dX_{t_2}$.

Using these integrals, it is possible to develop an analog of the multivariable calculus which in k-dimensions ($k \geq 2$) leads to stochastic differential forms and equations. Here we merely indicate a few possible avenues for research, since the whole area is being studied with much interest. (More details of the following account can be found in Walsh [2].)

For the ensuing work, and to simplify the discussion, we consider the particular $L^{2,2}$-bounded process, namely, the Wiener-Brownian motion $X = \{X_t, \mathcal{F}_t, t \in T\}$, which in the two parameter case is simply termed a *Brownian sheet.* Here the filtration is again $\mathcal{F}_t = \sigma(X_s, s \preceq t)$ so that the quadratic variation of X is $\langle X \rangle_t = t_1 t_2, t = (t_1, t_2)$, and we wish to study martingales $Y = \{Y_t, \mathcal{F}_t, t \in T\}$. Such a Y is often called a *Brownian functional.* A representation of these processes, due to Wong and Zakai [1], will be given as it leads to several other developments.

Let $\mathcal{L}_0^2$ be the set of processes Y, as above, which are (jointly) measurable when regarded as mappings $Y : \Omega' = T \times \Omega \to \mathbb{R}$ relative to $\mathcal{B}(T) \otimes \Sigma$, and $\|Y\|^2 = \int_{\Omega'} |Y|^2 dtdP < \infty$. Then one can verify (after some computation) that $\{\mathcal{L}_0^2, \|\cdot\|\}$ is a Hilbert space and (predictable) simple functions are dense in it. Consequently, using (21) first for simple Y and then for all $Y \in \mathcal{L}_0^2$, one has $\tau(Y)_t = \int_0^t Y(s)dX_s$, to be defined and is a strong martingale.

Similarly, let $\mathcal{L}_{00}^2$ be the class of all processes $\tilde{Y} : \Omega'' = T \times T \times \Omega \to \mathbb{R}$, which satisfy: (i) measurable relative to $\mathcal{B}(T \times T) \times \Sigma$ and $\tilde{Y}(t,t')$ is $\mathcal{F}_{t \vee t'}$-adapted for each $(t,t') \in T \times T$,(ii)$\tilde{Y}(t,t') = 0$ unless the points $t = (t_1,t_2), t' = (t_1',t_2')$ obey the order relation $t_1 < t_1'$ and $t_2 > t_2'$ (i.e., t is to the left of t' and is above it), and (iii)$\|\tilde{Y}\|'^2 = \int_{\Omega''} |\tilde{Y}|^2 dtdt'dP < \infty$. Then by analogous computations, one shows that $\{\mathcal{L}_{00}^2, \|\cdot\|'\}$ is a Hilbert space with the linear span of $\tilde{Y}_{A,B}(t,t') = a\chi_A(t)\chi_B(t')$ as a dense subspace, where a is a bounded $\mathcal{F}_{t \vee t'}$-adapted random variable and $A, B \in \mathcal{B}(T \times T)$ such that $t \in A, t' \in B$ iff t, t' satisfy the order relation of (ii) above. It can be verified that $X(\cdot) \times X(\cdot)$ satisfies the basic $L^{2,2}$-boundedness condition of Bochner's, since for the above $\tilde{Y}_{A,B}(u,u')$ with $A = (s,s'], B = (t,t']$ we have

$$E(\tilde{Y}_{A,B}^2(u,u')) = E(a^2)m(A \cap (0,u])m(B \cap (0,u']), \quad u,u' \in T$$

where $m(\cdot)$ is the planar Lebesgue measure. Then the integral

$$\tau(\tilde{Y})(t,t') = \int_0^t (\int_0^{t'} \tilde{Y}(u,v)dX(u))dX(v), \tag{22}$$

is defined first for simple processes and then for all $\tilde{Y} \in \mathcal{L}_{00}^2$. Also one can verify that the integrals in (22) have the "Fubini property", i.e.,

$$\int_0^t (\int_0^{t'} \tilde{Y}(u,v)dX(u))dX(v) = \int_0^{t'} (\int_0^t \tilde{Y}(u,v)dX(v))dX(u), \ \tilde{Y} \in \mathcal{L}_{00}^2. \tag{23}$$

We can now present the Wong-Zakai representation as:

11. Theorem. *Let $X = \{X_t, \mathcal{F}_t, t \in T\}$ be a Brownian sheet with its natural filtration and $M = \{M_t, \mathcal{F}_t, t \in T\}$ be an integrable martingale.*

Then there exist unique $Y \in \mathcal{L}_0^2, \tilde{Y} \in \mathcal{L}_{00}^2$ such that

$$M_t - M_0 = \int_0^t Y(s)dX(s) + \int_0^t \int_0^t \tilde{Y}(u,v)dX(u)dX(v), \text{ a.e.} \quad (23')$$

The result admits generalizations for an adapted M which is a square integrable i-martingale. [The generality is that M need not be a martingale.] An interesting consequence of the latter is the following if we restrict to strong martingales. Let $\mathcal{M}_S^2$ be the subset of strong martingales of $\mathcal{M}^2$, the space of all square integrable (two-parameter) martingales with the same stochastic base as in the theorem. Both are Hilbert spaces, as in Theorem V.3.28, and let $\mathcal{N}_S^2$ be the orthogonal complement of $\mathcal{M}_S^2$ in $\mathcal{M}^2$. Then, one shows that if $M \in \mathcal{M}_S^2$, the second term of (23′) vanishes and if $M \in \mathcal{N}_S^2$, the first term of (23′) vanishes. This is recorded as:

12. Proposition. *If $M \in \mathcal{M}_S^2$ and $N \in \mathcal{N}_S^2$, then we have*

$$(i) M_t - M_0 = \int_0^t Y(s)dX_s; \; (ii) N_t - N_0 = \int_0^t \int_0^t \tilde{Y}(u,v)dX_u dX_v. \quad (24)$$

for unique $Y \in \mathcal{L}_0^2$ and $\tilde{Y} \in \mathcal{L}_{00}^2$.

These ideas allow us to define stochastic differential forms, and then (real) holomorphic processes as distinct from the complex (one-parameter) class discussed in Definition 1.7 and Proposition 1.8. For this let C be a continuous curve starting at a and ending at b, in T, which parametrically is representable as $\gamma : [0,1] \to T$ where $\gamma(0) = a, \gamma(b) = b$. It is termed increasing (decreasing) if for $0 < u < v < 1, \gamma(u) \prec \gamma(v), (\gamma(u) \succ \gamma(v))$. Thus if C is as above, its reverse oriented curve $\hat{C} = \{t = \hat{\gamma}(u) = \gamma(1-u) : 0 \leq u \leq 1\}$ is in T and both are monotone curves. If $M (\in \mathcal{M}^2)$ is given, then $\tilde{M}_u = M_{\gamma(u)}$ defines a one-parameter square integrable martingale relative to the filtration $\{\mathcal{F}_{\gamma(u)}, u \in [0,1]\}$ if γ is increasing. A similar argument holds for $\bar{M}_u = M_{\hat{\gamma}(u)}$. We then define, when $\gamma(\hat{\gamma})$ is monotone, two processes for a given Brownian sheet X as follows. If $t \in T$, consider the horizontal (vertical) line $H_t(V_t)$ from t to the axis. Let $D_s^1(D_s^2)$ be the closed areas bounded by V_a and V_s (H_a and H_s), C and the axis for $s \in C, a = \gamma(0)$ and $b = \gamma(1)$, the end point of C. Now define the one-parameter processes $X_t^i = X(D_t^i), i = 1,2$. Then one verifies that $X^i = \{X_t^i, \mathcal{F}_t^i, t \in C\}$ is a continuous square integrable martingale and if C is increasing, X^1 and X^2 are also independent. Now for C monotone, we define a line integral for a measurable adapted process $Y = \{Y_t, \mathcal{F}_t^1, t \in C\}$ satisfying $P[\int_0^1 Y_{\gamma(u)}^2 t_2(u) dt_1(u) < \infty] = 1$ as

$$\int_C Y_t \partial_1 X_t = \int_0^1 Y_{\gamma(u)} dX^1(\gamma(u)), \tag{25}$$

where $t = (t_1(u), t_2(u)) = \gamma(u)$, since $\langle X^1 \rangle_b = \int_C t ds$ and $\langle X^2 \rangle_b = \int_C s dt$. Similarly we can define the line integral $\int_C Y_t \partial_2 X_t$ for X^2 with similar hypothesis. If Y_t is $\mathcal{F}_t$-adapted then it is $\mathcal{F}_t^i$-adapted, $i = 1, 2$, and both integrals exist so that one can define

$$\int_C Y_t \partial X_t = \int_C Y_t \partial_1 X_t + \int_C Y_t \partial_2 X_t, \tag{26}$$

and these definitions are consistent. Moreover they can be extended to curves which may have a finite number of such monotone pieces, and then $Y\partial X, Y\partial_1 X$ and $Y\partial_2 X$ are what may be appropriately called the *stochastic differential forms.*

These results can be used to define and study stochastic real holomorphic processes. Thus if $C \subset T$ is an increasing parametric curve, and $Y = \{Y_t, \mathcal{F}_t, t \in T\}$ is a measurable and adapted process as above such that $E(Y_s^2) \leq K_0 < \infty, s \in T$, (or it is replaced by $\mathbb{R}_+^2$, but bounded on compact sets), we call a process $Z = \{Z_t, t \in T\}$ *(real) holomorphic* whenever

$$Z_b = Z_a + \int_C Y_s \partial X_s, \tag{27}$$

and term Y_s, the (stochastic directional) *derivative* of Z relative to X along C and write $\partial Z_t = Y_t \partial X_t$. It can be shown that Y in (27) will also be (real) holomorphic so that Z is infinitely differentiable. After a detailed analysis (including the stochastic forms of the Green and Stokes theorems on the way to use in the proof) Cairoli and Walsh [1] established the following representation of all such processes.

13. Theorem. *If Z is a holomorphic process, then for each $t = (t_1, t_2) \in T$,*

$$Z_t = \sum_{n=0}^{\infty} Y_a^{(n)} H_n(X_t, t_1 t_2), \tag{28}$$

the series converging in $L^2(P)$. *Here* $Y^{(n)}$ *is the* n^{th} *holomorphic derivative of* Y *as in (27), and* $H_n(\cdot,\cdot)$ *is the* n^{th} *Hermite polynomial defined by*

$$H_n(x,u) = \frac{(-1)^n}{n!} e^{x^2/2u} \frac{\partial^n}{\partial x^n}(e^{-x^2/2u}),\ u > 0, x \in \mathbb{R}. \tag{29}$$

In particular, if for some $0 \prec t \in T, P[Z_t = 0] = 1$ *then* $Z \equiv 0$.

This result is given to show how the two parameter stochastic calculus contains several aspects of classical (deterministic) analysis. It is possible to extend parts of Cairoli-Walsh work for McShane processes, and perhaps eventually to the $L^{2,2}$-and even the L^{φ_1,φ_2}-bounded processes.

If one is interested in the two parameter stochastic integration relative to a Brownian sheet, then a direct procedure extending the classical Itô method can be given and then it may be employed to study the analogs of the stochastic differential equations of Chapter VI. This has been considered by Yeh [2]. Here a process $Y = \{Y_t, \mathcal{F}_t, t \in T\}$ is said to satisfy the (symbolic) differential equation:

$$dY_t = \alpha(t, Y_t)dX_t + \beta(t, Y_t)dt, \tag{30}$$

where $\alpha : T\times\mathbb{R} \to \mathbb{R}$, $\beta : T\times\mathbb{R} \to \mathbb{R}$ are some Borel functions satisfying suitable Lipschitz conditions, and $X = \{X_t, \mathcal{F}_t, t \in T\}$ is a Brownian sheet vanishing on the boundries $\tilde{\Gamma}_1$ and $\tilde{\Gamma}_2$ of T. Again (30) is to be understood in the integrated form. Thus if $t = (t_1, t_2)$, then Y is said to solve (30) iff it satisfies the following integral equation:

$$\begin{aligned} Y(t_1,t_2) &= Y(t_1,0) - Y(0,t_2) + Y(0,0) + \int_0^t \alpha(s, Y_s)dX_s \\ &\quad + \int_0^t \beta(s, Y_s)ds_1 ds_2,\ s = (s_1, s_2). \end{aligned} \tag{31}$$

Conditions for a solution of (31), extending the Picard method and related results of the one-parameter case, are discussed in detail by Yeh [2]. It appears difficult to apply these methods for a study of holomorphic processes (cf., Theorem 13), higher order equations of Section VI.4, or the line integrals. For these the general work presented above seems appropriate and essential.

Finally we touch another aspect of stochastic analysis to round out the present treatment.

7.4 Remarks on stochastic PDEs

The preceding discussion on stochastic differential equations of multiparameters suggests that one should consider stochastic partial differential equations (SPDE), and we introduce the concept essentially following Walsh [1].

Just as in the case of Langevin's equation (cf., Section VI.4), serving as a motivation for ordinary stochastic differential equations, we can consider the classical wave equation perturbed by random noise for SPDEs. Thus if $V(x,t), t \geq 0, x \in \mathbb{R}$, is the position of a vibrating string at time t and distance x from the start, disturbed by a noise X, then it satisfies the symbolic equation:

$$\begin{aligned} \frac{\partial^2 V}{\partial t^2}(x,t) &= \frac{\partial^2 V}{\partial x^2}(x,t) + \dot{X}(x,t),\ t > 0, x \in \mathbb{R}, \\ V(x,0) = 0 &= \frac{\partial V}{\partial t}(x,0),\ x \in \mathbb{R}. \end{aligned} \tag{1}$$

This makes sense only if we interpret it after multiplying by a smooth function (because $\dot{X}$ need not exist as a smooth function) $\varphi : \mathbb{R} \times [0,T] \to \mathbb{R}$ such that $\varphi(x,T) = 0 = \frac{\partial \varphi}{\partial t}(x,T)$, so that using integration by parts we get

$$\int_0^T \int_{\mathbb{R}} V(x,y)\Big(\frac{\partial^2 \varphi}{\partial t^2} - \frac{\partial^2 \varphi}{\partial x^2}\Big)(x,t)dxdt = \int_0^t \int_{\mathbb{R}} \varphi(x,t) X(dxdt). \tag{2}$$

This is well-defined and we say that V is a *weak solution* of the (formal) SPDE (1) if it satisfies (2) for all C^∞-functions φ with compact supports. Often V is also called a *distributional solution*, as in the classical PDE theory.

To indicate the flavor of the problem we present the following:

1. Theorem. *Under the above stated conditions, there is a unique continuous (weak) solution of (2), and it is given by*

$$V(x,t) = \frac{1}{2}\hat{X}\Big(\frac{t-x}{\sqrt{2}}, \frac{t+x}{\sqrt{2}}\Big), \tag{3}$$

where $\hat{X}$ is the modified Brownian sheet defined by $\hat{X}(s_1 s_2) = X(\hat{R}_{s_1 s_2})$, with $\hat{R}_{s_1 s_2} = D \cap ((-\infty, s_1] \times (-\infty, s_2]), D = \{(s_1,s_2) : s_1 + s_2 \geq 0\}$, so that $X(s_1 s_2) = \hat{X}(s_1 s_2) - \hat{X}(s_1, 0) - \hat{X}(0, s_2)$ for all $s_1, s_2 \geq 0$.

Proof. The uniqueness is easy. If V_1, V_2 are two solutions of (2), let $U = V_1 - V_2$ so that we have ($\varphi_{tt}, \varphi_{xx}$ denoting the partials)

$$\int_{\mathbb{R}} \int_0^T U(x,t)[\varphi_{tt} - \varphi_{xx}](x,t)dxdt = 0. \tag{4}$$

If $f \in C_c^\infty(\mathbb{R} \times [0,T])$, the space of compactly supported real C^∞-functions on $\mathbb{R} \times [0,T]$, then there exists a $\varphi \in C_c^\infty(\mathbb{R} \times [0,T])$ such that $\varphi(x,T) = 0 = \varphi_t(x,T)$ and $f = \varphi_{tt} - \varphi_{xx}$, and hence (4) gives $U = 0$ a.e.

As for the existence, let $u = (t-x)/\sqrt{2}, v = (t+x)/\sqrt{2}$, and define $\hat{\varphi}(u,v) = \varphi(x,t)$, $\hat{X}(dudv) = X(dxdt)$, and $\hat{R}(u,v;u_0,v_0) = \chi_{[u \le u_0, v \le v_0]}$. We find $\varphi_{tt} - \varphi_{xx} = 2\hat{\varphi}_{uv}$ and let $\hat{V}$ be defined by

$$\hat{V}(u,v) = \frac{1}{2} \iint_{[u'+v'>0]} \hat{R}(u',v';u,v)\hat{X}(du'dv').$$

This $\hat{V}$ satisfies (2) if the following expression vanishes:

$$\begin{aligned}
&\iint_{[u'+v'>0]} \Big[\iint_{[u+v>0]} \hat{R}(u,v;u',v')\hat{X}(dudv)\Big]\hat{\varphi}(u',v')du'dv' \\
&\quad - \iint_{[u+v>0]} \hat{\varphi}(u,v)\hat{X}(dudv) \\
&\quad = \iint_{[u+v>0]} \Big[\int_v^\infty \int_u^\infty \hat{\varphi}_{uv}(u',v')du'dv' - \varphi(u,v)\Big] \times \\
&\quad \hat{X}(dudv),
\end{aligned}$$

by (3.23). But by the fundamental theorem of calculus the expression in brackets vanishes, since $\hat{\varphi}$ has compact support. □

A more general equation is analogously obtained from the following parabolic PDE. Let $D \subset \mathbb{R}^k$ be a bounded domain with smooth boundary ∂D. If f, g are smooth functions in $C^\infty(\partial D)$, consider the (symbolic) equation

$$\begin{aligned}
&\frac{\partial V}{\partial t} = LV + J\dot{M}, \textit{ on } D \times [0,\infty) \\
&BV = 0, \textit{ on } \partial D \times [0,\infty), V(x,0) = 0 \textit{ on } D,
\end{aligned} \tag{5}$$

where L is a uniformly elliptic self-adjoint operator, $B = fD_N + g$, (D_N being the normal derivative on the boundary ∂D), and $J : f \mapsto$

$g \int_{\mathbb{R}} f(y)h(y)dy$, for suitable h. Here $(JM)_t(x) = g(x)M_t(h)$ where $M_t(h) = \int_{\mathbb{R}} hdM$ for a continuous square integrable martingale M. We interpret (5) in the integrated form, and V will be a distributional solution which satisfies

$$V_t(\varphi) = \int_0^t V_s(L\varphi)ds + \int_0^t \int_D J^*\varphi(x)M(dxds), \varphi \in S_B, \quad (6)$$

where $S_B = \{\varphi \in C_c^\infty(\bar{D}) : B\varphi = 0 \, on \, \partial D\}$, and J^* is the formal adjoint of J.

The existence and uniqueness of solutions of (6) involve several ideas from PDE along with stochastic analysis (cf. also Kunita [1], Chapter 6). But we observe that M satisfies the $L^{2,2}$-boundedness condition, and then by the generalized boundedness principle the second integral in (6) exists, defining a distribution valued process, i.e., a generalized random field. Under further conditions, it can be shown that $\{M_t(A), t \geq 0, A \in \mathcal{B}(\mathbb{R}^k)\}$ is a process which is a martingale for each A, and a stochastic measure on $\mathcal{B}(\mathbb{R}^k)$ for each t. Indeed the work uses the theory of Schwartz distributions. Thus the SPDE employs results from both functional analysis and probability theory with appropriate specializations. This shows how new areas in the subject are opened up for a separate and extensive treatment. With these indications, we conclude this account of the general theory at this point.

7.5 Complements and exercises

1. Let H be the class of all (complex) holomorphic processes relative to some fixed conformal basis $\mathcal{Z}$ of $\mathcal{M}+i\mathcal{M}$, as in Definition 1.7(c), and assume that each element of $\mathcal{Z}$ has continuous sample paths. Show that the set H is an algebra. (Use the fact that for each holomorphic function g on a disc $D \subset \mathbb{C}^N$ with $0 \in D, g(0) = 0$, and for $G^i \in$H$, 1 \leq i \leq N$, one has that $(g \circ G)_t = g(G_t^1, \ldots, G_t^N)$ is in H by the work of Section 1. See also Föllmer [2] in this regard.)

2. Let $X = \{X_t, \mathcal{F}_t, t \geq 0\}$ be a process with values in a smooth d-dimensional manifold M. If for each open $M_1 \subset M$, and a smooth $\varphi : M \to \mathbb{C}$ which is holomorphic on $M_1, \varphi \circ X$ is equivalent to a conformal martingale Y, provided for each $A(\in \Sigma \otimes \mathcal{B}(\mathbb{R}^+))$ we have that $\varphi \circ X$ is equivalent to Y on the set $A \cap X^{-1}(M_1)$, and $X(A) \subset M_1$. Suppose now that X is a semimartingale with values in M_1, a

submanifold. Show that X is equivalent to a conformal martingale Y on $A \subset [0, \infty] \times \Omega$, with $A(\omega)$ open for a.a. (ω) iff Y takes values in M. [This and several other extensions to complex analytic manifolds of these results have been extensively discussed in Schwartz [4].]

3. Consider a bidisc $D_2 = \{z = (z^1, z^2) \in \mathbb{C}^2 : |z^i| < 1, i = 1,2\}$. If D is the open unit disc in $\mathbb{C}$ with a Riemannian metric g defined by the differential form $ds^2 = |dz|^2(1 - |z|^2)^{-2}, z \in D$, let $Z(t) = \{(Z^1(t), Z^2(t)), t \geq 0\}$ be a two dimensional conformal martingale starting at $z_0 = (z_0^1, z_0^2) \in D_2$ where $\{Z^i(t), t \geq 0\}, i = 1,2$, are a pair of independent Brownian motions on (D, g). If $\widehat{\partial D_2} = \{z = (z^1, z^2) \in D_2 : |z^i| = 1, i = 1,2\}$ and $\partial D_2 = \{z \in D_2 : |z^1| = 1 \, and \, |z^2| \leq 1, \, or \, |z^1| \leq 1 \, and \, |z^2| \leq 1\}$, let $\varphi : \bar{D} = D_2 \cup \partial D_2 \to \mathbb{C}$ be continuous and (complex) holomorphic on D_2. Then using the fact that $\{\varphi(Z(t)), \mathcal{F}_t, t \geq 0\}$ is a uniformly bounded martingale, $\mathcal{F}_t$ being the natural filtration of the $Z(t)$, so $\varphi(Z(t)) \to \varphi(Z(\infty))$ a.e., show that $\max\{|\varphi(z)| : z \in \bar{D}\} = \max\{|\varphi(z)| : z \in \widehat{\partial D}\}$, so that $\widehat{\partial D}$ is a distinguished (or Šilov) boundary of D. [Cf., Ikeda-S.Watanabe [1] and Kaneko-Taniguchi [1] on this and related results.)

4. Let $X = \{X(t), \mathcal{F}_t, t \in T\}$ be a Brownian sheet, and define a new process $Y(t) = e^{-(t_1+t_2)} X(e^{2t_1}, e^{2t_2}), t = (t_1, t_2) \in \mathbb{R}_+^2$. Show that $Y = \{Y(t), \mathcal{F}_t, t \in T\}$ is a Gaussian process with mean zero, and covariance $e^{-|t_1 - t_1'| - |t_2 - t_2'|}$ where $t, t' \in T$. [The Y-process is a two parameter *Ornstein-Uhlenbeck process* (or sheet).] Verify that $Y(t_1, \cdot)$ takes values in $C[0,1])$, the space of continuous functions on [0,1] with uniform norm. (This process plays an important role in the Malliavin calculus.)

5. Let us call a process $X = \{X_t, \mathcal{F}_t, t \in T\} \subset L^1(P)$ an F-(or Fréchet) *quasimartingale* with $T = [0,1]$ if $K_F^x < \infty$ where

$$K_F^x = \sup\{E(|\sum_g a_i a_j E^{\mathcal{F}_t}(\Delta_g X)(t)|) : |a_i| \leq 1, g \, grid \, on \, T\}.$$

Verify that the class F contains both the V- and A-quasimartingales, and if $\mu_x(\cdot)$ is the associated additive set function as in (3.15), then one has $K_F^x = \|\mu_x\|(T \times \Omega)$, where

$$\|\mu_x\|(T \times \Omega) = \sup_{\|f_i\| \leq 1, i=1,2} \left| \int_T \int_\Omega f_1(t_1) f_2(t_2) \mu_x(dt_1, dt_2, \omega) dP(\omega) \right|.$$

6. The following is an extension of Cairoli's inequality (3.1) to V-quasimartingales. If $\Phi : \mathbb{R} \to \mathbb{R}^+$ is a symmetric nondecreasing function

$\Phi(0) = 0$, and $X = \{X_t, \mathcal{F}_t, t \in T\}$ is a right continuous process, T being the unit square, $E(\Phi(X_t)) < \infty$, then we say X is Φ – *bounded* if the following holds:

$$\sup\{E(\Phi(\sum_{t\in g} |E^{\mathcal{F}_t}(\Delta_g X)(t)|))\, g\ grid\ on\ T\} < \infty,$$

and on the right boundaries $\tilde{\Gamma}_j, j = 1, 2$, we have

$$\sup\{E(\Phi(\sum_{i=0}^{n-1} |E^{\mathcal{F}_{t_i}}(X_{t_{i+1}} - X_{t_i})|)) : t_i \prec t_{i+1}, t_i \in \tilde{\Gamma}_j\} < \infty.$$

These are the same as V- and A-processes if $\Phi(x) = |x|$. Generalizing the method of Theorem 3.1, show that if $\Phi(x) = |x| \lg^+ x$, then for each $\lambda > 0$,

$$P[\sup_{t\in T} |X_t| > \lambda] \leq \frac{1}{\lambda}(A + BE(\Phi(X_{1,1}))),$$

where $A > 0, B > 0$ are absolute constants. If $\Phi(x) = |x|^p, p > 1$, and X is of class D', then using the decomposition of X when it is a strong quasimartingale, obtain an analog of the second inequality of Theorem 3.1. If X is moreover an H-process, show with the above inequality that $\lim_{s\uparrow t} X_s$ exists a.e. and in $L^1(P)$[see also Theorem 3.9]. (The method of proof is an extension of the classical submartingale ideas, but is more involved; see Brennan [1] in this connection.)

7. In the SPDE study, one can replace the Brownian sheet by a more general family of functions, $\{M_t(\varphi), \mathcal{F}_t, t \in T, \varphi = \chi_A \in \mathcal{L}\}$ as follows. (a) Let $\mathcal{L}$ be replaced by $\mathcal{B}_k = \mathcal{B}(\mathbb{R}^k), k \geq 2$, and consider $M = \{M_t(A), \mathcal{F}_t, t \in T, A \in \mathcal{B}_k\}$ where $M : T \times \mathcal{B}_k \times \Omega \to \mathbb{R}$ satisfies (i)$M_t(A, \cdot)$ is $\mathcal{F}_t$-adapted, (ii)$M_0(A, \omega) = 0$, a.e., and (iii)$M_t(A, \cdot)$ is $L^{2,2}$-bounded for all $A \in \mathcal{B}_k, t \in T$. [When (iii) is specialized to an $L^2(P)$-martingale, such an M is termed a *martingale measure* in Walsh [2].] By an analog of Proposition VI.2.12, the quadratic covariation $[M(A), M(B)]_t$ can be shown to exist. Verify that it is positive definite and is separately additive on $\mathcal{B}_k \times \mathcal{B}_k$. If $M_t(\cdot)$ is σ-additive in $L^2(P)$ for each $t \in T$, then the above covariation function will be σ-additive in each component defining a (vector) bimeasure, and yet *not* a (vector) measure. Show if it takes nonnegative values a.e., then it determines such a measure. (b) If $Q_M : \mathcal{B}_k \times \mathcal{B}_k \times \mathcal{B}([0, \alpha]) \times \Omega \to \mathbb{R}$ is defined by $Q_M(A, B, (0, t], \omega) = [M(A), M(B)]_t(\omega) - [M(A), M(B)]_s(\omega)$, verify

that Q is additive for a.a.(ω). Consider the σ-algebra generated by $\mathcal{B}_k \times \mathcal{B}_k \times \mathcal{B}((0,\alpha]) = \mathcal{B}(\mathbb{R}^{2k} \times (0,\alpha])$ and let $\mathcal{P}$ denote the predictable σ-algebra determined by this and the given filtration from Σ. If the variation $|Q|$ is dominated by a σ-finite measure $\mu : \mathcal{P} \times \Omega \to \bar{\mathbb{R}}^+$, for a.a.$(\omega)$, then verify that Q will be σ-additive, so that the integral

$$\int_{A\times B\times(0,t]} f(x,s)g(y,s)Q_M(dxdyds) = [f.M(A), g.M(B)]_t$$

is well-defined for all $\mathcal{P}$-measurable bounded f, g and where for step functions $f(x,s,\omega) = X(\omega)\chi_{(a,b]}(s)\chi_A(x)$, we define $f.M_t(B) = X(M_{t\wedge b}(A\cap B) - M_{t\wedge a}(A\cap B))$ and extend linearly. (Regarding this analysis see Walsh [1], and the general bimeasure integration theory is in Morse and Transue [1]. For its use in stochastic analysis with references, see Chang and Rao [1], and for the vector case Ylinen [1].)

Bibliographical remarks

One of the main purposes of this chapter is to highlight some of the recent advances of stochastic analysis in important parts of mathematics that have a great potential, giving new insights and results. The first steps in conformal martingales were taken by Getoor and Sharpe [1], and further extensions were made by Föllmer [2]. This constituted Section 1 here.

A deep analysis of a problem on hypoellipticity, first investigated by Hörmander [1], was studied (and extended) with probabilistic methods by Malliavin [1] who thereby created a stochastic calculus of variations. This new branch was further expanded by him, by Bismut [2], and by Stroock [1] among others. Schwartz [4] and Meyer [7] have then systematized the work and generalized the semimartingale theory to smooth manifolds. For a proper understanding of this, a certain amount of differential manifold theory is needed along with probability theory. A readable account of the former for beginners (especially for probabilists) can be found in Auslander and MacKenzie [1], Flanders [1], Guggenheimer [1], Goldberg [1], and Postnikov [1]. We have included only a minimal amount of the subject in Section 2 to explain the desired stochastic calculus. A more extensive discussion is given in Emery [1] which is devoted to the general theory of semimartingales in this set up.

Malliavin's calculus itself is being expanded in different directions. We should refer to Bell [1] for a quick overview and to Ikeda-S. Watanabe [1] for several extensions which we followed to a large extent. A vector valued and related analysis is presented in the Belopol'skaya-Dalecky [1] monograph. Here we have included some basic results of the subject in order to entice the reader for further research, and indicated how the $L^{2,2}$-boundedness principle can be helpful in this study.

All the preceding work deals with processes with a linear index set as is the case with most of the results in our treatment. But the multiparameter study with the corresponding (stochastic) geometric integration is also an important area that is developing contemporaneously. This work started with Cairoli [1], and has taken a leap with Cairoli and Walsh [1]. We have included some aspects of this theory following these authors, especially Walsh [2], and Brennan [1] as well as Wong and Zakai [1], restricting to two-parameter processes with componentwise ordering. In the case of $k(> 2)$ parameters, there is substantial notational complexity (although not the ideas). Such a k-parameter case has been included in Dozzi [1]. A general study with a partially ordered index, was considered with the corresponding integration by Hürzeler [1]. These works also lead to multiparameter stochastic differential equations. Yeh [2] has presented a detailed study of an aspect of the subject for two-parameter indexes, extending the classical treatments and solving some new problems. Many other results of stochastic analysis in this area have been detailed in Walsh's lecture notes refered to above. A new problem here is that there are (weak) martingales which are not stochastic integrators, but again the $L^{2,2}$-boundedness principle appears to be the right general condition for this integration for which we demand the validity of the dominated convergence criterion. Applications of the principle for V-quasimartingales are being studied by Green [1] in his thesis.

In applying the above integration to stochastic partial differential equations, we need to allow extra parameters in addition to the time as index. This leads to a somewhat different and more involved type of studies than the ordinary differential equations (of Chapter VI). Here one uses Schwartz distribution theory extensively, and hence the generalized random processes (or fields) come into play. We indicated the flavour in the last section. This work is also related to stochastic flows and has many applications. We refer to Walsh [1] and Kunita [1] for some of these developments, and a more detailed account of Freidlin [1] should also be consulted in this regard.

It is clear that the results discussed in this chapter show the new vigor with which stochastic analysis is growing into substantial areas and applications. We hope that this account shows the challenges and rewards to researchers in different and deeply interesting studies.

Bibliography

Alloin, C.

[1] "Processus prévisibles optimaux associés à un processus stochastique." Cahiers centre Etud. Rech. Opér. **11** (1969) 92–103.

Ambrose,W.

[1] "On measurable stochastic processes." Trans. Am. Math. Soc. **47**(1940) 66–79.

Andersen, E.S., and Jessen, B.

[1] "Some limit theorems on integrals in an abstract set." Danske Vid. Selsk. Mat.-Fys. Medd. **22**(14)(1946) 29pp.

[2] "On the introduction of measures in infinite product sets." *ibid.* **25**(4)(1948) 8pp.

Anderson, R.F.

[1] "Diffusions with second order boundary conditions, Parts I - II." Indiana Univ. Math. J. **25**(1976) 367-395; 403–441.

Auslander, L., and MacKenzie, R. E.

[1] Introduction to Differential Manifolds. McGraw-Hill, New York, 1963.

Austin, D. G.

[1] "A sample function property of martingales." Ann. Math. Statist. **37**(1966) 1396–1397.

Bartle, R. G.

[1] "A bilinear vector integral." Studia Math. **15**(1956) 337–352.

Bell, D. R.

[1] The Malliavin Calculus. Pitman Math. Mono. **34** London, 1987.

Belopol'skaya, Ya. I., and Dalecky, Yu. L.

[1] Stochastic Equations and Differential Geometry. Kluwer Acad. Publ. Boston, MA 1990.

Bichteler, K.

[1] "Stochastic integration and L^p-theory of semimartingales." Ann. Prob.**9**(1981) 49–89.

Bismut, J.-M.

[1] "Martingales, the Malliavin calculus, and hypoellipticity under general Hörmander conditions." Z. Wahrs. **56**(1981) 469–505.

[2] Large Deviations and the Malliavin Calculus. Birkhauser, 1984.

Blake, L. H.
[1] "A generalization of martingales and two consequent convergence theorems." Pacific J. Math. **35** (1970) 279–283.
Blumenthal, R. M., and Getoor, R. K.
[1] Markov Processes and Potential Theory. Academic press. New York, 1968.
Bochner, S.
[1] "Stochastic processes." Ann. Math. **48**(2)(1947) 1014–1061.
[2] Harmonic Analysis and the Theory of Probability. Univ. of Calif. Press, Berkeley, CA, 1955.
[3] "Partial ordering in the theory of martingales." Ann. Math. **62**(2) (1955) 162–169.
[4] "Stationarity, boundedness, almost periodicity of random valued functions." Proc. Third Berkley Symp. Math. Statist. Prob. **2**(1956) 7–27.
Bonami, A., Karoui, N., Roynette, B., and Reinhard, H.
[1] "Processus de diffusion associé à un operateur elliptique dégénéré." Ann. Inst. H. Poincaré, **7**(1971) 31–80.
Borchers, D. R.
[1] Second Order Stochastic Differential Equations and Related Itô processes. Ph.D. thesis, Carnegie-Mellon Univ. Pittsburgh, 1964.
Bourbaki, N.
[1] Élenents de Mathématique VI. Chapitre IX (also Chaps. 3–5). Hermann, Paris. 1969.
Brennan, M. D.
[1] "Planar semimartingales." J. Multivar. Anal. **9**(1979) 465–486.
[2] "Riemann-Stieltjes quasimartingale integration." J. Multivar. Anal. **10**(1980) 517–538.
Brooks, J. K., and Dinculeanu, N.
[1] "Stochastic integration in Banach spaces." Seminar on Stochastic Processes, Birkhauser, Basel, (1991) 27–115.
Burkholder, D. L.
[1] "Distribution function inequalities for martingales." Ann. Prob. **1**(1973) 19–42.
Burkholder, D. L., Davis, B. J., and Gundy, R. F.
[1] "Integral inequalities for convex functions of operators on martingales." Proc. Sixth Berkeley Symp. Math. Statist. Prob., Univ. of

Calif. Press, **2**(1972) 223–240.

Cabaña, E. M.

[1] "Stochastic integration in separable Hilbert spaces." Montevideo Publ. Inst. Mat. Estad. **4**(1966) 1–27.

Cairoli, R.

[1] "Une inéqualité pour martingales à indices multiples et ses applications." Lect. Notes Math. **124**(1970) 49–80.

[2] "Décomposition de processus à indices double." *ibid.* **191**(1971) 37–57.

Cairoli, R., and Walsh, J. B.

[1] "Stochastic integrals in the plane." Acta Math. **134**(1975) 111–183.

Carmona, R.A., and Nualart, D.

[1] Nonlinear Stochastic Integrators, Equations and Flows. Gordon and Breach, New York, 1990.

Cartan, H.

[1] Elementary Theory of Analytic Functions of One or Several Complex Variables. Addison-Wesley, Reading, MA 1963.

Cartier, P.

[1] "Introduction à l'étude des movements browniens à plusieurs paramèters." Lect. Notes Math. **191**(1971) 58–75.

Chacón, R.V.

[1] "A'stopped' proof of convergence." Adv. Math. **14**(1974) 365–368.

Chang, D. K., and Rao, M. M.

[1] "Bimeasures and nonstationary processes."in Real and Stochastic Analysis, Wiley, New York, (1986) 7–118.

Choksi, J. R.

[1] "Inverse limits of measure spaces." Proc. London Math. Soc. **8**(3) (1958) 321–342.

Choquet, G.

[1] "Ensembles $\mathcal{K}$-analytiques et $\mathcal{K}$-sousliniens, cas général et cas métrique."(et autrer articles) Ann. Inst. Fourier Grenoble, **9**(1959) 75–109.

Chow, Y. S.

[1] "Martingales in a σ-finite measure space indexed by directed sets." Trans. Am. Math. Soc. **97**(1960) 254–285.

[2] "A martingale inequality and the law of large numbers." Proc. Am. Math. Soc. **11**(1960) 107–111.

[3] "Convergence of sums of squares of martingale differences." Ann. Math. Statist. **39**(1968) 123–133.

[4] "Convergence theorems for martingales." Z. Wahrs. **1**(1963) 340–346.

Chung, K. L., and Doob, J. L.

[1] " Fields, optimality and measurability." Am. J. Math. **87**(1965) 397–424.

Clarkson, J. A., and Adams, C. R.

[1] "On definitions of bounded variation for functions of two variables." Trans. Am. Math. Soc. **35**(1933) 824–854.

Coddington, E.A. and Levinson, N.

[1] Theory of Ordinary Differential Equations. McGraw-Hill, New York, 1955.

Cornea, A., and Licea, G.

[1] "General optional sampling of super martingales." Rev. Roum. Math. Pures Appl. **10**(1965) 1379–1367.

Courrège, P., and Prioret, P.

[1] "Temps d'arrét d'une fonction aléatoire." Publ. Inst. Statist. Univ. Paris, **14**(1965) 245–274.

Cramér, H.

[1] Mathematical Methods of Statistics. Princeton Univ. Press, Princeton, NJ, 1946.

Cuculescu, I.

[1] "Spectral families and stochastic integrals." Rev. Roum. Math. Pures appl. **15**(1970) 201–221.

Dalecky, Yu. L.

[1] "Infinite dimensional elliptic operators and parabolic equations connected with them." Russian Math. Surveys, **22**(4)(1967) 1–53.

Dambis, K. E.

[1] "On the decomposition of continuous submartingales." Th. Prob. Appl. **10**(1965) 401–410.

Darling, R.W.R.

[1] Constructing Nonlinear Stochastic Flows. Memoirs Am. Math. Soc., **376**, (1987), 1–97.

Davis, B. J.

[1] "On the integrability of the martingale square function." Israel J. Math. **8**(1970) 187–190.

Davis, M.

[1] Applied Nonstand Analysis. Wiley-Interscience, New York, 1977.

de la Vallée Poussin, C.J.

[1] "Sur l'integrale de Lebesgue." Trans. Am. Math. Soc. **16** (1915) 435-501.

Dellacherie, C.

[1] Capacités et Processus Stochastiques. Springer-Verlag, Berlin, 1972.

[2] " Un survol de la théorie de l'integrale stochastique." Lect. Notes Math. **794**(1980) 368–395.

Dinculeanu, N.

[1] Vector Measures. Pergamon Press, London, 1967.

[2] "Conditional expectations for general measure spaces." J. Multivar. Anal. **1**(1971) 347–364.

[3] "Vector valued stochastic processes, I–V." I,J. Th. Prob. **1**(1988) 149–169; II, Proc. Am. Math. Soc. **102** 393–401; III, Sem. Stoch. Processes, Birkhauser (1988) 93–132; IV, J. Math. Anal. **142**(1989) 144–161; V, Proc. Am. Math. Soc. **104**(1988) 625–631.

[4] "Linear operators on spaces of totally measurable functions." Rev. Roum. Math. Pures Appl. **10**(1965) 1493–1524.

Dinculeanu, N., and Rao, M. M.

[1] "Contractive projections and conditional expectations." J. Multivar. Anal. **2**(1972) 362–381.

Doléans-Dade, C. (=Doléans, C.)

[1] "Variation quadratique des martingales continue à droit." Ann. Math. Statist. **40**(1969) 284–289.

[2] "Quelques applications de la formule de changement de variables pour les semimartingales." Z. Wahrs. **16**(1970) 181–194.

[3] "Existence du processus croissant natural associé à un potentiel de la class (D)." Z. Wahrs. **9**(1968) 309–314.

[4] "On the existence and unicity of solutions of stochastic integral equations." Z. Wahrs. **36**(1976) 93–101.

Doléans-Dade, C., and Meyer, P. A.

[1] "Intégrales stochastique par rapport aux martingales locales." Lect. Notes Math. **124**(1970) 77–104.

Doob, J. L.

[1] Stochastic Processes. Wiley, New York, 1953.

[2] "Notes on martingale theory." Proc. Fourth Berkley Symp. Math. Statist. Prob. **2**(1961) 95–102.

[3] Classical Potential Theory and its Probabilistic Counterpart. Springer-Verlag, Berlin, 1984.

Dozzi, M.

[1] Stochastic Processes with a Multidimensional Parameter. Res.Notes Math. Pitman, London, 1989.

Dubins, L. E.

[1] "Conditional probability distributions in the wide sense." Proc. Am. Math. Soc. **8**(1957) 1088–1092.

Dubins, L. E., and Schwarz, G.

[1] "On continuous martingales." Proc. Nat'l. Acad. Sci. **53**(1965) 913–916.

Dunford, N., and Schwartz, J. T.

[1] Linear Operators, Part I: General Theory. Wiley-Interscience, New York, 1958.

Dym, H.

[1] "Stationary measures for the flow of a linear differential equation driven by white noise." Trans. Am. Math. Soc. **123**(1966) 130–164.

Dynkin, E. B.

[1] Foundations of the Theory of Markov Process. Pergamon Press, London, 1960.

[2] Markov Process (2 Vols.), Springer-Verlag, New York, 1965.

Eberlein, W. F.

[1] "Abstract ergodic theorems and weak almost periodic functions." Trans. Am. Math. Soc. **67**(1949) 217–240.

Edgar, G. A., and Sucheston, L.

[1] "Amarts: a class of asymptotic martingales."J. Multivar. Anal. **6**(1976) 193–221, 572–591.

[2] Stopping Times and Directed Processes. Cambridge Univ. Press, London, 1993.

Emery, M.

[1] Stochastic Calculus in Manifolds. Springer-Verlag, New York, 1989.

Fefferman, C. L.

[1] "Characterization of bounded mean oscillation." Bull. Am. Math. Soc. **77**(1971) 587–588.

Fefferman, C. L., and Stein, E. M.

[1] "$\mathcal{H}^p$-spaces of several variables." Acta. Math. **129**(1972) 137–193.

Feldman, J.

[1] "Equivalence and perpendicularity of Gaussian processes." Pacific J. Math. **8**(1959) 699–708.

Feller, W.

[1] Theory of Probability and its Applications. Vol.2, Wiley, New York, 1966.

[2] "Non-Markovian processes with the semigroup property." Ann. Math. Statist. **30**(1959) 1252–1253.

[3] "The parabolic partial differential equations and the associated semigroup of transformations." Ann. Math. **55**(2)(1952) 468–519.

Fillmore, P. A.

[1] "On topology induced by measure." Proc. Am. Math. Soc. **17** 854–857.

Finlayson, H. C.

[1] "Measurability of norm proved by Haar functions." Proc. Am. Math. Soc. **53**(1975) 334–336.

[2] "Two classical examples of Gross' abstract Wiener measure." Proc. Am. Math. Soc. **53**(1975) 337–340.

[3] "Gross' abstract Wiener measure on $C[0,\infty]$." Proc. Am. Math. Soc. **57**(1976) 297–298.

Fisk, D. L.

[1] "Quasi-martingales." Trans. Am. Math. Soc. **120**(1965) 369–389.

[2] "Sample quadratic variation of sample continuous second order martingales." Z. Wahrs. **6**(1966) 273–278.

Flanders, H.

[1] Differential Forms with Applications to Physical Sciences. Dover, New York, 1989.

Föllmer, H.

[1] "On the representation of semimartingales." Ann. Prob. **1**(1973) 580–589.

[2] "Stochastic holomorphy." Math. Ann. **207**(1974) 245–255.

[3] "Quasimartingales à deux indices." C. R. Acad. Sci., Paris, Ser. A, **288**(1979) 61–64.

Freidlin, M.

[1] Functional Integration and Partial Differential Equations. Princeton Univ. Press, Princeton, NJ, 1985.

Gangolli, R.

[1] Positive definite kernels on homogeneous spaces and certain stochastic processes related to Lévy's Brownian motion." Ann. Inst. H. Poincré, **3**(1967) 121–225.

Garsia, A. M.

[1] "The Burgess Davis inequalities via Fefferman's inequality." Ark. Math. **11**(1973) 229–237.

[2] Martingale Inequalities. Benjaman, Inc., Reading, MA, 1973.

[3] "A convex function inequality for martingales." Ann. Prob. **1**(1973) 171–174.

Gel'fand, I. M., and Vilenkin, N. Ya.

[1] Generalized Functions, Vol. 4, Acad. Press, New York, 1964.

Getoor, R. M., and Sharpe, M. J.

[1] "Conformal martingales." Invent. Math. **16**(1972) 271–308.

Gikhman, I. I., and Skorokhod, A.V.

[1] Introduction to the Theory of Random Processes. Saunders, Philadelphia, PA, 1969.

[2] "On the densities of probability measures in function spaces." Russian Math. Surveys, **21**(6) 83–156.

Girsanov, I. V.

[1] "On transforming a certain class of stochastic processes by absolutely continuous substitution of measures." Th. Prob. Appl. **5** 285–301.

Girshick, M. A., and Savage, L. J.

[1] "Bayes and minimax estimates for quadratic loss." Proc. Second Berkely Symp. Math. Statist. Prob. (1951) 285–301.

Goldberg, S. I.

[1] Curvature and Homology. Dover, New York, 1982.

Goldstein, J. A.

[1] " Abstract evolution equations." Trans. Am. Math. Soc. **141**(1969) 159–185.

[2] "Second order Itô processes." Nagoya Math. J. **36**(1969) 27–63.

[3] "An existence theorem for linear stochastic differential equations." J. Diff. Eq. **3**(1967) 78–87.

Gould, G. G.

[1] "Integration over vector-valued measures." Proc. London Math. Soc. **15**(3)(1965) 193–225.

Green, M. L.

[1] Multiparameter Semimartingale Integrals and Boundedness Principles. Ph. D. thesis, Univ. Calif., Riverside, 1995.

Gross, L.

[1] "Measurable functions on a Hilbert space." Trans. Am. Math. Soc. **105**(1962) 372–390.

[2] "Abstract Wiener spaces." Proc. Fifth Berkeley Symp. Math. Statist. Prob. **2**(1967) 31-42.

[3] "Potential theory on Hilbert space." J. Funct. Anal. **1**(1967) 123–181.

Guggenheimer, H. W.

[1] Differential Geometry. Dover, New York, 1977.

Gundy, R. F.

[1] "A decomposition for L^1-bounded martingales." Ann. Math. Statist. **39**(1968) 134–138.

Hájek, J.

[1] "On a property of normal distribution of any stochastic process." Čech. Math, J, **8**(2)(1958) 610–618.

Hájek, J., and Rényi, A.

[1] "Generalization of an inequality of Kolmogorov." Acta Math. Acad. Sci., Hung. **6**(1955) 281–283.

Halmos, P. R.

[1] Measure Theory. Van Nostrand. Princeton, NJ, 1950.

Hardy, G. H., Littlewood, J. E., and Pólya, G.

[1] Inequalities. Cambridge Univ. Press, London, 1934.

Hays, C. A., and Pauc, C. Y.

[1] Derivation and Martingales. Springer-Verlag, Berlin. 1970.

Herz, C. S.

[1] "Bounded mean oscillation and related martingales." Trans. Am. Math. Soc. **193**(1974) 199–215.

Hida, T., and Nomoto, H.

[1] "Gaussian measures on the projective limit space of spheres." Proc. Japan Acad. **40**(1964) 301–304.

[2] "Finite dimensional approximation to band limited white noise." Nagoya Math. J. **29**(1967) 211–216.

Hille, E., and Phillips, R. S.

[1] Functional Analysis and Semi-Groups. Am. Math. Soc. Colloq. Publ., Providence, RI, 1958.

Hörmander, L.

[1] "Hypoelliptic second order differential equations." Acta Math. **119**(1967) 147–171.

Hunt, G. A.

[1] Martingales et Processus de Markov. Dunad, Paris. 1966.

Hürzeler, H. E.

[1] "Stochastic integration on partially ordered sets." J. Multivar. Anal. **17** (1985) 279-303.

Ikeda, N., and Watanabe, S.

[1] Stochastic Differential Equations and Diffusion Processes, North-Holland, Amsterdam, (2nd ed.) 1989.

Ionescu Tulcea, A., and Ionescu Tulcea, C.

[1] Topics in the Theory of Lifting. Springer-Verlag, Berlin. 1969.

Ionescu Tulcea, C.

[1] "Mesures dans les espaces produits." Att Acad. Naz. Lincei Rend. cl. Sci. Fis. Mat. Nat. **7**(8)(1949/50) 208–211.

Isaac, R.

[1] "A proof of the martingale convergence theorem." Proc. Am. Math. Soc. **16**(1965) 842–844.

Isaacson, D.

[1] "Stochastic integrals and derivatives." Ann. Math. Statist. **40**(1969) 1610–1616.

Itô, K.

[1] "On a formula concerning stochastic differentials." Nagoya math. J. **3** (1951) 55–66.

[2] "On stochastic differential equations." Mem. Am. Math. Soc. **4** (1951) 51pp.

[3] "Stochastic differentials." Appl. Math. Opt. **1** (1975) 374–381.

[4] "Multiple Wiener integral." J. Math. Soc. Japan, **3** (1951) 157–169.

Itô, K., and Watanabe, S.

[1] "Transformation of Markov processes by multiplicative functionals." Ann. Inst. Fourier Grenoble, **15**(1965) 13–30.

[2] "Introduction to stochastic differential equations." Proc. Internat. Symp. Stoch. Diff. Eq. Kyoto, (1978) i–xxv.

John, F., and Nirenberg, L.

[1] "On functions of bounded mean oscillation." Comm. Pure Appl. Math. **14**(1961) 785–799.

Johansen, S., and Karush, J.

[1] "On the semimartingale convergence." Ann. Math. Statist. **37** (1966) 680-694.

Johnson, G., and Helms, L. L.

[1] "Class (D) supermartingales." Bull. Am. Math. Soc. **69**(1963) 59–62.

Kailath, T., and Zakai, M.

[1] "Absolute continuity and Radon-Nikodým derivatives for certain measures relative to Wiener measure." Ann. Math. Statist. **42** (1971) 130–140.

Kakutani, S.

[1] "Concrete representation of abstract (M) spaces." Ann. Math. **42**(2)(1941) 994–1024.

Kampe de Fériet, J.

[1] "Sur un probléme d'algébre abstrait posé par la définition de la moyenne dans la théorie de la turbulence." An. Soc. Sci. Bruxelles, **63**(1949) 156–172.

[2] "Problémes mathématiques posés par la mécanique statistique de la turbulence." Proc. Internat. Cong. Math. **3**(1954) 237–242.

Kaneko, H., and Taniguchi, S.

[1] A stochastic approach to Šilov boundary." J. Functional Anal. **74** (1987) 415-429.

Karhunen, K.

[1] "Uber lineare Methoden in der Wahrscheinlichkeitsrechnung." Ann. Acad. Sci. Fenn. AI. Math. **37**(1947) 3–79.

Kazamaki, N.

[1] "Changes of time, stochastic integrals and weak martingales." Z. Wahrs. **22**(1972) 25–32.

[2] "Note on a stochastic integral equation." Lect. Notes Math. **258** (1972) 105–108.

Kingman, J. F. C.

[1] "Additive set functions and the theory of probability." Proc. Camb. Phil. Soc. **63**(1967) 767–775.

Knight, F. B.

[1] "A reduction of continuous square integrable martingales to Brownian motion." Lect. Notes Math. **190**(1971) 19–31.

Kolmogorov, A. N.

[1] Grundbegriffe der Wahrscheinlichkeitsrechnung. Springer-Verlag, Berlin. 1933. [Foundations of the Theory of Probability, (Translation) Chelsea Publishing Co. New York, 1956.]

Krasnosel'skiĭ, M. A., and Rutickiĭ, Ya. B.

[1] Convex Functions and Orlicz Spaces. Noordhoff, Groningen, The Netherlands, 1961.

Krinik, A.

[1] "Diffusion processes in Hilbert space and likelihood ratios."in Real and Stochastic Analysis, Wiley, New York, (1986)168–210.

Kunita, H.

[1] Stochastic Flows and Stochastic Differential Equations. Camb. Univ. Press, London, 1990.

Kunita, H., and Watanabe, S.

[1] "On square integrable martingales." Nagoya Math. J. **30**(1967) 209–245.

Kuo, H. H.

[1] "Stochastic integrals in abstract Wiener spaces." Pacific J. Math. **41**(1972) 469–483.

Lamb, C. W.

[1] "Representation of functions as limits of martingales." Trans. Am. Math. Soc. **188**(1974) 395–405.

Lévy, P.

[1] Processus Stochastiques et Mouvement Brownien.(2nd ed.) Gauthier- Villars, Paris, 1965.

Lindenstrauss, J., and Pelczyński, A.

[1] "Absolutely summing operators in $\mathcal{L}^p$-spaces and applications." Studia Math. **29**(1968) 275–326.

Lloyd, S. P.

[1] "Two lifting theorems." Proc. Am. Math. Soc. **42**(1974) 128-134.

Loève, M.

[1] Probability Theory. (3rd ed.) Van Nostrand, Princeton, NJ, 1963.

Loomis, L. H.

[1] Introduction to Abstract Harmonic Analysis. Van Nostrand, Princeton, NJ, 1953.

Maisonneuve, B.

[1] "Quelques martingales remarkables associées à une martingale continue." Publ. Inst. Statist., Univ. Paris, **17**(1968) 13–27.

Malliavin, P.

[1] "Stochastic calculus of variation and hypoelliptic operators." Proc. Intern. Symp. Stoch. Diff. Eq., Kyoto, (1978) 195-263.

Mallory, D. J., and Sion, M.

[1] "Limits of inverse systems of measures." Ann. Inst. Fourier Grenoble, **21**(1971) 25–57.

McKean,jr., H. P.

[1] Stochastic Integrals. Academic Press, New York, 1969.

McShane, E. J.

[1] Order Preserving Maps and Integration Processes. Ann. Math. Studies, **31**, Princeton Univ. Press, Princeton, NJ, 1953.

[2] "Families of measures and representations of algebras of operators." Trans. Am. Math. Soc. **102**(1962) 328–345.

[3] Stochastic Calculus and Stochastic Models. Academic Press, New York, 1974.

[4] "Stochastic differential equations." J. Multivar. Anal. **5** 121–177.

Mertens, J.-F.

[1] " Théorie des processus stochastiques généraux et surmartingales." Z. Wahrs. **22**(1972) 45–68.

Métivier, M.

[1] Semimartingales, W. de Guyter, Berlin, 1982.

Métivier, M. Pellaumail, J.

[1] "On Doléans-Föllmer measure for quasimartingales." Ill. J. Math. **19**(1975) 491–504.

[2] Stochastic Integrals, Academic Press, New York, 1980.

Meyer, P. A.

[1] Probability and Potentials. Blaisdell Co., Waltham, MA, 1966.

[2] Martingales and Stochastic Integrals. Lect. Notes Math. **284**(1972) 89pp.

[3] "A decomposition theorem for supermartingales." Ill. J. Math. **6**(1962) 193–205; "Uniqueness," *ibid.* **7**(1963) 1–17.

[4] "Stochastic integrals." Lect. Notes Math. **39**(1967) 72–162.

[5] Processus de Markov. Lect. Notes Math. **26** 190pp.

[6] "Une cours sur les intégrales stochastiques." Lect. Notes Math. **511**(1976) 245–400.

[7] "Geometrie differential stochastiques." Astérisque, **131**(1985) 107–113.

Millar, P. W.

[1] "Transforms of stochastic processes." Ann. Math. Statist. **39**(1968) 372–376.

[2] "Martingale integrals." Trans. Am. Math. Soc. **133**(1968) 145–166.

Millington, H., and Sion, M.

[1] "Inverse systems of group valued-measures." Pacific J. Math. **44** 637–650.

Minlos, R. A.

[1] "Generalized random processes and their extension to a measure." Trudy Moscow Mat. Obšč. **8**(1959) 497–518. [IMS-AMS Translation No. **3**(1963) 291–313.]

Molchan, G. M.

[1] "Some problems connected with the Brownian motion of Lévy." Th. Prob. Appl. **12**(1967) 682–690.

Morse, M., and Transue, W.

[1] "$\mathbb{C}$-bimeasures and their integral extensions." Ann. Math. **64**(2) (1956) 480–504.

Moy, S.-C.

[1] "Characterizations of conditional expectation as a transformation on function spaces." Pacific J. Math. **4**(1954) 47–64.

Nelson, E.

[1] "Regular probability measures on function spaces." Ann. Math. **69**(2)(1959) 630–643.

Neveu, J.

[1] Mathematical Foundations of the Calculus of Probabilities. Holden-Day, San Francisco, CA, 1965.

[2] Processus Aléatoires Gaussiens. Univ. Montreal Press, 1968.
[3] "Théorie des semi-groupes de Markov." Univ. Calif. Publ. Statist. **2**(1958) 319–394.

Norris, J.
[1] "Simplified Malliavin calculus." Lect. Notes Math. **1204**(1986) 101–130.

Nualart, D., and Pardoux, E.
[1] "Stochastic calculus with anticipating integrands." Prob. Th. Rel. Fields, **78**(1988) 535–581.

Olson, M. P.
[1] "A characterization of conditional probability." Pacific J. Math. **15**(1965) 971–983.

Orey, S.
[1] "Conditions for absolute continuity of two diffusions." Trans. Am. Math. Soc. **193**(1974) 413–426.
[2] "F-processes." Proc. Fifth Berkeley Symp. Math. Statist. Prob. **2**(1958) 301–313.

Paley, R. E. A. C., Wiener, N., and Zygmund, A.
[1] "Notes on random functions." Math. Z. **37**(1929) 647–668.

Pellaumail, J.
[1] "Sur l'intégrale stochastique et la décomposition de Doob-Meyer." Astérisque No. **9** (1973) 1–125.

Petersen, K. E.
[1] Brownian Motion, Hardy Spaces and Bounded Mean Oscillation. London Math. Soc. Lect. Notes **28**, 1977.

Pitcher, T. S.
[1] "Parameter estimation for stochastic processes." Acta Math. **112** (1964) 1–40.

Postnikov, M. M.
[1] The Variational Theory of Geodesics. (Translation) Dover, New York, 1983.

Prokhorov, Yu. V.
[1] "Convergence of random processes and limit theorems in probability." Th. Prob. Appl. **1**(1956) 157–214.
[2] "The method of characteristic functionals." Proc. Fourth Berkely Symp. Math. Statist. Prob. **2**(1961) 403–419.

Protter, P. E.

[1] "On the existence, uniqueness, convergence and explosions of solutions of stochastic integral equations." Ann. Prob. **5**(1977) 243–261.

[2] "Right continuous solutions of systems of stochastic integral equations." J. Multivar. Anal. **7**(1977) 204–214.

[3] Stochastic Integration and Differential Equations: A New Approach. Springer-Verlag, Berlin, 1990.

Radó, T.

[1] Subharmonic Functions. Chelsea, New York, 1949.

Rao, K. M.

[1] "On decomposition theorems of Meyer." Math. Scand. **24**(1969) 66–78.

[2] "Quasimartingales." *ibid.***24**(1969) 79–92.

[3] "On modification theorems." Trans. Am. Math. Soc. **167**(1972) 443–450.

Rao M. M.

[1] "Conditional expectations and closed projections." Proc. Acad. Sci., Amsterdam, Ser. A, **68**(1965) 100–112.

[2] "Inference in stochastic processes,I-VI." I, Th. Prob. Appl. **8**(1963) 282–298; II, Z. Wahrs. **5**(1966) 317–335; III, *ibid.***8**(1967) 49–72; IV, Sankhyā, Ser. A, **36**(1974) 63–120; V, *ibid.***37**(1975) 538–549; VI, Multivariate Analysis-IV(1977) 311–334.

[3] "Conditional measures and operators." J. Multivar. Anal. **5**(1975) 330–413.

[4] "Two characterizations of conditional probability." Proc. Am. Math. Soc. **59**(1976) 75–80.

[5] "Abstract nonlinear prediction and operator martingales." J. Multivar. Anal. **1**(1971) 129–157, and **9** 614.

[6] "Extensions of stochastic transformations."Trab. Estad. **26**(1975) 473–485.

[7] "Conjugate series, convergence and martingales." Rev. Roum. Math. Pures Appl. **22**(1977) 219–254.

[8] "Interpolation, ergodicity and martingales." J. Math. Mech. **16** (1966) 543–568.

[9] "Covariance analysis of nonstationary time series." Developments in Statist. Academic Press, New York,**1**(1978) 171–225.

[10] "Stochastic processes and cylindrical probabilities." Sankhyā, Ser. A, **43**(1981) 149–169.

[11] Measure Theory and Integration. Wiley-Interscience, New York, 1987.

[12] Probability Theory with Applications. Academic Press, New York, 1984.

[13] Conditional Measures and Applications. Marcel Dekker, New York, 1993.

[14] "Non L^1-bounded martingales." Lect. Notes Control Inf. Sci. **16** (1979) 527–538.

[15] "Stochastic integration: a unified approach." C. R. Acad. Sci. Paris, Sér I, **314**(1992) 629–633.

[16] "An approach to stochastic integration." in Multivariate Analysis: Future Directions, Elsivier, New York, (1993) 347–374.

[17] "$L^{2,2}$-boundedness, harmonizability and filtering." Stoch. Anal. Appl. **10**(1992) 323–342.

[18] Stochastic Processes and Integration. Sijthoff and Noordhoff, Alphen ann den Rijn, The Netherlands, 1979.

Rao, M. M., and Ren, Z. D.

[1] Theory of Orlicz Spaces. Marcel Dekker, New York, 1991.

Rao, M. M., and Sazonov, V. V.

[1] "A projective limit theorem for probability spaces and applications." Th. Prob. Appl. **38** 307–315.

Raoult J.-P.

[1] "Sur un généralization d'un théorèm d'Ionescu Tulcea." C. R. Acad. Sci., Ser. A, Paris, **259**(1964) 2769–2772.

[2] " Limites projectives de mesures σ-finites et probabilités conditionnelles." *ibid.* **260**(1965) 4893–4896.

Rényi, A.

[1] "On a new axiomatic theory of probability." Acta Math. Sci. Hung. **6**(1955) 285–333.

[2] Foundations of Probability. Holden-Day, San Francisco, CA, 1970.

Revuz, D., and Yor, M.

[1] Continuous Martingales and Brownian Motion. Springer-Verlag, Berlin, 1991.

Riesz, F., and Sz.-Nagy, B.

[1] Functional Analysis. F. Unger, New York, 1955.

Royden, H. L.

[1] Real Analysis. (2nd ed.) Macmillan and Co., New York, 1969.

Ryan, R.

[1] "Representative sets and direct sums." Proc. Am. Math. Soc. **15**(1964) 387–390.

Saks, S.

[1] Theory of the Integral. (2nd ed.) Hefner Publishing Co., New York, 1937.

Sazonov, V. V.

[1] "On perfect measures." Translations of Am. Math. Soc. **48**(2)(1965) 229–254.

Schatten, R.

[1] A Theory of Cross Spaces. Princeton Univ. Press, Princeton, NJ, 1959.

Schreiber, B. M., Sun, T. C., and Bharucha-Reid, A. T.

[1] "Algebraic models for probability measures associated with stochastic processes." Trans. Am. Math. Soc. **158**(1971) 93–105.

Schwartz, L.

[1] Radon Measures on Arbitrary Topological Spaces and Cylindrical Measures. Tata Institute, Bombay, 1973.

[2] "Prorobabilités cylindriques et applications radonifiantes." J. Fac. Sci. Univ. Tokyo, **18**(1971) 139–286.

[3] "Sur martingales régulière à valeurs mesures et désintégrations régulières d'une mesure." J. Anal. Math. **26**(1973) 1–168.

[4] Semi-Martingales sur des Variétés, et Martingales Conformes sur des Variétés Analytiques Complexes. Lect. Notes Math. **780** (1980) 132pp.

Segal, I. E.

[1] "Abstract probability spaces and a theorem of Kolmogoroff." Am. J. Math. **76**(1954) 721–732.

[2] "Equivalence of measure spaces." *ibid.***73** 275–313.

Shale, D.

[1] "Invariant integration over the infinite dimensional orthogonal group and related spaces." Trans. Am. Math. Soc. **124**(1966) 148–157.

Sierpiński, W.

[1] Theorie des Ensembles. Warsaw, 1951.

Sion, M.

[1] Introduction to the Methods of Real Analysis. Holt, Rinehart and Winston, New York, 1968.

[2] A Theory of Semi-Group Valued Measures. Lect. Notes Math. **355**(1973) 140pp.

Snell, J. L.

[1] "Applications of martingale system theorems." Trans. Am. Math. Soc. **73**(1952) 293–312.

Stein, E. M.

[1] Topics in Harmonic Analysis. Princeton Univ. Press, Princeton, NJ, 1970.

Strassen, V.

[1] "The existence of probability measures with given marginals." Ann. Math. Statist. **36**(1965) 423–439.

Stricker, C.

[1] "Mesure de Föllmer en théorie des quasi-martingales." Lect. Notes Math. **485**(1975) 408–419.

[2] "Une caracterisation des quasi-martingales." *ibid.* **485**(1975) 420–424.

Stroock, D. W.

[1] "Malliavin calculus: a functional analytic approach." J. Funct. Anal. **44**(1981) 212–257.

Stroock, D. W., and Varadhan, S. R. S.

[1] "Diffusion process with continuous coefficients." Comm. Pure Appl. Math. **22** 345–400, 479–530.

[2] Multidimensional Diffusion Processes. Springer-Verlag, Berlin, 1979.

Subrahmanian, R.

[1] "On a generalization of martingales due to Blake." Pacific J. Math. **48**(1973) 275–278.

Sudderth, W. D.

[1] "A 'Fatou equation' for randomly stopped variables." Ann. Math. Statist. **42**(1971) 2143–2146.

Taniguchi, S.

[1] "Malliavin's stochastic calculus of variations for manifold-valued Wiener functionals and its applications." Z. Wahrs. **65**(1983) 269–290.

Traynor, T.

[1] "An elementary proof of the lifting theorem." Pacific J. Math. **53**(1974) 267–272.

Walsh, J. B.

[1] "An introduction to stochastic partial differential equations." Lect. Notes Math. **1180**(1986) 265–439.

[2] "Martingales with a multidimensional parameter and stochastic integrals in the plane." Lect. Notes Math. **1215**(1986) 329–491.

Wiener, N.

[1] "Differential space." J. Math. Phy. MIT, **2** (1923) 131-174.

Wong, E., and Zakai, M.

[1] "Martingales and stochastic integrals for processes with a multi-dimensional parameter." Z. Wahrs. **29**(1974) 109–122.

Wright, J. D. M.

[1] "Stone algebra valued measures and integrals." Proc. London Math. Soc. **19**(3)(1969) 108–122.

Yaglom, A. M.

[1] "Some classes of random fields in n-dimensional space related to stationary random processes." Th. Prob. Appl. **2**(1957) 273–320.

Yamasaki, Y.

[1] "Projective limit of Haar measures on $\mathcal{O}(n)$." Publ. Res. Inst. Math. Sci., Kyoto Univ. **8**(1973) 141–149.

Yeh, J.

[1] "Wiener measure in a space of functions of two variables." Trans. Am. Math. Soc. **95**(1960) 433–450.

[2] "Two parameter stochastic differential equations." in Real and Stochastic Analysis, Wiley, New York, (1986) 249–344.

Ylinen, K.

[1] "On vector bimeasures." Ann. Mat. Pura Appl. **117**(4)(1978) 115–138.

Yor, M.

[1] "Existence et unicité de diffusion à valuers dans un space de Hilbert." Ann. Inst. H. Poincaré, **10**(1974) 55–88.

[2] "Sur quelques approximations d'integrales stochastiques." Lect. Notes Math. **581**(1977) 518–528.

Yosida, K., and Kakutani, S.

[1] "Operator theoretical treatment of Markoff's processes and mean ergodic theorem." Ann. Math. **42** (1941) 188-228.

Zaanen, A. C.

[1] "The Radon-Nikodým theorem." Proc. Acad. Sci., Amsterdam, Ser. A **64**(1961) 157–187.

Zygmund, A.

[1] Trigonometric Series. Cambrige Univ. Press, London, 1958.

Notation Index

Chapter I

(Ω, Σ, P)-Probability triple, 1 ($\mathbb{R}[\mathbb{C}]$ denotes real [complex] field)
$[a,b), (a,b], [a,b]$-the usual half-open and closed intervals
$E(X)$-expectation of X, 8
CBS=Cauchy-Buniakowski-Schwarz, 8
iff = if and only if, 9
$\mathbb{R}^T$-space of real functions on a set T, 11
$\mathcal{R}$-Borel σ-algebra of $\mathbb{R}$, 13
$\{(\Omega_\alpha, \Sigma_\alpha, P_\alpha, g_{\alpha\beta})_{\alpha<\beta} : \alpha, \beta \text{ in } D\}$-projective system, 17
$\mathcal{X}$- usually a Banach space, $\mathcal{X}^*$ its dual, $\mathcal{X}'$ its algebraic dual, 34
$(i, \mathcal{X}, B)$-abstract Wiener triple, 41

Chapter II

$E^{\mathcal{B}}(X)$-conditional expectation of X relative to a σ-algebra $\mathcal{B}$, 64
$P_{\mathcal{B}} = P|\mathcal{B}$, restriction of P to $\mathcal{B}$, 63
$P^{\mathcal{B}}$- conditional probability relative to $\mathcal{B}$, 64
$\Sigma(Y)$ or Σ_Y- σ-algebra generated by the given function Y, 75
$\vee(\wedge)$-maximum (minimum) operator, 105
$(\Omega, \Sigma, \mathcal{B}, P(\cdot|\cdot))$-Rényi's conditonal measure space, 104
$E(X_{t_{n+1}}|X_{t_1}, \ldots, X_{t_n})$-conditional expectation, 111
$L^p(\Omega, \mathcal{F}_s, P) = L^p(\mathcal{F}_s)$, 111
$\ell g^+ \lambda = \log^+ \lambda = \log \lambda$ for $\lambda > 1, = 0$ for $\lambda \leq 1$, 114

Chapter III

Chapter IV

Chapter V

Chapter VI

Chapter VII

Author Index

SUBJECT INDEX

A

B

C

D

E

F

G

H

Other *Mathematics and Its Applications* titles of interest:

P.M. Alberti and A. Uhlmann: *Stochasticity and Partial Order. Doubly Stochastic Maps and Unitary Mixing.* 1982, 128 pp. ISBN 90-277-1350-2

A.V. Skorohod: *Random Linear Operators.* 1983, 216 pp. ISBN 90-277-1669-2

I.M. Stancu-Minasian: *Stochastic Programming with Multiple Objective Functions.* 1985, 352 pp. ISBN 90-277-1714-1

L. Arnold and P. Kotelenez (eds.): *Stochastic Space-Time Models and Limit Theorems.* 1985, 280 pp. ISBN 90-277-2038-X

Y. Ben-Haim: *The Assay of Spatially Random Material.* 1985, 336 pp. ISBN 90-277-2066-5

A. Pazman: *Foundations of Optimum Experimental Design.* 1986, 248 pp. ISBN 90-277-1865-2

P. Kree and C. Soize: *Mathematics of Random Phenomena. Random Vibrations of Mechanical Structures.* 1986, 456 pp. ISBN 90-277-2355-9

Y. Sakamoto, M. Ishiguro and G. Kitagawa: *Akaike Information Criterion Statistics.* 1986, 312 pp. ISBN 90-277-2253-6

G.J. Szekely: *Paradoxes in Probability Theory and Mathematical Statistics.* 1987, 264 pp. ISBN 90-277-1899-7

O.I. Aven, E.G. Coffman (Jr.) and Y.A. Kogan: *Stochastic Analysis of Computer Storage.* 1987, 264 pp. ISBN 90-277-2515-2

N.N. Vakhania, V.I. Tarieladze and S.A. Chobanyan: *Probability Distributions on Banach Spaces.* 1987, 512 pp. ISBN 90-277-2496-2

A.V. Skorohod: *Stochastic Equations for Complex Systems.* 1987, 196 pp. ISBN 90-277-2408-3

S. Albeverio, Ph. Blanchard, M. Hazewinkel and L. Streit (eds.): *Stochastic Processes in Physics and Engineering.* 1988, 430 pp. ISBN 90-277-2659-0

A. Liemant, K. Matthes and A. Wakolbinger: *Equilibrium Distributions of Branching Processes.* 1988, 240 pp. ISBN 90-277-2774-0

G. Adomian: *Nonlinear Stochastic Systems Theory and Applications to Physics.* 1988, 244 pp. ISBN 90-277-2525-X

J. Stoyanov, O. Mirazchiiski, Z. Ignatov and M. Tanushev: *Exercise Manual in Probability Theory.* 1988, 368 pp. ISBN 90-277-2687-6

E.A. Nadaraya: *Nonparametric Estimation of Probability Densities and Regression Curves.* 1988, 224 pp. ISBN 90-277-2757-0

H. Akaike and T. Nakagawa: *Statistical Analysis and Control of Dynamic Systems.* 1998, 224 pp. ISBN 90-277-2786-4

Other *Mathematics and Its Applications* titles of interest:

A.V. Ivanov and N.N. Leonenko: *Statistical Analysis of Random Fields.* 1989, 256 pp. ISBN 90-277-2800-3

V. Paulauskas and A. Rackauskas: *Approximation Theory in the Central Limit Theorem. Exact Results in Banach Spaces.* 1989, 176 pp. ISBN 90-277-2825-9

R.Sh. Liptser and A.N. Shiryayev: *Theory of Martingales.* 1989, 808 pp. ISBN 0-7923-0395-4

S.M. Ermakov, V.V. Nekrutkin and A.S. Sipin: *Random Processes for Classical Equations of Mathematical Physics.* 1989, 304 pp. ISBN 0-7923-0036-X

G. Constantin and I. Istratescu: *Elements of Probabilistic Analysis and Applications.* 1989, 488 pp. ISBN 90-277-2838-0

S. Albeverio, Ph. Blanchard and D. Testard (eds.): *Stochastics, Algebra and Analysis in Classical and Quantum Dynamics.* 1990, 264 pp. ISBN 0-7923-0637-6

Ya.I. Belopolskaya and Yu.L. Dalecky: *Stochastic Equations and Differential Geometry.* 1990, 288 pp. ISBN 90-277-2807-0

A.V. Gheorghe: *Decision Processes in Dynamic Probabilistic Systems.* 1990, 372 pp. ISBN 0-7923-0544-2

V.L. Girko: *Theory of Random Determinants.* 1990, 702 pp. ISBN 0-7923-0233-8

S. Albeverio, PH. Blanchard and L. Streit: *Stochastic Processes and their Applications in Mathematics and Physics.* 1990, 416 pp. ISBN 0-9023-0894-8

B.L. Rozovskii: *Stochastic Evolution Systems. Linear Theory and Applications to Non-linear Filtering.* 1990, 330 pp. ISBN 0-7923-0037-8

A.D. Wentzell: *Limit Theorems on Large Deviations for Markov Stochastic Process.* 1990, 192 pp. ISBN 0-7923-0143-9

K. Sobczyk: *Stochastic Differential Equations. Applications in Physics, Engineering and Mechanics.* 1991, 410 pp. ISBN 0-7923-0339-3

G. Dallaglio, S. Kotz and G. Salinetti: *Distributions with Given Marginals.* 1991, 300 pp. ISBN 0-7923-1156-6

A.V. Skorohod: *Random Processes with Independent Increments.* 1991, 280 pp. ISBN 0-7923-0340-7

L. Saulis and V.A. Statulevicius: *Limit Theorems for Large Deviations.* 1991, 232 pp. ISBN 0-7923-1475-1

A.N. Shiryaev (ed.): *Selected Works of A.N. Kolmogorov, Vol. 2: Probability Theory and Mathematical Statistics.* 1992, 598 pp. ISBN 90-277-2795-X

Yu.I. Neimark and P.S. Landa: *Stochastic and Chaotic Oscillations.* 1992, 502 pp. ISBN 0-7923-1530-8

Other *Mathematics and Its Applications* titles of interest:

Y. Sakamoto: *Categorical Data Analysis by AIC*. 1992, 260 pp.
ISBN 0-7923-1429-8

Lin Zhengyan and Lu Zhuarong: *Strong Limit Theorems*. 1992, 200 pp.
ISBN 0-7923-1798-0

J. Galambos and I. Katai (eds.): *Probability Theory and Applications*. 1992, 350 pp.
ISBN 0-7923-1922-2

N. Bellomo, Z. Brzezniak and L.M. de Socio: *Nonlinear Stochastic Evolution Problems in Applied Sciences*. 1992, 220 pp. ISBN 0-7923-2042-5

A.K. Gupta and T. Varga: *Elliptically Contoured Models in Statistics*. 1993, 328 pp.
ISBN 0-7923-2115-4

B.E. Brodsky and B.S. Darkhovsky: *Nonparametric Methods in Change-Point Problems*. 1993, 210 pp. ISBN 0-7923-2122-7

V.G. Voinov and M.S. Nikulin: *Unbiased Estimators and Their Applications. Volume 1: Univariate Case*. 1993, 522 pp. ISBN 0-7923-2382-3

V.S. Koroljuk and Yu.V. Borovskich: *Theory of* U-*Statistics*. 1993, 552 pp.
ISBN 0-7923-2608-3

A.P. Godbole and S.G. Papastavridis (eds.): *Runs and Patterns in Probability: Selected Papers*. 1994, 358 pp. ISBN 0-7923-2834-5

Yu. Kutoyants: *Identification of Dynamical Systems with Small Noise*. 1994, 298 pp.
ISBN 0-7923-3053-6

M.A. Lifshits: *Gaussian Random Functions*. 1995, 346 pp. ISBN 0-7923-3385-3

M.M. Rao: *Stochastic Processes: General Theory*. 1995, 635 pp.
ISBN 0-7923-3725-5